STRUCTURAL HEALTH MONITORING

HOW TO ORDER THIS BOOK

BY PHONE: 800-233-9936 or 717-291-5609, 8AM–5PM Eastern Time

BY FAX: 717-295-4538

BY MAIL: Order Department
Technomic Publishing Company, Inc.
851 New Holland Avenue, Box 3535
Lancaster, PA 17604, U.S.A.

BY CREDIT CARD: American Express, VISA, MasterCard

BY WWW SITE: http://www.techpub.com

STRUCTURAL HEALTH MONITORING

Current Status and Perspectives

Edited by

Fu-Kuo Chang

Department of Aeronautics and Astronautics
Stanford University, Stanford, CA

Proceedings of the International Workshop on Structural Health Monitoring, Stanford University, Stanford, CA
September 18-20, 1997

Sponsors:
Air Force Office of Scientific Research
Army Research Office
National Science Foundation

LANCASTER · BASEL

Structural Health Monitoring
a **TECHNOMIC**® publication

Published in the Western Hemisphere by
Technomic Publishing Company, Inc.
851 New Holland Avenue, Box 3535
Lancaster, Pennsylvania 17604 U.S.A.

Distributed in the Rest of the World by
Technomic Publishing AG
Missionsstrasse 44
CH-4055 Basel, Switzerland

Printed in the United States of America
10 9 8 7 6 5 4 3 2 1

Main entry under title:
Structural Health Monitoring: Current Status and Perspectives

A Technomic Publishing Company book
Bibliography: p.
Includes index p. 799

Library of Congress Catalog Card No. 97-61700
ISBN No. 1-56676-605-2

Table of Contents

SESSION 3: MODELING AND DIAGNOSTIC METHODS I

KEYNOTE SPEAKERS

SESSION 4: MODELING AND DIAGNOSTIC METHODS II

KEYNOTE SPEAKERS

SESSION 13: MODELING AND DIAGNOSTIC METHODS V

SESSION 14: CIVIL INFRASTRUCTURES IV

Foreword

The purpose of the workshop was to assess the current state-of-the-art technologies in the field of structural health monitoring and to discuss and identify key and emerging issues in research and development that are critical and unique to structural health monitoring. The workshop is also intended to promote exchanges and cross-fertilization among many disciplines. Presentations were made by invited and selected distinguished speakers. Present and potential applications of the techniques to military and civilian structures were also discussed. This volume contains the papers that were presented at the workshop.

Nearly all in-service structures require some form of maintenance for monitoring their integrity and health condition to prolong their lifespan or to prevent catastrophic failure. Current schedule-driven inspection and maintenance techniques can be time-consuming, labor-intensive, and expensive. Recent advances in smart structure technologies and material/structural damage characterization combined with recent developments in sensors and actuators have resulted in a significant interest in developing new diagnostic technologies for the in-situ characterization of material properties in manufacturing, for monitoring their integrity, and for the detection of damage to both existing and new structures in real time with minimum human involvement.

The topics of the workshop fall into the following areas:

– Sensing Technology Development
  - Fiber optics, piezoelectrics
– Modeling and Diagnostic Methods
  - Traditional NDE techniques, neural network, fuzzy logic, probabilistic estimations, system identifications
– System Integration
  - Distributed sensors with controller
– Applications
  - Civil infrastructures: Bridges, highway systems, buildings, power plants, etc.

- Aircraft and missile structures: Helicopters, airplanes, engines, motor cases, etc.
- Space structures: Satellites, space stations, reusable launch vehicles, etc.
- Land/marine structures: Automobiles, submarines, ships, etc.

A total of sixty papers were presented by researchers from all over the world. This volume provides a most comprehensive presentation and discussion of the field of structural health. It contains not only the most up-to-date technology developments in structural health monitoring, but also provides information regarding the current and potential applications in the field.

The support for the Workshop of Capt. Brian Sanders of the Air Force Office of Scientific Research, Dr. Gary Anderson of the Army Research Office, and Dr. Ken Chong and Dr. C. C. Liu of the National Science Foundation for the workshop is gratefully acknowledged.

The Organizing Committee of the workshop consisted of the following individuals:

F. K. Chang (Stanford University–Chairman)
A. Guemes (Universidad Politecnica de Madrid, Spain–Co-Chairman)
E. Aktan (U. of Cincinnati)
G. Anderson (ARO)
K. Chong (NSF)
S. Donaldson (Air Force Wright-Patterson)
P. Fuhr (UVM)
T. Fududa (Osaka City University, Japan)
P. Hagedorn (U. of Darmstadt, Germany)
R. Ikegami (Boeing Defense)
D. Inman (VPI)
C. C. Liu (NSF)
P. Lagace (MIT)
C. C. Lin (Chung-Hsing U., Taiwan)
S. Masri (U. of S. California)
Y. Rajapakse (ONR)
J. Sirkis (U. of Maryland)
B. Sanders (AFOSR)
C. T. Sun (Purdue University)
J. Tani (Tohoku U., Japan)
M. L. Wang (U. of Illinois, Chicago)

Fu-Kuo Chang, the Organizing Chairman, would like to express his sincere appreciation to Mr. Shawn Beard, Mr. Mark Lin, Mr. Robert Seydel, Ms. Sandra L. Snyder, and Mr. Calvin Wang for their hard work and dedication in the preparation for the workshop.

FU-KUO CHANG
Department of Aeronautics and Astronautics
Stanford University, Stanford, CA 94305

KEYNOTE SPEAKERS

U. Goranson
D. Hyland

Jet Transport Structures Performance Monitoring

U. G. GORANSON

ABSTRACT

Structural safety is an evolutionary accomplishment, and attention to design features is key to its achievement. Acquisition and review of service data and other firsthand information from customer airlines is necessary to promote safe and economic operation of the worldwide Boeing fleet. Technology standards are used to ensure analysis commonality across different models, and operational loading spectra variations are monitored through sampling with digital flight data recorders, as well as traditional speed/acceleration/altitude (V-G-H) recordings. Structural performance validation of new aircraft includes full-scale static and fatigue testing. Similar tests and teardowns are conducted on older airframes retired from service, to gain better understanding of corrosion and/or accidental damage, as well as structural repairs and service-caused defects. Boeing has, in addition, conducted fleet surveys over the last 10 years to get firsthand knowledge of structural performance of aging jet transport systems and structures. This paper describes these structural health monitoring approaches.

OVERVIEW

Acquisition of structural performance data is necessary for continuous improvement of jet transport structures, which have achieved a credible safety record over the last four decades (figure 1). This has been achieved by diligent fleet monitoring, by inspections and preventive repair actions, and by continuous exchange of information between manufacturers, operators, and regulatory agencies.

Specific structural health monitoring devices can provide valuable insights of performance characteristics. Basic V-G-H recordings have provided the foundation for loading spectra used to determine limit load gust and maneuver criteria. Discrete and power spectral gust spectra have evolved from V-G-H recordings to characterize repeated load spectra for fatigue damage assessments and damage growth predictions. Early attempts to develop fatigue damage gages have met with limited success, primarily because of difficulty in predicting critical data. More sophisticated strain history recorders that complement V-G-H recorders on military airplanes have proved valuable for fleet usage variations and impact on damage accumulation. Recent advances in flight data recorders on commercial jet transports offer opportunities to monitor and record several key usage parameters more effectively. Such data is used to estimate the impact on fatigue damage

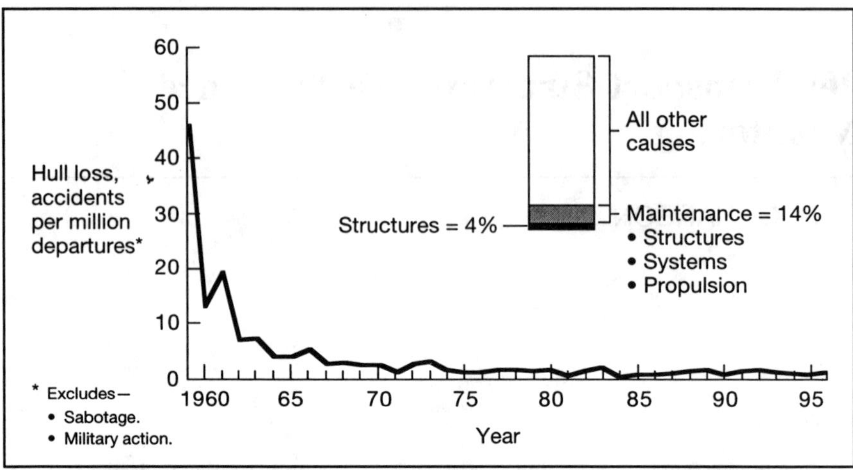

FIGURE 1. SAFETY RECORD—WORLDWIDE COMMERCIAL JET FLEET

accumulation and/or damage growth rates at several control points on wing, fuselage, and empennage structures. An intermediate monitoring approach involves parametric flight segment analyses and input of specific flight data parameters by the flightcrew for each mission. This approach has been employed for the E-4 Command Post and Air Force One, both 747 derivatives. Specific usage parameters then allow processing of mission data together with parametric flight profile data to derive commercial fleet inspection intervals adjustment factors.

Individual airplane tracking is occasionally performed for commercial airplanes. In the past, substantial V-G-H data were processed by NASA, and more recent activities are progressing with modern digital flight data recorders under FAA-sponsored research projects. The commercial jet transport fleet operates in a relatively stable operational environment although similar models may have substantially different flight lengths. Durability design criteria require independent assessments of short, medium, and long flight for a minimum of 20 years' operation to accommodate anticipated service usage variations. Inspection intervals are based on typical material spectrum data for the medium flight. This is sufficient for fleet monitoring of structure with no reported previous cracking. Once damage has been detected, preventive fleet inspections and repairs are focused on individual airplane safety monitoring, and adjustment of inspection thresholds and/or inspection intervals must reflect relevant usage parameters. In summary, planning of inspection programs based on past fleet monitoring usage can be relevant for inspection thresholds. Future usage should be the foundation for inspection interval selection, which is not always easy to predict from past usage recordings.

Major Boeing efforts have been focused on capturing structural performance based on fleet surveillance, testing, and teardowns of new and retired service airplanes. This type of structural health monitoring gradually leads to substantial performance improvements but requires diligence and determination beyond short-term sampling by special devices. Both approaches complement each other by providing validation of key parameters influencing structural performance projections for new and aging airplanes.

This paper provides fundamentals of the structural improvement process based on technology standards reflecting test and service performance. Examples of specific test and fleet monitoring approaches are also given.

STRUCTURAL PERFORMANCE VALIDATION

Accumulation of structural performance data is a major element of the structural improvement process. Such data comes from full-scale testing, teardown inspections of new and retired service airplane structure, fleet survey programs, and provides validation of the design for structural safety as well as for economic service performance.

TECHNOLOGY STANDARDS

Consistent application of durability and damage tolerance technology standards across airplane models is necessary for continuous structural performance improvements that reflect test validation and service experience. Interacting technologies employed in the evaluation of long-life jet transport structures have been available in various degrees for over 35 years but have generally been applied to aircraft analyses by a limited number of specialists using nonuniform methods. Performance demands, increasing structural complexity, aging fleets, and multiple airplane programs have required the development and documentation of disciplined Boeing technology standards (figure 2), with the following principal advantages:

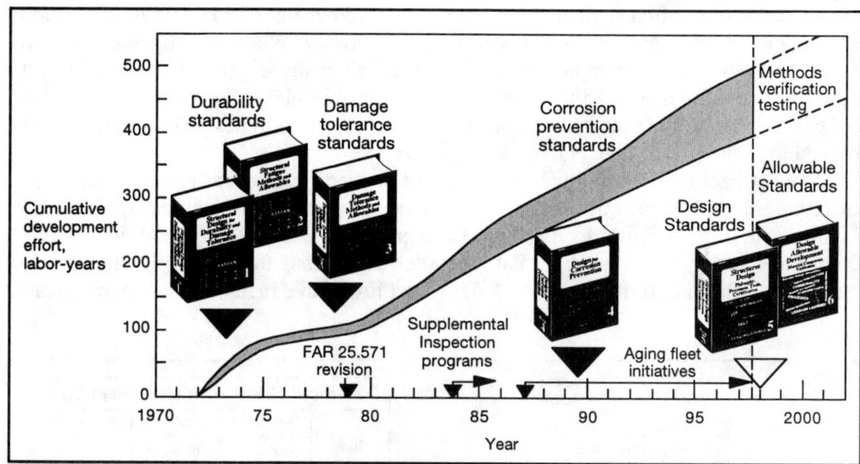

FIGURE 2. BOEING TECHNOLOGY STANDARDS DEVELOPMENT

- Application of common technology by large teams of structures engineers.
- Consistency between multiple airplane programs, sometimes involving great geographic distances.
- Visibility and control of key design and maintenance parameters.

Recent experience with design, validation, and certification of models 777 and 737-700 has prompted continued development of technology standards during 1996 and 1997. Significant amounts of service monitoring and testing have served as verification and validation of Boeing technology standards development. Durability standards were developed first, followed by damage tolerance standards. These technology standards were incorporated into the designs of the second generation of Boeing jet transports, the 757 and 767, and utilized in the certification of damage tolerant structures in compliance with the U.S. Federal Aviation Administration's (FAA) Regulations. Damage tolerance standards were also used in the development of the Supplemental Structural Inspection Documents for aging Boeing 707, 727, 737, and 747, as well as recent models 777 and 737-700.

Since the early 1970s, corrosion has been recognized as one of the dominant factors in the inspection and maintenance activities of airline operations. Boeing has devoted extensive resources to the technology standards development in the areas of corrosion prevention and control. Expanded corrosion coverage, as a result of the corrosion standards development, has been incorporated into the production lines of all current production airplanes as well as the aging fleet structures working groups. These groups include representatives from airframe manufacturers, airline operators, and regulatory agencies.

The Boeing commercial jet fleet can be visualized as a large group of major components loaded in real-life environments, and can therefore be used to obtain service-demonstrated fatigue performance. This fleet represents a database of over 6,000 delivered airplanes, with a total fleet experience exceeding 150 million flight cycles, and daily utilization exceeding 24,000 flights.

Where a statistically significant number of fatigue cracks have been reported in a fleet, maximum likelihood estimates of the Weibull shape and scale parameters are used to determine fleet-demonstrated fatigue lives. This provides a means of relating service experience for one model to other models with different utilization characteristics. Significant fleet findings, often augmented by extensive teardown inspections, are used to modify the fatigue methods and allowables described previously. When no fatigue cracks have been observed, a simpler approach based on the design shape parameter is used to estimate service-demonstrated lives. Such information provides a fundamental check and balance for the fatigue analysis system, and new design and/or redesign evaluations can be related to accumulated fleet performance (figure 3).

Standardization of operating load conditions between different airplane models is necessary to achieve the benefits of fleet performance monitoring and rational application of lessons learned. Operating load spectra have gradually evolved from traditional V-G-H recordings and by sporadic flight data recorder expansions to capture significant flight parameters. Recent efforts by CAA, NASA, and FAA have included more sophisticated

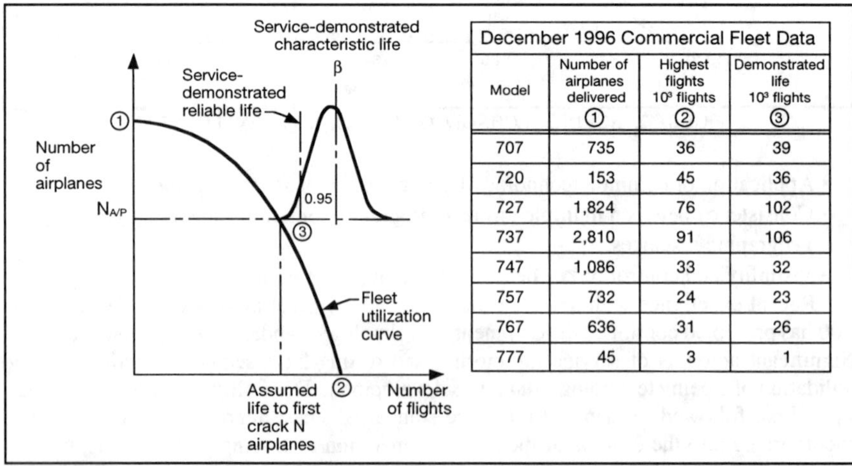

December 1996 Commercial Fleet Data

Model	Number of airplanes delivered ①	Highest flights 10^3 flights ②	Demonstrated life 10^3 flights ③
707	735	36	39
720	153	45	36
727	1,824	76	102
737	2,810	91	106
747	1,086	33	32
757	732	24	23
767	636	31	26
777	45	3	1

FIGURE 3. SERVICE-DEMONSTRATED FATIGUE LIVES

digital flight data recorders for validation of historical commercial jet transport load spectra. While these activities are progressing under FAA-sponsored research programs, several valuable lessons have been learned. Low-altitude turbulence intensity levels have typically increased relative to previous V-G-H recordings by up to 10% in terms of discrete gust velocity. Contrary, cruise altitude turbulence levels have decreased by up to 30%, most likely due to better airway traffic control and communication as well as weather surveillance capability improvements. Figure 4 shows some examples of discrete gust spectra changes based on digital flight data recordings.

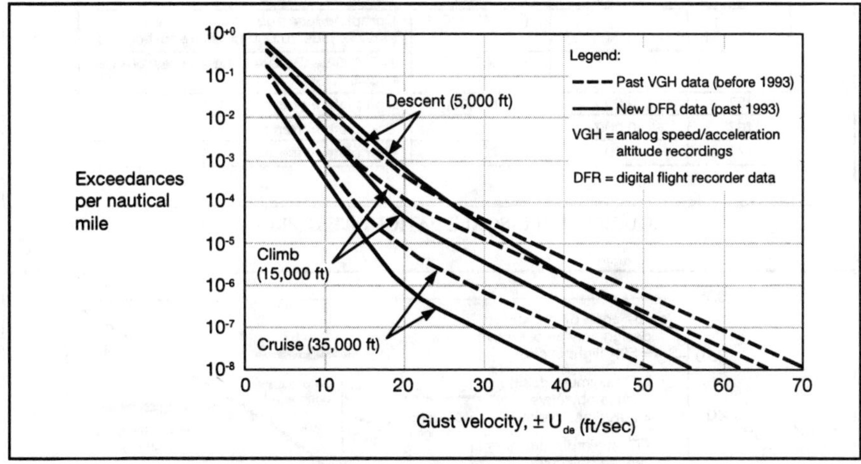

FIGURE 4. COMPARISON OF GUST EXCEEDANCES

FULL-SCALE FATIGUE TESTING

Full-scale fatigue testing of airplanes is a major part of Boeing structural performance acquisition. In addition to providing the validation of aircraft design concepts, full-scale fatigue testing is used to identify any preventive maintenance actions for the fleet. Figure 5 shows the Boeing airplane model, the minimum design service objective (DSO) in flight cycles, and the full-scale fatigue testing in flight cycles. Full-scale fatigue testing is generally accomplished to twice the minimum DSO.

The second-generation 757 and 767 were tested to twice their respective DSOs in flight cycles and improved technology allowed this testing to be completed in less time than it took to test the first-generation 747 to its one DSO in flight cycles. More significantly, the design changes identified in the 757 and the 767 fatigue testing in two DSO flight cycles are far fewer than the design changes identified during the 747 fatigue testing. This improvement was possible because durability technology standards were incorporated into the original designs of both the 757 and 767, (figure 6).The 777 full-scale fatigue test was recently completed with 120,000 cycles, equivalent to 60 years of service, applied around the clock for 26 months. Figure 7 further illustrates continued structural improvements by comparing fatigue test findings for models 767 and 777 after the equivalent of 40 years' simulated service usage. A reduction by approximately 60% was achieved for the model 777 relative to the model 767 certified over 15 years ago.

Test validation of the 777 horizontal stabilizer (similar in size to a model 737 wing)

Airplane	Minimum design service objective*	Fatigue test cycles	Remarks
707	20,000	50,000	Fuselage Hydro-fatigue test
727	60,000	(a) 60,000	Complete airframe
		(b) 170,000	Complete fuselage 47,000 cycles in service, plus 123,000 pressure test cycles
737	75,000	(a) 150,000	Fuselage section/pressure and shear
		(b) 129,000	Complete aft fuselage 59,000 cycles in service, plus 70,000 pressure test cycles
747	20,000	(a) 20,000	Complete airframe
		(b) 40,000	Complete fuselage 20,000 cycles in service, plus 20,000 pressure test cycles
		(c) 60,000	747-400 sections 41 and 42 pressure test cycles
757	50,000	100,000	Complete airframe
767	50,000	100,000	Complete airframe
777	40,000	120,000	Complete airframe

* 20 years

FIGURE 5. FULL-SCALE FATIGUE TEST PROGRAMS

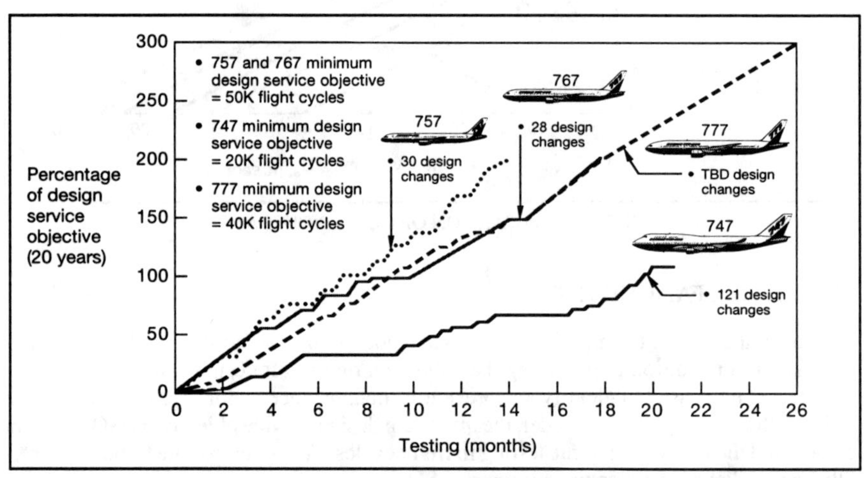

FIGURE 6. MAJOR AIRFRAME FATIGUE TESTS

made of carbon-fiber-reinforced plastic (CFRP) included fatigue testing to 120,000 cycles, followed by several ultimate load tests for several loading conditions. Acoustic emission (AE) technology was employed to ensure successful test sequence completion before the planned final destruction test.

Twenty-two channels of two Physical Acoustics Corporation (PAC) AE systems were used to monitor the stabilizer ultimate load and destruction tests. The Boeing-developed Prefailure Warning System was used to monitor the tests in real time. The second system, a commercially available PAC SPARTAN system, was used to acquire data that could be analyzed following the tests to extract additional information about any damage propagation. Up to 26 PACR15I, 150-kHz resonant AE sensors with integral preamplifiers were attached to the stabilizer with hot-melt adhesive. For some of the load conditions, sensors on opposite sides of the centerline splice were connected to the same channel, effectively

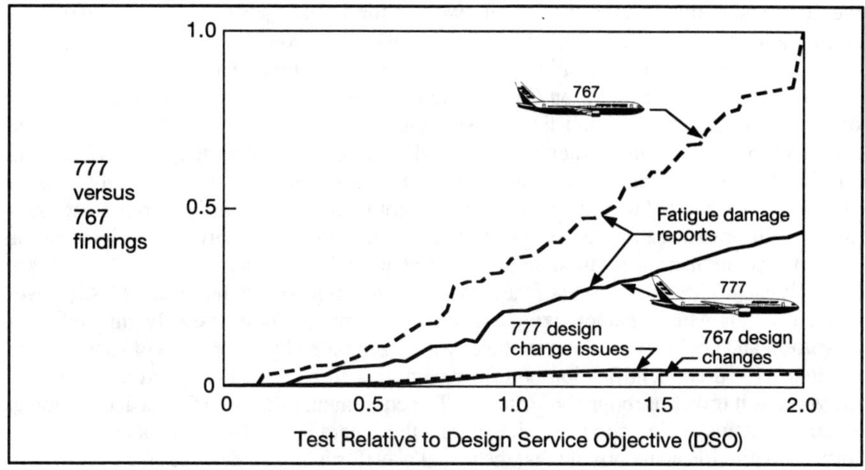

FIGURE 7. FATIGUE TEST PERFORMANCE COMPARISON

becoming one sensor monitoring AE activity in the vicinity of the splice without the capability of determining which side of the splice might be acoustically active. The sensor arrangements used for the 777 stabilizer ultimate load and destruction tests are shown in figure 8 for a specific load condition.

Acoustic emission data were acquired from the 777 stabilizer during three applications of design ultimate loads and the destruction load case. The acquired data were used to determine whether any subcritical damage propagation occurred during any of the load conditions and to help define the failure sequence. No significant activity was observed during the ultimate load conditions, indicating that the stabilizer was structurally sound and not damaged by the applied loads. During the destruction load condition, the stabilizer generated AE once design ultimate load was reached. Most of the AE activity was from the pivot fittings and center splice; this can be attributed to load redistribution around

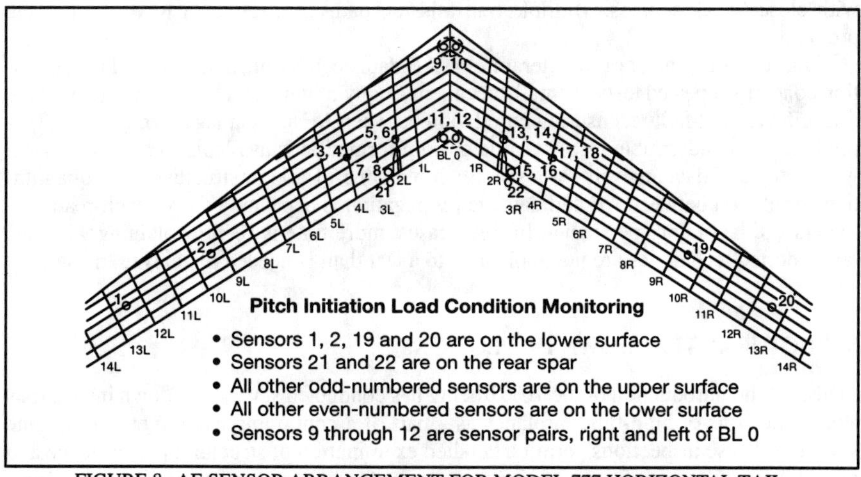

FIGURE 8. AE SENSOR ARRANGEMENT FOR MODEL 777 HORIZONTAL TAIL

the fasteners. Failure occurred at about 194% of limit load (or approximately 165%, accounting for temperature and humidity influences). There were no indications of damage propagation until just before failure, and these were concentrated around the failure site.

The successful use of AE on CRFP structures has not been paralleled in applications on aluminum structures. Acoustic Emission (AE) has the potential to detect and locate cracks in metal structures. Attempts to use AE in this capacity on the 727, 747 SR and 747-300 forward section pressure fatigue tests have been unsuccessful, primarily because possible AE generated by fatigue cracks could not be reliably separated from noise associated with pressurization, either pneumatic or mechanical. In part, this failure can be attributed to the instrumentation employed to acquire the AE data.

"Classical" AE data consist of various parameters intended to represent the AE waveform. The parameters characterizing a given waveform are not necessarily unique to that waveform and could be the same for other waves generated by other types of sources. The problem of source identification is further complicated by attenuation and dispersion of the wave as it travels through the structure. The equipment used for AE data acquisition at Boeing contains analog circuits which extract these parameters from the incoming waveforms and provide no record of what the actual waveform looked like, making quantitative analysis of the data very difficult.

This confirms that the application of AE technology for defining areas of aircraft structures that require either repairs or detailed inspection is not as straightforward as one would be led to believe by the vendor representatives. The variability of structural details (including sealant thickness, rivet tightness, bolt tension, presence or absence of repairs, and design modifications) contributes to the problem.

Acoustic emission analysis can be a valuable tool for determining the condition of a structure under load, as described above. However, on large complex structures, it has proved to be ineffective for precise AE source location or identification of damage mechanisms, for several reasons. Providing coverage of a large area with limited numbers of sensors forces the sensor array density to be just high enough to detect any reasonably high energy source, meaning that most of the AE sources will be seen by only one channel. This situation precludes calculating source locations using arrival times of the AE waves at multiple sensors. Furthermore, any analysis based solely on the characteristics of the waveform envelope is apt to confuse energetic sources far from the receiving sensor with closer weak sources. Multiple transmission paths and reflected waves add to the problem.

Unfortunately, most of the literature on AE data acquisition, analysis, and interpretation concerns repeated tests of small specimens, many of which were designed to produce a specific type of failure, ensuring correlation between AE and damage mechanisms. This produces two undesirable results. Little or no information is available to help with hardware setup and data analysis for tests of complex one-of-a-kind structures, and quantitative correlation between AE and damage propagation is expected by even well-read customers and laboratory personnel. In some cases, more time is spent explaining why certain reported techniques are not applicable to a test than is needed to analyze the data.

STRUCTURAL TEARDOWN INSPECTIONS

Since the introduction of the 707, Boeing has conducted several teardown inspections and evaluations of high-time airplanes as a part of a continuing assessment of airplane structure. These inspections permit a detailed examination of structural performance and provide much useful information for forecasting future structural maintenance requirements. Sophisticated inspection techniques, capable of finding smaller cracks than typically found during routine airline inspections, are used on the disassembled structure. Teardowns also provide an excellent database for calibrating analysis tools and developing structural modifications on future production airplanes, if required. Major teardown inspections supplementing normal fleet surveillance activities have been conducted on several Boeing models:

- 707 wing plus center section 1965
- 707 wing 1968
- 707 wing plus center section and fuselage 1973
- 707 empennage 1978
- 727 forward fuselage 1978
- 737 wing plus center section, forward fuselage, and empennage 1987
- 737 aft fuselage 1988
- 747 wing and empennage 1989
- 747 fuselage 1991
- 727 wing and empennage 1994
- 727 fuselage 1996
- 777 wing, fuselage and empennage 1997

Concerns related to an increased number of airplanes being used beyond their original design life objectives have spurred further activities to obtain airframes retired from service for teardown inspections. Boeing will continue to monitor the structural performance of the aging fleet to verify the effectiveness of preventive modifications incorporated as retrofit on older models and/or new model production improvements. Findings will be disseminated to operators by service bulletins, as required, and incorporated in maintenance recommendations issued by Boeing.

FLEET SURVEYS

Reported service data and other firsthand information from customer airlines are continually reviewed to promote safe and economic operation of the worldwide fleet. As a result of increasing costs and delivery times for airplane replacements, the active service usage for commercial airplanes has gradually been extended. The average age of the world airline jet transport fleet has increased from 8 years in 1980 to about 15 years today. This upward trend in airplane operating age is likely to continue at a slower pace. Boeing design goals have traditionally been established for a conservative number of flights over a 20-year period. This timeframe seemed more than adequate in the 1960s to carry these airplanes throughout their passenger-carrying years. Today there are more than 1,200 Boeing airplanes around the world that are over 20 years old. Approximately 150 airplanes in an active Boeing fleet exceeding 6,000 have reached their design flight-cycle goals in 1997.

Boeing implemented the fleet survey program 10 years ago to better understand the condition of older airplanes. The success of the initial program resulted in expansion to include new production airplane models such as the 737-300, 757, and 767, some of which have seen over 15 years of service. By reviewing the structural and systems performance of these airplanes during their early and maturing years, it is believed that Boeing and the operators can take the right actions to preclude the majority of aging fleet problems for these models.

The specific objectives of these surveys have been:

- Models 707, 727, 737, and 747:
  - Gain an engineering assessment of the condition of older airplanes, with emphasis on structures and systems.
  - Observe the effectiveness of Boeing corrosion prevention features and other corrosion control actions taken by operators.
  - Acquire additional fleet data to improve maintenance recommendations and design of new airplanes.
- Models 737-300, 757, and 767:
  - Avoid large-scale aging fleet problems beyond design service objectives.
  - Gain an early identification of problems experienced in service, particularly for new model features.

Boeing survey teams are dispatched to observe airplanes during scheduled heavy maintenance checks. Upto six experienced structures, systems, maintenance, and service engineers record their observations in survey documents covering up to 350 structures and 150 systems items. Typically, 70% of the items are surveyed because access is not available to all areas during the visit. Each survey includes a review of airline practices regarding airplane use, maintenance program, and dispatch reliability. The operators are briefed by the teams on their findings.

Airline acceptance of the program and cooperation with Boeing have been positive. Observing a significant number of airplanes in a variety of operating and climatic environments around the world provides a composite sample of each model and a better understanding of common and unique aspects within and between model fleets. As of July 1997, 195 airplanes owned by 100 operators have been visited in 49 countries around the world (figure 9). Although the number of airplanes observed is a small percentage of the total fleet, it does represent about 10% of the high-time airplanes.

All significant findings pertaining to a specific visit are recorded and assigned to appropriate organizations for necessary fleet action. The collected data have been pooled for fleet evaluations to determine if there are trends requiring additional actions by Boeing and/or the operators. A number of detailed action items have resulted from the surveys,

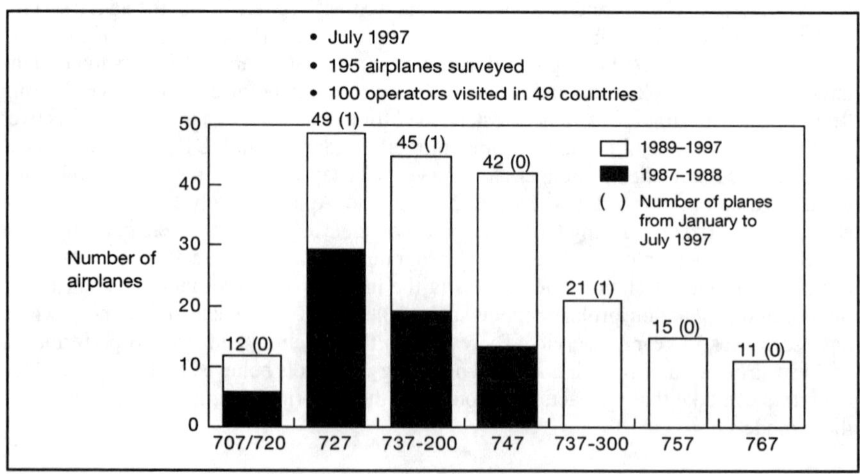

FIGURE 9. BOEING FLEET SURVEYS

and their applicability has been reviewed across all Boeing models. To ensure anonymity, all identifiable operator/airplane data are treated as confidential.

Findings over the last 5 years indicate that the condition of the airplanes generally is good and that they are receiving adequate and conscientious maintenance. Most discrepancies noted by the Boeing teams had already been recorded by the operator and corrective action was under way. Recent surveys have confirmed our belief that the aviation industry is more aware of the necessity of applying timely corrosion control and installing well-engineered repairs and, generally, has a more heightened awareness relative to sound maintenance practices.

SUPPLEMENTAL STRUCTURAL INSPECTION PROGRAMS (SSIP)

The continued structural airworthiness of aging transport airplanes has been the subject of considerable discussion among experts in the industry and airworthiness authorities. Attention has been focused on the adequacy of inspection programs for timely damage detection in support of fail-safe design practices during the last four decades.

CAA Airworthiness Notice 89 and FAA Advisory Circular 91-56 allow structural reassessments, including recommendations for any necessary supplemental structural inspections, as an alternative to imposing service life, operational, or inspection limitations. The resulting structural inspection program supplements existing operator maintenance programs that generally have been very effective in maintaining the inherent damage tolerance of fail-safe jet transport structures. Model-specific documents provide inspection options for selected structurally significant items (SSI) that will ensure timely detection of fatigue damage in older airplanes. This is achieved, where required, by use of more refined inspection techniques within the operator's existing scheduled maintenance program, or by supplemental inspections. These inspection requirements are applicable to those high-flight-cycle airplanes termed candidates because these are the most likely to experience the earliest fatigue cracking in the fleet (figure 10). When cracking is detected and reported, necessary action is taken to safeguard the total fleet. Reported information is used to establish a threshold; an inspection method; and repeat intervals to find all discrepancies in the fleet. Several structural item interim advisories (SIIA) and service

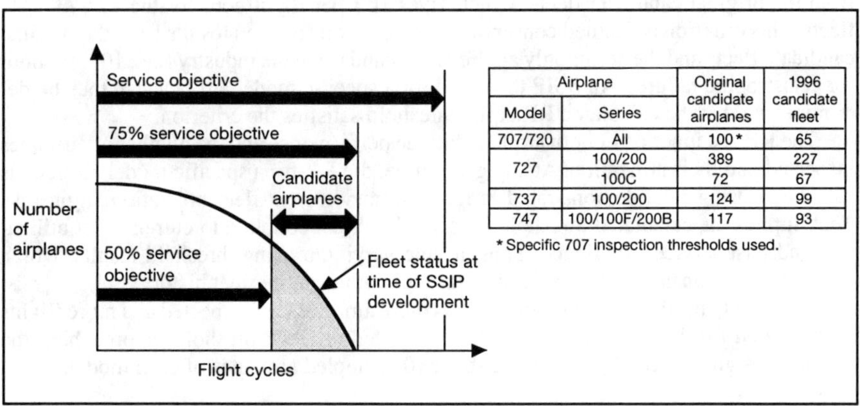

Airplane		Original candidate airplanes	1996 candidate fleet
Model	Series		
707/720	All	100*	65
727	100/200	389	227
	100C	72	67
737	100/200	124	99
747	100/100F/200B	117	93

* Specific 707 inspection thresholds used.

FIGURE 10. SUPPLEMENTAL INSPECTION CANDIDATE AIRPLANES

bulletins have been issued to ensure proper attention in the total fleet based on discrepancies that have been found. Supplemental structural inspections augment existing inspection programs and provide a valuable contribution toward maintaining fleet safety.

The supplemental structural inspection documents (SSID) for models 727, 737, and 747 were released in 1983 and contain usage procedures, lists of candidate airplanes, SSI information with example maintenance program requirements, and a discrepancy reporting procedure. The SSID for model 707 was first released in 1979 but is less sophisticated in terms of the inspection options that are provided. SSID inspections have been mandated on selected 727, 737, and 747 airplanes since 1984, and on selected 707 airplanes since 1985.

The SSIDs have been updated on a regular basis to reflect service experience and operator inputs. In the light of current aging fleet concerns, these inspection programs have recently been reviewed thoroughly to ensure adequate protection of the aging fleet. The major focus of these reviews was on:

- Inclusion/deletion of principal structural elements (PSE).
- Adequacy of the present fleet leader sampling.

The candidate airplanes were intended to be the sole group of airplanes covered by the SSIPs. However, through the years some airplanes have accumulated more flights than the candidate airplanes and have become fleet leaders while some candidate airplanes have been retired from service. This posed the necessity of replacing retired and low-utilization candidate airplanes with new fleet leaders in SSIPs for each model. Thus, a shift of paradigm has evolved from the candidate fleet concept to a threshold with full-fleet sampling.

To adapt to this change, Boeing has developed a new concept for deriving inspection thresholds that will ensure a high reliability for airplanes below the threshold while maintaining continuing airworthiness for airplanes exceeding the threshold through SSIP inspections. Normally, once cracks have been found in the fleet, fleetwide corrective actions will be taken that will safeguard the airplanes below the threshold as well. Therefore, the center of this new approach is to estimate the risk of having cracks in an airplane below the threshold while no cracks can be found in the SSIP-inspected airplanes. This risk estimate is calculated by the risk rate (RR) of having cracks at the threshold, which is an upper boundary for all uninspected airplanes.

The RR criterion for SSIP threshold has been derived based on service experiences from the original candidate fleets, which gives 10^{-7} per flight-hour for the 727/737/747 fleets. This criterion is deemed conservative in light of inspection results from the original candidate fleets and the commonly applied risk standard in the industry (i.e., 10^{-9} per hour for catastrophic failure). An SSIP threshold for a specific model airplane can thus be determined so that the calculated RR at the threshold satisfies the criterion.

The RR is a function of both airplane fatigue performance and the number of airplanes inspected and their utilization. At any given time, the RR for a specific model airplane is calculated using fleet-demonstrated fatigue life and current fleet utilization (figure 3). That implies the calculated threshold is robust by direct coupling to current fleet utilization and past service experience. Thus, a time varying/moving threshold results, which will provide a consistent fleet reliability throughout the entire fleet history.

In general, this threshold will increase as more airplanes are inspected and more flights accumulated for these airplanes. This has translated to 1997 thresholds approaching the original design service objectives and about 100 sampled airplanes of each model.

STRUCTURAL IMPROVEMENT PROCESS

The Boeing structural improvement process has three essential elements:

- Test and service experience.
- Technology standards developments.
- Implementation of design lessons learned.

The structural improvement process is a continuous living process at Boeing and, as improvements are identified, they are implemented in the next possible production-line number. The first generation of Boeing jet transports, designed in the 1950s and 1960s, includes the 707, 727, and the earlier production models of the 737 and 747. The second generation of Boeing jet transports comprises the 757, 767, and the current production models of the 737 and 747. The third generation of Boeing jet transports includes the 777 and the 737-NG (Next Generation). As design upgrades are developed, they are incorporated into the production of these airplanes, as appropriate. Figure 7 further illustrates continued structural improvements by comparing fatigue test findings for models 767 and 777, which shows approximately a 60% reduction in fatigue damage reports.

Another metric for the effectiveness of design improvements is "maintenance labor-hours per airplane to address corrosion and fatigue for the first 10 years of operation." In figure 11, this metric is compared for standard and widebody airplanes. This figure illustrates two points:

- The maintenance labor-hours to address corrosion and fatigue decrease with increase in production-line number.
- The second-generation 757 and 767 airplanes have much better corrosion and fatigue performance, compared with the first-generation airplanes.

Worldwide airplane fleet support is an important objective for Boeing. Structural design technology improvements have a vital role in accomplishing this objective. Current airplanes are in service longer than their original minimum DSO. It is expected that this trend will continue for some time in the future. Structural design technology improvements provide rational means to achieve this increased DSO for the operators by developing, under the auspices of the structures working groups, supplemental inspections, ag-

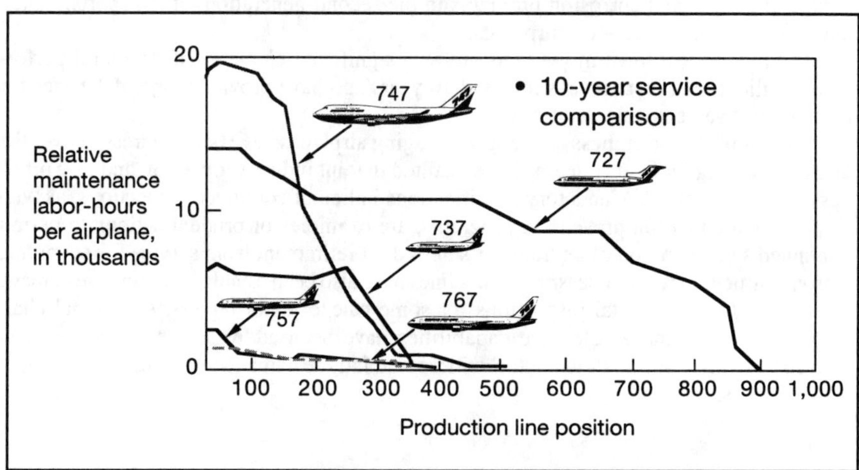

FIGURE 11. SERVICE BULLETIN LABOR-HOURS COMPARISON (STANDARD AND WIDE BODY)

gressive corrosion prevention and corrosion control programs, and any required structural modifications. A summary of Boeing structural design improvements in support of the airplane fleet is shown in figure 12.

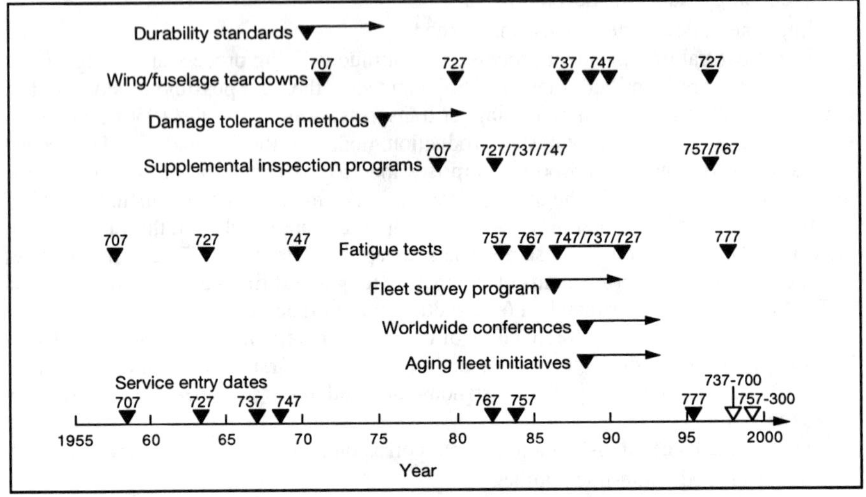

FIGURE 12. BOEING FLEET SUPPORT ACTIONS

CONCLUSIONS

Structural performance monitoring is a major element of design, manufacturing, and safe operation of commercial jet transports. Incorporation of lessons learned requires persistence and cross-model surveillance to ensure continuous improvements. Implementation of disciplined technology standards has resulted in order-of-magnitude reductions in structural fatigue and corrosion problems in the second-generation jet transports, which have also benefited derivative airplanes.

Testing and teardown inspections are also significant elements of structural performance validation. Fleet surveys initiated 10 years ago have provided large databases for design improvement considerations.

Continuing airworthiness challenges for aging airplanes have been addressed over the last 15 years. Aging fleet concerns have resulted in joint industry, operator, and airworthiness authority actions. Mandatory modifications in lieu of continued inspections, as well as mandated corrosion prevention programs, are examples of prudent actions to permit continued safe operation of jet transports until their retirement from service for economic reasons. Structural repair assessment guidelines have also been established to ensure damage detection by supplemental inspections for some categories of repairs. Additional challenges of local damage tolerance capabilities have been addressed in recent years to establish positive initiatives to control widespread fatigue damage affecting on continuing airworthiness.

ACKNOWLEDGMENTS

The author is greatly indebted to several Boeing structural specialists for their contributions. Special thanks are extended to the Structures Laboratories and the Technology Standards staff. The contributions by Ronald Slaminko, William T. Hardrath, Lisa A. Brunet, Cliff Chen, and J. Russ Horsfall with the compilation of illustrations and manuscript reviews are significant and truly appreciated. Coordination of graphics support by Wayne J. Brewer and Floyd A. Stewart made completion of the paper possible amidst other equally pressing business priorities.

REFERENCES

Goranson, U. G. "Structural Airworthiness Initiatives." Vice-President Al Gore's Commission on Aviation Safety and Security in the 21st Century. George Washington University, Washington, D.C. January 1997.

Connectionist Algorithms for Structural System Identification and Anomaly Detection

D. C. HYLAND[1]

ABSTRACT

This paper discusses a neural network architecture for system identification and adaptive control and its ramifications for structural system identification and anomaly detection. After reviewing the basic Adaptive Neural Control (ANC) architective, we discuss elaborations that permit rapid failure detection, structural damages location and autonomous control system recovery.

1. INTRODUCTION

The development of the Adaptive Neural Control (ANC) architecture by the author and associates over the past several years is part of an effort to develop neural network based controllers capable of self-optimization, on-line adaptation and autonomous fault detection and control recovery. The ANC architecture is a set of building blocks and rules for combining them so as to achieve massively parallel and decentralized processing for identification and adaptive control of dynamic systems. A sequence of papers[1-7] have outlined the ANC scheme and reported numerous simulation and experimental results.

First, we describe the more or less direct antecedents of ANC in the literature of 'connectionist' or 'artificial neural network' systems. Of course, the foundation of this work is the basic neuron model of McCulloch and Pitts[8], the early decentralized adaptation algorithm of Hebb[9] and the basic multi-layer feedforward structures defined by Rosenblatt[10-13]. Given this basis, the ANC approach utilizes gradient based learning schemes. Subsequent to the advances of Widrow and Hoff[14-16] wherein the gradient of the instantaneous error is used, the more general backpropagation algorithm was invented by Werbos[17] and independently developed by Parker[18], LeCun[19], Rumelhart, Hinton and Williams[20,21] and Rumelhart and McClelland[22]. It will be seen below that the ANC architecture involves a form of local backpropagation. There are various approachs to the application of backpropagation systems to temporal processing (of time series signals). These may be epitomized by the time-delay neural networks of Lang and Hinton[23] or Waibel et al.[24] and the finite-impulse response (FIR) multi-layer perceptrons of Wan[25]. ANC is more closely related to the latter approach. Associated with this scheme are several basic training algorithm approaches: "backpropagation through time" as explained by Werbos[17], Rumelhart et al[22] and Werbos[26], the "temporal backprop" of Wan[27,28] and the "real-time temporal learning" approach of Williams and Zipser[29]. The latter approach, to which ANC bears the closest resemblance, combines multilayer

[1] Department of Aerospace Engineering, The University of Michigan, Ann Arbor, MI

perceptrons with tapped delay lines and trains the network using the "temporal learning algorithm". This algorithm requires the use of auxiliary "sensitivity systems" to compute gradients. This scheme was applied by Narendra and Parthasarathy[30] to treat dynamic system identification and control. In contrast, the ANC approach decentralizes computations associated with both the forward signals and the backpropagated signals and adaptation occurs locally in each synapse. Because of the "local backpropagation" arrangement, no separate sensitivity systems are required.

In addition, as is discussed below, there are further novel features of ANC. One of the most important of these is the time varying rule for assigning the adaptive speed coefficient. The time-varying adaptive speed utilized in ANC greatly simplifies the investigation of convergence and allows a guarantee of convergence under certain broad conditions.

In the following section, we describe the hierarchy of ANC components and how they are used to build up more complex systems.

2. ANC ARCHITECTURE DESCRIPTION

The hierarchy of modular structures that compose the ANC architecture is depicted in Figure 1. In the order progressing from basic constituents to higher-level modules, the modular structures include individual neurons which are interconnected by synapses and dynamic arrays which are in turn, grouped into replicator networks, one or more of which may form an adaptive neural control system.

The lowest level of the hierarchy as shown in Figure 1 contains three devices: an individual *neuron*, an individual *synaptic connector* and a tapped delay line.

Unlike the one-way neurons usually postulated in neural net applications to adaptive control. (e.g. Williams and Zipser[29], Narendra[30]) the neuron defined here is inherently a two-way device with a *forward* signal flow path and a *backward* signal flow path. The neuron also receives a forward path bias input and a backward path input. Signals along the forward path are operated on by the neural function, which may be a sigmoid function. Signals along the backward path are multiplied by the derivative of the neural function. The forward and backward paths receive input signals from other neurons via synaptic connectors. On the forward path, during each computational cycle, the neuron sums the forward path inputs from the synaptic connectors and adds a forward path bias signal to form signal u_k. The bias signals are used to connect the system with the outside world, e.g. the forward bias might be a training stimulus injected to drive the learning process in the overall network. Signal u_k is then passed through the activation function linearity, $g(x)$, (called the *activation function*) to form the forward path output of the neuron, x_k.

Fully localized computational capability is provided by the backward path operations that are executed simultaneously with the forward path. The neuron sums the backward path bias inputs (which may represent error signals injected into the network) and the backward path inputs. This signal is then multiplied by $g'(u_k)$ to form the backward path output signal y_k. Thus, the neuron has a pair of inputs and a pair of outputs, each member of the pair being associated with signals flowing in opposite directions.

The capability to adapt over the time actually resides, not in the neuron, but in the *synaptic connectors* (or simply synapses) which are elements that serve to link several neurons together. A synaptic connector is also an inherently two way device with both forward and backward inputs and outputs. On a forward signal path the synapse merely multiplies the input x by a real number, $W(n)$ to produce an output y. $W(n)$ is the *synaptic weight* or, more simply, the "weight" of the connector. On a backward path, the same kind of operations occur: $\bar{x}$ is multiplied by $W(n)$ (the same weights as the forward path) to produce the backward path output.

Essentially the synapse is analogous to a bi-directional cable with the same resistance in both directions. In addition the synapse may also adjust its own weight. For example, the change in the weight from time n to time $n+1$ may be simply proportional to the product of the forward path input to the synapse and the backward path input to the synapse. This update rule, is purely decentralized or localized since it uses only information available to the individual synapse.

The constant of the weight change rule, $\mu(n)$, is called here the *adaptive speed* since it governs the rapidity of learning. As discussed below, the *adaptive speed* is not necessarily a constant and may be updated (using local information) so as to guarantee convergence in both system identification and adaptive control.

Systems built up of the neurons and synapses defined above are adequate to address *static* pattern classification and nonlinear mapping problems. By virtue of the backward signal path defined at the most fundamental level, backpropagation of error and adaptive learning capabilities are built in. Moreover, the learning capability is totally localized and decentralized — separate "sensitivity systems" and weight update computations are not needed. Each subdivision of a neural network composed of the elements defined above has a localized learning capability. This feature confers a high degree of fault tolerance — i.e. with a completely decentralized learning architecture, failure or damage to a subset of neurons will result in the remaining neurons eventually assuming the functions of the damaged components.

However, there remains the problem of applying networks of this type to tasks involving dynamic systems with time-varying input and output signals. This problem is addressed by the second member of the hierarchy shown in Figure 1. The key to applying the neurons and synapses defined here to dynamic system identification is to organize the neurons into larger building blocks — termed the dynamic arrays. Arrays may be connected by *Toeplitz synapses*. An array is a "stack" of neurons such that neurons within the same stack are not connected, but are (at most) only connected to neurons in some other stack. Basically, this organization into stacks introduces a temporal ordering into the network: the position of a neuron in the array indicates the "age" of the data (relative to the present instant) it is meant to process. A neuron that is further from the top of the stack represents a time instant that is further into the past.

In general, two such stacks of neurons can be interconnected with any number of connections. However, it is desirable to impose connectivity constraints such that the neuron k places from the top in one array receives (forward path) signals only from the neurons in another stack that are *k places or more* from the top. Any bundle of synaptic connectors obeying this constraint is termed a *Toeplitz synapse*. The weights of this bundle can be represented as an upper triangular matrix. Fundamentally, this upper triangular structure is designed to preserve temporal ordering and causality within higher order networks. In other words, by forbidding a top level neuron from one stack from feeding signals into a lower level neuron in another stack, the system is constrained to be capable of modelling only *causal* systems in which the future input signals cannot influence the past output signals.

Any network formed by an arbitrary number of arrays, linked by Toeplitz synapses in any arbitrary pattern (series connections, parallel, closed-loops or combinations thereof) is referred to herein as a *Teoplitz network*. The two top levels of the hierarchy in figure 1 are examples of Toeplitz networks, assuming Toeplitz synapses are being used. Considerable details on the construction of Toeplitz networks for various tasks in system identification and control are given in[1-7]. Here we discuss some of the general properties of Toeplitz networks that are relevant to all applications.

In a typical configuration for a Toeplitz network, an externally provided training signal, ξ is first passed through a tapped delay line so that $\bar{\xi}$ is the basic input to the network. The output of some designated array constitutes the output to the network

network. The output of some designated array constitutes the output to the network and this vector is compared with a reference signal $\overline{\eta}(n)$obtained by passing a reference signal $\eta(n)$ through a tapped delay line. $\eta(n)$ might be the output of some dynamic system with ξ as input. The difference, $\overline{\varepsilon}(n)$, is then injected into the backward signal path of the output array.

The formula for the update of any particular weight matrix can be obtained as a direct consequence of the definitions above. First, we introduce supporting notation. Let * denote the Hadamard product of matrices (element-by-element multiplication). U_0 is a particular instance of a synaptic constraint matrix. In the following, let U_n denote the matrix:

$$U_n \in \mathbf{R}^{L\times L} \qquad (U_n)_{kj} = \begin{cases} 0; j < k+n \\ 1; j \geq k+n \end{cases} \tag{1}$$

Furthermore, the symbols $\hat{g}(\cdot)$ and $\hat{g}'(\cdot)$ denote:

$$\begin{aligned} &\hat{g}(\overline{x}) \in \mathbf{R}^{L}: (\hat{g}(\overline{x}))_\kappa \triangleq g(\overline{x}_\kappa) \\ &\hat{g}'(u) \in \mathbf{R}^{L\times L}: \hat{g}(u) \triangleq \operatorname*{diag}_{\kappa=1\ldots L}\{g'(u_\kappa)\} \end{aligned} \tag{2}$$

where g is the sigmoidal nonlinearity of the neurons and g' is its derivative. Suppose that $W_{\kappa j}$ is the weight matrix of the bundle of synapses that feeds the output of neural array j into the inputs of array κ. Let the outputs of array κ in the forward and backward paths, respectively, be x_κ and X_κ. Then as a consequence of our definitions:

$$x_\kappa = \hat{g}(U_\kappa) \tag{3}$$

$$U_\kappa = \sum_j W_{\kappa j} x_j + A_\kappa \overline{\xi} \tag{4}$$

$$X_\kappa = \hat{g}'(U_\kappa)\left(\sum_j W_{j\kappa}^T X_j + \Omega_\kappa \overline{\in}\right) \tag{5}$$

where;

$$\begin{aligned} A_\kappa &= \begin{cases} 1, \text{ if array } \kappa \text{ is the input array} \\ 0, \text{ otherwise} \end{cases} \\ \Omega_\kappa &= \begin{cases} 1, \text{ if array } \kappa \text{ is the output array} \\ 0, \text{ otherwise} \end{cases} \end{aligned} \tag{6}$$

and where:

$$\overline{\in} = \overline{\eta} - \sum_\ell \Omega_\ell x_\ell \tag{7}$$

Equally obvious from our definitions (see Figure 3) and the above notation is that;

$$W_{\kappa j}(n+1) = W_{\kappa j}(n) + \mu(n) U_o * (X_\kappa x_j^T) \tag{8}$$

where μ(n)>0 (for all *n*) is the time varying adaptive speed. μ(n) is defined specifically below. However, before considering this matter we can readily show the following result:

Theorem 1

For the neural network defined by relations (1)-(8), if we assume that all the vectors x_r and X_r are uniquely determined, then:

$$\frac{1}{\mu(n)}(W_{\kappa j}(n+1)-W_{\kappa j}(n)) = -\frac{\partial}{\partial W_{\kappa j}}\|\in(n)\|^2 \tag{9}$$

Proof: Appendix A, Reference [31].
The above result indicates that the increment in each synaptic weight is proportional to the negative gradient of the square of the norm of the output error.

With regard to μ(n), we first define

$$\mu(n) = \alpha F(n), \quad \alpha > 0 \tag{10}$$

where we term α the *learning rate constant*. In gradient descent schemes, instability in the form of oscillatory divergence can result if the adaptive speeds are too large. In the present approach, it is desired to make μ(n) time varying so that such instability is avoided. In particular *F(n)* provides a scaling of μ such that stability and convergence requirements are reduced to a restriction on α. Under broad conditions, α is a "universal" constant in that it may be chosen once and for all (and built into the system) so as to secure desired convergence rates independently of the detailed characteristics of the training stimuli or of the systems to be identified and controlled. This confers the benefit of stable, autonomous performance.

The specific definition of *F(n)* in (10) is:

$$F(n) = \frac{P(n)}{A(n)} \tag{11}$$

P = "performance function"

$$= L\left(\frac{1}{2}\|\in(n)\|^2 - J^*\right) \tag{12}$$

where

$$L(x) \triangleq \begin{cases} x; & x \ge 0 \\ 0; & x < 0 \end{cases} \tag{13}$$

$$J^* = \text{desired mean-square error level} \tag{14}$$

and

$$A = \text{"neural activity level"}$$
$$= \sum_{\text{all synapses}} (X_\kappa x_\kappa)^2 \tag{15}$$

where x_κ and X_κ denote the inputs on the forward path and on the backward path, respectively, to synapse κ. In particular,

$$A = \sum_{\kappa,j} \left\| U_o * \left(* X_\kappa x_j^T \right) \right\|_F^2 \tag{16}$$

where $\|\cdot\|_F$ denotes the Frobenius norm. In the definition, (11), of $F(n)$, the numerator and denominator are derived from readily accessible signals. The numerator, P, signifies the performance requirement of the overall network in terms of the square of the norm of the output error, $\frac{1}{2}\|\in(n)\|^2$. Note that by virtue of the function chosen, the system is not required to achieve a minimum value of $\frac{1}{2}\|\in(n)\|^2$ but only to attain a value that falls below a desired level, J^*. Thus the network attempts to achieve "good-enough" performance, not necessarily optimal performance.

The above concludes our definition of the kind of modules that are used within the ANC architecture. The next section offers some simple results concerning convergence behavior of Toeplitz networks.

3. Some Convergence Properties of Toeplitz Networks

Here we consider some simple, introductory results on the convergence of Toeplitz networks of the described above and with the defining relations (1-8), (10-16). In particular, we trace the consequences for convergence of the time-varying adaptive speed adjustment, (11).

To begin, note that $\frac{1}{2}\|\in(n)\|^2$ is a function of all the weight matrices $W_{\kappa j}, \kappa, j = 1,\ldots,N_a$ (where N_a is the number of neural arrays) at time n and of the past history; $\{\xi(n-m), m = 0,1,\ldots\}$ of the training stimulus. Alternatively, if we stack all of the nonzero and independent elements of the $W_{\kappa j}$'s in a single vector, $W(n)$; then $\frac{1}{2}\|\in(n)\|^2$ is a function of $W(n)$ and $\{\xi(n-m), m = 0,1,\ldots\}$. We symbolize this by writing:

$$\frac{1}{2}\|\in(n)\|^2 = J(W(n), \{\xi(n-m); m = 0,1,\ldots\}) \tag{17}$$

or, more briefly, $\frac{1}{2}\|\in(n)\|^2 = J(W(n))$, when we wish to emphasize the functional dependence of $\|\in\|^2$on W. In (17) $W(n) \in \mathbf{R}^{N_s}$ where N_s is the is the total number of independent, nonzero, synaptic weights. Clearly, by (17), J is a nonnegative function of $W(n)$, i.e. $J(W(n)) \geq 0$ for all $W \in \mathbf{R}^{N_s}$.

For bounded and convergent response, the crux of the matter is the "shape" of $J(n)$ as a function of $W \in \mathbf{R}^{N_s}$. In order to state the simplest result, recall that $F(x_1, x_2, \ldots, x_N)$ is termed a *homogeneous function of degree M* if for $\beta \in \mathbf{R}$:

$$\begin{aligned} &F(\beta x_1, \beta x_2, \ldots, \beta x_N) \\ &= \beta^M F(x_1, x_2, \ldots, x_N) \end{aligned} \tag{18}$$

By differentiating both sides with respect to β and setting β equal to unity, we see that:

$$F = \frac{1}{M} \sum_{\kappa=1}^{N} x_\kappa \frac{\partial F}{\partial x_\kappa} \tag{19}$$

With the above terminology, we may define a slightly more general class of functions. We shall say that $J(W(n))$ is *bounded by a homogeneous function of degree M* if:

$$0 < \left((J - J_0) / \frac{1}{M} (W - W_0)^T \frac{\partial J}{\partial W} \right) \le 1$$
$$W_0 \in \mathbf{R}^{Ns}, \quad \forall (W - W_0) \neq 0_{Ns} \tag{20}$$

and J has a single global minimum at $W = W_0$.

Note that the network can succeed in accomplishing its task only when there is a minimum of J that is smaller than the stipulated level, J^*, of acceptable error. Thus, we shall only consider the case in which $J(W_0) \le J^*$. The following preliminary result must be stated:

<u>Lemma 1</u>

If, together with the assumptions of Theorem 1, $J(W(n))$ is bounded by a homogeneous function of degree M and $J(W_0) \le J^*$ then, defining:

$$\tilde{W}(n) \triangleq W(n) - W_0 \tag{21}$$

the vector $\tilde{W}$ satisfies:

$$\tilde{W}(n+1) = \left[I_{Ns} - \frac{\alpha}{M} \delta(n) \Psi(n) \Psi^T(n) \right] \tilde{W}(n) \tag{22}$$

where:

$$\Psi(n) \triangleq \frac{\partial J(n)}{\partial W(n)} / \left\| \frac{\partial J(n)}{\partial W(n)} \right\|$$
$$(\therefore \|\Psi\| = 1) \tag{23}$$

$$\delta(n) = \frac{L(J - J^*)}{J - J_0} \frac{J(n) - J_0}{\left(\frac{1}{M} \tilde{W}^T \frac{\partial J}{\partial W} \right)} \in [0,1] \forall n \tag{24}$$

<u>Proof</u>: Appendix B, Reference [31].

Equation (22) is similar to an equation arising frequently in linear adaptive control. A complicating factor here is that the unit vector Ψ may also depend upon W. Nevertheless, it is easy to derive a sufficient condition for the boundedness of the time series $\{ W(n) \in \mathbf{R}^{Ns}; n = 0, 1, \ldots \}$. We have:

Theorem 2

Under the assumptions of Lemma 1:

$$0 < \alpha < 2M \Rightarrow \left\|\tilde{W}(n+1)\right\| \le \left\|\tilde{W}(n)\right\| \forall n \tag{25}$$

Proof: Reference [31]

Theorem 2 shows a simple condition on the constant α that guarantees boundedness of the sequence of synaptic weights. However, the result does not prove convergence of the network output to some desirable level of error. We complete our investigation of the simple case assuming condition (20) with the following result.

Theorem 3

Under the assumptions of Lemma 1 and considering $\alpha < 2M$:

$$\lim_{n\to\infty} J(n) \le J^* \tag{26}$$

Proof: Reference [31].

4. Rapid Anomaly Detection via a Competitive Model Principle.

Toeplity networks as described above, are used to form system identifiers of various forms (termed "replicator networks" in Fig. 1), the form depending on the underlying model form (impulse response, ARMA, etc.) selected a priori combined, in turn, to form adaptive control systems.

A considerable degree of fault tolerance is already implicit in the basic ANC architecture. ANC provide on-line solution of the optimization problem, dealing with complex constraints (e.g., from sensor and actuator limitations) implicitly. If sensor, actuator or structure hardware fails, ANC merely solves the optimization problem under new constraints — achieving convergence to a new controller that is optimal under the new set of hardware limitations. This has been demonstrated experimentally, see [7]. If an abrupt system change occurs however, the controller can recover only at a rate comparable to its convergence when starting with a "blank slate". For typical aerospace control problems, involving tens of structural modes, this recovery period is some fraction of an hour — too long to prevent serious performance degradation in the interim.

Let us consider some basic limitation to the learning rate constants and convergence speed using, as a simple example, the replicator structure for a linear system (M=2). With W denoting a vector containing the unconstrained synaptic weights minus their optimal values, we absolutely must require step-by-step nondivergence — i.e.: the right hand inequality of (25). If this condition is violated, sustained or at least momentary divergence can result. A necessary and sufficient condition for satisfaction of is that the learning rate constant, α, be chosen in accordance with Theorem 2:

$$0 \le \alpha \le 4. \tag{27}$$

There can be no benefit to the use of larger values of α than 4. With this restriction, $\|W(n+1)\|/\|W(n)\|$ is a random variable in the range:

$$\frac{\|W(n+1)\|}{\|W(n)\|} \in \left[\sqrt{1-\alpha+\frac{1}{4}\alpha^2},\ 1\right] \tag{28}$$

We typically take $\alpha = 2$ so (28) promises a best case value of 0 for $\|W(n+1)\|/\|W(n)\|$ — i.e., one step convergence. However the probability of this best case is vanishingly small.

Practical measures of convergence are achieved by considering the expected value $E[W]$. Here we have:

$$\frac{\|E[W(n+1)\|}{\|W[W(n)]\|} \simeq \left(1-\frac{1}{\eta^2}(\alpha-\frac{1}{4}\alpha^2)\right)^{1/2}$$

$$\eta \equiv \text{number of synaptic weights being updated} \tag{29}$$

An analogous formula applies to the ANC in this entirety, where η can be of order 10^2. Hence, actually observed convergence is much slower than the best case suggested in (28) and the speed decreases as the number of synaptic connections increases. Apparently, the situation cannot be easily remedied since α *must* be chosen below 4 to avoid the possibility of divergence.

The solution, for the replication example, is to find a way to adjust a very small number of synaptic weights *first* and fine-tune the rest later. Specifically, in place of a single replicator, we set up a bank of N replicators. As indicated in Fig. 2, the outputs of the replicators are fed, via Toeplitz synapses constrained proportional to identity, into an output array. Each of the replicators is, essentially, an alternative model. The total output is a linear combination of these models, with coefficients $\Theta_1,\ldots,\Theta_N$. If $\Theta_1^*,\ldots,\Theta_N^*$ denote the least squares optimal values, we have analogously to (29):

$$\frac{\left\|E\left[\Theta_\kappa(n+1)-\Theta_\kappa^*\right]\right\|}{\left\|E\left[\Theta_\kappa(n)-\Theta_\kappa^*\right]\right\|} \simeq \left(1-\frac{1}{N^2}(\alpha-\frac{1}{4}\alpha^2)\right)^{1/2} \tag{30}$$

Thus, if N is of modest size, Θ_κ's converge very rapidly. Usually, that Θ_κ associated with the replicator that happens to be closest to the unknown system to be replicated is largest. This replicator then becomes fine tuned faster than all the others. Ultimately only on of the Θ_κ's is nonzero.

In essence, this scheme sets up a "model competition" where the "winner" is determined quickly and then becomes fine-tuned at a slower rate. All the losers are ultimately disconnected from the output.

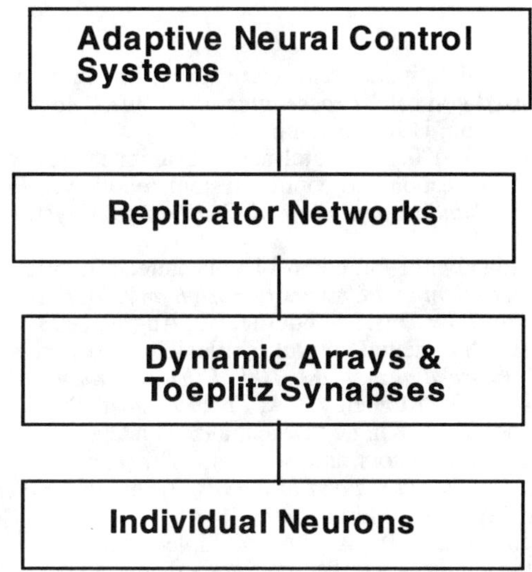

Fig. 1 Hierachy Of Modular Neural Structures Progressing From Basic Constituents To Higher-Level Modules.

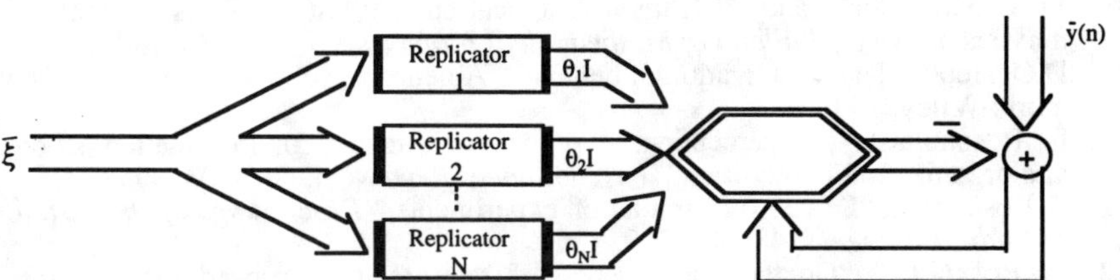

Fig. 2 The Competitive Model Scheme For System Replication Combines Several Replicator Units (Representing Alternative Models) To Form The Total Output Through Rapidly Adaptive Coefficients Of Combination. The System Quickly Switches To The Most Satisfactory Replicator Unit, Disconnecting All The Others.

5. Summary And Conclusions

In this paper, we outlined a modular/hierarchical organization of artificial neurons for the construction of neural networks capable of autonomous system identification and of simultaneous structural identification and on-line control optimization for sampled-data systems. Extensions to the basic scheme were given to achieve rapid fault detection and control system reconfiguration. These extensions involved the introduction of levels of neural redundancy through the competitive neural model approach. Applied to spacecraft structures and control systems, these developments open the possibility of greater spacecraft autonomy and, indeed, of *self reliance*, in the sense that the spacecraft is capable of rapidly and effectively working around equipment failures and other unforeseen conditions.

References

1. D.C. Hyland, "Neural network architecture for on-line system identification and adaptively optimized control," *Proceedings of the IEEE conference on Decision and Control.*, Brighton, U.K., December 1991.
2. D.C. Hyland and J.A. King, "Neural network architectures for stable adaptive control, rapid fault detection and control system recovery," *Proceedings of the 15th Annual AAS Guidance and Control Conference*, Keystone, CO, February 1992.
3. D.C. Hyland, "Adaptive neural control architecture: a tutorial," *Proceedings of the Industry, Government and University Forum on Adaptive Neural Control for Aerospace Structural Systems*, Melbourne, FL, August 1993.
4. D.C. Hyland, "Adaptive neural control for flexible aerospace systems: Progress and prospects," *Proceedings of the 10th IEEE International Symposium on Intelligent Control 1995*, Monterey, CA, 27-29 August 1995.
5. D.C. Hyland and J.A. King, "Decentralized adaptive neural control for distributed mesoscale actuators and sensors," *Proceedings of the 1995 SPIE Conference on Smart Structures and Materials, Technical Conference 2442*, San Diego, CA, 26 February - 3 March, 1995.
6. D.C. Hyland and L.D. Davis, "A Multiple-input, multiple-output neural architecture for the suppression of a multi-tone disturbance," *19th Annual AAS Guidance and Control Conference*, Paper No. AAS 96-067, Breckenridge, Colorado, February 7-11, 1996.
7. D.C. Hyland, L.D. Davis, A. Das and G. Yen, "Antonomous neural control for structure vibration suppression," AIAA Guidance, Navigation and Control Conference, Paper No. AIAA-96-3923, San Diego, CA, July 29-31, 1996.
8. W.S. McCulloch, and W. Pitts, "A logical calculus of the ideas immanent in nervous activity," *Bulletin of Mathematical Biophysics* 5, 115-133, 1943.
9. D.O. Hebb, "The organization of behavior: A neurophsychological theory," New York: Wiley, 1949.
10. F. Rosenblatt, "The perception: A probabilistic model for information storage and organization in the brain," *Psychological Review* 65, 386-408, 1958.
11. F. Rosenblatt, "Perceptron simulation experiments," *Proceedings of the Institute of Radio Engineers*, 48, 301-309, 1960.
12. F. Rosenblatt, "On the convergence of reinforcement procedures in simple perceptrons," Report VG-1196-G-4. Cornell Aeronautical Laboratory, Buffalo, NY, 1960.
13. F. Rosenblatt, *Principles of Neurodynamics*, Washington, DC: Spartan Books, 1962.
14. B. Widrow and M.E. Hoff, Jr., "Adaptive switching circuits," *IRE WESCON Convention Record*, 96-104, 1960.
15. B. Widrow and M.A. Lehr, 1990, "30 years of adaptive neural networks: Perceptron, madaline, and backprogagation," *Proceedings of the IEEE* 78, 1415-1442, 1990.
16. B. Widrow and S.D. Stearns, *Adaptive Signal Processing*, Englewood Cliffs, NJ: Prentice-Hall, 1985.
17. P.J. Werbos "Beyond regression: New tools for prediction and analysis in the behavioral sciences." Ph.D. Thesis, Harvard University, Cambridge, MA, 1974.
18. D.B. Parker "Learning-logic: Casting the cortex of the human brain in silicon," Technical Report TR-47. Center for Computational Research in Economics and Management Science, MIT, Cambridge, MA, 1985.
19. Y. LeCun. "Une procedure d'apprentissage pour reseau a seuil assymetrique," Cognitiva 85, 599-604, 1985.

20. D.E. Rumelhart, G.E. Hinton and R.J. Williams, "Learning representations by back-propagating errors," *Nature (London)*, 323, 533-536, 1986.
21. D.E. Rumelhart, G.E. Hinton and R.J. Williams, "Learning internal representations by error propagation," in *Parallel Distributed Processing: Explorations in the Microstructure of Cognition* (D.E. Rumelhart and J.L. McClelland, eds.) Vol. 1, chapter 8, Cambridge, MA: MIT Press, 1986.
22. D.E. Rumelhart and L.J. McClelland, eds., *Parallel distributed processing: Explorations in the Microstructure of Cognition*, Vol. 1 Cambridge, MA: MIT Press, 1986.
23. K.J. Lang and G.E. Hinton, "The development of the time-delay neural network architecture for speech recognition," Technical Report CMU-CS-88-152. Carnegie-Mellon University, Pittsburgh, PA, 1988.
24. A. Waibel, T. Hanazawa, G. Hinton, K. Shikano and K.J. Lang, "Phoneme recognition using time-delay neural networks," *IEEE Transactions on Acoustics, Speech, and Signal Processing* ASSP-37, 328-339, 1989.
25. E.A. Wan, "Time series prediction by using a connectionist network with internal delay lines," *In Time Series Prediction: forecasting the Future and Understanding the Past* (A.S. Weigend and N.A. Gershenfeld, eds), 195-217. Reading, MA: Addison-Wesley, 1994.
26. P.J. Werbos "Backpropagation through time: What it does and how to do it," *Proceedings of the IEEE*, 78, 1550-1560, 1990.
27. E.A. Wan, "Temporal backpropagation for FIR neural networks," *IEEE International Joint Conference on Neural Networks*, Vol. 1, 575-580, San Diego, CA, 1990.
28. E.A. Wan "Temporal backpropagation: An efficient algorithm for finite impulse response neural networks," *In Proceedings of the 1990 Connectionist Models Summer School* (D.S. Touretzky, J.L. Elman, T.J. Sejnowski, and G.E. Hinton, eds.), 131-140, San Mateo, CA: Morgan Kaufmann, 1990.
29. R.J. Williams and D. Zipser "A learning algorithm for continually running fully recurrent neural networks," *Neural Computation* 1, 270-280, 1989.
30. K.S. Narendra and K. Parthasarathy "Identification and control of dynamical systems using neural networks," *IEEE Transactions on Neural Networks* 1, 4-27, 1990.
31. D.C. Hyland, "Connectionist Algorithms for identification and control: System structure and convergenceanalysis," Paper No. AIAA 97-0686, 35th Aerospace Sciences Meeting Reno, NV, January 1997.

SESSION 1

SENSING TECHNOLOGY DEVELOPMENT I

Use of Insitu Dielectric Sensing for Intelligent Processing and Health Monitoring

D. KRANBUEHL, W. NEWBY-MAHLER, D. HOOD,
S. CASE, K. REIFSNIDER and A. LOOS

ABSTRACT

Frequency dependent dielectric measurements (FDEMS) provide a sensitive, automated insitu sensing technique for intelligent processing and an insitu means for monitoring durability, that is degradation of polymers during use. FDEMS insitu sensing can be designed and calibrated to monitor changes in processing properties during fabrication and changes in mechanical service life properties of polymer materials during use in the field environment. With a proper understanding of the type of polymer and the use environment, the sensor output can be related to changes such as in modulus, maximum load and elongation at break. The FDEMS technique has advantages over other monitoring techniques, nondestructive; accuracy/reproductivity; sensitivity; insitu capability; remote sensing; automated.

This report will discuss recent work on the use of FDEMS sensing to monitor aging in thermoplastics during use in the field, particularly those Nylon 11, PVDF and PPS being used for flexible composite pipe to transport oil-gas in the offshore subsea environment. Current life monitoring work focuses on characterizing the chemical and physical processes occurring during aging, using FDEMS sensing to monitor the aging rate and state of the polymer and integrating the sensor output with a model for predicting the remaining service life and state of the structure. The model is periodically updated through the insitu online sensing measurements. This report will also discuss work on developing a fundamental understanding of the relationship of the sensor measurement of the ionic and dipolar mobility to the macroscopic properties of polymeric materials during synthesis-fabrication and sensor driven intelligent automated process control.

INTRODUCTION

With the continuing expansion in the use of polymeric materials in extended use structures such as airplanes, bridges and pipelines where the expected lifetimes are 20 to 40 years and where failure can be catastrophic, there is a clear need to develop high quality processing and an insitu health monitoring sensor capability. This paper describes progress toward the development of a frequency dependent dielectric measurement sensor (FDEMS) which is capable of detecting the change in the physical and chemical state of a polymeric material insitu during processing-fabrication and insitu during use in the field environment.

Essential to the development of reliable, automated, intelligent processing is the ability to monitor continuously the changing state of the polymeric resin insitu in the fabrication tool during processing. This sensing capability is essential because 1) resin processing properties vary with age, batch and handling: 2) tool heat transfer properties vary; 3) significant differences exist between one autoclave, press, oven, pultruder and 4) operators differ. In addition, there is the need to quickly develop and optimize cure procedures for new tool designs, new resins etc. as well as a need to verify quality during processing while reducing post fabrication testing costs.

Health monitoring involves monitoring the degradation of a polymeric materials performance properties. In one sense, health monitoring is the reversal of monitoring cure during which there is a buildup in properties.

This report will discuss recent work on the use of FDEMS sensing to monitor aging in thermoplastics during use in the field, particularly those used in flexible composite pipe to transport oil-gas in the offshore subsea environment. Current life monitoring work focuses on characterizing the chemical and physical processes occurring during aging, using FDEMS sensing to monitor the aging rate along with state of the polymer and integrating the sensor output with a model for predicting the remaining service life and health of the structure. The model is periodically updated through the insitu online sensing measurements. This report will also discuss work on developing a fundamental understanding of the relationship of the sensor measurement of the ionic and dipolar mobility to the macroscopic properties of polymeric materials during synthesis-fabrication and aging-use. It will briefly describe the use of FDEMS sensing for intelligent automated cure using resin transfer modeling as one example.

BACKGROUND

Frequency dependent dielectric measurement, made over many decades of frequency, Hz-MHz, are a sensitive, convenient automated means for characterizing the processing and performance properties of thermosets and thermoplastics.[Kranbuehl, 1989] Using a planar wafer thin sensor, measurements can be made insitu in almost any environment. Through the frequency dependence of the impedance, this sensing technique is able to monitor changes in the molecular mobility of ions and dipoles. These changes in molecular mobility are then related to chemical and physical property changes which occur during use or during processing. The FDEMS techniques have the advantage that measurements can be made both in the laboratory, insitu in the fabrication tool, and insitu during use. Few laboratory measurement techniques have the advantage of being able to make measurements in a processing tool and in the field in a composite, in an adhesive bond line, of a thin film or a coating. It can be used at temperatures exceeding 400°C and at pressures of 60 atm, with an accuracy of 0.1% and a range in magnitude of over 10 decades. It is difficult for most other in the field techniques attain this level of sensitivity in harsh environments.

At the heart of dielectric sensing is the ability to measure the changes at the molecular level in the translational mobility of ions and changes in the rotational mobility of dipoles in the presence of a force created by an electric field. Mechanical properties reflect the response in displacement on a macroscopic level due to a mechanical force acting on the whole sample. The reason why dielectric sensing is quite sensitive is rooted in the fact that changes on the macroscopic level originate from changes in force displacement relationships on a molecular level. Indeed, it is these molecular changes in force-displacement relationships which dielectric sensing measures as the resin cures and ages. They are the origin of the resin's macroscopic changes in mechanical performance properties, during use and processing properties during procedure.

INSTRUMENTATION

Frequency dependent complex dielectric measurements are made using an Impedance Analyzer controlled by a microcomputer.[Kranbuehl, 1989; Senturia, 1986] In the work discussed here, measurements at frequencies form Hz to MHz are taken continuously throughout the entire cure process at regular intervals and converted to the complex permittivity $\epsilon^*=\epsilon'-i\epsilon''$. The measurements are made with a geometry independent DekDyne micro sensor which has been patented and is now commercially available and a DekDyne automated dielectric measurement system. This system is used with either commercially available impedance bridges or specially built marine environmental bridges designed for use on offshore oil platforms. The system permits multiplexed measurement of several sensors. The sensor itself is planar, 2.5 cm x 1.25 cm area and 5 mm thick. This single sensor-bridge microcomputer assembly is able to make continuous uninterrupted measurements of both ϵ' and ϵ'' over decades in magnitude at all frequencies. The sensor is inert and has been used at temperatures exceeding 400°C and over 60 atm pressure.

THEORY

Frequency dependent measurements of the materials' dielectric impedance as characterized by its equivalent capacitance, C, and conductance, G, are used to calculate the complex permittivity, $\epsilon^*=\epsilon'-i\epsilon''$, where $\omega=2\pi f$, f is the measurement frequency and C_o is the equivalent air replacement capacitance of the sensor.

$$\epsilon'(\omega) = \frac{C(\omega)\ material}{C_o} \qquad \epsilon''(\omega) = \frac{G(\omega)\ material}{\omega C_o} \tag{1}$$

This calculation is possible when using the sensor whose geometry is invariant over all measurement conditions. Both the real and the imaginary parts of ϵ^* can have a dipolar and ionic-charge polarization components;

$$\epsilon' = \epsilon_d' + \epsilon_i' \qquad \epsilon'' = \epsilon_d'' + \epsilon_i'' \tag{2}$$

Plots of the product of frequency (ω) multiplied by the imaginary component of the complex permittivity $\epsilon''(\omega)$ make it relatively easy to visually determine when the low frequency magnitude of ϵ'' is dominated by the mobility of ions in the resin and when at higher frequencies the rotational mobility of bound charge dominates ϵ''.

LIFE MONITORING - EXPERIMENTAL

FDEMS life monitoring results are being conducted on three material systems;

a polyamide nylon-11, polyvinyl diflouride (PVDF) and polyphenlyl sulfide (PPS).[Kranbuehl, 1996]. The first two materials are already in use in flexible subsea oil-gas transmission pipes on oil platforms throughout the world. The nylon-11 system is used as the oil-gas retention barrier in pipes with operating temperatures up to 80°C. The PVDF is a higher temperature polymer barrier used for oil-gas temperatures up to 130°C. The PPS material is being considered with graphite as an unidirectional tape for use as the axial wrap. It will replace steel bands and create a lighter, higher performance flexible pipe for extremely deep water and arctic environments. As the outer layer of the pipe, the PPS graphic tape is exposed to seawater. The nylon-11 and PVDF, on the other hand, serve as the inner fluid gas barrier and are exposed to the acidic H_2S, water, oil mixture coming up from the ground.

FDEMS sensors were embedded in all three material systems by heating the polymer to its glass transition temperature, placing the sensor between 2 pieces of the polymer and encapsulating the sensor in the center with pressure. The resulting material sensor system was approximately 1 cm thick. The embedded sensor material system and mechanical test dog bones of ASTM D638 specifications were then placed in the following aging environments. Nylon-11: 95% oil, 5% water, pH 4.6, 70° and 105°C. PVDF: 95% oil, 5% water, pH 4.6 130°C. PPS: 100% simulated sea water 90° and 120°C. Periodically FDEMS sensor data was taken and pieces were removed for mechanical testing.

LIFE MONITORING - RESULTS

For nylon-11, Figures 1-2, display the value of ϵ'' multiplied by the frequency at 100, 120, 10^3, 10^4 and 10^5 Hz versus time in the 105° and 70° oil-water acidic aging environment.[Kranbuehl, 1996] The results are on a log scale. They show a rapid very large rise, 10^4, during the initial days as water diffuses into the polymer. This process occurs over 30 days at 105°. It occurs over 150 days at 70°, although a large fraction of this change occurs in the initial 30 days. After water impregnation has occurred at 105° there is a gradual decline in ϵ''. Other experimental work in our laboratory and recently reported by others has shown that the nylon is degrading due to hydrolysis. This is accompanied by embrittlement, as shown in Figure 3. By making molecular weight measurements of the sensor at various times and of the mechanical tested dog bones, a calibration plot relating the normalized sensor output to the load and elongation at break was prepared. This is shown in Figures 4 and 5. The normalized value of ϵ is defined as the absolute value of ϵ at each time divided by the maximum value achieved around day 30 due to water impregnation.

Using the calibration, it is possible to monitor the mechanical properties of the critical polymer oil-gas barrier with time insitu in the field. This was done over a 400 day period on an oil platform in the British sector of the North sea in 1995-6. The sensor data suggested the pipe was in a failure state. The pipe did fail, was retrieved and the sensor data was verified.

Similar work is being conducted on PVDF. Figure 6 shows the results of the sensor output over a 300 day period at 130°. Figure 7 shows the change in mechanical properties. Again for this material system the sensor is able to monitor the gradual embrittlement of the polymer oil-gas barrier. The aging process in PVDF is not primarily molecular weight degradation as in nylon. Aging involves changes in polymer

Nylon Sensor 2L in 105C 5%ASTM/95% H2O
Log e"*w vs Day

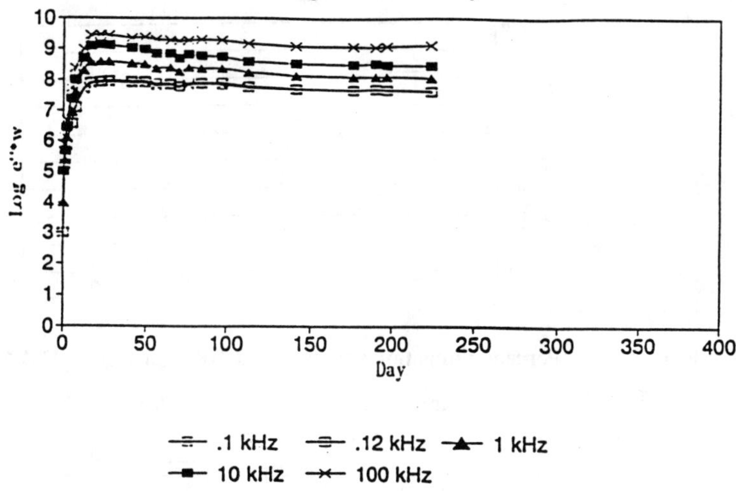

Figure 1 Log(ϵ″*ω) versus day for nylon-11 at 105°C in oil water pH 4.68 frequencies from top to bottom 100, 10, 1, .12, .10 kHz.

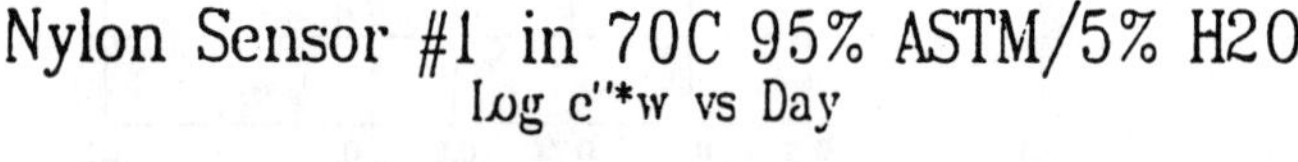

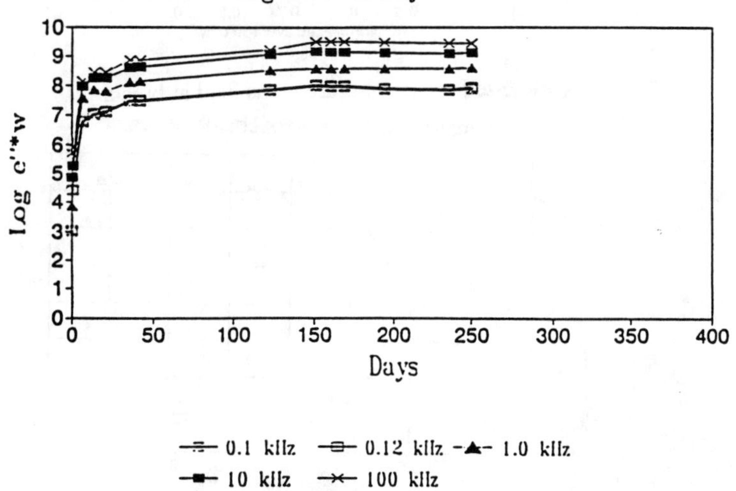

Figure 2 Log (ϵ″*ω) versus day 70°C.

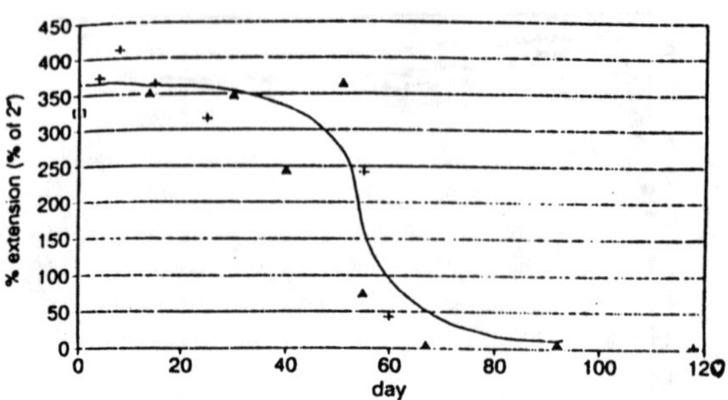

Figure 3 Percent elongation versus day at 105° oil-water pH 4.68.

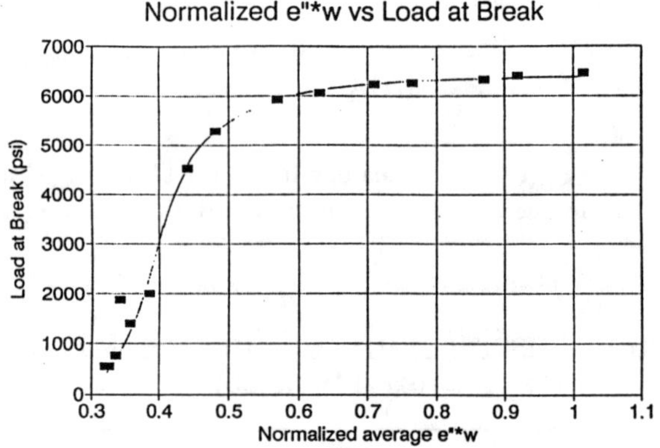

Figure 4 Normalized sensor output versus load at break.

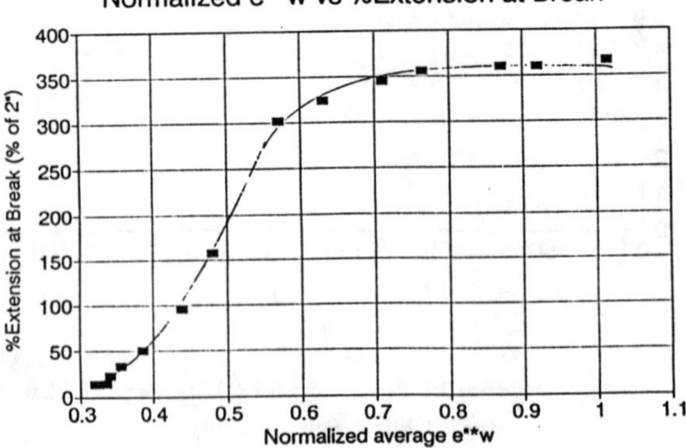

Figure 5 Normalized sensor output versus elongation at break.

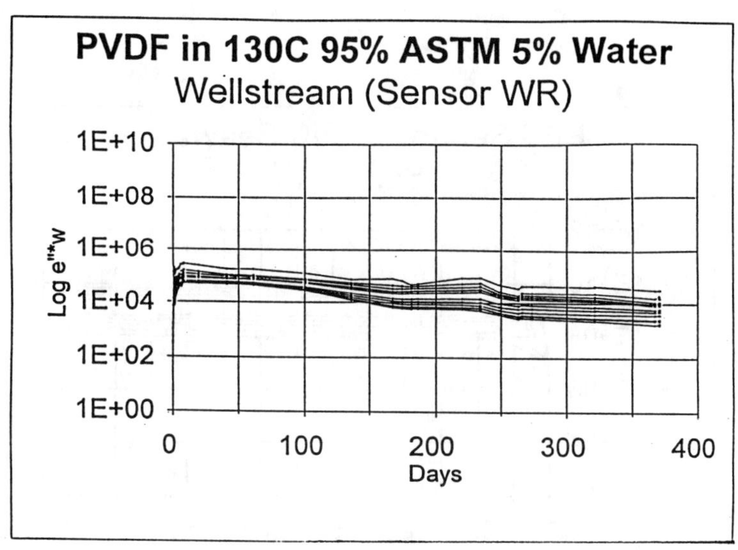

Figure 6 Output of sensor embedded in PVDF while at 130°C, oil-water pH 4.68.

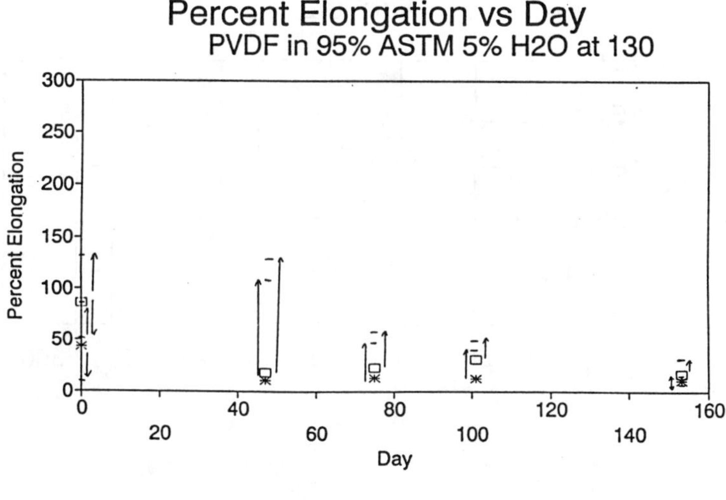

Figure 7 Displacement versus day.

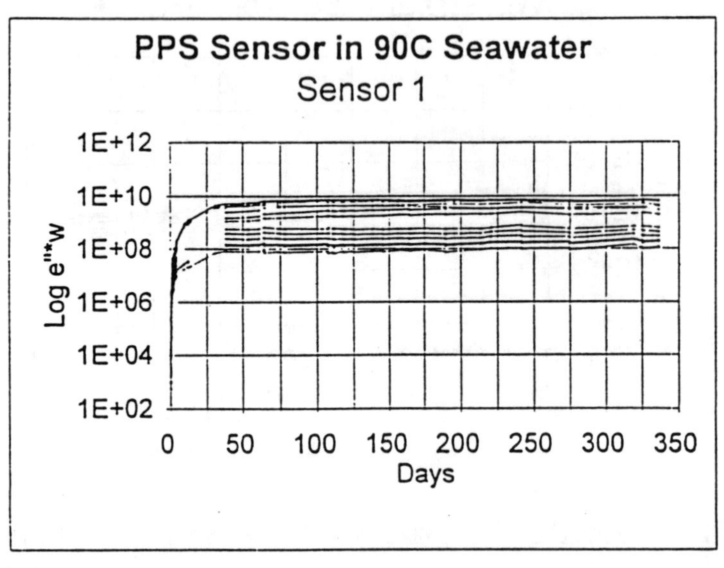

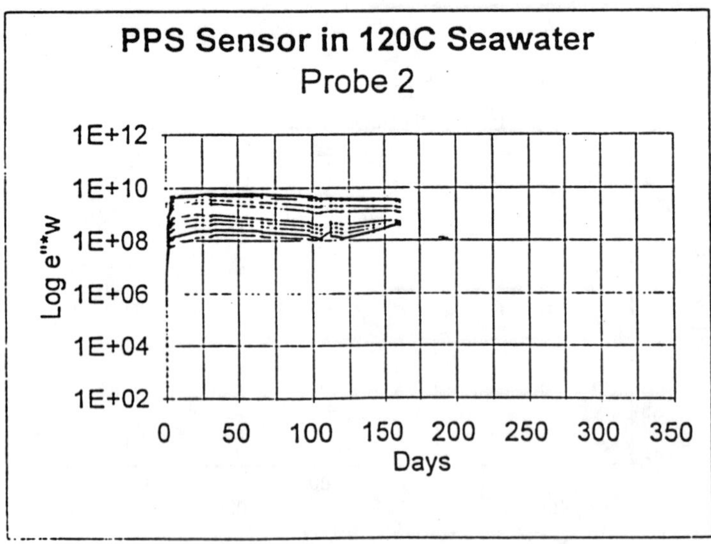

Figure 8 Output of sensor embedded in PPS while at 90° (a) and 120° (b).

morphology in both the crystalline and amorphous regions and is currently under study.

The PPS graphite unidirectional tape was studied in its sea water use environment at 90° and 120° as shown in Figures 8a and 8b. Little to no change is observed in the output over 330 days at 90°. A slight drop is observed at 120°. The mechanical load and elongation at break show similarly consistent results with no change in properties over the 150 days of aging.

INSITU INTELLIGENT PROCESS CONTROL - RESULTS

An insitu intelligent control system is based on monitoring the actual *state* of the resin, not on a predetermined time temperature procedure.[Kranbuehl, 1989, 1994, 1995; Loos, 1987; Ciriscioli, 1990] Thus the temperature scheme the part experiences is not based on time, but on the achievement of molecular landmarks in the curing process. Because these critical points may not occur at the exact same time for each part being made, time is not wasted holding a part at some stage when the cure process could be advanced. This gives the smart, closed-loop process a flexibility to adjust to variations in resin cure and heat transfer properties. Overall an intelligent sensor closed-loop system produces more consistent parts because the advancement to the next stage of cure is based on the resin achieving a certain molecular state and not on time. The advancement of the viscosity and degree of cure is monitored. Final cure is defined by a universal degree of cure.

An insitu sensor closed-loop system composed of a press, FDEMS sensors able to gather data insitu from the composite part, and smart software that collects the sensor information, interprets it, and then makes decisions that control the fabrication device was assembled.[Kranbuehl, 1994] The smart software combines both the predictions of the processing model and the insight of operator experience. The expert system was constructed to optimize and control the achievement of six critical stages during the resin film infusion cure process. These were: 1) achieve a low resin viscosity; 2) maintain a low viscosity until impregnation is complete; 3) advance the resin reaction at an intermediate temperature to a particular degree of cure that avoids an excessive exothermic effect (this value varies with part thickness); 4) ramp to final cure temperature once reaction has advanced; 5) monitor degree of cure during final hold; 6) determine achievement of proper degree of cure, which is related to attainment of ultimate Tg, and desired use properties and then 7) turn the process cycle off.

Figure 9 shows the sensor output for the smart automated sensor expert system-controlled run. The resin reached the center sensor at 37 min. The viscosity is maintained at a low value by permitting slow increases in the temperature. At 60 min/fabric impregnation was complete. The resin was advanced during a 121°C hold to a predetermined value of degree of cure of 0.35, based on the Loos model's predictions of the extent of the exothermic effect. This value is clearly dependent on panel thickness. At 130 min., the ramp to 177°C was begun. Achievement of an acceptable complete degree of cure was determined by the sensor at 190 min. Then the cure process was shut down.

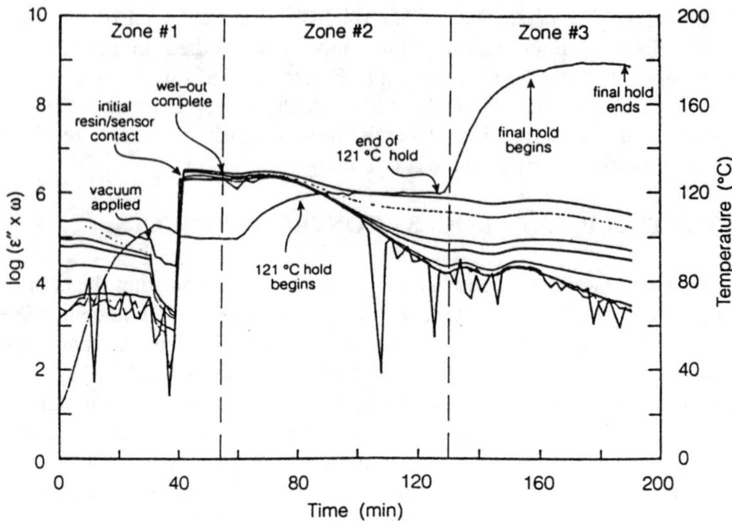

Figure 9 **Sensor output during automated sensor expert system controlled RTM run.**

CONCLUSIONS

FDEMS sensing is an attractive method for monitoring of the changing state of a polymeric structure either to monitor aging during use in the field or to monitor the changing processing properties in a tool during fabrication. Through calibration in the laboratory, this molecular force displacement sensing technique can monitor continuously and insitu the changes in macroscopic structural properties such as load and elongation at break or processing properties such as viscosity, Tg and degree of cure.

ACKNOWLEDGMENT

David E. Kranbuehl appreciates partial support form the NSF Science and Technology Center at Virginia Polytechnic and State University under Contract #DMR91-2004; a NASA Langley grant NAGI-23; Exxon, Elf, Shell, BP, Amerada Hess, Statoil, Norsk Hydro and Robit, and a NIST STP grant with Wellstream Corporation.

REFERENCES

1. Ciriscíolo, P., Springer, G., 1990, “Smart Autoclave Cure” *Technomic*, Lancaster, PA.

2. Kranbuehl, D., Hood, D., McCullough, L., Aandahl, H., Haralampus, N., Newby, W., Eriksen, M., 1996, "Frequency Dependant Electromagnetic Sensing for Life Monitoring of Polymers in Structural Composites During Use", Progress in Durability Analysis of Composite Systems, ed. A.H. Cordon, Balkema, Rotterdam, 53-62.
3. Kranbuehl, D., Hood, D., Rogozinski, J., Barksdale, R., Loos, A., MacRae, D., 1995, "FDEMS Sensing for Automated Intelligent Processing of Polyimides", Proc. ASMW International Mechanical Engineering Congress H1041A Vol 2 pp 1017-1030.
4. Kranbuehl, D., 1989, "Dielectric Cure Monitoring", *Encyclopedia of Composites,* ed., Stuart M. Lee, VCH Publishers, New York, pp. 531-43.
5. Kranbuehl, D., Kingsley, P., Hart, S., Hasko, G., Dexter, B., Loos, A.C., 1994, "Control of Composite Cure, *Polymer Composites*, 15 (4), 297-305.
6. Loos, A.C., Kranbuehl, D. E. and Freeman, W. T., 1987,"Modeling and Measuring the Cure of a Composite", *Intelligent Processing of Materials and Advanced Sensors,* ed., H.N.G. Wadley, et al. The Metallurgical Society, Inc., Warrendale, PA, pp. 197-211.
7. Senturia, S., Sheppard, S., 1986, *Applied Polymer Science*, **80** 1-48.

Damage Detection in CFRP by Resistance Measurements at the Boundary

R. SCHUELER, S. P. JOSHI and K. SCHULTE

ABSTRACT

Carbon Fiber Reinforced Polymer (CFRP) composites derive the excellent mechanical strength, stiffness and its conductivity from carbon fibers. The mechanical deformation and electrical resistance are coupled in these fibers that make them inherently sensors. Therefore, Carbon Fiber Reinforced Polymer (CFRP) composites can be considered a self-monitoring material without any need for additional sensing elements. However for this to become reality the conductivity map of the entire structure need to be constructed and relationship between the conductivity and various usage and damage related variables need to be established. Experimental results demonstrate that internal damage, such as fiber fracture and delamination, decreases the conductivity of the composite laminates. In general, the information about the damage size and position can be obtained by utilizing Electrical Impedance Tomography (EIT) and using an iterative reconstruction algorithm. But the traditional EIT is not capable of extracting this information when the medium possesses highly anisotropic electrical conductivity. Above a certain level of anisotropy it is advantageous to modify the traditional EIT. This paper presents a method to extract the damage size and position for highly orthotropic (unidirectional) CFRP without the need of complex calculations, thus enabling damage detection in real-time. These results indicate that a practical EIT has a potential of being a cost effective Health and Usage Monitoring Technique (HUMT) for CFRPs.

1. INTRODUCTION

Carbon fiber reinforced polymers (CFRP) consist of electrically conductive carbon fibers and a polymeric matrix, which is an insulator. The carbon fibers are responsible for both the strength and the conductivity of the composite material. Thus the stiffness and strength as well as the conductivity is much higher in fiber direction than in transverse direction. Many investigators have suggested optical fibers and other sensors as smart sensing constituents for health and usage monitoring of CFRP laminates. Instead of introducing additional devices into the laminate, our approach suggest that the load carrying carbon fibers themselves be utilized for self-monitoring of the CFRP laminates. It is possible to extract the stress/strain field as well as the damage state by mapping the specimen electrical resistivity (or impedance) information.

The EIT-method gained wide recognition in the 1920's by geophysicists who placed arrays of electrodes into the ground. By injecting current through pairs of electrodes and measuring the resulting voltage reply at all others, information about oil-bearing rocks under the surface could be acquired. In our case we want to determine the resistivity distribution of a laminate sheet. A possible experimental setup is shown in Figure 1. Various electrodes are connected to the edges of the sample. An electrical current is injected via two electrodes and the potential difference between all other neighboring electrodes is measured. By taking various combinations of current injecting electrodes and repeating the potential difference measurements at the remaining electrodes, a wealth of information can be obtained. This information can be utilized to extract resistivity distribution inside the sample using a reconstruction algorithm based on fundamental electrodynamic equations.

Once refined, the EIT has practical applications in monitoring CFRP laminated structures. The technological and manufacturing hurdles do not appear to be barrier in implementation of the technique. For example, by placing electrodes at a one mm spacing along grid lines on a array of size 50 mm by 50 mm, a detectable damage size is about 5 mm. See Figure 2. In existing laminated composite structures which are often joined together by metal rivets, these rivets can be used as electrodes to measure the resistivity distribution inside a structure. In these structures, the inter rivet distance is typically 1" to 2". From this riveting configuration one could expect to detect a damage of size 1".

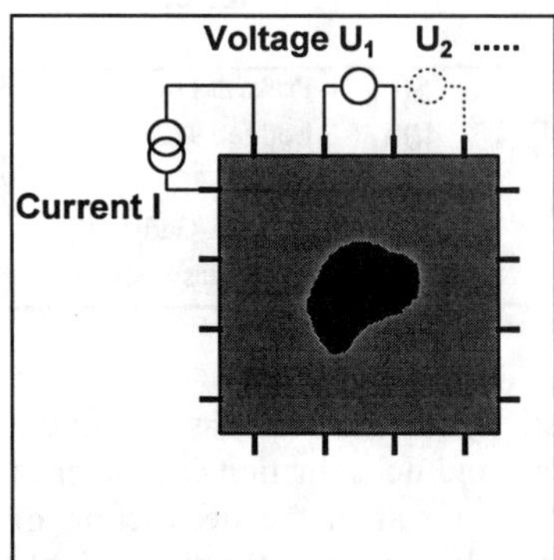

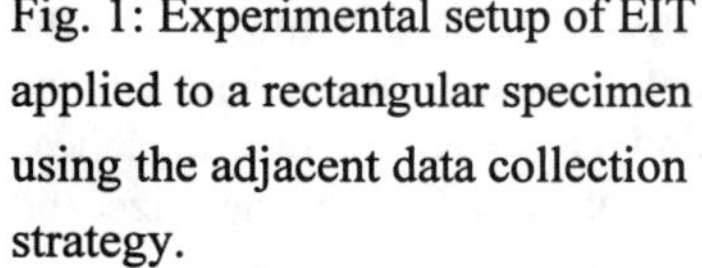

Fig. 1: Experimental setup of EIT applied to a rectangular specimen using the adjacent data collection strategy.

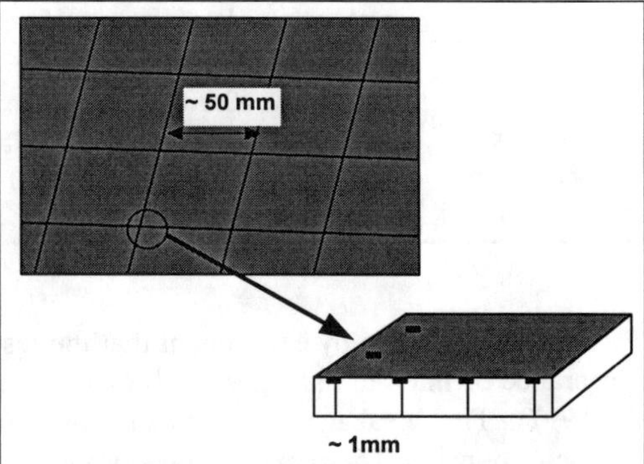

Fig. 2: Proposed application of EIT to monitor CFRP laminated structures. Sensing electrodes are placed inside the laminate along grid lines on an array of size 50 mm by 50 mm.

2. CONDUCTIVITY AND DAMAGE IN CFRP

2.1 ELECTRICAL PROPERTIES OF CFRP

For the most part the conductivity paths follow individual carbon fibers, since fibers are surrounded by insulating polymer. Only at inter-fiber contact points the current can switch to an adjacent fiber. The electrical resistance of the specimen is an assemblage of the resistance of all fibers. Due to its internal structure, electrical properties of CFRP are anisotropic. This is especially true for unidirectional laminates. In fiber direction (0°-direction) the conductivity of the composite is dominated by the single fiber conductivity and the fiber volume fraction and can be calculated by the rule of mixtures. Perpendicular to the fibers (90°-direction) the conductivity is relatively small. Here the current transport depends on the existence and number of inter-fiber contact points. This number is related with the fiber volume fraction via a percolation process, the waviness of fibers and fiber bundles and the amount of misaligned fibers in an nominal unidirectional laminate. Because of the variation in manufacturing procedures, many different values have been reported. A few of them are reported in Table 1.

As can be seen, there is a huge anisotropy in the electrical resistivity of CFRP. In literature, $\rho_{90°}/\rho_{0°}$ electrical resistivity ratio of 50 to about 500 are found. Our own measurements on AS4/3501-6 yielded an even higher value of 2000. As will be shown later, this high anisotropy makes the applicability of the traditional experimental EIT-setup very difficult.

TABLE I: RESISTIVITY OF CFRP.

Material	Fiber content	$\rho_{0°}$ [Ωcm]	$\rho_{90°}$ [Ωcm]	$\rho_{90°}/\rho_{0°}$	Reference
Carbon/epoxy	55 vol%	0.08	0.4	50	Prakash/Owsten 1976
Carbon/epoxy	65 vol%	0.0025	1	400	Lodge 1982
Carbon/epoxy	-	0.0021	0.95	450	ERA 1979
Graphite/epoxy	-	0.0625	40	640	Scruggs/Gadja 1977
AS4/3501-6	-	0.003	6	2000	own measurements

It has been shown by experiment that the resistivity of CFRP increases with the appearance of internal damages, such as fiber break and delamination (Schueler et al., 1997). The breakage of fibers or fiber bundles result in the destruction of conductivity paths. These experiments prove that the electrical resistivity is a viable manifestation of damage in CFRP. A natural expansion of this method is the determination of damage class, e.g. fiber breakage or delamination, and damage size and position. However additional information is needed to quantify these aspects. A promising method for the evaluation of at least the damage size and position is the Electrical Impedance Tomography (EIT).

3. ELECTRICAL IMPEDANCE TOMOGRAPHY

There are different strategies for collecting experimental data for EIT. One of them is the adjacent strategy, where current injection is done at two adjacent electrodes and the voltage is measured between all other adjacent electrodes. This gives N^2 measurements, where N is the number of electrodes, from which N(N-1)/2 are independent. The use of the current electrodes is often omitted due to possible contact resistance at these points which could be a source of an error. This reduces the number of independent measurements to N(N-3)/2. For 16 electrodes this is equal 104.

The desired conductivity distribution $\sigma(x,y)$ have to be derived from the following fundamental equation. In a 2-dimensional case and if the principle axis of the conductivity tensor coincide with coordinate axis $\sigma(x,y)$ has to fulfill the equation:

$$\frac{\partial}{\partial x}\left(\sigma_x \cdot \frac{\partial \Phi}{\partial x}\right) + \frac{\partial}{\partial y}\left(\sigma_y \cdot \frac{\partial \Phi}{\partial y}\right) = 0 \quad (1)$$

with the boundary conditions:

$$\left(\sigma_x \cdot \frac{\partial \Phi}{\partial x}\right) + \left(\sigma_y \cdot \frac{\partial \Phi}{\partial y}\right) = j_s \quad (2a)$$

$$\Phi = \Phi_s \quad (2b)$$

Φ : electric potential; σ : conductivity; j_S : current density applied at boundary; Φ_S : potential applied at boundary.

If the conductivity distribution and the boundary conditions j_S and Φ_S are given, the internal potential and current densities can be obtained straightforward using equation (1) and (2) and a Finite Element Method, FEM-program. This is called a forward problem. In EIT instead, we measure the potential and current distributions on the electrodes placed at the boundary of a specimen to determine the internal conductivity distribution. This so called inverse problem requires many iterative, numerical solutions of a forward problem.

An often used reconstruction algorithm to solve the inverse problem is the Newton-Raphson method. This iterative method minimizes the error F between the measured, $\mathbf{V}_{meas}$, and estimated potential distribution, $\mathbf{V}_{est}(\sigma)$.

$$F = \frac{1}{2}\left(\mathbf{V}_{est}(\sigma) - \mathbf{V}_{meas}\right)^T \cdot \left(\mathbf{V}_{est}(\sigma) - \mathbf{V}_{meas}\right) \quad (3)$$

Starting with an initial guess, σ_0, the estimated potential distribution is calculated using FEM. If the error F is more than a criterion error, the conductivity distribution is altered by $\Delta\sigma$, which is a function of $\mathbf{V}_{meas}$, $\mathbf{V}_{est}(\sigma)$, and the derivative $\mathbf{V}_{est}(\sigma)'$. Again, to obtain $\mathbf{V}_{est}(\sigma)'$ FEM has to be used. This is the most time-consuming step, since for each different current electrode pair and each conductivity value the calculation of the voltage response have to be done. With the improved conductivity distribution the potential distribution and the error are calculated. These steps are repeated until the convergence criterion is met.

Usually EIT is applied to materials with an isotropic conductivity. CFRP has a highly anisotropic conductivity. Therefore, it is necessary to assign at least two conductivity values to each element. The spatial resolution, i.e. the element size, is determined by the maximal number of equations which is equal the number of independent measurements.

To give an example: with 16 electrodes we get N(N-3)/2 = 104 independent measurements. Dividing a quadratic specimen in i x i elements give 2 x i^2 unknown conductivity values. This yields i ≈ 7 which means that each specimen side is divided into 7 sections. Doubling the number of electrodes (N=32) gives i ≈ 15.

4. APPLICATION OF EIT TO CFRP

In the case of low anisotropy the conventional EIT can be applied to obtain a resistivity distribution of specimen. To do this only a minor problem has to be solved. Until now only materials with isotropic, i.e. not direction dependent but position dependent, conductivity are examined using this technique. That means reconstruction algorithm made for the isotropic case has to be expanded to account for anisotropy.

At high ρ-ratios the situation is different as illustrated in figure 3. An injected current flows first along the fibers which are connected to the electrodes into and out of the specimen. Between these connected fiber bundles (light area in Fig. 3) the current flow is almost homogeneously. This means, only voltages measured at electrodes contacted to the light area give some information about the resistivity distribution inside the light area. In the rest of the specimen the current flow is negligible resulting in a constant potential in each of these areas (painted dark in Fig. 3). No information about the resistivity distribution inside these areas can be obtained. Furthermore it becomes clear that the traditional equidistant positioning of the electrodes (see Fig. 2) around the specimen edges has to be adjusted. Measurements at the left and right specimen edges (see Fig. 3) reveal only information about the small fiber bundles at each edge, i.e. they give no information from inside the sample and are therefore useless. Instead, it is advantageous to place the electrodes only along the sides that are perpendicular to the fiber direction.

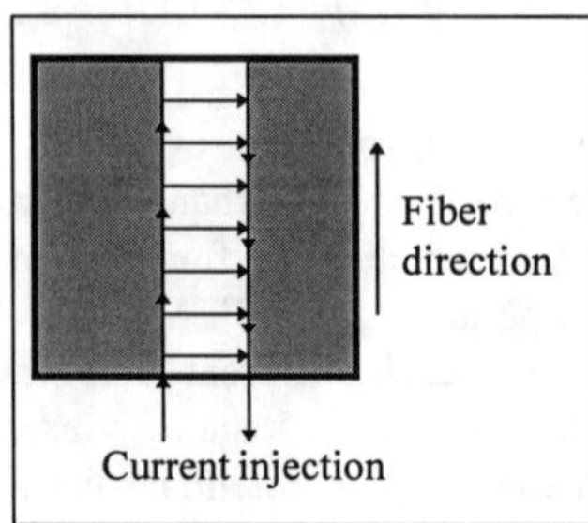

Figure 3: Current flow in case of high ρ-ratio.

In Figure 5 numerical results for a specimen with a hole in the center (hole 1, see Fig. 4) are shown. Here the traditional setup with equidistant electrodes are used. For clarity, only the difference in potential between the damaged and the intact specimen is presented. Only one significant peak is found. This is, when current is applied between the electrodes 2 and 3 and the voltage is measured between 10 and 11. This peak for hole 1 reveals that the entire hole is between x=1.5 and x=2.5 and thus determine the hole position in transverse direction. The very same peak is found for hole 3, determing only the hole position in transverse direction. Thus hole 1 and 3 are undistinguishable. For hole 2 the peak shift to the current electrodes 3-4 and potential electrodes 9-10 indicating that entire hole 2 is between x=2.5 and x=3.5. Experiments agree well with these FEM-results (Schueler et al., 1997).

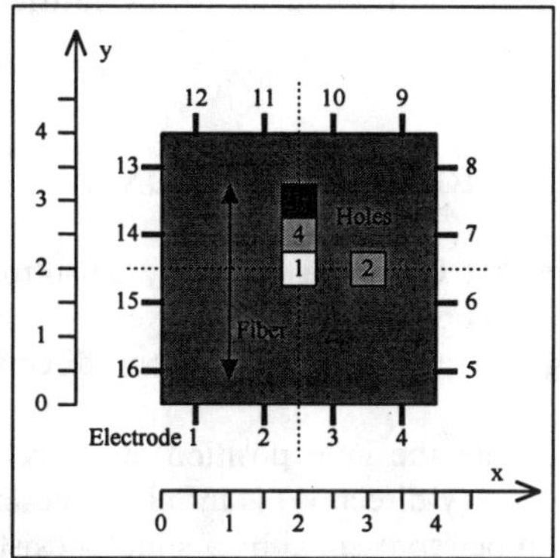

Fig. 4: 2D-Model used for finite element calculations. Holes have edge length of 0.5 cm.

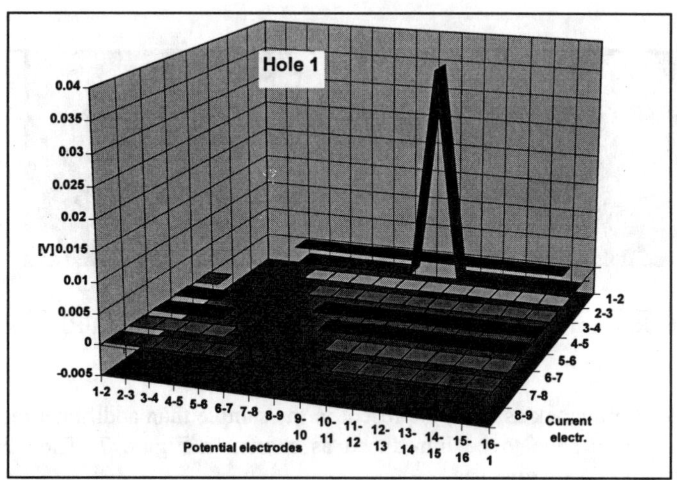

Fig. 5: Numerical results for specimen with hole in the center (hole 1).

The peak height provides information about the amount of conductivity drop in the 'examined specimen section'. For instance, by applying current at the electrodes 2 and 3 and measuring the voltage between the electrodes 10 and 11 this section would be the rectangle spanned between these corners (2-3-10-11). For the intact specimen the voltage is 0.482 compared to 0.518 with hole 1. That means the conductivity decreased by 7%. In fact we reduced the conductive area by the hole size, which is about 6%.

Using more electrodes with smaller distances the hole position in transverse direction could be easily determined more accurately. More electrodes between the current electrodes and on the opposite side allows additionally a finer resolution of the only peak. Keeping electrode 2 and 3 as current electrodes the comparison of the voltage distribution at the electrode side with the opposite side reveals the y-position of the hole, too. But this is only possible, if the exact shape of the hole is known. For experimental detection of a hole this restriction is not a realistic constraint.

5. RESISTOR NETWORK REPRESENTATION

It is mentioned above that CFRP has a highly anisotropic conductivity. It is much higher in fiber direction than in transverse direction. Therefore the approximation of setting the conductivity in fiber direction to infinity seems feasible.

The procedure to calculate the hole position and size by assuming that the conductivity in fiber direction (y-direction) is infinite is described below. With this assumption, the sample can be represented by a simple resistor network, where the resistance between two arbitrary electrodes at position x_i and x_{i+1} is given as

$$R_{i,i+1} = \rho_x*(x_{i+1}-x_i)/a. \qquad (4)$$

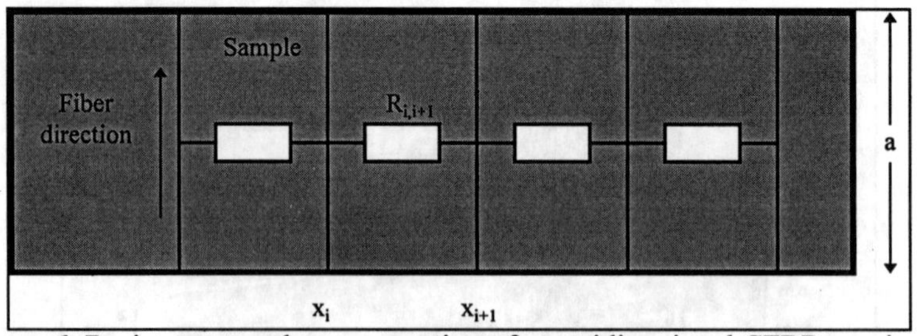

Figure 6: Resistor-network representation of an unidirectional CFRP-specimen.

If a hole (or a crack as well) is present in the sample than additional resistors at the hole location have to be introduced as shown in Figure 7. Then the inner resistors R_3 and R_4 are given as

$$R_{3/4}=\rho_x*d/y_{u/l} \quad (5a)$$

where ρ_x=resistivity in x-direction, perpendicular to fibers; d=electrode distance; y_u, y_l =distance between sample edge and hole edge on upper and lower side, respectively

The outer resistors R_0,R_7 are given as:

$$R_0=\rho_x*(x_L-x_0)/a \quad (5b)$$

$$R_7= \rho_x*(x_3-x_R)/a \quad (5c)$$

where x_L,x_R=x-position of left and right hole edge; x_i=x-position of electrode i; a=sample height.

The other resistors are defined by:

$$R_{1/2}= \rho_x*(x_1-x_L)/y_{u/l} \quad (5d)$$

$$R_{5/6}= \rho_x*(x_R-x_3)/y_{u/l} \quad (5e)$$

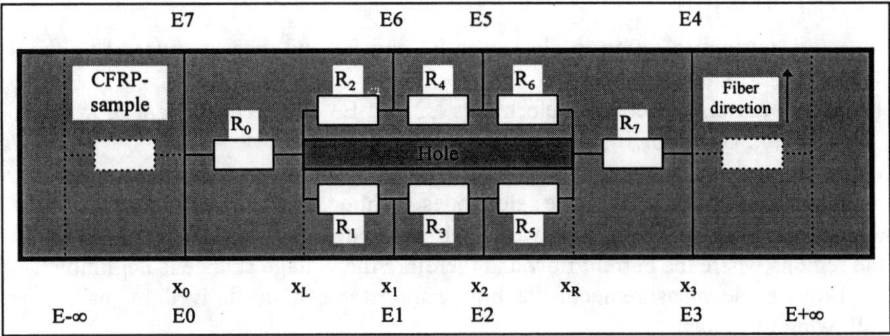

Figure 7: Resistor-network representation of an unidirectional CFRP-specimen with a hole.

In this model the hole is represented as a rectangle. It is defined by its center position (x_h, y_h) and its width in longitudinal (h_y) and in transverse direction (h_x) or equivalently by the position of the left and right edge, x_L and x_R, and distance between sample edge and hole edge on upper, y_u, and lower side, y_l. These parameters can be obtained by applying current at four different electrode pairs and measuring the voltage response at certain electrodes, which are named below.

In experiment one seeks to avoid the usage of current electrodes for voltage measurements due to the unknown contact resistance. If only one electrode is at the hole position, then only insufficient information can be gained to calculate the values of the 6 resistors. That means, to calculate the position of the hole in y-direction at least two electrode position (x_1 and x_2) have to be at hole position. Since we examine only symmetric setups (electrodes at both sides have same x-positions) as shown in Figure 7, overall 8 local electrodes (E0 ... E7) and two at the outer sample edges (E-∞ and E+∞) have to be considered. The following calculation/procedure can be easily expanded to more electrodes.

1. Apply current near left and right edge and measure voltage response at both sides of the intact sample.

2. Apply current near left and right edge and measure voltage response at both sides of the sample containing a hole. In comparison with data obtained from the intact sample, the position of the hole in the transverse to fiber direction can be derived as follows. A voltage increase between an adjacent pair of electrodes indicates that at least a portion of the hole lies with in the area which is between fiber direction lines aligned with this pair of electrodes. These pairs of electrodes are utilized for obtaining an accurate estimate of the hole size and position. Figure 8 depicts these electrodes as E0 -- E7.

3. Performing three additional sets of measurement using these electrodes:
   i. current injection in electrodes E0 and E1 and measurement of voltages $U_\infty^{0,1}$, $U_{23}^{0,1}$, $U_{45}^{0,1}$, $U_{56}^{0,1}$;
   ii. current injection in electrodes E1 and E2 and measurement of voltages $U_\infty^{1,2}$, $U_{56}^{1,2}$;
   iii. current injection in electrodes E2 and E3 and measurement of voltages $U_\infty^{2,3}$, $U_{01}^{2,3}$, $U_{56}^{2,3}$, $U_{67}^{2,3}$;

Subscripts indicate voltage electrodes, while superscripts indicate current electrodes. U_∞ is the voltage between two electrodes, E-∞ and E+∞, which are in the regions where the current flow and therefore the voltage change is negligible.

From these measurements the hole parameter can be derived by using the following formulas:

$$\Delta U_\infty^{1,2} = \rho_x I \frac{d}{a - h_y} \Leftrightarrow h_y = a - \frac{\rho_x I d}{\Delta U_\infty^{1,2}} \tag{6}$$

$$\Delta U_\infty^{0,1} = \rho_x I\left(\frac{x_1 - x_L}{a - h_y} + \frac{x_L - x_0}{a}\right) \Leftrightarrow x_L = \frac{\dfrac{\Delta U_\infty^{0,1}}{\rho_x I} - \dfrac{x_1}{a - h_y} + \dfrac{x_0}{a}}{\dfrac{1}{a} - \dfrac{1}{a - h_y}} \tag{7}$$

$$\Delta U_\infty^{2,3} = \rho_x I\left(\frac{x_R - x_2}{a} + \frac{x_3 - x_R}{a - h_y}\right) \Leftrightarrow x_R = \frac{\dfrac{\Delta U_\infty^{2,3}}{\rho_x I} + \dfrac{x_2}{a - h_y} + \dfrac{x_3}{a}}{\dfrac{1}{a - h_y} - \dfrac{1}{a}} \tag{8}$$

The formula to determine the hole position in longitudinal direction is:

$$y_h = \frac{1}{2}\left[a + \rho_x d\left(\frac{1}{R_3} - \frac{1}{R_4}\right)\right], \tag{9a}$$

$$\text{where } R_3 = \frac{U_{56}^{1,2}}{I}\left(\frac{U_{01}^{2,3}}{U_{56}^{2,3}} + \frac{U_{67}^{2,3}}{U_{56}^{2,3}} + \frac{U_{56}^{1,2}}{U_{56}^{2,3}}\frac{U_{23}^{0,1}}{U_{56}^{0,1}} + 1 + \frac{U_{23}^{0,1}}{U_{56}^{0,1}} + \frac{U_{45}^{0,1}}{U_{56}^{0,1}}\right) \quad (9b)$$

$$\text{and} \qquad R_4 = R_3 \frac{U_{56}^{1,2}}{U_{56}^{2,3}}\frac{U_{23}^{0,1}}{U_{56}^{0,1}}. \quad (9c)$$

The advantage of this outlined procedure to obtain the hole position and size is that no time-consuming numerical calculations and iterations are necessary like in the Newton-Raphson-method mentioned in Section 3. Thus the damage detection in real time is possible without super-fast and expensive computers.

5.1 NUMERICAL SIMULATION

A 2-dimensional, orthotropic sheet with uniform anisotropic conductivity is examined. Its size was chosen to be 4 cm x 4 cm representing a unidirectional ply. Corresponding to our experimental findings of $\rho_{0°}= 3x10^{-3}$ Ωcm and $\rho_{90°}= 6$ Ωcm and a ply thickness of 0.015 cm, the resistivity in 0° and 90°-direction were set to 0.2 Ω and 400 Ω respectively, i.e. ρ-ratio = 2000. To parametrically study the effect of various ρ-ratios the resistivity in 90°-direction was decreased.

The ANSYS 5.3 program was used for the finite element calculation. Element type was thermal-electric solid (PLANE 67) with 4 nodes. The element size was 0.03125 cm x 0.125 cm near the electrode and hole position and 0.125 cm x 0.125 cm elsewhere. Damages to the specimen were modeled by setting the resistivity of different specimen regions to 10^{20}. The potential at the origin of the coordinate system is set to 0 and a current of $5x10^{-3}$ A is applied.

5.2 RESULTS

To check the accuracy of our model, the data from FEM-calculations were used. First a square hole in orthotropic plates with resistivity-ratios of 2000, 500, 100 and 20, respectively is analyzed. Second the influence of hole shape and the electrode position was examined.

The center of the hole was arbitrarily chosen to be x_h=1.5 and y_h=2.5. The width and height was set to h_x=h_y=0.5. That means the hole lies between x=1.25 and x=1.75. The electrodes were placed at x=1.0625, 1.3125, 1.4626, and 1.8125. The electrodes are not placed symmetric with respect to the hole center x=1.5. A symmetric arrangement of electrodes is a special case that will always give the exact x-location of the hole.

Table II shows that down to a ρ-ratio of 100 the calculated parameters are close to the actual values used in the FEM-calculation. With further decreasing ratios, the hole height h_y tends to decrease while the hole width h_x increases. This results in a relatively constant hole size with only about 2% difference from the actual value of 0.25. Below a ratio of 100 the error, especially in longitudinal position y_h, increases significantly. Here the assumption of an infinite conductivity in longitudinal direction is not valid anymore.

Until now only rectangular holes are considered. In reality one has to deal with holes of an aspect ratio bigger than 1, and holes that are not aligned either to the y- or x-axis. To account for such situations the hole shape shown in Figure 8 is used for the FEM-calculation. The ρ-ratio is again 2000. Two different sets of electrodes are examined. First the same set as used before (x_0,x_1,x_2,x_3 in Figure 8) and one shifted 0.125 cm to the right (x_0',x_1',x_2',x_3'). Thus the influence of the electrode position is evaluated.

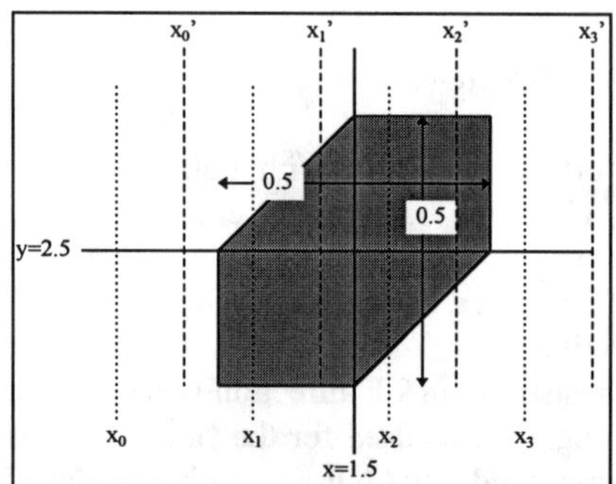

Figure 8: Alternative hole shape used for FEM-calculations.

TABLE III: DEPENDENCE OF CALCULATED HOLE PARAMETER ON HOLE SHAPE AND ELECTRODE POSITION (ρ-RATIO=2000)

	x_h	y_h	h_x	h_y	area
actual values	1.5	2.5	-	-	0.1875
Electrode set					
x_0,x_1,x_2,x_3	1.442	2.475	0.418	0.454	0.1897
x_0',x_1',x_2',x_3'	1.439	2.536	0.539	0.359	0.1935

In both cases the calculated hole center position and the hole size are in good agreement with the actual values, as shown in Table III. The center is slightly shifted to the left, which is probably due to the fact that the current electrodes are on the lower sample side and the influence on the result of the lower left hole corner is higher than that of the upper right corner, which is further away from the electrodes.

Inspite of the relatively simple resistor model the calculation of hole position and size is accurate.

6. CONCLUSION

It was shown that damage location in CFRP can be performed with the electrical impedance tomography. Depending on the anisotropy of CFRP, electrodes have to be positioned either at all specimen edges or only at two opposing sides. For the case of high anisotropy a method was outlined to determine the damage size and position in real-time.

These preliminary results show that EIT is a promising method for structural health and usage monitoring. However, sustained research efforts are needed before this promising technique can become a viable technology.

ACKNOWLEDGMENTS

Dr. R. Schueler is supported by Deutsche Forschungsgemeinschaft.

REFERENCES

Electrical Research Association, 1979, "Measurements of R.F. Properties of Carbon Fibre Composite Materials (CFC).", No. 3320/R/2

Lodge, K.J., 1982, "The electrical properties of joints in carbon fibre composites." *Composites* ,13 , pp. 305-310

Prakash, R., Owston, C.N., 1976, "Eddy-current method for the determination of lay-up order in cross-plied cfrp laminates." *Composites,* 7, pp. 88-92

Schueler, R., Joshi, S.P., Schulte, K., 1997, "Conductivity of CFRP as a tool for health and usage monitoring.", Proceedings of SPIE, Vol. 3041, Smart Structures and Materials, 2-6 March 1997, San Diego, Smart Structures and Integrated Systems, pp. 417-426

Scruggs, L.A., Gadja, W.J., 1977, "Low frequency conductivity of unidirectional graphite/epoxy composite samples." Proc. IEEE-EMC Symp., 2.-4. August 1977, Seattle, WA, pp. 396-402

Fault Identification in Vibrating Structures Using a Scanning Laser Doppler Vibrometer

A. B. STANBRIDGE, A. Z. KHAN and D. J. EWINS

ABSTRACT

A Continuously-Scanning Laser Doppler Vibrometer (CSLDV) can be used for a high-resolution measurement of a structure's vibrating mode shape, measured along a straight or circular scan line so that, if the scan crosses a defect line, this may be detected as a mode shape discontinuity. The process has been illustrated with reference to a vibration mode of three flat plates: one defect-free, one with a saw-cut and one with a fatigue crack. The defects were clearly evident in the demodulated CSLDV output, and in the altered sideband structure of its output spectrum. Further trials, and improvements in technique are continuing.

INTRODUCTION

Methods of identifying faults and flaws in engineering structures and components can be divided into two categories:

(i) localised searches, such as ultrasonic pulse-reflection techniques, X-Ray measurements, fluorescent oil crack detection, etc.; and

(ii) measurement of changes in the overall properties of a structure such as, and of particular interest in the context of this paper, vibration modal properties: the natural frequencies, modal damping constants and the mode-shapes, as defined at specific measurement locations on the structure. Despite considerable effort and ingenuity, changes in frequency response, natural frequency and damping have generally been found to be somewhat insensitive and unreliable in establishing the presence of weaknesses and flaws in practical structures. Commonly, measurements have been compared on a basis of repeated measurements on a structure in service, or comparisons made with finite-element models - a variation on "model updating".

More detailed mode-shape measurements might be expected to offer a better method of fault identification. If a structure has, for example, a crack extending to an accessible surface, or a loose bolt or rivet, one might expect this to have a noticeable *local* effect on vibration mode-shapes, provided they are such as to flex the defect. The problem has been that, for instance, using accelerometers the accuracy and repeatability of measurement is insufficient to give a reliable warning of a mode-shape discontinuity, even if such a finely-spaced survey is in practice

---

Anthony B. Stanbridge, Arshad Z. Khan and David J. Ewins, Imperial College, Dept. of Mechanical Engineering, London, SW7 2BX, UK

feasible. The Laser Doppler Vibrometer (LDV) provides a far more flexible measurement system, the point of aim of the measuring laser beam being quite simply directed to any required point using moveable mirrors. Mode shapes have been measured by this means at hundreds or thousands of points with, in these days of rapid and large computer data stores, relatively little trouble, (Meza et al., 1997) for example. Unfortunately, a step-scanned LDV is liable to occasional faulty measurements due to an unavoidable optical noise problem, speckle noise drop-out. Partly for this reason, mode shapes acquired by this means have usually been subjected to a smoothing process which, unfortunately, is also likely to mask the evidence of the mode shape discontinuities being sought.

A number of methods have been described (Barker, 1992; Sriram et al., 1990; Stanbridge and Ewins, 1995, 1996a) for using *continuously-scanning* Laser Doppler Vibrometers (CSLDVs). The present authors' technique for vibration measurement, (Stanbridge and Ewins, 1995, 1996a), uses either straight-line or circular scans and simple spectrum analysis to define mode shapes. There are still speckle drop-outs but they are not coherent with the vibration frequency and appear as very sharp spikes, on digital sampling usually as single-point drop-outs. On Fourier transformation they leave a broad-band noise floor, typically giving a 40dB signal/noise ratio (i.e. 1%). There is therefore a potential for very high spatial resolution of vibration mode shapes, preserving any discontinuities which might be indicative of structural faults.

It should perhaps be pointed out that the term "mode shape" as applied here means the as-measured deflected form, not the separated contribution due to an individual natural mode, derived by "Modal Analysis".

Other optical techniques, using holography or speckle-pattern interferography, can give high spatial resolution, but they are essentially laboratory-based whereas the CSLDV is capable of use in an in-field situation.

THE LASER DOPPLER VIBROMETER

The LDV measures vibration with an interferometer, detecting the instantaneous difference in wavelength of laser light directed at, and scattered back from, a moving surface. Commercially-manufactured units have been available for at least 10 years and only a brief outline of their operation is warranted here.

By the Doppler principle, the change in wavelength of the returned light is proportional to the velocity of the moving surface. With an LDV, therefore, a beam of coherent, monochromatic light is focused as a spot on a vibrating, diffusing surface. Some of the scattered light is collected in the instrument, mixed in an interferometer with incident light diverted by a beam splitter, and directed at a detector which detects the interference fringes as a measure of the target's vibrational velocity, in the direction of the incident beam. Means, such as Bragg-cell wavelength-shifters, have to be adopted to discriminate negative-going velocities. Some LDVs incorporate x and y axis deflecting mirrors so that the point addressed can easily be varied - a facility which is a requirement in the scanning LDV. There is no need for the target surface to be polished. Indeed it is necessary that it is diffusing, and the measurement does not have to be perpendicular to the vibrating surface. For the CSLDV used in the investigation described here, at a range of around 2m, most unpolished surfaces were satisfactory, though a coat of matt white paint was used to avoid specular reflection.

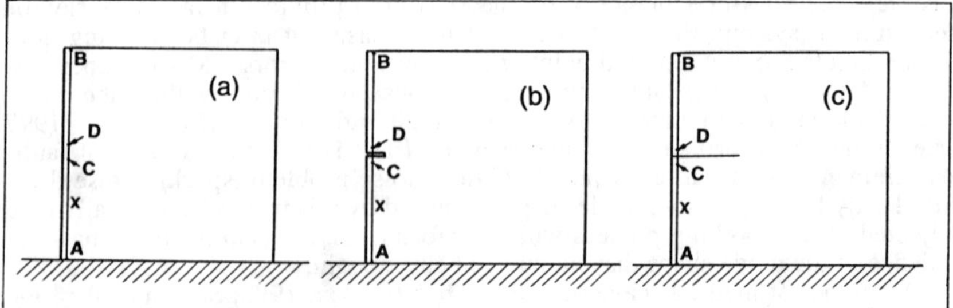

Figure 1 Cantilever-Plate Test Specimens

SPECKLE NOISE

Any diffusing surface illuminated by coherent laser light has a speckled appearance which is caused by interference effects between discrete, random-phased light sources, localised near the plane of the surface. The boundaries between the speckles are defined by a network of lines along which the return light is totally extinguished. The number of 'speckles' covered by a focused LDV beam depends on the size of the illuminated spot: if it is too large there can be so many speckles that phase coherence is almost totally destroyed. With a scanning LDV there must be occasional drop-outs, where the scan passes through a null-line, although these have an extremely short time-scale and may disappear as a result of filtering or sampling, in digital data acquisition.

TEST PIECES

CSLDV test methods for defect detection are explored here using the plates shown in Figure 1. Each of these plates was 175 x 175 x1mm , cantilevered at its lower edge by clamping it firmly between two massive steel blocks. The plate in Figure 1(a) was uniform and effectively "defect-free"; that shown in 1(b) had a saw cut half way up one side, as a gross "defect"; and 1(c) had a fatigue crack artificially induced where shown.

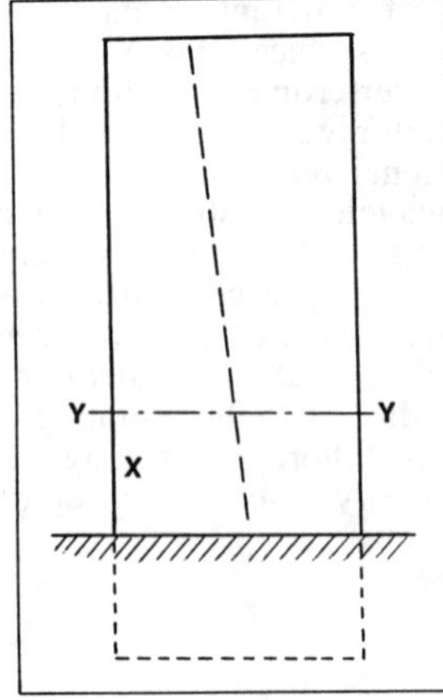

Figure 2 Plate Clamped so as to produce Fatigue Cracks at the Clamp Line.

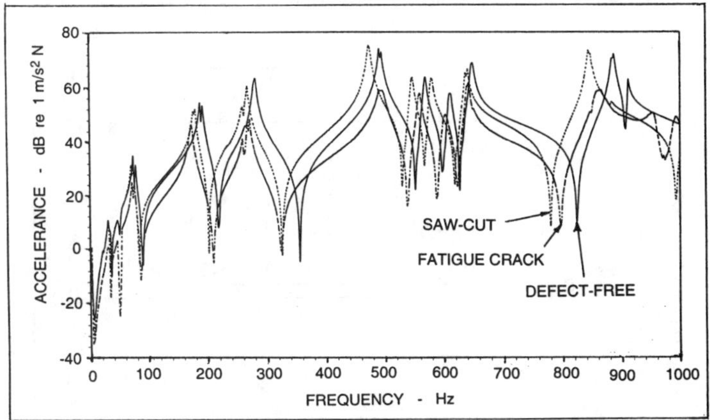

Figure 3 Drive-Point Frequency Response Functions of Test Plates

The saw-cut in the second plate was in practice backed by a piece of thin adhesive tape so as to prevent the scanning LDV beam passing through. Without it the scanned mode shapes would have had unwanted zero-amplitude pulses corresponding to the gap.

Fatigue damage to the third plate was actually produced by exciting the plate in a twisting mode, clamped as shown in Figure 2. A tip amplitude of 20mm was produced in a resonance exhibiting, incidentally, considerable 'stiffening stiffness' non-linearity. The resulting fatigue crack was produced, in practice, just below the apparent clamping line, and was not seen until the plate was removed. The plate was then reclamped, gripping the top part of the plate in Figure 2 up to line Y-Y, so that the crack was at half-height, as indicated in Figure 1.

Figure 3 displays point FRFs for each plate, i.e. Frequency Response Functions with excitation applied at point X (in Figure 1) and response measured at the same point. Response was measured perpendicular to the plate, with a non-scanning LDV and the LDV signal differentiated, in this case, to produce a conventional Accelerance FRF. The plates had a number of natural frequencies below 1000Hz. The defects had a noticeable effect on the FRFs, as well as having a natural frequency shift, the cracked plate exhibiting noticeable damping and non-linearity. In fact, difficulty was found, with this plate, because the damage appeared to extend during the testing, from initial surface cracks on either face, to the point where the plate was cracked right through along the line shown in Figure 1(c).

All the illustrations included here relate to a resonance at about 870 Hz. Some measurements were made at other frequencies and the results were basically similar.

MODE-SHAPE MEASUREMENT AND ANALYSIS

MODE SHAPE DEMODULATION

(Stanbridge and Ewins, 1995, 1996a) describe measurements of vibratory mode shapes using heterodyne demodulation, and this technique has been used in Figure 4 to illustrate these mode-shape discontinuities, using a straight-line scan

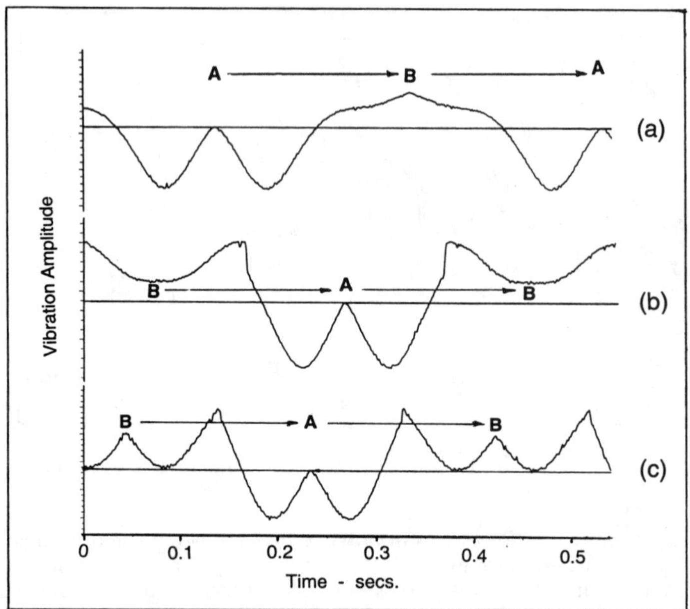

Figure 4 Mode shapes measured from Demodulated CSLDV. Full-Length Uniform Straight-Line Scans: (a) on Defect-Free Plate, (b) on Plate with Saw-Cut, (c) on Plate wlth Fatigue Crack.

perpendicular to the defect. (Briefly, the LDV signal was mixed with a reference signal at the excitation frequency, followed by a low-pass filter, the phase of the reference signal being adjusted to maximise the mixer output.)

Figure 4 shows the time-signals recorded for scans along the lines A-B in Figure 1. The CSLDV was scanned at about 2Hz, with a triangular wave input to the y-axis mirror drive, so that, in the demodulated time-signal plots, the mode shape was scanned at a uniform rate, successively in opposite directions. The locations of the faults can be clearly seen, producing discontinuities in addition to those at A and B, where the scan reversed.

The demodulated mode shape results include speckle noise interference which, though noticeable, does not visually mask the evidence of mode-shape discontinuity.

SPECTRUM ANALYSIS

A spatial mode shape defined along a scan line, can be defined by a set of spatial Fourier coefficients. Any discontinuity in the mode shape will result in increased magnitudes of the higher-order terms but these might be masked, with a long scan such as that in Figure 4, by the coefficients describing the full mode shape. This problem may be avoided by using a short scan for which the defect-free mode shape is nearly a straight line, as in Figure 5 - which is a demodulated mode shape for the defect-free plate, with the same conditions as Figure 4, but with a scan, (C-D in Figure 1) across the defect line, only 30 mm in length. This is an improvement, but discontinuities still remain, at the ends of the scan.

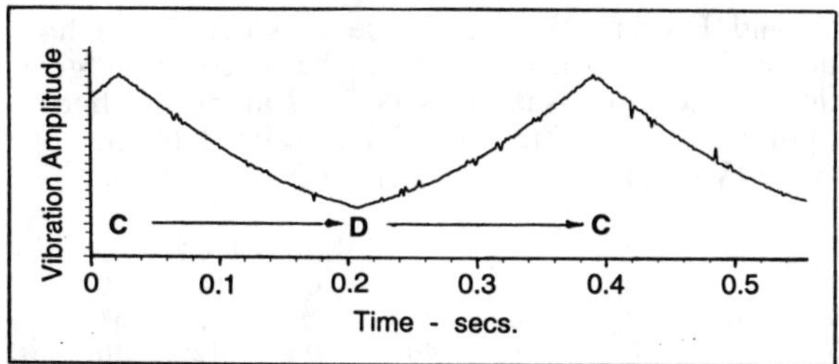

Figure 5 Demodulated Mode Shapes of Defect-Free Plate.
Short, Uniform-Rate Straight-Line Scan.

If, on the other hand, the scan is sinusoidal, this difficulty disappears (it also puts less strain on the mirror drive system). Figure 6 illustrates demodulated CSLDV traces obtained with such a sine-scan, directly comparable with Figure 4. The only discontinuities in these traces (apart from speckle noise) are those due to the "defects", which should therefore be easily identifiable from the corresponding Fourier spectra.

There is no need to demodulate a CSLVD signal before Fourier transformation, the spectrum of a CSLDV signal consists of a component at the vibration frequency together with a symmetrical sideband structure with components spaced at multiples of the scan frequency.

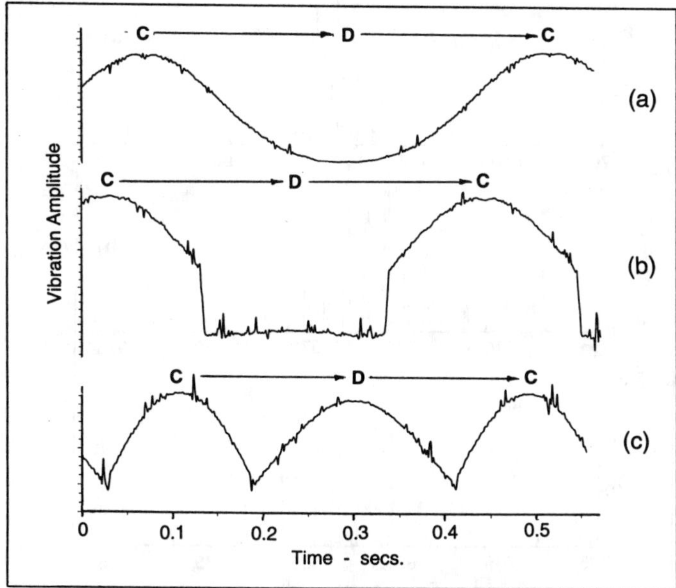

Figure 6 Demodulated Mode Shapes of: (a) Defect-Free Plate,
(b) Plate with Saw-Cut, (c) Plate with Fatigue Crack.
Short, Sinusoidal Straight-Line Scan.

(Stanbridge and Ewins, 1995) describe in some detail how the Fourier transform of such a sine-scan LDV signal may be processed to give a polynomial series description of the mode-shape, as defined along the line. The column vector of the polynomial coefficients **{V}** is related to that of the sideband amplitudes, **{A}**, by **{V} = [T] {A}** where (for the first ten terms):

$$[T] = \begin{bmatrix} 1 & 0 & -2 & 0 & 2 & 0 & -2 & 0 & 2 & 0 & -2 \\ 0 & 2 & 0 & -6 & 0 & 10 & 0 & -14 & 0 & 18 & 0 \\ 0 & 0 & 4 & 0 & -16 & 0 & 36 & 0 & -64 & 0 & 100 \\ 0 & 0 & 0 & 8 & 0 & -40 & 0 & 112 & 0 & -240 & 0 \\ 0 & 0 & 0 & 0 & 16 & 0 & -96 & 0 & 320 & 0 & -800 \\ 0 & 0 & 0 & 0 & 0 & 32 & 0 & -224 & 0 & 864 & 0 \\ 0 & 0 & 0 & 0 & 0 & 0 & 64 & 0 & -512 & 0 & 2240 \\ 0 & 0 & 0 & 0 & 0 & 0 & 0 & 128 & 0 & -1152 & 0 \\ 0 & 0 & 0 & 0 & 0 & 0 & 0 & 0 & 256 & 0 & -2560 \\ 0 & 0 & 0 & 0 & 0 & 0 & 0 & 0 & 0 & 512 & 0 \\ 0 & 0 & 0 & 0 & 0 & 0 & 0 & 0 & 0 & 0 & 1024 \end{bmatrix}$$

and the vibration v, a function of time, t, and distance, x, along the scan line, measured from the scan centre, is:

$$v(x,t) = (V_0 + V_1x + V_2x^2 + V_3x^3 + V_4x^4 \text{........}) \sin \omega t$$

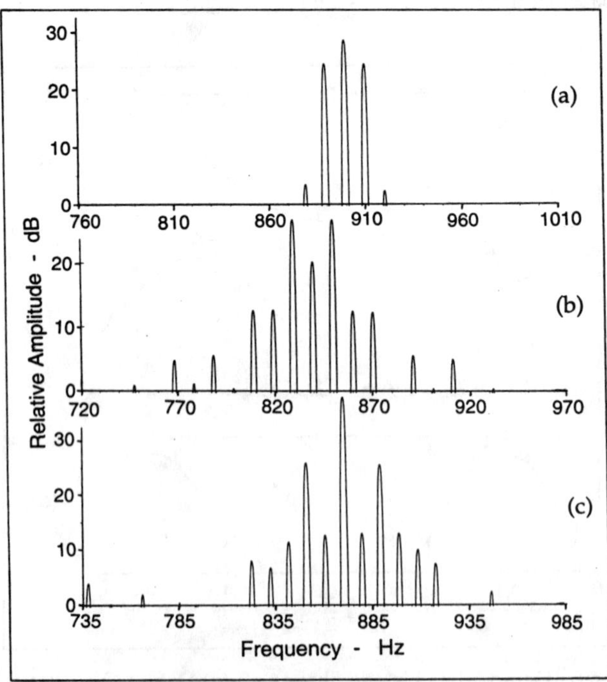

Figure 7 CSLDV Short Straight-Line Sinusoidal Scan Spectra: (a) on Defect-Free Plate, (b) on Plate with Saw-Cut, (c) on Plate with Fatigue Crack.

In order to derive mode shapes in this way it is necessary additionally to consider the CSLDV output *phase* spectrum, which enables real and imaginary mode shape components to be identified, and gives signs to the polynomial coefficients. (Stanbridge and Ewins, 1995, 1996a) give details of a method for doing this but, in the present context, there is no need to curve-fit the mode shapes, no phase data are necessary and the extra complication in the measurement equipment is unwarranted.

It will be seen from the form of **[T]** above that (i): any sideband of order n contains information on polynomial coefficients $V_{(n+2)}$, $V_{(n+4)}$, etc., as well as V_n, and (ii): the sensitivity to higher order components decreases rapidly.

Figure 7 shows spectra corresponding to the short sine-scan spectra shown in Figure 6; the additional sideband components due to the discontinuities are clearly evident. A Zoom FFT analysis was used with a 'Flat-Top' window to minimise peak amplitude errors. The sideband spacing is equal to the scan rate, 10 Hz. A restricted (log) amplitude range has been used, so that the speckle and other noise components are largely supressed.

(For a near-planar vibration, as detected by a short sine scan over a defect-free surface, the spectrum consists predominantly of components at the vibration frequency, corresponding to translational movement, and first-order sidebands corresponding to angular rotation - a feature exploited in (Stanbridge and Ewins, 1996b) as a means of measuring these.)

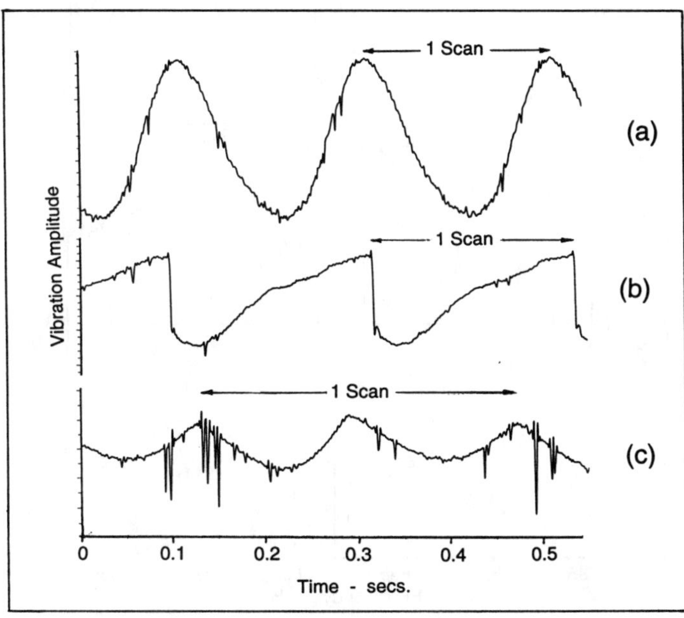

Figure 8 Mode Shapes measured from Demodulated CSLDV Circular Scans: (a) on Defect-Free Plate, (b) on Plate with Saw-Cut, (c) on Plate with Fatigue Crack.

CIRCULAR SCANS

A potential problem with the above method is that a straight-line scan would be unlikely to give a good indication of any defect which happened to run along the line of the scan. This can be overcome by using a circular scan, obtained by applying a sine input to the LDV x-axis scan mirror and a cosine wave to the y-axis mirror. Whatever its orientation, a defect is then bound to cross the scan line at right angles, if the scan centre is suitably positioned.

In this situation the scan is at a uniform rate around the circle, and the spectrum sidebands give spatial Fourier coefficients for the mode shape directly. As before, phase data are necessary for a full mode shape description, but again this is unnecessary in this application.

Figure 8 shows demodulated mode shapes derived from a 10 mm diameter scan for each of the three plates, centred on C-D as a diameter (in Figure 1). The scan can be seen to have crossed the sawcut once and the crack twice. For some reason speckle noise was more pronounced in the circular scan on the cracked plate (Figure 8(c)).

Figure 9 shows CSLDV spectra corresponding to the plots in Figure 8, the wide-ranging spectrum for the saw-cut plate is particularly striking.

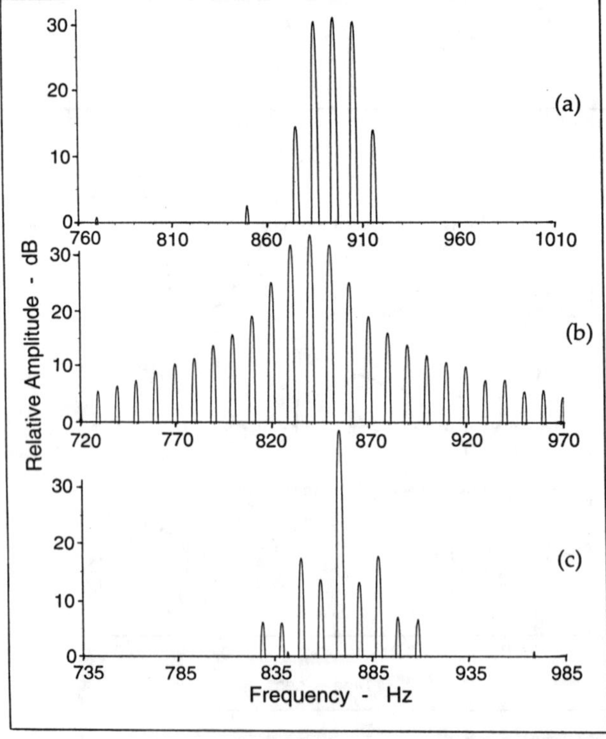

Figure 9 CSLDV Circular Scan Spectra: (a) on Defect-Free Plate, (b) on Plate with Saw-Cut, (c) on Plate with Fatigue Crack.

CONCLUDING REMARKS

The somewhat-artificial experiments described above have demonstrated a clear potential for the continuously-scanning laser Doppler vibrometer as a means of damage detection and location in structures, if vibration to flex the defects can be induced, and if the defects are such as to produce localised mode shape distortions. A small-diameter circular scan appeared most likely to give a reliable and unambiguous defect indication. To survey a structure's surface the circle would have to be scanned in turn over the whole area.

The paper reports a research project which is only part-completed. Spatial Fourier spectra have been explored as a means of providing a quantifiable defect indicator, but these were not quite as effective as had been hoped and further work is required on this aspect.

Additional, more realistic and less extreme failure samples, including cracked concrete specimens, are to be investigated.

ACKNOWLEDGEMENTS

The authors wish to acknowledge the support given by BRITE-EURAM 'MARS' Project No. 5464, and their partners in that Project, and currently by EPSRC.

REFERENCES

Barker, A. J., 1992, "Rapid Full Field Vibration Pattern Imaging using a Laser Doppler Vibration Sensor.", *Proc. IMAC 10*, San Diego, pp. 650-655.

Meza, R., C. J. Carrasco, R.A. Osegueda, G. James and N. Robinson, 1997. "Damage Detection in a DC-9 Fuselage using Laser Doppler Velocimetry.", *Proc. IMAC 15*, Orlando, pp. 1779-1785.

Sriram, P., S. Hanagud, J. Craig and N. M. Komerath, 1990, "Scanning Laser Doppler Technique for Velocity Profile Sensing on a Moving Surface.", *Applied Optics* 29, pp. 2409-2417.

Stanbridge, A. B. and D. J. Ewins, 1995, "Structural Modal Analysis using a Scanning Laser Doppler Vibrometer", *R. Ae. Soc. Forum on Aeroelasticity and Structural Dynamics*, Manchester, pp. 85.1-85.7.

Stanbridge, A. B. and D. J. Ewins, 1996a, "Using a Continuously-Scanning Laser Doppler Vibrometer for Modal Testing.", *Proc. IMAC 14*, Dearborn, pp. 816-822.

Stanbridge, A. B. and D. J. Ewins, 1996b, "Measurement of Translational and Angular Vibration using a Scanning Laser Doppler Vibrometer", *Shock and Vibration*, 3 (2), pp. 141-152.

Nondestructive Evaluation of Damage in Stitched Composites Using Laser Based Ultrasound

R. F. ANASTASI, A. D. FRIEDMAN, M. K. HINDERS and E. I. MADARAS

ABSTRACT

Composite material applications in civil transport aircraft primary structures such as wings are currently being planned. In a building block approach, sub components of the wing structure are tested to understand the structural response and failure characteristics. Nondestructive evaluation is an important element which can provide quantitative information on the damage initiation, propagation, and final failure modes for the composite structural components. Although ultrasound is generally accepted as a test method, the use of conventional ultrasound in situ health monitoring of damage during a test is not practical. The use of lasers to both generate and detect ultrasound extends the application of ultrasound to in situ sensing of damage in a deformed structure remotely and in a non-contact manner. The goal of the present research is to utilize this technology to monitor damage progression during testing.

The present paper describes the application of laser based ultrasound to quantify damage in thick stitched composite structural elements to demonstrate the method. This method involves using a Q-switched laser to generate a rapid, local linear thermal strain on the surface of the structure. This causes the generation of ultrasonic waves into the material. A second laser illuminates the surface and a Fabry-Perot interferometer measures the surface deflections from the reflected light. The use of fiber optics provides a convenient method of delivering the laser over long distances to the specimens. The material for these structural elements is composed of several stacks of composite material assembled together by stitching through the laminate thickness. The specimen thickness varies from 0.5 to 1.0 inch.

R.F. Anastasi, U.S. Army Vehicle Technology Center, ARL, Nondestructive Evaluation Sciences Branch, NASA Langley Research Center, Hampton, VA 23681

A.D. Friedman and M.K. Hinders, Dept. of Applied Science, The College of William and Mary, Williamsburg, VA 23187

E.I. Madaras, Nondestructive Evaluation Sciences Branch, NASA Langley Research Center, Hampton, VA 23681

The specimens used for these nondestructive evaluation studies had either impact damage or skin/stiffener interlaminar failure. Although no visible surface damage existed internal damage was detected by laser based ultrasound.

INTRODUCTION

Nondestructive evaluation is a standard component of aviation safety programs, where it is used to provide quantitative information on the presence and degree of damage in aviation structures. Engineers are also recognizing the value of nondestructive evaluation as a tool that could be used in laboratories during the development phases of new materials and structures. The goal is to use such laboratory testing to continuously monitor a structure's health and identify damage initiation, propagation, and final failure modes. The demands on these types of nondestructive evaluation techniques are different in a laboratory environment because of the potential for dynamic testing. This is leading to system improvements that also represent opportunities for the aviation industry.

Ultrasonics is an example of such an opportunity. Ultrasound is a well accepted test methodology used in the aviation industry, but the use of conventional ultrasound for in situ health monitoring of damage evolution during a laboratory test is not currently practical. A system improvement for ultrasonics is the use of lasers to both generate and detect ultrasound. This extends the application of ultrasound to in situ sensing of damage in a deformed structure remotely and in a non-contact manner. Using this technology in a laboratory setting will allow the testing of sub-components, components and then full scale test articles. It will also generate the knowledge base for application of this unique technology to new commercial aviation structures.

Laser ultrasonic generation and interferometric detection methods have been described in detail by other authors, (Scruby and Drain, 1990; Monchalin and Heon, 1986; Wagner, 1990; Hutchins, 1988; Monchalin et al. 1989). Ultrasound is generated by laser pulses rapidly heating a thin material surface layer. This small heated volume expands creating stresses that act as a source of elastic waves in the material. The resulting surface displacement, u, in an isotropic material, using a simple one-dimensional model is given by (Scruby and Drain, 1990),

$$u = \frac{(1+\nu)}{(1-\nu)} \frac{\alpha \delta E}{\rho C} \tag{1}$$

where ν is the Poisson's ratio, α is coefficient of linear expansion, δE is absorbed laser pulse energy assumed to have a uniform energy distribution, ρ is the material density, and C is the specific thermal capacity of the material. More recently, authors have developed analytical and numerical models that take into account thermal conductivity (Telschow and Conant, 1990) and laser beam intensity distributions in orthotropic material (Dubois et al., 1994). These more advanced models may help understand the generation of ultrasound by non-uniform laser beam intensity distributions in anisotropic composite materials. Two additional mechanisms for generating ultrasound occur when material is melted or ablated by the laser, and when that material surface is constrained by paint or other thin layer. The amplitude of the elastic waves by both these methods can be 100 times greater then those generated by thermoelastic expansion. However, the ablation process damages the material and painting may not always be possible.

Typical ultrasonic vibrations generated by normal piezoelectric or laser sources have a 1 to 50 nanometer amplitude and 100 kHz to 10 MHz frequency range. Backscatter of a probe laser beam is used to detect this ultrasonic vibration and convert it to an electronic signal. The vibrating surface causes the backscatter laser wavelength to be modulated by the vibrating surface. The Doppler effect describes how the wavelength of light reflected from the vibrating surface is shifted by an amount proportional to the surface velocity. This shifted wavelength, λ_s is given by

$$\lambda_s = \left(1 - 2\frac{\mathrm{v}}{c}\right)^{-1} \lambda_0 \tag{2}$$

where v is the velocity of the vibrating surface, c is the speed of light and λ_0 is the unshifted laser wavelength. Two interferometric methods are capable of detecting this wavelength modulation: optical heterodyning and velocity interferometry. In the first method a reference beam is made to interfere with an ultrasonic encoded beam. In the second method, the ultrasound encoded beam is made to interfere with a split portion from itself which is time delayed. The first method is limited by its low light gathering efficiency caused by backscatter from optically rough surfaces. The Fabry-Perot interferometer, a multiple wave velocity interferometer, over comes this light gathering inefficiency.

The classic Fabry-Perot interferometer is composed of two plane partially reflecting mirrors separated to form a cavity. When light from the vibrating surface enters the cavity, multiple reflection are generated between the partial mirrors with some light transmitted and reflected out of the cavity. The relation governing the intensity of light transmitted, I_t , given by (Born and Wolf, 1964) is

$$I_t = \frac{(1-R)^2}{(1-R)^2 + 4R\sin^2(\phi)} I_i \tag{3}$$

where I_i is the intensity of incident light, R is the reflectivity of the mirrors, and $\phi = (2\pi nL/\lambda_i)\cos\theta$ is the phase difference due to the mirror separation, in which n is the index of refraction of the cavity, L is cavity length or mirror separation, λ_i is the incident optical wavelength, and θ is the angle of incidence of light upon the mirror.

The connection between the Fabry-Perot interferometer and modulated backscatter can be seen in the phase term of equation (3). When ultrasonic vibrations move the surface of a structure they cause a Doppler shift in the wavelength of light backscattered from the probe laser beam. Thus, substituting the Doppler shifted wavelength λ_s, in equation (2) for λ_i in equation (3), illustrates how the Fabry-Perot interferometer signal will be modulated by the ultrasonic vibration.

SYSTEM DESCRIPTION

Figure 1, shows major components of the laser based ultrasonic system including the generation and detection lasers, Fabry-Perot interferometer, and some optical components. The complete system consists of a PC computer, 100 MHz, 8 bit analog to digital board, encoder board, translation stage controller, and translation stage with axis encoders and cable drivers. The generation and detection lasers are Nd:YAG lasers, one for ultrasonic generation and the second for detection. The generation laser is a pulsed laser with a wavelength of 1064 nm, 6.5 mm beam diameter, and a variable pulse energy up to 200 mJ with a 12 nsec pulse duration. The detection laser is a continuous wave laser with a 532 nm wavelength, 0.32 mm beam diameter and 400 mW maximum output power.

The system is operated using a PC-based virtual instrument controller and data acquisition package that stores ultrasonic A-scan signals. The main control screen has control buttons and graphics that allow the user to see scanner motion, signals, and an image of the scanned area. Ultrasonic signals are digitized to eight bit resolution at sampling rates of 6.25, 12.5, 25.0, 50.0 or 100 MHz. These are chosen by the user and can have up to 4000 points in length. The scanning area, also user selectable, can have a width of 24 inches, length of 36 inches, and can be scanned with a spatial resolution of 0.05 inches to 1.00 inch.

The system is being developed for inspecting large samples in a structural testing laboratory. The large open laboratory area raises concerns for eye safety. This system incorporates fiber optics to deliver generation and detection laser beams and to transmit the laser reflections back to the interferometer. Fiber optic cables allow the laser sources to be remote and contained. The inspection end of the fiber optic cables are attached to the scanning stage that translates the cables and thus the laser beams across the sample surface in a raster pattern. It can be enclosed to contain laser beam reflections, making the system eye safe. The system is currently being assembled and tested in an optics laboratory with three fiber optic cables to deliver and detect laser energy. Reducing the number of fibers is presently being pursued by using partial mirrors, polarizers, and filters to control laser energy delivery to and from the sample.

Figure 1. Laser Ultrasonic System Lasers, Fabry-Perot Interferometer, and Optics.

TEST SPECIMENS AND TESTING

Specimens evaluated in this study were obtained from a composite wing stub box that had been structurally tested to failure. The stub box is shown in Figure 2, between the extension box and a transition box. The skin consists of ten stacks of $[45/-45/0_2/90/0_2/-45/45]$ AS-4-3501-6 graphite epoxy material stitched together, resulting in a 0.58 inch total thickness. Stitched reinforcements of kevlar are in rows spaced 0.2 to 0.5 inches apart with stitch spacing approximately 0.1 inches.

A flat plate specimen was impact damaged in a drop weight test machine to create internal delaminations. A one inch diameter impact head weighing 25.25 lb. was dropped from a height of 3 ft 11.5 inches to generate an impact energy of 100 ft-lb. This produced a dent approximately 1 mm in depth and small localized cracks on the back surface of the panel. A second specimen has bi-directional T-stiffeners attached to the skin with through-the-thickness stitches. Stiffeners are made of the same material as the skin. During wing box structural testing this joint failed in regions along the stiffener/skin interface. These samples (see Figure 3) show no damage from the top surface.

Figure 2. The Composite Wing Stub Box (above the person in photograph) is located between a metal extension box (show center to right side of the photo) and a metal transition box attached to the wall. The metal boxes were used to form realistic loads in the Wing Stub Box.

These samples were laser ultrasonically scanned in the laboratory by placing them 8-10 inches from the system optics. The generation and detection beams at the sample surface were approximately 5 mm and 2 mm in diameter, respectively. The scanned area on the impacted plate was 2.5 inches by 2.5 inches with a 0.05 inch spatial resolution. Scanning on the stiffened plate was 6.0 inches by 6.0 inches with a 0.10 inch spatial resolution. Amplitude C-scans or density plots were generated during the scanning process by gating the A-scan signal and plotting the amplitude signal under the gate as a color or gray level. A-scans were stored to file and were post-processed for evaluation. Contour plots of the same data helped highlight the contrast between stitching rows and damaged and undamaged regions.

To improve imaging of internal damage, stored A-scans were processed using methods similar to those described by (Gammell, 1981) and (Smith et al. 1989, 1991). In the first processing step, signals were time gain amplified to account for reduced echo amplification due to material attenuation. To remove some amplified noise, generated by this operation, signals were band-pass filtered in the frequency domain and then transformed back into the time domain from which the magnitude of the analytic signal was calculated. This signal processing resulted in a smooth positive value function which is an envelope of the filtered signal. Signal bandwidth was in the range of 1 to 5 MHz.

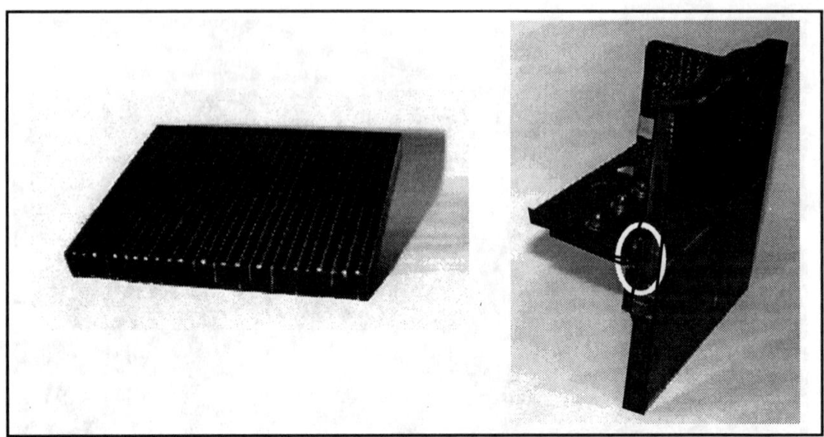

Figure 3. Thick Stitched Composite Samples, (left) Impacted Plate with dimensions 4.5" x 5.5" x 0.58" and (right) Bi-directional Skin/Stiffened Plate with interlamina failure (circled in photograph) with dimensions 7.0" x 7.0" x 0.58".

RESULTS AND DISCUSSION

Through-the-thickness stitches in composites produce several effects on ultrasonic imaging. One is caused by increased interlaminar shear strength which moderates the damage and subsequent image. In conventional composites, impact damage image would present itself as an irregular shape about the impact site, preferentially oriented along ply directions in the lamina. In stitched composites, the impact site tends to be circular because through-the-thickness stitches increase interlaminar shear strength which slows and contains delamination progression. Secondly, these through the thickness stitches also perturb ultrasound transmission (Long, et al 1990) because the sound velocity is greater in the stitching than the surrounding material. These perturbations cause a phase-distorted wavefront at the material surface or point of ultrasonic detection and produce a reduced signal amplitude. Both these effects are visible in Figure 4, as a dark circular area and light and dark bands. This figure was generated by measuring the amplitude of the back surface ultrasonic echo. Spacing between the bands is approximately 0.2 inches and corresponds to stitch spacing in the sample. Representative A-scan signals from these areas are shown in Figure 5. A conventional ultrasonic signal with a back surface echo is shown in Figure 5a, which represents a signal from light band regions. The back wall echo occurs approximately 10 microseconds after the front wall signal. A representative signal from the dark band region, shown in Figure 5b, lacks a back surface echo and has reduced interior echoes. A signal from the dark circular area, shown in Figure 5c, also lacks a back surface echo but has large echoes occurring during the first five microseconds from internal damage near the front surface. This circular region is approximately 2.25 inches in diameter and results from the one inch diameter impactor that created only a millimeter depth indentation on the sample surface. A third interesting feature in Figure 4 is seen at the center of the damage, where there is evidence that sound penetrates to the back wall at certain points in spite of the damage. The stitches may be partially responsible for the transmission of ultrasound within this area of impact. If the through-the-thickness stitches do not break during the impact, it is possible that they act as waveguides for the sound to travel to the back wall and back. While in other location in the damage area, sound travel to the back wall is blocked. Therefore, at these locations no phase cancellation occurs and some waveguided signals are detected.

The second stitched composite sample, studied using laser based ultrasound has a complex structure consisting of a horizontal and a vertical stiffener. It is

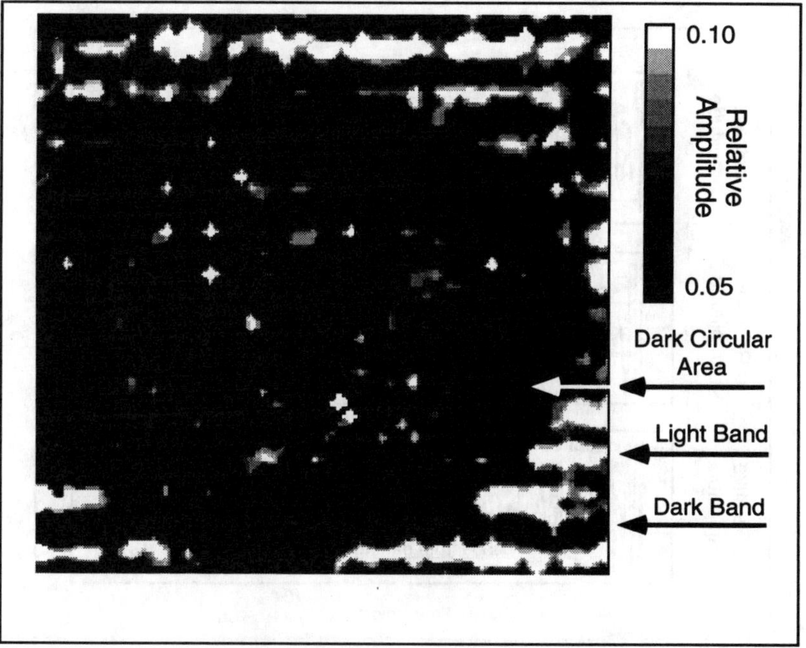

Figure 4. Peak detected back wall C-scan contour plot of impact damaged plate. Scan area is 2.5" by 2.5" and the scan spatial resolution is 0.05". The dark circular area is approximately 2.25" in diameter and horizontal lines correspond to stitching rows.

shown in Figure 3 and its horizontal and vertical profiles of varying thickness are shown along with scan results in Figure 6 and Figure 7. To highlight this structure ultrasonically, a back surface echo from a particular thickness was isolated and measured. Two resulting scans are shown in Figure 6. The first figure, Figure 6a, measures the echo from the skin or thinnest section of the sample. This scan shows dark horizontal and vertical lines that correspond directly to stitched rows as well as a wide dark vertical band, greater than three inches thick, that corresponds to the thicker stiffener section. This region is dark because its back surface echo is much later due to a thickness increase and is therefore ignored in the figure. The result of shifting the sampling point to detect the thicker material section is shown in Figure 6b. Here a dark horizontal and vertical band can be seen with the vertical band being about 0.7 inches thick at its widest point. These bands correspond to the horizontal and vertical webs of the stiffener, the thickness of the vertical web is approximately 0.6 inches. With a finer scanning spatial resolution these thickness values would be approximately equal. The vertical white bands represent the back wall of the thicker web and also show evidence of stitching. Taken together these show underlying structure of the sample but no substantial evidence of damage.

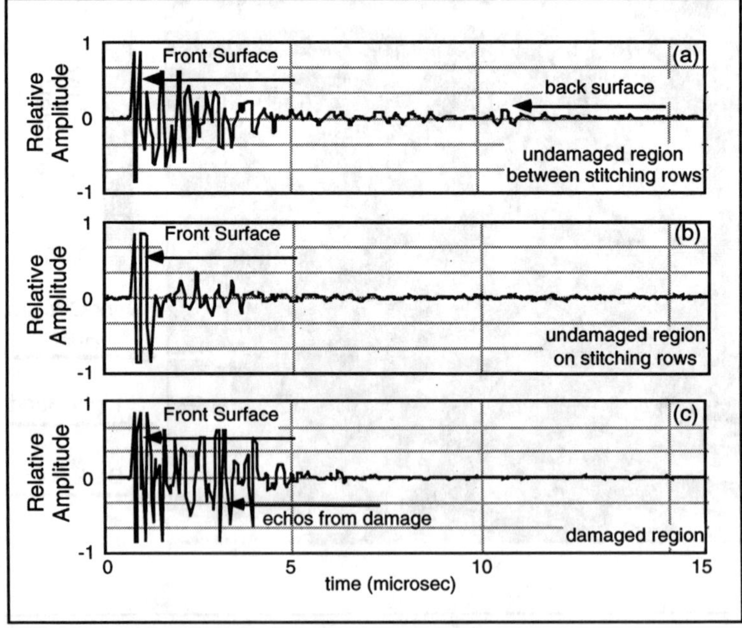

Figure 5. Representative A-scan signals from damaged and undamaged regions of impacted plate (a) form light band, (b) from dark band, and (c) from dark circular area.

To highlight internal damage, a third image was created by plotting the location in time of the back wall echo instead of the amplitude. Damage would be highlighted if the echo occurred earlier than anticipated. The result in Figure 7, shows evidence of damage located above the horizontal stiffener region, on the left and right sides. The circled region on the scan corresponds to the left side of the sample and is circled and outlined in the picture. This delamination occurs about the same depth as the thinnest back wall section. Thus the time-of-flight to the delamination is similar to the time-of-flight to the back wall of the skin, seen in the four corners of the scan. If the sample was undamaged, these regions would be shaded gray instead of black. In an intact wing structure, this stiffener flaw would be internal to the structure and could be over looked if visual examination of back and side surfaces were not possible.

These scans demonstrate that laser generated and detected ultrasound via fiber optic cable is capable of detecting damage in thick stitched composite material in situ. Fiber optic cable provides a means of delivering eye hazardous laser energy to the inspection area in order to generate and detect ultrasound in a non-contact manner to monitor structural health and ultrasonically map internal damage and underlying structure.

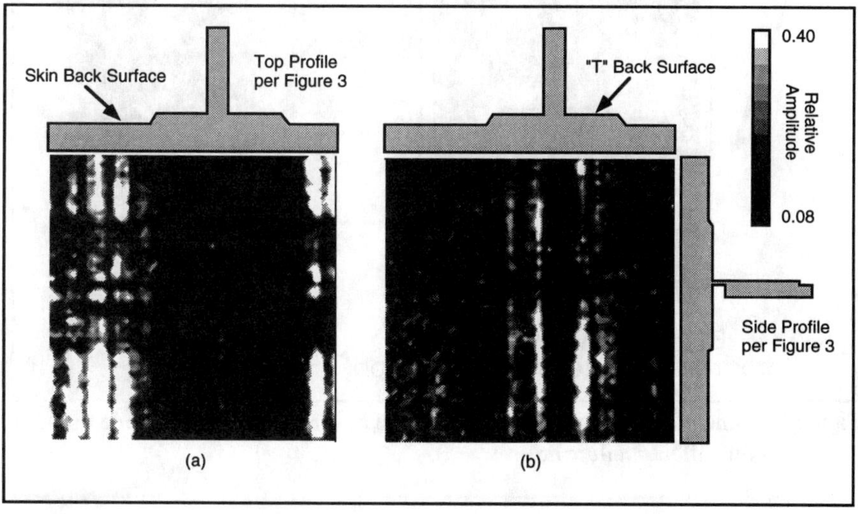

Figure 6. Peak detected back wall C-scan contour plots of bi-directional skin/stiffened sample with; (a) gate over skin back surface and (b) gate over thicker "T" back surface.

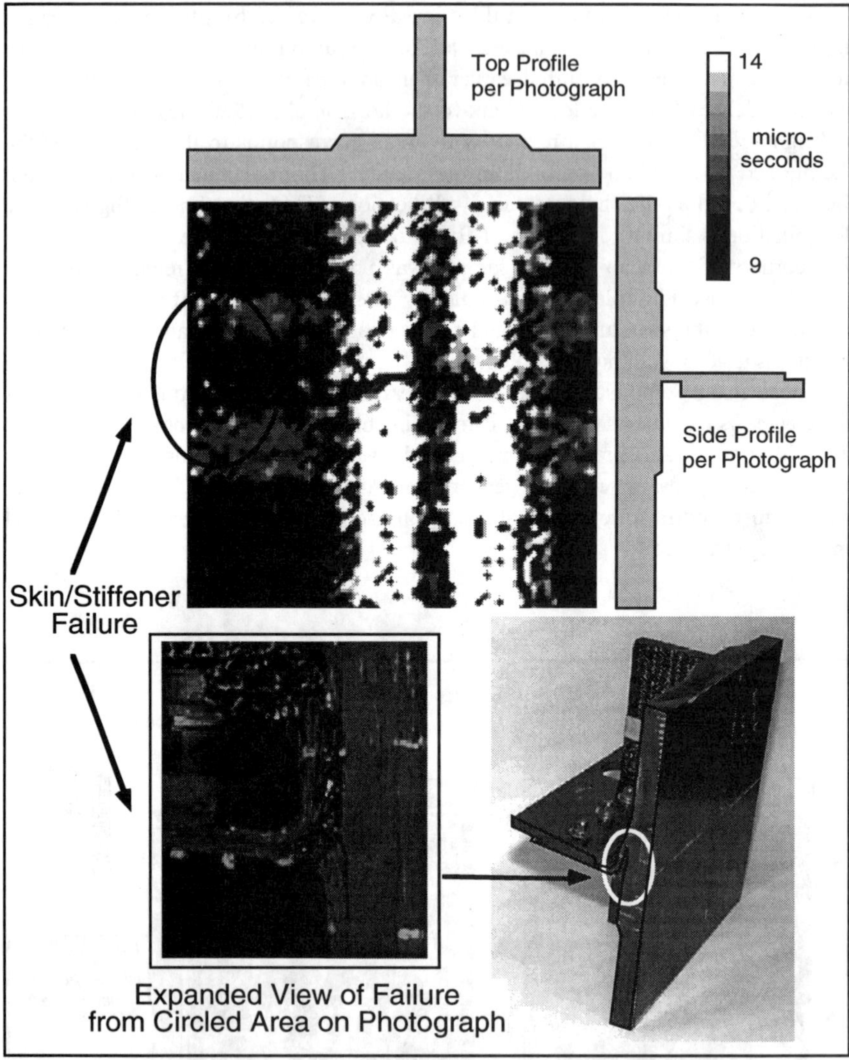

Figure 7. Time-of-Flight C-scan contour plot of bi-directional skin/stiffened sample showing skin/stiffener failure areas.

REFERENCES

Born, M. and E.Wolf, *Principles of Optics*, Pergamon Press, New York, 1964.

Dubois, M., F. Enguehard, M. Choque, J.-P. Monchalin, and L. Bertrand, "Numerical Modeling of Laser Generation of Ultrasound in Orthotropic Materials," in *Review of Progress in Quantitative*

Nondestructive Evaluation, Vol. 13, Edited by D.O. Thompson and D.E. Chimenti, Plenum Press, New York, 1994.

Gammell, P.M., "Improved Ultrasonic Detection using the Analytic Signal Magnitude," *Ultrasonics*, Vol. 19, March 1981.

Hutchins, D.A., *Ultrasonic Generation by Pulsed Lasers*, Physical Acoustics. Vol. 18, 1988.

Long, E.R., S.M. Kullerd, P.H. Johnston, and E.M. Madaras, "Ultrasonic Detection and Identification of Fabrication Defects in Composites," Presented at the First NASA Advanced Composite Technology Conference, Seattle, Washington, NASA Technical Report #CP3104, October 29 - November 1, 1990.

Monchalin, J.-P., "Optical Detection of Ultrasound," *IEEE Transactions on Ultrasonics, Ferroelectrics, and Frequency Control*, Vol. UFFC-33, No. 5, September 1986.

Monchalin, J.-P. and R. Heon, "Laser Ultrasonic Generation and Optical Detection with a Confocal Fabry-Perot Interferometer," *Materials Evaluation*, Vol. 44, September 1986.

Monchalin, J.-P., R. Heon, P. Bouchard, and C. Padioleau, "Broadband Optical Detection of Ultrasound by Optical Sideband Stripping with a Confocal Fabry-Perot," *Applied Physics Letters*, 55(16), October 1989.

Scruby, C.B., L.E. Drain., *Laser Ultrasonics: Techniques and Applications*, Adam Hilger, New York, 1990.

Smith, B.T., J.S. Heyman, A.M. Buoncristiani, E.D. Blodgett, J.G. Miller, and S.M. Freeman, "Correlation of the Deply Technique with Ultrasonic Imaging of Impacted Damage in Graphite-Epoxy Composites," *Materials Evaluation*, Vol. 47, No. 12, December 1989.

Smith, B.T., G.L. Farley, J. Maiden, D. Coogan, J.G. Moore, "Damage Assessment and residual Compression Strength of Thick Composite Plates with Through-The-Thickness Reinforcements," in *Review of Progress in Quantitative Nondestructive Evaluation*, Vol. 10B, Edited by D.O. Thompson and D.E. Chimenti, Plenum Press, New York, 1991.

Telschow, K.L. and R.J. Conant, "Optical and Thermal Parameters Effect on Laser-Generated Ultrasound," *J. Acoust. Soc. Am.*, 88(3), September 1990.

Wagner, J.W., *Optical Detection of Ultrasound*, Physical Acoustics, Vol. XIX, Academic Press, 1990.

SESSION 2

CIVIL INFRASTRUCTURES I

Control of Bridges Using Magneto-Rheological Fluid Dampers

B. C. HANSEN,[1] F. GORDANINEJAD,[2]
M. SAIIDI[3] and F.-K. CHANG[4]

ABSTRACT

Supplemental energy-dissipating systems for bridge structures have been proposed to mitigate harmful effects of earthquakes and severe storms. Controllable devices are among these systems and have attracted significant attention because they potentially provide immediate vibration control for structures subjected to destructive seismic motions. In this work, theoretical and experimental studies are performed to evaluate the response of a scaled model bridge using magneto-rheological fluid (MRF) dampers. The scaled model bridge is investigated for its dynamic response to various inputs when MRF dampers are utilized to control the structure. The results demonstrate that MRF dampers can proved to significantly improve the response of the scaled model bridge.

INTRODUCTION

A variety of energy-dissipating devices and systems have been proposed for mitigating the harmful effects of earthquakes. Passive, active, and semi-active systems are considered for controlling structures and systems [1]. Semi-active control systems are receiving significant attention because they provide much-needed technologies that bridge the gap between the active and passive structural control systems. Examples of such devices include variable orifice dampers [2-3], variable stiffness devices [4], friction controllable isolators and braces [5-9], and controllable fluid devices.

Passive control systems have been used to control motions of structures during seismic or wind vibrations. The term "passive" is used to indicate that these systems do not require external power to operate. Moreover, the motion of the structure produces relative motion within the damping device, which, in turn, dissipates energy. Such passive devices include: yielding of mild steel, viscoelastic action in rubber-like materials, sloshing of fluid, shearing of viscous fluid, orificing

[1] Graduate Student, Mechanical Engineering Dept., University of Nevada, Reno, NV 89557, USA
[2] Professor, Mechanical Engineering Department, University of Nevada, Reno, NV 89557, USA
[3] Professor, Civil Engineering Department, University of Nevada, Reno, NV 89557, USA
[4] Associate Professor, Department of Aeronautics and Astronautics, Stanford University, Stanford, CA 94305, USA

of fluid, and sliding friction [1]. Active systems, which typically consist of electro-hydraulic actuators and system feedback control loops, require large amounts of power (on the order of tens of kilowatts).

A semi-active system may consist of a variable actuator, but requires low power to operate. Semi-active systems have the control flexibility of active systems and capabilities of passive systems. Magneto-rheological fluid dampers are a type of semi-active devices that may potentially provide safe and reliable vibration control to structures under seismic activities. MRF dampers are a very promising class of semi-active devices because of their mechanical simplicity, environmental robustness, and demonstrated potential for full-scale applications [10]. The present work addresses the control capabilities of MRF dampers in a two-span scaled model bridge, both theoretically and experimentally.

THEORETICAL ANALYSIS

A finite element analysis is performed to estimate the response of a two-span 1/12 scaled model bridge using the steel column when subjected to two different sinusoidal motions. A Direct Linear Transient Analysis is conducted with the exact dimensions of the experimental scaled model bridge. Two sinusoidal inputs were examined. The first has a peak displacement of 0.3 cm (0.11 in) with a frequency of 2 Hz. The second input has a peak displacement of 0.1 cm (0.37 in) with a frequency of 4 Hz. The response of the scaled model bridge is analyzed for three different damping conditions: the case where 1) no damper is used, 2) the damping coefficient is equal to that of the MRF dampers when connected but not activated with any electric current, C_0, and 3) the damping coefficient is equal to that of the MRF dampers when connected and activated with 2.0 Amps of current input, C_2. The equivalent damping coefficients, C_0 and C_2 are determined experimentally for a sinusoidal motion at 2 and 4 Hz each with 1.0 cm peak displacement.

Structural damping is also a parameter needed for the finite element analysis. Logarithmic decrement method along with the snap test are used to determine the percentage of the critical damping of the scaled model bridge. The results of the snap test yielded a 2% critical damping value with a natural frequency of 3.0 Hz for the steel column.

EXPERIMENTAL STUDY

A brief description of the experimental set-up is to follow. A horizontal shake table, which can carry a load of 8,880 N (2,000 lbf), supports the bridge and the base of the two abutments. The abutments are made from 1.27 cm (1/2 in) thick steel plate and have adjustable mounts that extend under the deck. Four roller bearings are placed in grooves cut in the top surface of each of the extended mounts and interface with the underside of the deck. The deck is a solid steel plate 30.48 cm (12 in) wide and 60.96 cm (24 in) in length. A steel box filled with lead is mounted on the top of the deck and has a total combined mass of 680.4 Kg (1,500 lb.). A steel column is joined via solid circular steel inserts with a flat base to the center of the deck and to the area between the bases of the two abutments. The base of one solid circular insert bolts to the magnesium plate between the abutments while the solid circular protrusion is pressed into the end of the column. The base of the other solid circular insert bolts directly to the underside of the deck while its circular protrusion is pressed into the other end of the column. Two MRF dampers are mounted on the underside of the bridge deck and attached to one of the abutments. Figure 1 is a photograph of the scaled model bridge.

Snap tests are performed on the steel column to determine its natural frequency and percent of critical damping. The MRF dampers are not connected for these snap tests. Two sinusoidal tests are performed on the structure and analyzed for five damping scenarios: 1) no damper is used and 2) each MRF damper is connected to the abutment and activated with 0.0, 0.5, 1.0, and 2.0 Amps of current input. The final testing phase of the project involves random testing using a scaled version of the El Centro earthquake. The Power Spectrum Density (PSD) profile of the scaled earthquake is used in the experiment and can be seen in Figure 2. These tests are performed and used to analyze the capabilities and effects of the MRF dampers.

RESULTS AND DISCUSSION

Theoretical and experimental results for the two-span scaled model bridge tests when subjected to two sinusoidal motions and an earthquake input are presented. The relative displacement between the deck and the abutment is analyzed for each input with various MRF damping situations. Four snap tests are performed to determine the natural frequency of the model and the percent of structural damping. The natural frequency of the column is obtained by taking the average for four tests. The logarithmic decrement method is used on the average of the peak values for the four tests to determine the percent of structural damping. The measured natural frequency is 3.0 Hz with a structural damping coefficient of 2.0%.

Relative displacements obtained from the finite element analysis are generated by subtracting the two absolute displacements of the bridge deck and the abutment. This process is repeated for each damping scenario. The results of the 2 Hz input for all three damping cases can be seen in Figure 3. A 40.2% reduction in peak motion is obtained by changing the damping coefficient from 0.0 (no damper) to the C_0 equivalent damping coefficient (connected but not activated) for the MRF damper. When the damping coefficient is changed from 0.0 to the C_2 equivalent damping coefficient (activated with 2.0 Amps), the relative displacement reduces 52.8%. The results of the 4 Hz data for all three damping cases can be seen in Figure 4. A 36.7% reduction in peak displacement is obtained by changing the damping coefficient from 0.0 to the C_0. When the damping coefficient is changed from 0.0 to the C_2, the maximum relative displacement reduces 54.6%.

Experimental response of the system is first analyzed for a sinusoidal input of 2 Hz without the MRF dampers connected. The result of the test yields a maximum relative displacement between the deck and the abutment of 0.132 cm. The dampers are then connected and the test is repeated for 0.0, 0.5, 1.0, and 2.0

Amps per damper. When the dampers are connected, but not activated, the maximum relative displacement of the deck with respect to the abutment reduces to 0.071 cm, a reduction of 46.2%. When the MRF dampers are activated with 2.0 Amps, the maximum relative displacement of the deck with respect to the abutment further reduces to 0.058 cm, an additional reduction of 17.7%. The relative displacement of the two-span scaled model bridge under the 2 Hz sinusoidal test with the MRF dampers not present and present, activated with 0.0 and 2.0 amps are shown in Figure 5. Damping greater than 20% of critical typically does not yield additional significant reductions in relative displacement between the deck and the abutment because the system damping capabilities become saturated beyond this point. Hence, the two-span scaled model bridge is at its 20% of critical damping value when 1.0 Amp is applied to the MRF dampers.

The response of the system is then analyzed for a 4 Hz sinusoidal input without the MRF dampers connected. The result of the test yields a maximum relative displacement between the deck and the abutment of 0.278 cm. The dampers are then connected and the test is repeated for 0.0, 0.5, 1.0, and 2.0 Amps per damper. When the dampers are connected, but not activated, the maximum relative displacement of the deck with respect to the abutment reduces to 0.155 cm, a reduction of 44.2%. When the MRF dampers are activated with 2.0 amps, the maximum relative displacement of the deck with respect to the abutment further reduces to 0.1175 cm, an additional reduction of 24.2%. The results are shown in Figure 6. The saturated 20% of critical damping has not been reached by the MRF dampers for this test. This is due to the lower damping equivalent of the MRF dampers at higher frequencies. Nonetheless, the relative displacement between the deck and the abutments are reduced by 23.9% from when the MRF dampers are not activated to when they are activated at 2.0 amps.

Results from the theoretical finite element analysis are compared with the results obtained experimentally for the two-span scaled model bridge tests and show very good agreement. For each sinusoidal input test, the finite element analysis and experimental results for the steel column with the MRF dampers activated with 0.0, 1.0, and 2.0 amps are compared. The largest error for any one of the six sinusoidal input tests is 7.24 %. Table 1 shows the comparison of the theoretical and experimental relative displacements of the two-span scaled model bridge deck for all damping cases.

The two-span scaled model bridge is subjected to a scaled El Centro 1940 earthquake input and analyzed for its dynamic response with the MRF dampers in a passive mode (0.0 amps) and in an activated 1.0 and 2.0 Amp mode. A sample of the displacement response history can be seen in Figure 7. Due to shake table limitations, the random earthquake could not be perfectly reproduced as it is run from a PSD. The El Centro earthquake has two "power" sections: 1) the initial acceleration spike, and 2) the smaller secondary acceleration section. The responses for relative displacement, acceleration, and strain are compared for each test in these two sections and referred to as Section I and Section II, respectively.

The maximum peak-to-peak relative displacement for Section I and II with the MRF dampers connected, but not activated, are 0.636 cm and 0.308 cm, respectively. When the MRF dampers are activated with 1.0 amps, the maximum peak-to-peak relative displacement decreases to 0.405 cm and 0.174 cm, approximately a 40% ± 4% reduction. The maximum peak-to-peak relative displacement further decreases to 0.255 cm and 0.170 cm, when 2.0 amps are supplied to the MRF dampers.

SUMMARY AND CONCLUSIONS

A theoretical model of the two-span 1/12 scaled model bridge using a steel column with various damping coefficients representing the damping capabilities of the MRF dampers is constructed and analyzed for two sinusoidal inputs. The results of which are compared to and match the response of the experimental two-span scaled model bridge specimen.

A 1/12 two-span model bridge is constructed, fitted with the MRF dampers, and tested for its dynamic response for two different sinusoidal motions and a random motion equivalent to a scaled El Centro earthquake. The relative displacement between the deck and the abutment is analyzed for the various damping scenarios.

MRF dampers proved to substantially reduce the relative displacement of the scaled model bridge when subjected to various input motions. The application of MRF dampers to bridge structures has proven to be a success in a scaled environment and perhaps should be explored for full-scale applications.

ACKNOWLEDGMENTS

This project is funded by the National Science Foundation Grant Number CMS-9402535. The authors are thankful for the constant encouragement by Dr. S. C. Liu, the Program Director.

REFERENCES

1. Akbay, Z. and Aktan, H. M., "Intelligent Energy Dissipation Devices," *Proc. of the Fourth U.S. National Conference on Earthquake Engineering,* Vol. 3, No. 4. Pp. 427-435, 1990.

2. Patten, W. N., He, Q., Kuo, C. C., Liu, L., Sack, R. L., "Suppression Of Vehicle Induced Bridge Vibration Via Hydraulic Semi-Active Vibration Dampers (SAVD)," *First World Conference on Structural Control,* Los Angeles, California, USA, pp. FA1-30 -38, 1994.

3. Kurata, N., Kobori, T., Takahashi, M. Niwa, N. and Kurino, H., "Shaking Table Experiments of Active Variable Damping System," *Proceedings of the First World Conference on Structural Control,* Vol. 2 pp. TP2-107 through TP2-108, 1994.

4. Inaudi, J. A. and Kelly, J. M., "Experiments on Tuned Mass Dampers Using Viscoelastic, Frictional, and Shape-Memory Alloy Materials," *Proceedings of*

the First World Conference on Structural Control, Vol. 2 pp. TP3-107, 1994.

5. Kawashima, K., Unjoh, S. and Shimizu, K., "Experiments on Dynamics Characteristics of Variable Damper," *Proc. of the Japan National Symposium on Structural Response Control,* Tokyo, Japan, p. 121, 1992.

6. Akabay, X. and Aktan, H. M., "Actively Regulated Friction Slip Devices," *Proc. of the 6th Canadian Conference on Earthquake Engineering,* pp. 367-374, 1991.

7. Dowdell, D. J. and Cherry, S., "Semi-active Friction Dampers for Seismic Response Control of Structures," *Proc. of the Fifth U.S. National Conference on Earthquake Engineering,* Vol. 1, pp. 819-828, 1994.

8. Cherry, S., "Research on Friction Damping at the University of British Columbia," *Proceedings of the International Workshop on Structural Control,* USC Publication Number CE-9311, pp. 84-91, 1994.

9. Feng, Q. and Shinozuka, M., "Use of a Variable Damper for Hybrid Control of Bridge Response Under Earthquake," *Proc. of the U.S. National Workshop on Structural Control Research.* USC Publication No. CE-9013, pp. 107-112, 1990.

10. Dyke, S. J. Spencer, Jr., B. F., Sain, M. K., and Carlson J. D., "Seismic Response Reduction Using Magnetorheological Dampers," *Proceedings of the 13th Triennial World Congress, International Federation of Automatic Control,* Vol. L, pp. 145-150, 1996.

Table 1. Comparison of Theoretical and Experimental Results of the Maximum Relative Displacements (cm) of the Deck for Various Inputs.

Input	Amps	Theoretical	Experimental	% Difference
2 Hz / 0.3 cm Peak Displacement	No Damper	0.127	0.132	3.79
	0	0.076	0.0705	7.24
	2	0.060	0.058	3.33
4 Hz / 0.1 cm Peak Displacement	No Damper	0.251	0.269	6.69
	0	0.159	0.149	6.29
	2	0.114	0.119	4.20

Figure 1. Photograph of the two-span scaled model-bridge.

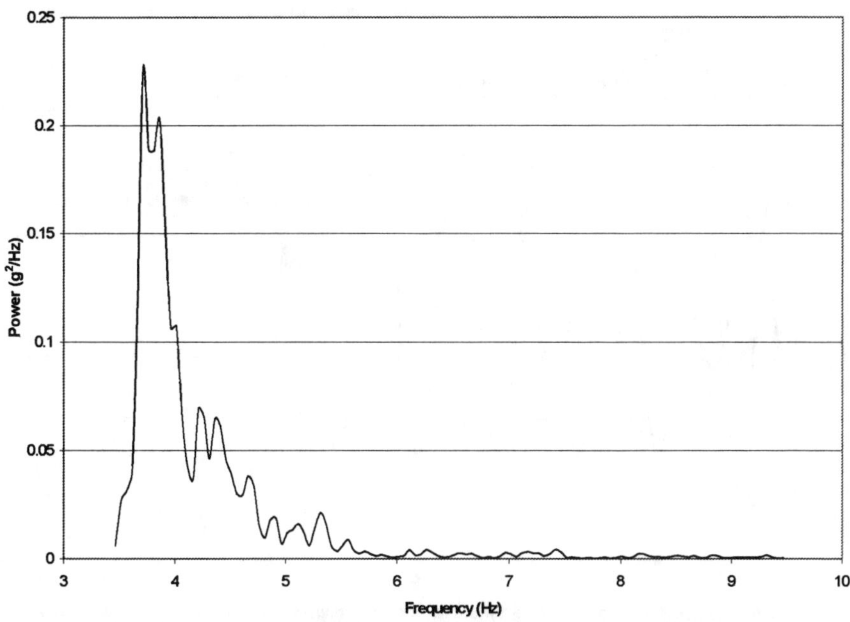

Figure 2. Power Spectrum of the El Centro earthquake of 1940 at 1/12 scale.

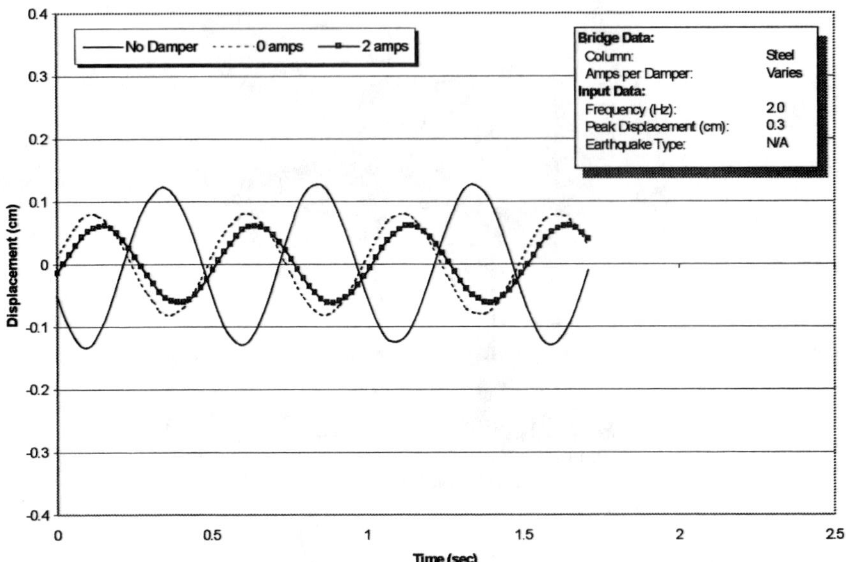

Figure 3. **Theoretical relative displacement of the finite element bridge model under sinusoidal motion using the steel column with MRF dampers not present and present, activated with 0.0 and 2.0 amps.**

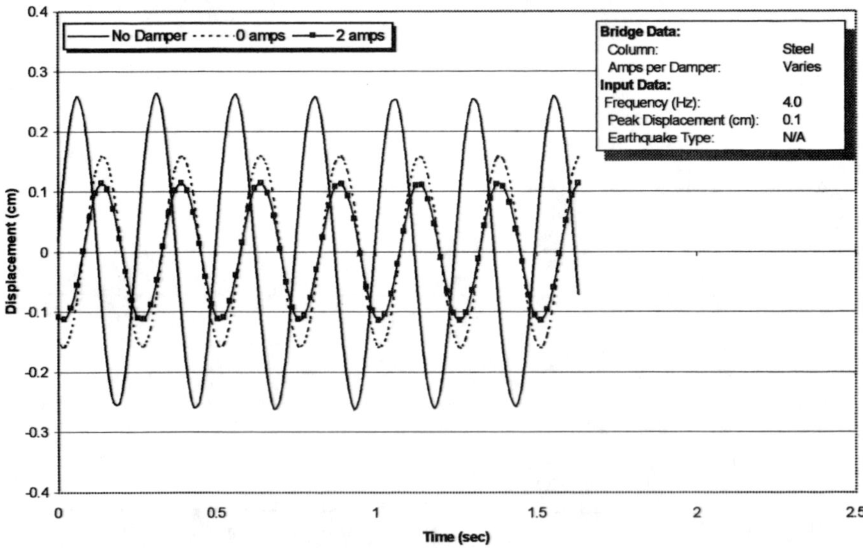

Figure 4. **Theoretical relative displacement for the finite element bridge model under sinusoidal motion using the steel column with MRF dampers not present and present, activated with 0.0 and 2.0 amps.**

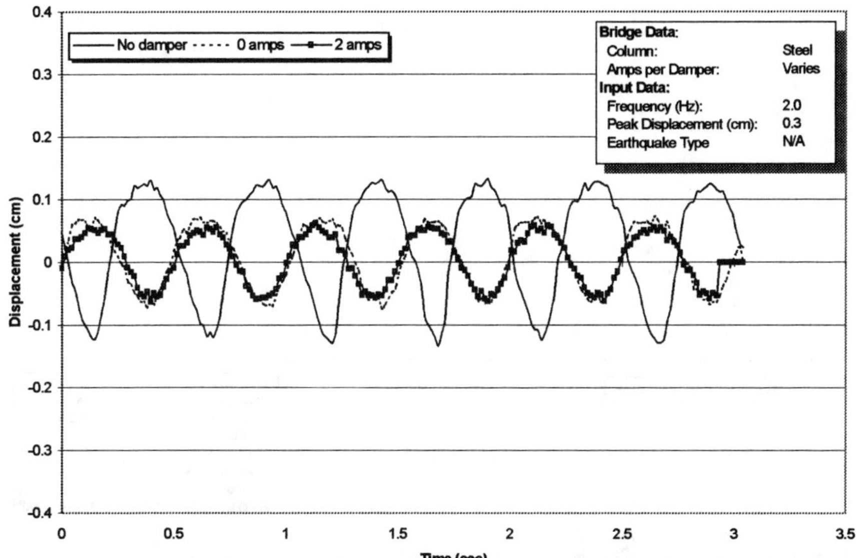

Figure 5. Experimental results of the relative displacement of the two-span scaled model-bridge deck under sinusoidal motion using the steel column with MRF dampers not connected, connected, and activated with 2.0 amps.

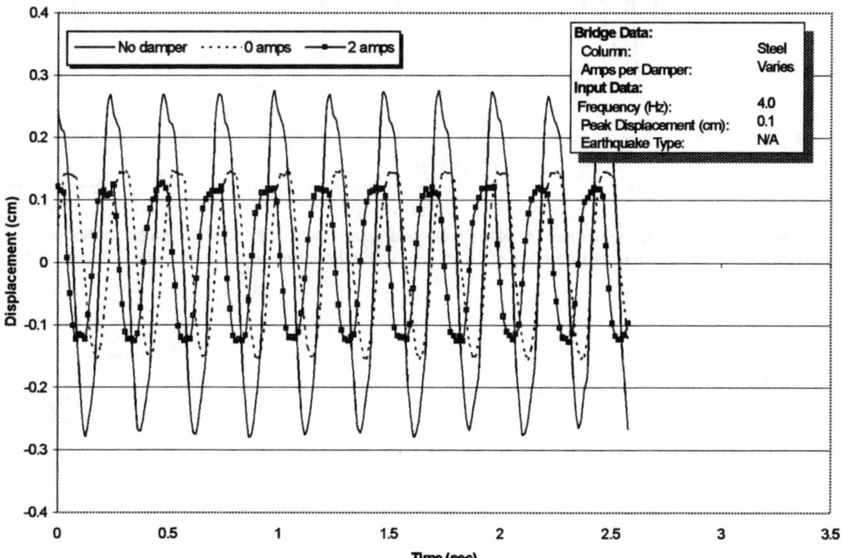

Figure 6. Experimental results of the relative displacement of the two-span scaled model bridge deck under sinusoidal motion using the steel column with MRF dampers not connected, connected, and activated with 2.0 amps.

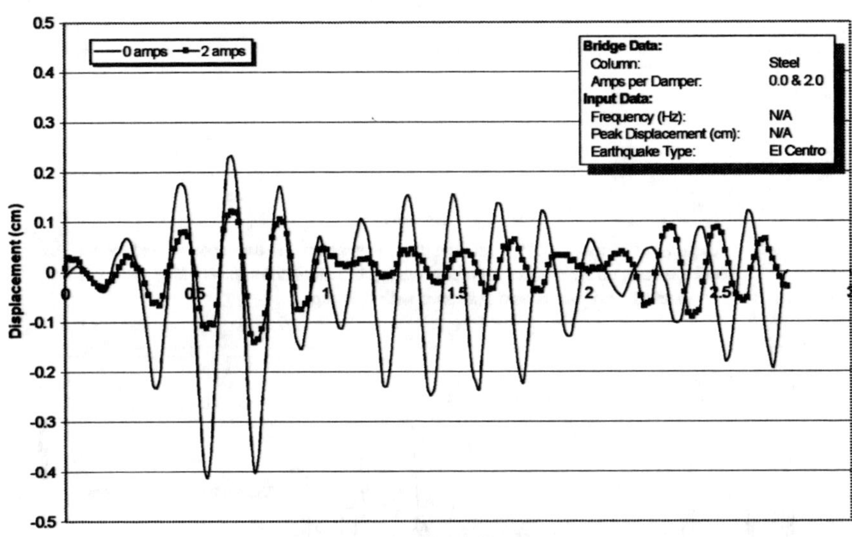

Figure 7. **Experimental results of the relative displacement of the two-span scaled model-bridge deck using the steel column with MRF dampers activated with 0.0 and 2.0 amps.**

Bridge Deck Evaluation with Ground Penetrating Radar

D. HUSTON, P. FUHR, C. ADAM, W. WEEDON and K. MASER

ABSTRACT

Highway bridge decks must survive decades of repeated loading and environmental stress that can cause cracking, delamination, and corrosion. Efficiently maintaining and rehabilitating bridge decks requires accurate deck health assessments. Bridge deck inspection is often hampered by large portions of the damage occurring initially inside of deck and not appearing on the surface until it is at an advanced stage. Hence the need for non-destructive methods of evaluating the internal state of the deck. This paper describes the results of an ongoing study into the use of ground penetrating radar (GPR) for the inspection of bridge decks. GPR assumes that the interactions of electromagnetic waves with a material depends on the properties of the material. Spatial variations of the dielectric constants of the material will cause reflected waves that can be measured. Analyzing the reflection of an electromagnetic wave can be used to assess the material properties of the deck. Key features in the bridge deck that will affect GPR wave propagation are air and water voids, saturation, chloride ions, asphalt overlays, and rebars. This study is a multifaceted approach to the problem of using GPR for bridge deck inspection. Results of a numerical analysis of GPR interactions with bridge decks, laboratory studies of GPR, and antenna design, will be reported. The radar frequencies range from 0.5 to 6.0 GHz.

BACKGROUND

The bridge infrastructure of the United States is distressed. The United States Department of Transportation estimated in early 1994 that as many as 40% of the total number of bridges in this country are structurally deficient, with repairs estimated in the billions of dollars (Smith, 1995; Halabe et al., 1995). The

D. Huston, P. Fuhr and C. Adam, College of Engineering, Univ. of Vermont, Burl. VT 05405
W. Weedon, Applied Radar Analysis, 14 Union St., Watertown, MA 02172
K. Maser, Infrasense, Inc., 14 Kensington Rd. Arlington, MA 02174

leading cause of this structural deficiency is the deterioration of the bridge deck. To further complicate the problem, bridge deterioration including rebar corrosion, delaminations and disintegrating concrete is often hidden under an asphalt overlay, rendering it invisible to the naked eye (Maser and Kim Roddis, 1990). In addition to hiding defects in the bridge deck until they are well advanced, the asphalt overlay often makes it difficult to distinguish between deterioration in the concrete deck slab and debonding of the overlay (Manning and Holt, 1984). These difficulties in turn lead to a deviation from budget and time estimates somewhere between 100 and 200 percent (Maser, 1989). It therefore becomes necessary to recognize the potential damage before it occurs in order to improve budget and time estimates, to reserve the structural strength and to reduce its life cycle cost (Al-Qadi et al., 1995).

Because it is not economically feasible to replace all decks which have a low condition rating, a more efficient approach is to gather information on the condition of the bridge decks and to develop a comprehensive program of maintenance, rehabilitation and replacement (Halabe et al., 1995; Manning and Holt, 1984; Al-Qadi et al., 1995). This approach involves periodic deterioration assessments leading to the subsequent development of strategies to retard deterioration. Current methods of deterioration assessment, including chain drag and coring, are limited by the fact that they are slow, labor intensive, intrusive to traffic and inaccurate (Maser, 1990). In comparison, non-destructive techniques are usually quicker, can cover a larger area, and the integrity of the interior of the structure can be studied without disturbing the structure itself (Al-Qadi et al., 1995; Zoughi et al., 1991). Besides the obvious advantages of non-destructive techniques, the proper implementation of such methods would increase the accuracy in deck deterioration prediction, which would in turn lead to improved allocation of limited maintenance and rehabilitation funds (Maser, 1989). Since non-destructive testing (NDT) seems to be an invaluable tool in assessing the performance and characteristics of constructed facilities, the development and implementation of an accurate and widely accepted non-destructive method of assessment is necessary (Al-Qadi et al., 1996).

When compared to other non-destructive methods of evaluation, such as infra-red thermography, ground penetrating radar (GPR) has been favored from a research perspective because it is less sensitive to ambient conditions and because it is equally effective with or without asphalt overlays (Maser and Kim Roddis, 1990). GPR also has the ability to evaluate the properties of materials composed of a mixture of several constituents, and to collect data at highway speeds (Maser, 1990; Sansalone and Carino, 1989). In field testing, radar survey equipment has proved to be a functional and reliable field device capable of determining the condition of concrete pavements and structures (Cantor and Knetter, 1982).

The principle of operation is based on reflections of a high frequency electromagnetic wave caused by changes in the electromagnetic properties of the material being probed. Whenever a transmitted wave encounters an interface in the medium where the dielectric properties change, such as material boundaries (e.g. asphalt-concrete interface) and metallic inclusions (e.g. rebar), it is partially reflected and the reflected pulses are detected by a receiving antenna (Maser and

Kim Roddis, 1990; Joyce, 1984). The shape of the resulting composite waveform is characterized by the amplitude and time of flight of each reflected pulse. These parameters are, in turn, dependent upon the nature of the reflecting interfaces and the material involved. Theoretically, then, the shape of the composite waveform recorded when electromagnetic pulses are directed through an overlaid deck is influenced by the condition of the deck, and therefore provides a qualitative picture of that condition (Clemena, 1984). In fact, alterations in the composite waveform have been correlated with deterioration and delamination conditions (Maser, 1989). For a sound deck, the waveform will show echoes from the top and bottom surfaces of the deck, with an additional peak or peaks if the deck contains layers of rebar. Faults such as delaminations or fractures produce echoes, which are more pronounced the more severe the defect, occurring after the surface signal but before the echo from the bottom boundary of the deck (Joyce, 1984; Clemena, 1984). Because the amplitude and wave shape of return signals for delaminated concrete are different from those of sound concrete, areas containing voids and deteriorated concrete are identifiable (Joyce, 1984). In fact, Cantor and Kneeter (1982) were able to produce a 90 percent correlation between radar evaluation predictions and pavement condition. Based upon this accuracy, combined with the fact that the inspection would not be hampered by normal surface conditions, with the exception of standing water (Joyce, 1984), GPR can be a viable method for the evaluation of bridge decks.

One obstacle which stands in the way of implementing a program of non-destructive testing using GPR is data processing. Present analysis and interpretation of radar data requires a research engineer and manual processing, which constitutes a time consuming procedure. By developing a computerized, automated system of data interpretation, manual data processing would be eliminated, thus making GPR a practical NDT tool for use by field technicians (Cantor and Kneeter, 1982).

A second barrier lies in the inherent limitations of commercially available radar used for bridge deck and other civil engineering applications. Maser (1989) reported that because these radar systems operate at wavelengths on the order of 10cm (frequency of 1GHz), small (~1mm) air-filled delaminations are not directly detectable by radar. However, it has been found that these cracks are detectable if they and the adjacent concrete are filled with moisture, or if the properties of the concrete abruptly change in going from above to below the delamination (Maser, 1989). Since most delaminations are small air filled cracks, this problem presents a dilemma which is as challenging as that of data processing, and must be over come in order to create an accurate means of bridge inspection.

ONE DIMENSIONAL LAYERED MEDIA MODEL

A theoretical model for the synthesis and analysis of radar waveforms was developed Halabe et al. (Halabe et al., 1993, 1995a, 1995b; Bhandarkar, 1993). This model idealizes the bridge deck or pavement as a multi-layered medium,

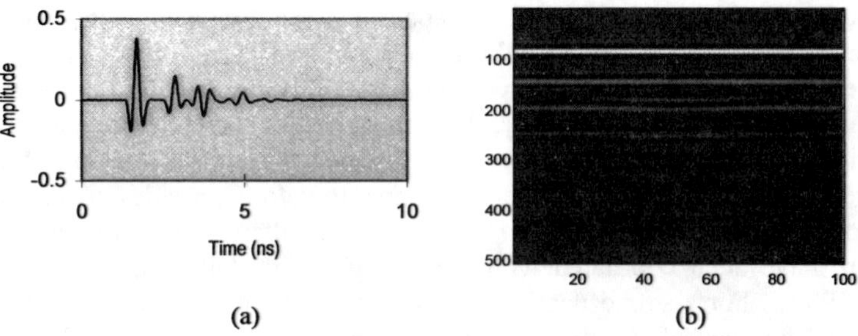

Figure 1- (a) Simulated waveform for a bridge deck containing a 1mm water filled crack. (b) Color band representation of simulated bridge deck containing a water filled crack of varying thickness.

each of which is considered homogeneous, with cylindrical inclusions (rebars). It takes into account the effect of several parameters such as porosity, saturation, chloride content, temperature and frequency on the complex dielectric permittivity of concrete (Halabe et al., 1995a). In this model the reflection coefficient for each layer is computed recursively, working from the bottom interface and proceeding upwards. These computations are performed in the frequency domain to account for the multiple reflections from all of the interfaces (Halabe et al., 1995a, 1995b). Reflections from the longitudinal rebars are accounted for in the time domain, because metals are perfect reflectors with a reflection coefficient of -1. It should be noted that this model ignores the reflections from the transverse rebars since they are usually perpendicular to the polarization direction of the radar antenna's electric field in a typical bridge deck survey (Halabe et al., 1995a, 1995b).

The Halabe model was used to simulate data obtained while monitoring a bridge deck containing a delamination. Using this program three different bridge deck simulations were generated using a 2GHz source pulse. Each simulation contained one hundred waveforms, of which the first and last twenty five waveforms were created with a no-delamination condition and the middle fifty contained a symmetrical delamination. This delamination gradually increased from 0.1 to 1.0mm and then back to 0.1mm in a symmetric manner. The simulations differed by the amount of water that was contained in the delaminated layer. These included a water-filled delamination, an air-filled delamination, and a delamination which contained fifty percent water and fifty percent air. A single waveform generated for a 1mm water filled crack can be seen in Figure 1(a). It should be noted that all other parameters were held constant throughout the simulations.

The data for each simulation were plotted using Matlab. Two-dimensional plots were generated in which time was represented by the y axis, waveform number, or distance along the deck, was represented along the x axis and

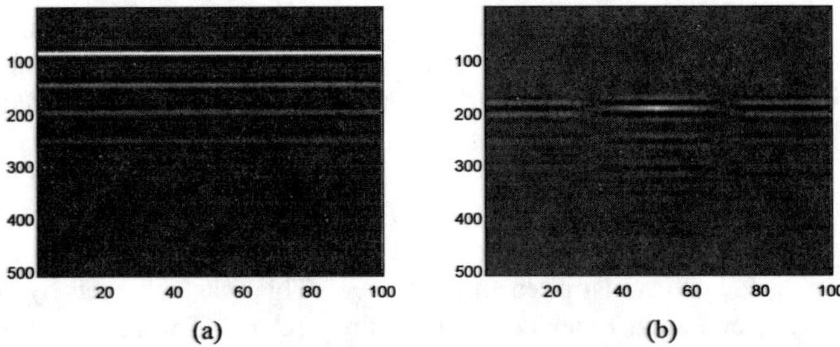

Figure 2 - Color band plots for 1mm air filled crack. Time step is represented on the y-axis and distance along deck is represented by the x-axis. (a) unfiltered (b) filtered.

amplitude was represented by a color. The resulting plots, which can be seen in Figures 1(b) and 2(a), showed various attributes contained within the deck as stripes of color, or lightened bands in these black and white versions. The top surface of the deck, the asphalt- concrete interface, and the rebars could all be seen clearly, however the water filled delamination was the only one of the three which was visible in this format. Because the air filled delamination presented the greatest detectability problem, attention was centered around locating this defect. In an attempt to find the air filled crack, two filtering techniques were used. The first technique subtracted a waveform containing the no-delamination condition from each of the one hundred waveforms contained in the simulation. This resulted in a zero amplitude for the waveforms which did not contain a delamination and caused the air filled delamination to become clearly defined. The second filtering technique, seen in Figure 2(b), subtracted an average waveform from each of the one hundred waveforms contained within the simulation. Again the result clearly showed the location of the air filled delamination. In addition to showing the delamination, these filtering techniques also demonstrated the ability of this model to simulate the resonance effects caused by the cracks.

These simulations illustrate the possibility of using a filtering technique to detect smaller amplitude reflections caused by cracks on the order of 1mm in thickness. Of the two filtering techniques which were examined, the method of subtracting an average waveform seems to be more applicable to actual field data.

TWO DIMENSIONAL FINITE DIFFERENCE MODEL

A two-dimensional model was developed by Weedon and Chew (Weedon, 1994). It is based upon the Finite difference-Time domain (FDTD) algorithm for solving initial boundary value problems involving Maxwell's equations in isotropic media (Yee, 1966). The original Yee FDTD algorithm applies second-order accurate central difference approximations for the space and time derivatives of the electric and magnetic fields, and can be used to calculate either scattered fields or total fields (Shlager and Schneider, 1995; Taflove, 1988). This code also includes a perfectly matched boundary layer which allows Maxwell's equations to be solved for problems involving theoretically infinite domains (Berenger, 1994) by the absorption of electro-magnetic waves at the boundaries of the computational domain.

Initial work with the software involved examining the reflections from rebar, with a co-located transmitter and receiver. This was achieved by running two simulations, one with rebar and one without rebar. The results were then subtracted to obtain the reflected field from the rebar. Ongoing research using this program includes examining different transmitter and receiver locations, the inclusion of delaminations, and the effect of corroded rebar.

LABORATORY STUDIES

CONCRETE SLABS

Thus far three slabs, 1175mm(46.25")x813mm(32")x152.5mm(6"), have been cast using Sakrete concrete mix. The third, and most recent slab was cast on 2 June 1997, and has not as of yet been tested. Each of the slabs has been cast with simulated defects. The first two contain 6.35mm (1/4"), 12.7mm (1/2"), and 19.05mm (3/4") layers of styrofoam, along with several film canisters. The third slab contains 6.35mm (1/4") and 12.7mm (1/2") layers of styrofoam, a sheet of packing foam approximately 2mm thick, and a piece of 3.81cm (1 1/2") diameter PVC piping. The piping was a last minute addition to the third slab, and is oriented perpendicular to the rebar layers. The locations of the various defects were documented on individual drawings of the slabs for later comparison with experimental data.

EQUIPMENT

A GPR system is composed of four essential components, namely the transmitter, the antenna, the receiver, and the signal processor. For this project a Hewlett Packard 8753D Network Analyzer with a 30kHz-6GHz frequency range and time domain capability is being used as the transmitter, receiver, and signal processor. Initial experiments have been performed using a pair of Vivaldi antennas. In addition to these antennas a pair of TEM horn antennas have been designed and built.

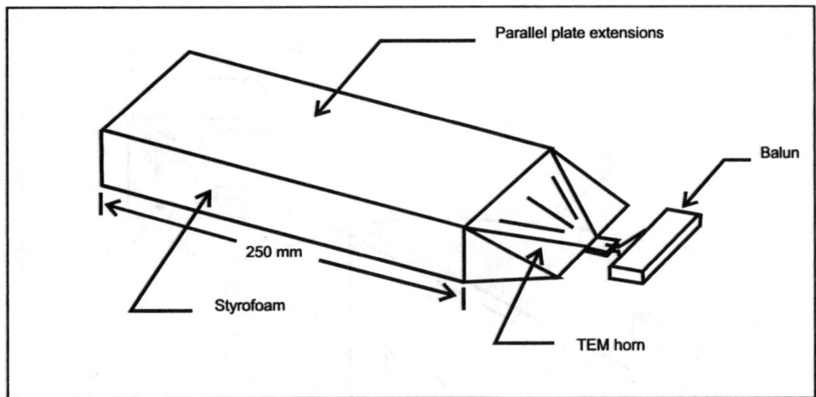

Figure 3 - Diagram of a TEM horn antenna.

TEM HORN ANTENNA DESIGN

In order to preserve the shape of a time waveform, an antenna operating in the time domain should have a narrow impulse response, which in turn implies very broadband operation (Al-Qadi et al., 1996), and a linear phase characteristic. One antenna with this property is the TEM horn antenna. The design which was used for this project follows the design guidelines developed at NIST (Ondrejka et al., 1995). A diagram of one of the TEM horn antennas built for this project can be seen in Figure 3.

To date two antennas have been designed and built. Testing of the horn antennas is continuing in an attempt to optimize performance. To this end several tests on the antennas themselves have been performed. These tests included an S_{11} measurement to examine the internal reflections of the horns, an S_{21} measurement between the two horns, an S_{21} measurement between a horn antenna and a Vivaldi antenna, and an S_{21} between the Vivaldi antennas. The purpose of the S_{21} measurements was to examine the transmitted pulse between antennas.

The S_{11} measurement was performed using a step response in order to examine the internal reflections from the TEM horns. The results of these measurements showed that there was a small reflection resulting from the feed point, and a larger reflection from the end of the antenna. This reflection was a result of the impedance mismatch between the antenna (100Ω) and air (377Ω). At this point several design modifications are being examined in order to match more closely the antenna impedance to that of air.

The S_{21} measurement between the horns resulted in a waveform which showed a reflection occurring approximately 3ns after the main impulse. It is believed that this reflection is caused by low frequency moding, which can be eliminated through resistive loading of the parallel plate extensions (Ondrejka et al., 1995). In addition to the S_{21} measurement which was conducted between the

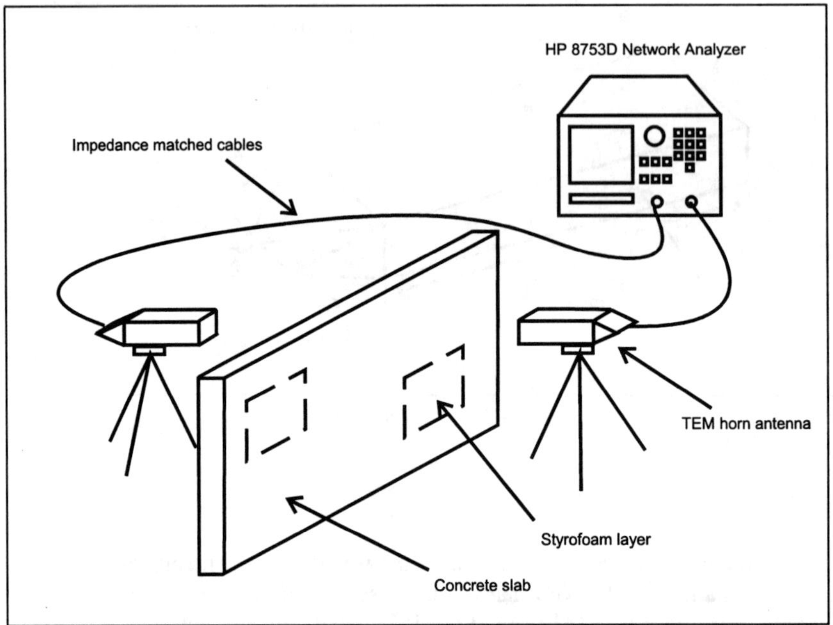

Figure 4 - Typical setup for transmission measurement using TEM horn antennas.

two horn antennas, S_{21} measurements were taken between a horn and a Vivaldi antenna and two Vivaldi antennas. The purpose of these measurements was to compare the transmitted signal from the horn to that of the Vivaldi antenna. The results of these measurements showed that the transmitted signal from the Vivaldi antennas was more powerful than that of the horn antennas. The lower power output generated by the horn antennas is caused by the internal reflection at the end of the horn as evidenced by the S_{11} measurement. As previously mentioned, antenna characterization and tuning is still ongoing.

TRANSMISSION MEASUREMENTS

Initial laboratory experimentation has involved measuring the attenuation and time shift of the initial pulse using the time domain capability of the HP8753D. The signal which has been used is a stepped frequency broad band pulse ranging from 500 MHz to 6 GHz. The first set of measurements were taken using an identical pair of Vivaldi antennas. The second set of experiments were taken using a set of identical TEM horn antennas. For all measurements, the antennas were set up three inches from the surface of the concrete slabs. A diagram of the basic laboratory set up can be seen in Figure 4.

Using the Vivaldi antennas transmission measurements were taken through slab #1 in areas of solid concrete and in areas where styrofoam was known to be located. It should be noted at this time that these measurements were taken between the rebar layers so as to limit the number of variables affecting the measurements. Based upon the velocity change in going from concrete (velocity~4"/ns) to styrofoam (velocity~12"/ns), and the time shift in the transmitted peaks caused by the presence of the styrofoam, it was possible to determine the approximate thickness of the air void (styrofoam). The measurements which were taken confirmed the presence of the styrofoam layers, however these measurements indicated that the thickness' contained in slab #1 were 12.7mm (1/2") and 19.05mm (3/4"), as opposed to the 12.7mm (1/2") and 6.35mm (1/4") thickness' as indicated on the diagrams.

Transmission measurements were taken through slab #2 in a manner similar to those taken for slab #1. In this case, the thickness' which were measured were 12.7mm (1/2") and 6.35mm (1/4") as opposed to the 12.7mm (1/2") and 19.05mm (3/4") which were indicated on the diagrams.

In addition to the transmission measurements which were taken using the Vivaldi antennas, a transmission measurement was taken using the TEM horn antennas. The results of this measurement showed the presence of a 6.35mm (1/4") piece of styrofoam in slab #2, and supported the results obtained using the Vivaldi antennas.

CHANGE IN DIELECTRIC CONSTANT WITH CURING

For a constant antenna separation, it is possible to determine a single representative value of dielectric constant, ε_r, for a sample of known thickness using the following formula (Aurand, 1996), which assumes minimal dispersion through the layer

$$\varepsilon_r \cong \left[1+\frac{\Delta\tau}{\tau_o}\right]^2 = \left[1+\frac{\Delta\tau}{d/c}\right]^2 \qquad (1)$$

$\Delta\tau$ is the time shift in the transmitted pulse caused by placing the sample between the antennas, d is the sample thickness, and c is the speed of light.

In order to measure the change in dielectric constant during the curing cycle, three test slabs, 228.6mm(9")x279.4mm(11")x76.2mm(3"), were cast using Sakrete mix. The slabs were allowed to cure for four hours before the first measurement was taken in order to ensure that they would not slump. The TEM

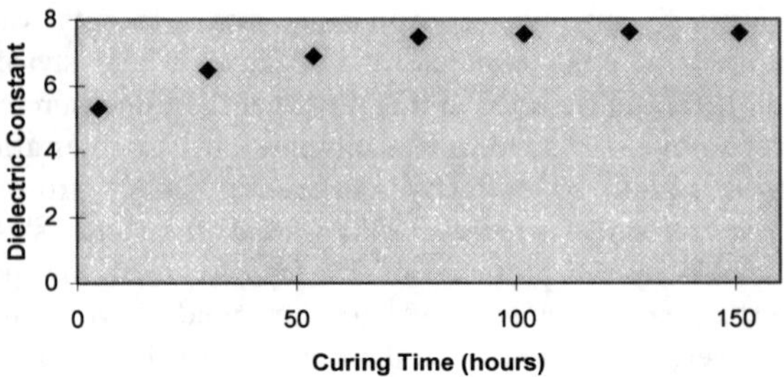

Figure 5 - Typical dielectric constant curve as a function of curing time for Sakrete.

horn antennas were used for these measurements, and a stand was built in order to ensure a constant antenna separation. After the initial measurement, additional measurements were taken every twenty four hours over the next seven days. A typical curve for the change in dielectric constant over the first seven days of curing can be seen in Figure 5. During the curing cycle the slabs were kept moist with damp layers of cloth.

CONCLUSIONS AND FUTURE WORK

The results of the transmission measurements indicate that the current radar set up has the ability to detect and differentiate between layers of styrofoam of varying thickness. The results also indicate the presence of layers which are 6.35mm (1/4"), 12.7mm (1/2"), and 19.05mm (3/4") thick, however the locations differ from those on the drawings. It is possible that the drawings are not accurate. A third slab has subsequently been cast in order to quantify the results obtained for the first two slabs. This third slab contains 6.35mm (1/4") and 12.7mm (1/2") layers of styrofoam, whose location have been carefully documented. In addition a 1mm piece of packing styrofoam, which has been doubled over, was included to simulate a more realistic cracking situation. This piece of styrofoam was not laid flat, but rather it was given some degree of contour. A fourth slab is currently being designed. This slab will include removable rebar, and a void which can be filled with water in order to simulate a water filled crack. Future work will include testing the additional slabs, as well as working with reflection measurements. Because reflection measurements offer a more practical placement of equipment, this is the method which would be used in field testing.

The curing measurements indicate that there is a detectable change in nominal dielectric constant over the frequency range of 500MHz-6GHz. This change was constant for each of the three specimens which were prepared. The dielectric constant leveled out at around 7.5 for all three samples after a period of seven days. Future work will include a repetition of these experiments using the improved antenna designs. To this end antenna design and enhancement is continuing.

ACKNOWLEDGMENT

This research has been sponsored by the New England Transportation Consortium. Peter Bergendahl assisted in the preparation of the laboratory tests.

REFERENCES

I.L. Al-Qadi, S.M. Riad, R. Mostafa, W. Su, "Design and Evaluation of a Coaxial Transmission Line Fixture to Characterize Portland Cement Concrete", Proc. of 6th International Structural Faults and Repair, Vol. 2, London, UK, pp.337-348, July 1995.

I.L. Al-Qadi, R. Mostafa, W. Su, S.M. Riad, "Measuring Dielectric properties of Portland Cement Concrete: New Methods", Accepted by the Structural Materials Technology and Nondestructive Testing Conference, San Diego, CA, Feb. 20-23, 1996.

J.F. Aurand, "Measurements of Transient Electromagnetic propagation Through Concrete and Sand", Sandia National Laboratories Technical Report, Albuquerque, NM, Sept. 1996.

J.P. Berenger, "A Perfectly Matched Layer for the Absorption of Electromagnetic Waves", *J. Comp. Phys.*, Vol. 114, pp.185-200, 1994.

V.A. Bhandarkar, "Detection of Sub-Surface Anomalies in Concrete Bridge Decks Using Ground Penetrating Radar", Masters Thesis, Dept. of Civil Engineering, West Virginia University, 1993.

T.R. Cantor, C.P. Kneeter, "Radar as Applied to the Evaluation of Bridge Decks", Transportation Research Board Record No. 853, January 1982.

C.R. Carter, T. Chung, F.B. Holt, D.G. Manning, "An Automated Signal Processing System for the Signature Analysis of Radar Waveforms From Bridge Decks", *Canadian Electrical Engineering Journal*, Vol. 11, No. 3, pp. 128-137, 1986.

G.G. Clemena, "Non-destructive Inspection of Overlaid Bridge Decks with Ground- Penetrating Radar", Transportation Research Board Record No. 899, 1984.

U.B. Halabe, A. Sotoodehnia, K.R. Maser, E.A. Kausel, "Modeling the Electromagnetic Properties of Concrete", *ACI Materials Journal*, Vol.90, No.6, Nov./Dec., 1993, pp. 552-563.

U.B. Halabe, H.L. Chen, V. Bhandarkar, Z. Sami, "Detection of Sub-Surface Anomalies in Concrete Bridge Decks Using Ground Penetrating Radar", Accepted by *ACI Materials Journal*, August 1995.

U.B. Halabe, K.R. Maser, E.A. Kausel, "Condition Assessment of Reinforced Concrete Structures Using Electromagnetic Waves", *ACI Materials Journal*, Vol. 92, No. 5, Sept./Oct., 1995,pp. 511-523.

R.P. Joyce, "Rapid Non-Destructive Delamination Detection", Federal Highway Administration FHWA RD-84/076, November 1984.

D.G. Manning, F.B. Holt, "Detecting Deterioration in Asphalt-Covered Bridge Decks", Transportation Research Board Record No. 899, 1984.

K.R. Maser, "New Technology for Bridge Deck Assessment", New England Transportation Consortium Phase I Final Report, Center for Transportation Studies, MIT, October 1989.

K.R. Maser, "New Technology for Bridge Deck Assessment", New England Transportation Consortium Phase II Final Report, Center for Transportation Studies, MIT, January 1990.

K.R. Maser, W.M. Kim Roddis, "Principles of Thermography and Radar for Bridge Deck Assessment", *ASCE Journal of Transportation Engineering*, Vol.116 No.5, pp. 583-601, Sept./Oct. 1990.

A.R. Ondrejka, J.M. Ladbury, H.W. Medley, "TEM Horn Antenna Design Guide", NIST Technical Report, Boulder, CO, 1995.

M. Sansalone, N.J. Carino, "Detecting Delaminations in Concrete Slabs with and without Overlays Using the Impact-Echo Method", *ACI Materials Journal*, Vol.86, No.2, March-April 1989, pp.175-184.

S.S. Smith, "Detecting Pavement Deterioration with Subsurface Interface Radar", *Sensors*, pp.29-40, September 1995.

W.J. Steinway, J.D. Echard, C.M. Luke, "Locating Voids Beneath Pavement Using Pulsed Electromagnetic Waves", National Research Council, November 1981.

W.H. Weedon, Broadband Microwave Inverse Scattering: Theory and Experiment, Ph.D. Dissertation, University of Illinois at Urbana-Champaign, 1994.

K.S. Yee, "Numerical Solution of Initial Boundary Value Problems Involving Maxwell's Equations in Isotropic Media", *IEEE Trans. on Antennas and Propagation*, Vol. AP-14, No. 3, pp. 302-307, 1966.

R. Zoughi, G.L. Cone, P.S. Nowak, "Microwave Non-destructive Detection of Rebars in Concrete Slabs", *Materials Evaluation*, Vol.49 No.11, Nov.1991, pp.1385-1388.

The Application of Artificial Neural Networks on the Health Monitoring of Bridges

P.-L. LIU[1] and S.-C. SUN[2]

ABSTRACT

This study develops a neural network system to monitor the safety of a bridge structure. A truck of constant mass is driven at constant speed through the target bridge. Then, the maximal and minimal values of the bridge elongations are processed by a monitoring system to evaluate the current condition of the bridge. The monitoring system is composed of several backpropagation neural networks. Each neural network monitors a part of the bridge. The neural networks are trained using simulation data. The numerical example shows that the monitoring system is effective in the damage detection of the bridge.

INTRODUCTION

Bridges play an indispensable role in land transportation. The collapse of a bridge often results in tremendous losses. Therefore, the safety assessment of bridges is an important and challenging task for civil engineers. The conventional way of evaluating the integrity of a bridge is by visual inspection. However, bridge structures are usually large. Therefore, such inspection is costly and time-consuming. Furthermore, the inspection may not be complete because the structure is not totally accessible.

This paper proposes an intelligent inspection system for bridges. A truck of constant mass is driven at constant speed through the target bridge. The response of the bridge is then processed by artificial neural networks to evaluate the condition of the bridge automatically.

[1] Professor, Institute of Applied Mechanics, National Taiwan University, Taipei, Taiwan.
[2] Graduate student, Institute of Applied Mechanics, National Taiwan University, Taipei, Taiwan.

ARTIFICIAL NEURAL NETWORK

The artificial neural network is a powerful tool in the areas of pattern recognition, signal processing, control, and so on. Recently, several researchers have applied artificial neural networks in the damage assessment of structures, for example, Wu et al. (1990), Elkordy and Chang (1992), and Tsou and Shen (1994).

The artificial neural networks adopted in this study are multi-layered backpropagation networks (Zurada, 1992). The backpropagation network constitutes of an input layer, an output layer, and several hidden layers. Each layer is composed of several neurons, and each neuron is linked to every neuron of the adjacent layers, as shown in Fig. 1.

The input layer reads the input data and passes it to the first hidden layer. The neurons in the first hidden layer take the weighted sums of the input data and evaluate the activation function f (a sigmoid function in this study):

$$y_k = \mathbf{f}\left(\sum_{j=1}^{p} W_{kj} x_j\right) \tag{1}$$

where x_j is the input data, y_k is the output of the activation function, and W_{kj} is the weight of a link. Then, y_k is passed forward to the next layer. The neurons in the next layer again take the weighted sums of the incoming data, evaluate the activation function, and pass the function values forward. The same process is repeated until the output layer sends out the results.

The weights of the links, which characterize the network, are determined by training. To train a backpropagation network, a number of samples must be provided. Suppose the target output of a sample is $\mathbf{d} = \begin{bmatrix} d_1 & d_2 & \cdots & d_r \end{bmatrix}^T$ and the network output is $\mathbf{o} = \begin{bmatrix} o_1 & o_2 & \cdots & o_r \end{bmatrix}^T$. A sum-squared error can be defined as follows:

$$E = \frac{1}{2}\sum_{k=1}^{r}\left(d_k - o_k\right)^2 \tag{2}$$

In each training cycle, the weights are modified along the direction of negative gradient of the error function. The weight corrections are performed layer by layer, starting from the output layer backward to the input layer.

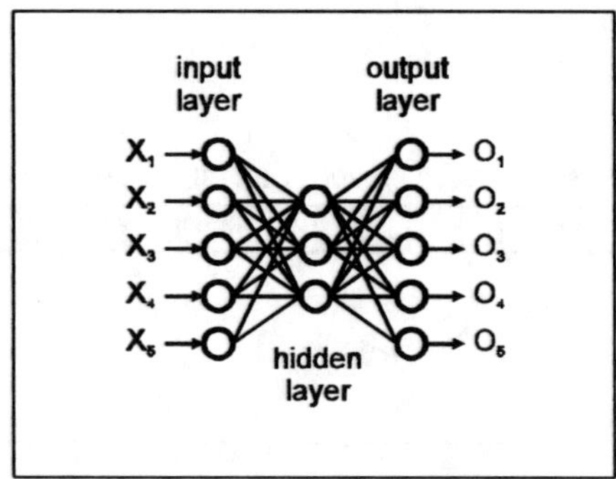

Figure 1 Multi-Layered Backpropagation Neural Network

DYNAMIC ANALYSIS OF A BRIDGE

A truck moving at constant speed V across a bridge can be modeled by a moving mass on a continuous beam. The dynamic response of a beam subjected to a moving mass is as follows:

$$\frac{\partial^2}{\partial x^2}\left(EI\frac{\partial^2 w(x,t)}{\partial x^2}\right)+\rho A\frac{\partial^2 w(x,t)}{\partial t^2}+M\delta(x-\eta)\frac{\partial^2 w(\eta,t)}{\partial t^2}=Mg\delta(x-\eta) \quad (3)$$

where w is the transverse displacement of the beam, EI is the flexural rigidity, ρA is the mass density per unit length, M is the mass of the truck, g is the gravitational acceleration, δ is the Dirac delta function, and $\eta = Vt$ is the position of the truck.

Finite element formulation of the above equation leads to

$$\mathbf{KU}+\mathbf{M\ddot{U}}=\mathbf{F} \quad (4)$$

where **K**, **M**, **U**, and **F** are the stiffness matrix, mass matrix, nodal displacement vector, and external load vector, respectively. Notice that **M** and **F** are dependent on the truck position. The equation can be integrated by the Newmark scheme to obtain the bridge response.

THE EXAMPLE BRIDGE

Consider a simply supported, three-spanned bridge. Each span is 60m long. The original flexural rigidity EI of the bridge is 7.34×10^{10} N·m^2 and the mass density ρA is 1570 kg/m. The mass of the test truck is 20000 kg, and the speed is 60 km/hr. In the finite element analysis, the bridge is discretized into 30 uniform beam elements.

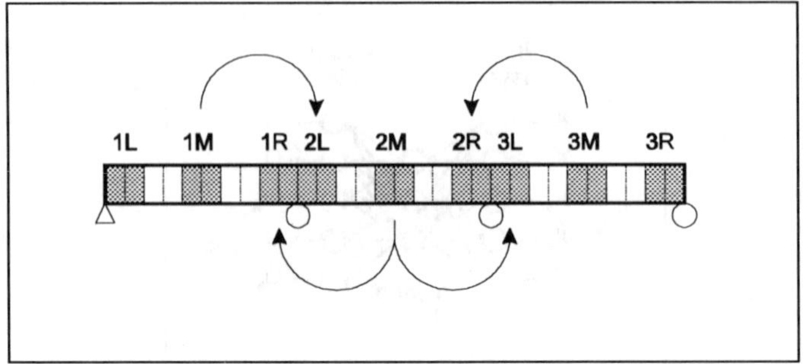

Figure 2 The Example Bridge and Its Damage Zones

Nine damage zones are considered for the bridge, namely, 1L, 1M, 1R, 2L, 2M, 2R, 3L, 3M, and 3R, where 1, 2, and 3 denote the span number, and L, M, and R denote the left, middle, and right of the span, as shown in Fig. 2. Each damage zone constitutes of two elements. For example, 1L constitutes of elements 1 and 2, 2M constitutes of elements 15 and 16, and 3R constitutes of elements 29 and 30.

In this study, damage of the bridge is modeled by stiffness reduction. Define the stiffness ratio as

$$r_i = \frac{EI_i'}{EI_i} \tag{5}$$

where EI_i and EI_i' are the original and current flexural rigidities of zone i. Take zone 2M for example. $r_{2M} = 0.8$ means that the flexural rigidities of elements 15 and 16 are reduced to 80% of the original value.

STRUCTURE OF NEURAL NETWORK

While designing an artificial neural network, the most crucial step is the selection of input parameters. The input parameters must be chosen such that they are representative of the output. Hence, the input vector must be able to reflect the current condition of the bridge in this application.

Observing the response curves of the bridge, one can find that the longitudinal elongation of each zone is highly correlated to the condition of that zone. Furthermore, one can use the maximal and minimal values of the elongation curves to estimate the damage level accurately. Therefore, the maxima and minima of the 9 elongation curves are used as the input features of the neural networks.

Further examination of these extreme values shows that

1. The extreme values of zone 1L change notably only when the zone itself is damaged. Damage in other zones has almost no influence on these values. This is also true for zone 3R.
2. The extreme values of zone 1R are influenced by the damage of all 9 zones.

extremes increase. Similarly, the extreme values of zone 3L are influenced by the damage of all 9 zones, but only when zone 3L or 2M is damaged do the absolute values of both extremes increase.

3. The extreme values of zone 2L and 2R are influenced by the damage of all 9 zones. However, only when zone 2L or 1M is damaged do the absolute values of both extremes of 2L increase. Similarly, only when zone 2R or 3M is damaged do the absolute values of both extremes of 2R increase.

The above relations are shown in Fig. 2. The same phenomenon is also observed in a five-spanned bridge, that is, when the middle of a span is damaged, the absolute values of extremes of the zones in the adjacent spans near the support will increase simultaneously.

Based on above observations, the 9 damage zones are divided into five groups: (1L), (1M, 2L), (1R, 2M, 3L), (2R, 3M), and (3R). The damage and features of a zone are dependent of those of the zones in the same group, but independent of those of the zones in other groups.

According to such grouping, the bridge monitoring system is constructed using five parallel neural networks, namely, network 1L, network 1M-2L, network 1R-2M-3L, network 2R-3M, and network 3R. Network 1L processes the features of zone 1L to monitor the health of zone 1L; network 1M-2L processes the features of zones 1M and 2L to monitor the health of zones 1M and 2L; and so on.

The output parameters of the networks are the stiffness ratios of the monitored zones. The output ratios are between 0 and 1.

Only one hidden layer is used in each of the 5 networks. The numbers of neurons in the hidden layers, ranging from 2 to 6, are determined by trial runs.

TRAINING AND TEST OF THE MONITORING SYSTEM

As mentioned before, one must provide a number of samples to train the neural networks. Hence, finite element analysis was carried out for various damage conditions. In the simulations, 5 damage levels are considered at each zone, the stiffness ratios being 0.1, 0.3, 0.5, 0.7, and 0.9. The training database for the five networks contains the following samples:

1. Network 1L and 3R:
   46 cases = the undamaged case + 45 single-zone-damaged cases
2. Networks 1M-2L and 2R-3M :
   71 cases = the undamaged case + 45 single-zone-damaged cases
   + 25 1M-2L-damaged cases or 2R-3M-damaged cases
3. Network 1R-2M-3L:
   121 cases = the undamaged case + 45 single-zone-damaged cases
   + 25 1R-2M-damaged cases + 25 1R-3L-damaged cases
   + 25 2M-3L-damaged cases

After the training of the neural networks is complete, more damage cases were simulated by finite element analysis. Then, the new samples are used to test the effectiveness of the monitoring system. The test results are as following:

1. Single zone damaged

There are 163 test cases, including the undamaged case and 162 single-zone-damaged cases. Eighteen stiffness ratios are considered for each damaged zone, namely, 0.95, 0.9, 0.85, ..., 0.15, 0.1. Using the parallel networks to analyze these cases, the damaged zone can be located accurately in all cases. In most cases, the errors of the output ratios are less than 5%, and the maximum error is 8.2%.

Since noise always presents in practice, 20% random noise was superposed to the test data to see if the system is vulnerable to noise. A thousand Monte Carlo simulations were performed. It turned out that the biases of the output ratio are less than 15%, and the standard deviations are less than 19%. Both the biases and standard deviations do not exceed the noise level. Apparently, this system is insensitive to noise.

2. Two zones damaged

Twelve combinations of damaged zones are considered in the test samples, namely, 1M-2L, 2R-3M, 1R-2M, 1R-3L, 2M-3L, 1L-1M, 1L-1R, 1L-3R, 1M-1R, 1M-3M, 2L-2R, and 2R-3L. In each combination, 9 stiffness ratios (0.9, 0.8, ..., 0.2, and 0.1) are assumed for each zone. Hence, there are totally 972 test samples.

First, the noiseless data were input to the system. It turned out that most output errors are less than 10%, and the maximum error is 15.9%. Next, tests were performed using the noisy data. The noise level was 20%. The maximum bias of the output ratios is 16%, and the maximum standard deviation is 19%. Again, both the biases and standard deviations do not exceed the noise level. This shows that system is insensitive to noise in the detection of two damage zones.

3. Multiple zones damaged

Fifty three-zones-damaged cases were randomly chosen to examine the ability of the monitoring system to detect multiple damage zones. Although the three-zones-damaged cases are not included in the training database, the damaged locations still can be identified successfully in all these cases. Furthermore, the output errors are all less than 15%.

4. Zones partially damaged

In all the aforementioned cases, damage of a zone refers to the stiffness reduction of both elements in the zone. However, this is rarely the case for real damages. Hence, the cases in which a zone is partially damaged are also tested. The partially damaged zones considered here are 1M (elements 5 and 6), 1R (elements 9 and 10), and 2M (elements 15 and 16). The stiffness ratios for the damaged elements are identically 0.5.

TABLE 1 SYSTEM OUTPUT — ZONES PARTIALLY DAMAGED

damaged elements	1L	1M	1R	2L	2M	2R	3L	3M	3R
5	0.99	0.63	1.00	0.99	1.00	1.00	1.00	0.99	0.99
6	0.99	0.65	1.01	0.99	1.00	1.00	1.00	1.00	1.00
5, 6	0.99	0.50	1.01	0.98	0.99	1.00	1.00	1.00	1.00
9	0.99	1.00	0.66	0.98	0.99	0.99	1.00	1.00	1.00
10	0.99	0.99	0.63	1.01	0.99	1.00	0.99	1.00	0.99
9, 10	0.99	0.99	0.50	1.00	0.99	0.99	1.00	1.00	0.99
15	0.99	0.99	0.96	1.01	0.68	1.00	1.01	1.00	1.00
16	0.99	1.00	1.00	1.00	0.65	1.01	0.99	0.99	1.00
15, 16	0.99	0.99	1.00	1.01	0.51	1.01	0.99	1.00	1.00

The output ratios are shown in Table 1. It is seen that the results are satisfactory. The damaged locations can be identified without difficulty. Furthermore, when only one element of the zone is damaged, the output ratio lies between the true stiffness ratios of the two elements. Therefore, even when the zone is partially damaged, the monitoring system still can identify the damage location and damage level successfully.

CONCLUSIONS

An intelligent monitoring system for a bridge is developed in this paper. The backpropagation neural networks are adopted in the system for data processing. Several conclusions can be drawn from the numerical tests:

1. The longitudinal elongations of a bridge subjected to a moving truck can reflect the damage condition of the bridge.
2. The damage zones of the bridge can be divided into several groups. Each group contains at most three zones: the middle of a span and the zones next to the supports in the adjacent spans. The extremes of elongation of a zone are only influenced by the damage of zones in the same group.
3. According to the zone grouping, the bridge monitoring system can be constructed using several neural networks. Each neural network monitors the health of a certain group of zones. Such parallel structure can reduce the number of training samples and training time.
4. The monitoring system developed in this study can detect the locations and levels of damages accurately. The system is effective even in the case of multiple damages or partial damage.
5. The monitoring system is insensitive to noise.

ACKNOWLEDGMENT

This work was supported by the National Science Council of the Republic of China under Grant NSC85-2211-E-002-017.

REFERENCES

1. Elkordy, M. F. and Chang, K. C., 1993. "Neural networks trained by analytical simulated damage state." *Journal of Computing in Civil Engineering*, 7(2), 130-145.
2. Tsou, P. and Shen, M.-H. H., 1994. "Structure damage detection and identification using neural networks." *AIAA Journal*, **32**(1), 176-183.
3. Wu, X., Ghaboussi, J., and Garrett, J. H., 1992. "Use of neural networks in detection of structure damage." *Computer and Structures*, **42**(4), 649-659.
4. Yao, G. C., Chang, K. C., and Lee G. C., 1992. "Damage Diagnosis of Steel Frames Using Vibrational Signature Analysis." *Journal of Engineering*, ASCE, **118**(9), 1949-1961.
5. Zurada, M. J., 1992. *Introduction to Artificial Neural Systems*, Info Access & Distribution Pte Ltd., Singapore.

Substructural Identification of a Bridge—A FFT-Based Spectral Analysis

C.-H. LOH and Z.-K. LEE

ABSTRACT

The purpose of this paper is to use the seismic response data of a 5-span continuous box-girder bridge to identify the dynamic characteristics of the system. The spectral finite element method was used by considering the flexural wave in beam to form the dynamic stiffness matrix of the bridge substructure. Through matrix condensation of the substructure system the interior nodal displacements can be expressed in terms of the boundary nodal displacements. The identification was performed by using the recorded nodal displacement of the bridge substructure to identify the EI-value and damping factor η of the system. From this analysis the bending moment and shear force time histories at the nodal points can be retrieved after the identification. The wavelet analysis was also used to identify the response differences between weak motion and strong motion of ground excitation.

INTRODUCTION

The transient response of civil engineering structures can be obtained by using the methodology of finite elements combined with a suitable scheme to solve the equation of motion. An approach similar in style to that of the finite element method (FEM) of analysis is the spectral representation. The spectral formulation begins with the equation of motion of the beam including the inertia term [Doyle, 1989]. The spectral approach to dynamic problems temporarily removes time from the description that the effect of damping can easily be incorporated by changing the spectrum relation. This dynamic stiffness of the beam element can be established in spectral representation. Application of the spectral finite element method to the parametric identification of frames using transient data had been employed through simulation

C. H. Loh, Professor, Department of Civil Engineering, National Taiwan University, Taipei, Taiwan, R. O. C.

Z. K. Lee, Graduate student, Department of Civil Engineering, National Taiwan University, Taipei, Taiwan, R. O. C.

[Chen and Liu, 1995]. Except the spectral finite element method, there has been many noticeable research works using the system identification techniques to study the dynamic response of structures subjected to earthquake excitations. Most of them were concentrated on the methodology, such as Kalman Filtering Techniques (Shinozuka *et al.*, 1982; Hoshiya *et al.*, 1984) and time domain recursive identification (Safak, 1988). The purpose of this paper is to use the spectral finite element method to form the dynamic stiffness of the substructure system and perform the identification of bridge dynamic characteristics from seismic response data. Research work will be concentrated on the substructural identification and the discussion on the result of identification.

DYNAMIC STIFFNESS OF BEAM ELEMENT (TRANSFER MATRIX)

Consider a beam element with uniform cross-section to have constant properties along its length. The equation governing $\hat{v}_k$, the Fourier transform of the displacement within a typical uniform section, say section k, is as follows:

$$(EI)_k \frac{d^4 \hat{v}_k}{dx^4} - \left(\omega^2 m_k - i\omega c_k\right)\hat{v}_k = 0 \tag{1}$$

The external load is assumed to be absent from this section and the rotational inertia term $\rho I \ddot{\phi}$ is neglected. Let

$$\sigma_k^4 = \frac{\omega^2 m_k - i\omega c_k}{(EI)_k} \tag{2}$$

The general solution for the deflection curve is written:

$$\hat{v}_k = \frac{\Lambda_0}{2}\left(\cosh \sigma_k x_k + \cos \sigma_k x_k\right) + \frac{\Lambda_1}{2\sigma_k}\left(\sinh \sigma_k x_k + \sin \sigma_k x_k\right)$$
$$+\frac{\Lambda_2}{2\sigma_k^2}\left(\cosh \sigma_k x_k - \cos \sigma_k x_k\right) + \frac{\Lambda_3}{2\sigma_k^3}\left(\sinh \sigma_k x_k - \sin \sigma_k x_k\right) \tag{3}$$

The symbols Λ_0, Λ_1, Λ_2, Λ_3, are $\hat{v}_k$, $\hat{v}_k'$, $\hat{v}_k''$, $\hat{v}_k'''$ evaluated at $x_k = 0$, respectively. Define $\overline{Z}(x_k = 0) = \{\Lambda_0,\ \Lambda_1,\ EI\,\Lambda_2,\ EI\,\Lambda_3\}$, then we are able to establish the following relationship between two ends of the beam element:

$$\overline{Z}(x_k) = \mathbf{F}_k(x_k)\,\overline{Z}(x_k = 0) \tag{4}$$

where

$$\mathbf{F}_k(x_k) = \begin{bmatrix} C_0 & S_1 & \frac{C_2}{EI} & \frac{S_3}{EI} \\ \sigma^4 S_3 & C_0 & \frac{S_1}{EI} & \frac{C_2}{EI} \\ \sigma^4 EI\, C_2 & \sigma^4 EI\, S_3 & C_0 & S_1 \\ \sigma^4 EI\, S_1 & \sigma^4 EI\, C_2 & \sigma^4 S_3 & C_0 \end{bmatrix}_k \tag{4a}$$

and

$$
C_0 = \frac{\cosh \sigma_k x_k + \cos \sigma_k x_k}{2}, \qquad C_2 = \frac{\cosh \sigma_k x_k - \cos \sigma_k x_k}{2\,\sigma_k^2}
$$
$$
S_1 = \frac{\sinh \sigma_k x_k + \sin \sigma_k x_k}{2\,\sigma_k}, \qquad S_3 = \frac{\sinh \sigma_k x_k - \sin \sigma_k x_k}{2\,\sigma_k^3} \tag{4b}
$$

The subscript k associated with the square matrix on the right-hand side of Eq. (4) indicates that $EI = (EI)_k$, $\sigma = \sigma_k$, etc. The matrix $\mathbf{F}_k(x_k)$ transfer the state vector $\overline{Z}$ from $x_k = 0$ to x_k within a uniform section. The element dynamic stiffness can be obtained through the rearrangement of Eq. (4):

$$
\left\{ \begin{array}{c} \left\{ \begin{array}{c} EI\,\Lambda_2 \\ EI\,\Lambda_3 \end{array} \right\}_0 \\ \left\{ \begin{array}{c} EI\,\Lambda_2 \\ EI\,\Lambda_3 \end{array} \right\}_k \end{array} \right\} = \Big[\, K \,\Big]_{\mathrm{dyn}} \left\{ \begin{array}{c} \left\{ \begin{array}{c} \Lambda_0 \\ \Lambda_1 \end{array} \right\}_0 \\ \left\{ \begin{array}{c} \Lambda_0 \\ \Lambda_1 \end{array} \right\}_k \end{array} \right\} \tag{5}
$$

where $[K]_{\mathrm{dyn}}$ is the dynamic stiffness matrix of the beam element. The subscript 0 and k associated with the vector on the right hand site of Eq. (5) indicate the both end of the beam element. Assume that there are no external loads applied between two ends at node 1 and node 2 (as shown in Fig. 1). At each node, there are two essential beam actions, namely, the bending moment and shear force. The corresponding nodal degree of freedom are the rotation $\phi(x,t)$ and the vertical displacement $v(x,t)$. The stiffness matrix is of order 4×4 because there must be two degree-of-freedom at each node. The dynamic stiffness of each beam element can be used to assemble the stiffness of the structural system in frequency domain.

SUBSTRUCTURE FORMULATION AND IDENTIFICATION

Consider a pier-girder substructure system (T-section), as shown in Fig. 2. Five beam elements were used to construct the substructure. The dynamic stiffness matrix of the substructure system for motion in transverse direction can be constructed with order (13×13). The nodal forces and displacements relationship is shown as follows:

$$
\left\{ \begin{array}{c} \hat{V}_1 \\ \hat{M}_1 \\ \hat{V}_2 \\ \hat{M}_2 \\ \hat{V}_3 \\ \hat{M}_3 \\ \hat{V}_4 \\ \hat{M}_4 \\ \hat{V}_5 \\ \hat{M}_5 \\ \hat{M}_6 \\ \hat{V}_7 \\ \hat{M}_7 \end{array} \right\}
=
\begin{bmatrix}
K_{11} & K_{12} & K_{13} & K_{14} & & & & & & & & & \\
K_{21} & K_{22} & K_{23} & K_{24} & & & & & & & & & \\
K_{31} & K_{32} & K_{33}^{*} & K_{34}^{*} & K_{35} & K_{36} & & & & & & & \\
K_{41} & K_{42} & K_{43}^{*} & K_{44}^{*} & K_{45} & K_{46} & & & & & & & \\
 & & K_{53} & K_{54} & \overline{K}_{55}^{*} & K_{56}^{*} & K_{57} & K_{58} & & & K_{511} & K_{512} & K_{513} \\
 & & K_{63} & K_{64} & K_{65}^{*} & K_{66}^{*} & K_{67} & K_{68} & & & & & \\
 & & & & K_{75} & K_{76} & K_{77}^{*} & K_{78}^{*} & K_{79} & K_{7\,10} & & & \\
 & & & & K_{85} & K_{86} & K_{87}^{*} & K_{88}^{*} & K_{89} & K_{8\,10} & & & \\
 & & & & & & K_{97} & K_{98} & K_{99} & K_{9\,10} & & & \\
 & & & & & & K_{10\,7} & K_{10\,8} & K_{10\,9} & K_{10\,10} & & & \\
 & & & & K_{11\,5} & & & & & & K_{11\,11} & K_{11\,12} & K_{11\,13} \\
 & & & & K_{12\,5} & & & & & & K_{12\,11} & K_{12\,12} & K_{12\,13} \\
 & & & & K_{13\,5} & & & & & & K_{13\,11} & K_{13\,12} & K_{13\,13}
\end{bmatrix}
\left\{ \begin{array}{c} \hat{u}_1 \\ \hat{\phi}_1 \\ \hat{u}_2 \\ \hat{\phi}_2 \\ \hat{u}_3 \\ \hat{\phi}_3 \\ \hat{u}_4 \\ \hat{\phi}_4 \\ \hat{u}_5 \\ \hat{\phi}_5 \\ \hat{\phi}_6 \\ \hat{u}_7 \\ \hat{\phi}_7 \end{array} \right\} \tag{6}
$$

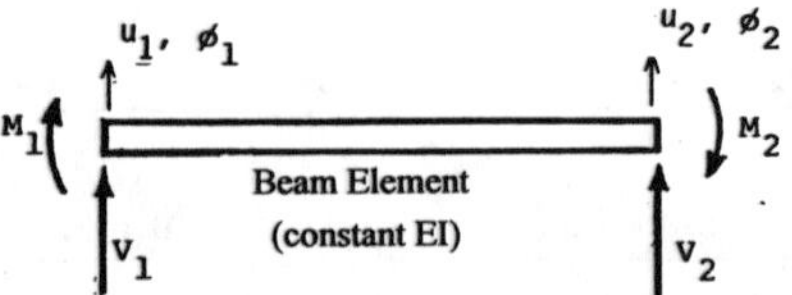

Fig. 1: Nodal loads and degree of freedom of beam element (consider only flexural deformayion).

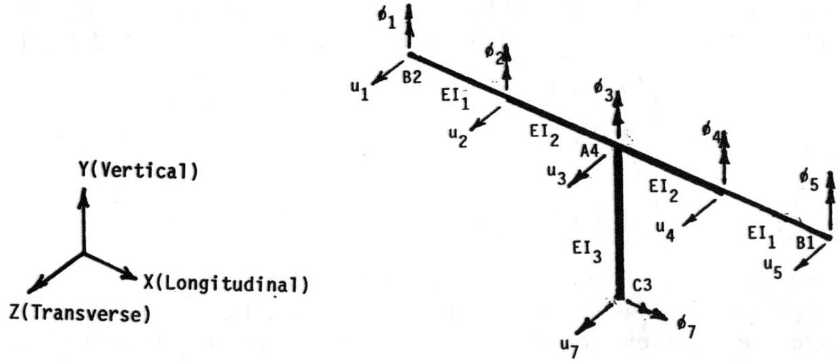

Fig.2: Pier-Girder substructure system (B2-A4-B1) with three different EI-values. The nodal displacements are indicated.

It is clear that $\hat{u}_1$, $\hat{\phi}_1$, $\hat{u}_2$, $\hat{\phi}_2$, $\cdots$, denote the nodal displacement and $\hat{V}_1$, $\hat{M}_1$, $\hat{V}_2$, $\hat{M}_2$, $\cdots$, denote the nodal force of the specified substructure, respectively. Equation (6) can be rearranged between interior nodal points and boundary nodal point in the following form:

$$\begin{array}{c}\text{interior}\\\text{nodal forces}\\ \\ \text{boundary}\\\text{nodal forces}\end{array}\left\{\begin{array}{c}\hat{V}_2\\\hat{M}_2\\\hat{V}_3\\\hat{M}_3\\\hat{V}_4\\\hat{M}_4\\\hat{M}_6\\\hline\hat{V}_1\\\hat{M}_1\\\hat{V}_5\\\hat{M}_5\\\hat{V}_7\\\hat{M}_7\end{array}\right\}=\left[\begin{array}{c:c}\underset{(7\times7)}{B} & \underset{(7\times6)}{A}\\\hdashline\underset{(6\times7)}{C} & \underset{(6\times6)}{D}\end{array}\right]\left\{\begin{array}{c}\hat{u}_2\\\hat{\phi}_2\\\hat{u}_3\\\hat{\phi}_3\\\hat{u}_4\\\hat{\phi}_4\\\hat{\phi}_6\\\hline\hat{u}_1\\\hat{\phi}_1\\\hat{u}_5\\\hat{\phi}_5\\\hat{u}_7\\\hat{\phi}_7\end{array}\right\}\begin{array}{c}\text{interior}\\\text{nodal displacement}\\ \\ \text{boundary}\\\text{nodal displacement}\end{array} \quad (7)$$

Because there is no external loads were applied at the interior nodal points of the substructure, condensation was performed by describing the internal nodal displacement in terms of the boundary nodal displacement. Then

$$\begin{Bmatrix} \hat{V}_2 \\ \hat{M}_2 \\ \hat{V}_3 \\ \hat{M}_3 \\ \hat{V}_4 \\ \hat{M}_4 \\ \hat{M}_6 \end{Bmatrix} = \begin{bmatrix} B \\ (7 \times 7) \end{bmatrix} \begin{Bmatrix} \hat{u}_2 \\ \hat{\phi}_2 \\ \hat{u}_3 \\ \hat{\phi}_3 \\ \hat{u}_4 \\ \hat{\phi}_4 \\ \hat{\phi}_6 \end{Bmatrix} + \begin{bmatrix} A \\ (7 \times 6) \end{bmatrix} \begin{Bmatrix} \hat{u}_1 \\ \hat{\phi}_1 \\ \hat{u}_5 \\ \hat{\phi}_5 \\ \hat{u}_7 \\ \hat{\phi}_7 \end{Bmatrix} = \begin{Bmatrix} 0 \end{Bmatrix} \tag{8}$$

Based on Eq. (8), the interior nodal displacement can be expressed from the boundary nodal displacements:

$$\begin{Bmatrix} \hat{u}_2 \\ \hat{\phi}_2 \\ \hat{u}_3 \\ \hat{\phi}_3 \\ \hat{u}_4 \\ \hat{\phi}_4 \\ \hat{\phi}_6 \end{Bmatrix} = -\begin{bmatrix} B \end{bmatrix}^{-1} \begin{bmatrix} A \end{bmatrix} \begin{Bmatrix} \hat{u}_1 \\ \hat{\phi}_1 \\ \hat{u}_5 \\ \hat{\phi}_5 \\ \hat{u}_7 \\ \hat{\phi}_7 \end{Bmatrix} = \begin{bmatrix} G \\ (7 \times 6) \end{bmatrix} \begin{Bmatrix} \hat{u}_1 \\ \hat{\phi}_1 \\ \hat{u}_5 \\ \hat{\phi}_5 \\ \hat{u}_7 \\ \hat{\phi}_7 \end{Bmatrix} \tag{9}$$

The $\begin{bmatrix} A \end{bmatrix}$ and $\begin{bmatrix} B \end{bmatrix}$ matrices contain unknown parameters of the spectral element properties, EI values and η value. The deformation vector on the right hand side of Eq. (9) is the boundary nodal displacement vector. One can select this deformation vector as known response (which can be obtained from measurement). If one of the interior nodal displacement on the left-hand side of Eq. (8) is also a known response from measurement then the identification problem can be formulated. By defining an objective function J which minimize the difference between the interior nodal displacement, $\hat{u}_3(\omega)$, and the boundary nodal displacements, i.e.,

$$\text{Minimize } J = \int_0^\infty \left| \hat{u}_3(\omega) - \left(G_{31}\,\hat{u}_1 + G_{32}\,\hat{\phi}_1 + G_{33}\,\hat{u}_5 + G_{34}\,\hat{\phi}_5 + G_{35}\,\hat{u}_7 + G_{36}\,\hat{\phi}_7 \right) \right|^2 d\omega \tag{10}$$

where G_{ij} is the element in $\begin{bmatrix} G \end{bmatrix}$ matrix. It is function of unknown modal parameters $(EI)_i$ and η_i. " ^ " denote the Fourier transform of the measurement nodal displacement. In Eq. (10), the boundary nodal rotation $\hat{\phi}_i$ may be retrieved from the vibration data of bridge deck in transverse direction or may be assumed as zero if $\hat{\phi}_i$ is small.

APPLICATION TO BRIDGE SEISMIC RESPONSE DATA

The New-Lian River Bridge (NLRB), a continuous five-span prestressed box-girder bridge, located at the northeast coast of Taiwan. This bridge was instrumented in November 1994 by Central Weather Bureau. This instrumentation comprises single-axis, two-axes and triaxial force-balanced accelerometers with 16-bit resolution which makes the apparatus capable of recording-high-resolution ground motion and structural response within $\pm 2g$ over a nominal frequency range of 0 to 100Hz and with a pre-event and post-event memory. A total of 24 strong motion accelerometers along its deck, at its abutments and a nearby free-field location (as shown in Fig. 3). Most of these instruments were triggered during Feb. 23, 1995 and June 25, 1995 earthquakes, providing one of the most extensive array of strong motion measurements. Figure 4 shows some recorded acceleration from these three events. Preliminary analysis on the seismic response data had been studied using time domain identification scheme (Loh & Lee, 1997). Using lumped mass model the dynamic characteristics of the structure was analyzed.

To illustrate the application of spectral finite element method, the pier-girder substructure of the bridge (the T-section includes two over hang girders and the pier system) will be analyzed, as shown in Fig. 2. Table 1 shows the identified EI-value and η value in transverse direction between points B2 and A4, between C4 and A4, and between A4 and B1. The identified EI-value for each element are quite similar even from different seismic event. The η-value is larger from the result of 1995-6-25 earthquake. This is quite

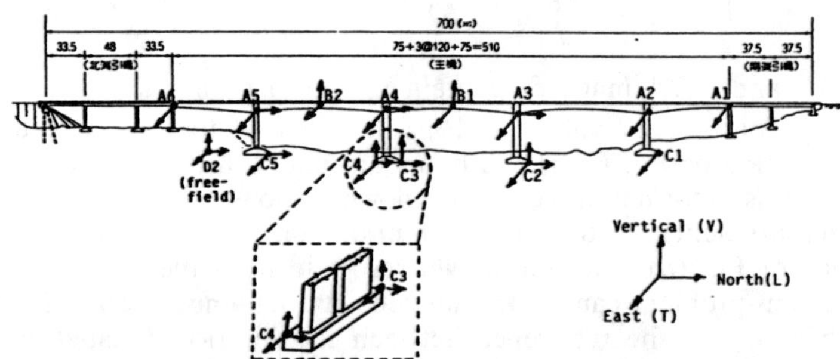

Fig.3: Instrumentation layout at New-Lian River Bridge (Taiwan). A total of 28 channels can be operated.

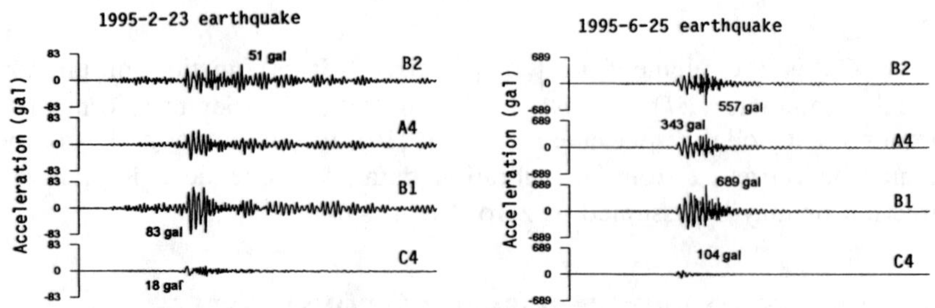

Fig.4: Part of the recorded accelerations from station B2, A4, B1 and C4 for 1995-2-3 earthquake and 1995-6-25 earthquake.

Table 1: Identified model parameter (EI value and damping) of the pier-girder substructure.

	1995-2-23 earthquake	1995-6-25 earthquake
EI_1	6.75×10^9	6.78×10^9
EI_2	1.01×10^{10}	1.02×10^{10}
EI_3	2.59×10^{10}	2.64×10^{10}
η_1	44.	118.
η_2	68.	180.
η_3	91.	243.
Note: EI ($KN\text{-}m^2$) η (kpa-sec)		

Table 2: Identified maximum bending moment and shear force at the nodal point of the pier-girder substructure (B2-A4-B1).

	Maximum Shear Force (KN) ---		Maximum Bending Moment	
	1995-2-23 earthquake	1995-6-25 earthquake	1995-2-23 earthquake	1995-6-25 earthquake
a	519	3310	17562	116630
b	346	2980	11929	78080
c	798	6190	19558	148830
d	1570	10640	19558	148830
e	613	5250	14900	102100
f	614	5030	18718	169840
g	2318	16382	170030	1124400
h	3870	24661	0	0

reasonable because larger structural responses were measured. From Eq. (7), the boundary nodal forces (bending moment and shear force) can also be retrieved from the following equation:

$$\begin{Bmatrix} \hat{V}_1 \\ \hat{M}_1 \\ \hat{V}_5 \\ \hat{M}_5 \\ \hat{M}_7 \\ \hat{V}_7 \end{Bmatrix} = \begin{bmatrix} C \\ (6\times7) \end{bmatrix} \begin{Bmatrix} \hat{u}_2 \\ \hat{\phi}_2 \\ \hat{u}_3 \\ \hat{\phi}_3 \\ \hat{u}_4 \\ \hat{\phi}_4 \\ \hat{\phi}_6 \end{Bmatrix} + \begin{bmatrix} D \\ (6\times6) \end{bmatrix} \begin{Bmatrix} \hat{u}_1 \\ \hat{\phi}_1 \\ \hat{u}_5 \\ \hat{\phi}_5 \\ \hat{\phi}_7 \\ \hat{u}_7 \end{Bmatrix} \tag{11}$$

Table 2 shows the retrieved maximum bending moment and shear force at each nodal point of the substructure (T-section). It is found that significant bending moment was identified at node "g". Figure 5 shows the retrieved time history of shear force at the nodal points of A4 from different beam elements. It has to be pointed out that the identified shear force at the juncture of pier and girder in transverse direction can be balanced [such as 6190kN (point c) + 10640kN (point d) $\doteq$ 16382kN (point g)]. These identified

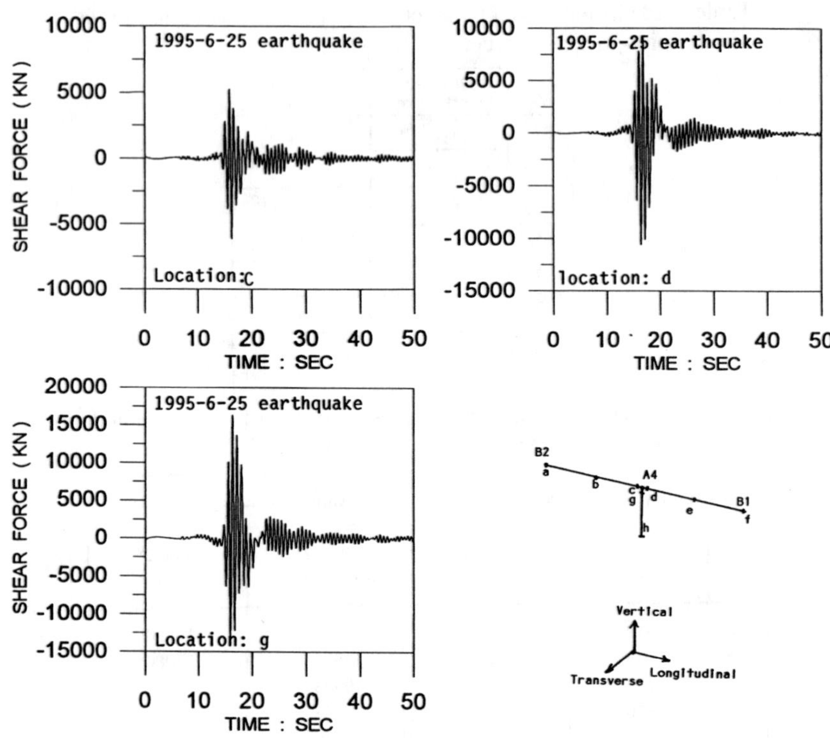

Fig.5: Retrieved time history of shear force (kN) at nodal points of "c', "d' and "g" for the pier-girder substructure (B2-A4-B1) during the 1995-6-25 earthquake.

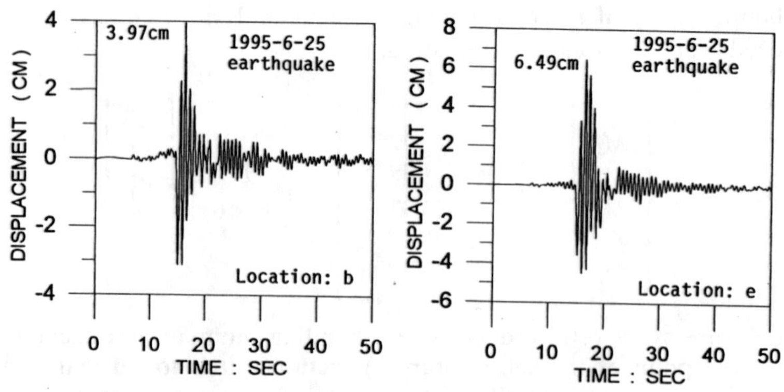

Fig.6: Retrieved displacement at location "b" and "e" from the spectral element method for 1995-6-25 earthquake.

shear force and bending moment can be used to check with the design value for safety evaluation. Figure 6 also shows the retrieved displacement at interior nodal displacements "b" and "e".

DISCUSSIONS AND CONCLUSIONS

For the identification of seismic response data of bridge structure the spectral finite element method was used by considering the flexural waves in beam. The dynamic stiffness of beam element was developed first. Then dynamic structural stiffness matrix of a selected bridge substructure is assembled in a completely analogous way to that used for the static analysis. EI-value and η-value of each beam element are assumed as modal parameters. Matrix operation is performed in such a way that the interior nodal displacements of substructure can be expressed in terms of the boundary nodal displacements (it is a known displacement vector). By matching one of the interior nodal displacement with the recorded displacement at that particular node to that of the boundary nodal displacement, the modal parameters can be estimated. Identification was performed in frequency domain through such a displacement relationship. From the analysis of the seismic response data of a 5-span continuous box-girder prestressed bridge, the following conclusions are drawn:

1. Wavelet analysis was performed on the seismic response data of some measuring points, as shown in Figs. 7a and 7b. Difference on the frequency-time domain analysis between weak motion and strong motion response is obvious. For the seismic response data of 1995-6-25 earthquake the rocking motion of footing plays an important role to the contribution on the transverse motion of location B1. The frequency-time distribution of energy is quite significant for the rocking motion of footing for 1995-6-25 earthquake. This rocking of footing induced the propergation of high frequency signals to the transverse motion of location B1.

2. In the analysis of 1995-6-25 seismic response data of the bridge, because of significant rocking of the footing (at peri C4-A4), we assumed $M_7 = 0$ for this particular case. The result of identification under such assumption shows consistent on the identified EI-value as compare to other set of seismic response data.

3. The substructural formulation provides a methodology in which the interior nodal displacement can be expressed as a function of exterior (or boundary) nodal displacement. The EI-value and η-value of the flexural beam element can be identified. Some of the unknown nodal displacement can also be retrieved after the identification.

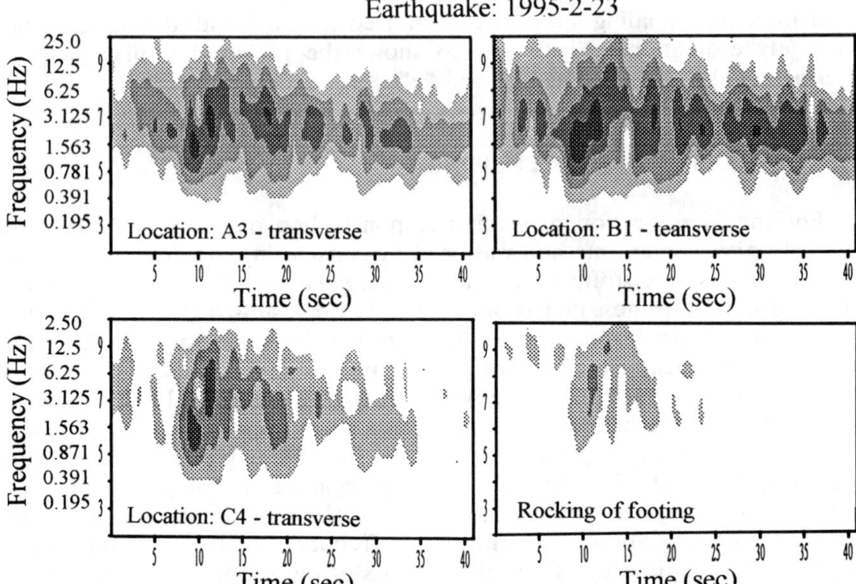

Fig.7a: (above) Wavelet analysis on the recored acceleration at location A3, B1, C4 and rocking of footing for 1995-2-23 earthquake in transverse direction.

Fig.7b: (below) Wavelet analysis on the recorded acceleration at location A3, B1, C4 and rocking of footing for 1995-6-25 earthquake in transverse direction.

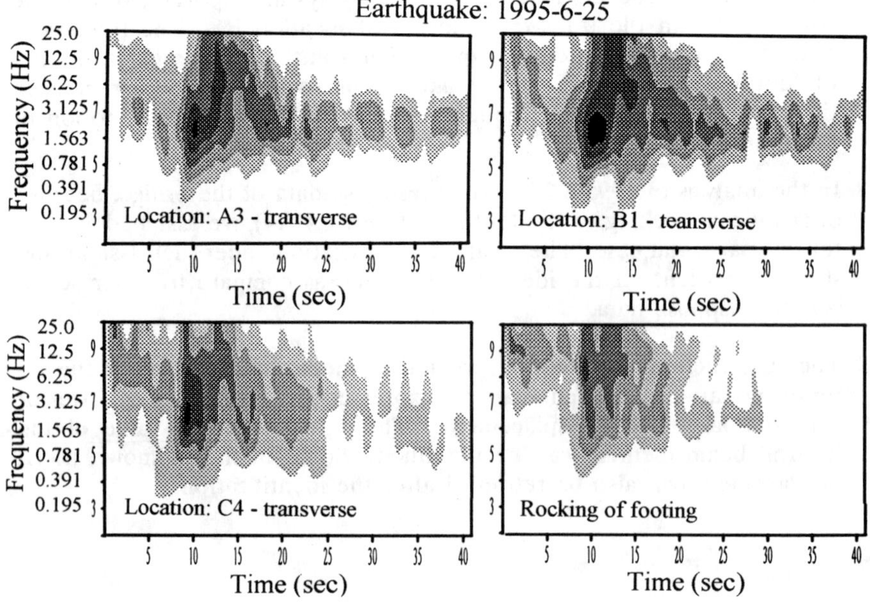

ACKNOWLEDGEMENTS

This research was supported by National Science Council under Grants No. NSC86-2621-P-002-033. The authors wish to express their thanks for this support.

REFERENCES

1. Doyle, James F., *Wave Propagation in Structures: An FFT-Based Spectral Analysis Methodology*, Springer-Verlag, New York, 1989.
2. Chen C. C. and Liu, P. L., "Parametric Identification of Frames Using Transient Response," *Proceedings of the 8-th Asia-Pacific Conference on Nondestructive Testing*, Taipei, December 1995, pp. 575–582.
3. Shinozuka, M., Yun, C. B. and Imai, H., "Identification of Linear Structural Dynamic Systems," *ASCE, J. of Engineering Mechanics*, Vol. 108, No. EM6, 1982, pp. 1371–1390.
4. Hoshiya, M. and Saito, E., "Structural Identification by Extended Kalman Filter," *J. of Engineering Mechanics*, ASCE, Vol. 110, No. 12, 1984, pp. 1757–1770.
5. Safak, E., "Analysis of Recordings in Structural Engineering: Adaptive Filtering, Prediction and Control," U.S. Geological Survey, Open-file Report 88-646, Oct. 1988.
6. Loh, C. H. and Lee, Z. K., "Seismic Monitoring of a Bridge: Assessing Dynamic Characteristics from Both Weak and Strong Ground Excitations," *Earthquake Engineering and Structural Dynamics*, 26, 1997, pp. 269–288.

SESSION 3

MODELING AND DIAGNOSTIC METHODS I

Structural Damage Detection Using Localized Flexibilities

K. C. PARK, G. W. REICH and K. F. ALVIN

ABSTRACT

The present paper offers structural damage detection methods based on the relative changes in localized flexibility properties. The localized flexibility matrices are obtained either by applying a decomposition procedure to experimentally determined global flexibility matrix or by processing the output signals on a substructure-by-substructure manner. Three methods are evaluated in the present paper: a free-free substructural flexibility method, a deformation-based flexibility method, and a strain-basis flexibility method. These methods are applied to a model ten-story building problem, an experimental bridge damage data, and a hybrid experimental-simulated engine structure. The present methods are found to correctly identify damage locations.

1. INTRODUCTION

This paper addresses model-based structural damage detection. Specifically, we focus on the use of localized flexibility properties that can be deduced either from the experimentally determined global flexibility matrix or from a localized structural identification procedure (Alvin and Park, 1996; Reich and Park, 1997; Park, 1997). For various damage detection methods based on the global stiffness matrix or its sensitivities, the reader is referred to recent excellent survey publications, e.g., (Doebling et al, 1996; James III, 1996). To this end, first we present the underlying theory that can be viewed a generalized flexibility formulation in three different generalized coordinates, viz., localized or substructural displacement, elemental deformation-basis, and elemental strain basis.

It is shown that the three different-basis substructural flexibilities can be related to the measured global flexibility matrix via a Ricatti-like equation. The changes in the substructural flexibilities of nominal and damaged cases are used as localized damage indicators. The present methods are then applied to a model ten-story building, I-40 bridge damage experiment data, and an engine mount ladder.

2. A THEORY FOR LOCALIZED FLEXIBILITY

The discrete energy functional Π for a linear damped structure can be expressed as

$$\Pi(\mathbf{u}_g) = \mathbf{u}_g^T\,(\tfrac{1}{2}\mathbf{K}_g\,\mathbf{u}_g - \mathbf{f}_g^D), \qquad \mathbf{K}_g = \mathbf{L}^T\,\mathbf{K}\,\mathbf{L}$$

$$\mathbf{K} = \begin{bmatrix} \mathbf{K}^1 & & & & \\ & \mathbf{K}^2 & & & \\ & & \mathbf{K}^s & & \\ & & & \cdot & \\ & & & & \mathbf{K}^{n_s} \end{bmatrix} \tag{1}$$

$$\mathbf{f}_g^D = \mathbf{f}_g - \mathbf{M}_g\ddot{\mathbf{u}}_g - \mathbf{C}_g\,\dot{\mathbf{u}}_g$$

where $\mathbf{u}_g$ is the displacement vector of the assembled structure; $\mathbf{f}_g^D$ is D'Alembert's force vector that consists of the applied force vector $\mathbf{f}_g$, the resisting inertia force $\mathbf{M}_g\ddot{\mathbf{u}}_g$, and the dissipating force $\mathbf{C}_g\,\dot{\mathbf{u}}_g$; $\mathbf{K}_g$ is the assembled stiffness matrix; $\mathbf{C}_g$ is the assembled damping matrix; $\mathbf{M}_g$ is the mass matrix of the assembled structure; $\mathbf{K}$ is the block diagonal collection of unassembled substructural stiffness matrices $\mathbf{K}^s$; and, $\mathbf{L}$ is the assembly Boolean matrix that relates the global and substructural displacements; the superscript ($\dot{}$) designates time differentiation, and the subscript (**g**) designates 'an assembled global structure' to distinguish from 'partitioned substructures.' .

The discrete damped, time-invariant linear equations of motion for vibrating structures can be obtained from the stationary value of the preceding discrete energy functional, viz., $\delta\Pi = 0$:

$$\mathbf{M}_g\,\ddot{\mathbf{u}}_g + \mathbf{C}_g\,\dot{\mathbf{u}}_g + \mathbf{K}_g\,\mathbf{u}_g = \mathbf{f}_g \tag{2}$$

The input-output relation, referred to *frequency response function* (FRF), is obtained by substituting a harmonic form of the input-output vectors as

$$\begin{bmatrix} \mathbf{u}_g \\ \mathbf{f}_g \end{bmatrix} = e^{j\omega t} \begin{bmatrix} \bar{\mathbf{u}}_g \\ \bar{\mathbf{f}}_g \end{bmatrix} \tag{3}$$

and solving for the frequency-domain output $\bar{\mathbf{u}}_g$:

$$\begin{aligned} \bar{\mathbf{u}}_g &= \mathbf{H}(\omega)_g\,\bar{\mathbf{f}}_g \\ \mathbf{H}(\omega)_g &= (\mathbf{K}_g + j\omega\,\mathbf{C}_g - \omega^2\,\mathbf{M}_g)^{-1} \end{aligned} \tag{4}$$

where $\mathbf{H}(\omega)_g$ will be called the 'global' FRF.

The mass-normalized 'global' flexibility matrix $\mathbf{F}_g$ is obtained as

$$\mathbf{F}_g = \mathbf{\Phi}\,\mathbf{\Lambda}^{-1}\,\mathbf{\Phi}^T, \quad \mathbf{\Phi}^T\,\mathbf{M}_g\,\mathbf{\Phi} = \mathbf{I}, \quad \mathbf{\Phi}^T\,\mathbf{K}_g\mathbf{\Phi} = \mathbf{\Lambda} \tag{5}$$

In practice, the size of the experimentally identified global flexibility matrix is given by

$$\mathbf{F}_g = \mathbf{\Phi}_m\,\mathbf{\Lambda}_m^{-1}\,\mathbf{\Phi}_m^T \tag{6}$$

where the subscript m denotes the measured modes and mode shapes which are substantially smaller than analytical mode size.

2.1 Substructural Flexibility Equation

In order to derive the localized flexibilities, we employ a general partitioned flexibility method presented in (Park and Felippa, 1996). First, a structure that is in equilibrium under applied forces is partitioned into substructures or elements. Hence, each of the substructures formed after partitioning would be subject to the corresponding applied forces plus the Lagrange multipliers acting along the substructural partition interfaces. Consider a structure partitioned into substructures as shown in Fig. 1.

In order to maintain the kinematical compatibility along the partitioned boundaries, the displacement vector of the partitioned substructures $\mathbf{u}$ must satisfy the following relation:

$$\mathbf{u} - \mathbf{L}\mathbf{u}_g = 0 \tag{7}$$

where $\mathbf{u}$ are the collections of all the substructural displacements, and $\mathbf{L}$ is the compatibility Boolean operator that is introduced in assembling the global stiffness matrix from the block diagonal substructural stiffness matrices given in (1).

The force vector that is conjugate with the substructural displacement is given by

$$\mathbf{L}^T\,\mathbf{f} = \mathbf{f}_g \tag{8}$$

so that the desired FRF must assume the following form:

$$\bar{\mathbf{u}} \;=\; \mathbf{H}(\omega)\,\bar{\mathbf{f}} \tag{9}$$

Observe that (9) relates the input $\mathbf{f}$ that are acting on each substructure defined in (8) to the output of individual substructural displacements $\mathbf{u}$ introduced in (7).

Therefore, by introducing a Lagrange multiplier vector $\boldsymbol{\lambda}_b$ to enforce the kinematic compatibility condition (7), the system energy functional expressed in terms of the global nodal quantities (1) is transformed into a three-variable functional

$$\Pi(\mathbf{u},\,,\boldsymbol{\lambda}_b,\mathbf{u}_g) = \mathbf{u}^T\;(\tfrac{1}{2}\,\mathbf{K}\,\mathbf{u} - \mathbf{f} + \mathbf{M}\ddot{\mathbf{u}} + \mathbf{C}\dot{\mathbf{u}}) - \boldsymbol{\lambda}_b^T\mathbf{B}^T\;(\mathbf{u} - \mathbf{L}\,\mathbf{u}_g) \tag{10}$$

where and $\mathbf{M}$ and $\mathbf{C}$ are the substructural mass amd damping matrices, and $\mathbf{B}$ is a constraint projection matrix to be determined.

The stationary value of the above functional leads to the following equations of motion for partitioned structures:

$$\begin{bmatrix} \frac{d^2}{dt^2}\mathbf{M} + \frac{d}{dt}\mathbf{C} + \mathbf{K} & -\mathbf{B} & \mathbf{0} \\ -\mathbf{B}^T & \mathbf{0} & \mathbf{L}_b \\ \mathbf{0} & \mathbf{L}_b^T & \mathbf{0} \end{bmatrix} \begin{Bmatrix} \mathbf{u} \\ \boldsymbol{\lambda}_b \\ \mathbf{u}_g \end{Bmatrix} = \begin{Bmatrix} \mathbf{f} \\ 0 \\ 0 \end{Bmatrix}, \qquad \mathbf{L}_b = \mathbf{B}^T\mathbf{L} \tag{11}$$

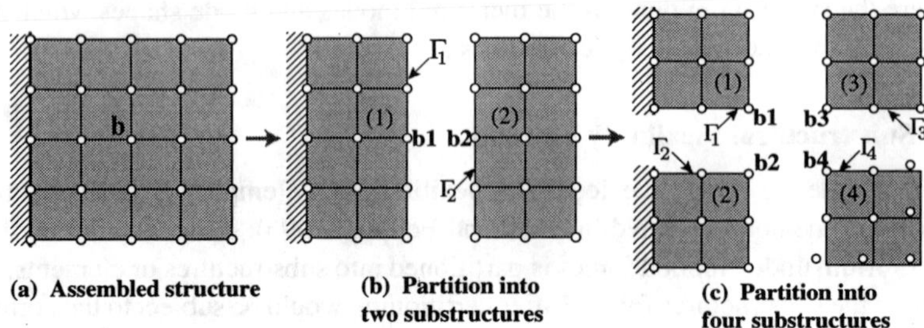

Fig. 1 Example of partitioning into 4 substructures

The substructure-by-substructure FRF that is sought after can now be obtained from (11) as

$$
\begin{aligned}
\mathbf{H}(\omega) &= \mathbf{H}(\omega)_S\,[\mathbf{I} - \mathbf{B}\,\bar{\mathbf{K}}_B(\omega)\,[\mathbf{I} - \mathbf{L}_b\,\mathbf{H}(\omega)_L\,\mathbf{L}_b^T\,\bar{\mathbf{K}}(\omega)_B]\,\mathbf{B}^T\,\mathbf{H}(\omega)_S] \\
\mathbf{H}(\omega)_S &= (\mathbf{K} + j\omega\,\mathbf{C} - \omega^2\,\mathbf{M})^{-1} \\
\bar{\mathbf{K}}(\omega)_B &= (\mathbf{B}^T\,\mathbf{H}(\omega)_u\,\mathbf{B})^{-1} \\
\mathbf{H}(\omega)_L &= (\mathbf{L}_b^T\,\bar{\mathbf{K}}(\omega)_B\,\mathbf{L}_b)^{-1}
\end{aligned} \tag{12}
$$

It can be shown that the substructural FRF $\mathbf{H}(\omega)$ can also be obtained in terms of the global FRF $\mathbf{H}(\omega)_g$, viz.,

$$
\begin{aligned}
\mathbf{H}(\omega) &= \mathbf{L}\,\mathbf{H}(\omega)_g\,\mathbf{L}^T \\
&= \mathbf{L}\,(\mathbf{K}_g + j\omega\,\mathbf{C}_g - \omega^2\,\mathbf{M}_g)^{-1}\,\mathbf{L}^T
\end{aligned} \tag{13}
$$

Observe that the substructural FRF given by (12) reveals the substructural compositions consisting of the substructural mass, damping and stiffness matrices. On the other hand, the same FRF obtained by the input/output invariance property given by (13) masks the substructural compositions, showing only the global quantities. It is their combined use that leads to an important application in detecting damages as shown below.

For the quasistatic limit, i.e., ($\omega \rightarrow \mathbf{0}$), the two expressions provide the following Ricatti-like equation:

$$
\begin{aligned}
\mathbf{L}\,\mathbf{F}_g\,\mathbf{L}^T &= \mathbf{F}\,[\mathbf{I} - \mathbf{B}\,\mathbf{F}_B^{-1}\,[\mathbf{I} - \mathbf{L}_b\,\mathbf{F}_L\,\mathbf{L}_b^T\,\mathbf{F}_B]\,\mathbf{B}^T\,\mathbf{F}] \\
\mathbf{F} = \mathbf{K}^+, \quad \mathbf{F}_B &= (\mathbf{B}^T\,\mathbf{F}\,\mathbf{B}), \quad \mathbf{F}_L = (\mathbf{L}_b^T\,\mathbf{F}_B^{-1}\,\mathbf{L}_b)^{-1}
\end{aligned} \tag{14}
$$

where the symbol ($^+$) designates a generalized inverse and $\mathbf{F}$ is called the substructural flexibility matrix of block diagonal form.

Equation (14) was used by Alvin and Park (1996) for determining the substructural flexibility via an iterative solution procedure. This will be further assessed in Applications section.

2.2 Deformation-Basis Flexibilities

Let us decompose the substructural displacements $\mathbf{u}$ further into deformational and rigid parts:

$$\mathbf{u} = \mathbf{d} + \mathbf{R}\boldsymbol{\alpha}, \tag{15}$$

where $\mathbf{d}$ is the deformational part, $\mathbf{R}$ is the substructural rigid-mode amplitudes, and $\boldsymbol{\alpha}$ are the rigid-mode amplitudes.

In order to eliminate the rigid-body motions from the substructural displacements, we partition (15) according to:

$$\begin{Bmatrix} \mathbf{u}_c \\ \mathbf{u}_f \end{Bmatrix} = \begin{Bmatrix} \mathbf{d}_c \\ \mathbf{d}_f \end{Bmatrix} + \begin{Bmatrix} \mathbf{R}_c \\ \mathbf{R}_f \end{Bmatrix} \boldsymbol{\alpha}, \quad \mathbf{R} = \begin{bmatrix} \mathbf{R}^1 & & & & \\ & \mathbf{R}^2 & & & \\ & & \cdot & & \\ & & & \cdot & \\ & & & & \mathbf{R}^{n_s} \end{bmatrix} \tag{16}$$

where $\mathbf{R}_c$ is a square invertible submatrix corresponding to 'temporarily constrained nodes' and $\mathbf{R}_f$ refers to unconstrained nodes. For example, for a plane beam each free-free substructure would have three rigid-body modes. Hence, $\mathbf{R}_c$ would be a (3×3) matrix. Similarly, the constrained-node rigid-body amplitudes $\mathbf{R}_c$ for other substructural cases can be chosen.

Solving for $\boldsymbol{\alpha}$ from the first row of (16), we obtain

$$\boldsymbol{\alpha} = \mathbf{R}_c^{-1} \, (\mathbf{u}_c - \mathbf{d}_c) \tag{17}$$

We now introduce a deformation measure that represents deformations at nodes f with respect to nodes c:

$$\begin{aligned} \mathbf{v} &= \mathbf{d}_f - \mathbf{R}_f \mathbf{R}_c^{-1} \, \mathbf{d}_c = \mathbf{T}\,\mathbf{u} = \mathbf{T}\,\mathbf{L}\mathbf{u}_g \\ \mathbf{T} &= [\,-\bar{\mathbf{R}} \quad \mathbf{I}\,], \quad \bar{\mathbf{R}} = \mathbf{R}_f \mathbf{R}_c^{-1} \end{aligned} \tag{18}$$

Using this relation, plus its conjugate force vectors given by

$$\mathbf{T}^T \mathbf{f}_v = \mathbf{f} \quad \Rightarrow \quad \mathbf{L}^T \mathbf{T}^T \mathbf{f}_v = \mathbf{L}^T \mathbf{f} = \mathbf{f}_g, \quad \mathbf{f}_\lambda = \mathbf{B}\boldsymbol{\lambda}_b = \mathbf{T}^T \mathbf{B}_v \boldsymbol{\lambda}_v, \tag{19}$$

the variational functional (10) expressed in terms of the substructural displacement $\mathbf{u}$ is transformed into that of the deformation variable $\mathbf{v}$:

$$\Pi(\mathbf{v}, \boldsymbol{\lambda}_v, \mathbf{u}_g) = \mathbf{v}^T \, (\tfrac{1}{2} \, \mathbf{K}_v \, \mathbf{v} + \mathbf{M}_v \ddot{\mathbf{v}} + \mathbf{C}_v \dot{\mathbf{v}} - \mathbf{f}_v) - \boldsymbol{\lambda}_v^T \mathbf{B}_v^T \, (\mathbf{v} - \mathbf{T}\mathbf{L} \, \mathbf{u}_g) \tag{20}$$

A simple expression is obtained if the contraint operator $\mathbf{B}_v$ is chosen as a nullspace of **TL**, viz.,

$$\mathbf{B}_v^T\ (\mathbf{TL}) = \mathbf{0} \tag{21}$$

so that the stationarity condition $\delta\Pi = 0$ of (20) yields

$$\begin{bmatrix} \mathbf{M}_v\frac{d^2}{dt^2} + \mathbf{C}_v\frac{d}{dt} + \mathbf{K}_v & -\mathbf{B}_v \\ -\mathbf{B}_v^T & \mathbf{0} \end{bmatrix} \begin{Bmatrix} \mathbf{v} \\ \boldsymbol{\lambda}_v \end{Bmatrix} = \begin{Bmatrix} \mathbf{f}_v \\ 0 \end{Bmatrix}, \qquad \mathbf{f}_v = \mathbf{L}^T\mathbf{T}^{T+}\ \mathbf{f}_g \tag{22}$$

which will be called the deformational equations of motion for partitioned structures.

The frequency response function that relates the forcing function $\mathbf{f}_v$ to the deformation vector **v** can be obtained in a similar manner as for the substructural displacement case (9) and(12):

$$\begin{aligned} \bar{\mathbf{v}}\ &= \mathbf{H}(\omega)_v\ \bar{\mathbf{f}}_v \\ \mathbf{H}(\omega)_v\ &= \bar{\mathbf{F}}_v\ \{\mathbf{I} - \mathbf{B}_v\ [\mathbf{B}_v^T\ \bar{\mathbf{F}}_v\ \mathbf{B}_v]^{-1}\mathbf{B}_v^T\ \bar{\mathbf{F}}_v\ \} \\ \bar{\mathbf{F}}_v &= (\mathbf{K}_v + j\omega\mathbf{C}_v - \omega^2\mathbf{M}_v)^{-1} \end{aligned} \tag{23}$$

where $\mathbf{H}(\omega)_v$ is the frequency response function in terms of the deformation-basis vector **v** introduced in (18) and its corresponding force vector (19).

The deformation-basis flexibility equation is thus obtained for the quastatic limit, i.e., ($\omega \to 0;\ \ \bar{\mathbf{v}} \to \mathbf{v}$) from the above eqution:

$$\mathbf{TLF}_g\mathbf{L}^T\mathbf{T}^T = \mathbf{P}_v\ \mathbf{F}_v\ \mathbf{P}_v{}^T, \quad \mathbf{P}_v = \mathbf{I} - \mathbf{F}_v\mathbf{B}_v\ [\mathbf{B}_v^T\ \mathbf{F}_v\ \mathbf{B}_v]^{-1}\mathbf{B}_v^T \tag{24}$$

2.3 Strain-Basis Flexibilities

Let us assume that the strain output **s** can be related to the substructural displacement **u** according to

$$\mathbf{s}\ =\ \mathbf{D}\,\mathbf{u}\ =\ \mathbf{D}\,\mathbf{L}\,\mathbf{u}_g \tag{25}$$

where **D** is the discrete strain-displacement relation matrix that can be derived in a variety of ways, e.g., by relying on the finite element shape functions. With this change of basis, one arrives at the following strain-basis functional and the corresponding equations of motion:

$$\begin{aligned} &\Pi(\mathbf{s}, \boldsymbol{\lambda}_s) = \mathbf{s}^T\ (\tfrac{1}{2}\ \mathbf{K}_s\ \mathbf{s} + \mathbf{M}_s\ddot{\mathbf{s}} + \mathbf{C}_s\dot{\mathbf{s}} - \mathbf{f}_s) - \boldsymbol{\lambda}_s^T\mathbf{B}_s^T\ \mathbf{s} \\ &\Downarrow \\ &\begin{bmatrix} \mathbf{M}_s\frac{d^2}{dt^2} + \mathbf{C}_s\frac{d}{dt} + \mathbf{K}_s & -\mathbf{B}_s \\ -\mathbf{B}_s^T & \mathbf{0} \end{bmatrix} \begin{Bmatrix} \mathbf{s} \\ \boldsymbol{\lambda}_s \end{Bmatrix} = \begin{Bmatrix} \mathbf{f}_s \\ 0 \end{Bmatrix} \\ &\mathbf{B}_s^T\ (\mathbf{DL}) = 0 \\ &\mathbf{L}^T\mathbf{D}^T\ \mathbf{f}_s = \mathbf{f}_g, \quad \mathbf{f}_\lambda = \mathbf{B}\boldsymbol{\lambda}_b = \mathbf{D}^T\mathbf{B}_s\boldsymbol{\lambda}_s \end{aligned} \tag{26}$$

The input-output relation between the strain-basis input $\mathbf{f}_s$ and the strain $\mathbf{s}$ can now be expressed in the frequency domain as

$$\begin{aligned} \bar{\mathbf{s}} &= \mathbf{H}(\omega)_s\, \bar{\mathbf{f}}_s \\ \mathbf{H}(\omega)_s &= \bar{\mathbf{F}}_s\, \{\mathbf{I} - \mathbf{B}_s\, [\mathbf{B}_s^T\, \bar{\mathbf{F}}_s\, \mathbf{B}_s]^{-1} \mathbf{B}_s^T\, \bar{\mathbf{F}}_s\, \} \\ \bar{\mathbf{F}}_s &= [\mathbf{D}^T (\mathbf{K}_s + j\omega \mathbf{C}_s - \omega^2 \mathbf{M}_s) \mathbf{D}]^{-1} \end{aligned} \tag{27}$$

where $\mathbf{H}(\omega)_s$ is the frequency response function in terms of the strain $\mathbf{s}$ introduced in (25). The global flexibility $\mathbf{F}_g$ is then related to the strain-basis substructural flexibility $\mathbf{F}_s$ according to:

$$\mathbf{L}\,\mathbf{D}\,\mathbf{F}_g\,\mathbf{D}^T\,\mathbf{L}^T = \mathbf{F}_s\,\{\mathbf{I} - \mathbf{B}_s\,[\mathbf{B}_s^T\,\mathbf{F}_s\,\mathbf{B}_s]^{-1}\mathbf{B}_s^T\,\mathbf{F}_s\,\} \tag{28}$$

which can be used for localized damage detection.

As analytical examples for the substructural flexibility, deformation-bais substructural flexibility are given in (Alvin and Park, 1996; Reich and Park, 1997), we will present the strain-basis flexibility. Consider a free-free plane beam stiffness matrix given by

$$\mathbf{K} = \frac{EI}{\ell^3} \begin{bmatrix} 12 & 6\ell & -12 & 6\ell \\ 6\ell & 4\ell^2 & -6\ell & 2\ell^2 \\ -12 & -6\ell & 12 & -6\ell \\ 6\ell & 2\ell^2 & -6\ell & 4\ell^2 \end{bmatrix}, \tag{29}$$

where EI is the bending rigidity and ℓ the elemental beam length.

It can be shown that the strain-displacement relation for each element is given by

$$\begin{aligned} \mathbf{s} &= \mathbf{D}\mathbf{u} \\ \mathbf{D} &= \frac{1}{\ell^2} \begin{bmatrix} -2\sqrt{3} & -\ell(1+\sqrt{3}) & 2\sqrt{3} & \ell(1-\sqrt{3}) \\ 2\sqrt{3} & \ell(-1+\sqrt{3}) & -2\sqrt{3} & \ell(1+\sqrt{3}) \end{bmatrix} \\ \mathbf{u}^T &= [\,w_1 \quad \theta_1 \quad w_2 \quad \theta_2\,] \end{aligned} \tag{30}$$

where $\mathbf{s}$ represents the bending strains at the two Barlow points; $(w\ \ \theta)$ are nodal displacement and rotation; and, subscripts (1,2) refer to element nodal points. Therefore, the strain-basis element flexibility matrix $\mathbf{F}_s$ can be derived as

$$\mathbf{F}_s = \frac{2}{EI\,\ell} \begin{bmatrix} 1 & 0 \\ 0 & 1 \end{bmatrix} \tag{31}$$

whose diagonal terms correspond to the two strain sampling points. This implies that the strain-based FRF given by (27) may be substantially more diagonally dominant than both the free-free substructural flexibility $\mathbf{F}$ and deformation-basis flexibility $\mathbf{F}_v$. The diagonal dominancy feature may offer sharper damage indication than otherwise possible.

3. APPLICATIONS

3.1 Analytical Assessment Using Free-Free Ladder

Before we demonstrate the three damage detection methods derived in Section 2

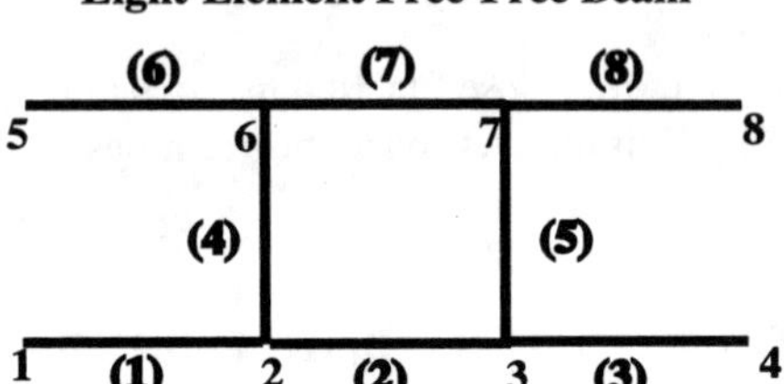

Fig. 2 Free-free indeterminate beam structure

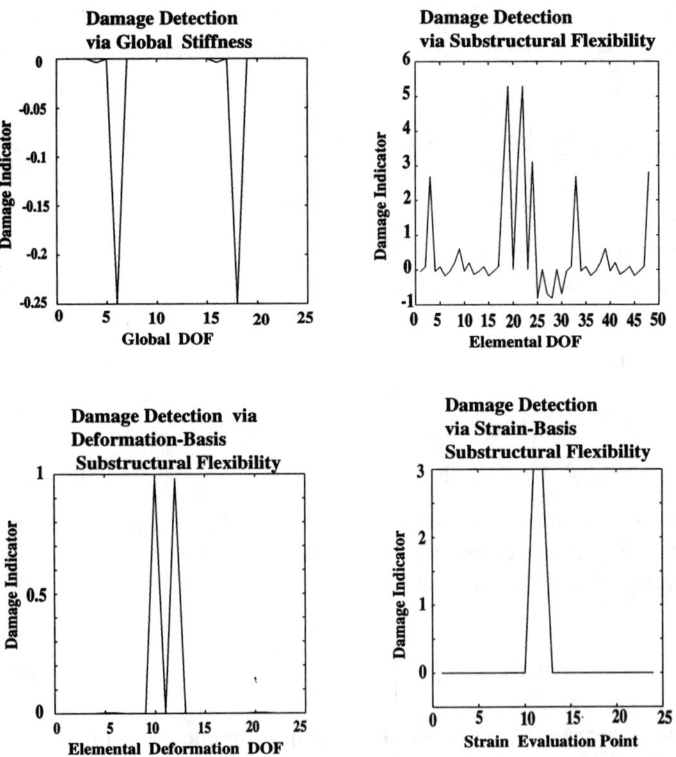

Fig. 3 Damage Detection by Various Methods

as applied to experimental data, we will use an ideal problem in order to offer their relative performance. Figure 2 shows a plane ladder consisting of eight beam elements, with damage invoked in the bending rigidity of element (4). Its bending rigidity is reduced by 1/4. For comparison purposes, the damage indication due to the global diagonal stiffness changes is also included. Figure 3 summarizes the four damage detection methods. For this ideal case, the global method predicts the damage has occurred at nodes 2 and 6, which gives rise to ambiguities as to the location of damages. One plausible choice is certainly the element (4). But it could also involve elements (1), (2), (6) and (7), as nodes 1 and 2 are shared by these elements. It should be noted that, if one were to use the global flexibility changes for this case, the resulting damage indicator becomes completely erraneous.

As for the damge detection based on free-free elemental flexibility, damages are indicated at the elemental degrees of freedom 18 and 24, which belong to element (4). About half of that damage level is also detected for rotational DOFs of elements (1), (2), (6) and (7). This is erraneous although the levels are much lower than the actual damage DOFs.

The damage indicator using the deformation-basis elemental flexibility correctly predicts the damaged DOFs 10 and 12, which correspond to the transverse and rotational DOFs of element (4). Finally, the strain-basis method not only correctly predicts the damage at the two bending strain evaluations points of element (4), but also yields the closest damage level compared to analytical value of 4.

3.2 A Beam Model of a Ten-Story Building

The first example is a ten-story building modeled as plane beam subjected

to ground excitations as shown in Fig. 4 where sample time history and its Fourier transform are also shown. The output consists of 12 horizontal and 12 rotational accelerations. As no distinct input is available, modifications were made to view the building subject to a base motion. System modal identification using the wavelet procedure (Robertson, Park and Alvin, 1995a,b) yielded three high-confidence modes. Using these three modes, the strain-basis localized flexibility was obtained for the nominal and damaged cases. Figure 5 shows the strain evaluation point vs. the relative strain-basis elemental flexibility changes. Note that the structure is damaged between the strain evaluation points 3 and 4.

After checking with the data source it was confirmed that the indicated damage location was in fact the damaged node.

4.3 Damage Identification of I-40 Bridge Experiment

Figure 6 shows the schematic of the I-40 bridge test model that consist of 6 modes. Using the identified 6 modes (Farrar et al, 1994), the present strain-basis localized flexibility changes predicts the damage location coincident with the experimentally induced damage location. It should be noted that the strain evaluation point 16 corresponds to the damaged location N_7 in Fig. 6.

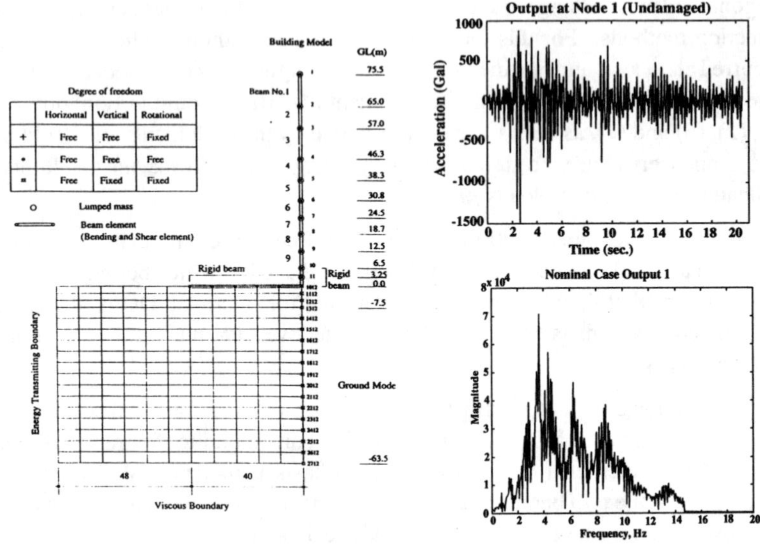

Fig. 4 A Beam Model of Ten-Story Building

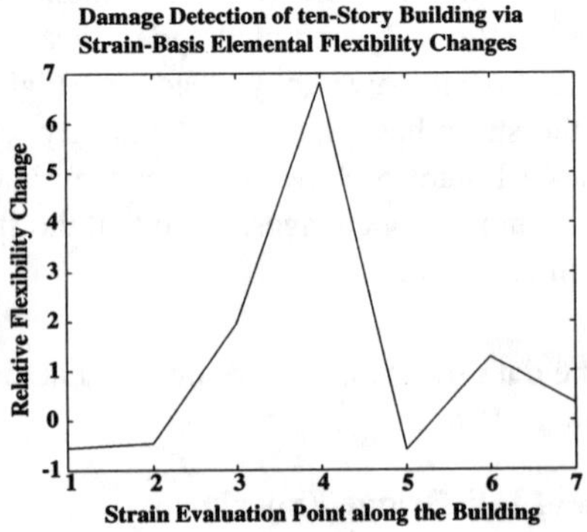

Fig. 5 Damage Detection of Ten-Story Building

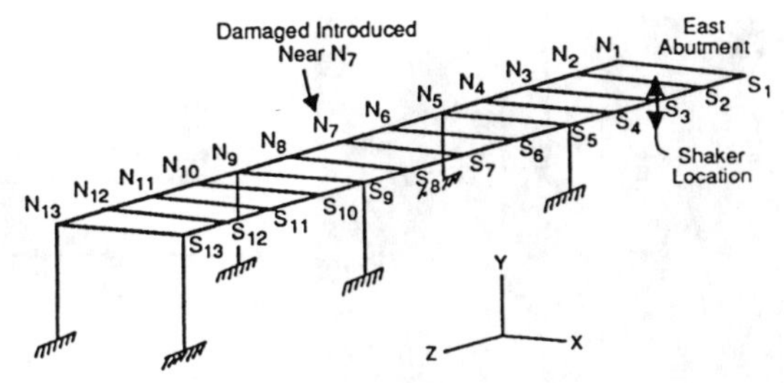

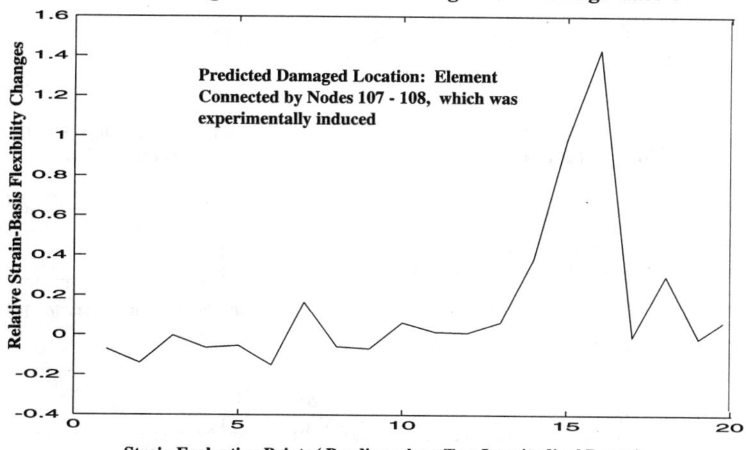

Fig. 6 Schematic and Damage Detection of I-40 Bridge

4.4 Damage Identification of Engine Mounting Ladder

The final example is a ladder-like structure which was carefully tested and modeled at Sandia National Laboratories. A total of 96 usable output sensors was instrumented, whose vibration records were used to obtain 19 global modes. An elaborate finite element model was constructed, which was subsequently modified to match the experimental modes and mode shapes (Redhorse and Alvin, 1996) using updating algorithms of Farhat and Hemez (1993) and Alvin (1996). Using the experimentally validated finite element model, damage was introduced at element 5 shown in Fig. 7. Figure 8a shows damage detection by the changes in the diagonal entries of the

Key
Global Nodes
Elements
Global DOF #s = 6*(Global Node # - 1) + 1:6
Substructure DOF #'= 12*(Element # - 1) + 1:12

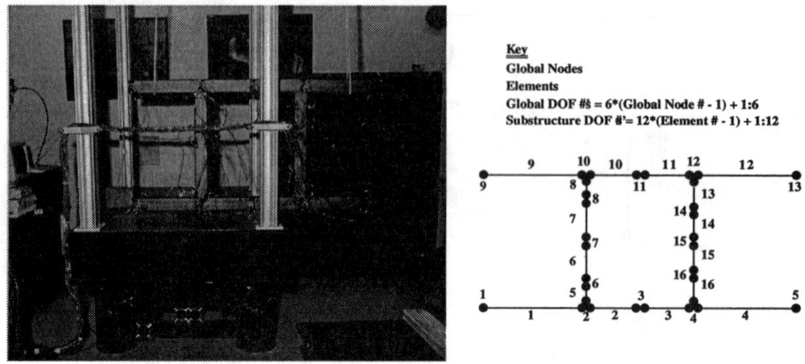

Fig. 7 Ladder and Output Sensor Instrumentation

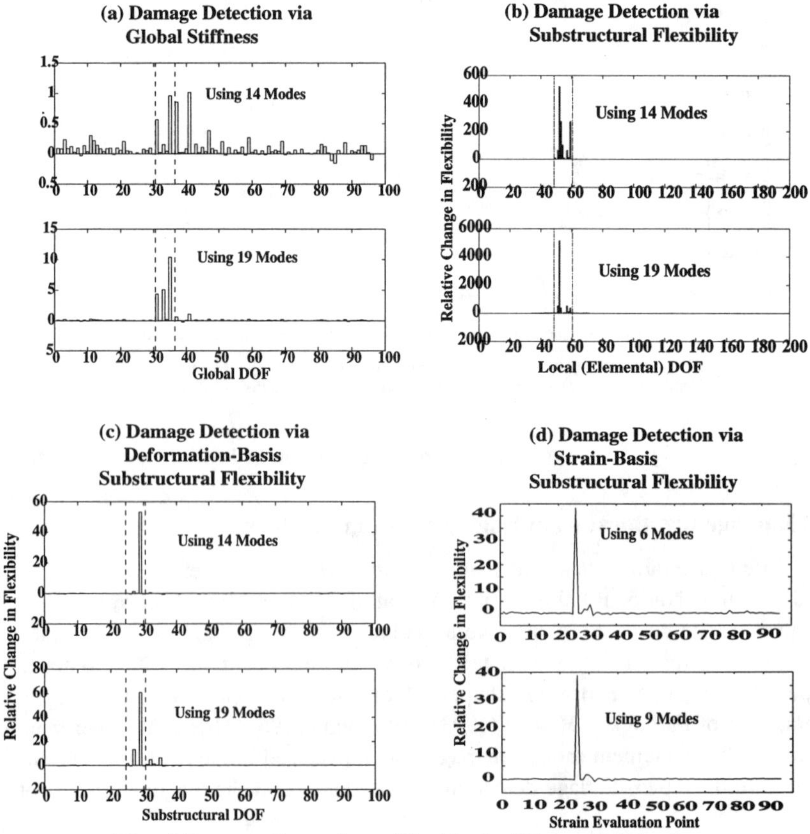

(a) Damage Detection via Global Stiffness
Using 14 Modes
Using 19 Modes
Global DOF
(b) Damage Detection via Substructural Flexibility
Using 14 Modes
Using 19 Modes
Relative Change in Flexibility
Local (Elemental) DOF
(c) Damage Detection via Deformation-Basis Substructural Flexibility
Using 14 Modes
Using 19 Modes
Relative Change in Flexibility
Substructural DOF
(d) Damage Detection via Strain-Basis Substructural Flexibility
Using 6 Modes
Using 9 Modes
Relative Change in Flexibility
Strain Evaluation Point

Fig. 8 Damage Detection of Ladder by Various Methods

global flexibility matrices using 14 and 19 modes. Note that the damage location is ambiguous when using 14 modes and continues to indicate three damage locations.

Figure 8b shows the results by using the substructural flexibility method of Alvin and Park (1996). When 14 modes are used, it predicts two possible damage locations whereas the correct location is identified with 19 modes. Damage prediction using the deformation-basis substructural flexibility is shown in Fig. 8c. Observe that it predicts the damage location correctly using both 14 and 19 modes. The strain-basis damage prediction is finally shown in Fig. 8d. Note that it can predicts the damage location correctly with only 6 modes.

5. DISCUSSIONS

Methods for localized damage detection is presented, which utilize substructural flexibility determined from the partitioned variational formulation of structural mechanics problems. Based on the applictions discussed in the paper we offer the following observations.

1. As far as damage locations are concerned, all of the three damage indicators can predict damage locations with high confidence.
2. Among the three, the strain-basis substructural flexibility method appears to be the most desirable.
3. The initial structural properties such as the bending rigidity, torsional spring constant, the axial stiffness, etc. may be used as initial parameters for model update search algorithms. It should be noted that, when using the three Ricatti-like equations (14), (24) and (28), accurate converged solutions were not necessary for damage location determination. Hence, accurate solutions of the three flexibilities would greatly improve the initial structural parameters to be used for subsequent model updating optimization. This remains an open research subject.
4. It should be noted that the present work reported herein has made no attempt to assess the damage levels. To do this, first further research is required to improve the model construction methodologies such as accurate determination of system models involving nonproportional damping (Alvin, Park and Peterson (1997), residual flexibility (Doebling, Peterson and Alvin, 1996), and signal noise filtering (Robertson and Park, 1997). Second, robust ways of obtaining the initial structural parameters are needed so that subsequent model updating search procedures would yield reasonably reliable parameter determination.
4. Finally, it should be noted that all of the three methods presented can be implemented for localized on-line health monitoring. An initial report on our on-line health monitoring strategy was reported in Reich and Park(1997).

Acknowledgments

It is a pleasure to acknowledge support by Sandia National Laboratories (Contract AP-1461) and a donation from Shimizu Corporation. We thank Dr. Haru Namba of Shimizu for the model ten-story building time response data, and Drs. Scott Doebling and Chuck Farrar of Los Alamos National Laboratories for their assistance in using the I-40 bridge test data.

References

1. Alvin, K. F. (1997), "Finite Element Model Update via Baysian Estimation Estimation and Minimization of Dynamic Residuals," *AIAA Journal*, Vol.35, No. 5, 879-886.

2. Alvin, K. F. and Park, K. C.(1996), Extraction of substructural flexibility from measured global modes and mode shapes, *Proc. 1996 AIAA SDM Conference*, Paper No. AIAA 96-1297, Salt Lake City, Utah, April 1996; submitted to *AIAA Journal.*

3. Doebling, S. W., Peterson, L. D. and Alvin, K. F. (1996), " Estimation of Reciprocal Residual Flexibility from Experimental Modal Data," *AIAA Journal*, Vol. 34, No.8, 1678-1685.

4. Doebling, S. W. et al, (1996), "Damage Identification and health Monitoring of Structural and Mechanical Systems from Changes in Their Vibration Characteristics: A Literature Survey," Los Alamos National Laboratory, Report No. LA-12767-MS, Los Alamos, NM.

5. Farhat, C. and Hemez, F. (1993), "Updating Finite Element Dynamic Models Using an Element-by-Element Sensitivity Methodology," *AIAA Journal*, Vol. 31, No. 9, 1702-1711.

6. Farrar, C. R. et al (1994), "Dynamic characterization and Damage Detection in the I-40 Bridge over the Rio Grande," Los Alamos National Laboratory, Report No. LA-13070-MS, Los Alamos, NM.

7. Games III, G. H. (1996), "Development of Structural Health Monitoring Techniques Using Dynamics Testing," Report No. SAND96-0810.UC-706, Sandia National Laboratories, Albuquerque, NM.

8. Park, K. C. and Felippa, C. A. (1996), A variational framework for solution method developments in structural mechanics, Center for Aerospace Structures, Report No. CU-CAS-96-22, University of Colorado, 1996; to appear in *ASME Journal of Applied Mechanics*

9. Park, K. C., " A Theory for Strain Output-Based Structural System Identification," Center for Aerospace Structures, Report No. CU-CAS-97-02, University of Colorado, February 1997.

10. Redhorse, J. R. and Alvin, K. F., "Finite Element Modeling, Reconciliation, and Evaluation of Predictive Accuracy: A Case Study," *Proc. the 1996 AIAA/ASME Adaptive Structures Forum*, Paper No. AIAA 96-1296, Salt Lake City, UT, April 1996.

11. Robertson, A.N., Park, K.C., and Alvin, K.F. (1996a), "Extraction of Impulse Response Data via Wavelet Transform for Structural System Identification," *To appear in ASME Journal of Vibration and Acoustics.*

12. Robertson, A.N., Park, K.C., and Alvin, K.F. (1996b), "Identification of Structural Dynamics Models Using Wavelet-Generated Impulse Response Data," *To appear in ASME Journal of Vibration and Acoustics.*

13. Robertson, A. N. and Park, K. C. (1997), " Filtering of vibration records and markov parameters via waelet and Fourier methods," *Proc. 1997 AIAA SDM Conference*, Paper No. AIAA 97-1213, April 7-10 1997, Kissimmee, FL.

14. Reich, G. W. and Park, K. C. (1997), " Localized system identification and structural health monitoring from vibration test data," *Proc. 1997 AIAA SDM Conference*, Paper No. AIAA 97-1318, April 7-10 1997, Kissimmee, FL.

A Bayesian Probabilistic Approach to Structural Health Monitoring

M. W. VANIK and J. L. BECK

ABSTRACT

Some general issues associated with on–line structural health monitoring are discussed. In order to address the problem of determining the existence and location of damage in the presence of uncertainties, a global model–based structural health monitoring method which utilizes Bayesian probabilistic inference is developed. The results of tests using simulated data are described.

INTRODUCTION

Structural health monitoring, or SHM for short, is the process of establishing some knowledge of the current condition of a structure. The ultimate goal is to determine the existence, location, and degree of damage in a structure if damage occurs. The need for this type of tool is demonstrated by the failure of structures due to gradual fatigue and corrosion type damage, such as the Mianus river bridge in Connecticut (Levy and Salvadori , 1992) and Aloha Airlines flight 737 (Ott and O'lone , 1988), as well as loss of structural integrity due to severe loading events, as seen in steel frame structures during the Northridge earthquake in southern California in 1994 (EERI , 1996)

A great deal of research in the past thirty years has been aimed at establishing effective local and global methods for health monitoring in civil, mechanical, and aerospace structures. An extensive survey of global methods which use vibration characteristics to perform SHM is presented in Doebling et al. (1996). One typical global approach involves comparing structural models identified using sets of modal data (i.e. frequencies and modeshapes) from a structure before and after damage has occurred. This *model–based* SHM approach relies on structural model updating methodologies to solve the inverse problem of determining the parameters of a structural model given some modal data. Farhat and Hemez (1993) and Kim and Bartkowicz (1993) present examples of this type of SHM method. The critical assumption

M. W. Vanik and J. L. Beck, Division of Engineering and Applied Science, California Institute of Technology, MS 104–44, Pasadena, CA 91125

is made that changes in the parameters of the structural model imply changes in the parts of the real structure associated with the model parameters.

There are some inherent features of structural model updating which lead to difficulties in the model–based SHM approach. The process of identifying the model parameters from the modal data is generally ill–conditioned. Thus, small changes in the modal data lead to proportionally larger changes in the model parameters. Model error and variations in the modal data due to noise, when combined with the ill–conditioning, can lead to large variations in the identified model parameters which are not due to true changes in the structure. Thus, there is uncertainty in whether changes in identified model parameters reflect damage in the structure.

Many of the available papers in the SHM literature do not consider this uncertainty. When the structure under consideration is well–characterized by the analytical model, many controlled measurements can be taken, and the measurements have very low noise levels, no significant uncertainty may be present, and ignoring it may not lead to any problems. However, in many cases, such as with civil structures, these assumptions do not apply. The analytical models rarely capture the full behavior of the structure. For instance, the model may not account for effects such as thermally-induced diurnal variations and excitation amplitude dependence of the modal parameters. Further, the available measured information is restricted by limits on the amount of instrumentation and the fact that only a few of the lower modes of a civil structure can generally be determined with confidence. Finally, measured modal data tends to show significant variation from one measurement to the next. Any SHM method applied in the case of civil structures should therefore account for the substantial uncertainty which arises in the identified structural model parameters.

In addition to neglecting the uncertainties, most methods tend to look for damage using only one set of data from the undamaged structure and another from the structure in a damaged state. Such approaches may attempt an ad hoc treatment of the uncertainty by using average modal data sets over a series of modal measurements as the undamaged and damaged data sets. In situations where the structure is only measured during infrequent periodic inspections or following a severe loading event which had given cause to suspect the structure was damaged, such methods are potentially useful. However, treating the problem in this fashion ignores the long–range "monitoring" goal of SHM. This goal is to continually monitor a structure so that gradual deterioration, as well as damage from severe events such as earthquakes, can be detected. Few, if any, methods explicitly consider this "monitoring" aspect of SHM.

There are a number of advantages to treating SHM as a continual process. First, the effects of noise in the data can potentially be mitigated by using multiple modal measurements. Also, by observing the structure continually, systematic changes may be separated from random fluctuations. Those systematic variations which are not due to damage, such as diurnal effects, can be included in the model to decrease model error. In this manner, effects of gradual damage, such as that due to fatigue and corrosion, can be detected.

A BAYESIAN PROBABILISTIC SHM METHODOLOGY

This paper introduces a *continual on–line* Bayesian probabilistic SHM technique which addresses the ill–conditioning inherent in the inverse problem. The approach requires a linear structural model whose stiffness matrix is parameterized to develop a class of possible models. The parameterization involves grouping the elements of the structural model into substructures. Modal data (i.e. frequencies and incomplete modeshapes) measured from a structure are used to identify the model substructure "stiffness" parameters. In a deterministic SHM scheme, differences in the "best" parameters identified from different modal data sets would be used as indicators of damage. However, rather than consider only single "best" models for each modal data set, the probabilistic method takes uncertainties in the identified model into account by treating the problem within a framework of plausible inference. Bayes' theorem is invoked to develop a probability density function (PDF) for the model stiffness parameters conditional on measured modal data and the class of possible models. Using conditional PDFs derived from sets of modal data determined at different times, a probabilistic damage measure is developed. The probabilistic damage measure arises in answer to the question: *Based on the available modal data and acknowledging the uncertainty, what is the probability that the current model stiffness parameters are less than the corresponding undamaged stiffness parameters?* A simple graphical representation of the damage measure and the interpretation of this measure for use in SHM will be discussed. Before presenting the SHM method, a few terms are first defined.

MODAL DATA

The identified modal parameters consist of N_m frequencies, $\hat{\omega}_r$, and N_m generally incomplete modeshapes, $\hat{\psi}_r \in \mathbb{R}^{N_o}$, where N_o represents the number of observed degrees of freedom. These modal parameters can be identified from measured time–domain data using any of the standard modal parameter identification methods. The modal parameters identified from the nth time–domain data set are referred to by $\hat{\Upsilon}_n$. A grouping of modal parameter data sets taken at different times is called *data*, $\mathcal{D}$. The data from a structure in a known undamaged state is *undamaged data*, $\mathcal{D}_{ud}$, while data from the same structure in an unknown state is called *potentially damaged data*, $\mathcal{D}_{pd}$.

STRUCTURAL MODEL CLASS

A set of N_d degree–of–freedom (DOF) deterministic models, $\mathcal{M}_{N_d}$, which have dynamic behavior characterized by the equation of motion, $M\ddot{x} + C(\theta)\dot{x} + K(\theta)x = f(t)$, with $f, x \in \mathbb{R}^{N_d}$, and $M, C, K \in \mathbb{R}^{N_d \times N_d}$ is used as the model class. The mass matrix is assumed known with sufficient accuracy from structural drawing, and the damping matrix is assumed to be of a form so that the models possess classical normal modes. The models in $\mathcal{M}_{N_d}$ are parameterized by the structural parameters, $\theta \in \mathbb{R}^{N_\theta}$, which define $K(\theta)$ in terms of a linear combination of substructure stiffness matrices, K_i:

$$K(\theta) = K_0 + \sum_{i=1}^{N_\theta} \theta_i K_i. \tag{1}$$

The substructure stiffness matrices model the contributions of a portion of the structure to the overall stiffness matrix. The nominal model, which corresponds to simply summing the K_i, is given by $\theta_F = [1, \ldots, 1]^T$. Expressing the dependence of the stiffness matrix on the structural parameters in the form given by (1) is convenient for mathematical analysis. However, the method which is developed does not preclude using a more general parameterization. For the purpose of health monitoring, the linear parameterization given is sufficient to enable determination of the existence and location of damage.

PROBABILISTIC FRAMEWORK

In the absence of noise and model error, a single model in the model class could be used to match the modal data. Differences in models identified from undamaged and potentially damaged data would then indicate changes in the structure. However, due to the aforementioned uncertainties, such a direct approach to SHM is not possible. The problem is therefore treated in a probabilistic context.

The method presented in this paper extends work done by Beck (1989, 1996) in probabilistic system identification. Beck used a system of plausible inference based on work by Cox (1961) and Jaynes (1978). In this framework, conditional probabilities are used as measures of the plausibility of certain statements given other statements. Bayes' theorem is used to express the conditional probabilities of the model parameters given the data:

$$p(\theta|\mathcal{D}, \mathcal{M}_{N_d}) = k p(\mathcal{D}|\theta, \mathcal{M}_{N_d}) p(\theta|\mathcal{M}_{N_d}). \tag{2}$$

Here, k is a normalizing constant. The distribution $p(\theta|\mathcal{M}_{N_d})$ is the conditional initial distribution on the model parameters based on engineering and modeling judgment. The data distribution, $p(\mathcal{D}|\theta, \mathcal{M}_{N_d})$, is the conditional distribution of the modal parameters given the model parameters. Unless required to maintain a conditional form for a PDF, the explicit dependence of distributions on the model class, $\mathcal{M}_{N_d}$, is dropped in future expressions in order to simplify the notation.

Suppose that $\mathcal{D}$ comprises N_s data sets such that $\mathcal{D} = \mathcal{D}_{N_s} = \{\hat{\Upsilon}_1, \ldots, \hat{\Upsilon}_{N_s}\}$. Using the axioms of probability, $p(\mathcal{D}|\theta)$ can be written as

$$p(\mathcal{D}_{N_s}|\theta) = p(\hat{\Upsilon}_{N_s}|\mathcal{D}_{N_s-1}, \theta)\, p(\mathcal{D}_{N_s-1}|\theta) = \cdots = \prod_{n=1}^{N_s} p(\hat{\Upsilon}_n|\mathcal{D}_{n-1}, \theta), \tag{3}$$

where $p(\hat{\Upsilon}_1|\mathcal{D}_0,\theta) = p(\hat{\Upsilon}_1|\theta)$. The assumption is made that $p(\hat{\Upsilon}_n|\mathcal{D}_{n-1},\theta)$ is independent of $\mathcal{D}_{n-1}$. This assumption reflects that the user's uncertainty in the nth modal parameters, when a structural model is given, is not influenced by the previous modal data. Thus, (2) becomes

$$p(\theta|\mathcal{D}_{N_s}) = kp(\theta|\mathcal{M}_{N_d})\prod_{n=1}^{N_s} p(\hat{\Upsilon}_n|\theta). \tag{4}$$

This result shows that within the Bayesian framework, new data can be incorporated into the PDF for the model parameters in a systematic and consistent fashion by simply extending the product by one term. The details of forming $p(\hat{\Upsilon}_n|\theta,\mathcal{M}_{N_d})$ and $p(\theta|\mathcal{M}_{N_d})$ appear in Vanik (1997). Once the necessary choices are made, the final form of the PDF for θ conditional on $\mathcal{D}$ is given as:

$$p(\theta|\mathcal{D}) = k\exp\left[-\frac{1}{2}J(\theta)\right] \tag{5}$$

where the *overall measure of fit* (MOF), $J(\theta)$, is

$$J(\theta) = (\theta-\theta_F)^T S^{-1}(\theta-\theta_F) + \sum_{n=1}^{N_m} J_r(\theta), \tag{6}$$

and the *modal measure of fit* (MMOF), $J_r(\theta)$, is

$$J_r(\theta) = \sum_{n=1}^{N_s}\left[\frac{\left\|(K-\hat{\omega}_r^2(n)M)\hat{\phi}_r(\theta)\right\|_{M^{-1}}^2}{\sigma_{\omega_r^2}^2\|\hat{\phi}_r(\theta)\|_M^2} + \frac{\hat{\phi}_r(\theta)^T\Gamma^T\left(I-\psi_r(n)\psi_r^T(n)\right)\Gamma\hat{\phi}_r(\theta)}{\epsilon_{\psi_r}^2\|\Gamma\hat{\phi}_r(\theta)\|^2}\right]. \tag{7}$$

The matrix S is a diagonal matrix of variances which reflect the initial level of uncertainty in the analytical model based on engineering and modeling judgment. The parameter $\sigma_{\omega_r^2}$ is the experimentally determined standard deviation for the rth measured frequency. The parameter $\epsilon_{\psi_r}^2$ is an experimentally determined measure of the variation in the rth modeshape. The vectors $\hat{\phi}_r(\theta) \in \mathbb{R}^{N_d}$ are *optimally expanded modeshapes*, which minimize the MOF with respect to ϕ_r for a fixed θ. The matrix Γ picks the observed degrees of freedom from $\hat{\phi}_r$. Although the PDF in (5) gives a non–zero probability for non–physical negative stiffness values, the amount of probability volume less than zero is generally negligible, so truncation of the PDF for negative values followed by a re–normalization is not necessary.

The distribution on θ is integrated over all parameters but one to get the marginal distribution for a single parameter, θ_i, giving

$$\tilde{p}(\theta_i|\mathcal{D}) = \int_{\Theta_i} p(\theta_i,\theta^i|\mathcal{D})d\theta^i \tag{8}$$

where $\theta^i = [\theta_1,\ldots,\theta_{i-1},\theta_{i+1},\ldots,\theta_{N_\theta}]^T$ and Θ^i is the set where θ^i varies across its full possible range. This process is continued for each parameter so that all the individual marginal distributions are found.

The integration cannot be performed analytically. Therefore, the integral is evaluated using an asymptotic expansion technique based on Laplace's method (Papadimitriou, Beck, and Katafygiotis , 1997). The final result gives

$$\tilde{p}(\theta_i|\mathcal{D}) = \varphi\left[\frac{\theta_i - \hat{\theta}_i}{\sqrt{e_i^T L(\hat{\theta})^{-1} e_i}}\right], \tag{9}$$

where φ is the Gaussian PDF with zero mean and unit variance, $\hat{\theta}$ maximizes $p(\theta|\mathcal{D})$, $L(\hat{\theta})$ is the Hessian matrix of $J(\theta)$ evaluated at $\hat{\theta}$, and e_i is a vector with the ith component equal to one, and all other components zero. Using the marginal distributions, the question which was posed at the beginning of this section can now be answered.

PROBABILITY DAMAGE MEASURE

The marginal PDF for the ith model parameters derived using only the undamaged data, $\mathcal{D}_{ud}$, and potentially damaged data, $\mathcal{D}_{pd}$, are $\tilde{p}^{ud}(\theta_i^{ud}|\mathcal{D}_{ud})$ and $\tilde{p}^{pd}(\theta_i^{pd}|\mathcal{D}_{pd})$ respectively. The probability that the ith potentially damaged parameter is less than the ith undamaged parameter is

$$P_i^{dam} = P\{\theta_i^{pd} < \theta_i^{ud}|\mathcal{D}_{ud}, \mathcal{D}_{pd}\}. \tag{10}$$

The quantity P_i^{dam} is called the *probability damage measure*, or simply the *damage measure*, since it is a probabilistic measure of the variation of θ_i^{pd} from θ_i^{ud}. Using the marginal distributions, P_i^{dam} can be shown to be

$$P_i^{dam} = \int_{-\infty}^{\infty} (1 - C^{ud}(\theta_i^{pd}))\tilde{p}_i^{pd}(\theta_i^{pd})d\,\theta_i^{pd}, \tag{11}$$

where $C^{ud}(\cdot)$ is the cumulative distribution for θ_i^{ud}. The damage measure is calculated and monitored for each model parameter in order to perform SHM. Changes in P_i^{dam}, rather than changes in the θ_i, are studied to detect structural damage.

USING P^{DAM} FOR SHM

In a continual on–line monitoring scenario, many modal data measurements will be available. Those from the structure in a known undamaged state are grouped together into $\mathcal{D}_{ud}$ and used to form the undamaged marginal PDFs. Modal parameters taken afterward fall into the potentially damaged data, $\mathcal{D}_{pd}$. A question arises as to which data from $\mathcal{D}_{pd}$ should be used to form the potentially damaged marginal PDFs and calculate P_i^{dam}. If only the most recently measured modal parameter set is used, then damage could be detected as soon as it occurs. However, the damage measure would be sensitive to noise in the modal parameters and could therefore give rise to misleading conclusions. If many sets of data are used, the effects of noise will be mitigated. Thus, smaller levels of damage will be detectable. Unfortunately, if damage occurs, the marginal PDF, and thus P_i^{dam}, will be strongly biased by the undamaged data already in $\mathcal{D}_{pd}$, so many additional measurements must be made before the damaged data can overcome the bias to indicate the existence of damage. In between these extremes are choices which trade off between sensitivity, noise mitigation, and bias. Rather than consider only one such choice, P_i^{dam} is calculated for a range of modal parameter sets from $\mathcal{D}_{pd}$. Thus, P_i^{dam} is calculated for the most recently measured data set, the two most recently measured data sets, and so forth until a limiting number of previous data sets, N_{win}. Each time a modal measurement is performed, P_i^{dam} is recalculated for all of the different subgroupings of modal parameter sets. The number of modal measurements performed is the current *monitoring cycle*, t_{mon}. The *monitoring cycle window*, k, indicates how many previous data sets from $\mathcal{D}_{pd}$, starting at t_{mon}, are used to calculate P_i^{dam}. Thus, P_i^{dam} is a function of t_{mon} and k.

In existing SHM techniques, the damage measure would be calculated from only the current modal parameter set (i.e. $k = 1$), or by simply using a modal parameter set averaged from multiple measurements. Calculating P_i^{dam} in the fashion mentioned above, as both t_{mon} and k vary, leads to the novel concept of monitoring the fluctuation of the damage measure as a function of time and the amount of data used. In testing with simulated data, this approach has shown promise as a means of detecting and locating a wide range of levels of damage.

The measure of the likely variation in P_i^{dam} for a given k due to noise in the modal parameters when no damage is present is characterized using $P_i^{alarm}(k)$, where

$$P_i^{alarm}(k) = \bar{P}_i^{dam}(k) + \gamma\sigma_i(k). \tag{12}$$

The first term is the mean value of $P_i^{dam}(t_{mon}, k)$ for a fixed k as t_{mon} varies, and $\sigma_i(k)$ is the standard deviation. The parameter γ determines the relative frequency of indicating damage when there is none (i.e. a *false alarm*) and missing damage when it exists (i.e. a *missed alarm*). For a given monitoring cycle, if $P_i^{dam}(t_{mon}, k)$ exceeds $P_i^{alarm}(k)$ for any k, an alarm is sounded that the substructure for which the excessive variation occurred may be "damaged":

$$\begin{aligned} P_i^{dam}(t_{mon}, k) &\geq P_i^{alarm}(k) \text{ for any } k \Rightarrow \text{“Damage Alarm”} \\ &< P_i^{alarm}(k) \text{ for all } k \Rightarrow \text{“No Damage Alarm.”} \end{aligned} \tag{13}$$

Thus, smaller values of γ increase the chance of false alarms while larger values increase missed alarms. In testing with simulated data, a value of γ of around 2.5 has proved to be a good choice.

Since P_i^{alarm} is a soft level rather that a hard level, additional decisions should be made when the damage alarm is set in order to determine whether the level was exceeded due to damage, or an extreme data set. In addition to the damage criterion in (13), therefore, other criteria should be considered in order to determine the difference between false and true alarms. In practice with simulated data, $P_i^{dam}(t_{mon}, k)$ is found to have characteristic behaviors depending on whether the structure is undamaged or damaged. Therefore, the behavior of $P_i^{dam}(t_{mon}, k)$ can be studied by a human or an expert system as a function of t_{mon} and k in order to determine the state of damage. During the description of the simulation testing in the next section, the behaviors of $P_i^{dam}(t_{mon}, k)$ for a simple structure are discussed, and some guidelines for use in determining a true state of damage from a false one are given.

APPLYING THE SHM APPROACH

In practice, the proposed method could be applied as an on–line automated structural health monitoring system through the following procedure. This procedure assumes that the structure under consideration has already been instrumented, and the model class and form of the PDFs for the application have already been determined.

- During an *initialization phase*, take as many measurements from the structure in the undamaged state as are needed to find a stable reference undamaged PDF and reasonable bounds on the variations in P_i^{dam} for each substructure when there is no damage in the structure.
- Start the *monitoring phase* wherein the structure is measured periodically and $P_i^{dam}(t_{mon}, k)$ in each substructure is calculated after each measurement.
- If $P_i^{dam}(t_{mon}, k)$ does not exceed $P_i^{alert}(k)$, wait for the next set of data.
- If P_i^{dam} exceeds $P_i^{alert}(k)$, consult an expert system (human or computer encoded logic) about how to proceed. The expert system decides whether the alarm appears false, true, or uncertain based on criteria establish through previous investigations. If the alarm is determine to be true, the appropriate safety measures are implemented immediately. For false or uncertain alarms, the rate at which data is taken is increased and the additional data used to further determine the validity of the alarm.

TESTING THE SHM METHOD

This section reports on the results of testing the method. During the testing, 2–DOF and 10–DOF shear structure models are used to explore the various features, strengths, and limitations of the probabilistic SHM approach. A typical 2–DOF model is pictured in Figure 1. The lumped masses are $m_1 = 2e4$ kg and $m_2 = 1e4$ kg. The substructure stiffness matrices are

$$K_1 = \begin{bmatrix} k_1 & 0 \\ 0 & 0 \end{bmatrix} \quad \text{and} \quad K_2 = \begin{bmatrix} k_2 & -k_2 \\ -k_2 & k_2 \end{bmatrix}, \tag{14}$$

where the "undamaged" model has $k_1 = 1200$ kN/m and $k_2 = 1000$ kN/m. The model parameters, θ_1 and θ_2, therefore scale the "undamaged" interstory stiffnesses, so that $K(\theta) = \theta_1 K_1 + \theta_2 K_2$. The 10–DOF shear structure model is of the same form with $m_1, m_2, m_3 = 3e4$ kg, $m_4, \ldots, m_9 = 2e4$ kg, $m_{10} = 1e4$ kg, $k_1, k_2, k_3 = 48000$ kN/m, $k_4, \ldots, k_8 = 44000$ kN/m, and $k_9, k_{10} = 40000$ kN/m.

Noisy "measured" modal parameters are generated by adding random values chosen from zero–mean Gaussian distributions to the modal parameters of a model

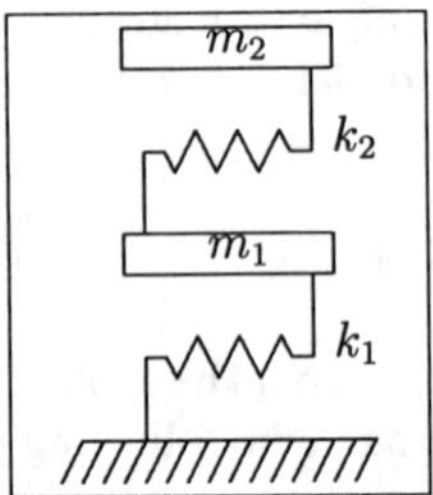

Figure 1 This is a representation of a two DOF shear structure model. The spring elements represent resistance to lateral motion between the "floors." The masses are assumed lumped at the "floors."

from the model class. A coefficient of variation of 0.02 is used for the frequencies and the modeshapes. In practice, the noise levels generally vary from mode to mode, and worsen for higher modes. Also, for a given application, the relationship between the frequency and modeshape noise levels may not be the same used in this testing. Adding these levels of complexity is not considered necessary in this exploratory study.

The monitoring procedure suggested at the end of the previous section is implemented using the synthetically generated modal data. Figures 2(a) to 2(d) show results from the 2–DOF example where the modal data comprises two modes with full modeshapes. The figures show $P_1^{dam}(t_{mon}, k)$ as the monitoring cycle and window vary for different levels of damage in the first substructure. Recall, an increase in the monitoring cycle, t_{mon}, indicates that another modal parameter set has been measured, and an increase in the window parameter, k, indicates that more of the previous potentially damaged data is being used to calculate P_1^{dam}. Figure 2(a) shows the results with no damage. Figures 2(b) to 2(d) show the results with 2%, 5%, and 10% reduction in stiffness k_1 while holding k_2 fixed. Damage is present during the first depicted monitoring cycle in each of Figures 2(b) to 2(d). When damage is present, there is a marked increase in P_1^{dam} over successive monitoring cycles. The damage measure for the undamaged substructure, P_2^{dam}, behaves similarly to the undamaged measure in Figure 2(a). For these examples, therefore, there is no "smearing" of the damage into adjacent substructures.

Figures 3(a) and 3(b) show P_4^{dam} and P_5^{dam} for the 10–DOF example with a 15% reduction in stiffness in the fifth story. The modal data consists of only two modes with full modeshapes in each mode. The time sequence of plots for the fifth substructure damage measure clearly indicate change in that substructure, while those for the fourth substructure damage measure indicate no change in that substructure. Other cases were run with only partial modeshape information and more modes. Damage could still be detected in some cases. In a few cases, however, potential damage was indicated in undamaged substructures. This "smearing" effect in SHM has also been observed by other researchers (Farhat and Hemez , 1993). Investigation is continuing into how to handle the smearing, and determining many modal parameters per measurement are needed in order to successfully apply the proposed SHM approach.

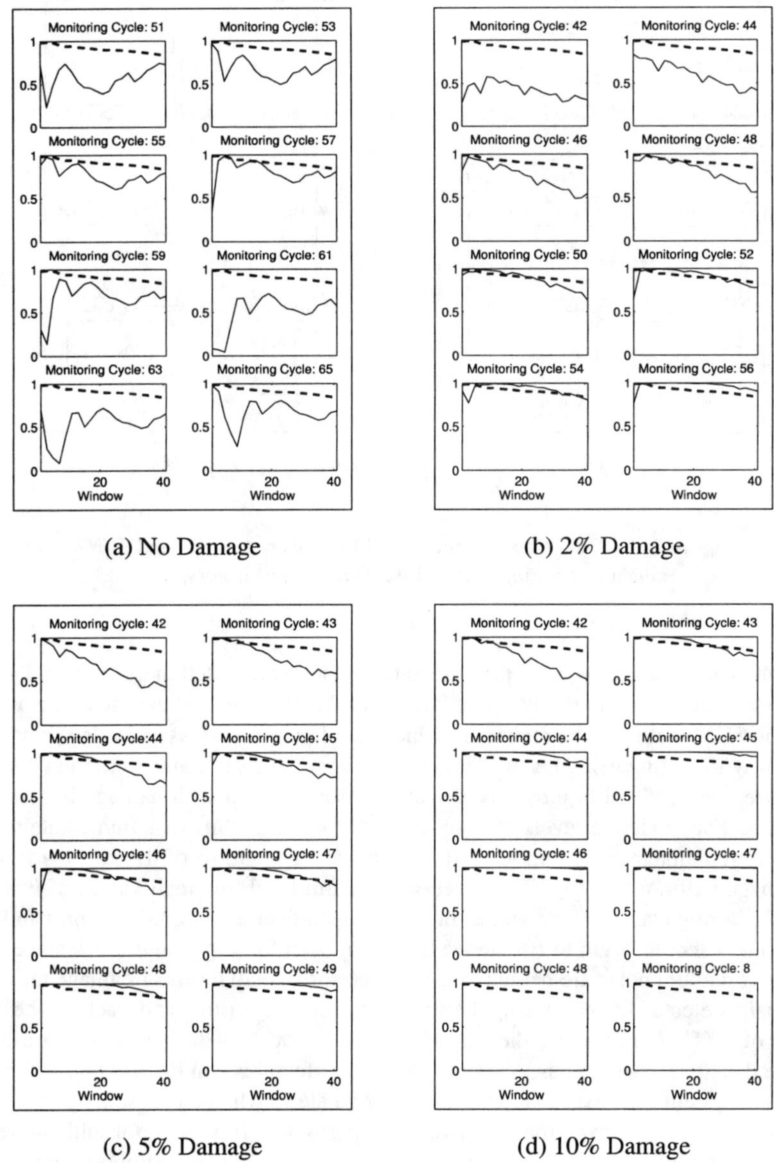

(a) No Damage

(b) 2% Damage

(c) 5% Damage

(d) 10% Damage

Figure 2 2–DOF Case, 2 modes, full modeshapes, P_1^{dam} over sixteen ((a) and (b)) or eight ((c) and (d)) monitoring cycles, (a) undamaged, (b) 2%, (c) 5%, and (d) 10% damage in first story, dashed line is the alarm function for $\gamma = 2$.

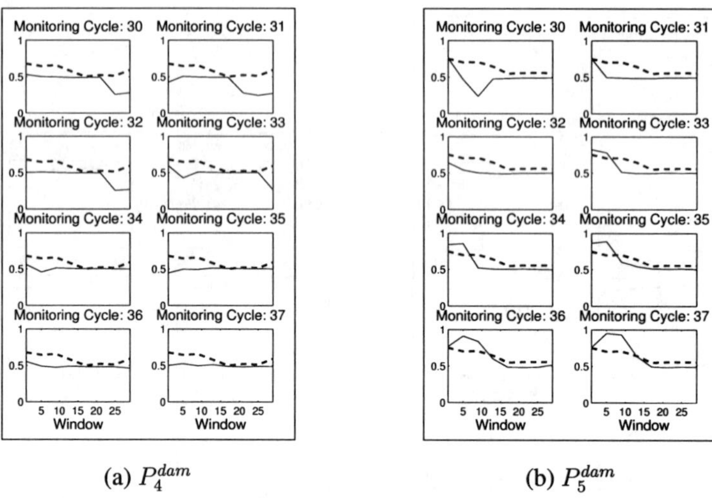

(a) P_4^{dam} (b) P_5^{dam}

Figure 3 10–DOF Case, 2 modes, full modeshapes, P_4^{dam} and P_5^{dam} over eight monitoring cycles, 15% damage in fifth story, $\gamma = 2.8$.

Many more analyses were performed than can be presented in this paper. These few are depicted in order to give an indication of the behavior and the use of the probabilistic SHM method. The manner in which the alarm function is exceeded provides some way to distinguish types of damage. For large levels of damage in the ith substructure, the P_i^{dam} will quickly be driven to 1 for all k, so such damage is quickly detected. For moderate levels of damage, P_i^{dam} will not shift to 1 immediately, as in the large damage case, but will still tend toward 1 for most of the k. Low levels of damage will not cause P_i^{dam} to exceed the alarm level for small values of k with few monitoring cycles. However, as more damaged data is acquired, the probability of variation should begin to rise above the alarm level for large values of k since the effects of random noise are being reduced. Therefore, small levels of damage may be eventually detected by monitoring the structure over longer times and tracking the behavior of P^{dam}. Alarms when there is no damage do not show the behaviors described for the damaged cases. Thus, recognition of these features can be programmed into an expert system to assist in the verification of an alarm when one is set.

Note that considering the data over long periods of time will not mitigate regularly persistent variations such as those due to diurnal changes. Suppose, however, that such effects can be observed while the structure is in its undamaged state, and different sets of "undamaged" data can be associated with different environmental conditions. Then, different undamaged PDFs formed from these data sets can be used as the reference PDF depending on the measured conditions. This is one way in which this type of model error can be accounted for in the proposed SHM framework.

CONCLUSIONS AND FUTURE WORK

This work has discussed the issues associated with uncertainty in applying SHM to real structures, and presented a method for continual on–line implementation which takes those factors into consideration. A novel approach to monitoring which involves studying the variation in time of a probabilistic damage measure was introduced. This approach may enable small levels of damage to be detected through monitoring of the structure over long times. Some preliminary results of testing on simulated data were shown, but additional work is required in a number of areas. The method must be tested on more complex simulated structures, and with data from real structures. Also, the variation of P^{dam} with and without damage must be further characterized so that a set of rules for use by an expert system or end–user can be established. Finally, implementation in an automated fashion must be developed to provide real–time monitoring.

REFERENCES

Beck, J. L. (1989, August). Statistical system identification of structures. In *Proceedings of the Fifth International Conference on Structural Safety and Reliability*, New York, pp. 1395–1402. ASCE.

Beck, J. L. (1996). System identification methods applied to measured seismic response. In *Proceedings Eleventh World Conference on Earthquake Engineering*, Acapulco, Mexico.

Cox, R. T. (1961). *The Algebra of Probable Inference*. Baltimore: Johns Hopkins Press.

Doebling, S. W., C. R. Farrar, M. B. Prime, and D. W. Shevitz (1996). Damage identification and health monitoring of structural and mechanical systems from changes in their vibrations characteristics: A literature review. Technical Report LA–13070–MS, Los Alamos National Laboratory.

EERI (1996). Supplement C to Volume 11, January 1996 Northridge Earthquake Renonnaissance Report. In W. T. Holmes and P. Somers (Eds.), *Earthquake Spectra*, Volume 2. Oakland, CA: EERI.

Farhat, C. and F. M. Hemez (1993). Updating finite element dynamics models using an element–by–element sensitivity methodology. *AIAA Journal 31*(9), 1702–1711.

Jaynes, E. T. (1978). Where do we stand on maximum entropy? In R. D. Levine and M. Tribus (Eds.), *The Maximum Entropy Formalism*. Cambridge, MA: MIT Press.

Kim, H. M. and T. J. Bartkowicz (1993, June). Damage detection and health monitoring of large space structures. *Sound and Vibration 27*(6), 12–17.

Levy, M. and M. Salvadori (1992). *Why Buildings Fall Down*. New York, NY: W. W. Norton and Company.

Ott, J. and R. G. O'lone (1988, May 9). 737 fuselage separation spurs review of safeguards. *Aviation Week and Space Technology*, 92–95.

Papadimitriou, C., J. L. Beck, and L. Katafygiotis (1997). Asymptotic expansions for reliability and moments of uncertain systems. *Journal of Engineering Mechanics, ASCE*.

Vanik, M. W. (1997). *A Bayesian Probabilistic Approach to Structural Health Monitoring*. Ph. D. thesis, California Institute of Technology.

A Probabilistic Approach to Structural Health Monitoring Using Dynamic Data

L. S. KATAFYGIOTIS and H. F. LAM

ABSTRACT

Global health monitoring of a structure is approached by detecting any significant changes in its stiffness distribution through continual updating of its model using vibration measurements. To continually update the structural model a Bayesian probabilistic formulation is followed which allows for the explicit treatment of the uncertainties arising from measurement and modeling errors, and an inherent nonuniqueness common in this inverse problem. Following this formulation one can obtain the updated probability distributions of the stiffness parameters before and after damage. Asymptotic results can be used to obtain simplified approximations of these distributions. Using the updated probability distributions of the stiffness parameters one can obtain the probability of some damage index parameters selected to measure damage within the structure. This approach allows for a probabilistic assessment of damage based on the measured data and any prior information. The methodology is demonstrated with an example using laboratory experimental data.

INTRODUCTION

Increasing interest has been shown in using system identification approaches to develop a global means of structural health monitoring using vibration measurements from a structure (Doebling et al., 1996; Lam, 1997; Katafygiotis and Beck, 1991). The basic idea is to use dynamic test data to continually update the stiffness distribution of a model of the structure, and to use any observed local decrease in the stiffness to indicate the location and severity of possible damage. Even if the mass distribution is assumed to be modeled accurately, the inverse problem of using vibration data to determine the stiffness distribution based on a prescribed class of structural models commonly leads to nonunique solutions. Additional difficulties are caused by measurement noise in the data,

L. S. Katafygiotis and H. F. Lam, Department of Civil & Structural Engineering, Hong Kong University of Science & Technology, Clear Water Bay, Hong Kong

and because any mathematical model is only an approximation of the dynamics of a real structure. For these reasons, a reliable methodology for global health monitoring of real structures has not yet been demonstrated.

This paper is based on a general Bayesian probabilistic methodology for model updating applicable to both linear and nonlinear models presented in Beck and Katafygiotis (1997). The interest here is in using this approach with linear models to detect changes in stiffness in a structure which may be due to damage. This would be useful, for example, if applied to small-amplitude ambient or forced vibration data to continually monitor an off-shore platform for possible fatigue damage caused by cyclic wave loading, or, following a severe loading event such as an earthquake, to use "before" and "after" vibration data from a structure to check for possible event-induced damage. If actual recorded strong earthquake motions are used from a building, care must be taken in interpreting the results since, in addition to the expected nonlinear behavior, past studies have shown that substantial changes in stiffness occur during an earthquake which are not related to structural damage. These changes are possibly due to such effects as temporary loss of stiffness from loosened nonstructural components, micro-cracking in concrete components, and softening in the foundations.

In the following we first review the formulation of the model updating methodology presented in Beck and Katafygiotis (1997). We also review some asymptotic results valid in an identifiable case for approximating the updated probability density function (PDF) of the model parameters; such a case usually occurs when the number of the "free" model parameters to be updated is relatively small, there exists relatively low measurement noise and modeling error and there are adequate measured data available. Extensions of these asymptotic results to nonidentifiable cases were presented in Katafygiotis et al. (1997) and are also briefly reviewed.

Next, the application of the above model updating methodologies for damage detection is discussed. It is shown that instead of utilizing separately "before" and "after" data to obtain the updated PDFs of the stiffness parameters of the undamaged and damaged structure one should consider these data jointly. It is shown that the joint consideration of these data reduces the uncertainty in the updated parameters. A set of damage index parameters is introduced to allow for quantification of damage. It is shown how the probability distribution of these damage indices can be calculated and how it can be utilized for probabilistic assessment of damage throughout the structure.

Finally, the methodology is demonstrated with examples using simulated as well as laboratory experimental data.

STATISTICAL SYSTEM IDENTIFICATION

A short overview of the Bayesian statistical system identification framework (Beck, 1989; Beck, 1996; Beck and Katafygiotis, 1997) and the related asymptotic results for the identifiable case are next presented. Structural model updating is viewed as a system identification problem which involves choosing the "best" (optimal) model from a specified class of structural models $\mathcal{M}$ defined by prescribing a functional relationship $\mathbf{q}(n; \mathbf{a}, Z_1^n, \mathcal{M})$ for the input-output behavior of a structure, where $\mathbf{a} \in S(\mathbf{a}) \subset R^{N_a}$ are the "free" parameters which need to be assigned values from a region $S(\mathbf{a})$ in order to choose a particular model $M(\mathbf{a})$ in $\mathcal{M}$; $\mathbf{q}(n; \mathbf{a}) = \mathbf{q}(n; \mathbf{a}, Z_1^n, \mathcal{M}) \in R^{N_d}$ is the *model* output vector at N_d DOFs (degrees of freedom) at time $t_n = n\Delta t$ where Δt is a prescribed sampling interval; and $Z_1^n = \{\mathbf{z}(m) \in R^{N_I} : m = 1, 2, \ldots, n\}$ is the *system* input up to this time.

A Bayesian statistical framework is utilized by imbedding the class of deterministic structural models $\mathcal{M}$ in a class of probability models $\mathcal{M}_P$ for the system output which is created by choosing a prediction-error probability model $P(\sigma)$ from a class of maximum-entropy probability models $\mathcal{P}$ where σ is a prediction-error model parameter.

The class of probability models $\mathcal{M}_P$, which is defined by the selection of the classes $\mathcal{M}$ and $\mathcal{P}$, is parameterized by $\boldsymbol{\alpha} = [\mathbf{a}^T, \sigma]^T \in S(\boldsymbol{\alpha}) \subset R^{N_\alpha}$ where $N_\alpha = N_a + 1$. Each value of $\boldsymbol{\alpha}$ specifies a function g_M giving the PDF for the system output sequences at the observed and unobserved DOFs, $Y_1^M = \{\mathbf{y}(n) \in R^{N_o} : n = 1, 2, \ldots, M\}$ and $X_1^M = \{\mathbf{x}(n) \in R^{N_d - N_o} : n = 1, 2, \ldots, M\}$ respectively:

$$
\begin{aligned}
&p(Y_1^M, X_1^M \mid \boldsymbol{\alpha}, Z_1^M, \mathcal{M}_P) = \\
&g_M(\mathbf{y}(1), \ldots, \mathbf{y}(M), \mathbf{x}(1), \ldots, \mathbf{x}(M); \boldsymbol{\alpha}, Z_1^M) = \\
&\frac{1}{(\sqrt{2\pi}\sigma)^{MN_d}} \exp\left[-\frac{1}{2\sigma^2} \sum_{n=1}^{M} \|\mathbf{y}(n) - S_o \mathbf{q}(n; \mathbf{a})\|^2\right] \\
&\times \exp\left[-\frac{1}{2\sigma^2} \sum_{n=1}^{M} \|\mathbf{x}(n) - S_u \mathbf{q}(n; \mathbf{a})\|^2\right]
\end{aligned}
\tag{1}
$$

where $\|.\|$ denotes the standard Euclidean 2-norm and S_o and S_u are selection matrices for the observed and unobserved DOF respectively, which have only one non-zero element, equal to unity, in each row and column. In order to account for the uncertainty in the values for the parameters $\boldsymbol{\alpha}$, the specification of $\mathcal{M}_P$ also involves choosing an initial ("prior") PDF $\pi(\boldsymbol{\alpha})$ over the set $S(\boldsymbol{\alpha})$ of possible parameter values, that is:

$$
p(\boldsymbol{\alpha} \mid \mathcal{M}_P) = \pi(\boldsymbol{\alpha}) \tag{2}
$$

The choice for $\pi(\boldsymbol{\alpha})$ allows engineering judgement regarding the plausibilities of the different models to be incorporated. It can be chosen as a smooth slowly-varying PDF which is mathematically convenient and roughly reflects the engineer's judgement regarding the relative plausibilities of the different values of the parameters $\boldsymbol{\alpha}$.

An updated PDF for the system parameters can be obtained for given dynamic test data $\mathcal{D}_N$ consisting of sampled input and output time histories, $\hat{Z}_1^N = \{\hat{\mathbf{z}}(n) \in R^{N_I} : n = 1, 2, \ldots, N\}$ and $\hat{Y}_1^N = \{\hat{\mathbf{y}}(n) \in R^{N_o} : n = 1, 2, \ldots, N\}$ from the structure. Applying Bayes Theorem one obtains the updated ("posterior")

PDF from the initial PDF:

$$p(\boldsymbol{\alpha} \mid \mathcal{D}_N, \mathcal{M}_P) = c f_N(\hat{Y}_1^N; \boldsymbol{\alpha}, \hat{Z}_1^N)\; \pi(\boldsymbol{\alpha}) \tag{3}$$

where

$$f_N(\hat{Y}_1^N; \boldsymbol{\alpha}, \hat{Z}_1^N) = \frac{1}{(\sqrt{2\pi}\sigma)^{NN_o}} \exp\left[-\frac{1}{2\sigma^2}\sum_{n=1}^{N} \|\hat{\mathbf{y}}(n) - S_o\mathbf{q}(n;\mathbf{a})\|^2\right] \tag{4}$$

and c is a normalizing constant such that the integral of the right side of (3) over the space $S(\boldsymbol{\alpha})$ is equal to unity.

The posterior PDF of the structural model parameters $\mathbf{a}$ can be obtained from:

$$p(\mathbf{a} \mid \mathcal{D}_N, \mathcal{M}_P) \;=\; \int_0^\infty c f_N(\hat{Y}_1^N; \boldsymbol{\alpha}, \hat{Z}_1^N)\; \pi(\boldsymbol{\alpha}) d\sigma \tag{5}$$

Assuming that the prior distribution $\pi(\boldsymbol{\alpha})$ is a slowly varying function of σ and that the number N of observed data is large it can be shown that the integrand in the right hand side of (5) is very peaked at the value

$$\hat{\sigma}^2(\mathbf{a}) = \frac{1}{N_o N}\sum_{n=1}^{N} \|\hat{\mathbf{y}}(n) - S_o\mathbf{q}(n;\mathbf{a})\|^2 = J(\mathbf{a}) \tag{6}$$

The integral in (5) can be calculated using the asymptotic approximation described in Papadimitriou et al. (1997) to obtain:

$$p(\mathbf{a} \mid \mathcal{D}_N, \mathcal{M}_P) = c_1 J(\mathbf{a})^{-N_J} \pi(\mathbf{a}, \hat{\sigma}(\mathbf{a})) \tag{7}$$

where $N_J = (NN_0 - 1)/2$, c_1 is a normalizing constant, and $\hat{\sigma}(\mathbf{a})$, $J(\mathbf{a})$ are defined in (6). Note that for a large number of observed data points the exponent N_J in (7) is a large number and therefore, the relative posterior probabilities of the various structural model parameters are very sensitive to the corresponding values of $J(\mathbf{a})$.

Beck and Katafygiotis (1997) presented an asymptotic approximation for the updated PDF $p(\mathbf{a} \mid \mathcal{D}_N, \mathcal{M}_P)$ which is valid for a large number N of sampling points and when the number N_a of parameters for identification is relatively small. According to this result the updated PDF of the structural model parameters can be approximated as a weighted sum of K multivariate Gaussian distributions. Each such distribution $k = 1, ..., K$ has weighting w_k, mean $\hat{\mathbf{a}}^{(k)}$ and covariance matrix $A_N^{-1}(\hat{\mathbf{a}}^{(k)})$, where $A_N(\hat{\mathbf{a}}^{(k)})$ is the Hessian of the function $N_J \ln J(\mathbf{a})$ calculated at the point $\mathbf{a} = \hat{\mathbf{a}}^{(k)}$ (Beck, 1989). The *optimal* structural model parameters $\hat{\mathbf{a}}^{(k)}, k = 1, \ldots, K$ are the values of $\mathbf{a}$ which *globally* minimize $J(\mathbf{a})$ given in (6), that is:

$$J(\hat{\mathbf{a}}^{(k)}) = \min_{\mathbf{a}\in S(\mathbf{a})} J(\mathbf{a}) \tag{8}$$

For a large number of measured data each of these normal distributions is very peaked and, therefore, the updated PDF $p(\mathbf{a} \mid \mathcal{D}_N, \mathcal{M}_P)$ can be effectively viewed as a weighted sum of Dirac delta functions centered at the optimal values of the parameters. The weighting coefficients w_k account for the total probability in the neighborhood of each optimal point and an expression for their calculation is given in Beck and Katafygiotis (1997).

The difference between an identifiable and a non-identifiable case is that while in the identifiable case the updated PDF is concentrated in small neighborhoods of a finite number of optimal points, in the non-identifiable case the updated PDF is concentrated in the neighborhood of a manifold $\mathfrak{S}$ in the parameter space (Katafygiotis et al., 1997). The dimension of this manifold is $N_{\mathfrak{S}} < N_a$. The identifiable case can be considered as a special case of this general case where the dimension of the manifold $\mathfrak{S}$ is zero. There is an increased challenge in handling model updating in a non-identifiable case mainly because of the extreme computational difficulty in calculating this manifold. Katafygiotis et al. (1997) have presented an algorithm for searching the parameter space in an efficient self-expanding manner to find a finite set $S_{\mathfrak{S}}$ comprised of almost uniformly distributed points on $\mathfrak{S}$. For practical model updating and damage detection purposes the updated PDF of the model parameters can be approximated as a weighted sum of Dirac delta functions each centered at one of the points of $S_{\mathfrak{S}}$. The set $S_{\mathfrak{S}}$ plays therefore a role similar to that of the set of optimal points $\{\hat{\mathbf{a}}^{(k)}, k = 1, \ldots, K\}$ in the identifiable case. An expression for the corresponding weightings in the non-identifiable case is given in Katafygiotis et al. (1997). In order to present the proposed damage detection formulation proposed in the next section in a unified manner, that is independent of whether the model parameters are identifiable or not, we will write in both cases $S_{\mathfrak{S}} = \{\hat{\mathbf{a}}^{(k)}, k = 1, \ldots, K\}$ and we will denote the weighting corresponding to the kth point by w_k. This enables us to write the following general approximation for the updated PDF:

$$p(\mathbf{a} \mid \mathcal{D}_N, \mathcal{M}_P) \simeq \sum_{k=1}^{K} w_k \delta(\mathbf{a} - \hat{\mathbf{a}}^{(k)}) \tag{9}$$

where $\delta(\mathbf{a} - \hat{\mathbf{a}}^{(k)})$ denotes the Dirac delta function centered at the point $\hat{\mathbf{a}}^{(k)}$.

DAMAGE DETECTION FORMULATION

In this section we discuss the application of the previously presented model updating methodology for damage detection. Consider an undamaged structure and a class of models $\mathcal{M}_P$ chosen to describe the input-output behavior of the structure. Let $\mathbf{a}^u$ denote the set of structural model parameters and $\mathcal{D}^u$ denote a set of measured dynamic data corresponding to the undamaged structure. Using the methodology described in the previous section one may obtain the updated

PDF $p(\mathbf{a}^u \mid \mathcal{D}^u, \mathcal{M}_P)$ which can be approximated according to (9) as:

$$p(\mathbf{a}^u \mid \mathcal{D}^u, \mathcal{M}_P) \simeq \sum_{k=1}^{K^u} w_k^u \delta(\mathbf{a}^u - \hat{\mathbf{a}}^{(k),u}) \tag{10}$$

The weightings w_k^u depend on the data $\mathcal{D}^u$ and the assigned prior PDF $\pi(\mathbf{a}^u)$. It must be noted that significant prior information may be available from the structural drawings of the undamaged structure which may justify the selection of an appropriate non-uniform prior PDF $\pi(\mathbf{a}^u)$. This non-uniform prior may have a significant effect on the calculated weightings and may significantly reduce the uncertainty in the updated model and lead to a reduced number of significant terms K^u to be included in the right hand side of (10).

Next, consider the structure at a damaged stage and assume that the class $\mathcal{M}_P$ is still appropriate for modeling its structural behavior. Let $\mathbf{a}^d$ denote the set of structural model parameters and $\mathcal{D}^d$ denote a set of measured dynamic data corresponding to the damaged structure. In analogy to equation (10) one obtains:

$$p(\mathbf{a}^d \mid \mathcal{D}^d, \mathcal{M}_P) \simeq \sum_{k=1}^{K^d} w_k^d \delta(\mathbf{a}^d - \hat{\mathbf{a}}^{(k),d}) \tag{11}$$

As in the undamaged case, the weightings w_k^d depend on the data $\mathcal{D}^d$ and the assigned prior PDF $\pi(\mathbf{a}^d)$. However, in the damaged case there is not as much prior information available since there is no prior knowledge about the location and extent of the incurred damage. Therefore, in the damaged case one needs to adopt a relatively uniform (non-informative) prior. Thus, the resulting uncertainty of $\mathbf{a}^d$ is usually larger than that of $\mathbf{a}^u$.

Let $\boldsymbol{\theta} = [\theta_1, ..., \theta_{N_\theta}]^T$ be a vector of uncertain non-dimensional positive stiffness parameters chosen to parameterize the stiffness matrix in the form:

$$K = K_0 + \sum_{i=1}^{N_\theta} K_i \theta_i \tag{12}$$

where $K_i, i = 1, ..., N_\theta$ are known symmetric positive semi-definite matrices corresponding to the expected contribution of various members or substructures to the global stiffness matrix. The remaining parameters in $\mathbf{a}$ are all other non-stiffness uncertain model parameters. For example, in applications where the mass is treated as deterministic and the damping and stiffness as uncertain one may select $\mathbf{a} = [\boldsymbol{\theta}^T, \boldsymbol{\zeta}^T]^T$, where $\boldsymbol{\zeta}$ represents the vector of uncertain damping ratios.

For damage detection the interest lies in detecting decreases in stiffness. From (10) and (11) one may obtain the updated PDFs of the stiffness parameters $\boldsymbol{\theta}^u$ and $\boldsymbol{\theta}^d$ before and after damage, respectively, as follows:

$$\begin{aligned} p(\boldsymbol{\theta}^u \mid \mathcal{D}^u, \mathcal{M}_P) &\simeq \sum_{k=1}^{K^u} w_k^u \delta(\boldsymbol{\theta}^u - \hat{\boldsymbol{\theta}}^{(k),u}), \\ p(\boldsymbol{\theta}^d \mid \mathcal{D}^d, \mathcal{M}_P) &\simeq \sum_{k=1}^{K^d} w_k^d \delta(\boldsymbol{\theta}^d - \hat{\boldsymbol{\theta}}^{(k),d}) \end{aligned} \tag{13}$$

where $\hat{\boldsymbol{\theta}}^{(k),u}$ and $\hat{\boldsymbol{\theta}}^{(k),d}$ denote the stiffness parameters in $\hat{\mathbf{a}}^{(k),u}$ and $\hat{\mathbf{a}}^{(k),d}$, respectively. Equation (13) gives the updated PDF of the stiffness parameters $\boldsymbol{\theta}^u$ ($\boldsymbol{\theta}^d$) of the undamaged (damaged) structure conditional only on its own data $\mathcal{D}^u$ ($\mathcal{D}^d$) and its own prior information.

It can be shown that the uncertainty in the updated models of both the undamaged and damaged structure can be reduced if all available information for both structures is considered simultaneously, and if the additional assertion is made:

$$\mathcal{A} = \{\boldsymbol{\theta}^u \geq \boldsymbol{\theta}^d\} = \{\theta_i^u \geq \theta_i^d, i = 1, ..., N_\theta\} \tag{14}$$

Equation (14) states that after damage the value of any stiffness parameter may either remain unchanged or decrease, but may not increase. The expression for the updated PDF of the parameters $\boldsymbol{\theta}^u$ given all the available data and the assertion $\mathcal{A}$ is:

$$p(\boldsymbol{\theta}^u \mid \mathcal{D}^u, \mathcal{D}^d, \mathcal{A}, \mathcal{M}_P) = \sum_{k=1}^{K^u} w_k^{*,u} \delta(\boldsymbol{\theta}^u - \hat{\boldsymbol{\theta}}^{(k),u}) \tag{15}$$

which is similar to the expression for $p(\boldsymbol{\theta}^u \mid \mathcal{D}^u, \mathcal{M}_P)$ given by (13) except that it involves modified weightings $w_k^{*,u}$ which are given by:

$$w_k^{*,u} = \frac{\sum_{l=1}^{K^d} w_k^u w_l^d f(k,l)}{\sum_{k=1}^{K^u} \sum_{l=1}^{K^d} w_k^u w_l^d f(k,l)} \tag{16}$$

where $f(k,l)$ is a step function such that $f(k,l) = 0$ if $\hat{\boldsymbol{\theta}}^{(k),u} < \hat{\boldsymbol{\theta}}^{(l),d}$, and $= 1$ otherwise. An expression analogous to (15) can be derived for the updated PDF $p(\boldsymbol{\theta}^d \mid \mathcal{D}^u, \mathcal{D}^d, \mathcal{A}, \mathcal{M}_P)$.

In order to quantify the damage in a structure we introduce a set of damage index parameters $\mathbf{D} = [D_1, ..., D_{N_\theta}]^T$. The damage index D_i is used to measure the extent of damage in the ith substructure and is defined as follows: for given values of θ_i^u and θ_i^d the ratio $d_i = \theta_i^d/\theta_i^u$ is calculated; if $d_i \geq d_i^*$, where d_i^* is a specified threshold value, then $D_i = 1.0$; if $d_i < d_i^*$ then $D_i = 1.0 - d_i/d_i^*$. This definition is schematically illustrated in Figure 1. The threshold values $d_i^*, i = 1, ..., N_\theta$ are established during the continual monitoring of the undamaged structure by determining the "normal" variations in the undamaged stiffness parameters. These variations are established by calculating the updated PDFs $p(\boldsymbol{\theta}^u \mid \mathcal{D}^u, \mathcal{M}_P)$ for several sets of "undamaged" data $\mathcal{D}^u$.

The proposed damage detection methodology is based on calculating the PDF of the damage index vector $\mathbf{D}$ given all available information:

$$p(\mathbf{D}|\mathcal{D}^u, \mathcal{D}^d, \mathcal{A}, \mathcal{M}_P) = c \sum_{k=1}^{K^u} \sum_{l=1}^{K^d} w_k w_l f(k,l) \delta(\mathbf{D} - \mathbf{D}_{kl}) \tag{17}$$

where $c^{-1} = \sum_{k=1}^{K^u} \sum_{l=1}^{K^d} w_k w_l f(k,l)$. In equation (17) $\mathbf{D}_{kl}$ denotes the value of the damage index obtained for $[\boldsymbol{\theta}^u, \boldsymbol{\theta}^d] = [\hat{\boldsymbol{\theta}}^{(k),u}, \hat{\boldsymbol{\theta}}^{(l),d}]$. Once the PDF in

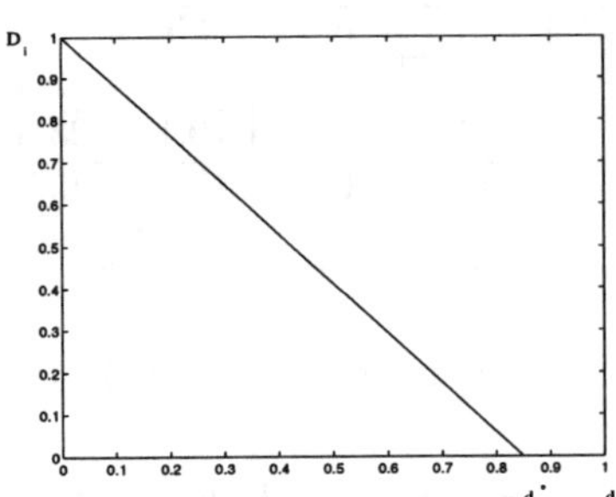

Figure 1: Definition of damage index D_i

Figure 2: Two-storey building model

equation (17) is calculated one may directly use it for a probabilistic assessment of damage.

EXPERIMENTAL CASE STUDY

A two-storey building model shown in Figure 2 was built and tested in the structures laboratory of HKUST. The floors of the building model are made of aluminum plates with dimension 303 × 303 × 8 mm. Each floor is supported by four columns and each column is made of two identical aluminum bars with dimension 38 × 4.5 × 274 mm. The components are connected by eight angles and a number of bolts and nuts.

White noise generated by shaking table was employed to vibrate the building model. The acceleration response histories at both floors as well as the ground acceleration of the building were measured with the help of accelerometers. The three channels of acceleration were digitized with 800 Hz sampling frequency per channel and stored into a computer. The recorded acceleration time histories were about 60 seconds.

Damage in a storey was simulated by removing the aluminum bars from some of its columns. As there are 8 aluminum bars for each storey, removing one aluminum bar corresponds to a theoretical 12.5% reduction in the storey lateral stiffness. In order to keep the motion planar and to avoid torsional effects, two aluminum bars were removed at a time thus preserving the symmetry of the building model with respect to the direction of motion.

Six damage cases were simulated with different locations and extents of damage. All damage cases considered are summarized in Table I, where R_1 and R_2 represent the theoretical percentage reduction in stiffness at the first and second storey, respectively.

Next, an analytical linear finite element model of a classically damped two-storey lumped mass shear building with masses m_1 and m_2, interstorey stiffnesses k_1 and k_2 and damping ratios ζ_1 and ζ_2 was constructed. The masses were

Damage Case	R_1	R_2	$\hat{\theta}_1$	$\hat{\theta}_2$	$\hat{\zeta}_1(\%)$	$\hat{\zeta}_2(\%)$	D_1	D_2
UD	0.00	0.00	0.494	0.762	1.0	0.5	0.00	0.00
DG01	0.00	25.00	0.483	0.620	0.8	0.5	2.31	18.59
DG02	0.00	50.00	0.449	0.492	0.7	0.3	9.03	35.42
DG03	25.00	50.00	0.390	0.439	0.4	0.2	21.06	42.42
DG04	25.00	25.00	0.400	0.580	0.6	0.4	19.13	23.81
DG05	25.00	0.00	0.406	0.713	0.7	0.5	17.78	6.33
DG06	50.00	0.00	0.307	0.691	0.4	0.5	37.82	9.27

Table I: Optimal parameters (using the response of both floors)

calculated from the structural drawings as $m_1 = 4.4482$ kg and $m_2 = 3.9562$ kg and were treated as deterministic, while the stiffness and damping parameters were considered as uncertain. The following four "free" (uncertain) parameters were considered:

1. two stiffness parameters θ_1 and θ_2 scaling the interstorey stiffnesses:

$$k_i = \theta_i k_i^o, \; i = 1, 2$$

   where $k_1^o = k_2^o = 2.3694 \times 10^5 \; N/m$ are the a priori best estimates of the interstorey stiffnesses calculated from the structural drawings.

2. the two damping ratios ζ_1 and ζ_2.

Thus, the uncertain parameter vector is: $\mathbf{a} = \{\theta_1, \theta_2, \zeta_1, \zeta_2\}$. These four parameters were used to update the analytical model. Following the methodology presented earlier in this paper the updated PDF $p(\mathbf{a}|\mathcal{D})$ of these parameters given the data was calculated. As the initial conditions of motion were not measured, the first 30 seconds of the measured data was not used in order to ensure that the effect of the initial conditions had faded away. It was found that $p(\mathbf{a}|\mathcal{D})$ is peaked at one or two optimal points, depending on the number of floor responses assumed to have been measured. If both floor responses are assumed to have been measured only one optimal solution was found, while two optimal solutions were found when only the second floor was assumed to have been measured. Tables I and II display the optimal solutions found in each damage case considered when both floor responses and when only the response of the second floor, respectively, were used in the identification.

We first discuss the results obtained when the response at both floors is assumed to have been measured. According to Table I, the updated first and second interstorey stiffnesses for the undamaged structure are only 49% and 76%, respectively, of the originally estimated values. The main reason for the large discrepancy between the updated and the originally estimated stiffnesses is modeling error. In the estimation of k_i^o, $i = 1, 2$ it was assumed that the joints connecting the columns to the floors were perfectly rigid. However, this assumption is unrealistic; as found the flexibility of the connections results in significantly reduced lateral interstorey stiffnesses.

The percentage stiffness reductions of the first and second storey were calculated and are summarized in columns D_1 and D_2 respectively of Table I. The values of the damage indices D_1 and D_2 in Table I were calculated assuming threshold values $d_1^* = d_2^* = 1$. By comparing in Table I the columns R_1 and R_2 with the columns D_1 and D_2, respectively, it can be seen that the identified damage for all damage cases is in general agreement with the simulated damage. The discrepancies are mainly caused by modeling error. As mentioned before, the connections are not perfectly rigid but have finite stiffness. Therefore, the total flexibility F_t of a storey is due to both the flexibility of its columns F_c and the flexibility F_j of its non-rigid joints, that is:

$$F_t = F_c + F_j \tag{18}$$

Removing some of the column bars in the experimental simulation of damage increases only the flexibility F_c but not F_j. For example, in the damage case DG02 removing one bar from each column of the second storey causes the term F_c to double. However, as it can be seen from equation (18) the effect on F_t is for it to less than double. Therefore, the effect on the second interstorey stiffness is for it to be reduced by less than 50% of its undamaged value. This is the reason why the identified stiffness index D_2 is only 35.42%, instead of the theoretically expected value of 50%. Note that the calculated stiffness reduction of the first storey listed under column D_1 is 9.03%. This is despite the fact that no column was removed from the first floor, and thus no damage was induced in this floor. Thus, we conclude that an identified stiffness reduction may be solely introduced by modeling error and its direct physical interpretation as damage may be erroneous.

Next, in order to address the problem of incomplete data often encountered in practice due to limitations in the number of measured DOFs, we discuss the results obtained when only the response of the second floor is used in the model updating and the damage detection. In this case, two different optimal points were found for every damage case considered and are shown in Table II. Both optimal solutions correspond to global minima of $J(\mathbf{a})$ and therefore fit equally well the measured data. Furthermore, it can be shown that these two optimal solution are output-equivalent and therefore they can be obtained using the methodology for resolving model identifiablity presented by Katafygiotis and Beck (1997). Both solutions are found to have equal weighting assuming a non-informative prior distribution. Following the approach described earlier in the paper it is found that based on the measured data two damage scenarios are possible in each damage case. For example, in the damage case DG02, it can be easily seen that out of the four possible combinations of "before" and "after" damage parameters $\{\boldsymbol{\theta}^u, \boldsymbol{\theta}^d\}$ only the following two are probable:

$$\{[0.496, 0.761]^T, [0.450, 0.491]^T\} \quad \text{and} \quad \{[1.616, 0.233]^T, [1.043, 0.212]^T\}$$

because they are the only ones corresponding to a non-increasing stiffness event. Furthermore, it can be shown that both combinations have equal weighting. Thus, assuming no prior information two equally probable damage scenarios exist:

$$[D_1, D_2] = [9.26\%, 35.44\%] \quad \text{and} \quad [D_1, D_2] = [35.45\%, 9.30\%]$$

Damage Case	$\hat{\theta}_1$	$\hat{\theta}_2$	$\hat{\zeta}_1$ (%)	$\hat{\zeta}_2$ (%)
UD	0.496	0.761	0.9	0.5
	1.616	0.233		
DG01	0.483	0.620	0.7	0.5
	1.317	0.227		
DG02	0.450	0.491	0.6	0.3
	1.043	0.212		
DG03	0.390	0.438	0.4	0.2
	0.931	0.184		
DG04	0.400	0.580	0.6	0.5
	1.232	0.188		
DG05	0.407	0.713	0.6	0.6
	1.514	0.191		
DG06	0.307	0.691	0.3	0.6
	1.467	0.145		

Table II: Optimal parameters (using only the second floor response)

Unless additional measurement and/or additional prior information is used, one is unable to satisfactorily detect damage in this case.

CONCLUSIONS

A probabilistic framework for damage detection has been presented. The approach utilizes a Bayesian statistical system identification methodology to update the structural model of the undamaged and damaged structure. Simultaneous consideration of both "undamaged" and "damaged" data in the model updating, as well as use of any available prior information, results in updated structural parameters which are less uncertain. A set of damage indices is introduced to quantify damage. These indices are treated as random variables and their PDF can be estimated allowing for a probabilistic assessment of damage. Following the proposed formulation, damage is established in a discrete probabilistic manner as a finite set of possible damage scenarios which can be ranked according to their relative plausibilities. However, one must be aware that the direct interpretation of these indices as damage may not be very accurate, especially in the case where the modeling error is significant.

ACKNOWLEDGMENT

This paper is based upon work partly supported by the Hong Kong Research Grant Council under grant HKUST 639/95E. This support is gratefully acknowledged.

REFERENCES

1. Beck, J. L., 1989. "Statistical system identification of structures." *Proceedings Fifth International Conference on Structural Safety and Reliability*, ICOSSAR'89, 1395-1402, ASCE, New York.

2. Beck, J. L., 1996. "System identification methods applied to measured seismic response." *Proceedings* 11th *World Conference on Earthquake Engineering*, Acapulco, Mexico.

3. Beck, J. L. and L. S. Katafygiotis, 1997. "Updating structural dynamic models and their uncertainties: statistical system identification." *Journal of Engineering Mechanics*, ASCE, in print.

4. Doebling, S. W., C. R. Farrar, M. B. Prime and D. W. Shevitz, 1996. "Damage identification and health monitoring of structural and mechanical systems from changes in their vibration characteristics: a literature review." Los Alamos National Laboratory Report LA-13070-MS

5. Katafygiotis, L. S. and J. L. Beck, 1991. "Probabilistic approach to structural health monitoring from dynamic testing." *Proc. Intl. Workshop on Technology for Hong Kong's Infrastructure Development*, Hong Kong, 291-305, Dec. 1991.

6. Katafygiotis, L. S. and J. L. Beck, 1997. "Updating structural dynamic models and their uncertainties: model identifiability." *Journal of Engineering Mechanics*, ASCE, in print.

7. Katafygiotis, L. S., C. Papadimitriou and H. F. Lam, 1997. "Probabilistic model updating using dynamic data", *Proc. Seventh International Conference on Structural Safety and Reliability,* ICOSSAR'97, Kyoto, Japan, Nov. 1997.

8. Lam, H.F., 1997. "Damage detection based on vibration data: a literature review," Report 97-01, Dept. of Civil & Structural Engineering, HKUST, Hong Kong.

9. Papadimitriou, C., J. L. Beck and L. S. Katafygiotis, 1997. "Asymptotic expansions for reliabilities and moments of uncertain dynamic systems", *Journal of Engineering Mechanics*, ASCE, in print.

Health Monitoring Systems of Linear Structures Using Wavelet Analysis

A. AL-KHALIDY,[1] M. NOORI,[2] Z. HOU,[2]
S. YAMAMOTO,[3] A. MASUDA[3] and A. SONE[3]

ABSTRACT

Detection of damage rate in structures caused by low cycle fatigue under dynamic loading such as seismic or wind excitations is of great concern in the structural community. Wavelet transform has been implemented for detecting the fatigue damage which is modeled as a random impulse in the input signal. In this work, a linear single-degree-of-freedom oscillator is used as a model for the structure. The occurrence time of the impulses has been detected for several signal-to-noise ratios. The detectability of the impulse signal was observed to be affected by the sampling rate, the signal-to-noise ratio, and the vanishing moments of the wavelets used. Additional noise was introduced in the output response of the system and it turned out that the detectability in general was greatly deteriorated for even very small noise amplitudes specially at high sampling frequencies and was almost lost when the noise level was 10 percent of the output signal. Sampling frequency plays an important and reverse role in the detection of impulsive singularities in the presence or absence of output measurement noise. There is a tradeoff between the sampling rate and detectability depending on the level of signal noise. For low noise level, detectability is improved while increasing the sampling frequency. However, in the case of high output noise, a low sampling rate is suggested.

INTRODUCTION

Wavelet theory was initially proposed by J. Morlet, a geophysicist, and A. Grossmann, a theoretical physicist. Together with their fellow Frenchman, Y. Meyer, their 'French school' developed the mathematical foundations of wavelets. Later, two America-based researchers Daubechies and Mallat defined the connection between wavelets and digital signal processing (Bentley and McDonnell, 1994).

Wavelets have been applied to a number of areas, including data compression, image processing and time-frequency spectral estimation. There have been several

[1] Department of Civil Engineering, Cornell University, Hollister Hall, Ithaca, NY 14853
[2] Department of Mechanical Engineering, Worcester Polytechnic Istitute, Worcester, MA 01609
[3] Department of Mechanical Engineering, Kyoto Institute of Technology, Kyoto, Japan

traditional tools for time frequency analysis of discrete signals such as the Fourier transform (FT) and the modified short-time Fourier transform (STFT). Advantages of the wavelet transform, which is the main concern of this work, can be seen by studying the limitations of these methods.

FOURIER TRANSFORM

In signal analysis few, if any, tools are as universal as the Fourier transform. It is used as the keystone of modern signal processing. The Fourier transform and its inverse are defined as follows:

$$F(\omega) = \int_{-\infty}^{\infty} f(t)\exp(-i\omega t)dt \tag{1}$$

$$f(t) = \frac{1}{2\pi}\int_{-\infty}^{\infty} F(\omega)\exp(i\omega t)d\omega \tag{2}$$

where $F(\omega)$ is the Fourier transform of a signal f(t). The inverse transform can be described in terms of sine and cosine functions rather than a complex exponential function:

$$f(t) = \frac{1}{2\pi}\int_{-\infty}^{\infty} F(\omega)(\cos\omega t + i\sin\omega t)d\omega \tag{3}$$

For many signals , Fourier analysis is extremely useful because the signal's frequency content is of great importance. Yet Fourier transform has a serious drawback. In transforming to the frequency domain, time information is completely lost (Misiti et al., 1996). It is impossible to know when certain events happened by just looking at the Fourier transform of a signal. Generally, it is not a problem for stationary signals where little change occurs over time. However, most interesting signals contain numerous non-stationary or transient characteristics of interests, including drift, trends, abrupt changes, and discontinuities.

SHORT-TIME FOURIER ANALYSIS

This method is developed mainly to correct the deficiency of FT when dealing with non-stationary signals. This is achieved by the *windowing* technique, i.e. modifying the FT and making it to analyze only a small section of the signal at a time. The STFT of a signal s(t) is defined by Gabor (Bentley and McDonnell, 1994):

$$STFT(\tau,\omega) = \int s(t)g(t-\tau)\exp(-i\omega t)dt \tag{4}$$

The result can be interpreted as the Fourier transform of the signal s(t) windowed by a function g(t) around time τ. As the window function is shifted in time over the whole signal and consecutive overlapped transforms are performed, a description of the evolution of the signal spectrum with time is achieved. This method assumes signal stationarity over the limited window g(t). A drawback of STFT is that size of the time window chosen is the same for all frequencies. Analysis of many signals requires a more flexible approach such as one with variable window sizes to obtain more accurate information about time and/or frequency of an event (Misiti et al., 1996).

WAVELET TRANSFORM

Wavelet analysis represents the next logical step: a windowing technique with variable-sized regions. Wavelet analysis allows the use of longer time intervals for low frequency information and shorter intervals for high frequency information. Wavelet analysis is capable of revealing aspects of data that other signal analysis techniques miss, such as trends, breakdown points, discontinuities in higher derivatives, and self similarity. Indeed, in their brief history within the field of signal processing, wavelet analysis has already proven itself to be an indispensable addition to the analyst's collection of tools and continue to enjoy a burgeoning popularity today.

WAVELETS

A wavelet is a waveform of effectively limited duration that has an average value of zero. This is often described as compact support in time. The sine waves which are the basis of Fourier transform extend from minus infinity to plus infinity and thus they have infinite support. The sine waves are smooth and predictable while wavelets tend to be irregular and asymmetric. Fourier analysis consists of breaking up a signal into sine waves of various frequencies and phases. Similarly, wavelet analysis is a breaking up of a signal into shifted and scaled versions of the original (or mother) wavelet (Misiti et al., 1996).

CONTINUOUS WAVELET TRANSFORM(CWT)

The continuous wavelet transform of a function f(t) is defined as:

$$(Wf)(a,b) = \frac{1}{\sqrt{a}} \int_{-\infty}^{\infty} f(t)\overline{\Psi}\left(\frac{t-b}{a}\right)dt \tag{5}$$

where $a>0$ and $b \in \Re$ are the dilation and translation parameters, respectively. The function $\Psi(t)$ is called the analyzing wavelet and the overbar indicates the complex conjugate. The analyzing wavelet should satisfy the *admissibility condition*:

$$C_{\Psi} = \int_{-\infty}^{\infty} \frac{|\Psi(\omega)|^2}{|\omega|} d\omega < \infty \tag{6}$$

where $\Psi(\omega)$ denotes the Fourier transform of $\Psi(t)$. Thus, the wavelet expansion of f(t) is defined by the following equation:

$$f(t) = \frac{1}{C_\Psi} \int_{-\infty}^{\infty} \int_{-\infty}^{\infty} (Wf)(a,b) \Psi\left(\frac{t-b}{a}\right) \frac{1}{a^2} \, da \, db \tag{7}$$

DISCRETE WAVELET TRANSFORM (DWT)

In the wavelet transform, a function or a signal is expanded by a basis which is compactly supported in both time and frequency domain. For the actual computation of wavelet transform, the dilation parameter a and the translation parameter b should be discretized. This is called the Discrete Wavelet Transform (DWT). It turns out that analysis will be much more efficient and just as accurate if dyadic scales and positions, i.e. scales and positions are chosen based on powers of two, are employed. For some special choices of Ψ, the discretized wavelets $\{\Psi_{j,k}\}$ constitute an orthonormal basis for $L^2(R)$. By using the orthonormal bases, the wavelet expansion of a function x(t) and the coefficients of wavelet expansion are defined as

$$x(t) = \sum_j \sum_k \alpha_{j,k} \Psi_{j,k}(t) \tag{8}$$

and

$$\alpha_{j,k} = \int_{-\infty}^{\infty} x(t) \overline{\Psi}_{j,k}(t) dt = \langle x(t), \Psi_{j,k}(t) \rangle \tag{9}$$

where $\alpha_{j,k}$ is the coefficient of the wavelet expansion of x(t) and $\Psi_{j,k}$ is the discrete basis generated by dilating and translating an analyzing wavelet Ψ. Integers j and k $\in Z$ are the dilation and translation parameters, respectively. The discretized wavelet can be written as

$$\Psi_{j,k}(t) = 2^{j/2} \Psi(2^j t - k) \tag{10}$$

Mallat (1988) developed an efficient algorithm using filters which is equivalent to the Fast Fourier Transform and is sometimes referred as the Fast Wavelet transform: a box into which a signal passes, and out of which wavelet coefficients quickly emerge.

For many signals, the low-frequency content is the most important part, which gives a signal its identity. The high-frequency content, on the other hand, imparts flavor and contains noise and possibly discontinuities. In wavelet analysis, a signal may be represented by its *approximations* and *details*. The approximations are the high-scale, low-frequency components of the signal. The details are the low-scale, high-frequency components. By selecting different dyadic scales, a signal can be broken down into many lower-resolution components, referred as the *wavelet decomposition tree*. The downsampling technique is used to efficiently reduce the data size in the wavelet decomposition tree.

Daubechies Wavelets

The analyzing wavelets chosen for this work are the family of Daubechies wavelets which were invented by I. Daubechies. This family of wavelets is a set of compactly supported orthonormal wavelets that are practical for discrete wavelet analysis. The first nine wavelets in the set are shown in Figure 1.

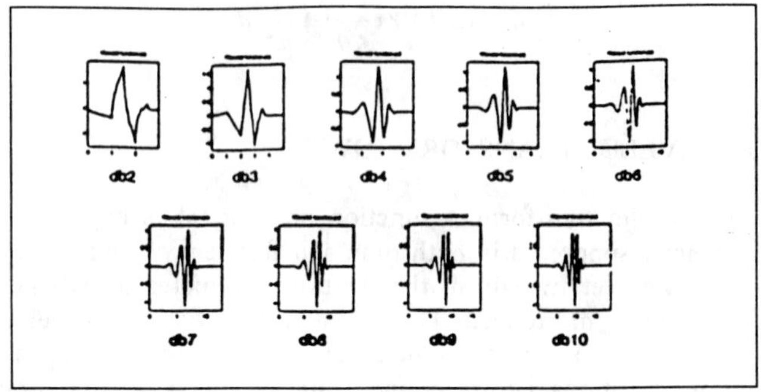

Figure 1 Family of Daubechies wavelets(db2 - db10)

HEALTH MONITORING

Developing health monitoring systems for large structures is of prime concern in the structural community. Several methods were proposed for detecting signals which indicate the development of fatigue damage under different loading. The main objective is to develop an on-line service to monitor the damage rate of a structure to assure its safety during severe environmental loading. The wavelet analysis were utilized to detect fatigue signals which are modeled as impulses superimposed to the input data (Masuda et al., 1993). It was shown that the moments when these impulses were applied can be clearly observed by spikes in the wavelet representations of the response signal.

In the present study, the approach given in (Masuda et al., 1993) is adopted to investigate the damage detection of a building structure subjected to an earthquake-like excitation. The building is modeled as a linear single-degree-of-freedom system. The seismic excitation is modeled as a filtered white noise. The fatigue damage is modeled as an impulse in the input data. The Daubechies wavelets are used to detect the damage signal from the simulated response data. Considering that both input and output signals are usually noisy due to various reasons, detection of the fatigue signal under noisy conditions is specially addressed.

STRUCTURAL MODEL

A simple mass-spring-dashpot building model, as shown in Figure 2, is considered in the present study. As a preliminary phase of the study, only linear system is considered.

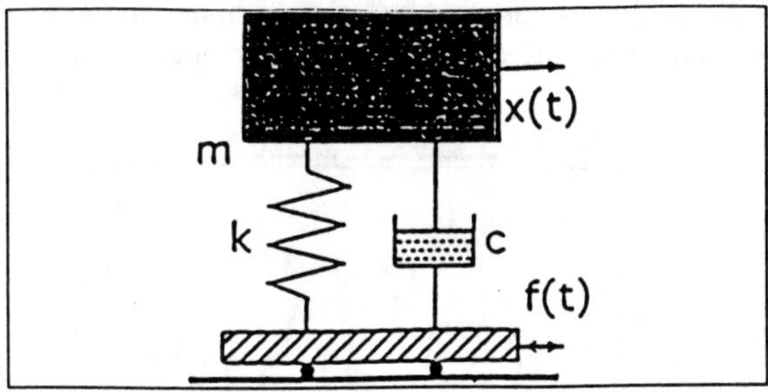

Figure 2 A simple single-degree-of-freedom structural model

The governing differential equation of motion of the system is given by:

$$m\frac{d^2}{dt^2}x(t)+c\frac{d}{dt}x(t)+kx(t)=f(t)=y(t)+\sum_{i=1}^{n}S\delta(t-\tau_i) \qquad (11)$$

where m, c, and k are mass, damping coefficient, spring constant of the system, respectively. The excitation, f(t), is assumed to be a sum of the seismic excitation, y(t), and a sequence of impulses occurring at random times. The former is modeled as a filtered white noise process as defined by:

$$m_f\frac{d^2}{dt^2}y(t)+c_f\frac{d}{dt}y(t)+k_f y(t)=\zeta(t) \qquad (12)$$

in which m_f, c_f, and k_f are the filter parameters and $\zeta(t)$ is a stationary white noise with zero mean. The random impulses model the fatigue damages occurring at random times τ_i (i=1,2,...n) with a magnitude S. The signal-to-noise ratio S/N is taken to be the ratio of S to the root mean square of y(t). The natural frequency of the system is taken to be $\omega_o = (k/m)^{1/2} = \pi$ rad/sec and the damping ratio $\xi_o = c/2(mk)^{1/2} = 0.05$, while the natural frequency of the filter is $\alpha_f = \pi/4$ rad/sec and the damping ratio $\xi_f = 0.5$ (Masuda et al., 1995).

SIMULATION RESULTS

Two sets of data used in this study have sampling rates of 8 Hz and 100 Hz, respectively; the former was also employed by (Masuda et al., 1993). A Gaussian white noise was generated by *Matlab*™ which was fed into the filter and the output was added to a series of three pulses at the times: 20.00 sec, 43.00 sec, and 65.00sec. The S/N ratio was managed by a proper choice of S. Seven different ratios were used. They are: 0.10, 0.20, 0.50, 1.0, 3.0, 5.0, and 7.0. Figure 3 illustrates a sample input of a filtered white noise with the addition of three impulses of S/N=1.0. Figures 4 and 5 show the time history and its Fast Fourier Transform (FFT) of the simulated response signal of the system. Note that while the three impulses are clearly observed in the input signal data, it is impossible to visually detect the abnormality of the response data. The FFT result shows that natural frequency of the system is being excited at $\omega = \pi$ rd/sec. A numerical test was also done by using the filtered white noise without adding the impulses and the FFT plot showed no visual difference from that in Figure 5.

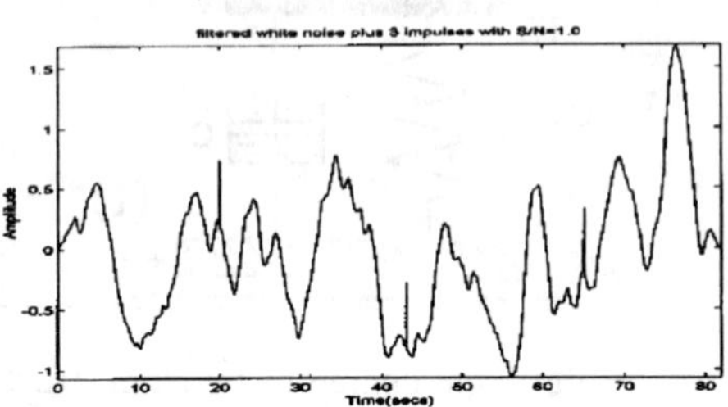

Figure 3 A filtered white noise signal with three impulses

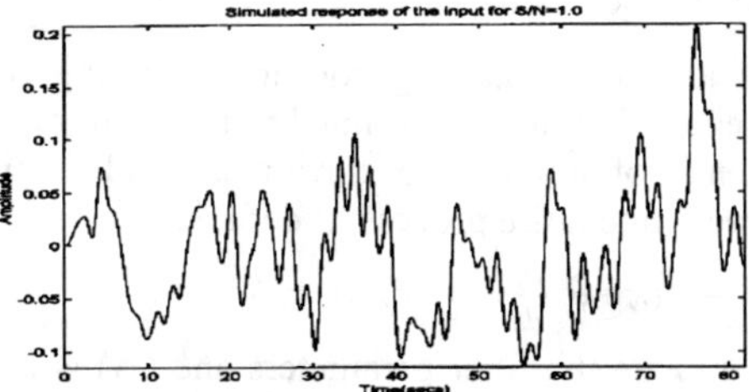

Figure 4 Simulated response history of the system

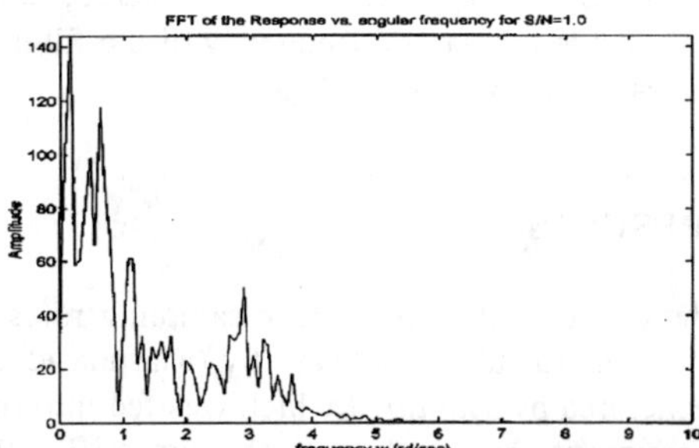

Figure 5 FFT of the simulated response

WAVELET ANALYSIS

The Wavelet Toolbox of Matlab was used to analyze the response signals with varying S/N ratios. The estimation of the occurrence time of impulses depends on both the S/N ratio and the wavelet used. The wavelet transform clearly picked up the exact time of impulse occurrence in the input, especially when wavelets with high order moments were used like db4 and higher as observed below in Figure 6. Here db4 stands for Daubechies wavelet number 4.

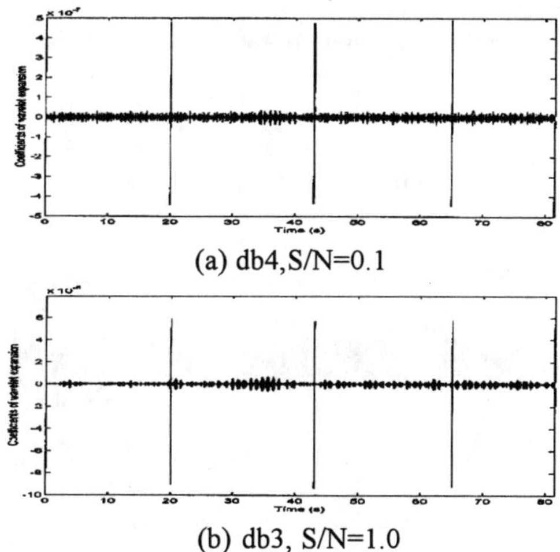

(a) db4,S/N=0.1

(b) db3, S/N=1.0

Figure 6 Wavelet analysis results

MEASUREMENT NOISE AND ROBUSTNESS OF RESULTS

Although the wavelet analysis was successful in detecting the occurrence times of the impulses in the input, as observed in the previous results, the amplitudes of coefficients of the wavelet expansion were small relative to the output response itself. Since most real-time sensors introduce noise in their measurement, the robustness of the wavelet transform results under noisy conditions of the output signal needs to be investigated. The root mean square of the output signal was calculated; a small white noise component was then added. Finally, wavelet analysis was performed for those contaminated data for three different values of S_o/N_o, which is now a measure of the standard deviation of the output signal to the added noise. The ratios used were: 10, 100, and 1000. The illustrated results are obtained for a sampling frequency of 100 Hz and are shown in Figure 7.

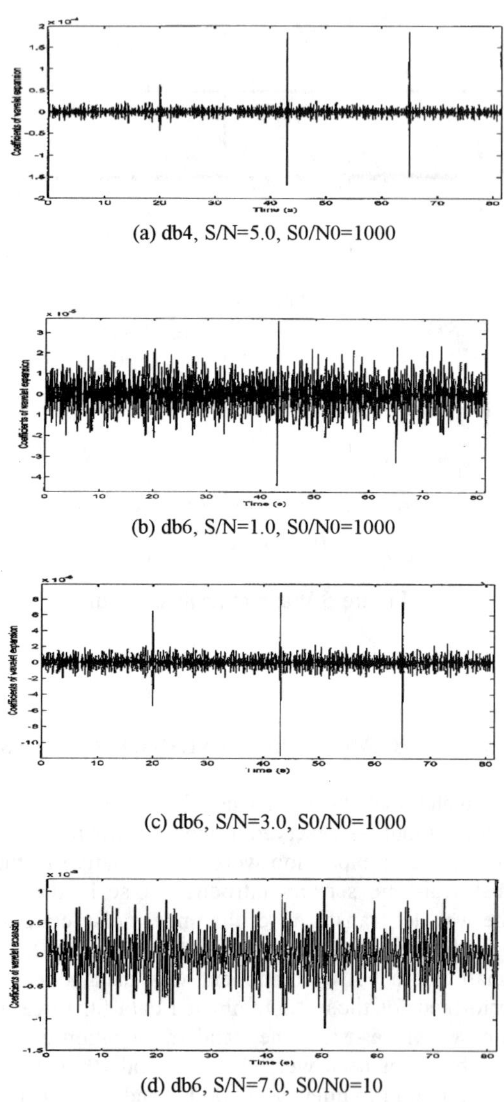

(a) db4, S/N=5.0, S0/N0=1000

(b) db6, S/N=1.0, S0/N0=1000

(c) db6, S/N=3.0, S0/N0=1000

(d) db6, S/N=7.0, S0/N0=10

Figure 7 Wavelet coefficients in the presence of output noise

DISCUSSION OF RESULTS

As observed in the previous wavelet analysis results, there are several factors that played a role in the successful detection of the impulses in the input force such as sampling rate, wavelet regularity, and signal-to-noise ratio. It is also noticed that by increasing the sampling rate from 8Hz to 100Hz, detection of impulses becomes easier. The reason behind it could be due to the fact that with higher sampling rate, the impulse will get sharper in the time domain and thus a more pronounced impulse response is expected. Also, the impulse signal is observed after being filtered by 2nd order system G(s), while the white noise is filtered first by the 2nd order filter F(s) and then by the system G(s) and then observed. So the power spectrum of the observed signal is equal to that of the system G, while the power spectrum of the observed noise is equal to that of GF. Since G and F are 2nd order low pass systems, GF has lower amplitude than G and the rate $|G|/|GF|$ increases in the higher frequency. This means that the higher the sampling rate is, the more the signal will overcome the noise in the frequency band $[f_N/2, f_N]$, where f_N stands for the Nyquist frequency. However, there is a drawback to using higher sampling rates. The magnitude of wavelet coefficients $\alpha_{j,k}$ decrease 2 to 3 orders of magnitude when increasing the sampling rate from 8 to 100Hz due to the low pass characteristics of the system itself that has been explained above.

The effect of the wavelet vanishing moment on the detection of impulsive singularities is explained by (Ashino et al., 1995). The system under study is a single-degree-of-freedom system which is described by a second order differential equation. The goal is to detect the singularities in the response which are due to impulses in the input force f(t). The required regularity of the wavelet to be used to detect the impulses in the input force is assumed to be D^0 class for f(t) because wavelet transform is performed directly on f(t). On the other hand, for the case of x(t), they are assumed to be D^2 class, because the output signal is obtained by integrating twice the f(t), then the wavelets used in the wavelet transform of x(t) have to travel the system inversely by getting differentiated twice.

For the Daubechies wavelets, the regularity of the wavelet is given by $D^{\lambda(N)}$ where $\lambda(N)= 0.3485N$ and N is the Daubechies wavelet number. Therefore, for our case N has to be 5.739 which is rounded up to 6. In case a fourth order system is concerned, the regularity needed would be 4 and this makes N=11.48 (Ashino et al., 1995). Of course, this condition is conservative in the sense that even if $N<6$ for the 2nd order system, there is still a good chance of detecting the impulse signal if the S/N ratio is large or if the sampling rate is high enough as seen earlier. It was observed that one may still detect impulses when Daubechies wavelet number was 4 and higher for S/N ratios as small as 0.1. For S/N ratios above 0.5, the impulses were detected by using db3.

ROBUSTNESS OF RESULTS

Deterioration in the detection of the signal was well observed when a certain level of noise was added to the output. The reason is that the coefficients of wavelet expansion $\alpha_{j,k}$ were relatively small when compared with the initial response x(t). This is especially true when high sampling frequency is used. Therefore, the addition of white noise would completely engulf the impulse signature even for a S_o/N_o ratio of 100. Several S_o/N_o were tested and it was found that detection is possible for ratios which are 100 or higher for a sampling rate of 100Hz. If the impulse amplitude is relatively large, better detection is expected. It was noticed that detection was not possible for S/N ratios less than 1.0 even for S_o/N_o as high as 1000. The results may be improved by using lower sampling frequencies. This is due to the fact that the coefficients of wavelet expansion are larger for the lower sampling rate by at least 2 orders of magnitude. This means that better robustness is attained at low sampling frequency.

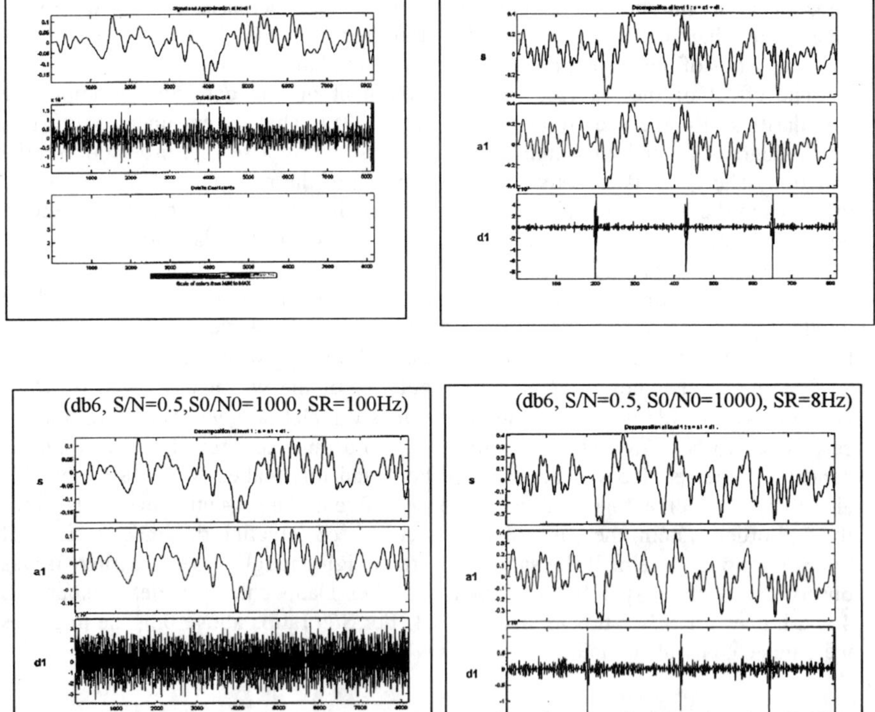

Figure 8 Effect of Sampling rate on detection

CONCLUSION

In this work, a health monitoring system was designed and implemented by using wavelet analysis. Detection of fatigue signals was seen to be affected by several factors such as sampling rate, signal-to-noise ratio, and the vanishing moments of the wavelets used. Robustness of wavelet analysis to external noise was investigated and was found out that it greatly deteriorated the ability to detect the impulsive signals even for very low noise levels. There is a tradeoff between sampling frequency and detection in the presence or absence of an output noise. Better detection is possible at a high sampling frequency if output noise level is small. However, if noise level is not negligible, a low sampling frequency is suggested.

REFERENCES

1. Ashino, R., Masuda, A., Nakaoka, A., Sone, A., & Yamamoto, S., 1995, "Regularity Between Orthonormal wavelet transform of Output Signal and Input signal of Transmitting System", *ICES*.
2. Bentley, P.M., and McDonnell, J.T.E., 1994, "Wavelet transforms: an introduction", *Electronics & Communication Engineering Journal*, pp. 175-186.
3. Inoue, H., Kishimoto, K., & Shibuya, T.,1995, "Experimental Wavelet Analysis of Flexural waves in Beams", *Experimental Mechanics*, pp 212-217.
4. Masuda, A., Nakaoka, A., Sone, A., & Yamamoto, S., 1993, " Health monitoring system of Building by using wavelet analysis", *IIWCEE*.
5. Masuda, A., Nakaoka, A., Sone, A., & Yamamoto, S., 1995, " Health monitoring system of structures based on orthonormal wavelet Transform", *Seismic Engineering, Transactions of ASME*, Vol 312, pp.161-167.
6. Misiti, M., Misiti, Y., Oppenheim, G., & Poggi, J.M., 1996, *Wavelet Toolbox*, The Math Works, Inc., Natick, MA.

KEYNOTE SPEAKERS

J. Achenbach
T. Vandiver

Techniques and Instrumentation for Structural Diagnostics

J. D. ACHENBACH, B. MORAN and A. ZULFIQAR

ABSTRACT

This paper discusses principal aspects of condition monitoring for structural reliability. For specificity the paper focuses on the development of cracks. In the first part techniques and instrumentation for flaw detection are discussed. The topics include quantitative non-destructive evaluation, measurement models, laser based ultrasonics, neural networks and integrated microsensors. The second part of the paper discusses the use of condition monitoring in the assessment of residual structural reliability. The topics in this part include probabilistic fatigue methods and fatigue reliability. The beneficial effect of inspections with a prescribed probability of detection on the probability of failure is demonstrated for the example of cyclic loading of a component containing an edge crack.

INTRODUCTION

This country's civilian and military infrastructure and its transportation equipment are becoming old. At this point it is expected that many of the systems that were built twenty, thirty or more years ago will in the near future, not be replaced by next generation structures and equipment. Consequently these aging systems will have to be kept in service well into the next century. This is quite feasible, provided that adequate measures are taken to prevent poor performance, inadequate safety and increasingly expensive maintenance. The need for extended life expectancy does, however, raise a substantial technical challenge.

J. D. Achenbach, B. Moran and A. Zulfiqar, Northwestern University, Center for Quality Engineering and Failure Prevention, 2137 Sheridan Road, Evanston, IL 60208

To respond to this challenge, defects must be detected before failures occur. Flaws and defects are introduced into materials during processing and they develop in structural systems during service. The presence of defects can severely compromise the structural integrity and effective performance of systems. As aging systems have to last longer, it will become more and more important to monitor the structural integrity. As time goes on, there will be deterioration of materials, metals as well as composite materials, and there will be fatigue and corrosion. The degree to which this deterioration can be allowed has to be determined by appropriate studies of material behavior and failure analysis, and by the judicious use of nondestructive evaluation. A good program of life extension of aging structures can obviously save any organization millions of dollars. It will lead to the prevention of undesirable failure and it will be very important for in-service performance of structures.

The recommended approach may be labeled as condition monitoring for structural reliability. The techniques for condition monitoring belong in the general area of quantitative non-destructive evaluation. The associated reliability analysis uses established techniques of failure prediction and reliability assessment. For specificity this paper focuses on the development of cracks. We briefly consider both relevant aspects of quantitative non-destructive evaluation (QNDE), particularly ultrasonics, and fracture mechanics.

In QNDE, a combined theoretical/experimental approach is recommended. In the theoretical work quantitative models, referred to as measurement models, should be developed for the measurement processes of the ultrasonic inspection techniques. These models can be used in the design of experiments and in the interpretation of the experimental data. Immersion transducers, contact transducers and laser-based ultrasonics may be used for the experimental work. The observed damage should be related to strength considerations by the use of fracture mechanics.

Fracture mechanics is the fundamental technology underlying failure analysis and life prediction based on the presence of cracks. Given the fracture toughness of the material and the load sustained by the component, fracture mechanics enables an engineer to calculate the critical size of a crack at a given location. A component is judged to be safe if the crack is smaller than a critical size and is not expected to grow to a critical size prior to the next inspection. Cracks which are judged acceptable may still grow, but at a rate which in principle is predictable. This leads to the "damage tolerant" philosophy: a component containing a macroscopic crack or flaw is acceptable if it can be shown that at the predicted stress levels, the flaw has almost zero probability of growing to critical size prior to the next inspection.

This systematic approach to quantitative non-destructive evaluation for a metallic structural component containing a crack is illustrated in Figure 1. As can be seen from the figure, the underlying concept in developing accept/reject criteria for a component is based on detecting and characterizing a defect and evaluating it in

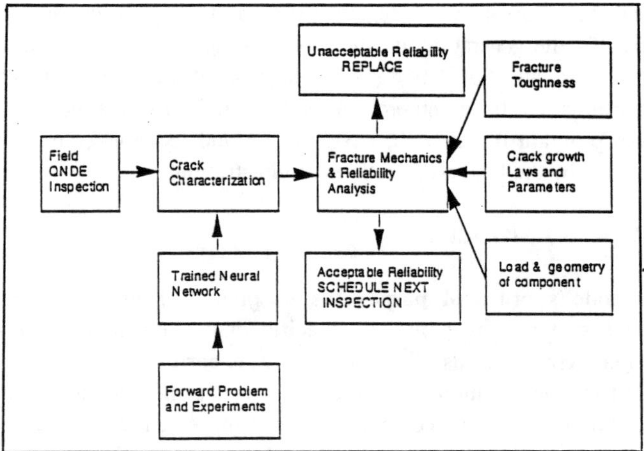

Figure 1. General procedure for QNDE and reliability assessment

terms of fracture mechanics and a fatigue crack growth law. The aim is to determine whether a crack in a structure is sufficiently small so that failure can be precluded with a high degree of certainty.

A program in the general area of condition-based monitoring for Structural Reliability should have the following components:

1. *QNDE:* Sensors and Techniques (smallest detectable flaw, POD), Measurement Models, Imaging Methods, Neural Networks, Flaw Characterization.
2. *Structural Reliability*: Critical Failure Mechanisms, Nucleation of Flaws (to determine need for first inspection), Flaw Growth Rates (for frequency of inspection), Reliability Analysis.

The application of QNDE techniques can be either on a fixed inspection interval basis, or it can be as a continuous process, where the latter is generally referred to as condition monitoring. The monitoring approach is generally achieved with permanently installed sensors, which sound an alarm when as undesirable defect situation is reached. Structures that can be monitored this way are considered to belong to the general category of smart structures.

Occasionally the world is reminded that aging structures may fail catastrophically if appropriate measures are not taken. For example, the potential perils of flying aging aircraft jumped into the public's consciousness in April 1988 when an Aloha Airlines 737 lost the upper front part of its fuselage, twenty thousand feet above the Pacific A cabin attendant, who was standing in the aisle, lost her life, but the pilot managed to land the seriously damaged airplane. This event was traumatic, although the OEM's and the airlines had known for some years that many aircraft were exceeding their original design lives, and a number of actions had been taken with regard to aircraft inspection and maintenance procedures. To the general public the advanced age of many in-service airplanes came as a surprise, and April 1988 marks a significant increase of the public interest in the airworthiness of the commercial aircraft fleet. The Aloha incident gave rise to an extensive program of research, development and technology transfer for aging aircraft, funded by the Federal Aviation Administration.

MEASUREMENT MODELS

A model's principal purpose is to predict, from first principles, the measurement system's response to specific anomalies in a given material or structure, (e.g., cracks, voids, distributed damage, corrosion, deviations in material properties from specifications and others). Thus, a measurement model includes the configuration of probe and component being inspected, as well as a description of the generation, propagation and reception of the interrogating energy.

The availability of a measurement model has many benefits. Numerical results based on a reliable model are very helpful in the design and optimization of efficient testing configurations. A good model is also indispensable in the interpretation of experimental data and the recognition of characteristic signal features. The relative ease of parametrical studies based on a measurement model facilitates an assessment of the probability of detection of anomalies. A measurement model is a virtual requirement for the development of an inverse technique based on quantitative data. Last, but not least, a measurement whose accuracy has been tested by comparison with experimental data provides a practical way of generating a training set for a neural network or a knowledge base for an expert system. Examples of measurement models for ultrasonic techniques have been discussed by Achenbach (1992).

LASER-BASED ULTRASONICS

Laser-based ultrasonics (LBU), involves the generation of ultrasound by laser illumination and the detection of ultrasonic signals by laser interferometric techniques. LBU has many advantages for applications to quantitative nondestructive evaluation including non-contact generation and detection, remote placement of equipment using fiber-optics, easy scanning, absolute displacement calibration, both broad band and narrow band signal generation, wide frequency band measurement, and applicability to curved surfaces.

Laser generation of ultrasound and the detection of the ultrasonic waves using laser interferometry are areas of active research. In earlier papers, the present author and co-workers have discussed an LBU system which employs a diffraction grating for illumination of a line-array to generate narrow-band surface waves and Lamb waves, see Huang, Krishnaswamy and Achenbach (1992), and a fiberized heterodyne dual-probe laser interferometer to measure the ultrasonic signals, Huang and Achenbach (1991). In this paper, we draw attention to the use of optical fibers to remotely generate and detect ultrasound with the acoustic energy focused into a selected narrow frequency band. The generation system uses a binary diffraction grating to separate a signal laser beam into 10 equal but spatially separated laser beams which are focused into 10 individual fibers. Results have been obtained over a range of frequencies from 2.7 to 7.7 MHz.

A recent paper by Fomitchov et al. (1997) reports progress towards the development of a robust low cost fiberized Sagnac laser interferometer suitable for field applications. It presents a low noise system using a low cost, long coherence HeNe laser that has better intensity noise characteristics than typically used laser diodes. A scheme for elimination of parasitic interference utilizing a frequency shifting technique has been developed. The primary advantage of the Sagnac interferometer is that it is exactly path matched and as such requires no heterodyning or static path compensation for sensor stabilization. The Sagnac interferometer is suitable for the measurement of ultrasonic surface waves arising from laser- or PZT-generated sources or from acoustic emissions. The general configuration is displayed in Figure 2.

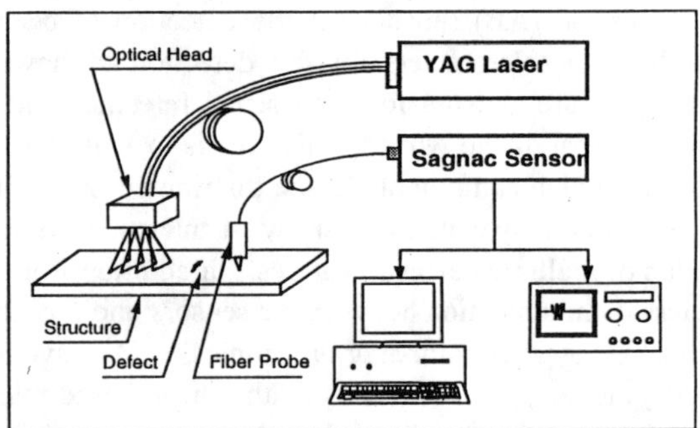

Figure 2. General configuration of LBU system

NEURAL NETWORKS

Whatever sensor, data-collecting and data-processing methods are used, it is to be expected that the volume of the data will be large and that it will be difficult to discriminate between ultrasonic signals generated by flaws and harmless noise. As a consequence a human operator will not be effective in extracting relevant information for decision making purposes from an overwhelming flow of data. Neural networks can play a major role in the recognition of sets of data that are related to damage or failure phenomena in the component or structure that is being monitored.

Neural networks must, however, be applied in an intelligent manner A brute force application of a neural network will require a very large set of training data (usually not available) and a large computer, and then might still not give very satisfactory results.

Specifically a neural network with an analog output has been worked out to estimate the crack-depth from ultrasonic signals backscattered from a surface-breaking crack in an aluminum plate. The network has only one response unit and this unit directly reports the crack depth as an analog value from the measured signals. A completely synthetic data set, spot-checked by comparison with experimental results, has been utilized for the training of the network. The synthetic data set was obtained by solving the boundary integral equations that formulate the problem by the boundary element method. A Gaussian modulated sinusoid was utilized as an incident signal. Results have been presented by Takadoya, et al., (1996).

INTEGRATED MICROSENSORS

An interesting new development in the sensing area brings together solid-state microsensors and advanced signal processing methods. The microsensors include acoustic emission (AE) sensors for the detection of crack nucleation and generation, microbeam accelerometers for the detection of unwanted mechanical vibrations and temperature sensors for both signal reference and identification of excessive heating. A goal of the work in this area is to realize inexpensive silicon chips, or "coupons", which can potentially be distributed over critical regions of a structure. Such a coupon may contain a variety of microsensors for detections and cross-configuration of multiple defect signatures. In addition it may be possible to implement wireless communication between the sensors and a central location. By also integrating analog signal conditioning electronics, A/D converters, digital signal processing circuits, memory, and telemetry with silicon based microsensors, smart microsystems can be formed. The silicon-based concept as envisioned by Polla and Francis (1996) is shown in Figure 3. The dimensions of the coupons are 1 x 1 cm.

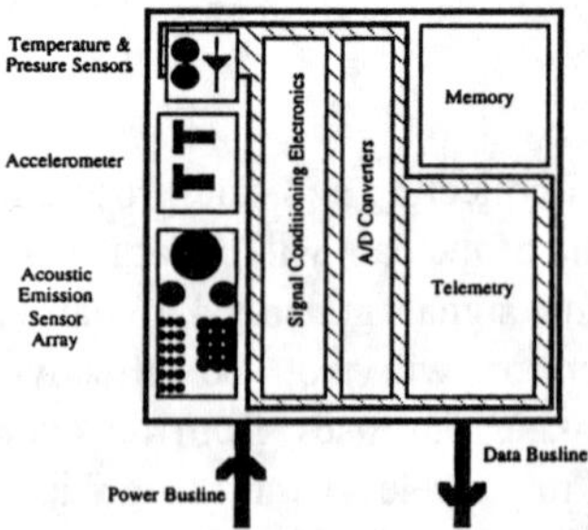

Figure 3. Integrated diagnostic coupon containing microsensors, signal conditioning electronics, and telemetry circuitry. From Polla and Francis (1996).

PROBABILISTIC FATIGUE METHODS

Probabilistic fatigue methods are often applied when critical structural components are subjected to non-destructive evaluation (NDE) techniques, so that cracked components can be identified and repaired or replaced. These inspections can significantly reduce the probability of fatigue failure of structures. Quantified measures of reliability (provided by probabilistic methods) allow maximization of inspection benefits through optimization of the inspection schedule. They also allow comparison of effectiveness for various inspection methods.

A risk analysis methodology for the assessment of structural integrity of aircraft structures has been outlined by Berens *et. al.* (1991). This methodology, which is based on the direct integration of the probability of failure, works well when the number of random variables is relatively small and a single parameter characterization of crack size is adequate. However, when it is desirable to characterize crack growth in detail, other modeling techniques such as Monte Carlo simulations (MCS) and the first order reliability method (FORM) are needed for calculating probabilities of failure (see Harkness *et. al.* (1992) and references therein). An alternative approach is given in Fleming *et. al.* (1995), where the first order reliability method is augmented to account for the effects of the inspections so that the crack size distribution need only be characterized at an initial state.

In the following section, the fatigue reliability problem is outlined. The role of NDE inspections, which enters through the probability of detection (POD) curve, is illustrated. A numerical example is given to explore the influence of POD on fatigue reliability.

FATIGUE RELIABILITY

Fatigue reliability analysis requires a mathematical model derived from the principles of mechanics and experimental data that relates various input random parameters for a specific performance criterion of interest. An example is the Paris relation (Paris and Erdogan, 1963):

$$\frac{da}{dN} = D(\Delta K)^m \tag{1}$$

where N is the number of cycles, da/dN is the rate of crack growth, D and m are material parameters, and ΔK is the amplitude of the stress intensity factor. Consider a simple failure criterion

$$a(N) > a_f \tag{2}$$

where a_f is a deterministic critical crack size. The fail-safe conditions are as follows:

$$\begin{array}{ll} G(\boldsymbol{r}) < 0 & \text{fail} \\ G(\boldsymbol{r}) = 0 & \text{limit state} \\ G(\boldsymbol{r}) > 0 & \text{safe} \end{array} \tag{3}$$

where

$$G(\boldsymbol{r}) = a_f - a(N) \ . \tag{4}$$

The limit case of $G(\boldsymbol{r}) = 0$ is often included in the failure domain. Now, the probability of failure is defined as

$$P_f = P[G(\boldsymbol{r}) \leq 0], \tag{5}$$

which is given by

$$P_f = \int_{\Omega_f^r} f_X(\boldsymbol{r}) d\boldsymbol{r}, \tag{6}$$

where Ω_f^r is the failure space ($G(\boldsymbol{r}) \leq 0$), and $f_R(\boldsymbol{r})$ is the joint probability density function for realizations in the space of random variables $\boldsymbol{r}$. For example, the initial crack size or the Paris relation parameters might be taken as random.

To incorporate in-service NDE inspections in a fatigue reliability analysis, the probability density function for realizations with undetected cracks after the first I inspections, is given by

$$f_U(\boldsymbol{r}) = f_R(\boldsymbol{r}) P_{nd}(\boldsymbol{r}, N), \tag{7}$$

where, $P_{nd}(\boldsymbol{r}, N)$, which is the probability that cracks associated with the realization $\boldsymbol{r}$ are not detected in all of the inspections prior to the current cycle, N, can be written as

$$P_{nd}(\boldsymbol{r}, N) = \prod_{i=1}^{I} \{1 - POD[a(\boldsymbol{r}, N_i)]\}, \tag{8}$$

where $POD[a(\boldsymbol{r},N_i)]$ is the probability of detection for the inspection method, and $a(\boldsymbol{r},N_i)$ is the crack length upon the ith inspection for the realization $\boldsymbol{r}$. The probability of failure is therefore given by

$$P_f(N) = \int_{\Omega_f^r} f_R(\boldsymbol{r}) P_{nd}(\boldsymbol{r}, N) d\boldsymbol{r}, \tag{9}$$

where Ω_f^r includes all $\boldsymbol{r}$ such that $G \leq 0$.

DIRECT INTEGRATION METHOD

Sobczyk and Spencer (1992) have given a detailed description of the representation of random variable fatigue models by Liouville equations. These equations govern the evolution of the probability density function and form the basis of the Direct Integration Method, which can be interpreted in terms of probability mass conservation which is depicted below. An efficient discretization of the random variable space and a selective integration scheme are incorporated in this method to make it computationally efficient (Moran *et. al.*, 1996). This method is also capable of incorporating NDE techniques through the use of probability of detection (POD) curves.

The concept of probability mass conservation is illustrated in Figure 4. Consider the area 'a' under the initial crack size distribution a_i (equivalent initial flaw size distribution). The crack size distribution evolves with time due to fatigue cycling. According to probability mass conservation the area *'a'* = area *'b'*. The area *'a'* can be mapped onto the final crack size a_f line. By mapping the whole distribution in this manner, the failure probability density function (p_f) as a function of fatigue cycles is obtained. The probability of failure (P_f) at time N is given by

$$P_f = \int_0^N p_f(N)dN. \tag{10}$$

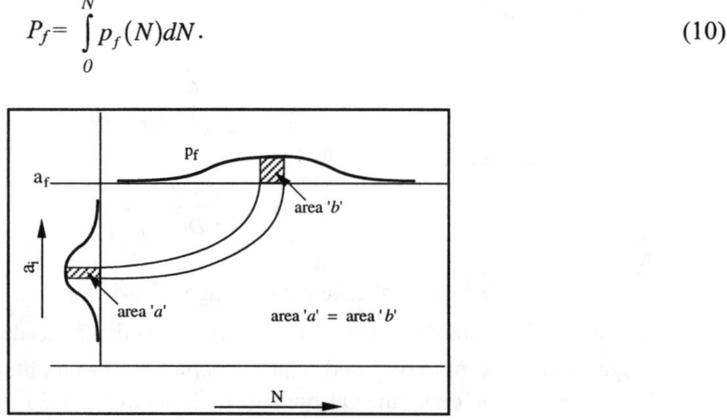

Figure 4: Mapping of crack size distribution on the final crack line

NDE inspections help to reduce the risk of failure. Let us assume that a component is inspected after N_1 cycles, and that all cracks which are detected are repaired. The effect of inspection is shown schematically in Figure 5. By inspecting at suitable intervals, the probability of failure can be kept below a specified value during the component service life. The effect of inspection on probability of failure will be illustrated in Figure 7.

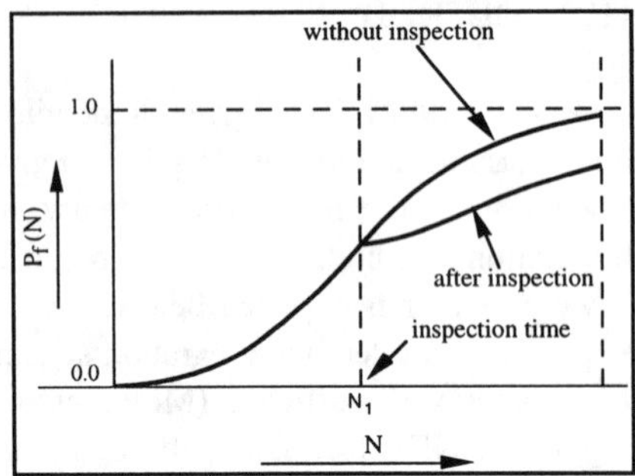

Figure 5: Change of probability of failure (P_f) due to inspection

NUMERICAL EXAMPLES

We consider the fatigue of an edge crack in a semi-infinite plate. It is assumed that the cracks propagate according to the Paris relation as shown in Eq. (1). The stress intensity factor for an edge crack of length a, is given by $\Delta K = 1.12s(pa)^{1/2}$ where s is the amplitude of the applied stress. Using this relation, Eq. (1) can be integrated to obtain the number of cycles for a crack of initial length a_i to grow to a crack of length a_f

$$N_f = \frac{a_f^n - a_i^n}{Dn(1.12\sigma\sqrt{\pi})^m} \qquad \text{for } m \neq 2 \tag{11}$$

where $n = (2-m)/2$

$$N_f = \frac{\ln(a_f / a_i)}{\pi D(1.12\sigma)^2} \qquad \text{for } m=2. \tag{12}$$

The material is taken to be ingot 304 stainless steel with fracture toughness K_{Ic} = 48MPa $\sqrt{\text{m}}$. The crack is cyclically loaded in tension-tension fatigue with an R -ratio of 0 and remote applied stress amplitude of s = 250 MPa. Motivated by the experimental findings of McGuire (1993)*, the initial crack size distribution is taken to be lognormal with mean 0.1 mm and standard deviation 0.033 mm. The quantity m-1 is also taken to be lognormal with mean 2.67 and standard deviation 0.75. The coefficient D and the exponent m are also found to be functionally related as log$D$=-1.50m -7.29. The units of D are (m/cycle)/(MPa$\sqrt{\text{m}}$)^m.

* The experimental data collected by McGuire (1993) are for fatigue crack growth from a hole in a tension-loaded bar. The statistical data obtained for this configuration are not directly transferrable to the edge crack configuration, but do provide a reasonable representation of fatigue crack growth in this material.

It is assumed that the probability of detecting an existing crack of length a upon inspection is (Palmberg, et al., 1987)

$$POD(a) = \frac{\alpha a^{\beta}}{1 + \alpha a^{\beta}}, \tag{13}$$

where the parameters a and b depend on the inspection technique. POD curves for different values of a and b are shown in Figure 6, with curve A representing a relatively poor inspection technique and curve C representing a relatively good one.

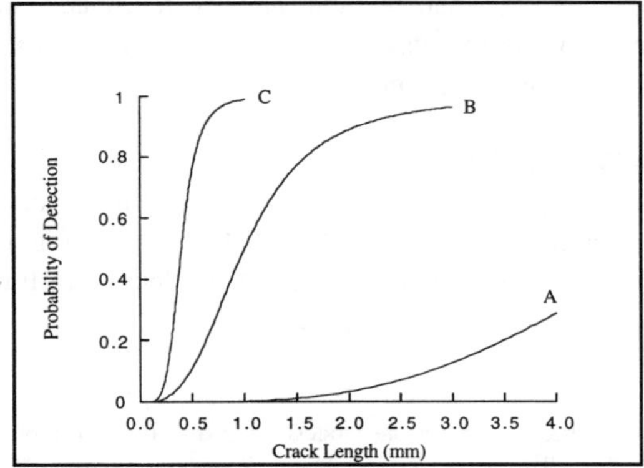

Figure 6: Probability of detection curves
(A: a =0.0032 mm^{-b}, b =3.5 ; B: a =1 mm^{-b}, b =3; C: a =100 mm^{-b}, b =5)

The probability of failure can be kept below a desired level by improving the probability of detection through better NDE inspection techniques. In Figure 7 the cumulative probability of failure ($P_f(N)$) is illustrated for the edge crack problem

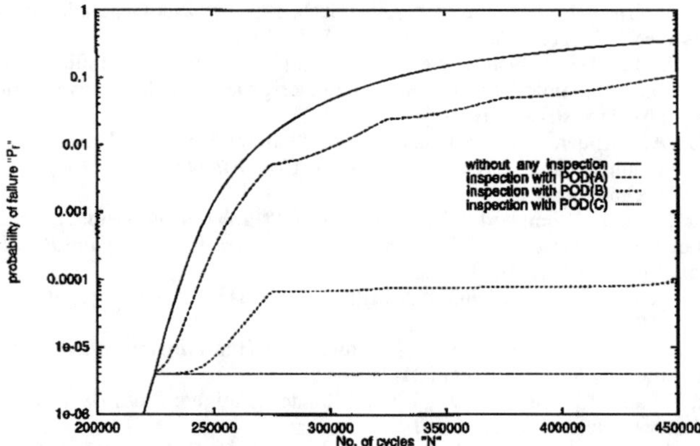

Figure 7: Effect of different POD on probability of failure (P_f)

and for evenly space inspections at $2.25\text{x}10^5$, $2.75\text{x}10^5$, $3.25\text{x}10^5$ and $3.75\text{x}10^5$ cycles. This figure shows results for different inspection techniques represented by POD curves A, B and C (see Figure 6). The computations were carried out using the Direct Integration Method. Clearly POD has a significant influence on the fatigue reliability with the curve C giving a dramatic reduction in probability of failures compared to curves A and B.

CONCLUDING COMMENT

The complementary roles of quantitative non-destructive evaluation and probabilistic reliability assessment methods have been discussed in this paper. As shown, the reduction of the probability of failure due to scheduled inspections or continuous monitoring can be determined analytically.

ACKNOWLEDGMENT

This paper was prepared in the course of research supported by the Office of Naval Research under the M-URI Integrated Diagnostics Program through the Multi University Center for Integrated Diagnostics, Georgia Institute of Technology, ONR Grant N0014-95-1-0539.

REFERENCES

Achenbach, J. D., 1992. "Measurement Models for Quantitative Ultrasonics." *J. Sound and Vibration, 159*: 385-401.

Berens, A. P., J. G. Burns, and J. L. Rudd, 1991. "Risk Analysis for Aging Aircraft Fleets." In *Structural Integrity of Aging Airplanes*. Eds. S. N. Atluri, S. G. Sampath, and P. Tong. Springer-Verlag, New York.

Fleming, M., H. Harkness, B. Moran, and T. B. Belytschko, 1995. "Fatigue Reliability Method with In-Service Inspections." Final report for U.S. DOT/FAA/CT-94/88.

Fomitchov, P., S. Krishnaswamy, and J. D. Achenbach, 1997. "Compact Phase Shifted Sagnac Interferometer for Ultrasound Detection." *Optics and Laser Technology*, in press.

Huang, J., and J. D. Achenbach, 1991. "Dual-Probe Laser Interferometer." *J. Acoust. Soc. Am*, 90(3): 1269-1274.

Huang, J., S. Krishnaswamy, and J. D. Achenbach, 1992. "Laser Generation of Narrow-Band Surface Waves." *J. Acoust. Soc. Am.* 92(5): 2527-2531.

McGuire, S. M., 1993. "Quantitative Measurement of Fatigue Crack Initiation and Propagation in 304 Stainless Steel as Related to Design and Nondestructive Evaluation." *Ph.D. Diss*., Northwestern University.

Moran, B., A. Zulfiqar, and N. Sukumar, 1996. "Failure Prediction Methodology using the Direct Integration Method." *Technical report for Center for Integrated Diagnostics*, Georgia Institute of Technology.

Palmberg, B., A. F. Blom, and S. Eggwertz, 1987. "Probabilistic Damage Tolerance Analysis of Aircraft Structures." In *Probabilistic Fracture Mechanics and Reliability*, Ed. J. W. Provan. Martinus Nijhoff, Boston.

Paris, P. C., and F. Erdogan, 1963. "A Critical Analysis of Crack Propagation Laws." *J. Bas Engng.*, 85, 528-534.

Polla, D. L., and L. F. Francis, 1996. "Ferroelectric Thin Films in MEMS Applications." *Materials Research Society Bulletin*, 21(7):59-65.

Sobczyk, K., and B. F. Spencer, Jr., 1992. "Random Fatigue: From Data to Theory." Academic Press, New York, pp. 167-177.

Takadoya, M., Y. Yabe, M. Kitahara, J. D. Achenbach, Q. C. Quo, and M. L. Peterson, 1996."An Artificial Intelligence Technique to Characterize Surface-Breaking Cracks." *Rev. of Progress in QNDE*, Ed. by D. O. Thompson and D. E. Chimenti, Plenum Press, New York, pp. 771-778.

Health Monitoring of U.S. Army Missile Systems

T. L. VANDIVER

ABSTRACT

Health monitoring of U.S. Army Missile systems is of paramount importance in this time of Army downsizing and limited new missile systems coming on line. The emphasis has been shifted from new missile systems to maintaining our existing ones through new means of assessing damage and extending the shelf-life of missile components. Missiles, ground support equipment, auxiliary hardware, tanks, and military vehicles are all areas of concern in the area of health monitoring and shelf-life extension. The same technology used for health monitoring could also be beneficial in the areas of design optimization, testing, and instrumentation of missile system components.

INTRODUCTION

Health monitoring for damage detection, shelf-life extension, design optimization, testing, and instrumentation for missile systems is of major concern to the U.S. Army Missile Command. Due to the current world events, downsizing, and the end of the Cold War, emphasis has been placed on maintenance of existing missile systems rather than development of new ones. One of the most widely used techniques to monitor the structural health of rocket motorcases and propellant depend on either taking the missile out of stock and conducting static firings or actually dissecting the motorcases, nozzles, and propellant and conducting structural integrity tests. The tests include dog bone tension test specimens and specially configured nozzle test specimens. This approach for determining the structural health and shelf-life of missile systems is very time consuming, expensive, and represents only a small sample of the inventory. Ideally, non-obtrusive techniques could be developed that can determine damage and access the structural health of a missile system components. Of special interest in missiles would be a system that could determine if propellant has separated, or sagged from the bondline, or decomposed.

U.S. Army Missile Command, AMSMI-RD-ST-CM, Redstone Arsenal, AL 35898-5247

SENSOR TYPES

Fiber optics, piezoelectrics, interferometrics, ultrasonics, acoustics, thermography, and micro electromechanical systems (MEMS) offer a host of possibilities in U.S. Army missile systems to assess damage and monitor the health of the component structures. Currently, there is a great deal of research and development in the areas of embedding optical sensors in filament wound composite rocket motorcases. The fiber optical sensors could be used for cure monitoring, component testing, damage assessment, and shelf-life assessment. Piezoelectrics utilize, both active and passive sensors, and have a great deal of potential in the structural health monitoring of missile structures to include rocket motorcases, launchers, and ground support equipment. Interferometric techniques require contact and non-contact techniques, and are somewhat limited to laboratory usage and surface strain measurements. Ultrasonics is used extensively as a non-destructive evaluation technique that proves useful in damage assessment and could prove useful in health monitoring of missile structures. Acoustics are being used to assess the health of certain structures in the stockpile of missile systems. Thermography has advanced considerably in the past few years with the advent of advanced thermal imaging cameras and computer imaging. MEMS systems offer promise to monitor the health life of missile system components. The most desirable techniques are those that are the least intrusive or that require the least amount of excitation of the missile structure. The installation of health monitoring devices create special challenges. The ingress and egress of fiber optic embedded sensing devices, plus the embedded sensor, can pose difficulties on composite rocket motorcases and launchers that may have case walls as thin as 0.050 inch (Foedinger, Sirkis, and Vandiver, 1997). Externally bonded sensors for retrofit can be effective, but special attention must be placed on the adhesives and mounting techniques. Internal sensors have the greatest potential for health monitoring, but also offer the greatest manufacturing, data link, and instrumentation challenges.

FIBER OPTIC SENSORS

Technology Developments Associates (TDA) in a PHASE I SBIR entitled "Embedded Fiber Optic Sensors for Filament Wound Composite Structures," (Foedinger, 1996) investigated techniques for embedding fiber optic sensors in filament wound composite structures. The effort investigated the effects of different optical fiber types, coating methods, and sensor locations on the structural integrity of missile structures. TDA, Pressure Technology, and the University of Maryland are currently in a Phase II SBIR effort utilizing Bragg Gratings and other optical sensors to further the research of fiber optics in missile structures (Foedinger and Sirkis, 1997). The fiber optic sensors are being considered for monitoring the cure cycle of composite rocket motorcases, dynamic testing of motorcases, and structural health monitoring. The cure monitoring could result in optimized cure cycles that may result in energy savings and a more efficient part with minimized residual stresses. Testing a static motor instrumented with fiber optic sensors to determine the strain fields along the motorcase and in the nozzle area could optimize a motorcase design. Production Products Manufacturing and Sales Company, and Blue Road Research, Inc. are currently performing embedded fiber optic sensors efforts that can monitor liquid molded composite armor structures for ballistic damage and structural health throughout the components service life (Corona and Udd, 1996). Astro Technology recently demonstrated the strain measurement of a solid propellant rocket motor static firing using fiber optic sensor technology (Brower, 1996). Production Products Manufacturing and Sales, Inc. used a Bragg grating sensor system based on a scanning Fabry-Perot etalon to determine tensile and compressive loads in carbon fiber composites wrapped on concrete cylinders which were tested to failure. The research investigated whether the composite overwrap could significantly enhance the compressive strength and earthquake resistance of concrete piers, columns, and other structural members. The strains were determined by embedded fiber grating arrays in both the axial and circumferential directions of the structure under compressive loading conditions. The embedded gratings survived compressive strains in excess of 3.8 percent strain, (Corona, Kersey, and Slattery, 1996). Fiber optics technology has great potential to assess damage and predict service life of Army missile structures.

PIEZOELECTRIC SENSORS

Research at Stanford University is being conducted by both active and passive sensing diagnosis using surface mounted and embedded piezoceramics to detect impact damage of composite structures. Experiments were conducted with both composite plates and cylinders. The technique was demonstrated by applying impact damage to the composite structures, confirming the impact event by the passive sensing diagnosis process, and generating the appropriate diagnostic signals with the active sensing diagnostic from built-in actuators to identify damage size resulting from the impact. Lamb waves with narrow-banded frequencies were selected as diagnostic signals for the impact damage detection. Test results show that impact damage in composites could be detected by selecting appropriate Lamb waves generated from built in piezoceramics (Chang, F., 1996). It is suggested that the sensor-only passive sensing diagnostics be used to estimate location of damage and the sensor/actuator diagnostics be used to obtain unlimited information for diagnosis of the damage.

INTERFEROMETRY

Interferometric techniques such as electronic interferometry, laser speckle interferometry, acoustical interferometry, and holographic interferometry have promise; but many of the techniques are still at the laboratory stage and are currently not ready for field use. Residual stress measurement and strain measurement are being made on missile structures by the University of Tuskegee and the University of Florida using electronic interferometry. The structure must be excited via a loading condition. Interferometric techniques are somewhat limited since they measure surface conditions which limit the information received from thick wall structures. Also, the inherent rough surfaces and circular geometry of missile motorcases and launchers offer further challenges due to a limited region of observation. These methods are currently in the laboratory stage with aspirations of building portable units for field use.

ULTRASONICS

Ultrasonic testing is used extensively as a non-destructive evaluation tool for missile structures. Ultrasonics can be used for characterizing voids, cracks, inclusions, and delamination in metals, ceramics, plastics, and composite materials. However, ultrasonics is better suited for homogeneous type structures rather than composite structures which are non-homogenous in nature. Ultrasonic C-scanning can observe a structure layer by layer with the proper transducers, power, and couplants. Los Alamos National Laboratory (LANL) has successfully used a technique called Ultrasonic Computed Tomography; but the system has not progressed into a technique that can be used outside the laboratory. The areas to be inspected are typically small and require an array of transducers to be less time consuming. NASA Langley has developed a state-of-the-art system that utilizes a computer controlled phased array system consisting of 32 transducers.

ACOUSTICS

Acoustical testing techniques have been used to determine bond line structural health of the Kevlar attachment thread ring to the radome forebody of the PATRIOT missile. Acoustical techniques have been used to determine the structural integrity of composite pressure vessels during the proof test cycle to determine if the acoustical signatures could be analyzed to reduce the level of pressure used during the proof cycle. Acoustical holographic interferometry has been used by MICOM to determine internal displacements of flawed regions of a three dimensional body. LANL uses Acoustic Resonant Spectroscopy for chemical weapon munition identification and surveillance. The system can compare known acoustic signatures to measured signatures and determine their structural health by their natural modes of vibration of the munition.

THERMOGRAPHY

Thermographic techniques have evolved considerably in the past few years as an effective nondestructive testing technique. Advanced infrared thermal imaging cameras and data acquisition equipment have made thermography a viable source of damage assessment, and health monitoring. Thermal imaging is more effective with thin wall structures as opposed to thicker wall specimens. A joint effort with MICOM and Army Research Lab on damage assessment of composite missile structures involved impacting missile structures, conducting thermographic nondestructive evaluation, and developing finite element analysis (FEA) codes to correlate a certain amount of impact energy versus damage (Triplett, Patterson, and Zalameda, 1997). A very low level of heat was applied to impacted composite cylinders at varying energy levels, infrared thermographic images taken, and the composite cylinders hydroburst to determine residual strength. The FEA code predictions were in agreement with the residual strength predictions. If thermographic imaging techniques can be perfected for easy field use with the backup of a set of signature images, a structural damage and health monitoring system for missile structures is plausible.

MICRO ELECTRO-MECHANICAL SYSTEMS (MEMS)

MEMS sensor technology is being considered for applications to U.S. Army missile systems for damage detection, health monitoring, wing/fin deployment, and drive units for sensors and gimbals. In early 1996, Lockheed Martin and System Excelerator personnel participated in a MICOM health monitoring prototype development program called Missile Advanced Remote Monitoring System. A spent PATRIOT launcher was instrumented with three axis MEMS accelerometers, a humidity sensor, and an ambient temperature sensor connected to a commercial remote monitoring unit. The output data was transmitted over a cellular telephone to a remote location. Numerous shock, vibration, humidity, and temperature data were transmitted from Orlando, FL to MICOM offices in Huntsville, AL. With the miniaturization of electromechanical components and advancements in micropackaging, and instrumentation, MEMS will certainly play a major role in the future of health monitoring of missile structures.

SUMMARY

In summary, there are numerous techniques under development for determining damage level, monitoring, and evaluating structural health of missile structures. Many of the techniques are improvements and advancements of the conventional nondestructive evaluation methods. However, improvements in MEMS, miniaturization of components, sensor technology, instrumentation, and computerization are meeting the challenges of extending shelf-life of missile systems. The challenge of structural health monitoring of missile components from the factory, in storage, and to the soldier in the field can be met with a combination of theory, experimental data, and application by government, industry, and academia working together. The technological advances in damage assessment and health monitoring of missile systems is of paramount importance to the U.S. Army Missile Command and will certainly contribute to a strong national defense.

REFERENCES

Brower, D., 1996, "Embedded Optical Fiber Sensors in Solid Rocket Motorcases," MICOM SBIR Phase I Contract DAAH01-96-C-R071.

Chang, F., 1996, "Passive and active Sensing Diagnostics for Damage Detection in Composite Structures," Workshop on Fiber Optics for Missile Applications, Redstone Arsenal, AL.

Corona, K., Kersey, A., and Slattery, 1996, "Fiber Optics in Filament Wound Structures," Workshop on Fiber Optics for Missile Applications, Redstone Arsenal, AL.

Corona, K. and Udd, A., October 1996, "Liquid Molded Composite Armor Smart Structures Using Embedded Sensors," Army Research Laboratory Contract DAAH01-97-C-0034.

Foedinger, R., 1996, "Embedded Fiber Optic Sensors for Filament Wound Composite Structures," MICOM SBIR Phase I Contract DAAH01-96-C-R060.

Foedinger, R. March 1997, "Embedded Fiber Optic Sensors for Filament Wound Composite Structures," MICOM SBIR Phase II Contract DAAH01-97-C-R165.

Foedinger, R, Sirkis, J. and Vandiver, T., 1997, "Structural Integrity of Filament Wound Composite Structures with Embedded Fiber Optic Sensors," proceedings of 42nd International SAMPE Symposium.

Triplett, M., Patterson, J., and Zalameda, April 1997, "Impact Damage Evaluation of Graphite Epoxy Cylinders," 38th Annual AIAA Structures, Structural Dynamics, and Materials Conference, Kissimmee, FL.

SESSION 4

MODELING AND DIAGNOSTIC METHODS II

Active Composite Materials and Damage Monitoring

P. BLANAS, M. P. WENGER, R. J. SHUFORD and
D. K. DAS-GUPTA

ABSTRACT

Health monitoring of composite structures is essential as these types of structures find wider applications. Piezoelectric composite materials can provide the means by which sensing and actuating devices become integral parts of the composite structural system. In this study the suitability of piezoelectric composite materials as in-situ multipurpose sensors for the nondestructive monitoring of fiber reinforced composites has been investigated. Results obtained from embedded composite bimorph sensors will be presented in this paper.

INTRODUCTION

Smart materials can be defined as those possessing the capability to respond in an advantageous manner to their environment. As such these materials must in general have the ability to sense a stimulus, compute a reaction and optimally respond by actuating the structural system. Smart composite structures could contain embedded sensors, such as piezoelectric elements or optical fibers, to determine in-process and in-service parameters and responses to loads, such as deflections and local strains. The incorporation of "active" actuator and sensor elements into an otherwise "passive" structure could allow for such capabilities as damage detection and vibration reduction or shape control. Such smart composite structures offer significant potential for extending the effective operations envelope and improving the performance of many Army systems. For example, improving the performance of a composite structure can be thought as the ability of such a system to sense and evaluate service or battle induced damage. Towards that goal, research efforts have been undertaken to develop polymer based active materials

Panagiotis Blanas and Richard J. Shuford, US Army Research Laboratory, Weapons and Materials Research Directorate, Materials Division, Aberdeen Proving Ground, MD, 21005-5069, USA
Matthew P. Wenger and Dilip K. Das-Gupta, University of Wales, Bangor, School of Electronic Engineering and Computer Systems, Bangor, Gwynedd, LL57 1UT, UK

that can provide sensing capabilities for integrated smart composite structural systems. As part of these efforts, ferroelectric ceramic/polymer composites have been developed and evaluated as embedded sensing devices in laminated composite structures (Blanas et al., 1995).

Diphasic piezoelectric composite materials consisting of ferroelectric ceramics within a polymer matrix can take advantage of the favorable properties of their constituents while reducing the detrimental ones. The high piezoelectric and electromechanical properties of the ceramic phase combine with the mechanical strength and low dielectric permittivity of the polymer matrix to enhance the overall performance of the composites (Chilton, 1991). Thus, the electroactive as well as the mechanical properties of the composite materials can be tailored to provide specific advantages, over single phase materials, for specific applications. A film made out of piezoelectric composites, referred to here as a monomorph, can be used as a multipurpose sensor. It can be incorporated into a surface mounted acoustic emission (AE) transducer (Dias and Das-Gupta, 1993), or be embedded into a fiber reinforced composite structure (Wenger et al., 1996a). Two or more monomorphs can be incorporated into one sensor to form a bimorph or multimorph respectively. The present work is concerned with the sensing properties of bimorphs constructed from ferroelectric ceramic/polymer composites embedded into laminated plate structures. Bimorph sensors differ from monomorphs in their ability to work in two modes of operation. A study on the ability of embedded monomorphs to detect simulated acoustic emission signals has also been conducted and reported in a previous paper (Wenger et al., 1996b). This paper will report the results obtained from a study on the ability of the embedded bimorph sensors to detect and distinguish simulated AE sources and a comparison will be made between the sensing properties of the monomorphs and the bimorphs.

PIEZOELECTRIC COMPOSITES

In this study mixed connectivity composites made of ceramic powder dispersed in a polymer matrix have been developed and utilized. Calcium modified lead titanate (PTCa), supplied by GEC-Marconi Materials Technology (UK), has been used as the filler ceramic with an average grain size of 10-30μm. Two composites have been developed using two different polymers as the matrix phase of the composites. These polymers are the polar copolymer of polyvinylidene with trifluoroethylene (P(VDF-TrFE)), 75/25 molar percentage, supplied by Solvay in fine powder form, and a non-polar epoxy resin, Epon 828 (Shell Resins), a thermoset.

The PTCa/P(VDF-TrFE) composites were fabricated using a solvent casting technique. Weighted amounts of the ceramic and polymer powders were dry mixed before a certain amount of methyl-ethyl ketone was added to the powder mix so as to fully dissolve the polymer. Stirring of the mixture with a mortar and pestle enabled the solvent to evaporate leaving behind the composite material. Thin films (~100μm) of the composite were then formed by pressing the material, at the softening temperature (~170°C) of the copolymer, in a mechanical press. The composites of PTCa/Epoxy were fabricated (Wenger et al., 1994) by gradually mixing the ceramic powder into the epoxy resin and hardener mix. Once the desired amount of ceramic had been mixed, a degassing procedure ensued to remove gases trapped in the epoxy phase during the mixing process. Afterwards, thin films (~100μm to 150μm) of the composite were cast by pressing the material in a temperature controlled mechanical press.

Aluminum electrodes lcm in diameter and ~1000Å thick were vacuum deposited onto both sides of the film. Poling of the films was accomplished by applying an electric field, up to a maximum of 25 MV/m, across the thickness of each sample in a silicone oil bath. Poling temperatures were 100°C, for the composites of PTCa/P(VDF-TrFE), and 80°C for the composites of PTCa/Epoxy. The samples were poled for approximately 30 minutes and were allowed to cool in the presence of the field. During poling samples were submersed in a silicone oil bath to sustain sample temperature and prevent electrical breakdowns.

The concept of connectivity in a composite describes the manner in which the phases of the composite are arranged and is designated by two numbers (Newnham et al., 1978). These numbers indicate the number of dimensions each phase in the composite is self-connected. In the present case, the first digit indicates the connectivity of the ceramic while the second that of the polymer. A 0-3 connectivity is expected for composites obtained from a ceramic powder dispersed into a polymer matrix. However, at high ceramic loadings or for thin films, where film thickness is comparable to the size of the ceramic grains, certain 1-3 connectivity is observed where the ceramic is connected to itself through the thickness. The composite films produced for this work show both 0-3 and 1-3 connectivity patterns and thus posses what has been termed a mixed connectivity pattern (Dias and Das-Gupta, 1996).

Bimorphs were obtained by adhering two of the composite monomorphs together. Care was taken so as to minimize the thickness of the adhesive layer. This was accomplished by coating one surface of one monomorph with the epoxy resin while the corresponding surface of the other monomorph was coated with the hardener. Excess resin and hardener were removed with a blade before the two monomorphs were placed together and the epoxy was allowed to cure under light pressure (~lPa). Depending on how the monomorphs were joined together, series or parallel connected bimorphs were obtained. Bimorphs were connected in series when the directions of polarization of the forming monomorphs were opposing one another as shown in figure 1. On the other hand, monomorph components of parallel connected bimorphs had their directions of polarization aligned. The bimorphs were constructed in such a way so that electrical contacts could be made to all three electrodes with the two center electrodes acting as one.

A bimorph transducer can be configured four different ways depending on monomorph polarization and electrode arrangements Shown in figure 1, are the two possible configurations for a series connected bimorph along with their equivalent capacitor circuits. Two more configurations can be obtained for bimorphs connected in parallel with similar electrode arrangements.

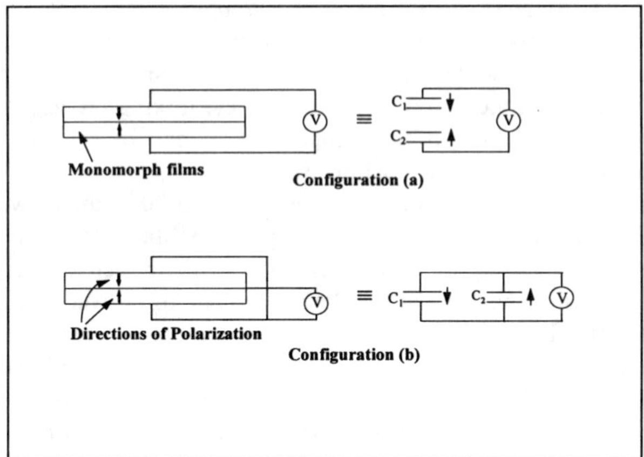

Figure 1. Series Connected bimorph in two configurations

When a bimorph is used as a sensor, there are essentially two modes in which it can operate. These modes will be called here the thickness and bending mode of operation respectively. In configuration (a) of figure 1 the application of a normal stress to the film surface will result in cancellation of signals arising from the two monomorphs. In this, the thickness mode of operation, the signals are due to the piezoelectric g_{33} coefficient and will be of opposite signs because of the series connected bimorph. The piezoelectric g_{ij} coefficients relate the electric field developed in the electroactive material to the applied stress, and the subscript indices i and j indicate the direction of the response and the applied stimulus respectively. When a bending moment is applied to the bimorph, the signals obtained from the two monomorphs are due to the piezoelectric g_{31} coefficient and compliment each other. In this, the bending mode of operation of the bimorph, the top monomorph will be under extension (or compression) while the monomorph making up the lower half of the bimorph will be under compression (or extension). Thus, under bending the signals produced by the two monomorpohs will be of the same polarity. In this configuration therefore, the bimorph will only be sensing in the bending mode as the thickness mode will produce a null response. In configuration (b), where the two outer electrodes of the bimorph are connected together and thus the signal is sensed between the inner electrode and the outer electrodes, the opposite will be true. With this configuration the bimorph will be working in the thickness mode as here the bending mode will produce the null result. Therefore, for a given bimorph connected either in series or in parallel, the sensing mode can be defined by the electrode arrangement thus providing the bimorph with a dual mode of operation. The preceding discussion has assumed that the properties of the individual monomorphs are the same and the thickness of the glue layer is negligible.

The embedded bimorph sensors have been characterized for their response to simulated acoustic emission sources. Simulated AE sources can be generated by pencil lead breaks on the surface of the test structure, Hsu-Nielsen method,

(ASTM, 1993). This method is often used to simulate AE events because of the transient nature of the lead breaking which is similar to the transient nature of real AE sources. The elastic wave produced by such a method can be efficiently reproduced by always ensuring the lead is broken at the same point on the plate and at the same angle. The output of the sensors were recorded directly on a digital storage oscilloscope for analysis.

ELASTIC WAVE DETECTION IN COMPOSITE PLATES

Accumulation of damage in a structure can be thought as the generation and growth of local instabilities. The rapid localized change of the elastic state at the vicinity of damage mechanisms within a body, causes the release of transient elastic waves which propagate throughout the body. Acoustic emission can be defined as the transient elastic waves generated by damage events. Such transient elastic waves or acoustic emissions can be generated within a composite structures by a number of damage mechanisms. These damage mechanisms are then referred to as acoustic emission sources and include matrix cracking, fiber brakes and delaminations. Organic composites, when deformed, produce extensive amounts of acoustic emission. These waveforms contain information about the damage source such as location, type and size. If these information can then be extracted by capturing and analyzing the waveforms, then they can be used to determine the extent of damage in a structure. Monitoring of acoustic emissions in composites can then be utilized as a passive, nondestructive and noninterfering method for damage detection and evaluation.

When dealing with thin plate structures, elastic waves generated by acoustic emission sources would propagate as Lamb or plate waves. (Lamb, 1917). The two plate modes of acoustic emission waves, observed in an AE signal, are the symmetrical or extensional and antisymmetric or flexural mode. Extensional plate waves contains higher frequency components while flexural waves are higher amplitude lower frequency waves. Furthermore, extensional waves are non-dispersive in contrast to the flexural waves which are dispersive in nature. It has been suggested that plate waves can be utilized to detect damage in composites due to their ability to travel long distances and because source orientation affects the relative amplitudes of the two plate wave modes. (Gorman and Prosser, 1990).

Acoustic emission wave detection results reported in this paper were obtained from a composite plate specimen with an embedded sensor array. The plate was fabricated from prepreg of S-Glass continuous reinforcing fibers in an Epoxy (NCT-301) matrix supplied by Newport Composites. The laminate consisted of 25 plies with a stacking sequence of $[0]_{25}$ and a nominal thickness of 0.3175 cm (0.125 in). The in plane dimensions of the unidirectional laminate were 56cm x 56cm (22in x 22in). Four sensors, two monomorphs and two bimorphs, were embedded in a 10cm square array at the center of the plate, away from the edges to avoid signal reflection problems, and two layers down from the surface. The bimorphs in the array were constructed from monomorphs of PTCa/P(VDF-TrFE) 65/35vol%

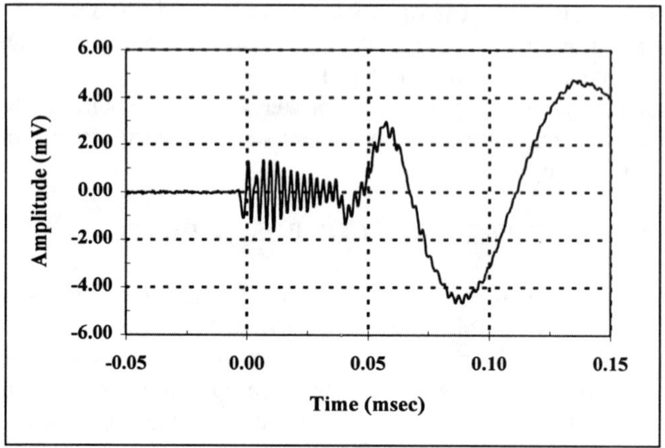

Figure 2. Response of PTCa/P(VDF-TrFE) bimorph to a simulated AE source

and PTCa/P(VDF-TrFE) 60/40vol% connected in series. The two monomorphs used in the array were also obtained from the same piezoelectric composite films, i.e. PTCa/P(VDF-TrFE) with 65 and 60 vol% ceramic respectively. After the sensors had been embedded, the plates were cured in an autoclave at 90°C and 0.55 MPa for 6hrs according to manufacturer's specifications. A low temperature cure cycle was used in this case because it was desired to stay below the poling temperature of the composite films, 100°C, thus avoiding depoling problems. The composite plates so obtained had 60 % fiber volume and a tensile modulus of 56.5 GPa (8.2 Msi) (Newport Composites Product Data Sheet).

Typical responses of the embedded bimorph sensors to simulated acoustic emission sources can be seen in figures 2 and 4. Figure 2 shows the response of an

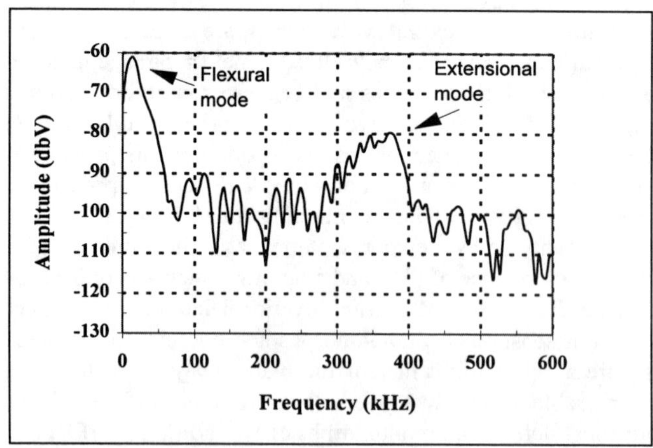

Figure 3. Frequency response of time signal in figure 2

embedded PTCa/P(VDF-TrFE) bimorph connected in series, with electrodes arranged in configuration (b) of figure 1, to a simulated acoustic emission source 10cm away and along the fiber direction. The two plate wave modes of propagation can be clearly seen in this figure. The oscilloscope was triggered on the arrival of the extensional mode, which is seen as the higher frequency component of the signal, arriving before the lower frequency component due to the flexural mode. The frequency spectrum of the signal can be seen in figure 3 and it was obtained from a Fast Fourier Transform (FFT) of the time signal. From this figure the two distinct frequency components, associated with the two types of wave propagation, are clearly distinguished. Although not shown here, this signal is equivalent to a signal given by a single monomorph in response to the same input (Wenger et al., 1996b). Figure 4 shows the response of the same bimorph with electrode configuration (a), and its FFT frequency spectrum is shown in figure 5. The immediately noticeable difference between the two responses is that of attenuation. The latter response showing a greatly attenuated signal to that of the former. Upon examination of the corresponding frequency spectra it can be seen that the flexural modes have been attenuated more than the extensional modes, by a factor of 27dBV as opposed to 6dBV. A slight shift in peak frequencies was also observed between the two responses, but this is believed to be due to the resolution of the FFT and not an artifact of the response.

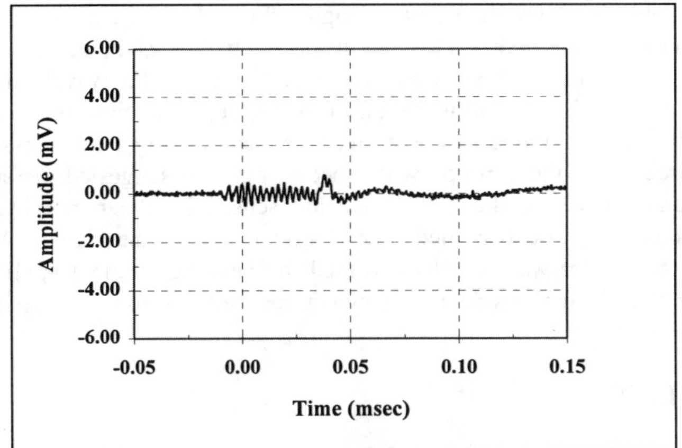

Figure 4. Response of PTCa/P(VDF-TrFE) bimorph to simulated AE source

Similar responses were seen from bimorph constructed from PTCa/Epoxy composites but relative signal attenuations were 18dBV and 8dBV for the flexural and extensional modes respectively. The difference in signal attenuation between the two types of bimorphs can be understood when considering their respective coupling coefficients. The measured piezoelectric d_{31} coefficients are 4.5 and 0.4 pCN^{-1} while the measured dielectric permittivities are 45 and 26 for the composites

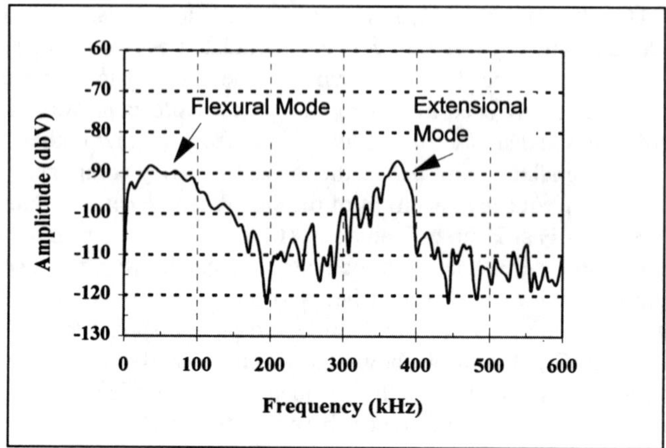

Figure 5. Frequency response of time signal in figure 4

of PTCa/P(VDF-TrFE) and PTCa/Epoxy respectively. These numbers would indicate an approximate 20dBV difference in a signal arising from the g_{31} coefficients as would be the case for a bimorph sensing in its bending mode.

From wave velocity measurements along the fiber direction on a unidirectional composite plate, the approximate wavelengths found for the extensional and the flexural modes were 1.4 and 11.0 cm respectively. It is noted here that the dimensions of the bimorph transducers are close to those of the wavelengths of the extensional modes. Under configuration (a) of figure 1, the bimorph will only be sensing in its bending mode of operation, as signals arising from stresses in the thickness direction of the bimorph will cancel. It can be suggested therefore, that the extensional modes of the plate wave are sensed by the bimorph due to a sufficient bending moment applied onto it giving rise to the signal. However, because of the greater wavelengths involved the bending moment applied to the bimorph by the flexural mode is somewhat reduced resulting in signals being greatly attenuated.

CONCLUSIONS

It follows from the preceding discussion that bimorph sensors constructed from ferroelectric ceramic/polymer composite materials, can successfully be employed as embedded sensing devices for the detection of elastic waves in laminated plate structures. Furthermore, due to their dual mode of operation the bimorphs can essentially differentiate between the two plate modes of acoustic emission waves.

REFERENCES

1. P. Blanas, M. Wenger, R. J. Shuford, and D.K. Das-Gupta, September 1995. "Smart Ceramic/Polymer Composite Materials for Sensors", *Proceedings of the Fourth International Symposium on Advanced Composites*, pp.

2. J.A. Chilton, 1991. Electroactive Composites, *GEC Review*, **6**(3):156.

3. C. Dias, D. K. Das-Gupta, Y. Hinton and R. J. Shuford, 1993. "Polymer/Ceramic Composites for Piezoelectric Sensors", *Sensors and Actuators A*, 37-38:343.

4. M. Wenger, P. Blanas, C.J. Dias, R. J. Shuford, and D.K. Das-Gupta, 1996. "Ferroelectric Ceramic/Polymer Composites and Their Applications", *Ferroelectrics*, 187:75.

5. M. Wenger, P. Blanas, R.J. Shuford and D.K. Das-Gupta, 1996. "Acoustic Emission Signal Detection by Ceramic/Polymer Composite Piezoelectrets Embedded in Glass-Epoxy Laminates", *Polymer Engineering and Science*. 36(24):2945.

6. M. Wenger, P. Blanas, R. J. Shuford, and D.K. Das-Gupta, 1994. "Ferroelectric Ceramic and Epoxy Composite Films as Pyroelectric Detectors", *Proceedings of the 2nd European Conference on Smart Structures and Materials*, 358.

7. R.E. Newnham, D.P. Skinner and L.E. Cross, 1978. "Connectivity and Piezoelectric-Pyroelectric Composites, *Mater. Res. Bull.,* 13:525.

8. C.J. Dias and D.K. Das-Gupta, 1996. "Inorganic Ceramic/Polymer Ferroelectric Composite Electrets, *IEEE Trans. Dielectr. and Electr. Insul.*, 3(5):706.

9. ASTM, E 976, 1993. "Standard Guide for Determining the Reproducibility of Acoustic Emission Sensor Response, ASTM Book of Standards, 03-03.

10. H. Lamb, 1917. "On Waves in an Elastic Plate", *Proc. Roy. Soc. London,* Series A, 93:114.

11. M.R. Gorman and W.H. Prosser, 1990. "Acoustic Emission Source Orientation by Plate Wave Analysis, *J. Acoustic Emission,* 9(4):283.

Nonlinear Active Damping of Adaptive Space Structures

L. GAUL and D. SACHAU

ABSTRACT

An improved simulation backed by active damping control for the dynamics of flexible lightweight structures is provided. Large deployable space systems often consist of trusses with interconnections at mechanical joints. Microslip and macroslip in the joint contact surfaces are the dominating dissipation mechanisms as compared to material damping and environmental damping if no additional damping measures are applied. So far only control of operational modes by actuators in the truss element is realized. The present paper aims to control the nonlinear transfer behaviour of joints by adapting the contact pressure. This is achieved by piezoelectric elements in bolted connections. Active joint description by ODE with internal variables backed by experimental data is implemented in the hybrid multibody system (MBS) of the assembled truss structure. The structural response is decomposed in large rigid body motion and superimposed small elastic deformations. The equations of motion are linearized in the deformation coordinates. The flexibility of the MBS is treated by superposition of structural modes calculated by FEM in the sense of a Ritz approximation. Simulation of free as well as forced vibrations of the structure with active joints in a closed control loop underline the gain of damping performance compared to the associated passive system.

INTRODUCTION

Various fields of technology where Mechanical Engineering plays a central role such as aerospace (air- and spacecraft), transportation (road and rail vehicles), automation (robotics and machine tools) show similar present tendencies: In order to save weight and energy drastic reduction of material is a major request in all fields, *lightweight constructions* have highest priority in all major design groups. As a conse-

Lothar Gaul, Institute A of Mechanics, University Stuttgart, D-70550 Stuttgart, Germany.
Delf Sachau, Institute for Robotics and System Dynamics,German Aerospace Research Establishment, D-82330 Wessling, Germany.

quence, the elastic deformations and structural vibrations become of focal interest, even in situations where formerly a rigid body modelling was more than appropriate.

Two major options are available to analyse flexible mechanical systems: The finite element method (FEM) and the multibody system (MBS) approach. A simulation based on finite element models is – despite of the labour for setting up the model data – straightforward, the corresponding codes are well developed, and include linear and nonlinear theory of elasticity. But there is a disadvantage: a dynamic analysis with FEM-codes is very time consuming. In many applications one is confronted with system models, in which the deformations of the flexible bodies are small but superimposed on a large reference- or rigid-body-motion. In multibody system simulation one exploits this fact to reduce the computational burden for such applications by linearizing the equations of motion assuming small deflections. Using relative variables to represent the reference motion and applying O(N)-formalisms (Schwertassek and Rulka, 1991), MBS-codes provide an efficient alternative for system analysis via simulation. Following such arguments, flexible multibody formalisms have been used frequently (Man and Sirlin, 1989); however, not all of the mathematical models are correct (Kane, Ryan and Banerjee, 1987). Because of neglecting so-called "geometric stiffening terms" some codes did yield wrong results. Geometric stiffening is a well known effect in dynamics and its importance for multibody system simulation has been pointed out clearly, in (Wallrapp and Schwertassek, 1991). The modelling of flexible bodies in the MBS-program SIMPACK has been described at various places (Wallrapp and Sachau, 1994), (Wallrapp, Rulka and Maurer, 1991) including the problem of how to compute geometric stiffening terms for beams and arbitrary structures. Recent investigations (Sachau, 1996) demonstrate how to increase computational efficiency by neglecting terms which are not required to describe the body deformations with the accuracy guaranteed by a linear order of approximation.

The ability of an assembled structure to dissipate vibration energy is of importance. In built up structures bolted or riveted joints are the primary source of damping if no special damping treatment is added to the structure. The actual normal contact pressure distribution in a dynamically loaded lap joint is not constant in the interface. Depending on the transmitted load, the contact interface is divided into stick and slip zones. Before gross slip behaviour occurs, the so called macroslip, only local slip exists, the so called microslip. In constructions joints are transmitting static or dynamic loads and they fail to do this if macroslip occurs. But proper design of joints can lead to microslip which contributes significantly to the damping capacity of a structure. At implemented active joints, the joint pressure can be changed by a piezoelectric disc. This allows to adapt the bolted joint connection in a control loop to improve the overall system behaviour. Control parameters are adapted via simulation and the improvement compared to the passive system can be shown in the design phase of a structure.

BASIC EQUATIONS FOR AN ELASTIC MULTIBODY SYSTEM

The starting point for the development of the $O(N)$-formalism in SIMPACK is a set of equations of motion of the flexible bodies. The flexible body data required to generate the MBS-equations are those, which are needed to compute the elements of the mass matrix, the gyroscopic terms and the stiffness terms. To compute such data, preprocessors have been developed. After reading the flexible body data and the remainder of the system data, SIMPACK can simulate the MBS-motion.

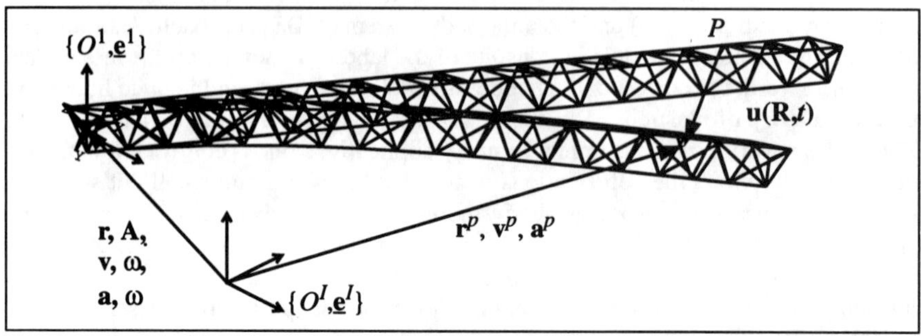

Figure 1: Representation of Motion, First Eigenmode, 2 Hz, Bending

The coordinate systems shown in Figure 1 will be used to describe the body's motion. They are the inertial frame $\{O^I, \underline{e}^I\}$ and the body reference frame $\{O^1, \underline{e}^1\}$ located at one end of the body in the reference configuration. In general all vectors will be resolved in the reference frame $\{O^1, \underline{e}^1\}$, e.g. $\underline{r} = \underline{e}^{1^T}\mathbf{r}$. The motion of a representative material point P with respect to inertial space is described as:

$$\mathbf{r}^P(\mathbf{R}, t) = \mathbf{r}(t) + \mathbf{R} + \mathbf{u}(\mathbf{R}, t). \tag{1}$$

In this representation the body motion is given in terms of the reference or rigid body motion as described by $\mathbf{r}(t)$ and $\mathbf{A}(t)$ and in terms of small displacements $\mathbf{u}$, i.e.

$$\mathbf{u} = \bar{\mathbf{u}}(\mathbf{R}^u, \mathbf{y}(\mathbf{R}^y, t)), \left(\mathbf{R} = \begin{bmatrix} \mathbf{R}^u \\ \mathbf{R}^y \end{bmatrix}\right). \tag{2}$$

The displacements are expressed in terms of deformation variables $\mathbf{y}$ (Bremer and Pfeiffer, 1992), e.g. for beams the motion of the neutral axis. This representation of body motion is suitable for incorporating the assumption of small deformations. The coordinates of the absolute velocity and acceleration of mass element dm at point P with respect to reference frame are derived from (1)

$$\mathbf{v}^P = \mathbf{v} + \tilde{\omega}(\mathbf{R} + \mathbf{u}) + \dot{\mathbf{u}}, \qquad \mathbf{v} = \dot{\mathbf{r}} + \tilde{\omega}\mathbf{r} \tag{3}$$

$$\mathbf{a}^P = \mathbf{a} + \tilde{\omega}\tilde{\omega}(\mathbf{R} + \mathbf{u}) + \dot{\tilde{\omega}}(\mathbf{R} + \mathbf{u}) + 2\tilde{\omega}\dot{\mathbf{u}} + \ddot{\mathbf{u}} \tag{4}$$

where $\mathbf{v}$ and $\mathbf{a}$ are the reference point velocity and acceleration, ω and $\dot{\omega}$ are the angular velocity and acceleration of the reference frame respectively. The tilde operator represents the cross product of vectors. Differentiation of displacements $\mathbf{u}$ with respect to time yields

$$\dot{\mathbf{u}} = \mathbf{J}(\mathbf{y})\dot{\mathbf{y}} \tag{5}$$

with the Jacobian

$$\mathbf{J} = [J_{\alpha i}] = \left[\frac{\partial u_\alpha}{\partial y_i}\right] \tag{6}$$

containing the derivatives of **u** w.r.t. **y**. Deformations are measured by the symmetric Green-Lagrange strain tensor. A common matrix form is

$$\varepsilon = [\varepsilon_{11}, \varepsilon_{22}, \varepsilon_{33}, 2\varepsilon_{12}, 2\varepsilon_{23}, 2\varepsilon_{31}]^T = \left[\mathbf{L}^0 + \frac{1}{2}\mathbf{L}^1(\mathbf{u})\right] \tag{7}$$

where operator matrix $\mathbf{L}^0$, which is well known from the linear theory of elasticity (Washizu, 1982), contains partial derivatives with respect to the material coordinates $\partial_\alpha = \partial(\,)/\partial\mathbf{R}_\alpha$. Products of ∂_α and $\mathbf{u}_{\alpha,\beta}$, so-called bilinear terms, are collected in $\mathbf{L}^1$. They result from the nonlinear terms of the strain-displacement relation (7). Stresses are represented here by the symmetric second Piola-Kirchhoff stress tensor. For a linear material law, the stresses are related to the strains by

$$\sigma = [\sigma_{11}, \sigma_{22}, \sigma_{33}, \sigma_{12}, \sigma_{23}, \sigma_{31}] = \mathbf{H}\varepsilon \tag{8}$$

with the 6×6 Hookean matrix **H**. Its elements are given in terms of two independent elasticity constants Young's modulus E and Poisson's ratio ν of the isotropic material. A Ritz-approximation of the deformation variables **y**

$$\mathbf{y}(\mathbf{R}^y, t) = \Phi(\mathbf{R}^y)\mathbf{q}(t) \tag{9}$$

is used with the n_q unknown nodal coordinates $\mathbf{q}(t)$ and finite elements with interpolation functions Φ. Introducing (9) into Hamilton's principle and using the fundamental lemma of variational calculus yields the equations of motion

$$\mathbf{M}(\mathbf{q})\begin{bmatrix}\mathbf{a}\\ \dot{\omega}\\ \ddot{\mathbf{q}}\end{bmatrix} + \mathbf{k}(\omega, \mathbf{q}, \dot{\mathbf{q}}) + \begin{bmatrix}\mathbf{0}\\ \mathbf{0}\\ \mathbf{K}(\mathbf{q})\mathbf{q}\end{bmatrix} = \mathbf{h}(\mathbf{r}, \mathbf{A}, \mathbf{q}, \ldots) \tag{10}$$

where matrix **M** contains the inertia terms, **k** the gyroscopic terms, **K** the stiffness terms and **h** the applied forces.

The number of coordinates is reduced by modal transformation. This is performed by the preprocessor FEMBS (FEMBS, 1996) using finite element results from the FE-program ANSYS. The antenna with a mass of 35 kg is attached to the spacecraft by a truss structure as shown in Figure 9. The truss has a length of 20.3 m and a mass of 95 kg. The pre-computations required for a MBS-analysis of the manoeuvre are as follows:

I. Set up the Finite-Element-Model of the truss structure, 108 rods are interconnected at 57 nodes.
II. Compute mass- and stiffness-matrices plus eigenmodes – boundary condition: structure is clamped at x=0, see Figure 1 and Figure 9.
III. Compute stress stiffening matrix caused by a unit load in longitudinal direction and loads resulting from centrifugal forces due to a unit angular velocity, using the geometrically nonlinear modelling options of a FE-code (here ANSYS).
IV. Read FE-results and compute flexible body data by the preprozessor FEMBS.

INVESTIGATION OF ISOLATED JOINT

The damping of lightweight flexible structures is often dominated by bolted joint connections. The nonlinear transfer behaviour of an isolated bolted joint connection has been experimentally investigated (Lenz and Gaul, 1995). Figure 2 shows the experimental setup. A lap joint is implemented between two lumped masses suspended by Nylon cords such that the system can be driven as a resonator. The system is excited by a shaker. The joint normal pressure is measured by a strain gauge mounted on the joint bolt. The accelerations and the driving force are measured by piezoelectric pickups. The hysteresis loop of transmitted force F versus relative displacement u requires double integration of the acceleration signals to obtain displacements. Figure 3 shows the hysteresis for different excitation force amplitudes F_e in the range from 20 N to 310 N at 467 Hz. During all these measurements the joint normal pressure was $P = 0.25\text{N/mm}^2$.

The measured behaviour can be well represented by the so called Valanis model (Valanis, 1971). For simulations a well suited form is the ODE (32) (Gaul et al., 1994). The change of the generalized force in time is given by

$$\dot{F} = \frac{E_0\dot{u}\left(1 + \frac{\lambda}{E_0}\operatorname{sgn}(\dot{u})(E_t u - F)\right)}{1 + \kappa\frac{\lambda}{E_0}\operatorname{sgn}(\dot{u})(E_t u - F)} \tag{11}$$

as a function of the generalized force F, the relative displacement u and the velocity $\dot{u}$. The dimensionless parameter λ is defined by

$$\lambda = \frac{E_0}{\sigma_0\left(1 - \kappa\frac{E_t}{E_0}\right)}. \tag{12}$$

The four parameters E_0, E_t, σ_0 and κ in (32) and (33) can be identified easily from a measured hysteresis as shown in Figure 4. For the resonator investigated here, these parameters are found from Figure 3 to be $E_0 = 1.28 \cdot 10^8$ N/m, $E_t = 1.36 \cdot 10^7$ N/m, $\lambda = 8 \cdot 10^5$ 1/m and $\kappa = 0.01$.

Figure 2: Experimental Resonator with Bolted Lap Joint

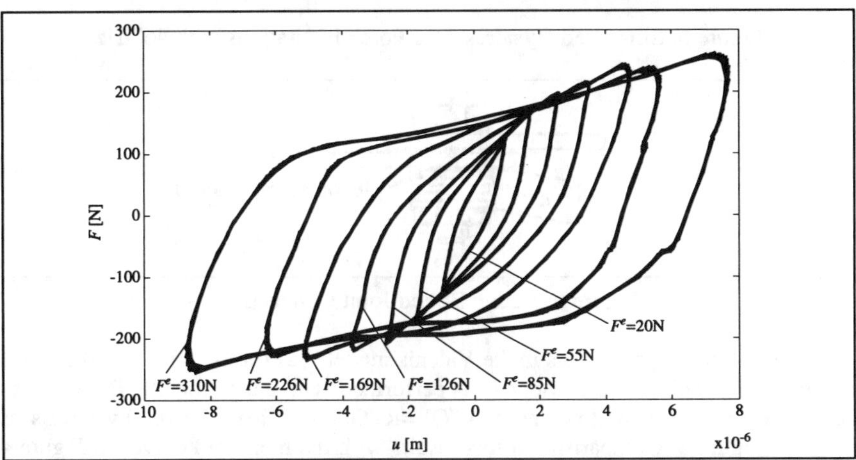

Figure 3: Measured Hysteresis of Resonator with $P = 0.25$ N/mm²

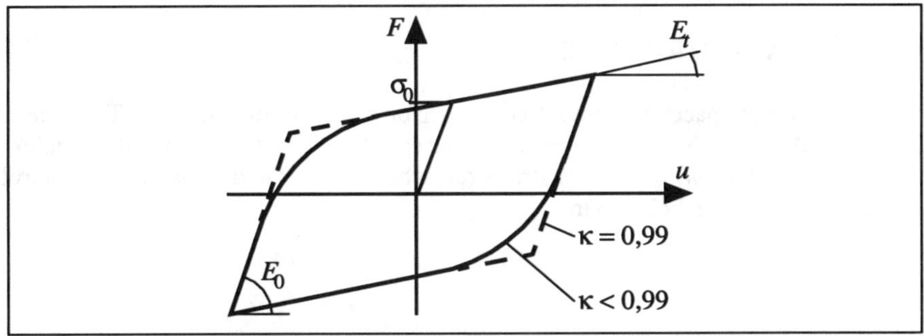

Figure 4: Closed Hysteresis of the Valanis Model of Joint

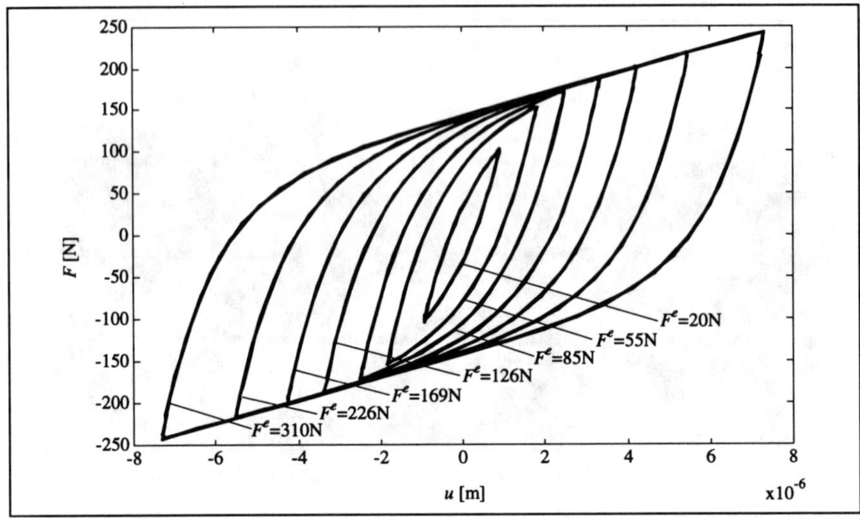

Figure 5: Simulated Hysteresis for Force Excitations with 467 Hz

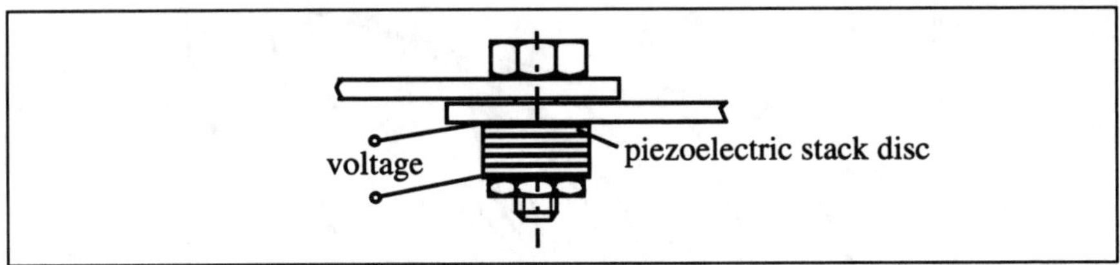

Figure 6: Active Bolted Joint Connection

By using these parameters in the Valanis model (32) for the transmitted force, a multibody simulation of the resonator is performed. The integrator LSODE with absolute and relative error tolerances ATOL=RTOL=10^{-8} leads to the hysteresis as shown in Figure 5. Comparison of this results with the measured curves in Figure 4 shows a good fit of the Valanis model with the measured joint behaviour.

ACTIVE SPACE STRUCTURE

Lightweight space structures have to perform precise manoeuvres. To achieve this goal, vibrations due to disturbances of the structure have to be damped out quickly. This can be done with adaptive structures, where sensors and actuators in a control loop adapt the system behaviour.

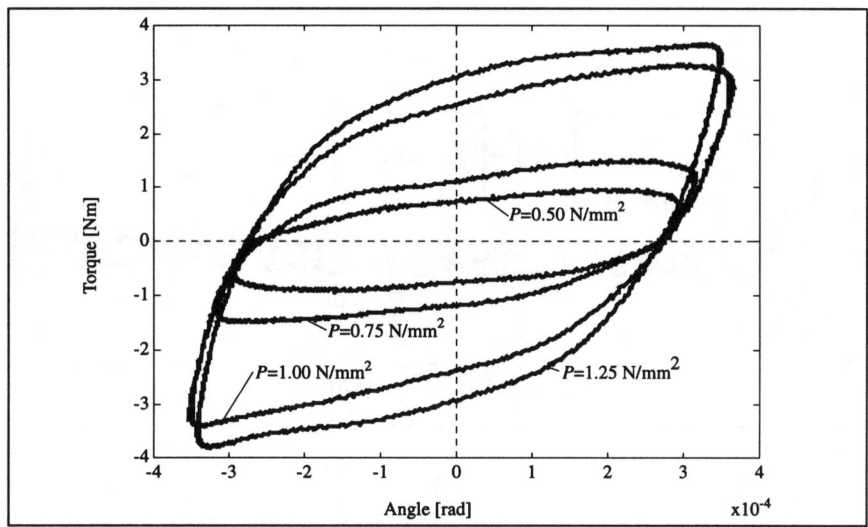

Figure 7: Hysteresis Plot of Lap Joint With Varying Normal Pressure

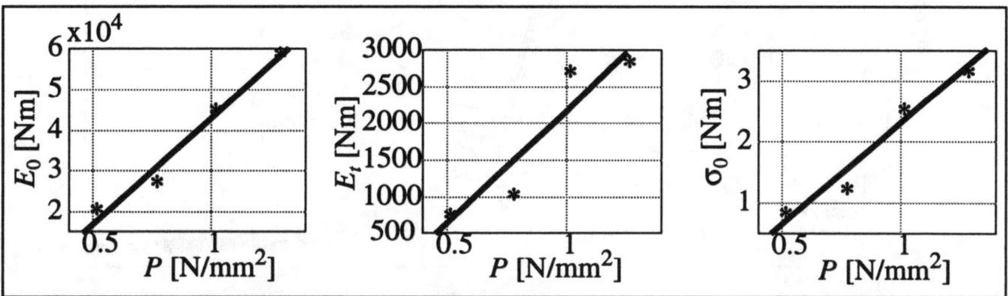

Figure 8: Dependency of the Valanis Parameters on the Joint Normal Pressure *P*

Here a single input single output control system (SISO) is proposed to control the system shown in Figure 9. The antenna displacement is measured directly with a sensor (e.g. an optical device). A new actuator design is proposed: an active bolted joint connection in the connecting rod of the truss structure to the spacecraft. Figure 6 shows the active bolted joint connection. A piezoelectric disc is used to modify the joint normal pressure *P*. The transfer behaviour of the joint connection depends strongly on the normal pressure *P*. Variation of the hysteresis with different contact pressures is depicted in Figure 7 for a transmitted torque versus relative angle of rotation. The corresponding parameters are marked as ' * ' in Figure 8. The dependency of the parameters E_0, E_t and σ_0 on the normal pressure *P* can be approximated by linear functions, plotted by lines in Figure 8. The selected parameter κ is approximately 0.01 for all cases.

Figure 9 shows the control system block diagram. The controller calculates the adapted variable *P* for the bolted joint connection depending on the control deviation $u_A = u_a^d - u_a^a$. The control algorithm most widely used in standard controller design features is a PID feedback. In the present case the integrator (I-element) is omitted.

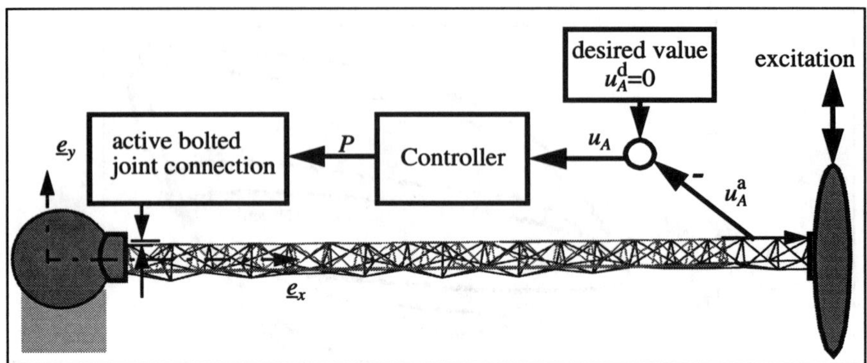

Figure 9: Active Space Structure

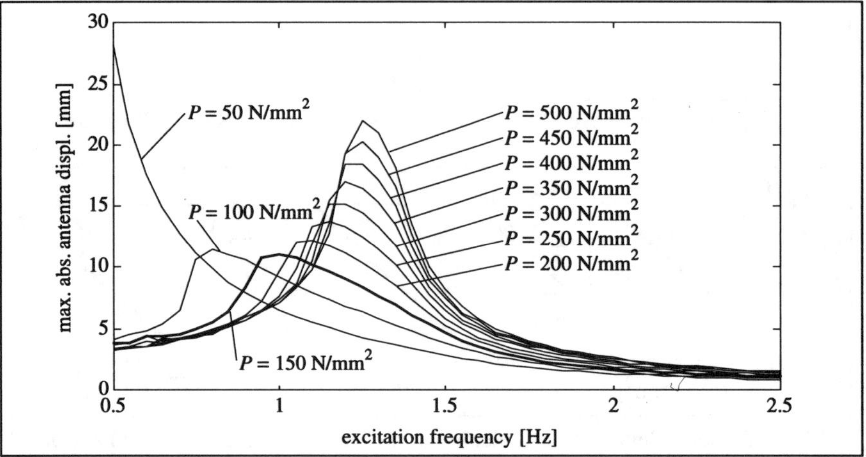

Figure 10: Maximal Absolute Antenna Displacement for Different Constant Joint Normal Pressure P

For simulation purposes the controller has to be described in time domain. The state space form is given with the variable x as

$$\dot{x} = -N_0 x + u_A, \; x(0) = 0, \; P = (Z_0 - Z_1 N_0)x + Z_1 u_A. \tag{13}$$

Herewith all portions of the controlled space structure are expressed in the form of nonlinear ODE, ready for time integration within the multibody simulation. Due to the nonlinear behaviour of the bolted joint connection, a linearization of the equations of motion is inappropriate. Therefore the dynamic behaviour of the system has to be investigated in time domain via simulation.

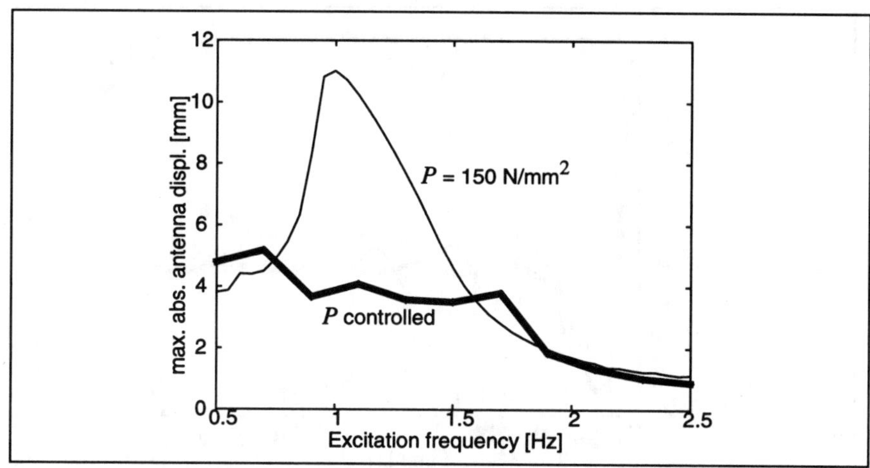

Figure 11: Maximal Absolute Antenna Displacement: a) Controlled with $Z_0 = 2{\cdot}10^5$ N/(m^{3}s), $Z_1 = 7{\cdot}10^5$ N/m^3 b) Constant with $P = 150$ N/mm^2

The space station is considered to be fixed in inertial space. The system reaction with respect to an oscillating force in *y*-direction at the antenna is investigated. The excitation acts with an amplitude of 10 N. The system behaviour near the lowest eigenfrequency (1.36 Hz) of the truss structure with fixed bolted joint connection is investigated. Here the frequency band from 0.5 Hz up to 2.5 Hz is considered.

For discrete excitation frequencies in this range, the maximal antenna deflection has been calculated and plotted versus the excitation frequency in Figure 10 for different constant joint normal pressure P in the range of 50 up to 500 N/mm^2 in steps of 50 N/mm^2. The best passive system behaviour which means the smallest antenna excursion is achieved by choosing $P_1 = 150$ N/mm^2.

Now the PD-controller is used. Suitable controller parameters Z_0 and Z_1 are found from systematic parameter variation where the parameters are varied from 0 up to 10^6 in steps of 10^5 N/(m^{3}s) respectively N/m^3. The parameter N_0 is fixed to $N_0 = 100$ 1/s. The large number of simulations is performed with the MBS program SIMPACK (Rulka, 1990). Figure 11 shows the achieved enhancement of the system behaviour, after selecting the control parameters $Z_0 = 2{\cdot}10^5$ N/(m^{3}s) and $Z_1 = 7{\cdot}10^5$ N/m^3. Especially the amplitudes in the frequency range from 0.9 Hz up to 1.5 Hz are much smaller compared with the uncontrolled passive system.

The achieved improvement of the system behaviour can as well be seen by the free vibration response, Figure 12. An initial antenna tip displacement of 60 mm attenuates after a few oscillations down to 7 mm for the passive system with pressure $P_1 = 150$ N/mm^2. Due to microslip the amplitude reduction is small during the following oscillations. Much more efficient is the controlled system behaviour. Here the deflection reduces fast to a much lower level of about 2 mm.

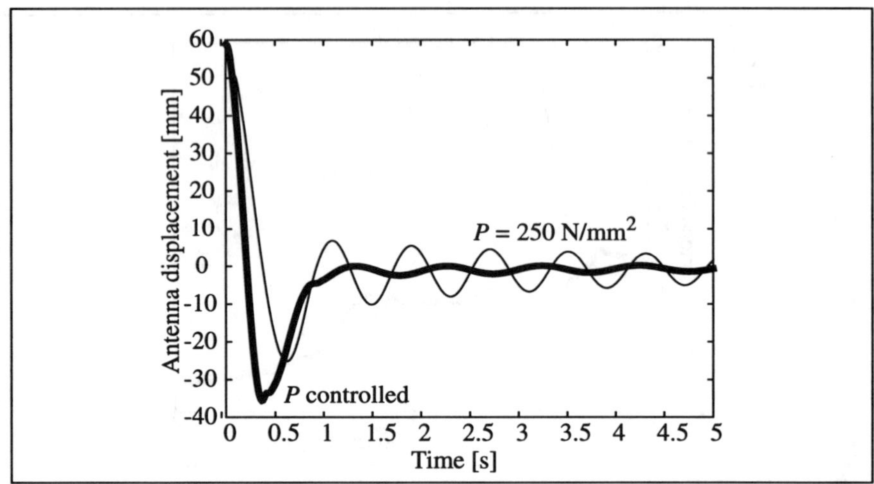

Figure 12: Antenna Deflection During Free Vibration

CONCLUDING REMARKS

The present paper aims to improve the prediction of structural response of lightweight space structures. Linear restoring and inertia forces are well understood in the design phase and the description of material damping has made progress. Still there are gaps in modelling of structural damping e.g. by bolted, riveted or clamped joints.

This is why the paper firstly provides a geometrically nonlinear description of flexible body deformation within the frame of multibody dynamics such that the degree of approximation is consistent within the order of the associated perturbation approach.

Secondly the paper presents an effective evolution equation for the simulation of nonlinear joint transfer behaviour. Experimental results of an isolated lap joint as part of a resonator validate the joint simulation, which combines nonlinear stiffness and damping description in micro- as well as macroslip regime with a suitable formulation for implementation in MBS programs.

Thirdly the paper present the innovative idea of an adaptive joint, (Gaul, 1997). Variation of contact pressure, thereby joint hysteresis and associated dissipation is achieved by voltage controlled deformation of a stack of piezoelectric discs on the bolt. The damping capacity of structures with such and similar active joints associated with optimized location in a control loop turns out to be significantly increased as compared to passive damping with fixed contact pressure.

ACKNOWLEDGEMENTS

Support by the German Research Society (DFG) under Grants Ga 209, Schw 443 is gratefully acknowledged.

REFERENCES

Bremer, H. and F. Pfeiffer, Elastische Mehrkörpersysteme. Stuttgart: Teubner,1992.

FEMBS, Users Manual, INTEC GmbH, Münchnerstr. 20, D-82230 Wessling, Germany, 1996.

Gaul, L., U. Nackenhorst, K. Willner and J. Lenz, Nonlinear Vibration Damping of Structures With Bolted Joints. in 12th International Modal Analysis Conference. Hawaii, p.875-881, 1994.

Gaul, L., Aktive Beeinflussung von Fügestellen in mechanischen Konstruktionselementen und Strukturen, German Patent, DE 19702518 A1, 1997.

Kane, T.R., R.R. Ryan, and A.K. Banerjee, Dynamics of a Cantilever Beam Attached to a Moving Base. Journal of Guidance, 10: p. 139-151, 1987.

Lenz, J. and L. Gaul. The Influence of Microslip on the Dynamic Behaviour of Bolted Joints. in 13th International Modal Analysis Conference. Nashville, Tennessee, p.248-254, 1995.

Man, G.K. and S.W. Sirlin. An Assessment of Multibody Simulation Tools for Articulated Spacecraft. in 3rd Annual Conference on Aerospace Computational Control. Pasadena, CA: NASA JPL Publ. p. 89-45, 1989.

Preumont, A., J.-P. Dufour and M. Sparavier. Active Damping of a Truss Structure Using Piezoelectric Actuators. in Dynamics of Flexible Structures in Space.. Cranfield, UK: Computational Mechanics Publications Southampton Boston, p. 369-382, 1990.

Rulka, W., SIMPACK - A Computer Program for Simulations of Large-Motion Multibody Systems, in Multibody System Handbook, W. Schiehlen, Editor. Springer-Verlag: Berlin. p. 265-284, 1990.

Sachau, D., Consideration of Flexible Bodies and Bolted Joint Connections in Multibody Systems for Simulation of Active Space Structures (in German). Ph.D. Thesis, Bericht aus dem Institut A für Mechanik 1/1996, Universität Stuttgart: Stuttgart, 1996.

Schwertassek, R. and W. Rulka, Aspects of Efficient and Reliable Multibody System Simulation, in Real-Time Integration Methods for Mechanical System Simulation, E.J. Haug and R.C. Deyo, Editors. Springer-Verlag: Berlin. p. 55-96, 1991.

Valanis, K.C., A Theory of Viscoplasticity Without a Yield Surface. Archive of Mechanics. 23(4): p. 517-551, 1971.

Wallrapp, O. and R. Schwertassek, Representation of Geometric Stiffening in Multibody System Simulation. Int. Journal for Numerical Methods in Engineering. 32: p. 1833-1850, 1991.

Wallrapp, O. and D. Sachau, Space Flight Dynamic Simulations Using Finite Element Analysis Results in Multibody System Codes, in 2nd Int. Conf. on Computational Structures Technology. Civil-Comp-Press: Athens, Greece, 1994.

Wallrapp, O., W. Rulka, and M. Maurer. Comparison of two Approaches to Incorporate Geometric Stiffness Terms in Flexible Multibody Dynamics. in 9th Modal Analysis Conference (IMAC).Firenze, Italy, April, p.31-37, 1991.

Washizu, K., Variational Methods in Elasticity & Plasticity. 3 ed. Oxford: Pergamon Press, 1982.

Fracture Mechanics Problem for a Shape Memory Alloy Plate

V. BIRMAN

ABSTRACT

The isothermal problem of a Mode I fracture crack in a shape memory alloy (SMA) plate is considered. Although the plate material is in the austenitic phase at a reference temperature, the martensitic transformation is triggered by high effective stresses at the tip of the crack. The phase transformation is accompanied by a reduction of the stiffness and strength that affects the extend of the plastic zone. The analysis is based on a two-dimensional version of the Tanaka constitutive theory for SMA. The goal is to evaluate the size of the plastic zone in front of the crack and the effect of the phase transformation on the stress intensity factor. This estimate can be used for a better understanding of damage tolerance of SMA plates.

INTRODUCTION

Development of smart materials and structures represents one of the most promising and challenging areas of applied mechanics. Recently, practical applications have been successfully pursued, notably in such high-tech fields as aerospace and biomedical engineering. Shape memory alloys represent a class of smart materials that can be employed in applications that require high recovery stresses or strains. The main feature of these alloys is related to the phase transformation from austenite to martensite and back that can be triggered by temperature or stresses. The martensite stiffness and strength are inferior compared to those of austenite. Therefore, the material can be plastically deformed in the martensitic phase with a relative ease. The material that was originally in the austenitic phase can be transformed into martensite as a result of either low temperature or external stresses and plastically deformed. However, it recovers the original shape if subject to the reverse transformation to austenite (shape memory).

Dr. Victor Birman, University of Missouri-Rolla, Engineering Education Center, 8001Natural Bridge Road, St. Louis, MO 63121

Significant strains or stresses can be generated in the regime of unconstrained or constrained recovery, respectively. A review of mechanics of SMA and their applications can be found in the forthcoming paper of Birman (1997).

The problem considered in this paper is related to fracture of SMA sensors or actuators. The Mode I fracture is analyzed for a SMA plate with a central crack. The edge or corner effects can be incorporated into the solution using the standard methods of fracture mechanics. The analysis is concerned with the presence of the plastic zone in front of the crack and its effect on the stress intensity factor. Therefore, the solution analyzes the influence of the martensitic phase transformation associated with a stress concentration around the tip of the crack on the size of the plastic zone. As follows from the analysis, the size of the plastic zone increases, as a result of the phase transformation. Accordingly, the stress intensity factor increases, as compared to the case of the austenitic plate where the phase transformation is disregarded. This phenomenon has to be accounted for when predicting structural health of SMA sensors or actuators, particularly, if the external stresses are comparable to the yield limit of martensite.

ANALYSIS

Consider a thin infinite SMA plate with a crack subjected to tensile stresses acting perpendicular to the crack axis (Mode I fracture). The temperature is higher than the martensite start value so that in the absence of external loads the material is in the austenitic phase. When the loads are applied, a plastic zone appears in front of the tip of the crack. The radius of this zone can be determined using the methods of Irwin (1960), Dugdale (1960) or Barenblatt (1962). Following Irwin (1960), this radius is determined based on the von Mises yield criterion. In the case of plane stress considered here the extend (diameter) of the plastic zone is

$$r_p = (K_I/\sigma_y)^2 / \pi \tag{1}$$

where K_I is the stress intensity factor and σ_y is the yield stress of the material.

The effect of the plastic zone on the linear elastic fracture mechanics solution can be accounted for, if its radius is added to the size of the crack in the calculation of the stress intensity factor (Anderson, 1991). Accordingly, in the case of Mode I fracture in an infinite plate

$$K_I = \sigma_0[\pi(a + r_p/2)]^{1/2} \tag{2}$$

where σ_0 is the external nominal stress and a is the size (half-length) of the crack. The effective stress intensity factor calculated according to eqn. (2) should be compared to the fracture toughness of the material to predict the failure.

In the case of a SMA plate at a temperature exceeding the martensite start level, a stress-induced martensitic transformation can be triggered by tensile stresses. Therefore, the approach outlined above has to be modified to account for formation of SIM (stress-induced martensite). If the material of the plate has been partially transformed into martensite, the region of pure martensite extends from the tip of the crack to the boundary of the region where the transformation is not complete. The radius of the region of pure martensite is denoted by r_M. The region of mixed austenite and martensite is located at a distance from the crack tip, within $r_M < r < r_A$ where r_A is the coordinate of the boundary of the outer region of pure austenite. The problem is to specify the size of the plastic zone. This zone may extend within the region of pure austenite. On the other hand, it may be encompassed by the region of pure martensite or its boundary may be located within the region of mixed austenite and martensite.

First of all, it is necessary to compare the size of the plastic zone to the size of the region of SIM formed at the tip of the crack. This requires us to specify a constitutive theory for SMA. The review of these theories can be found in the paper of Birman (1997). In the present paper, the theory of Tanaka (Tanaka and Sato, 1986) is adopted. While the original theory was developed for a one-dimensional case, it is generalized here for the plate in the state of plane stress (such generalization can be found in the papers of Boyd and Lagoudas, 1993 and Birman et al., 1996). According to the generalized theory, the martensitic fraction (ξ) is the following function of temperature and effective stresses:

$$\xi = 1 - \exp[b_M(M_S{}^0 - T) + (b_M/d_M)\sigma] \tag{3}$$

where σ is an effective stress, $M_S{}^0$ is the martensite transformation temperature corresponding to the stress-free state, T is a current temperature, and d_M is the slope of the martensite transformation temperature lines in the stress-temperature plane. It is assumed in this paper that the slope of the transformation temperature lines in the plane "effective stress - temperature," i.e. d_M, is identical to that in the one-dimensional case. The constant b_M is defined by

$$b_M = \ln 0.01/(M_S{}^0 - M_F{}^0) \tag{4}$$

where $M_F{}^0$ is the stress-free martensite finish temperature.

The effective stress in eqn. (3) is defined as in the theories of Liang and Rogers (1991) and Boyd and Lagoudas (1993):

$$\sigma = [(3/2)\sigma_{ij}'\sigma_{ij}']^{1/2} \tag{5}$$

where σ_{ij}' is a component of the deviatoric stress tensor.

Note that the effective stress should account for both the local as well as the uniform externally applied stresses. Therefore, in the present problem,

$$\sigma = (\sigma_{rr}{}^2 + \sigma_{\theta\theta}{}^2 - \sigma_{rr}\sigma_{\theta\theta} + \sigma_0{}^2 - \sigma_{rr}\sigma^0 + 2\sigma_{\theta\theta}\,\sigma_0)^{1/2} \tag{6}$$

If the stresses in front of the crack are considered, the linear elastic fracture mechanics solution corresponding to $\theta = 0$ yields

$$\sigma_{rr} = \sigma_{\theta\theta} = K_I/(2\pi r)^{1/2},\ \sigma_{r\theta} = 0 \tag{7}$$

Accordingly, in the vicinity of the tip of the crack where local stresses dominate, the effective stress is equal to the radial and circumferential stresses given by eqn. (7). The boundaries of the regions of pure martensite and pure austenite can now be determined from the nucleation criterion that specifies the effective stresses corresponding to the start and finish of the martensitic transformation:

$$d_M(T - M_S^0) < \sigma < d_M \ln 0.01/b_M + d_M(T - M_S^0) \tag{8}$$

Combining the effective stress with the right side of inequality (8) the radius of the martensitic zone in front of the crack tip is obtained as

$$r_M = K_I^2/\{2\pi[d_M \ln 0.01/b_M + d_M(T - M_S^0)]^2\} \tag{9}$$

Accordingly, the radius of the pure austenite region is obtained using the left side of inequality (8) as

$$r_A = K_I^2/\{2\pi[d_M(T - M_S^0)]^2\} \tag{10}$$

Note that eqns (9,10) are obtained by assumption that the stress distribution in front of the crack is not affected by the martensitic transformation. This assumption results in a violation of the compatibility condition in the region of mixed austenite and stress-induced martensite, i.e., for $r_M < r < r_A$. However, it is shown below that this approach may be justified if the region $r_M < r < r_A$ is subdivided into a number of small subregions so that both the modulus of elasticity and the martensitic fraction remain constant within each subregion (piece-wise constant functions).

The constitutive equations of the Tanaka theory corresponding to a superelastic martensitic transformation are

$$\sigma = \mathbf{C}\epsilon + \Omega\xi \tag{11}$$

where σ and ϵ are vectors of stresses and elastic strains, respectively, $\mathbf{C}$ is the matrix of instantaneous elastic stiffnesses, and Ω is the transformation tensor. The components of this tensor are selected proportional to the maximum transformation strain observed in uniaxial loading (Birman et al., 1996). Accordingly, if the martensitic fraction and the elastic moduli remain constant within a certain region, the second term in eqn. (11) is also constant within this region. The strain tensor can now be obtained from eqn. (11) as

$$\epsilon = \mathbf{C}^{-1}(\sigma - \Omega\xi) \tag{12}$$

where both $\mathbf{C}$ and $\Omega\xi$ are constant. Therefore, a correction to the tensor of strains due to the phase transformation is also constant. The solution of the plane fracture problem is obtained using the stress function that satisfies the equations of equilibrium. The compatibility equation employed to specify the stress function includes only derivatives of the components of the strain tensor. Therefore, this solution is not affected by SIM within the regions of constant martensitic fractions and elastic moduli. Reducing the width of these regions, it is possible to generate a solution that would satisfy the equilibrium and compatibility conditions. Of course, the strains calculated from eqn. (12) would exhibit discontinuities along the boundaries of the regions, but these discontinuities would decrease as the number of region increases.

Another issue that should be addressed in future investigation is applicability of eqn. (2) for the stress intensity factor when a phase transformation is present. In this paper, it is assumed that eqn. (2) remains valid, even when the external load is sufficient to trigger the transformation.

Now consider three possible scenarios. If the radius of the plastic zone calculated using the value of the yield stress corresponding to austenite is larger than r_A, plastic effects dominate the response of the material in the vicinity of the crack tip. In this case, fracture is predicted based on eqn. (2) and the value of fracture toughness corresponding to austenite. In the second scenario, the plastic zone is located within the region of pure martensite. In this case, the radius of this region should be calculated using the yield stress of martensite. Of course, this stress is much smaller than the austenite yield limit and the size of the plastic zone will increase accordingly.

The challenging task is to calculate the size of the plastic zone if its boundary is located between the regions of pure martensite and austenite, i.e., if $r_M < r_p < r_A$. In this case, it is necessary to specify the value of the yield stress corresponding to the martensitic fraction at $r = r_p$. There exists sufficient evidence to justify an assumption that the yield stress can be often evaluated as a linear function of the martensitic fraction. For example, a justification of this assumption is found in experimental results presented by Funakubo (1987) for three different NiTi systems and in the report of Jackson et al. (1972) for Nitinol-55 (alloy with about 55% weight contributed by nickel). Accordingly,

$$\sigma_y = \sigma_y^A + \xi(\sigma_y^M - \sigma_y^A) \tag{13}$$

where σ_y^A and σ_y^M denote the yield limits of austenite and martensite, respectively. Now for a given value of the external stress σ_0, the value of the stress intensity factor can be obtained from eqn. (2) where the plastic zone is disregarded. Subsequently, the effective stress and the martensitc fraction are obtained for each value of the radius within $r_M < r < r_A$ from eqns. (7) and (3). Finally, the yield stress $\sigma_y(r)$ is calculated from eqn. (13). This stress can be used in eqns. (1,2) to adjust the stress intensity factor and the process is repeated again until a satisfactory convergence has been achieved.

It is useful to specify temperatures corresponding to the situations when the boundaries of the plastic zone coincide with the boundaries of the region of pure austenite or the region pure martensite. Using equations (1, 9 and 10) one obtains

$$T > M_S^0 + \sigma_y^A/(2^{1/2} d_M) \tag{14}$$

and

$$T < M_F^0 + \sigma_y^M(2^{1/2}d_M) \tag{15}$$

Inequality (14) defines the values of temperature corresponding to the plastic zone expanding within the austenitic region. Temperature that results in the plastic zone within the martensitic region is specified by inequality (15). The range of temperatures between the values given by inequalities (14 and 15) corresponds to the case where the boundary of the plastic zone is located within the region of mixed austenite-martensite. Note that temperatures predicted by inequalities (14 and 15) are independent of the magnitude of the external stress. However, this stress will affect the actual size of the plastic zone.

Finally, it is necessary to indicate that the fracture toughness of austenite and martensite differ. Therefore, an extensive experimental program is needed to determine fracture toughness of SMA experiencing fracture at various temperatures.

NUMERICAL RESULTS

The material of the plate considered in the following examples was Nitinol-55, i.e., an alloy of nickel and titanium where the latter element constitutes about 45% of the total weight. The properties of the material were adopted from the work of Lei and Wu (1991) and White et al. (1993). In particular, according to Lei and Wu (1991), Ni-44.8wt%Ti annealed at 600^{0}C has $M_S^0 = 23^0$C, $M_F^0 = 5^0$C and d_M = 11.3 MPa/^{0}C. The yield limits of this material are taken as (White et al., 1993) σ_y^A = 276 MPa and σ_y^M = 96.5 MPa. Note that these limits, particularly in the case of austenite, differ from the values presented by Jackson et al. (1972). Such difference can be due to different technologies involved in manufacturing of the materials. It should also be noted that data presented by White et al. (1991) refers to two-way shape memory material. Nevertheless, the material data adopted here allows us to generate the results that reflect the qualitative features of fracture of SMA.

As follows from inequalities (14) and (15), the plastic zone in a Nitinol-55 plate subjected to Mode I fracture will expand into the austenitic region, if $T > 40^0$C. On the other hand, this zone will be within the martensitic region at $T < 11^0$C (note that this temperature is below the martensite start temperature).

Distributions of the martensitic fraction in front of a crack in the SMA plate at $T = 27^0$C are shown in Fig. 1 for three different values of external stresses. The martensitic fraction varies from $\xi = 1$ corresponding to pure martensite to $\xi = 0.01$ that is associated with pure austenite. As follows from this figure, the width of the region of mixed austenite-martensite drastically increases as a result of an increase of external tensile stresses.

Variations of the yield stress in the same plate are illustrated in Fig. 2. It is evident that the yield stress rapidly decreases in the vicinity of the tip of the crack due to formation of SIM. Note that in the cases considered in Figs. 1 and 2, the boundary of the plastic region corresponded to relatively small values of r/a and large values of the martensitic fraction. In particular, for the values of σ_0 equal

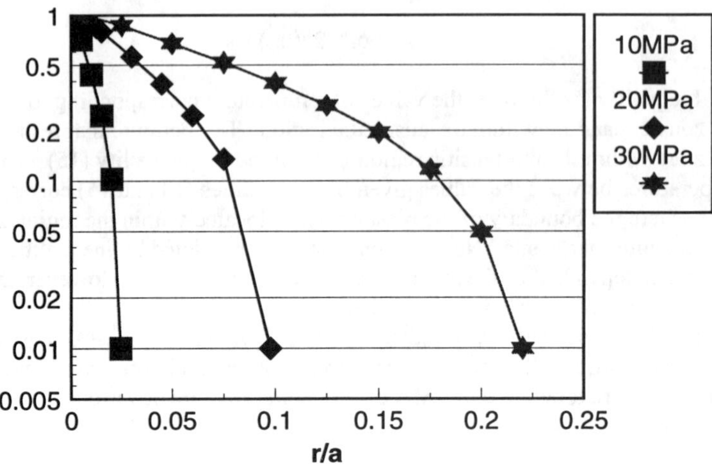

Fig. 1. Distribution of the martensitic fraction in a SMA plate as a function of external stresses.

to 10 MPa, 20 MPa and 30 MPa, r_p/a was equal to 0.008, 0.04 and 0.09, respectively. Obviously, the section of the corresponding curves in Figs. 1 and 2 located to the left of r_p/a should be disregarded.

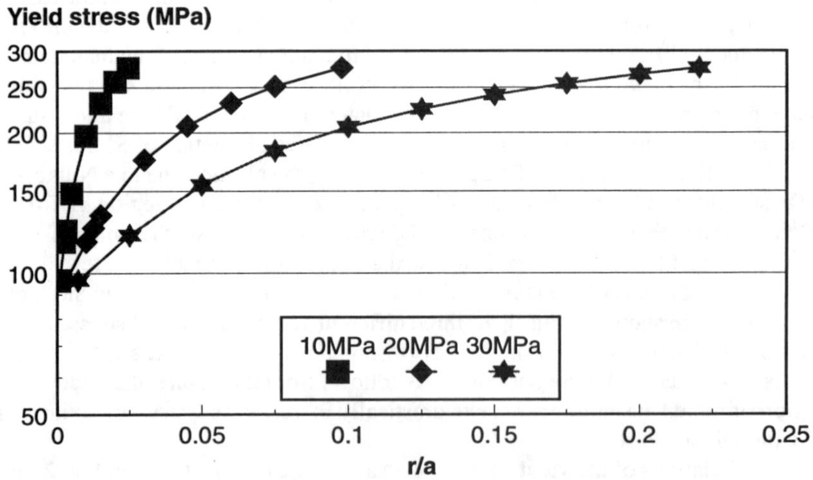

Fig. 2. Variations of the yield stress in front of the crack in a SMA plate.

It is necessary to estimate an effect of the phase transformation on the stress intensity factor. Combining eqns. (1 and 2) and neglecting the effect of the plastic zone on the stress intensity factor used to evaluate the radius of this zone, one obtains a simple formula for the ratio between the stress intensity factor affected by the phase transformation and this factor for the austenitic plate:

$$f = \{[2 + (\sigma_0/\sigma_y(\xi))^2]/[2 + (\sigma_0/\sigma_y^A)^2]\}^{1/2} \qquad (16)$$

Note that the actual yield stress cannot be less than the martensite yield stress. Therefore, a conservative prediction of the effect of the phase transformation on the ratio f can be obtained if $\sigma_y(\xi)$ is replaced with σ_y^M. The corresponding result is shown in Fig. 3 that implies a relatively small effect of the phase transformation on the stress intensity factor as long as the external load remains below the yield limit of martensite. In reality, the effect of the phase transformation is even smaller. For example, if the external stress is equal to 100 MPa, the actual result is f = 1.15, while the simplified solution obtained by (16) yields f = 1.20. Note that the results shown in Fig. 3 are not affected by temperature.

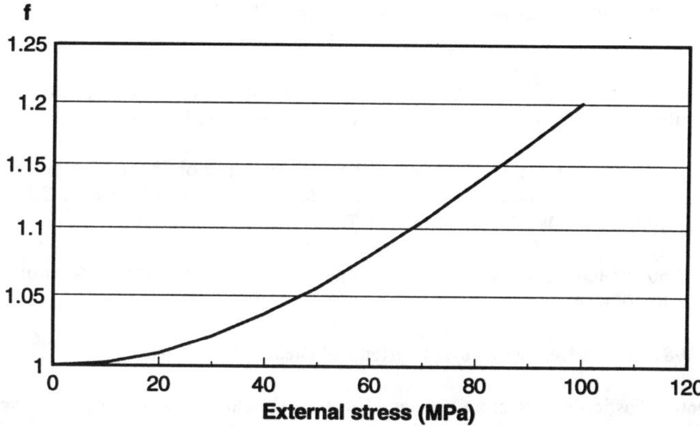

Fig. 3. Effect of external stresses on the stress intensity factor of SMA plates.

CONCLUSIONS

An approximate solution for the Mode I isothermal fracture problem of SMA plates is presented. The analysis concentrates on a stress-induced martensitic transformation in front of the crack and its effect on the stress intensity factor. The results generated in the paper indicate that the effect of the phase transformation on the stress intensity factor is relatively small. Accordingly, the conclusion from the paper is that the magnitude of the stress intensity factor may be evaluated based on the properties of austenite. Of course, the actual stress intensity factor is larger than its austenitic counterpart, but this difference becomes significant only at high external stresses approaching the martensite yield stress.

The principal problem in estimating the anticipated failure of a SMA component is related to the evaluation of fracture toughness. This parameter will also be affected by a stress-induced transformation in front of the crack and an extensive experimental investigation is needed to generate reliable estimates.

ACKNOWLEDGMENTS

This work was supported by the Office of Naval Research under the grant ONR N00014-94-1-1200 with Dr. Thomas M. McKenna as the Technical Monitor. A partial support from the Missouri Department of Economic Development is gratefully acknowledged.

REFERENCES

Anderson, T.L., 1991. *Fracture Mechanics*. CRC Press, Boca Raton.

Barenblatt, G.I., 1962. "The Mathematical Theory of Equilibrium Cracks in Brittle Fracture." *Advances in Applied Mechanics*, 7: 55-129.

Birman, V., 1997. "Review of Mechanics of Shape Memory Alloy Structures." *Applied Mechanics Reviews*. In press.

Birman, V., Saravanos, D.A. and Hopkins, D.A., 1996. "Micromechanics of Composites with Shape Memory Alloy Fibers in Uniform Thermal Fields." *AIAA Journal*, 34: 1905-1912.

Boyd, J.G. and Lagoudas, D.A., 1993. "Thermomechanical Response of Shape Memory Composites," *Smart Structures and Intelligent Systems*, Hagood, N.W. and Knowles, G.J., eds., SPIE, Vol. 1917, Bellingham, Washington, pp. 774-790.

Dugdale, D.S., 1960. "Yielding of Steel Sheets Containing Slits." *Journal of the Mechanics and Physics of Solids*, 8: 100-108.

Funakubo, H., 1987. *Shape Memory Alloys*. Gordon and Bleach, New York.

Irwin, G.P., 1960. "Plastic Zone Near a Crack and Fracture Toughness," *Proceedings of the Seventh Sagamore Ordnance Materials Conference*, New York, Syracuse University, Vol. IV, pp. 63-78.

Jackson, C.M., Wagner, H.J. and Wasilewski, R.J., 1972. "55-Nitinol - The Alloy with a Memory: Its Physical Metallurgy, Properties, and Applications." NASA-SP 5110.

Lei, C.Y. and Wu, M.H., 1991. "Thermomechanical Properties of NiTi-Base Shape Memory Alloys," *Smart Structures and Materials,* Haritos, G.K. and Srinivasan, A.V., eds., ASME, AD-Vol. 24 and AMD-Vol. 123, New York, pp. 73-77.

Tanaka, K. and Sato, Y., 1986. "Analysis of Superelastic Deformations during Isothermal Martensitic Transformation." *Res Mechanica*, 17: 241-252.

White, S.R., Whitlock, M.E., Ditman, J.B. and Hebda, D.A., 1993. "Manufacturing of Adaptive Graphite/Epoxy Structures with Embedded Nitinol Wires," *Adaptive Structures and Material Systems,* Carman, G.P. and Garcia, E., eds., ASME, AD-Vol. 35, New York, pp. 71-79.

Identification of External Force Acting on a Plate by Using Neural Networks

S. LIHUA and T. BAOQI

ABSTRACT

A neural network approach is used to solve the inverse problem of load identification on a plate structure. A smart structure system composed of glass fiber-epoxy composites and piezoelectric sensors was developed. The system monitors the structural response under the external force through several sensors in the structure. Neural network models are trained to approximate the relationship between load location, load amplitude and sensor output. Numerical simulations were first carried out to investigate the validity of the method. Results show the neural network can accurately identify either the load located on the training sites or the load located between each training sites. Experiments on a composite plate was also carried out. With four bonded PZT sensors, the identification accuracy of load location on the 1m×1m plate is within 10cm.

INTRODUCTION

The ability to sense the location and strength of external force acting on a structure is required in the health monitoring smart structures. The location and strength are usually determined from the displacements or strains measured by several sensors distributed on the structure. In order to do this, a system model is perversely needed to describe the relationship of the structural response and the force, then by solving the nonlinear least-

Shi Lihua, Research Division, Nanjing Engineering Institute, No.1 Haifuxiang, Nanjing 210007, China

Tao Baoqi, Smart Material and Structure Research Institute, Nanjing Univ. of Aero. & Astro., Nanjing 210016, China

square problem an optimal estimation of the force location and magnitude is obtained(D'Cruz et al, 1990; Choi and Chang, 1994). However, when the structure shape and boundary condition is complicated, it is less possible to establish a proper analytical model of the structure for force identification. So neural network approach was proposed to approximate the complex relationship of input force and sensor output(Shaw et al, 1995).

This paper also uses the neural network approach to determined the location and amplitude of external force. The location and magnitude of the load is identified separately. A normalizing method is used to remove the affect of load strength to the localization results. A neural network model is first trained to approximate the relationship of sensor output and load location. After the load location is decided, the magnitude is then identified by a similar network using the estimated coordinate of the load location and sensor output. By using a functional link structure, the neural network offers the advantages of fast learning speed, good convergency and simple structure. Numerical simulations of static load acting on a simply supported plate is carried out. Results show the neural network can accurately identify either the load located on the training sites or the load located between each training sites. An experimental system composed of a 1m×1m clamped glass fiber-epoxy composite plate bonded with PZT sensors and connected with external data processing system was developed. Experiments show the load localization error of the system is within 10cm.

RESPONSE OF THE PIEZOELECTRIC SENSOR ON THE PLATE

Considering an elastic rectangular plate simply supported along its four edges(Figure 1), the deflection at a point S(x, y) due to the static point load at (ξ,η) with magnitude of P_0 is

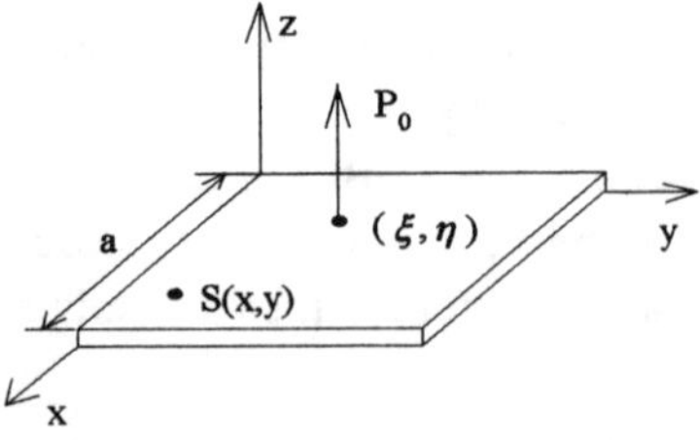

Figure 1 Plate under static load

$$w(x,y)=\frac{4P_0}{\pi^4 ab}\sum_{m=1}^{\infty}\sum_{n=1}^{\infty}\frac{\sin(m\pi\xi/a)\sin(n\pi\eta/b)\sin(m\pi x/a)\sin(n\pi y/b)}{D_1(m/a)^4+2D_3(m/a)^2(n/b)^2+D_2(n/b)^4} \tag{1}$$

where an orthogonal nonisotropic plate model is used, and

$$\begin{cases} D_1=\dfrac{E_1h^3}{12(1-\mu_{12}\mu_{21})} \\ D_2=\dfrac{E_2h^3}{12(1-\mu_{12}\mu_{21})} \\ D_3=D_1\mu_{21}+\dfrac{G_{12}h^3}{6} \end{cases}$$

E_1, E_2 is the Young's modulus in the direction of x and y, μ_{ij} is the Poisson ratio and G_{12} is the shear modulus. If a thin piezoelectric patch is bonded to point S, then output charge of the piezoelectric sensor is

$$q(t)=\iint_A (e_{31}S_1+e_{32}S_2+e_{33}S_3)dxdy \tag{2}$$

where A is the area covered by the electrode, $S_i(i=1,2,3)$ is strain, $e_{3i}(i=1,2,3)$ is the piezoelectric constant of piezoelectric materials. By substituting the strain-displacement relationship of elastic plate into equation (2), we have

$$q(t)=\frac{h}{2}\iint_A (e_{31}\frac{\partial^2 w}{\partial x^2}+e_{32}\frac{\partial^2 w}{\partial y^2})dxdy \tag{3}$$

where, h is the thickness of the plate. For simplicity, edge length of plate is assumed to be unity(a=b=1), sensor is assumed to be square with the coordinate of its four corner at (x_1,y_1), (x_1+c,y_1), (x_1+c,y_1+c) and(x_1,y_1+c). By substituting equation (1) into equation (3) and approximating infinite series with finite term, equation (3) becomes

$$q(x_1,y_1)=\frac{2P_0h}{\pi^4}\sum_{m=1}^{M}\sum_{n=1}^{N}\frac{\sin(m\pi\xi)\sin(n\pi\eta)A_{mn}(x_1,y_1)}{D_{mn}} \tag{4}$$

where

$$D_{mn} = D_1 m^4 + 2D_3 m^2 n^2 + D_2 n^4$$

$$A_{mn}(x_1, y_1) = \left\{ \frac{m}{n} e_{31} + \frac{n}{m} e_{32} \right\} \{\cos[m\pi(x_1 + c)] - \cos(m\pi x_1)\}$$
$$\{\cos[n\pi(y_1 + c)] - \cos(n\pi y_1)\}$$

Equation (4) is the response of piezoelectric sensor to the static point load acting on the plate.

IDENTIFICATION OF FORCE LOCATION AND MAGNITUDE

To solve the load location (ξ, η) and strength P_0 from sensor output q, at least three sensors at different sites are needed. Although we have equation (4) and it can be solved by least square estimation, when the problem is not for a simply supported square plate we assumed, the relationship of load location, load strength and sensor output is not so directly that can be expressed by a simple analytical equation. So a neural network is used to approximate the complicated relationship between load and sensor output. From the observing results of direct problem

$$q_i(x_i, y_i) = g(\xi, \eta, P_0) \tag{5}$$

a neural network is trained to solve the inverse problem

$$(\xi, \eta, P_0) = g^{-1}(q_i) \quad 6)$$

where g() is the relationship between load and sensor output, $g^{-1}()$ is the function that maps sensor output to load location and strength. Neural network is a kind of nonlinear approximation tool that fits for this problem.

The commonly used back-propagation (BP) neural network uses hidden layers to realize the nonlinear mapping function. However, the optimal number of hidden layer and the number of neuron in each hidden layer is difficult to define. Here we use a functional link neural network structure which needs no hidden layers to identify the load location from the output of sensors. Compared with BP network which realize complicated nonlinear mapping by hidden layers, the functional link net realize complicated mapping by an enhancement function Φ. Figure 2 is the structure of a functional link network.

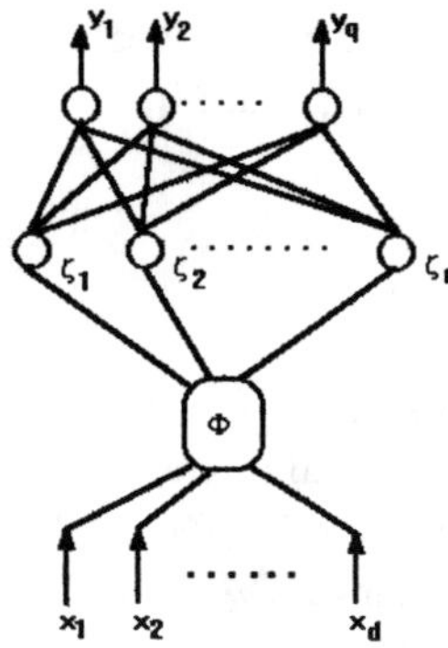

Figure 2 Functional link neural network

where $\mathbf{X}=[x_1\ x_2\ x_3\ \ldots\ x_d]$, is the input vector, $\Phi(\mathbf{X})$ is a nonlinear functional link, whose function is to transform a input vector **X** with lower dimension d to ζ with higher dimension r(r>d). Here, considering the physical model, we select a function spread model for $\Phi(\mathbf{X})$, with the transformation of $\Phi(\ )$, the enhanced vector ζ is set to be

$$\zeta= [\mathbf{X}\ \ \sin(\pi\mathbf{X})\ \cos(\pi\mathbf{X})\ \sin(2\pi\mathbf{X})\ \cos(2\pi\mathbf{X})\ \sin(3\pi\mathbf{X})\ \cos(3\pi\mathbf{X})]$$

The output of the network is

$$\mathbf{Y}=f(\mathbf{W}\zeta+\theta) \tag{7}$$

where **W** and θ is the weight and threshold to be learned through training, f( ) is the sigmoid function. Because the functional link network has no hidden layer, it has some advantages over the BP network, for example, it has fast learning speed and has no local minimum problem which is often encountered in the training of BP network(Pao, 1989).

In load identification, to improve the identification accuracy, load location and load strength is identified separately at two steps: first, peak values of four piezoelectric sensors are measured, they are denoted as q_1 q_2 q_3 and q_4. By dividing each with q1, the affect of strength is removed. The input to the neural network is $q_{21}=q_2/q_1$, $q_{31}=q_3/q_1$ and $q_{41}=q_4/q_1$. A neural network is trained to approximate the relationship between measured q_{i1}(i=2,3,4) and load location(ξ,η) by training data acquired through experiment. Then, after the location is decided, the load strength is estimated by similar methods using (ξ,η) and q_1.

NUMERICAL SIMULATION

A numerical simulation was carried out to investigate the validity of the proposed method. The model parameters are selected to be: a=b=1m; $\mu_{12} = \mu_{21} = 0.14$; h=0.005m; G=3.79GPa; $E_1=E_2=21.63$GPa, $e_{31}=54\times10^{-3}$ c/m^2, $e_{32}=24\times10^{-3}$c/m^2. Four sensors were arranged symmetrically near the corners of the plate(Figure 3). Sensor output when 1N load acting on 25 nodes of the grid shown in Figure 3 was simulated separately. The result are shown in Table 1.

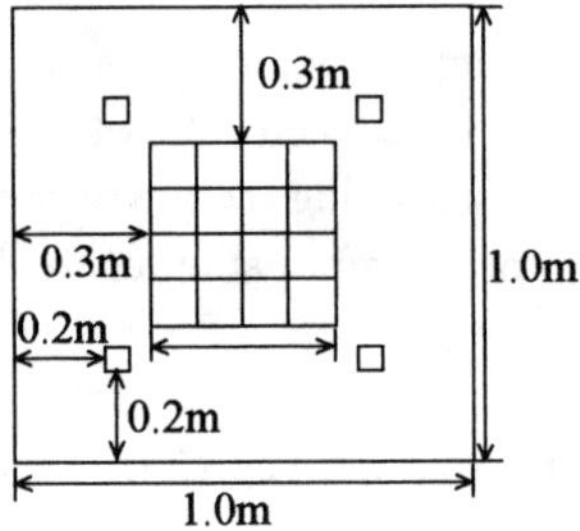

Figure 3 Geometry of the simulated structure

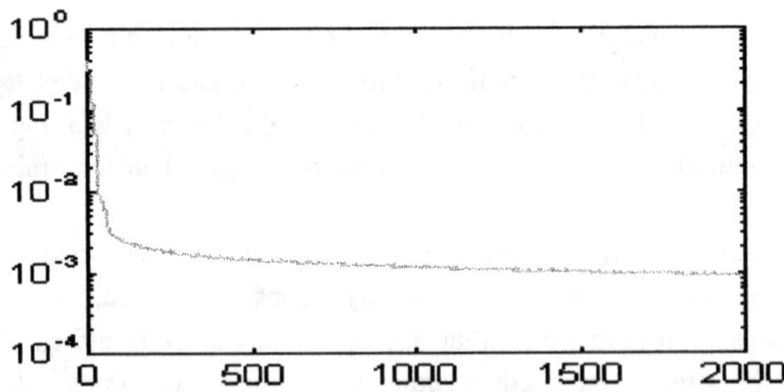

Figure 4 Error curve of the functional link network

A functional link neural network model for identification of load location is programmed in MATLAB environment. It is trained by these simulated results. Figure 4 is the error curve of the network with iteration number. It can be seen that the ultimate training error is controlled within 0.001. After training, the neural network was tested by four new simulated load. Table 2 is the identification results of the load locations. Although these new locations are different from the training data, the neural network can identify them accurately.

TABLE 1 SENSOR OUTPUT OF 25 LOAD LOCATIONS

No.	Location (ξ,η)		Normalized sensor output (q1, q2, q3, q4)			
1	0.30	0.30	0.9012	0.1971	0.1199	0.0726
2	0.30	0.40	0.6647	0.3056	0.1367	0.1075
3	0.30	0.50	0.4580	0.4580	0.1327	0.1327
4	0.30	0.60	0.3056	0.6647	0.1075	0.1367
5	0.30	0.70	0.1971	0.9012	0.0726	0.1199
6	0.40	0.30	0.4803	0.2124	0.1637	0.1351
7	0.40	0.40	0.4974	0.3077	0.2005	0.1890
8	0.40	0.50	0.4105	0.4105	0.2115	0.2115
9	0.40	0.60	0.3077	0.4974	0.1890	0.2005
10	0.40	0.70	0.2124	0.4803	0.1351	0.1637
11	0.50	0.30	0.2639	0.1988	0.1988	0.2639
12	0.50	0.40	0.3149	0.2640	0.2640	0.3149
13	0.50	0.50	0.3122	0.3122	0.3122	0.3122
14	0.50	0.60	0.2640	0.3149	0.3149	0.2640
15	0.50	0.70	0.1988	0.2639	0.2639	0.1988
16	0.60	0.30	0.1351	0.1637	0.2124	0.4803
17	0.60	0.40	0.1890	0.2005	0.3077	0.4974
18	0.60	0.50	0.2115	0.2115	0.4105	0.4105
19	0.60	0.60	0.2005	0.1890	0.4974	0.3077
20	0.60	0.70	0.1637	0.1351	0.4803	0.2124
21	0.70	0.30	0.0726	0.1199	0.1971	0.9012
22	0.70	0.40	0.1075	0.1367	0.3056	0.6647
23	0.70	0.50	0.1327	0.1327	0.4580	0.4580
24	0.70	0.60	0.1367	0.1075	0.6647	0.3056
25	0.70	0.70	0.1199	0.0726	0.9012	0.1971

TABLE 2 IDENTIFICATION RESULTS

No.	Actual location	Sensor output	Identified location
1	(0.50,0.45)	(0.3217 0.2912 0.2912 0.3217)	(0.49, 0.44)
2	(0.41,0.40)	(0.4767 0.3047 0.2072 0.1998)	(0.41, 0.36)
3	(0.39,0.49)	(0.4283 0.4077 0.2032 0.2019)	(0.43, 0.49)
4	(0.51,0.35)	(0.2788 0.2283 0.2372 0.3091)	(0.51, 0.36)

EXPERIMENTAL RESULTS

Experiments were also carried out on a 1.0m×1.0m clamped fiber-epoxy plate. Four PZT sensors were bonded to the plate on the sites similar to Figure 3. Sensor output is passed to multi-channel charge amplifier and then sampled by acquisition board in the computer. A computer program is developed to interface MATLAB with C^{++} for identification. Peak values of sensor output at 9 points were used to train a functional link neural network. Figure 5 gives the identification result of 30 load locations with the trained neural network. Where, '*' is the actual load location, 'o' is the identified location by the network. The average distance error is within 10cm. Figure 6 gives the identified load magnitude at point(0.5,0,5), the average error is within 20%.

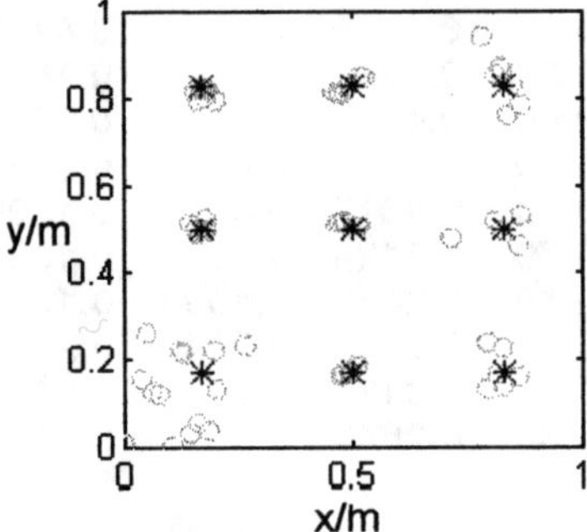

Figure 5 Identification results of load location

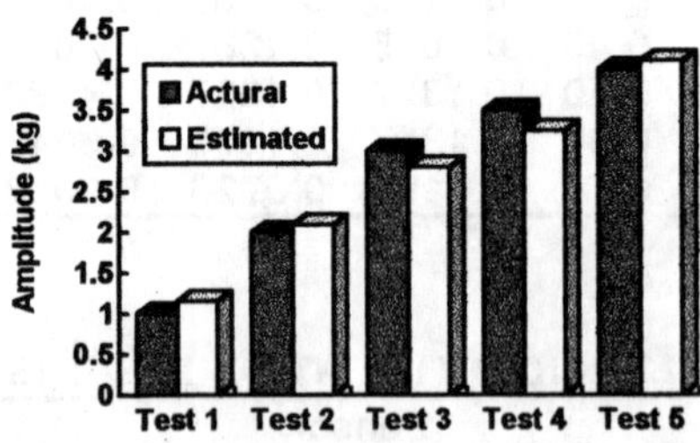

Figure 6 Identification results of load magnitude

CONCLUSION

A load identification approach using neural network and piezoelectric sensors was investigated. Numerical simulation and experiments all verify the validity of the proposed method. This paper only discussed the application of this method to static point load identification on square plate. This method can be directly used to other structures with different shape and boundary conditions. In our experiment, the applied force on the plate is not restricted to static load, for the dynamic load which does not change very quickly, this method also have good results.

REFERENCE

1. Choi, K. and Chang F.K., 1994. "Identification of Foreign Object Impact in Structures Using Distributed Sensors," J. of Intelligent Material Systems and Structures, 5(6):864-869

2. D'Cruz J, Crisp J.D.C., 1991. "Determining a Force Acting on a Plate-an Inverse Problem," J. AIAA, 29(3):464-470

3. Lee C.K. 1990. "Theory of Laminated Piezoelectric Plates for the Design of Distributed Sensors/Actuators. Part I: Governing Equations and Reciprocal Relationships," J. of Acoust. Soc. Am, 87(3): 1144-1158

4. Pao Y.H. 1989. Adaptive Pattern Recognition and Neural Networks. Addison-Wesley Publishing Company, Inc.

5. Shaw, J.K., Sirkis, J.S.,Friebel, E.J., etal, 1995. "Model of Transverse Plate Impact Dynamics for Design of Impact Detection Methodologies," J. AIAA, 33(7):1327-1334

SESSION 5

SYSTEM INTEGRATION

A Computer Toolbox for Damage Identification Based on Changes in Vibration Characteristics

S. W. DOEBLING,[1] C. R. FARRAR[2] and P. J. CORNWELL[3]

ABSTRACT

This paper introduces a new toolbox of graphical-interface software algorithms for the numerical simulation of vibration tests, analysis of modal data, finite element model correlation, and the comparison of both linear and nonlinear damage identification techniques. This toolbox is unique because it contains several different vibration-based damage identification algorithms, categorized as those which use only measured response and sensor location information ("non-model-based" techniques) and those which use finite element model correlation ("model-based" techniques). Another unique feature of this toolbox is the wide range of algorithms for experimental modal analysis. The toolbox also contains a unique capability that utilizes the measured coherence functions and Monte Carlo analysis to perform statistical uncertainty analysis on the modal parameters and damage identification results. This paper contains a detailed description of the various modal analysis, damage identification, and model correlation capabilities of toolbox, and also shows a sample application which uses the toolbox to analyze the statistical uncertainties on the results of a series of modal tests performed on a highway bridge.

INTRODUCTION

This paper introduces a new suite of graphical-interface software algorithms to numerically simulate vibration tests and to apply various modal analysis, damage identification, and finite element model refinement techniques to measured or simulated modal vibration data. This toolbox is known as DIAMOND (Damage Identification And MOdal aNalysis for Dummies), and has been developed at Los Alamos National Laboratory over the last year. DIAMOND is written in MATLAB

[1]Technical Staff Member, Engineering Sciences and Applications Division, Engineering Analysis Group (ESA-EA), M/S P946, (505) 667-6950, doebling@lanl.gov

[2]Materials Behavior Team Leader, Engineering Sciences and Applications Division, Engineering Analysis Group (ESA-EA)

[3]Associate Professor, Dept. of Mechanical Engineering

[1], a numerical matrix math application which is available on all major computer platforms. DIAMOND is unique in three primary ways:

1. DIAMOND contains several of the most widely used modal curve-fitting algorithms. Thus the user may analyze the data using more than one technique and compare the results directly. This modal identification capability is coupled with a numerical test-simulation capability that allows the user to directly explore the effects of various test conditions on the identified modal parameters.

2. The damage identification and finite element model refinement modules are graphically interactive, so the operation is intuitive and the results are displayed visually as well as numerically. This feature allows the user to easily interpret the results in terms of structural damage.

3. DIAMOND has statistical analysis capability built into all three major analysis modules: modal analysis, damage identification, and finite element model refinement. The statistical analysis capability allows the user to determine the magnitude of the uncertainties associated with the results. No other software package for modal analysis or damage identification has this capability.

The development of DIAMOND was motivated primarily by the lack of graphical implementation of modern damage identification and finite element model refinement algorithms. Also, the desire to have a variety of modal curve-fitting techniques available and the capability to generate numerical data with which to compare the results of each technique was a motivating factor. The authors are unaware of any commercial software package that integrates all of these features.

This paper is organized as follows: An overview of DIAMOND is provided, including an outline of each module (except for the numerical test simulator): experimental modal analysis and statistical analysis of modal data, damage identification, and finite element model updating. In each section, a flowchart of the menu structure of DIAMOND is presented. Following the overview section is an example of the application of DIAMOND to vibration data obtained from an actual highway bridge. This section contains a description of the testbed, data acquisition equipment, and testing procedure. It also contains a sample experimental modal analysis, complete with statistical analysis of the variability of the results with respect to the variations inherent in the data acquisition process as well as variations resulting from changes in the environmental conditions of the bridge.

OVERVIEW OF DIAMOND

DIAMOND is divided into four primary modules at the top level: numerical vibration test simulator, experimental modal curve fitting and statistical analysis, damage identification, and finite element model refinement. These four modules constitute the primary hierarchy in DIAMOND, as shown in Figure 1. In this paper, the three analysis-oriented modules (all but the test simulator) will be discussed in further detail.

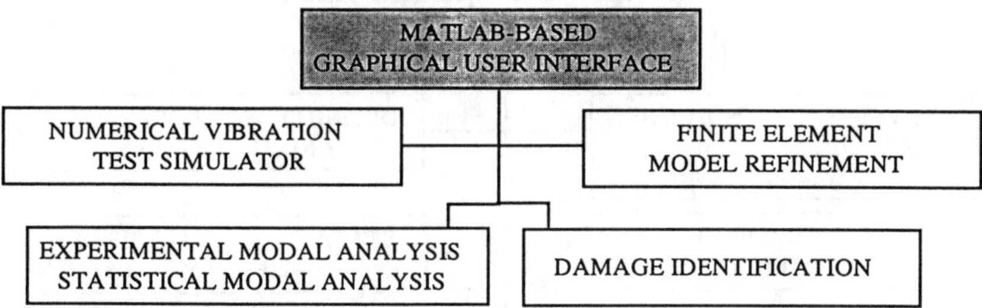

Figure 1: Flowchart of Top Level of DIAMOND

Experimental Modal Analysis / Statistical Analysis of Modal Data

The experimental modal analysis module provides a series of tools for plotting the data in various forms, plotting data indicator functions, defining sensor geometry, performing modal curve-fits, analyzing the results of modal curve fits, and analyzing the variance of identified modal parameters as a function of the noise in the measurements as defined by the measured coherence function. A flowchart of this module is shown in Figure 2.

The most important feature of the experimental modal analysis module is the variety of modal parameter identification algorithms which are available. These include:

1. Operating shapes (which is simply "peak picking," or "slicing" the FRF matrix at a particular frequency bin)

2. Eigensystem Realization Algorithm (ERA), [2] which is a low-order time domain modal parameter estimation algorithm.

3. Complex exponential algorithm, which is a high-order time domain modal parameter estimation algorithm. The specific algorithm implemented is the Polyreference Time Domain [3] approach.

4. Rational Polynomial Curve fit [4], which is a high-order frequency domain technique that uses orthogonal polynomials to estimate the coefficients of a rational polynomial representation of the frequency response function.

5. Nonlinear least squares fit, which uses a Levenberg-Marquardt nonlinear least-squares curve fitting routine [5] to estimate modal frequencies and modal damping ratios from the unfiltered Fourier spectral responses of a base-excited structure.

Any of these modal identification algorithms can be implemented in a statistical Monte Carlo [6] technique. In such an analysis, a series of perturbed data sets, based on the statistics of the measured FRFs as defined by the measured coherence functions, are generated and propagated through the selected algorithm. The statistics

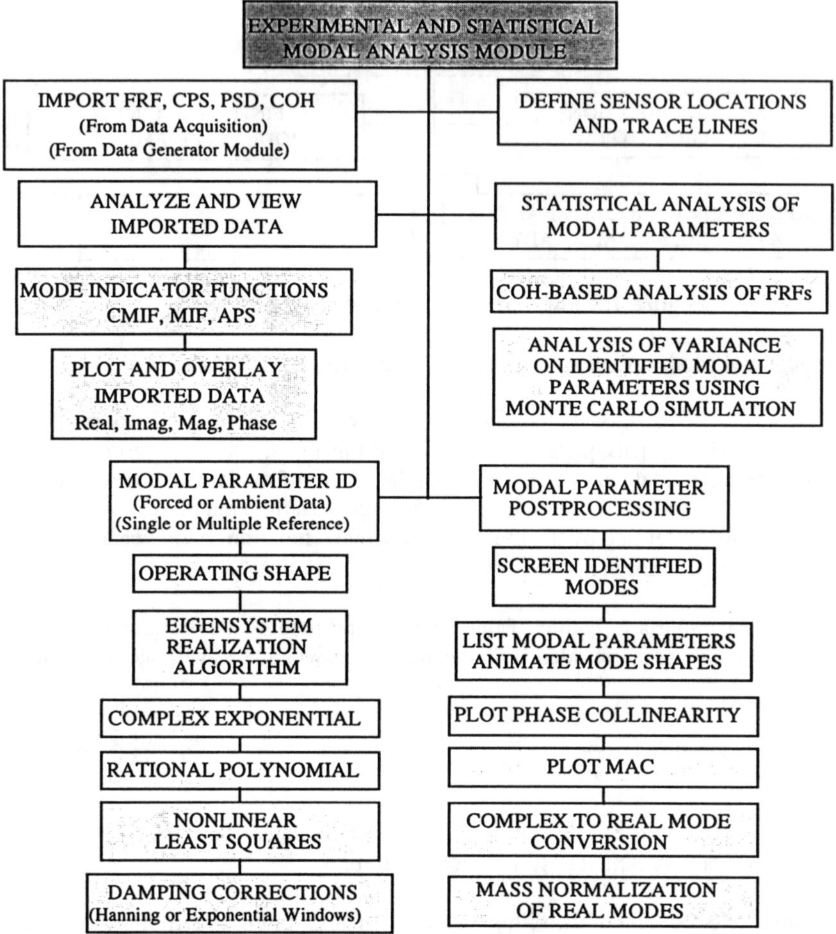

Figure 2: Flowchart of Experimental Modal Analysis / Statistical Modal Analysis Module

on the results are then used as uncertainty bounds on the identified modal parameters.

Damage Identification

The algorithms contained in the damage identification module of DIAMOND can be classified as modal-based, finite element refinement-based, or nonlinear. A flowchart of the damage identification module is shown in Figure 3.

The damage identification module presents a number of different algorithms:

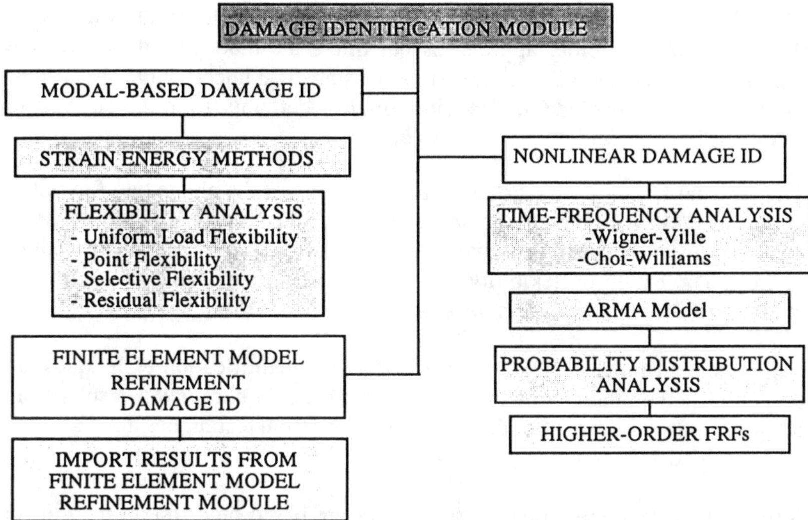

Figure 3: Flowchart of Damage Identification Module

1. Strain energy methods are based on the work of Stubbs [7], Cornwell [8], and others. The basic idea of these methods is the division of the structure into a series of beam or plate-like elements, and then the estimation of the strain energy stored in each element both before and after damage. The curvatures (second-derivatives with respect to space) of the mode shapes are used to approximate the strain energy content.

2. Flexibility methods all use some measure of the change in the modal flexibility matrix, estimated from the mass-normalized measured mode shapes, $[\Phi]$, and modal frequencies squared, $[\Lambda]$, as

$$[G] \approx [\Phi][\Lambda]^{-1}[\Phi]^T \tag{1}$$

The modal flexibility matrix is used to estimate the static displacements that the structure would undergo as a result of a specified loading pattern. The uniform load flexibility method [9] involves specifying a unit load at all measurement degrees of freedom (DOF), then comparing the change in the resulting displacement pattern before and after damage. The point flexibility method [10] specifies the application of a unit load at each measurement DOF one at a time, then looking for a change in the resulting displacements at the same point before and after damage.

The selective flexibility method, which is still under development, uses one of the above two flexibility approaches but filters the modes used to form the flexibility matrix according to their relative statistical uncertainty. The idea of this method is to exclude modes with a high uncertainty from the analysis to avoid biasing the results.

The residual flexibility method [11] also uses one of the above two flexibility approaches but includes the estimate of the residual flexibility, which is the contribution to the flexibility matrix from the modes above the bandwidth of interest. The resulting flexibility matrix is a closer approximation to the true static flexibility matrix than is the modal flexibility matrix.

3. Finite element model correlation-based damage identification techniques are based on the comparison of the finite element model correlation results from before damage to those after damage. The correlation techniques are discussed in the next section.

4. Nonlinear damage identification techniques are based on different theories of nonlinear signal processing. They are a widely varying group of methods and are reviewed and discussed in Ref. [12].

Finite Element Model Refinement

The finite element model refinement module consists of four options: pre-processing for update analysis, optimal matrix updating, sensitivity-based model update, and post-processing of update results. A flowchart of this module is shown in Figure 4.

The pre-processing phase of the model correlation analysis involves the selection of which modal parameters (i.e. modal frequencies and mode shapes) should be used in the correlation, as well as which finite element model parameters should be updated.

The optimal matrix update methods are based on the minimization of the error in the structural eigenproblem using a closed-form, direct solution. The minimum rank perturbation technique (MRPT) [13] is one such method which produces a minimum-rank perturbation of the structural stiffness, damping, and/or mass matrices reduced to the measurement degrees of freedom. The minimum rank element update (MREU) [14] is a similar technique which produces perturbations at the elemental, rather than the matrix, level. The Baruch updating technique [15] minimizes an error function of the eigenequation using a closed-form function of the mass and stiffness matrices.

The sensitivity-based model update methods also seek to minimize the error in the structural eigenequation, but do so using a Newton-Raphson-type technique based on solving for the perturbations such that the gradient of the error function is

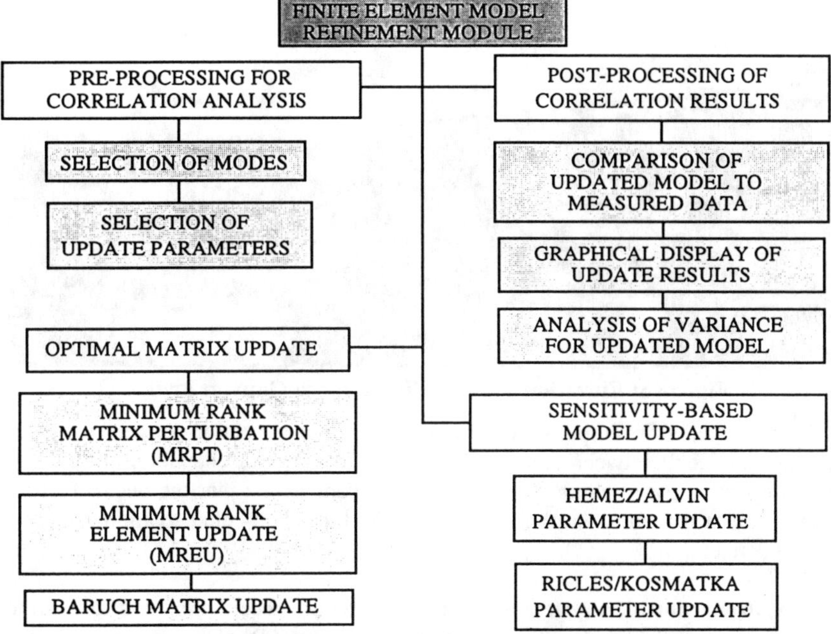

Figure 4: Flowchart of Finite Element Model Refinement Module

near zero. [6] Thus these methods require the computation of the sensitivity of the structural eigenproblem to the parameters which are to be updated. The Hemez/Alvin algorithm [16],[17] computes the sensitivities at the elemental level, then assembles them to produce the global sensitivity matrices. The Ricles/Kosmatka [18] algorithm computes a "hybrid" sensitivity matrix using both analytical and experimental sensitivities.

EXAMPLE APPLICATION: THE ALAMOSA CANYON BRIDGE

The following analysis of modal data from a series of tests performed on a highway bridge is intended to demonstrate the application of the unique capabilities of DIAMOND to data from an actual field test.

The Alamosa Canyon Bridge has seven independent spans with a common pier between successive spans. An elevation view of the bridge is shown in Figure 5. The bridge is located on a seldom-used frontage road parallel to Interstate 25 about 10 miles North of the town of Truth or Consequences, New Mexico, USA. Each span consists of a concrete deck supported by six W30x116 steel girders. The roadway in each span is approximately 7.3 m (24 ft.) wide and 15.2 (50 ft.) long. Integrally attached to the concrete deck is a concrete curb and concrete guard rail. Inspection of the bridge showed that the upper flanges of the beams are imbedded

Figure 5: Elevation View of the Alamosa Canyon Bridge

in the concrete, which implies the possibility of composite action between the girders and the deck. Between adjacent beams are four sets of channel section cross braces equally spaced along the length of the span. At the pier the beams rest on rollers, and at the abutment the beams are bolted to a half-roller to approximate a pinned connection.

The data acquisition system was set up to measure acceleration and force time histories and to calculate frequency response functions (FRFs), power spectral densities (PSDs), cross-power spectra and coherence functions. A modal sledge hammer was used as the impact excitation source. Accelerometers were used for the vibration measurements. More details regarding the instrumentation can be found in Ref. [19].

A total of 31 acceleration measurements were made on the concrete deck and on the girders below the bridge as shown in Figure 6. Five accelerometers were

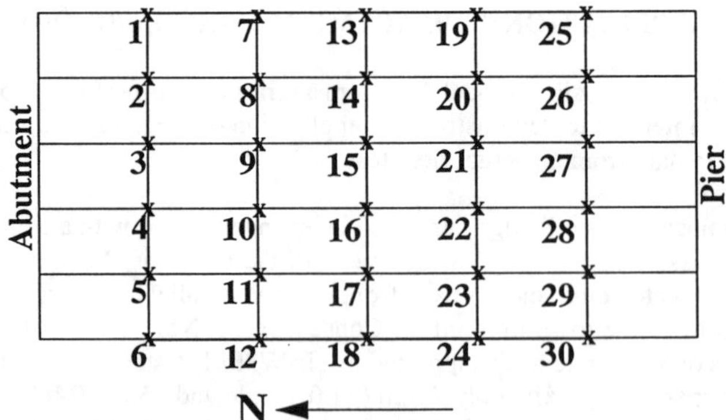

Figure 6: Accelerometer and Impact Locations

spaced along the length of each girder. Because of the limited number of data channels, measurements were not made on the girders at the abutment or at the pier. Two excitations points were located on the top of the concrete deck. Point 2 was used as the primary excitation location. Point 23 was used to perform a reciprocity check. The force-input and acceleration-response time histories obtained from each impact were subsequently transformed into the frequency domain so that estimates of the PSDs, FRFs, and coherence functions could be calculated. Thirty averages were typically used for these estimates. With the sampling parameters listed above and the overload reject specified, data acquisition for a specific test usually occurred over a time period of approximately 30 - 45 minutes. All of the results in this paper are from measurements made on span 1 of the bridge, which is located at the far North end.

A total of 52 data sets were collected over the course of the six days of testing. Temperature measurements were made at 9 locations around the bridge, both above and below the deck, to track the effects of ambient temperature changes on the test results. Reciprocity and linearity checks were conducted first. A series of modal tests was conducted over a 24 hour period (one test every 2 hours) to assess the change in modal properties as a result of variations in ambient environmental conditions, as discussed in Ref. [19]. A series of tests with various levels of attempted damage was also conducted, but the permitted alterations in the bridge did not cause a significant change in the measured modal properties. Specifically, the nuts on the bolted connections that hold the channel-section cross members to the girders, as shown in Figure 7 were removed. However the bolts could not be loosened sufficiently, and no relative motion could be induced at the interface under the loading of the modal excitation. For this reason, the damage cases presented in this paper are results from simulated stiffness reduction using a correlated FEM. The identified modal frequencies and mode shapes from the modal analysis of one of these data sets are shown in Figure 8.

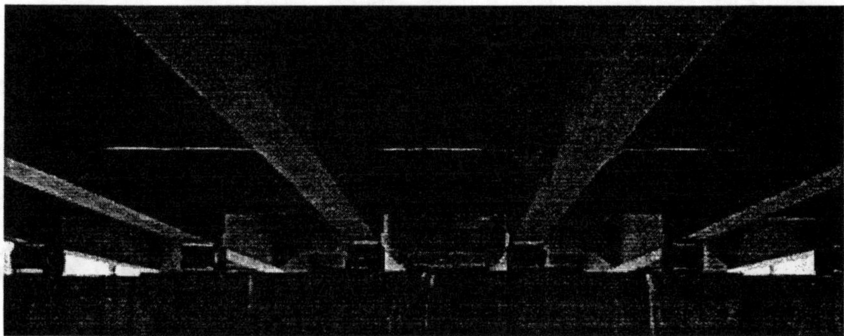

Figure 7: Bolted Connections of Cross-Members to Girders

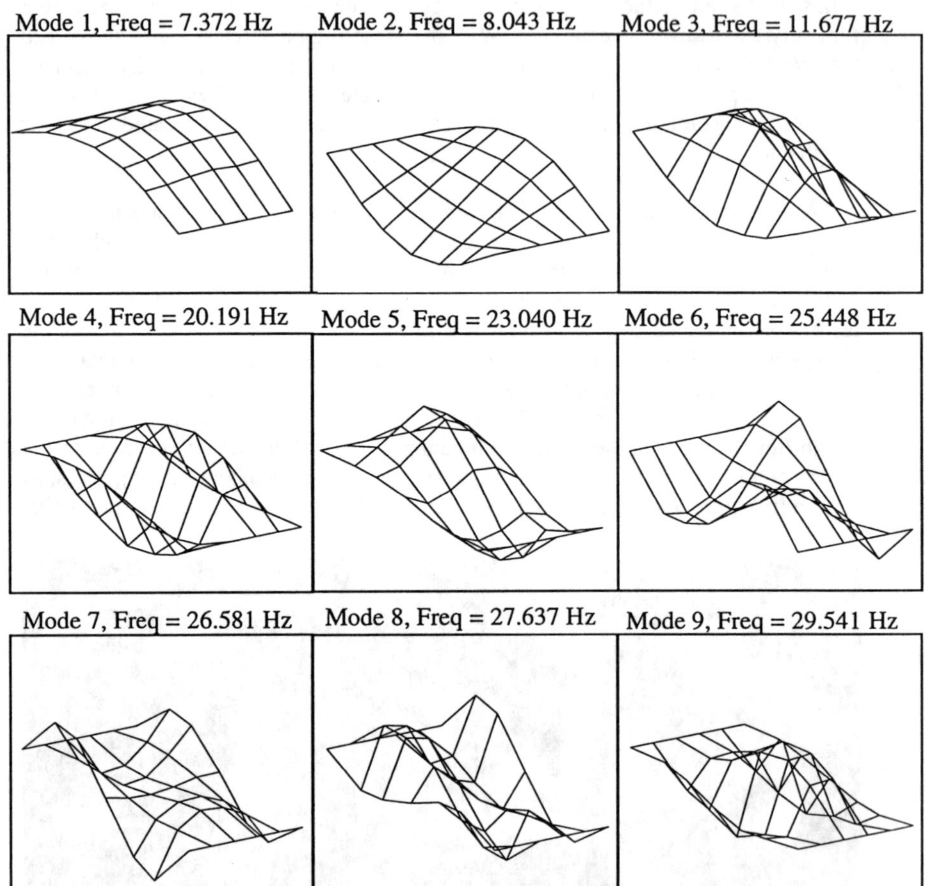

Figure 8: Identified Mode Shapes for Alamosa Canyon Bridge

Analysis of Uncertainty in Each Test

Statistical uncertainty bounds on the measured frequency response function magnitude and phase were computed from the measured coherence functions, assuming that the errors were distributed in a Gaussian manner, according to the following formulas from Bendat and Piersol [20]:

$$\sigma(|H(\omega)|) = \frac{\sqrt{1-\gamma^2(\omega)}}{|\gamma(\omega)|\sqrt{2n_d}}|H(\omega)|$$
$$\sigma(\angle H(\omega)) = \frac{\sqrt{1-\gamma^2(\omega)}}{|\gamma(\omega)|\sqrt{2n_d}}\angle H(\omega) \quad (2)$$

where $|H(\omega)|$ and $\angle H(\omega)$ are the magnitude and phase angle of the measured FRF, respectively, $\gamma^2(\omega)$ is the coherence function, n_d is the number of measurement averages, and $\sigma(\bullet)$ is the value of 1 standard deviation (68% uncertainty bound). These uncertainty bounds represent a statistical distribution of the FRF based on a realistic level of random noise on the measurement. Once the 1 standard deviation (68% uncertainty) bounds were known, 2 standard deviation (95% uncertainty) bounds were computed. Statistical uncertainty bounds on the identified modal parameters (frequencies, damping ratios, and mode shapes) were estimated using the uncertainty bounds on the FRFs via a Monte Carlo analysis [6]. The details of this procedure are shown in Ref. [21].

Effects of Uncertainties on Damage Identification

The changes in the bridge as a result of damage were predicted using the finite element model. The damage case modeled was the 100% failure of a connection between a cross-member and an interior girder. A comparison of the estimated 95% confidence bounds and the predicted changes as a result of damage for the modal frequencies are shown in Figure 9. The modal frequencies of modes 3, 4, 7, 8, and 9 undergo a change that is significantly larger than the corresponding 95% confidence bounds. The relative magnitudes of the changes indicate that the frequency changes of these modes could be used with confidence in a damage identification analysis. It should be noted from the y-axis scale of Figure 9 that the overall changes in frequency as a result of damage are quite small (< 1.2%), but as a consequence of the extremely low uncertainty bounds on the modal frequencies (many less than 0.2%), these small changes can be considered to be statistically significant.

The relative statistical significance of the changes in the various modes is one of the primary motivating factors for a "selective flexibility" approach. Using the selective flexibility approach, those modes where the frequency changes are statistically insignificant would be considered to be unchanged, while those modes with significant frequency change would be used in the flexibility analysis. This technique would prevent modes which are insensitive to the damage from masking the indications of damage from modes which are sensitive to the damage.

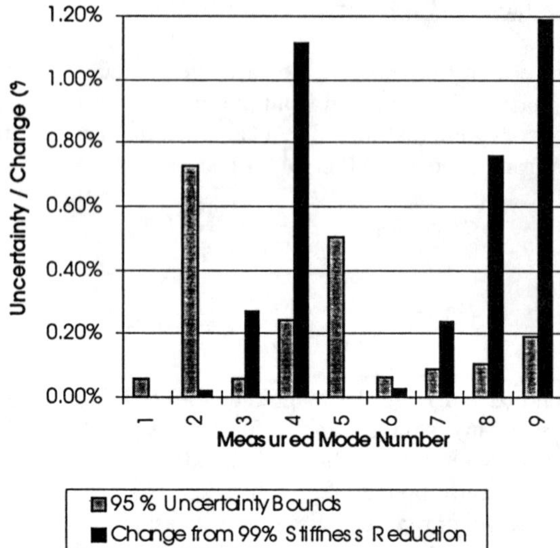

Figure 9: Comparison of Modal Frequency 95% Confidence Bounds to Changes Predicted as a Result of Damage

CONCLUSION

A new toolbox of graphical-interface software algorithms, known as DIAMOND, has been introduced and demonstrated. The toolbox provides the capability to simulate vibration tests, perform experimental modal analysis including statistical bounds, apply various damage identification techniques, and implement finite element refinement algorithms. The structure of the toolbox menus was described in detail in this paper, and a sample application to measured data from a highway bridge was presented.

ACKNOWLEDGMENTS

This work was supported by Los Alamos National Laboratory Directed Research and Development Project #95002, under the auspices of the United States Department of Energy. The authors wish to recognize the contributions of Mr. Erik G. Straser and Mr. A. Alex Barron of Stanford University.

REFERENCES

1. MATLAB, *Users Manual*, The Mathworks, Inc., 1994.

2. Juang, J.N. and Pappa, R.S., "An Eigensystem Realization Algorithm for Modal Parameter Identification and Model Reduction," *Journal of Guidance, Control and Dynamics*, Vol. 8, No. 5, pp. 620-627, 1985.

3. Vold, H., Kundrat, J., Rocklin, T., and Russell, R., "A Multi-Input Modal Estimation Algorithm for Mini-Computers," SAE Transactions, Vol. 91, No. 1, January 1982, pp. 815-821.

4. Richardson, M.H. and Formenti, D.L., "Parameter Estimation from Frequency Response Measurements using Rational Fraction Polynomials," *Structural Measurement Systems Technical Note 85-3*, 1985.

5. McVerry, G.H., "Structural Identification in the Frequency Domain from Earthquake Records," Earthquake Engineering and Structural Dynamics, Vol. 8, 1980, pp.161-180.

6. Press, W.H., Teukolsky, S.A., Vetterling, W.T., and Flannery, B.P., *Numerical Recipes in FORTRAN: The Art of Scientific Computing*, Second Edition, Cambridge Univ. Press, 1992, pp. 684-686.

7. Stubbs, N., J.-T. Kim, and C.R. Farrar, 1995, "Field Verification of a Nondestructive Damage Localization and Severity Estimation Algorithm," in *Proc. 13th International Modal Analysis Conference*, 210–218.

8. Cornwell, P., Doebling, S.W., and Farrar, C.R., "Application of the Strain Energy Damage Detection Method to Plate-Like Structures," to appear in *Proc. of the 15th International Modal Analysis Conference*, Orlando, FL, February, 1997.

9. Catbas, F.N., Lenett, M., Brown, D.L., Doebling, S.W., Farrar, C.R., and Turer, A., "Modal Analysis of Multi-Reference Impact Test Data for Steel Stringer Bridges," to appear in *Proc. of the 15th International Modal Analysis Conference*, Orlando, FL, February, 1997.

10. Robinson, N.A., L.D. Peterson, G.H. James, and S.W. Doebling, 1996, "Damage Detection in Aircraft Structures Using Dynamically Measured Static Flexibility Matrices," in *Proc. of the 14th International Modal Analysis Conference*, 857–865.

11. Doebling, S.W., Peterson, L.D., and Alvin, K.F., "Estimation of Reciprocal Residual Flexibility from Experimental Modal Data," AIAA Journal, Vol. 34, No. 8, pp. 1678–1685, August 1996.

12. Doebling, S.W., Farrar, C.R., Prime, M.B., and Shevitz, D.W., "Damage Identification and Health Monitoring of Structural and Mechanical Systems From Changes in Their Vibration Characteristics: A Literature Review," Los Alamos National Laboratory report LA-13070-MS.

13. Zimmerman, D.C. and M. Kaouk, 1994, "Structural Damage Detection Using a Minimum Rank Update Theory," *Journal of Vibration and Acoustics,* **116**, 222–230.

14. Doebling, S.W., "Damage Detection and Model Refinement Using Elemental Stiffness Perturbations with Constrained Connectivity," in *Proc. of the AIAA/ASME/AHS Adaptive Structures Forum*, pp. 360–370, AIAA-96-1307, 1996. Accepted and to appear in *AIAA Journal.*

15. Baruch, M., 1982, "Optimal Correction of Mass and Stiffness Matrices Using Measured Modes," *AIAA Journal*, **20**(11), 1623–1626.

16. Hemez, F.M. and C. Farhat, 1995, "Structural Damage Detection via a Finite Element Model Updating Methodology," *Modal Analysis: The International Journal of Analytical and Experimental Modal Analysis,* **10** (3), 152–166.

17. Alvin, K.F., 1996, "Finite Element Model Update via Bayesian Estimation and Minimization of Dynamic Residuals," in *Proc. of the 14th International Modal Analysis Conference*, Dearborn, MI, February 1996, 561–567.

18. Ricles, J.M. and J.B. Kosmatka, 1992, "Damage Detection in Elastic Structures Using Vibratory Residual Forces and Weighted Sensitivity," *AIAA Journal*, **30**, 2310–2316.

19. Farrar, C.R., Doebling, S.W., Cornwell, P.J., and Straser, E.G., "Variability Of Modal Parameters Measured On The Alamosa Canyon Bridge," in *Proc. of the 15th International Modal Analysis Conference*, Orlando, FL, February 1997. 257-263.

20. Bendat, J.S. and Piersol, A.G., *Engineering Applications of Correlation and Spectral Analysis*, John Wiley and Sons, New York, 1980, p. 274.

21. Doebling, S.W., Farrar, C.R., and Goodman, R.S., "Effects of Measurement Statistics on the Detection of Damage in the Alamosa Canyon Bridge," in *Proc. of the 15th International Modal Analysis Conference*, Orlando, FL, February 1997, 919-929.

Structural Control by Integrating Neural Network and Experimental Design

S. M. YANG and G. S. LEE

ABSTRACT

This paper developed a structural control design methodology so that the optimal network design parameters can be determined in a systematic way thereby avoiding the lengthy trial-and-error. The methodology combines the experimental design of quality engineering and the back-propagation neural network for their advantages in implementation feasibility and performance robustness. Vibration suppression experiments of a composite smart structure with embedded piezoelectric sensor/actuator validate that the methodology provides an efficient neural controller design, including the plant order, the number of hidden layer neurons, the number of training patterns, and the coefficients of adaptive learning rate.

INTRODUCTION

Artificial neural networks have been seen in the applications of system identification, structure damage detection, and structural control. Among many network models, the multilayer feedforward neural network is most representative for it can learn a mapping of any complexity. However, one of the major difficulties in applications is the selection of network parameters, including the network structure configuration and learning rule, for satisfactory system performance. Some of the previous work studied the number of hidden neurons required in a network configuration (Gutierrez and Grondin, 1989; Hellstrom and Kanal, 1990; Huang and Huang, 1991; and Wang and Hsu, 1991). Others proposed the algorithms to improve the convergence and accuracy of network training (Jacobs, 1988; Vogl et al., 1988). However, most of the network parameters were determined either by engineering experience or by trial-and-error. The influence of different parameter combinations on system performance is usually examined by

S. M. Yang, Professor., Institute of Aeronautics and Astronautics, National Cheng Kung University, Tainan 701, Taiwan, ROC.
G. S. Lee, Graduate research assistant.

varying one parameter at a time while keeping the remaining constant. This "one-parameter-at-a-time" method is impractical and likely fruitless in reaching an optimal neural network design.

A design methodology that combines the experimental design and classical PD control is recently developed by Yang et al. (1997) to calculate the optimal set of sensor/actuator location and feedback gains in minimizing the unbalanced vibration of a rotor system. Experimental design employs a set of matrix experiments, also called an orthogonal array, to study several design parameters concurrently for optimal design with minimum cost. Peterson et al. (1995) recently applied Taguchi method, similar to experimental design, to study the error sources of the neural network in function approximations. Khaw et al. (1995) also applied Taguchi method to determine the network structure parameters in meeting the training speed and accuracy requirements. These studies, however, were limited to numerical simulations. Few discussion on the parameters of network learning rule was given, nor was there any experimental verification in orthogonal array. This paper presents a structural control methodology that combines the experimental design in quality engineering and the back-propagation network with an adaptive learning rate such that their advantages in implementation feasibility and performance robustness can be integrated together.

BACKPROPAGATION NEURAL NETWORK

The multilayer feedforward neural network learning is accomplished by successively adjusting the weights based on a set of input patterns and the corresponding desired output patterns. In one of improvements of the back-propagation learning rule, Vogl et al. (1988) developed the technique by adding a momentum term and an adaptive learning rate to accelerate the convergence speed and improve the network accuracy. This momentum term can be added by making the weight change Δw equals to the sum of a fraction of the latest weight change and the new change defined by the backpropagation rule. The weight change is mediated by a momentum constant α and it can be expressed mathematically by

$$\Delta w(k) = -(1-\alpha)\eta(k)\frac{\partial J(k)}{\partial w(k)} + \alpha\,\Delta w(k-1) \tag{1}$$

where $\eta(k)$ is the learning rate at time step k and J is the sum-squared error function. In addition to the momentum term, an adaptive learning rate is employed to reduce the training time by keeping the learning rate as large as possible while being stable. If the new error at present step exceeds the previous one by more than a pre-defined error-ratio ξ, the new weight change is discarded and the learning rate is decreased by a factor β, $\beta \le 1$. Otherwise the weight change is kept and the learning rate is increased by a factor ϕ, $\phi \ge 1$.

In this paper a backpropagation network with one hidden layer in batch learning mode (Rumelhart et al., 1986) is employed to illustrate the neural controller design methodology. The transfer functions of the hidden/output neurons are sigmoid/linear function, respectively. One of the difficulties in neural network applications is to select the appropriate parameters in learning rule and network configuration so as to have best performance. Although there have been some guidelines on how to choose the undetermined parameters, most, if not all, are qualitative instead of quantitative. Analysis of the parameter influence on system performance has yet to be seen. For example, how many hidden neurons are required without decreasing network generalizability? How many training patterns are sufficient for successful learning? Some suggest that the initial weights ω should be small enough, but how small is small? As for the momentum constant α ($0 \leq \alpha < 1$), the error-ratio ξ ($\xi \geq 1$), the increasing and decreasing factors of the adaptive learning rate ($\phi \geq 1$ and $\beta \leq 1$), what are the optimal values and how do they affect neural controller performance? An effective neural network design is to select the design parameters by matrix experiments such that a combination of them yields the best possible performance.

STRUCTURAL CONTROL

Many advanced systems are often required to be stiffer, lighter, and have sufficient damping for high precision pointing accuracy. These requirements have motivated an emerging new technology: the smart structures, structure systems with built-in sensor and actuator that can adaptively sense and actively change their physical geometry and properties under external stimuli. A smart structure specimen of 265 x 40.5 x 1mm embedded with a piezoelectric sensor and two piezoelectric actuators is applied to study structural control. The details of fabrication process, material composition and mechanical properties can be found in Yang and Lee (1997). The structural control system is usually discretized either by finite element method or by normal mode analysis, the governing equation is written by

$$\boldsymbol{M}\ddot{q} + \boldsymbol{C}\dot{q} + \boldsymbol{K}q = \boldsymbol{B}u \tag{2}$$

where q is generalized modal coordinate vector. $\boldsymbol{M}$, $\boldsymbol{C}$, and $\boldsymbol{K}$ are the modal inertia, damping, and stiffness matrices. $\boldsymbol{B}$ is the actuator influence vector and u is the control input. The objective is to design a neural controller so as to suppress the vibration of the bending modes at 11 and 52.5 Hz by using the embedded piezoelectric (PZT) sensor and actuator.

It is known that the damping of a structural system can be increased by direct velocity feedback. The control input for active damping control is formulated by

$$u = -\boldsymbol{K}_V \boldsymbol{B}_V{}^{\mathrm{T}} \dot{q} \tag{3}$$

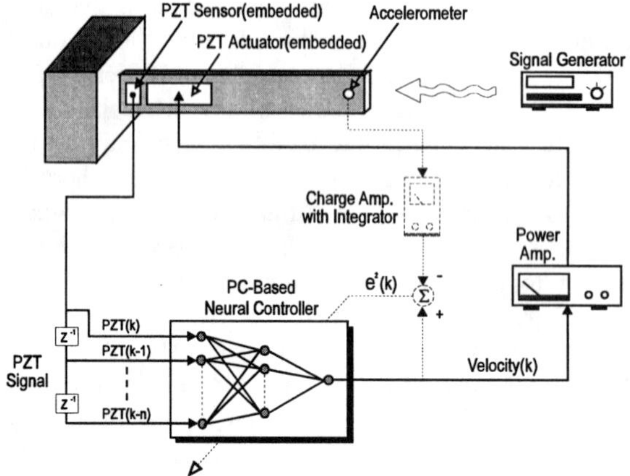

Figure 1 Vibration suppression experiment by using neural controller. The dashed line represents the neural controller training phase.

where $\boldsymbol{K}_V$ is the velocity feedback gain and $\boldsymbol{B}_V$ is the sensor measurement influence vector. Substituting Eq. (3) into Eq. (2), the closed loop system becomes

$$\boldsymbol{M}\,\ddot{q} + (\boldsymbol{C} + \boldsymbol{B}\boldsymbol{K}_V\boldsymbol{B}_V{}^{\mathrm{T}})\,\dot{q} + \boldsymbol{K}\,q = 0 \qquad (4)$$

where $\boldsymbol{B}\boldsymbol{K}_V\boldsymbol{B}_V{}^{\mathrm{T}}$ is the control-induced damping for vibration suppression. Engineering implementation of velocity signal often obtained by integrating the acceleration measurement from accelerometer. The neural network controller is to construct the mapping between the velocity signal and the piezoelectric sensor signal. The control input is sent to the embedded piezoelectric actuator for structural vibration suppression. The composite smart structure then becomes a "self-contained" system for the sensor and actuator are embedded.

Figure 1 shows the schematic diagram of the smart structure vibration suppression experiment. One piezoelectric actuator driven by a random signal uniformly distributed over ± 90V with an effective frequency range of 200 Hz is employed to excite the smart structure. An accelerometer located at tip measures the acceleration which is integrated to velocity. The embedded piezoelectric sensor located near the clamp end also measures the response at the sampling rate of 400 Hz simultaneously.

NEURAL CONTROLLER DESIGN

The neural controller design methodology is to determine a set of optimal design parameters for best vibration suppression. Consider the case where the error-ratio ξ is selected to have two levels and the remaining design parameters

TABLE I DESIGN PARAMETERS AND THEIR LEVELS

DESIGN PARAMETERS	LEVEL 1	LEVEL 2	LEVEL 3
Error-ratio (ξ)	1.02	1.08	-----
No. of input PZT steps (X)	4	6	8
No. of hidden neurons (Y)	4	8	12
No. of training patterns (Z)	1000	2500	4000
Momentum constant (α)	0.15	0.55	0.95
Magnitude of initial weight (ω)	0.001	0.1	1
Increasing factor of learning rate (ϕ)	1.02	1.1	1.5
Decreasing factor of learning rate (β)	0.3	0.6	0.9

have three levels. In addition to the parameters aforementioned, the number of input neurons that representing the number of steps of piezoelectric sensor signal required for estimating the velocity signal is also considered. In theory, at least four preceding steps of piezoelectric signal are required for controlling the vibration of the first two bending modes, two for each mode. In view of possible spillover in implementation, more steps of piezoelectric signal may be helpful in practice but the penalty is the increase of network complexity and training time. A total of eight design parameters and each in different levels are listed in Table I. In order to reduce the number of experiment and achieving a robust design, the orthogonal arrays of L_{12}, L_{18} and L_{36} (Phadka, 1989) are recommended for the interactions among design parameters are distributed to all columns. A standard $L_{18}(2^1 \times 3^7)$ array as listed in Table II is adopted for studying the neural controller design of one 2-level design parameter (ξ) and seven 3-level design parameters (X~β).

A quality loss index (QL), the percentage of vibration suppression under a constant feedback gain when the structure is in resonant condition at each of the two natural frequencies, is defined to investigate the system performance. The corresponding signal-to-noise ratio to be maximized is $\eta = 10 \cdot \log_{10}(1/QL)$. Since "log" is a monotonically increasing function, minimizing QL is equivalent to maximizing η. If the total quality loss of the structural control system can be formulated as a product of quality loss from different design parameters, then a reduced set of experiments can be formulated to cover all design parameters at all levels. The procedures of neural controller design are:

STEP 1: MATRIX EXPERIMENTS

A common technique in modal testing of a dynamic system is to measure the frequency response function(s) by the excitation of a broad spectrum. Conducting matrix experiment with several design parameters simultaneously is analogous to the use of multiple frequency input for frequency response function. Each experiment corresponds to each row of the L_{18} orthogonal array in Table II. In neural network training, the maximum number of training epochs is selected as N_{max} = 1000, and the sum-squared-error bound is SSE_{max} = ($0.53^2 \times Z$) for all

TABLE II $L_{18}(2^1 \times 3^7)$ ORTHOGONAL ARRAY AND THE LEVEL ASSIGNMENT

Exp. No.	(ξ)	(X)	(Y)	(Z)	(α)	(ω)	(ϕ)	(β)
1	1	1	1	1	1	1	1	1
2	1	1	2	2	2	2	2	2
3	1	1	3	3	3	3	3	3
4	1	2	1	1	2	2	3	3
5	1	2	2	2	3	3	1	1
6	1	2	3	3	1	1	2	2
7	1	3	1	2	1	3	2	3
8	1	3	2	3	2	1	3	1
9	1	3	3	1	3	2	1	2
10	2	1	1	3	3	2	2	1
11	2	1	2	1	1	3	3	2
12	2	1	3	2	2	1	1	3
13	2	2	1	2	3	1	3	2
14	2	2	2	3	1	2	1	3
15	2	2	3	1	2	3	2	1
16	2	3	1	3	2	3	1	2
17	2	3	2	1	3	1	2	3
18	2	3	3	2	1	2	3	1

experiments with an error tolerance of 5%. After conducting the 18 experiments, the signal-to-noise ratios are obtained by the vibration suppression experiment as shown in Fig. 1 and they are summarized in Table III. It should be noted that because of the control and/or observation spillover, the η of experiment 1, 2, 3, 10, 11 and 12 are lower than that of open-loop case. This implies that $X = 4$ (four preceding steps of piezoelectric sensor signal) may not be sufficient for controlling two modes effectively. The next step is analysis of means for estimating the effect of each design parameter on system performance.

STEP 2: ANALYSIS OF MEANS

The overall mean of η for the experiments defined in Table III is given by $m = \frac{1}{18}\sum_{i=1}^{18}\eta_i = -18.3307$. The average η of the experiments with error-ratio $\xi =$ 1.02 (level 1) is

$$m_{\xi 1} = \frac{1}{9}(\eta_1 + \eta_2 + \eta_3 + \eta_4 + \eta_5 + \eta_6 + \eta_7 + \eta_8 + \eta_9) = -18.3996 \qquad (5)$$

and the average η of the experiments with training patterns $Z = 4000$ (level 3) is

$$m_{Z3} = \frac{1}{6}(\eta_3 + \eta_6 + \eta_8 + \eta_{10} + \eta_{14} + \eta_{16}) = -18.5809. \qquad (6)$$

TABLE III EXPERIMENT RESULTS OF THE NEURAL CONTROLLER

Exp. No.	QL (%)	η
1	124.7222	-20.9594
2	122.2222	-20.8715
3	118.6667	-20.7433
4	66.8333	-18.2499
5	80.8889	-19.0789
6	94.7778	-19.7671
7	34.5000	-15.3782
8	34.5556	-15.3852
9	32.8333	-15.1631
10	115.0000	-20.6070
11	136.6667	-21.3566
12	117.2222	-20.6901
13	64.8889	-18.1217
14	63.0000	-17.9934
15	94.7222	-19.7645
16	33.2222	-15.2143
17	33.5000	-15.2504
18	34.3333	-15.3572
Optimal	30.5647	-14.8522
Open-Loop	100.0000	-20.0000

Table IV lists the mean effects of each design parameter in which the optimal level with the highest signal-to-noise ratio is marked. The matrix experiments show that the optimal design of neural controller for best vibration suppression is by selecting the error-ratio ξ = 1.08, the number of input layer neurons X = 8, hidden layer neurons Y = 4, training patterns Z = 2500, momentum constant α = 0.95, initial weight ω = 0.1 and the increasing/decreasing factor of the adaptive learning rate ϕ = 1.02 and β = 0.9, respectively. This optimal set is listed in the last column of Table IV. Note that it does not necessarily correspond to any row in the matrix experiments, nor should it be. The quality loss index and signal-to-noise ratio after implementing the optimal neural controller on the vibration suppression experiment are $QL^* = 30.5647$ and $\eta^* = -14.8522$, respectively. They are listed in the bottom rows of Table III. As was expected, the performance by using the optimal parameters is superior to all 18 experiments.

STEP 3: ANALYSIS OF VARIANCE

Before proceeding the verification of optimal design, analysis of variance should be conducted to study the relative importance of design parameters. The grand total sum of squares ($GTSS$) of η is defines by

$$GTSS = \sum_{i=1}^{18} \eta_i^2 = 6.1476 \times 10^3. \tag{7}$$

TABLE IV ANALYSIS OF MEANS FOR THE VIBRATION SUPPRESSION

Design Parameters	LEVEL 1	LEVEL 2	LEVEL 3	Optimal Value
Error-ratio (ξ)	-18.3996	-18.2617*	------	1.08
PZT steps (X)	-20.8713	-18.8293	-15.2914*	8
# of hid. neun. (Y)	-18.0884*	-18.3227	-18.5809	4
# of patterns (Z)	-18.4574	-18.2496*	-18.2850	2500
Momentum (α)	-18.4686	-18.3626	-18.1607*	0.95
Initial weight (ω)	-18.3623	-18.0404*	-18.5893	0.1
Inc. factor (ϕ)	-18.1832*	-18.6065	-18.2023	1.02
Dec. factor (β)	-18.5254	-18.4157	-18.0509*	0.9

* maximum in each row

It can be decomposed into two parts: the sum of squares (*SS*) due to mean and the total sum of squares (*TSS*). The former is analogous to the DC power of the signal while the latter is analogous to the AC part in Fourier analysis. The sum of squares of a design parameter is defined by the square deviation from the overall mean. For example, the *SS* of error-ratio ξ is

$$SS_{\xi} = \sum_{i=1}^{2} 9(m_{\xi i} - m)^2 = 0.0856. \tag{8}$$

The *SSs* of the other design parameters can be calculated in a similar way. The mean square of each parameter is then computed by dividing the sum of squares by the associated degree-of-freedom (usually equals to the number of levels minus one) as listed in Table V. Among all design parameters, the number of steps of piezoelectric sensor signal X is shown to be most sensitive (96.2189%) in vibration suppression. Parameter sensitivity is defined by the ratio of the sum of squares (*SS*) of a parameter to the total sum of squares (*TSS*). The higher the parameter sensitivity, the stronger its influence on performance. The neural controller design methodology not only identifies the optimal parameters but also provides their relative importance. These advantages are crucial in engineering applications.

STEP 4: VERIFICATION OF THE OPTIMAL DESIGN

In addition to the variance analysis, the optimal design parameters in step 2 have to be verified to see if the error caused by the interaction among the design parameters is within an acceptable tolerance. It is common in Fourier analysis to use the harmonics with higher power (sensitivity) as the dominant signals while leaving the rest as error. The pooled error in the matrix experiment is obtained by inspecting the variance ratio F which is defined by the ratio of the mean square of a design parameter to the error mean square. In general, only the parameters with $F \leq 2$ (with an asterisk in Table V) are included to calculate the pooled error. From

TABLE V VARIANCE ANALYSIS OF THE VIBRATION SUPPRESSION

Design Parameters	DOF	*SS*	*MS*	Sensitivity	*F*
Error-ratio (ξ)	1	* 0.0856	0.0856	0.0861 %	----
PZT steps (*X*)	2	95.6439	47.8219	96.2189 %	583.9
# of hid. neun. (*Y*)	2	0.7281	0.3641	0.7325 %	4.4
# of patterns (*Z*)	2	* 0.1482	0.0741	0.1491 %	----
Momentum (α)	2	* 0.2936	0.1468	0.2954 %	----
Initial weight (ω)	2	0.9131	0.4565	0.9185 %	5.6
Inc. factor (ϕ)	2	0.6856	0.3428	0.6898 %	4.2
Dec. factor (β)	2	0.7405	0.3702	0.7449 %	4.5
Error	2	* 0.1638	0.0819	0.1648 %	----
Total	17	99.4024	5.8472		
Pooled Error	7	0.6912	0.0987		

(DOF: degree-of-freedom, *SS*: sum of squares, *MS*: mean square, *F*: variance ratio)

Table V, the design parameters ξ, *Z*, and α are selected, and the pooled error of mean square is

$$\sigma_e^2 = \frac{(SS_\xi + SS_Z + SS_\alpha + SS_{error})}{(\text{Total DOFs of } \xi,\ Z,\ \alpha,\ \text{and } error)} = 0.0987. \tag{9}$$

Thus the variance of the prediction error σ_d^2 is defined by

$$\sigma_d^2 = \frac{1}{n_0}\sigma_e^2 + [\frac{1}{n} + (\frac{1}{n_{X3}} - \frac{1}{n}) + (\frac{1}{n_{Y1}} - \frac{1}{n}) + (\frac{1}{n_{\omega 2}} - \frac{1}{n}) + (\frac{1}{n_{\phi 1}} - \frac{1}{n}) + (\frac{1}{n_{\beta 3}} - \frac{1}{n})]\sigma_e^2 \tag{10}$$

where the number of confirmation experiment $n_0 = 5$, the number of total matrix experiments $n = 18$, and the number of matrix experiment of dominant design parameters at their level $n_{X3} = n_{Y1} = n_{\omega 2} = n_{\phi 1} = n_{\beta 3} = 6$. Equation (10) has two independent components: the first is the repetition error of the experiments, while the second is caused by the error in estimating the overall mean and the mean effect of the design parameters.

In statistical analysis, the two-sigma confidence is defined for the prediction error. The optimal set in Table IV can then be verified by the closeness between the optimal performance η^* and the performance from the dominant design parameters, denoted by η^+, the latter is determined by the five design parameters with higher mean square, *X*, *Y*, ω, ϕ, and β,

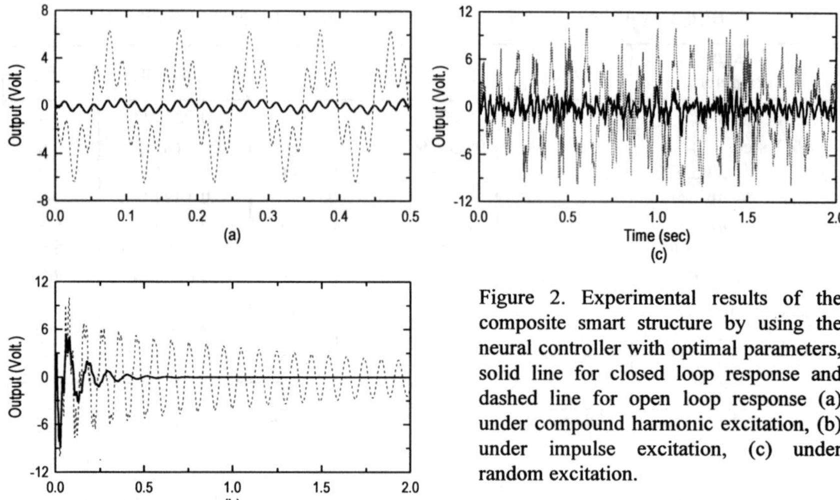

Figure 2. Experimental results of the composite smart structure by using the neural controller with optimal parameters, solid line for closed loop response and dashed line for open loop response (a) under compound harmonic excitation, (b) under impulse excitation, (c) under random excitation.

$$\eta^{+} = m + (m_{X3} - m) + (m_{Y1} - m) + (m_{\omega 2} - m) + (m_{\phi 1} - m) + (m_{\beta 3} - m) = -14.3316. \tag{11}$$

Prediction of the matrix experiments is valid if the discrepancy between η^{*} and η^{+} is bounded by the two-sigma confidence

$$|\,\eta^{*} - \eta^{+}| < 2\sigma_{d} \tag{12}$$

Because $\eta^{*} = -14.8522$, $\eta^{+} = -14.3316$, and $\sigma_{d} = 0.2830$, Eq. (12) is satisfied and the optimal design is hence validated. It worthies noting that not only is the performance optimal, but also the interactions among the parameters are negligible.

EXPERIMENTAL VERIFICATION

The optimal neural network design has been obtained in Tables IV. In addition, from analysis of variance in Table V, the number of steps of piezoelectric sensor signal (X) is found to be most significant with a largest sensitivity (96.2189%). As shown in Table IV, the more steps fed-in, hence more input neurons, the better the vibration suppression performance. Because of the number of input neurons (X) is heavily dominant, the other design parameters produce relatively minor influence on vibration suppression performance. For further investigating those parameters in detail, the neural controller can be trained with a fixed number of steps of piezoelectric sensor signal. For this study, the order of next four important design parameters are the magnitude of initial weights (ω), the decreasing factor of adaptive learning rate (β), number of hidden neurons (Y), and the increasing factor of adaptive learning rate (ϕ), respectively. The remaining are found to be trivial and their influences are almost negligible. The signal-to-noise ratio of optimal closed loop system experiment is listed in the bottom row of Table III. As aforementioned, the neural controller with the optimal parameter combination gets the maximal vibration suppression performance.

By using the optimal neural controller, Fig. 2(a) shows the open and closed loop piezoelectric sensor responses of the smart structure under the two-mode harmonic excitation. The steady state vibration amplitude is suppressed by about 90% with the control voltage limit of 100V. The settling time of the closed loop system under impulse excitation is also reduced considerably from over 2 seconds to about 0.4 second as shown in Fig. 2(b). In the case of random excitation, the peak-to-peak vibration amplitude is decreased to about 35% as shown in Fig. 2(c). These experimental results indicate that the neural controller is effective in vibration suppression and it is robust to unstructured uncertainties.

CONCLUSIONS

(1) A neural controller design methodology is developed for structural vibration control. The methodology combines the experimental design and the back-propagation neural network with an adaptive learning rate such that their advantages in implementation feasibility and performance robustness can be integrated together. This methodology facilitates the matrix experiment to search for the optimal neural network thus reducing the development cost and avoiding the lengthy trial-and-error. Moreover, the importance of the design parameters and their levels can be evaluated by analysis of variance. This is crucial in engineering applications.

(2) A glass fiber composite laminate structure with embedded piezoelectric sensor and actuators is employed to illustrate the design methodology. Active damping control is achieved by constructing a three-layer neural network to generate the mapping between the piezoelectric sensor signal and the velocity signal. The neural controller achieves the superior vibration suppression performance.

(3) The neural network design employs the orthogonal array to study eight design parameters in the network configuration and in the learning rule simultaneously. The network structure parameters include the number of input and hidden neurons (X and Y), the size of training pattern (Z), and the magnitude of the initial weight (ω). The parameters of learning rule include the error-ratio (ξ), momentum constant (α), and the increasing/decreasing factor (ϕ and β) of the adaptive learning rate. Analyses and experiments show that the number of steps of piezoelectric sensor signal (X) is most influential. All analyses are validated by laboratory experiment. Experimental results show that not only is the optimal neural controller effective in damping

control, it is also robust to unstructured (the unknown external noise or disturbance) uncertainties.

REFERENCES

Gutierrez, M., Wang, J., and Grondin, R., 1989, "Estimating Hidden Unit Number for Two-layer Perceptrons," *IJCNN*, vol. 1, pp. 677-681.

Huang, S. C. and Huang, Y. F., 1991, "Bounds on the Number of Hidden Neurons in Multilayer Perceptrons," *IEEE Trans. Neural Networks*, vol. 2, no. 1, pp. 47-55.

Hellstrom, B. J. and Kanal, L. N., 1990, "The Definition of Necessary Hidden Units in Neural Networks for Combinatorial Optimization," *IJCNN*, vol. 3, pp. 775-780.

Jacobs, R. A., 1988, "Increased Rates of Convergence through Learning Rate Adaptation," *Neural Networks*, vol. 1, pp. 295-307.

Khaw, J. F. C., Lim, B. S., and Lim, L. E. N., 1995, "Optimal Design of Neural Networks Using the Taguchi Method," *Neurocomputing*, vol. 7, pp. 225-245.

Peterson, G. E., St. Clair, D. C., Aylward, S. R., and Bond, W. E., 1995, "Using Taguchi's Method of Experimental Design to Control Errors in Layered Perceptrons," *IEEE Trans. Neural Networks*, vol. 6, no. 4, pp. 949-961.

Phadka, M. S., 1989, *Quality Engineering Using Robust Design*, Prentice-Hall.

Rumelhart, D. E., Hinton, G. E., and Williams, P. J., 1986, "Learning Internal Representation by Error Propagation," *in Parallel Distributed Processing,* D. E. Rumelhart and J. L. McClelland, Eds. Cambridge, MA: MIT Press.

Vogl, T. P., Mangis, J. K., Rigler, A. K., Zink, W. T. and Alkon, D. L., 1988, "Accelerating the Convergence of the Backpropagation Method," *Biological Cybernetics*, vol. 59, pp. 257-263.

Wang, S. D. and Hsu, C. H., 1991, "A Self Growing Learning Algorithm for Determining the Appropriate Number of Hidden Units," *IJCNN*, vol. 2, pp. 1098-1104.

Yang, S. M. and Lee, G. S., 1997, "Vibration Control of Smart Structures by Using Neural Networks," *J. of Dynamic Systems, Measurement, and Control*, vol. 119, no. 1, pp. 34-39.

Yang, S. M., Sheu, G. J., and Yang, C. D., 1997, "Vibration Control of Rotor Systems with Noncollocated Sensor/Actuator by Experimental Design," *J. of Vibration and Acoustics*, vol. 119, no. 2, pp. 200-207.

Problem-Specific Neural Networks for Detecting Structural Damage

H.-G. HERRMANN* and J. STRENG**

ABSTRACT

Problem specific neural networks (NN) are able to locate and quantify damages in plane truss structures. NN are well suited to this problem due to their ability to learn unknown, complex functional relations.

First, the behaviour of the undamaged structure as well as the behaviour of the structure including various states of damage are calculated or obtained from experimental results. The training patterns include characteristics of the damaged structure. It is shown that a trained neural network is able to detect and quantify existing damage in the structural system, based on the range of the presented training data. This holds under the assumption that the training patterns include a sufficient number of characteristics that distinguish between distinct states of the structural behaviour. Network training and its generalization capabilities can be considerably improved by applying dimensional analysis in order to reduce the amount of input and output patterns.

This paper evaluates an example in the field of two dimensional truss structures, different learning rules and network types, reduction (pruning) of fully connected initial neural network topologies and a particular problem specific data preprocessing (dimensional analysis).

Sensitive pruning algorithms minimize the network topology, which is necessary to comprehensively describe the given problem. Under the above mentioned limitations neural networks are able to compute inverse functional relations, which are virtually unsolvable using analytical methods and thereby assess the impairment of a damaged structural system with a minimized network topology.

KEYWORDS

Neural networks, structural damage, inverse functional relation, damage assessment, learning process, data preprocessing, dimensional analysis, network pruning.

* Hans-Georg Herrmann, Daimler-Benz, Research and Technology, Germany,
e-mail:Hans-Georg.Herrmann@Sifi.Mercedes-Benz.com

** Juergen Streng, Deutsche Forschungsanstalt fuer Luft- und Raumfahrt,
Institute for Structures and Design, Stuttgart, Germany, e-mail:juergen.streng@dlr.de

INTRODUCTION

The integrity of a complex structure has to be determined through a periodically performed inspection. Therefore a system for real time structural damage monitoring can be very useful. Hitherto, determining the place and amount of damage by means of the structure's behaviour is difficult and expensive. A principle assessment by a finite element analysis can only be applied if the damage is known.

Another possibility consists of the integration of sensors into engineering structures, since the increased variety and performance of sensors has made them highly attractive.

Optical fibres and piezoelectric materials are types of sensors which have recently been favoured for being integrated into or adapted onto structures and components. New sensor manufacturing techniques have improved the sensitivity of sensors and thus enabled sensing of effects which could hardly be monitored before. This is also due to developments arising from microsystems and micromechanics technology and has led to the wide field of micro electro mechanical systems (MEMS).

Depending on the size of the structure and the damage as well as the physical parameters to be monitored a larger number of sensors may be required. Neural networks have therefore been considered to be an interesting tool for handling the large number of sensors and the possibly complex signals generated by these sensors.

Neural networks are well suited to process a large number of sensors due to their capability of learning complex relationships.

NEURAL NETWORKS

For the first time in 1943 W.S. McCulloch et al. [MP43] suggested information processing by means of interconnected simplified neurons. After first enthusiasm (see D. Hebb [Heb49]) most of the researchers abandoned the field, when M. Minsky et al. [MP69] showed the great disadvantages of perceptron models in 1969. A renewed interest in neural networks emerged in the early eighties after some important theoretical results (like the generalized delta rule as a learning algorithm for multi-layered feed-forward networks inter alia by D.E. Rumelhart et al. [RM86] and J. McClelland et al. [MR86]) and new hardware developments provided increased processing capacities. Nowadays applications for neural networks can be found in robotics, image processing, data and speech recognition as well as in areas of mathematics and physics. A good summary about different neural network types is given by A. Zell [Zel94].

In this paper we focus on feed-forward networks for approximating real valued functions. During the learning process, the link weights of the interconnections between the layers are modified by a learning rule in order to minimize the error between the correct and actual output. This error is backpropagated through the network from output layer to input layer. The knowledge of a trained neural network can be understood as distributed information storage, which is integrated in its topology, link weights and transfer functions. Employing a neural network as an approximation tool, its operation can be interpreted as highly nonlinear mapping between input units as arguments and output units as function values of a given relation.

After a successful learning phase, the trained network is able to generalize in a way so as to compute output data from input patterns which have never been presented to it before. Yet, the network will succeed in this task only within the range of trained input patterns. Deviation of the presented data in supervised neural networks cannot be qualitatively assigned either to the experimental mean variation or to the network's limited ability for generalization due to its topology.
Neural networks are often employed as 'black boxes' that are simply applied without using any further available information about the problem. The unknown functional relation is approximated by the network by minimizing the difference between actual and teaching output according to the learning algorithm. Deficiencies mentioned above may be overcome by using further available knowledge about the problem applying dimensional analysis, in order to use a problem specific network. Network training for damage monitoring can be performed with data compressed by dimensional analysis and therefore without loss of any information about the given problem.

DIMENSIONAL ANALYSIS

Generally, every physical law or equation of the n variables x_i in implicit form

$$f(x_1,\ldots,x_n) = 0 \tag{1}$$

or explicit form

$$x_i = f(x_1,\ldots,x_{i-1},x_{i+1},\ldots,x_n) \tag{2}$$

has to comply with the principle of dimensional homogeneity, i.e. all $n-1$ arguments on the left hand side of equation (2) must have the same dimensions as x_i. This principle is assumed in all of physics and is part of dimensional analysis.
All further reflections on dimensional analysis are theoretically based on the *Product-Theorem* and the *Pi-Theorem*. For reasons of brevity only the main properties of these foundations are stated in this paper. We refer to P. BRIDGMAN [Bri32], J. PAWLOWSKY [Paw71], E. BUCKINGHAM [Buc14], H. GÖRTLER [Gör75], J. ZIEREP [Zie72] and S. KLINE [Kli86] for readers interested in proof and more detailed information.
By means of dimensional analysis a dimensionally homogeneous equation f of n physical variables x_i

$$f(x_1,\ldots,x_n) = 0 \tag{3}$$

can be written as a function G of $p = n - r$ dimensionless arguments π_j

$$G(\pi_1,\ldots,\pi_p) = 0 \ , \tag{4}$$

where r is the rank of the dimensional matrix containing dimensional variables and physical quantities.

The p dimensionless variables π_j are calculated according to

$$\pi_j \ = \ \prod_{i=1}^{n} x_i^{-k_{ji}} \tag{5}$$

with $j = 1, \ldots, p \in I\!N$ and $k_{ji} \in I\!R$ as constants.
Dimensional analysis turned out to be a versatile and efficient tool in other fields of research. For instance, dimensional analysis may considerably improve the learning behaviour and the ability for generalization of neural networks in structural mechanics (see H.-G. HERRMANN [Her96]).
The manner of using dimensional analysis for data preprocessing is explained hereafter and illustrated by an example taken from an engineering background.

BASIC PRINCIPLE

In this section the application of the above mentioned elements of the proposed methods concerning neural networks for damage monitoring are described.
By applying dimensional analysis, the number of dimensional input *and* output data per pattern is reduced in size. The reduction of the n dimensional input and output data to a smaller number $p = n - r$ of dimensionless numbers is carried out due to the rank r of the dimensional matrix, without losing any physical information about the problem. In general, each dimensionless number can be resolved into a desired dimensional argument by multiplying it with the remaining variables of which the number consists. The basic principle of the presented method is illustrated in figure 1.

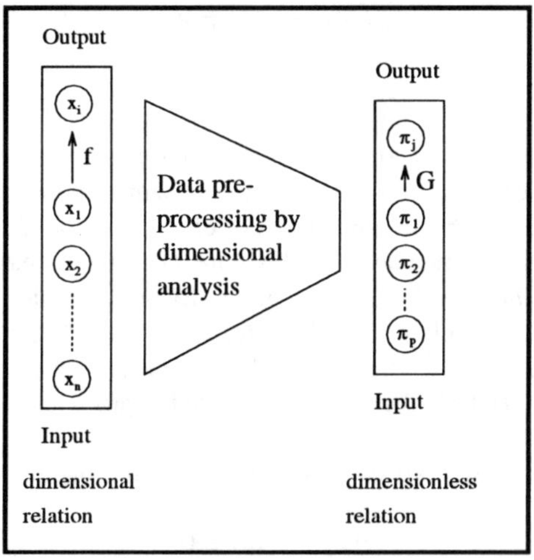

Figure 1: Basic principle by using dimensional analysis for data preprocessing

Because of the inherent property of each dimensionless number as similarity variable, one value of a dimensionless number includes a wide range of its dimensional arguments. Thus, a provable ability of generalization of dimensionless numbers or training data exists. Apart from the network's ability to generalize, its general performance properties are considerably improved.

Figure 2 illustrates the basic principle of applying neural networks for damage monitoring. Behaviour of the undamaged structure as well as bevaviour of the structure with various possible damage states are examined by a finite element analysis.

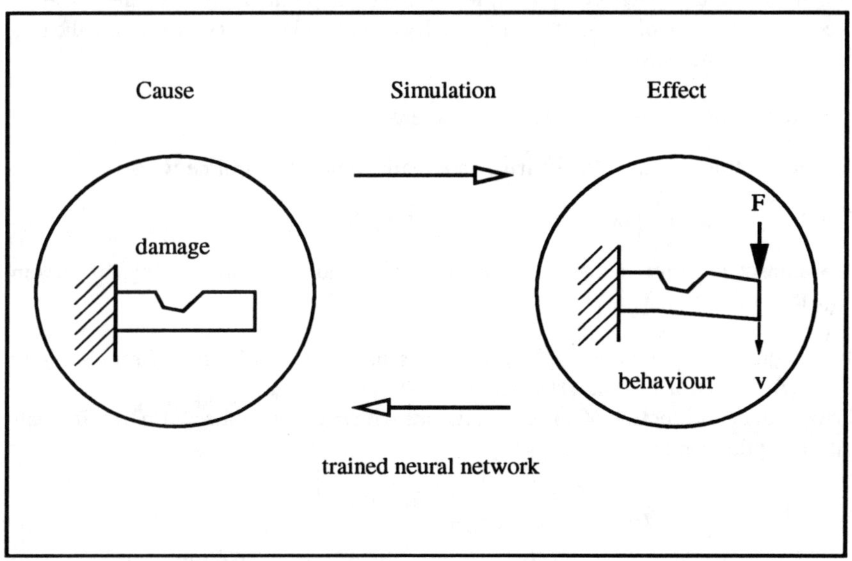

Figure 2: Damage monitoring using a neural network

Dimensional input and output data are preprocessed by dimensional analysis, in order to obtain dimensionless input and output data for network training. In this case, the input patterns consist of the dimensionless displacements describing the structural behaviour, whereas the dimensionless integrities of each element are used as output patterns. Thus, the network's task is to approximate the inverse above mentioned relation.

After a successful learning process the trained network should be able to detect and quantify any existing damage of a loaded structure. This prediction of structural damage in bar trusses is limited to conditions concerning neural networks that are shortly described below.

The inverse functional relation between the structural behaviour (dimensionless nodal displacements) and the dimensionless structural integrity of each element under a given load F is virtually unsolvable with analytical methods. A numerical solution is however possible under the condition that the training patterns include a sufficient number of characteristics that distinguish between distinct states of the structural behaviour. For example failures of different elements of the topology must not cause the same output pattern (nodal displacements) and thus lead to ambiguous results (see H.-G. HERRMANN ET AL. [Her96, HS96, HBS97]).

Additionally, the network topology can be considerably reduced by using pruning algorithms.

PRUNING

The main goal in machine learning is the consistent approximation of given training data with minimal system complexity and at the same point maximum generalization potential. Reduction of the system complexity of a neural network by pruning the network topology has several advantages:

- Reduction of training and storage expenses
- Reduction of expenses for hardware realizations of neural networks
- Simplification of complex network starting topologies
- Improved comprehension of data processing and important topological parts in the networks

The weight pruning method applied in this research is called Optimal Brain Surgeon (OBS) (B. HASSIBI ET EL. [HS93],[HSWW94]).
The overall error function of a neural network with m output units and p training pattern can be defined as

$$E = \sum_{i=1}^{p} \left[\frac{1}{2} \sum_{j=1}^{m} \left(t_{ij} - o_{ij}\right)^2 \right] . \tag{6}$$

In equation (6), t_{ij} denotes the teaching output and o_{ij} the actual output of the network pertinent to the output unit j and the training pattern i. Second derivative information of E is used to determine the weight which leads to the smalled increase of the E when eliminated. Pruning of a weight also deletes the corresponding link, pruning of all links that connect a unit to the rest of the topology also deletes that unit but was restricted to hidden units.
The approximation of equation (6) in a Taylor series leads to

$$\Delta E = \left(\frac{\partial E}{\partial \mathbf{w}}\right)^T \Delta\mathbf{w} + \frac{1}{2}\,\Delta\mathbf{w}^T\,\mathbf{H}\,\Delta\mathbf{w} + \mathbb{O}\left(\|\Delta\mathbf{w}\|^3\right) \tag{7}$$

with $\mathbf{w}$ as the vector of network weights and $\mathbf{H}$ the Hessian matrix of E with respect to $\mathbf{w}$. In equation (7), $\mathbb{O}$ denotes all third and higher order terms. The following assumptions

- Network is trained to a local minimum of E
- Neighbourhood of local minimum is quadratic

simplify equation (7) as they eliminate all linear, third and higher order terms. Pruning of the weight w_q is defined as

$$\mathbf{e}_q^T\,\Delta\mathbf{w} + w_q = 0 \tag{8}$$

where $\Delta \mathbf{w}_q$ is the change of weight q in direction of the unit vector $\mathbf{e}_q$ in the weight space $\mathbf{w}$. The goal now is to minimize the increase of error E while pruning w_q. This can be written as a restricted optimization problem

$$min_q \left[min_{\Delta \mathbf{w}} \left(\frac{1}{2} \Delta \mathbf{w}^T \, \mathbf{H} \, \Delta \mathbf{w} \right) \mid \mathbf{e}_q^T \, \Delta \mathbf{w} + w_q = 0 \right] . \tag{9}$$

and leads to optimal changes of all remaining network weights

$$\Delta \mathbf{w} = -\frac{w_q}{\left[\mathbf{H}^{-1}\right]_{qq}} \mathbf{H}^{-1} \, \mathbf{e}_q \, , \tag{10}$$

with $\left[\mathbf{H}^{-1}\right]_{qq} = \mathbf{e}_q^T \, \mathbf{H}^{-1} \, \mathbf{e}_q$. As a result the increase of overall error (saliency) follows as

$$L_q = \frac{1}{2} \frac{w_q^2}{\left[\mathbf{H}^{-1}\right]_{qq}} . \tag{11}$$

With this the pruning algorithm can be given as:

1. Training of problem specific network topology to minimum (local) of E
2. Compute $\mathbf{H}^{-1}$
3. Compute saliency L_q for all weights q
4. If smallest saliency L_q is much smaller than E delete weight q. Proceed to step 5 otherwise to step 6.
5. Adjust all remaining weights according to equation (10). Back to step 3.
6. Further pruning of weights increases E too much. Restoration of condition before last pruning. Retraining of network

OBS needs external criteria to terminate the pruning procedure. This algorithm and the efficient computation of the inverse Hessian matrix $\mathbf{H}^{-1}$ have been developed by B. HASSIBI ET AL. [HS93],[HSWW94].

SIX BAR TRUSS STRUCTURE

The investigated structural problem is a statically determinate two dimensional 6 bar truss structure subject to a horizontal F_x and a vertical F_y load in node 1 (see figure 3). The structural behaviour is characterized by horizontal u_i and vertical v_i displacements of the nodes $i = 1,2,3$. The Young's modulus E_{mod} is the same for all bars. Bars only transmit tractive and pressure forces and no momenta. The investigated truss structure has realistic dimensions, material properties and loads. Network A is a $[6-12-12-6]$ fully connected network topology with 6 input units, 2 layers of 12 hidden units each and 6 output units.

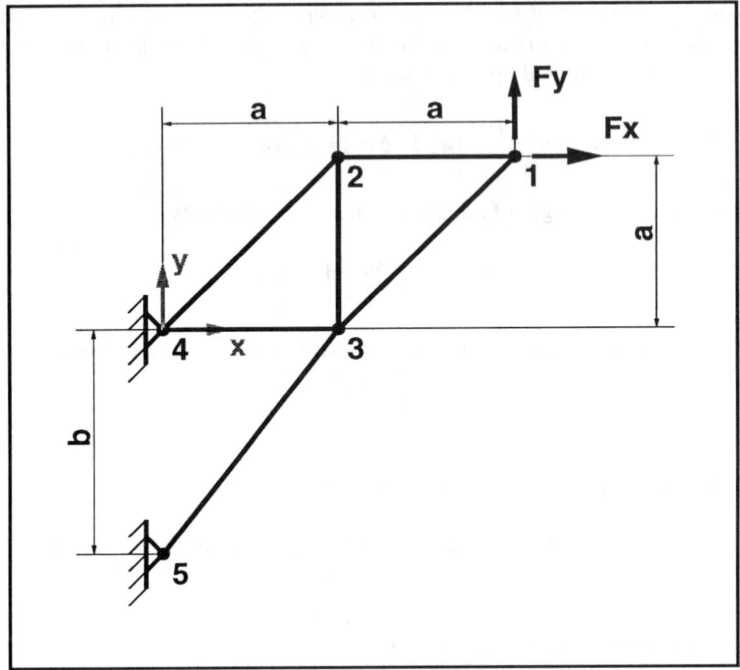

Figure 3: Six bar truss structure

DATA PREPROCESSING

The structural integrity coefficients a_k which denote the actual to nominal stiffness of the individual bars were chosen according to table 1. Consistent with that training and

Table 1: Integrity coefficients

data	a_k			bar diameter D_k in %		
training	0.64	0.81	1.0	80	90	100
test	0.7056	0.8464		84	92	

test data were generated using a finite element program. Since the loads F_x and F_y are constant they don't have to be considered as training data. Dimensional analysis (see table 2) as well as transformation of the training data into the interval $]0.05, 0.95[$ was performed prior to network training or testing.

The networks were trained with dimensionless training patterns and the trained or fully pruned network was then tested with unknown test patterns, which were up to that point not presented to the network. The training and test data for network A (see figure 4) consist of the dimensionless node displacemants $\tilde{u}_i$ and $\tilde{v}_i$ of the nodes

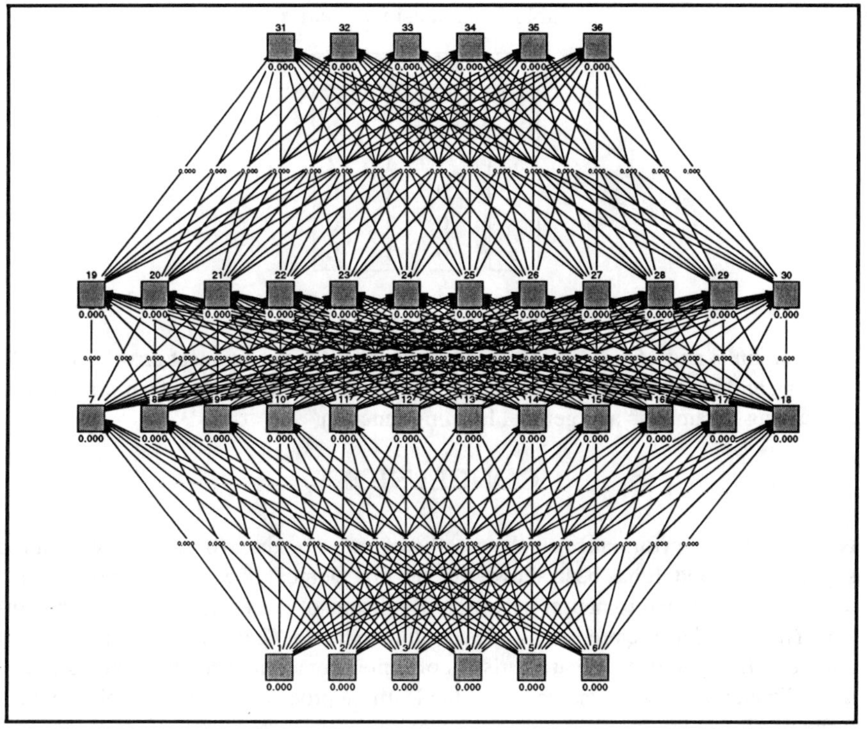

Figure 4: Network Topology A

$i = 1,2,3$ as input and the appertaining structural integrity coefficients a_k (see table 1) as teaching output data.

RESULTS

For simulation of the neural network behaviour SNNS (A. ZELL ET AL. [ZMVe95] and T. SCHREINER [Sch94]) was used.
Using OBS, the initial network topologies could be significantly reduced with respect to number of connection links and hidden units. This also led to a reduction of the test error, although few examples also showed a small increase.

Simple training of the initial network A (see figure 4) for 100000 epochs reduced the Sum Squared Error

$$SSE = \sum_{i=1}^{p} \sum_{j=1}^{m} \left(t_{ij} - o_{ij}\right)^2 \tag{12}$$

from 554.71 to 0.00465 with an average test error of 8.7 %. To illustrate the pruning results, figure 5 shows the pruned network topology A. Pruning with OBS reduced the initial value of SSE from 554.71 to 0.01075 while at the same time deleting 72.5

Table 2: Dimensional matrix

	$[M]$	$[L]$	$[T]$
S_1	1	1	-2
...	1	1	-2
S_6	1	1	-2
u_i	0	1	0
v_i	0	1	0
a	0	1	0

% of the links and 16.6 % of the hidden units with an average test error of 8.25 %. Training and pruning were conducted with the same training and test patterns as well as learning parameters and general learning function.

CONCLUSIONS

Neural network performance can be greatly enhanced by the two new elements of the suggested method. It has been shown that the first element, the dimensional analysis, can be easily applied to neural networks. By means of dimensional analysis dimensional training data are preprocessed according to the *Pi-Theorem* into dimensionless numbers. Because of the characteristics of dimensional analysis, the application of these dimensionless patterns improves the learning process. No relevant physical information is lost because of the inherent property of the dimensionless similarity variables.

The ability of the network to generalize has been considerably increased, in generalizing the problem parameters into dimensionless *Pi* quantities. In contrast to ordinary neural network training there is no practical limitation to the range of recall data, except for the underlying model limitations. This is a result of the similarity properties of the dimensionless numbers.

With the described example a successful application of neural networks for approximating inverse functional relations being virtually unsolvable with analytical methods is demonstrated. Neural networks are well suited in solving this task, provided that there are as many input as output neurons (i.e. the same number of sensors for e.g. displacements and integrities). This holds under the assumption that the training patterns include a sufficient number of characteristics that distinguish between distinct states of the structural behaviour. If these conditions are met, the network's ability for detecting, locating and quantifying damage with a minimized network topology in a given structure is shown.

It was shown that sensitive pruning algorithms have the ability to significantly reduce the complexity of an initial network topology. This led to improved test performance as well as to a reduction of further training expenses for the minimized network.

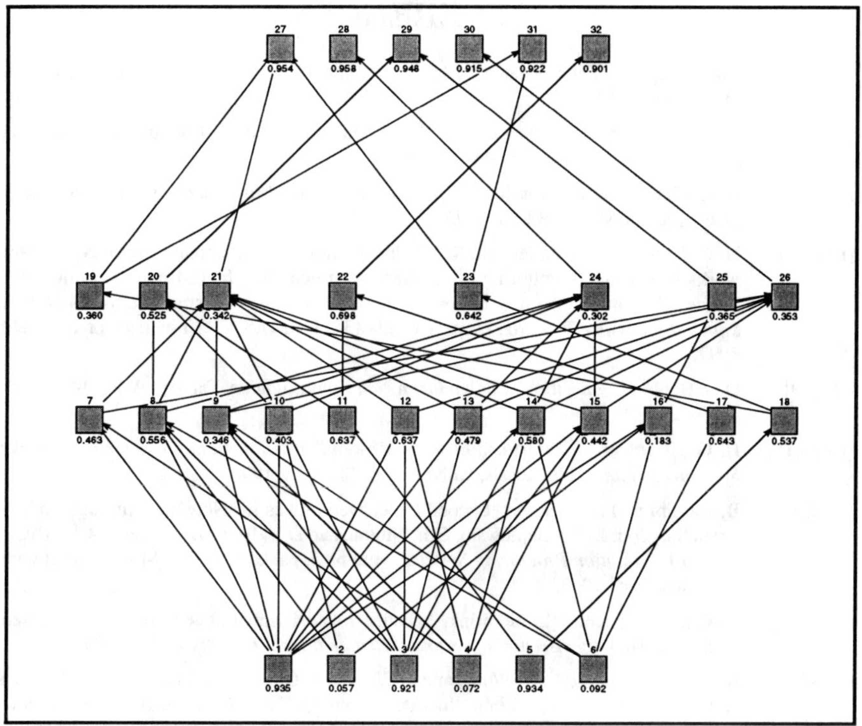

Figure 5: Pruned Network Topology A

APPLICATION POTENTIAL AND OUTLOOK

The suggested approach can be applied to all functional relations which can be described in terms of dimensionally homogeneous functions. In principle this includes all methods in damage monitoring for instance acoustic emission, lamb waves and vibration based methods.

Furthermore, application of this method to other types of neural networks like recurrent networks is conceivable. The proposed concept would also be applicable to neural networks used as classifiers, since the training data can be better assessed by using dimensional analysis and more easily assigned to different classes than with conventional approaches.

Minimized and problem specific network topologies in combination with dimensional analysis can improve the performance of the approach shown in this paper. Large systems can be handled by subdividing them into small subsystems (see H.-G. HERRMANN [Her96]). The correlation between *SSE* in training examples and the resulting test error of the corresponding pruned network will be subject to a more detailed research.

REFERENCES

[Bri32] P.W. Bridgman. *Theorie der Physikalischen Dimensionen.* Verlag B.G. Teubner Leipzig und Berlin, 1932.

[Buc14] E. Buckingham. On physically similar systems; illustration of the use of dimensional equations. *Phys. Review*, 4:345–376, 1914.

[Gör75] H. Görtler. *Dimensionsanalyse. Theorie der physikalischen Dimensionen mit Anwendungen.* Springer-Verlag, Berlin, 1975.

[HBS97] H.-G. Herrmann, C. Boller, and K. Stellbrink. Some aspects on the Use of Neural Networks in Damage monitoring of engineering structures. In J.M. Dulieu-Smith, W.J. Staszewski, and K. Worden, editors, *Structural Damage Assessment Using Advanced Signal Processing Procedures*, pages 121–134. DAMAS 97 University of Sheffield, 1997.

[Heb49] D.O. Hebb. *Organization of Behavior: A Neurophysiological Theory.* Wiley, New York, 1949.

[Her96] H.-G. Herrmann. *Untersuchungen zur Anwendbarkeit von neuronalen Netzen in der Strukturmechanik.* PhD thesis, University of Stuttgart, 1996.

[HS93] B. Hassibi and D.G. Stork. Second Order Derivatives for Network Pruning: Optimal Brain Surgeon. In T.J. Sejnowski, G.E. Hinton, and D.S. Touretzky, editors, *Advances in Neural Information Processing Systems*, volume 5, pages 164–171. Morgan Kaufmann Publishers Inc., 1993.

[HS96] H.-G. Herrmann and K. Stellbrink. Schadenslokalisierung mit neuronalen Netzwerken. *Zeitschrift für Flugwissenschaften und Weltraumforschung*, 20:163–168, 1996.

[HSWW94] B. Hassibi, D.G. Stork, G. Wolff, and T. Watanabe. Optimal Brain Surgeon: Extensions and Performance Comparisons. In J.D. Cowan, G. Tesauro, and J. Alspector, editors, *Advances in Neural Information Processing Systems*, volume 6, pages 263–270. Morgan Kaufmann Publishers Inc., 1994.

[Kli86] S.J. Kline. *Similitude and Approximation Theory.* Springer-Verlag, New York, 1986.

[MP43] W. S. McCulloch and W. Pitts. A logical calculus of the ideas immanent in nervous activity. *Bull. Math. Biophys.*, 5:115–133, 1943.

[MP69] M. Minsky and S. Papert. *Perceptrons: An Introduction to Computational Geometry.* MIT Press, 1969.

[MR86] J.L. McClelland and D.E. Rumelhart. *Parallel Distributed Processing: Explorations in the Microstructure of Cognition, Volume II: Psychological and Biological Models.* The MIT Press Cambridge, 1986.

[Paw71] J. Pawlowsky. *Die Ähnlichkeitstheorie in der physikalisch-technischen Forschung: Grundlagen und Anwendungen.* Springer-Verlag, Berlin, 1971.

[RM86] D.E. Rumelhart and J.L. McClelland. *Parallel Distributed Processing: Explorations in the Microstructure of Cognition, Volume I: Foundations.* The MIT Press Cambridge, 1986.

[Sch94] T. Schreiner. Ausdünnungsverfahren für Neuronale Netze. Master's thesis, Fakultät Informatik, Universität Stuttgart, Diplomarbeit Nr. 1140, Juli 1994.

[Zel94] A. Zell. *Simulation Neuronaler Netze.* Addison-Wesley, Bonn, 1994.

[Zie72] J. Zierep. *"Ahnlichkeitsgesetze und Modellregeln in der Strömungslehre.* Braun Verlag, Karlsruhe, 1972.

[ZMVe95] A. Zell, G. Mamier, M. Vogt, and N. Mache et.el. Stuttgart Neural Network Simulator, Version 4.1. Report 6/95, IPVR, University of Stuttgart, 1995.

Experimental Techniques and Analytical Model to Remove Geometric Effects in a Fatigue Crack Specimen

S. HURLEBAUS

ABSTRACT

The finite geometry of a laboratory specimen changes a measured acoustic emission waveform because of reflections, transmissions and mode conversions at the interfaces and boundaries of the specimen, thus making it difficult to interpret the measured signal. This paper develops a transfer function that removes the geometric effects from measured acoustic emission signals. This transfer function is developed using a repeatable, broad band acoustic emission source, a pulsed laser, and a broad band, high fidelity sensor, a laser interferometer. It can operate on an acoustic emission signal, measured in the same specimen, but caused by a different source. The accuracy of this transfer function is demonstrated by operating on different acoustic emission sources such as a double pulse, an ablation source and a pinducer.

Furthermore, an analytical model is developed in order to understand the behavior of the transfer function and to help interpret the experimental results. This analytical model is created by developing some artificial signals. With this analytical model, the influence of certain parameters is determined by varying their values and examining the effect of these changes. Such parameters are the separation distance between the source and receiver, different frequencies, repeated shapes, different amplitude ratios of waves, source size and noise.

INTRODUCTION

Acoustic emission techniques have great potential for diagnosing the structural integrity of in-service engineering components in real-time. This passive monitoring method "listens" for acoustic signals from internal "damage" events and attempts to quantify the amount of degradation in the structure. Unfortunately, this tool has failed to live up to its promise as a quantitative nondestructive evaluation technique. The chief reason for this shortcoming is acoustic emission's inability to identify and

---

Stefan Hurlebaus, Institute A of Mechanics, University of Stuttgart, Pfaffenwaldring 9, 70550 Stuttgart, Germany

separate signals caused by damage events from those due to ambient and extraneous sources. For acoustic emission testing to become more effective, new techniques must be developed that can accurately interpret and quantify the large amount of data obtained in a typical acoustic emission test. As a step in that direction, this research develops a methodology that removes extraneous information from a measured acoustic emission signal, and as a result, can help to isolate and identify the contribution from damage sources (Hurlebaus, 1996).

A measured acoustic emission signal depends upon three distinct factors: its source; component geometry and material properties; and receiving sensor and instrumentation. Each of these three elements must be understood in order to effectively identify and characterize measured acoustic emission waveforms. The source of an acoustic emission waveform determines its initial shape, amplitude and frequency content; these waveforms are transient and broad band in nature. The component geometry and material properties influence the wave propagation from the emission source to the receiver. Finally, the detection sensor and instrumentation can bias the signal by filtering certain frequencies or interfering with the process being measured.

The current research develops a transfer function that quantifies and removes these geometric effects. The steps in the development of the transfer function are as follows: an acoustic emission signal is generated and detected in a specimen whose geometry is being characterized; this experiment (same source) is repeated in a half-space. Then, the transfer function is obtained by (point-by-point) dividing the FFT of the half-space signal by the FFT of the specimen signal. Once the respective transfer function is developed, their robustness is tested by removing their geometric effects and isolating the source signal from three new sources: vaporized oil, a double laser pulse and a pinducer.

As a final step, an analytical model is developed to interpret the experimental results and help in understanding the behavior of the transfer functions. The advantage of the analytical model is that different parameters can be easily varied, enabling a systematic study of their influence on the transfer function.

EXPERIMENTAL SETUP

The experimental procedure consists of measuring a source function with a laser interferometer. The source function is produced with both a contact, piezoelectric pinducer that is driven with a short time duration pulse from a pulser and a non-contact optical source that is excited with a signal from a function generator.

Laser detection of these waveforms is accomplished with a heterodyne interferometer (Bruttomesso, Jacobs, and Costley, 1993). Briefly, single frequency light is split into two beams which are separated in frequency by 40 MHz using an acousto-optic modulator. These two beams are sent along two paths of the interferometer, one which contains the sample being monitored. The two beams are recombined at a photodetector and produce a beat frequency of 40 MHz. Frequency shifts in light reflected from the sample surface result in proportional frequency shifts in the 40 MHz beat signal. As a result, this 40 MHz signal acts as a carrier that is demodulated in real time with an FM-discriminator to obtain the time dependent surface velocity. The interferometer makes a high fidelity, absolute measurements of surface velocity over a bandwidth of 10 MHz.

For the optical generation of ultrasound a Q-switched Nd:YAG laser with a maximum output of 9 mJ and a pulse length of 15 ns is used (Hurlebaus, 1996). The maximum signal acoustic wave strength is obtained by focusing the laser beam as small as possible. A slightly different optical source (similar to ablation, but with an increased amplitude) is obtained by vaporizing a layer of oil on the surface.

The pinducer is a piezoelectric element with a tip diameter of approximately 2 mm that is designed for the detection of ultrasonic signals. However, in this study a pinducer is used for the broad band generation of ultrasound.

A fatigue specimen made out of 2024 aluminum is examined. The dimensions of the specimen and the location of the source and receiver are shown in Figure 1. The specimen surface is polished to increase the efficiency of optical detection.

EXPERIMENTAL RESULTS AND DISCUSSION

It should be noted that the wave scattering at a fatigue specimen is not simple. The reflected and transmitted Rayleigh waves do not contain all the energy that is contained in the original incident Rayleigh wave; some of the incident Rayleigh wave energy is mode-converted into longitudinal and shear waves that radiate into the body. However, it is not necessary to understand this scattering phenomena in order to develop the transfer function. An accurate experimental measurement of the phenomena is all that is necessary to make the transfer function. Figure 2 shows the specimen and the half-space waveforms, which are used to make the transfer function for the fatigue specimen. These signals are obtained in the ablation regime (without oil) and represent the average forty waves. Each of the signals in Fig. 2 is transformed into the frequency domain by taking the FFT of the entire waveform. The half-space signal is divided by the specimen signal to get the transfer function.

The validity of this transfer function is demonstrated by removing the geometric effects from signals created with different sources.

A different source is produced by vaporizing a thin layer of oil on the specimen surface. Figure 3 compares the oil source and ablation source signals; they show that the signal from the oil source is about ten times larger in amplitude than the ablation source, but they both have similar shapes. This oil signal represents the average of ten different laser shots. Since the oil vaporizes after each laser shot, the oil must be replaced before each subsequent measurement. A comparison of a signal

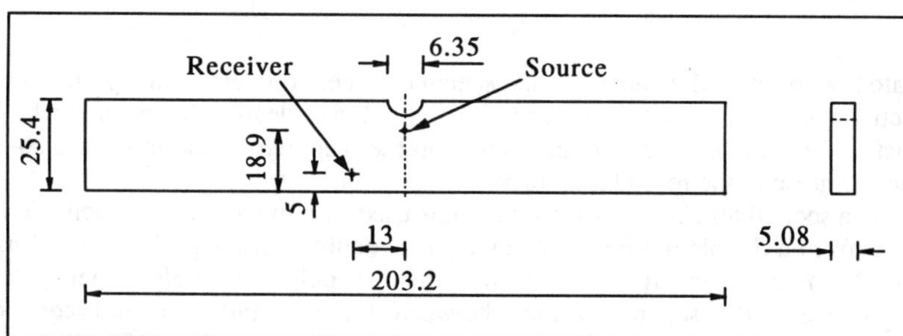

Figure 1: Dimensions of the Specimen (all dimensions in mm)

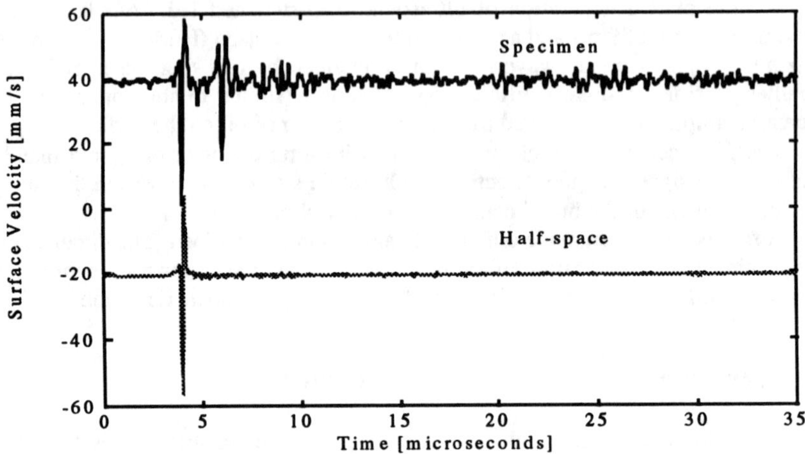

Figure 2: Specimen and Half-space Signals used for Transfer Function

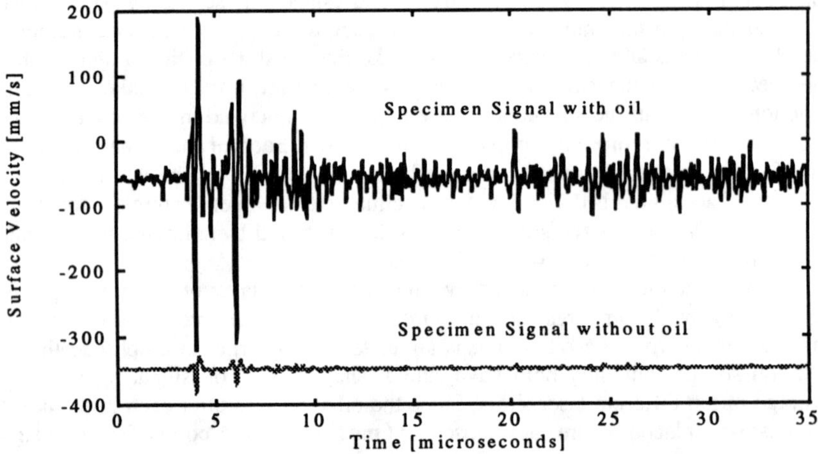

Figure 3: Specimen Signals Obtained with Oil and without Oil

created with the oil source in the specimen, then operated on by the transfer function, with an actual half-space signal (Fig. 4), clearly shows that while the transfer function is effective in removing the geometric features, it also adds spurious noise to the reproduced signal.

As a second test, the transfer function is used to remove the specimen geometry effects from a double pulse source signal. This double pulse signal is created by the same Nd:YAG laser; it produces two distinct pulses at higher energy levels. Unfortunately, the separation time between the two pulses is not completely repeatable, so it is only possible to average two signals. Figure 5 compares the double pulse signals in the specimen and the half-space (both in the ablation regime).

Note the increased noise in the double pulse signal caused by the small number of averages. Figure 6 compares the reproduced half-space with the actual half-space; the specimen geometry is again completely removed with the transfer function. However, the reproduced half-space signal contains more noise than the actual half-space signal.

Now the robustness of the transfer function is shown by removing the geometry from a completely different source, a pinducer. The pinducer source is different from the laser ablation source, because it is in contact with (and thus loads) the specimen and, it has a larger cross sectional area. In the following experiments, the pinducer is driven with a short time pulse. In order to get a reasonable SNR, all of

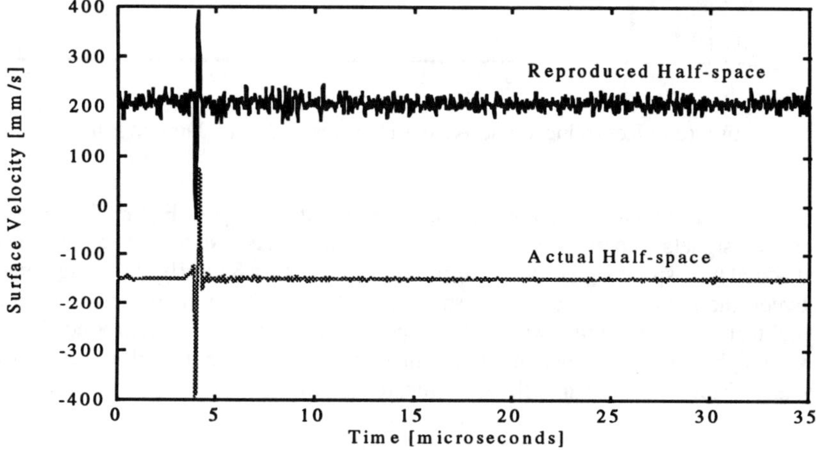

Figure 4: Reproduced and Actual Half-space Signal, Oil

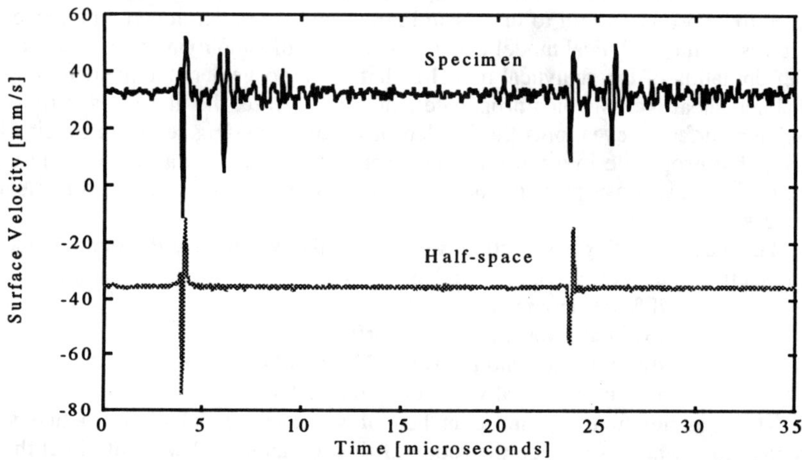

Figure 5: Double Pulse Signal: Specimen and Half-space

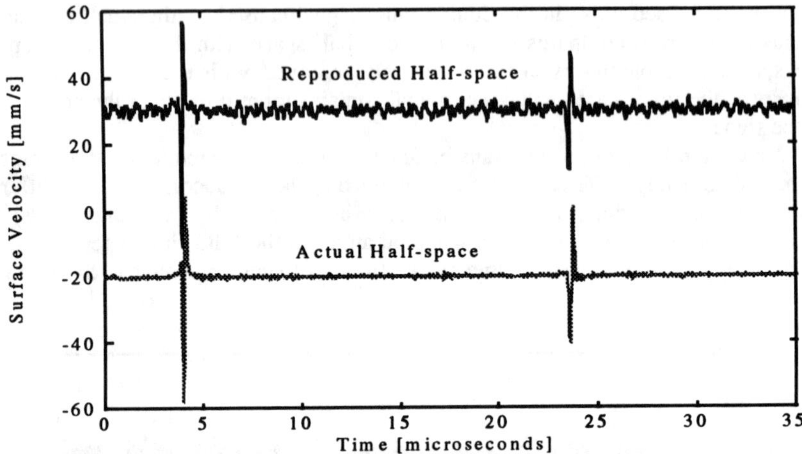

Figure 6: Reproduced and Actual Half-space Signal, Double Pulse

the following pinducer signals represent thousand averages. Figure 7 compares specimen signals created with a pinducer source and a laser source. The reproduced and actual half-space signals are contrasted in Fig. 8; Fig. 8 clearly shows agreement between the arrival time, the early shape and the amplitude of the most dominant signal feature: the Rayleigh wave. However, the noise level of the reproduced signal is very high and it is obvious that the transfer function does not work effectively for a vastly different source in such a complicated geometry.

DEVELOPMENT OF THE ANALYTICAL MODEL

The analytical model does not duplicate the experimental procedure; it is only meant to be used as a tool to understand trends and the effect of certain parameters. As a result, the analytical model contains a number of simplifying assumptions. The chief limitation of the analytical model is that it only considers the incident longitudinal and shear waves which propagate along the surface, the incident and reflected Rayleigh surface waves, plus the incident and corresponding reflected longitudinal waves that propagate into the specimen. This model is not an accurate representation of the actual wave propagation in a specimen; it does not consider a number of features.

The second limiting assumption is that all of the waves have the same frequency and that the input energy is distributed as follows:

70% Rayleigh wave
15% shear wave along the surface
10% reflected and transmitted longitudinal wave
5% longitudinal wave along the surface

These assumptions are made in lieu of an exact model of the actual wave propagation (which is outside the scope of this research) and will not effect the results. The specimen is assumed to be aluminum with velocities given as:

longitudinal wave: $c_L = 6420$ m/s
shear wave: $c_S = 3130$ m/s
Rayleigh wave: $c_R = 2930$ m/s

The ratio between the reflected and the incident Rayleigh waves at a 90° corner is taken as 0.4 (Li et al., 1992). The arrival times and amplitude ratios of the propagating waves are calculated using these assumptions, taking the initial Rayleigh wave amplitude equal to unity, and applying a spherical spreading model. Note that the only waves present in the half-space are the longitudinal and shear along the surface, plus the incident Rayleigh wave.

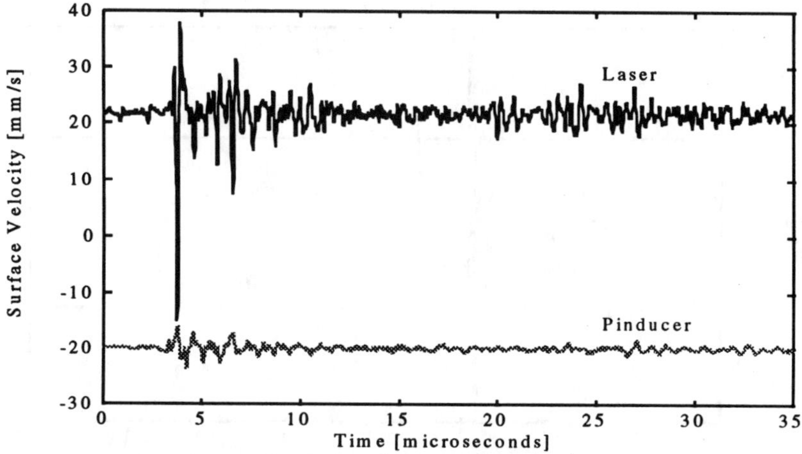

Figure 7: Specimen Signals with Laser and Pinducer (Pulse Input)

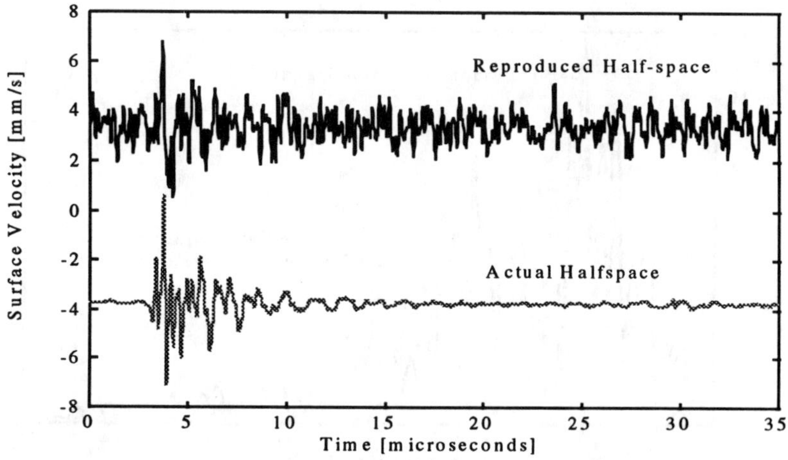

Figure 8: Reproduced and Actual Half-space Signal, Pinducer (Pulse Input)

APPLICATION OF THE ANALYTICAL MODEL

This analytical model is used to calculate artificial specimen and half-space signals that are shown in Figure 9. Each waveform is a single cycle of a sine wave. These signals are used to produce a transfer function by following the procedure presented earlier. This transfer function is used to determine the robustness of the proposed technique; it is used to remove the specimen geometry from new waves that contain different features from the original signals.

The first factor examined is the effect of frequency. A second set of waveforms are generated with a different frequency from the original signals, but all other para-

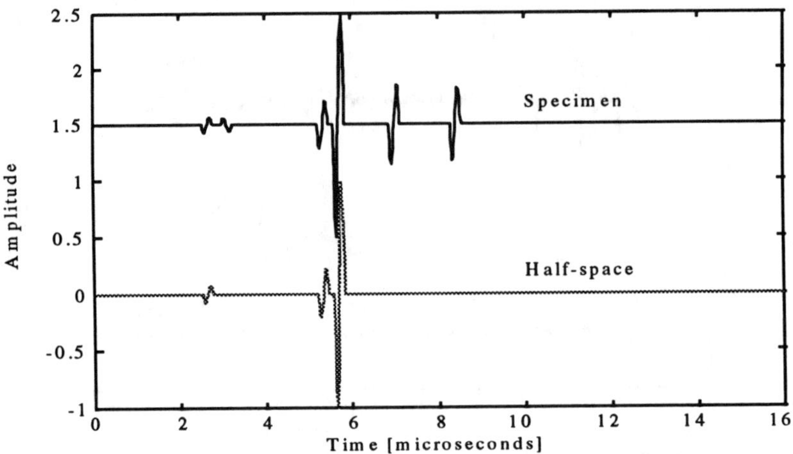

Figure 9: Specimen and Half-space Signal used for Transfer Function

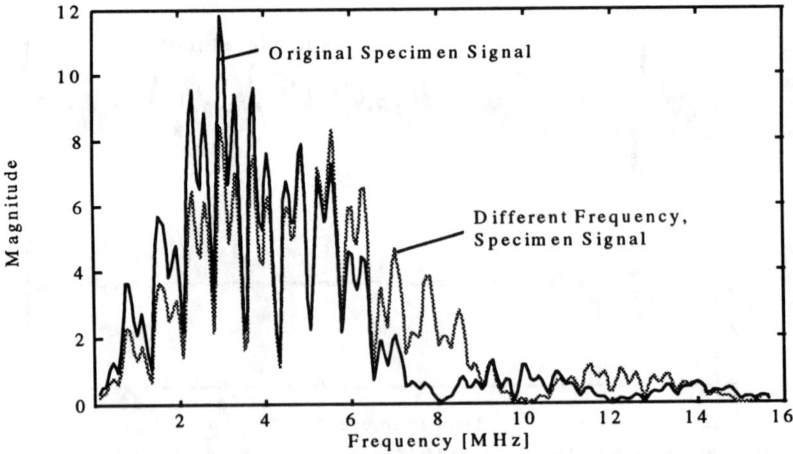

Figure 10: FFT's of the Specimen Signals with two Different Frequencies

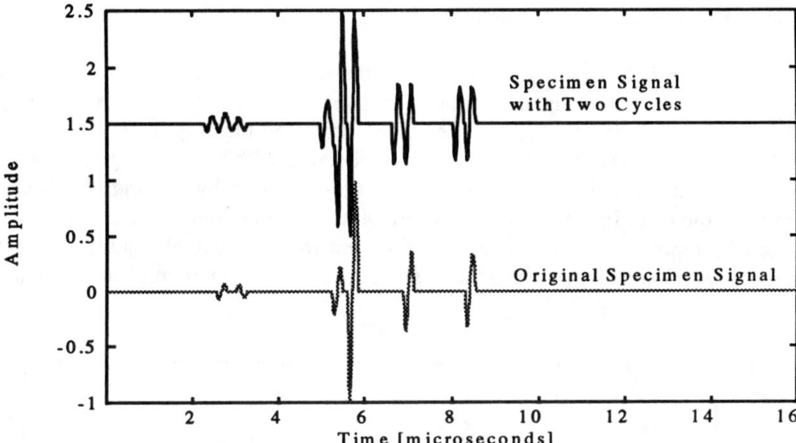

Figure 11: Specimen Signal with Two Cycle versus Original Specimen Signal

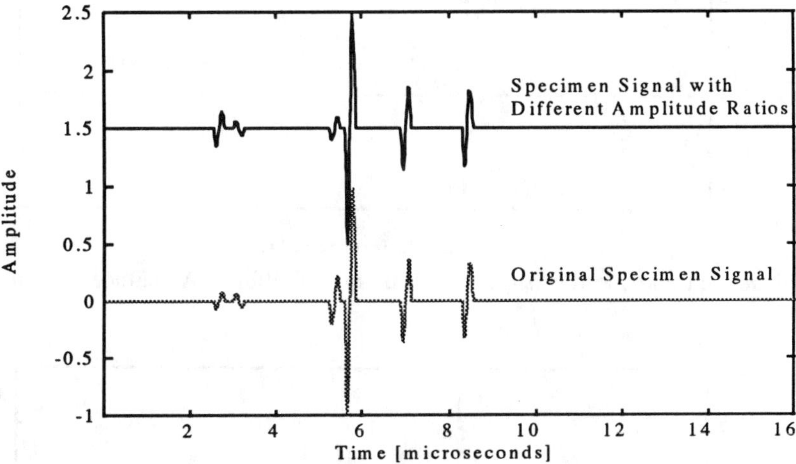

Figure 12: Signal with Different Amplitude Ratios versus Original Specimen Signal

meters are kept the same. A comparison of the frequency spectrum of the original and the new signal is shown in Fig. 10. The comparison of the reproduced half-space signal with the actual half-space signal shows that there is no difference between the two signal, thus the frequency content of a signal is not a critical factor.

The second factor investigated is the influence of the shape of the waveform; the original transfer function is used to operate on a signal that consists of two cycles of the same sine wave. Figure 11 shows the original specimen signal and the specimen signal with the two cycle sine wave. Again the transfer function exactly reproduces the two cycle sine wave signal in the half-space. These results show that multiples of the same signal (such as ringing) can be removed by the transfer function.

The next step is to investigate changes in the ratio of the different wave amplitudes. For example, a new signal is created by increasing the amplitude of the longitudinal wave propagating on the surface while decreasing the amplitude of the shear wave. Figure 12 is a comparison of the original signal with the new signal; both in the specimen. The new signal in the specimen is operated on by the original transfer function and the resulting signal is compared with the actual half-space signal in Fig. 13. Since the reproduced signal is clearly different from the actual signal, the original transfer function can not be used on a signal that has different amplitude ratios.

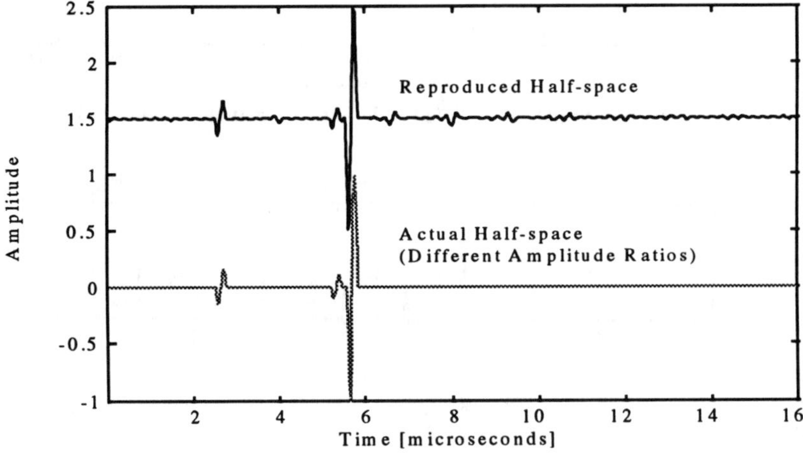

Figure 13: Reproduced versus Actual Half-space (Different Amplitude Ratios)

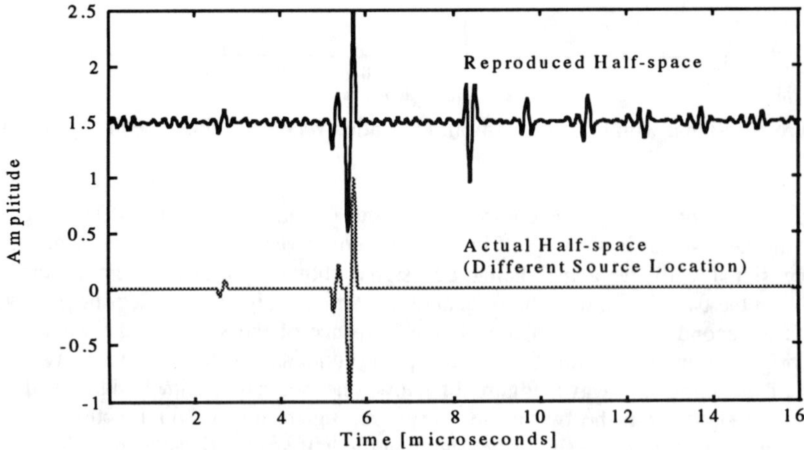

Figure 14: Reproduced versus Actual Half-space (Different Source Location)

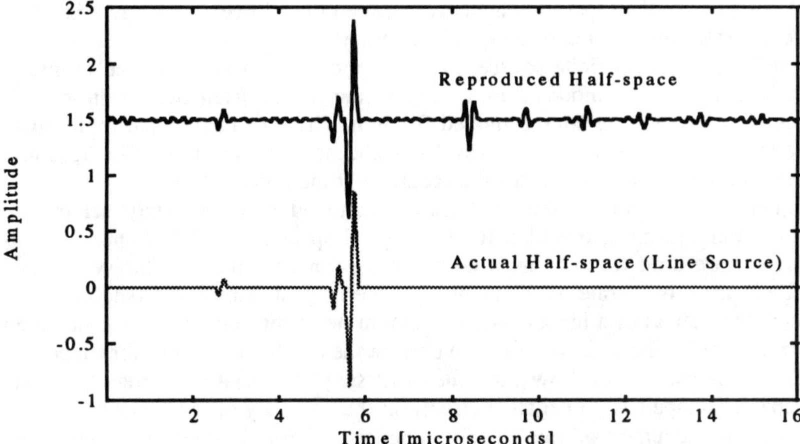

Figure 15: Reproduced Half-space versus Actual Half-space (Line Source)

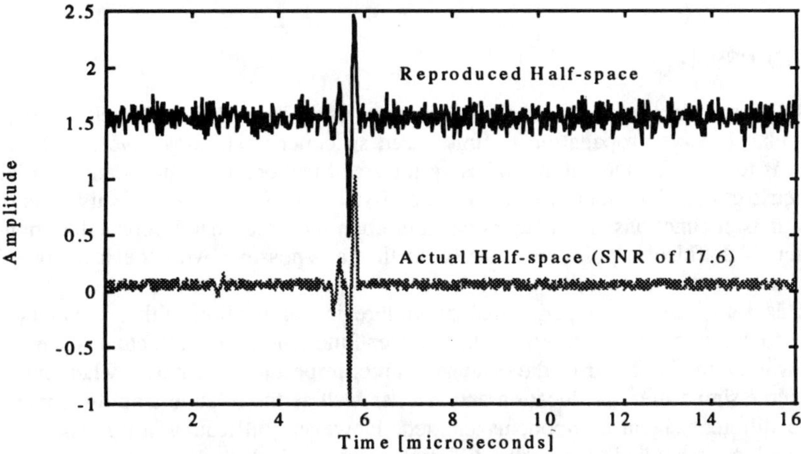

Figure 16: Reproduced Half-space versus Actual Half-space

The separation distance between the generation source and receiver is the next factor investigated with the analytical model. The distance between the original source location and the new source location is 0.1 mm. This case is important because small changes in source and receiver location are often introduced when an experiment is repeated. The new arrival times and wave amplitudes are calculated repeating the same procedure described earlier. There is no discernible change between the original specimen signal and the specimen signal with the new source. Figure 14 shows the signal with the slightly changed location that is operated on by the original transfer function compared with the actual half-space signal. It is clear that this small change in location has a great effect on the reproduced half-space signal. As a result, the separation distance between the source and receiver is a critical factor for the application of a transfer function.

The influence of a finite source size, as opposed to a point source is investigated. A line source is modeled by averaging the results from two locations: the original location and the source moved 0.1 mm. The averaged signal represents a line source. As expected from the results of the previous example, Fig. 15 shows that the line source greatly reduces the accuracy of the transfer function.

Signal-to-noise-ratio is the final factor examined with the analytical transfer function. Random noise is added to the original specimen and half-space signals. The signals are used to compute a new transfer function and this "noisy" transfer function is used to operate on a "noisy" specimen signal. This result (shown in Fig. 16) model the effect of a higher level of random noise throughout the measurement system. Figure 16 shows how noise can corrupt the results of the transfer function.

This analytical model shows, in spite of its simplicity, that a different frequency or a repeated shape has no negative effects on the accuracy of the transfer function. However, the accuracy of the transfer function is strongly dependent on several factors. These factors are: different amplitude ratios of waves; separation distance between the source and receiver; source size; and noise. As a result, these factors must be accounted for (or eliminated) in any application of the transfer function.

CONCLUSION

This paper demonstrates the effectiveness of using laser ultrasonic techniques to study elastic wave propagation in finite sized specimens. The objective of this research is to establish the validity of using transfer functions to quantify and remove geometric effects from measured acoustic emission waveforms. The effectiveness of these transfer functions is directly dependent upon the broad band, repeatable, non-contact, high fidelity, point measurements that are possible with the laser based techniques used in this work.

This study uses an experimental procedure in conjunction with an analytical model to show that the concept of the transfer function is most effectual when the unknown source is similar to the original source. Important parameters when defining source similarities are: location and size, as well as the relative amplitude ratios of the different signal components created. However, difficulties arise when the methodology is applied to a vastly different source or if there is excessive noise in the signal.

REFERENCES

1. Bruttomesso, D. A., L. J. Jacobs, and R. P. Costley, 1993. "Development of Interferometer for Acoustic Emission Testing." *Journal of Engineering Mechanics*, Vol. 119 (11), pp. 2303-2316.

2. Li, Z. L., J. D. Achenbach, I. Komsky, and Y. C. Lee, 1992. "Reflection and Transmission of Obliquely Incident Surface Waves by an Edge of a Quarter Space: Theory and Experiment." *Journal of Applied Mechanics*, Vol. 59, pp. 349-355.

3. Hurlebaus, S., 1996. "Laser Generation and Detection Techniques for Developing Transfer Functions to Remove the Effect of Geometry on Elastic Wave Propagation." *MS-Thesis*, School of Civil and Environmental Engineering, Georgia Institute of Technology, Atlanta, GA.

SESSION 6

AEROSPACE APPLICATIONS

Aircraft Structural Health Monitoring, Prospects for Smart Solutions from a European Viewpoint

G. BARTELDS

ABSTRACT

Structural health monitoring is an important safety factor in aviation that might benefit from advanced smart systems for damage sensing and signal processing. Current levels of structural safety and reliability do not present a particularly strong case for smart systems but cost considerations related to inspection and maintenance do. As an added benefit problems of poor accessibility and negative effects of human factors in inspection might be reduced.

The implementation of such system requires development and demonstration by dedicated and qualified multidisciplinary teams, acceptance by aircraft designers, manufacturers and operators and approval by the authorities. Current European collaborative schemes and the associated funding in conjunction with an apparent interest among potential end users provide excellent prospects for the realisation of smart solutions.

STRUCTURAL HEALTH AND USAGE MONITORING: WHY?

Structural health is directly related to structural performance and in this respect it is a governing parameter with regard to safety of operation. This aspect of structural health is particularly relevant to transportation systems including their infrastructural elements and in this connection structural health monitoring is a **safety issue**.

At the same time a change in structural health may affect structural performance to a degree that remedial actions become necessary. Structural repairs increase the cost of transportation in at least two ways. First, the design and implementation of repairs implies *direct costs*. Second, the execution of repairs generally requires the transportation system to be temporarily taken out of service and this induces

National Aerospace Laboratory, NLR. P.O. Box 153, 8300 AD Emmeloord, The Netherlands

indirect costs due to the loss of production volume or as a result of leasing a substitute system.

To reduce repair and maintenance cost one might attempt to repair at a very early stage of damage development to limit direct costs. Alternatively, it might be decided to postpone repair until the transportation system has to be taken out of service for scheduled major overhauls to reduce indirect costs. In this connection structural health monitoring becomes a **cost issue**.

In case of the latter option (relying on the delay measure) it may be necessary to adapt operational usage to limit or even stop damage growth. If sufficient knowledge exists to relate damage rates to mission types this can be achieved by **usage monitoring**.

In general usage monitoring can be viewed as a valuable addition to structural health monitoring. Prescribed maintenance schedules are based on an *estimated usage pattern*. Knowledge of the *actual utilization* can be translated into a severity parameter that can be compared to the value corresponding to the estimated loading spectrum. In this manner prescribed inspection intervals and times between overhauls can be tuned to actual needs.

It is worthy to note that there are substantial differences in damage development and as a consequence in the manner structural health will deteriorate with time between metal and composite structure. Whereas in metallic components cracking is a gradual and predictable process with a high probability of occurrence the wear-out of a composite component as a result after loading environment is much less pronounced but composites may suffer from discrete traumas due to accidental damage of a non-predictable random nature. The situation suggests that different health monitoring philosophies should be applied to the two families of structural components.

STRUCTURAL HEALTH MONITORING: HOW?

Structural health, or equivalently, the state of damage can be established either directly or indirectly. In the latter indirect approach structural performance or rather structural behaviour is measured and compared with the supposedly known global response characteristics of the undamaged structure. If the effect of certain damages on structural response characteristics is known this approach provides an *indirect measure* of damage and of structural health.

In a direct manner one checks for the damage type under consideration, like cracks, corrosion or delaminations, by applying an appropriate inspection technique. These techniques, based on physical phenomena, in fact sometimes amount to response measurements also but in this case they have a very *local and direct character*. The established inspection techniques vary from visual inspection by the naked eye to passing the structure through a fully automated inspection gantry.

Obviously in both the direct and indirect approaches the **sensitivity** and the **reliability** of inspection are important quantitative performance measures. They

are determined on the one hand by the laws of physics but on the other in practice also by the hardware and software quality of the inspection equipment and last but not least by the equipment operator: the inspector.
In this connection human factors like the loss of alertness in case of rare occurrences of damage and inspector fatigue in case of long and tedious inspections are important reasons to consider a smarter solution to inspection as an element of structural health monitoring.

OPTIONS FOR SMART SOLUTIONS

It is for both sensitivity and reliability that the particular features provided by **smart technologies** are considered.
Smart sensors could provide greater sensitivity provided that they are properly installed. This option is clearly related to specific inspections at precisely known critical locations that in addition may be poorly accessible. On the other hand, smart sensor systems with advanced data processing are relevant for inspecting larger areas for a variety of defects. If such systems function continously the time between inspections is effectively zero and then a moderate sensitivity might suffice.

In a more general sense smart system design and smart interpretation and use of data generated by the systems are desirable features in any solution and in this context it is necessary to define what is meant here by *smart solutions* to structural health monitoring requirements:

in the present paper smartness relates to either sensors for damage detection including their installation or to signal processing and presentation.

IS THERE A CASE FOR SMART SOLUTIONS IN AIRCRAFT?

In the first chapter of this paper structural health monitoring was identified first of all as a safety issue. Certainly in air transport where structural failures may lead to fatal accidents the safety of operation is of prime consideration. Continuous research in the areas of fatigue and corrosion of metallic aircraft structure including inspection techniques (sometimes spurred and accelerated by dramatic accidents or incidents) has helped to achieve a very high level of structural reliability. Design for damage tolerance is now widely applied. It relies on a very profound understanding of material behaviour, on a very accurate description of the loading environment (both external and internal) all of this in combination with advanced manufacturing techniques and, of course, proven and reliable inspection and maintenance procedures. And in situations where brittle material behaviour or poor accessibility with regard to inspection are in the way of a damage tolerant design approach detailed numerical analysis supported by advanced testing has produced low crack growth or safe life structure.

Any interest for automated integrated inspection systems could then result only from a need for greater **reliability of inspection**: the damage tolerance chain is only as strong as its weakest link which probably is inspection.

It is thought that from a safety of flight position there is not a strong case yet for smarter solutions. Only in special situations an integrated sensor system may provide greater reliability than current methods. However, if in view of the rapidly growing air transport volume, expressed in billions of passenger miles flown, a significant reduction in structural failure rates is needed smart solutions may become more relevant as a safety issue.

Another more important factor stimulating the development of smart systems, however, is the **cost of inspection**.

There is very little published data on the potential for cost reductions but the inspection efforts applied in current aircraft maintenance procedures are very considerable and moreover inspector training and motivation require continuous attention. It must be mentioned here that significant improvements have been achieved in traditional inspection equipment with regard to inspector friendliness and quantitative data presentation.

A recent study on inspection requirements for a modern fighter aircraft (featuring both metal and composite structure) revealed that an estimated 40 percent plus can be saved on inspection time by utilizing smart monitoring systems. The situation at hand is illustrated in the table below.

Inspection type	Current inspection time (% of total)	Estimated potential for smart systems	Time saved (% of total)
Flight line	16	.40	6.5
Scheduled	31	.45	14.0
Unscheduled	16	.10	1.5
Service instructions	37	.60	22.0
	100		44.0

Another estimate derived for a fully automated impact sensing system for a composite structure, based on the use of integrated distributed piezo sensors in combination with advanced signal processing software arrives at a 50 percent saving on regular inspection time again for a fighter aircraft.

Admittedly, these estimates are based on data derived from laboratory demonstrators. They provide a drive, however, for the development of full scale demonstrators of smart structural health monitoring systems. In fact a major programme, to be discussed in more detail further on, recently got underway on the basis of the assumption that up to 20 percent of current maintenance and inspection cost can be saved in civil and transportation by the use of integrated on-line damage monitoring systems.

So, the case for smart solutions to aircraft structural health monitoring requirements derives from cost considerations.

The development of integrated automated damage sensing systems relies on different research disciplines and in addition it affects design and manufacture as well as operation and maintenance. As primary flight systems such as the airframe, landing gear or engines are involved the airworthiness authorities will have to be involved. Obviously, the development risks of smart systems are considerable and at the same time a broad acceptance among all parties involved is necessary to achieve implementation.

These considerations have led, in Europe, to a number of initiatives aimed at setting up collaborative research and development projects. Not only countries that have significant aerospace programmes but also smaller nations with advanced system component expertise are involved in projects that are described in the next chapter.

EUROPEAN FRAMEWORKS FOR DEVELOPMENT OF SMART TECHNOLOGIES

The European Unions Directorate General for Research has funded so called Framework Programmes for research and technology development and demonstration since 1987.

Currently, the fourth Framework Programme is underway and in the four years' time frame between 1994 and 1998 the European Commission will provide 12 billion ECU split between different areas as shown below.

Information and Communication Technologies	28 percent
Energy	18 percent
Industrial Technologies	16 percent
Life Sciences and Technologies	13 percent
Socio-economic research, cooperation with third countries etc.	10 percent
Environment	9 percent
Training and mobility of researchers	6 percent

For each project the funding provided by the EC has to be supplement to the same amount by the contractors.

The programme on industrial (and material) technologies comprises an aeronautical chapter that addresses, among others, methods for improved operation and maintenance.

Under that heading a 4.7 MECU project was recently funded for the development and demonstration of on-line, integrated technology for operational reliability, MONITOR. A consortium led by British Aerospace and comprising all major

Airbus and AIR partners as well as research establishments in aeronautics and optics from seven different countries will develop and demonstrate integrated automated systems for damage detection and for load path monitoring. The systems will employ fibre optic sensors as well as the more traditional acoustic emission or lamb wave sensors and they will be implemented in two full scale ground based demonstrators (a composite and a metallic structure). Further the operational load path monitoring system will be flown also.

The project team interacts with the potential end user community consisting of aircraft manufacturers and operators (including the maintenance firms). Very early in the project the end users were invited to respond to a questionnaire clarifying the monitoring options considered in the project. The contacts established will be maintained during the project by performing interviews with the more engaged parties and by organizing workshops and demonstrations as the developments progress.

The response to a first attempt to capture the end user requirements by questionnaire already allows a ranking of inspection targets that might benefit from smart solutions (see diagram below).

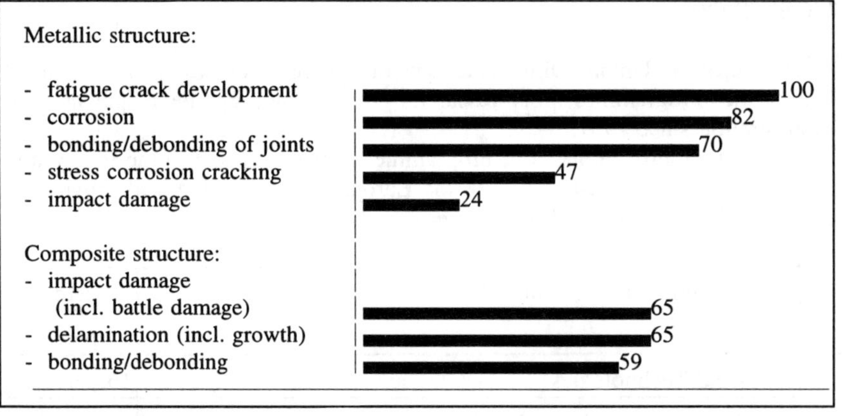

percentage of respondents with positive interest

There is a very strong support for automated integrated inspection concepts from all sides, but the interest is based not only on the cost reduction aspects but also on the potential of performing automated inspection in poorly accessible locations and on the prospect of reducing human factor effects on inspection reliability.

The Western European Armament Group (WEAG) comprising all European NATO countries also stimulates research and technology development, in principle for defence purposes but it pursues coordination with the civil oriented programmes such as the Framework Programme 5 now under consideration. Its efforts are organized under a framework programme called European Cooperation for the long term in Defence (EUCLID).

WEAG currently develops Research and Technology Projects (RTP's) in a number of Common European Priority Areas (CEPA's). In the CEPA devoted to

Advanced Structures and Materials there is considerable interest now in smart materials and structures. In 1995 an experts group with representation from five countries has been formed that is tasked with the identification of opportunities for smart applications to be developed in special RTP's.

The potential applications for smart materials and structures have been categorized as follows:

- active adaptive vibration control
- structural health and usage monitoring
- shape control of airfoils and antenna's.

Workshops are held on all three subjects and the subject of structural health and usage monitoring will be covered at a joint WEAG-NATO workshop at The Hague on 7 and 8 October of this year. US participation is welcomed to broaden the coverage and to establish opportunities for coordination.

RTP's are organized by WEAG nations that contribute equal-value shares to the project and provide for financial coverage of their share according to national rules and regulations. Funding for inidividual RTP's generally is in the 5 to 15 million ECU bracket.

Finally, an important mechanism for research coordination formed in 1973 is GARTEUR, the Group for Aeronautical Research and Technology in Europe. It aims to strengthen collaboration in aeronautical research and technology between European countries with major research and test capabilities and with Government funded programmes int his field. The group consists of six countries now and it is active in the following domains:

Aerodynamics
Structures and Materials
Flightmechanics, Systems and Integration
Helicopters
Propulsion Systems

An exploratory group has studied the current state of the art in smart structures and materials and has performed a cost-benefit evaluation of potential application in structural health monitoring. It has decided not to develop a GARTEUR activity in this field at this moment as the EC MONITOR project is running, but it will embark on an active adaptive vibration project as a precursor for a planned EUCLID project.

SUMMARIZING CONCLUSIONS

- Aircraft structural health monitoring is an essential element for continued safe operation. Current design capabilities and manufacturing and certification standards guarantee an extremely high level of structural reliability that can be maintained during the operational life of the aircraft provided that prescribed inspections are carried out, that data are processed and that remedial actions are taken when necessary. As a consequence, safety requirements do not contribute a strong case for advanced, smart, structural health monitoring systems with the possible exception of the requirement to limit the negative effects of human factors on inspection reliability.

- Both the direct costs of carrying out preventive inspections and the indirect costs associated with interrupted service, however, provide a strong stimulus for cost reduction programs. In this respect integrated damage sensing systems, advanced signal processing and maintenance oriented data presentation constitute smart solutions to inspection requirements that may reduce the cost of manpower for inspections and maintenance and at the same time increase reliability and enhance data presentation.

- Aircraft manufactures and operators have indicated that they would like to see more integrated automated inspection systems provided that they do offer a cost benefit and possibly are more reliable when compared to current inspection methods. They should not interfere with other flight systems and preferably be communicative to maintenance personnel.

The authorities will accept such smart systems as long as they do not adversely affect current safety levels.

- Current international programs for the development and demonstration of integrated damage sensing systems for aircraft structural health monitoring in Europe provide the opportunity to achieve a breakthrough for existing technology towards actual application. The broad participation representing all the different key expertise needed, the obvious interest among the potential end user community and the financial support by international bodies are important assets in the current efforts to demonstrate and exploit smart health monitoring systems.

Structural Health Monitoring of DC-XA LH_2 Tank Using Acoustic Emission

Q. HUANG and G. L. NISSEN

ABSTRACT

As part of integrated vehicle health management system developed by the Space System Division of Boeing North American, acoustic emission technology is taking an important role in the structural health monitoring. The Boeing North American's Acoustic Emission Flight Instrumentation System has achieved a world's first in recording structural acoustic emission during a flight of a rocket powered vehicle. For the first time, acoustic emission has been successfully monitored in a severe acoustic and temperature environment aboard the McDonnell Douglas' DC-XA under a national cooperative research agreement, during both static and flight testing. Some results from the DC-XA composite liquid hydrogen tank will be presented.

1

INTRODUCTION

The Delta Clipper-Advanced Experimental Vehicle (DC-XA), a prototype reusable launch vehicle, was developed by McDonnell Douglas for the NASA's single-stage-to-orbit program. It was an unmanned, rocket powered, and vertically takeoff/landing spacecraft (Figure 1a). As a flying technology testbed, one important technology tested on DC-XA was the polymer composite liquid hydrogen tank (Figure 1c) made of IM7 carbon fiber/epoxy. The composite tank was cylindrical in shape with elliptical domes, 8 ft in diameter and 16 ft tall. An internal thermal insulation was applied and bonded to the inside wall of the tank. Since it was the first of its size and material system to contain liquid hydrogen and to be tested under actual flight loads and conditions, the tank structural health conditions at all stages of the ground and flight tests became an important aspect of a series of tests.

The Boeing North American Acoustic Emission Flight Instrumentation System (AEFIS) monitored the integrity of the composite liquid hydrogen tank during both rocket powered static and flight testing. During two flights of the vertical

Qixin (Jerry) Huang and Gerry Nissen, Space Systems Division, Boeing North American, Inc., 12214 Lakewood Blvd., MS FC-97, Downey, CA 90242

takeoff/vertical land vehicle at White Sands Missile Range, New Mexico, the AE sensors provided vital data from the vehicle's cryogenic liquid hydrogen tank as it encountered flight stresses. The autonomous AEFIS system (Figure 1b) provided continuous monitoring during flight without degrading or interfering with other onboard systems.

Some key issues to be addressed during the test were to determine: (1) the feasibility and capability of AE technique for real-time monitoring of structural integrity during flight of reusable launch vehicle, (2) how in-flight vibro-acoustic environment effects sensor output, (3) how in-flight issues can be designed out quickly and inexpensively, (4) how the data acquisition performs in a spacecraft environment, and (5) overall test of AE system engineering prototype design.

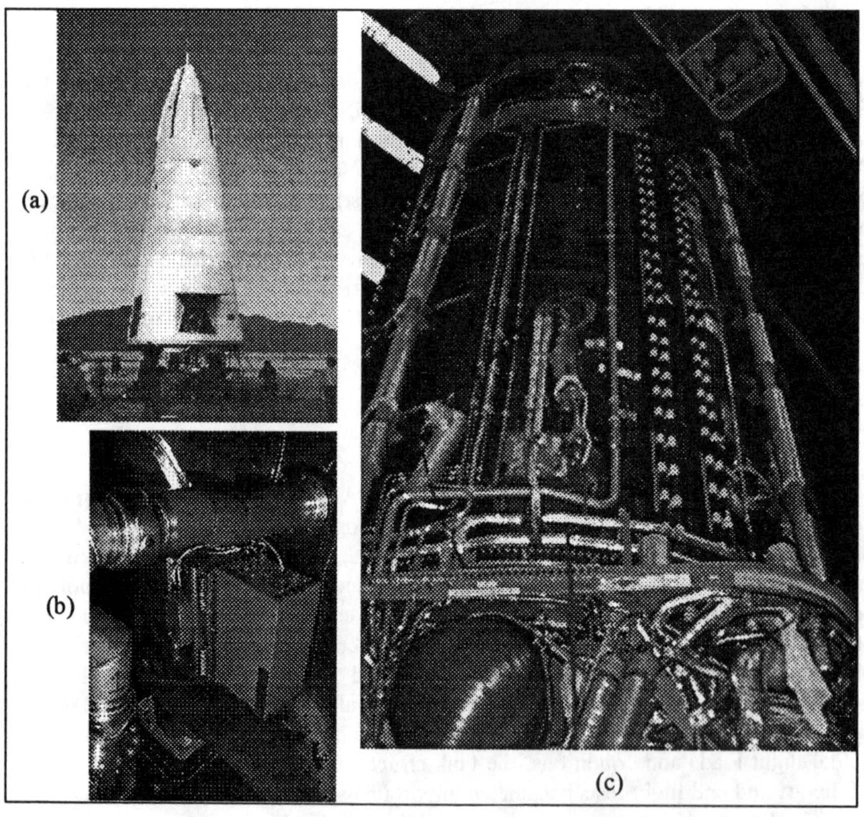

FIGURE 1. DC-XA vehicle, composite liquid hydrogen tank and AEFIS experiment installation.

EXPERIMENTAL

PREPARATION

To satisfy the flight requirements, the Boeing North American Space Systems Division and Physical Acoustics Corporation (PAC) developed a prototype Acoustic Emission Flight Instrumentation System (AEFIS). The system consists of a 4 channel AE system and a 486 computer inside a 6.5"x9.5"x15.5" chassis. It is EMI-shielded and shock/vibration resistant. Four AE sensors of three kinds, a Pico, a Mini-30 and two D9215 sensors (PAC), were used in the flight experiment. Hysol EA9394 epoxy was selected as the adhesive for bonding AE sensor to the composite surface. The sensor and adhesive selections were based on product information and a series of performance tests in the temperature range from the ambient temperature to -423°F.

SETUP

The AEFIS chassis was mounted on the No. 4 landing gear strut housing using a shock reduction device. Three sensors, one Pico and two D9215, were bonded in a 10" per side triangular array, and a Mini-30 sensor was bonded inside the array, close to one side of the triangle. The sensor array was just above the lower Y-joint of the skirt and the tank. Both of the traditional AE and waveform data were recorded during the flight testing. Waveform analysis helped characterizing features of the background noises. A real-time filtering method combining a high-pass filter and a software based digital filter was employed to eliminate the background acoustic and EMI noises. The noise sources were from onboard electromagnetic devices, engine firing and aeroshell acousto-vibration. The intensity of the noise from the rocket engine firing and aeroshell vibration during flight was approximately 160~170 dB. The detected AE data was stored in the onboard AEFIS unit during flight, and downloaded to a Laptop PC after landing. In the post data analysis, both the traditional AE parameter analysis and advanced waveform analysis were performed for evaluation of the structural condition, and failure mechanisms and locations if there is any structural failure.

RESULTS AND DISCUSSION

The AE experiment was highly successful. It met all the test requirements and objectives. The feasibility and capability of AE technique for real-time monitoring of structural integrity during flight of reusable launch vehicle have been demonstrated. Several technical barriers for AE application have been overcome. The results are discussed in two parts, validation of the AE technology and the captured anomaly during the flight test.

VALIDATION OF AE TECHNOLOGY

Would the AE data be corrupted by the extreme noise dominated environment was the most concerning question posed by this application of AE. To show the validation of the AE data, i.e., AE signals from the composite tank structure, some AE feature data and digitized waveforms are given in Figures 2~5.

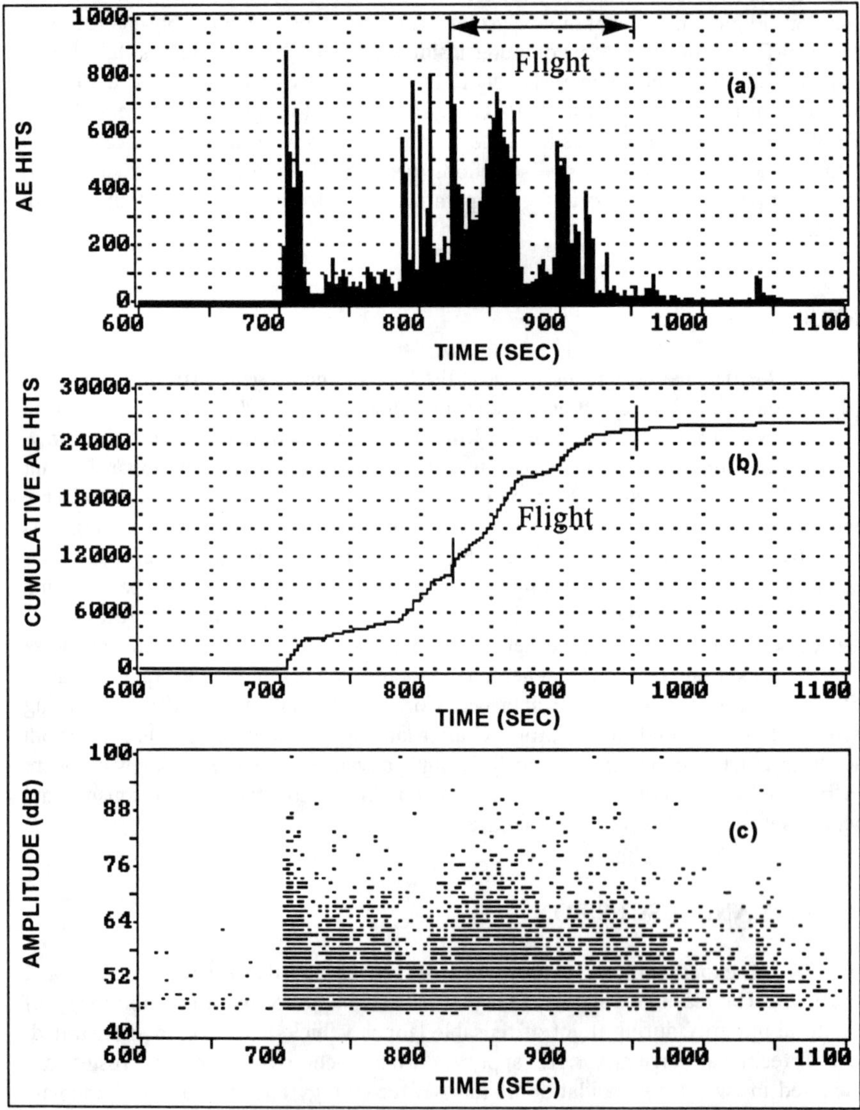

FIGURE 2. A general description of AE data recorded during a DC-XA flight test.

Though there are many ways to present a set of AE data in terms of various signal parameters and their combinations, we will use some commonly used AE parameters to show how the AE data response to activities associated to a flight process and stresses encountered by the composite liquid hydrogen tank. Figure 2 shows a general description the AE data recorded during a flight test where Fig. 1a is the acoustic emission history plot, Fig. 1b is the AE history plot also, but in a cumulative counts of AE hits, and Fig 1c is the distribution of the signal amplitude in the same time period as in Figures 2a and 2b. Figures 2a and 2b give a general idea about how AE responses to flight, and Fig. 2c gives amplitude distribution but also helps in AE system settings.

As we can see in these figures, the AE parameter profiles have well distinguishable features needed by the traditional analysis method based on AE parameters. This result is directly due to the effective filtering method applied in attempt to suppress the background noise. Without proper filtering, the recorded data from an earlier test has shown they were almost all noise signals, and the parameter profiles had no features. In that case, Figures 2a and 2c would appear like a band, and Fig. 2b would look like a increasing straight line. Another important finding is that the AE data recorded during another flight showed similar features as shown in Figure 2. Because such data analysis is almost real-time, it implies a quick evaluation based on a baseline data is possible. A deviation from the baseline data means a possible anomaly. We will discuss this issue later with a real anomaly event captured by AE monitoring during the test series.

The same AE data can be expressed in many different ways as mentioned earlier. For example, one can look at the AE energy vs. time and expand time scale for details in an interested time frame, as shown in Figure 3 for the same data as shown in Figure 1. The AE activities induced by vehicle takeoff and landing as well as other flight processes are clearly shown in Figure 3.

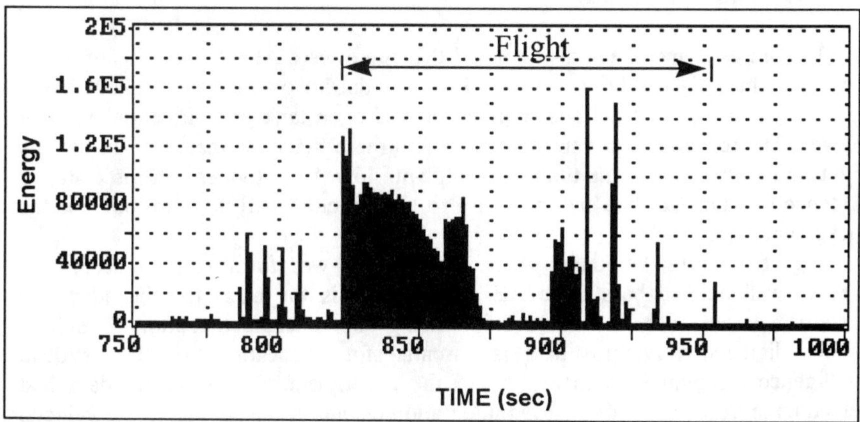

FIGURE 3. Alternative presentation of the same data as shown in FIGURE 1, in terms of signal energy and shorter time period to detail the part interested.

AE activities can be well correlated to the strain/stress data and the flight process such as the pressurization, launch, ascending/descending, acceleration /deceleration and landing.. By mapping Figures 2a, 3 and 4, and noting that there is a approximate 200 seconds time difference between Figure 4 and the rest of figures (subtract 200 seconds from Figure 4 to compare with other figures), we see that the AE activity peaks are corresponding to the strain increasing segments, and the valleys (low AE activities) or decreasing portions are matching the strain holding stages or decreasing portion.

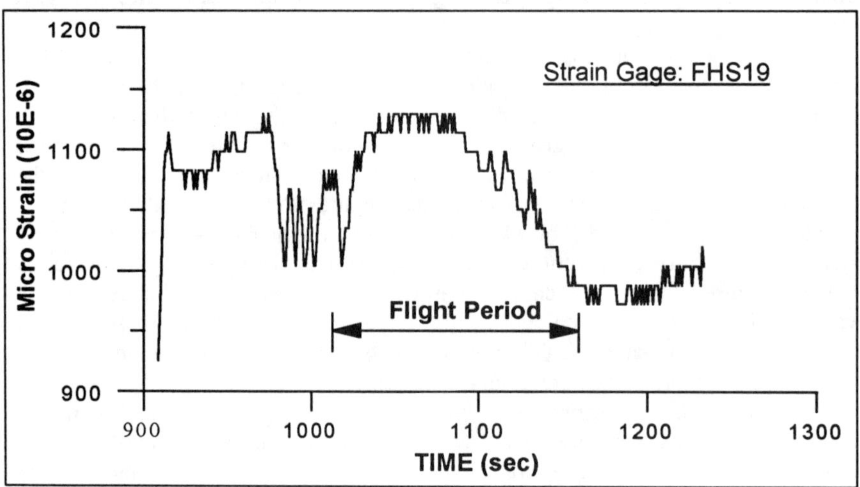

FIGURE 4. Micro strain measured at the location near the AE sensor array during the third flight testing.

In Figures 2 and 3, we also noticed that there were AE activities recorded in the pre-flight and after-flight periods. This was due to operations checkout being performed during these periods such as the hydrogen tank pressurization, venting the vehicle, etc.. The most important feature in these figures is that even during the flight period there are some time segments with low AE activities. This is a strong evidence of valid AE data that response to the structural stress/strain, but not the background noise.

Figure 5 shows the quality of the recorded AE waveform data. As seen, they are the well defined burst type AE signals. This is the basis for the advanced waveform based analysis method emphasizing source characterization. The high quality digitized waveform data is essential for applications of the artificial intelligence techniques. Actually, during the experiment, the waveform data had helped us in trouble shooting during integration testing of the AEFIS to the vehicle, identifying background noise and possible failure modes.

The detected AE signals were believed to be mainly from the micro-cracks in the composite material and the insulation system, as they encountered thermal stress at cryogenic temperature, pressurization and flight induced stresses. They were not harmful at that moment. Based on the data analysis, we concluded and reported to the flight control management that the tank was in a safe condition after each of the first three flights. The Kaiser and Felicity effects are observed in the flight data. They were used for data analysis and explanation, and should be taken into account in structural health evaluation.

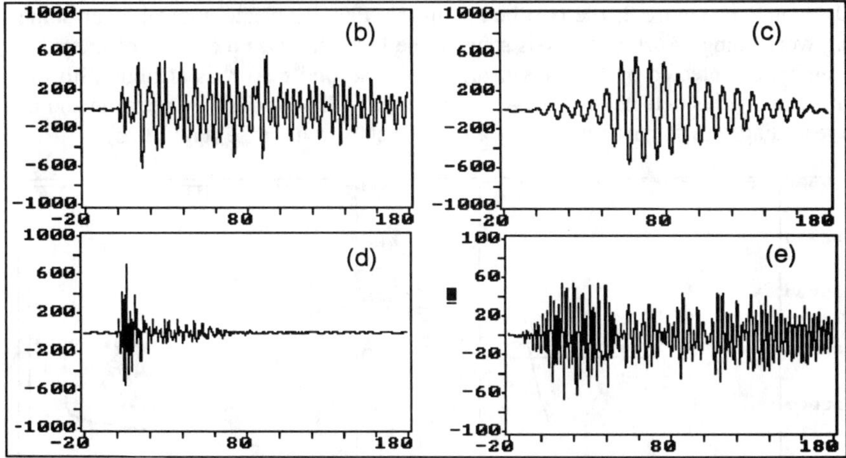

FIGURE 5. Some typical waveforms of AE signals recorded during DC-XA flight test. The horizontal axis in μs, and the vertical-axis in mV.

ANOMALY CAPTURED BY AE

During an attempted fourth flight, an anomaly was recorded by the AEFIS. Before the flight test was aborted, the tank was filled with liquid hydrogen at a temperature of -423 °F for several hours, much longer than the normal.

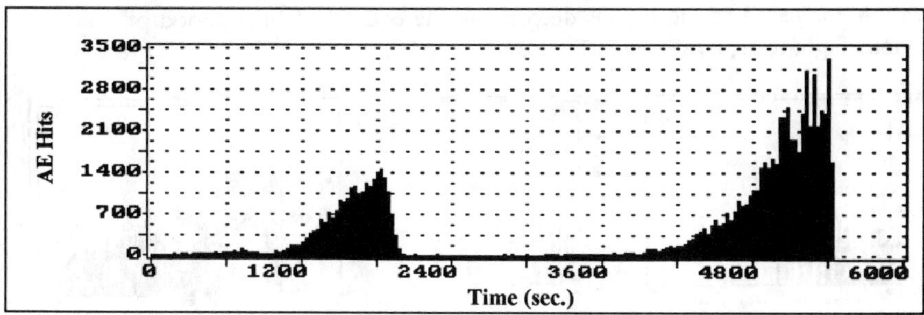

FIGURE 6. AE activities recorded before the fourth flight was aborted.

During this period, the tank was not pressurized and had no flight stresses. The stress condition in this situation was equivalent to that in the period before 700 seconds in Figure 2. In contrast to the low AE activity in that time period shown in Figure 2, approximately 85,000 signals were detected in 1 hour and 31 minutes (AEFIS was turned off at the time) in this aborted flight. Figure 6 shows the AE activities recorded during this period. Compared with the near zero AE hits before 700 seconds in Figure 2, the two bumps in this figure indicate anomalies or some thing were going. Further analysis showed the first peak was the result of intensive AE activities detected by the #1 sensor, and the second peak was attributed to the high AE activities detected by the #2 sensor (Figure 7a). At the time near the end of the recording, AE signals detected by #4 sensor start to increase (Figure 7b).

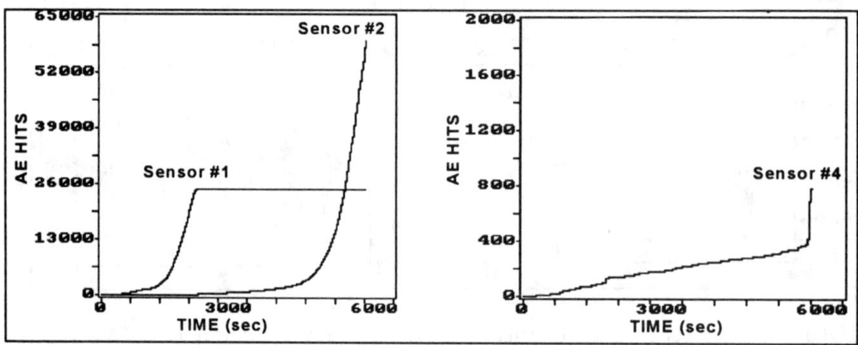

FIGURE 7. AE hits detected by different channels/sensors while the tank was filled with liquid hydrogen for a long period before this test aborted.

These three sensors were installed in 30 degree from the vertical direction with the #1 sensor on the top, #2 sensor at the middle and #4 sensor at the bottom. This suggests the AE sensors actually captured a AE activity movement and indicated the source was moving down from upper position and passing the sensor array. Analysis of the signal duration distribution showed these AE signals had much longer duration compared to the signals detected in the equivalent time period of the two previous flight tests, as shown in Figure 8.

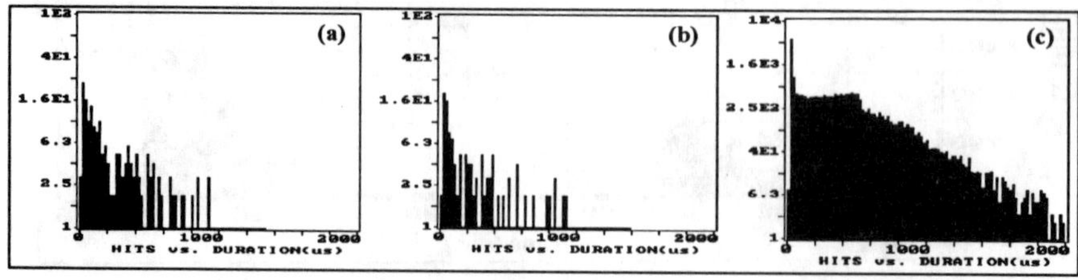

FIGURE 8. Duration distribution of AE signals detected during the pre-flight period. (a) the second flight test, (b) the third flight test, and (c) the aborted fourth flight test.

The long duration feature of the AE data recorded in this aborted flight test is highly similar to that found in the ground test of the same tank. This feature was characterized as the cryogenic insulation debonding and confirmed by inspection. The waveform analysis showed a large quantity of signals with features commonly characterizing the off-plane fracture phenomena. Based on our analysis, the anomaly was believed responsible to cryogenic insulation debonding caused by the thermal stress induced by thermal coefficient mismatch at the bondline between the composite material and the foam type insulation material. Another evidence of the insulation failure is the significant temperature drop recorded inside the vehicle in this aborted flight, compared to the temperature data recorded in the previous flight tests. The AEFIS chassis dropped well below freezing point in this test as shown in Figure 9 (even with a high system thermal load and no cooling fans). In the previous flight tests, the temperature inside the chassis always kept at between 60 °F and 80 °F. These evidences strongly suggest a failure of the tank insulation during the aborted fourth flight test. The resulted temperature drop below freezing point could cause problems to flight hardware inside the vehicle.

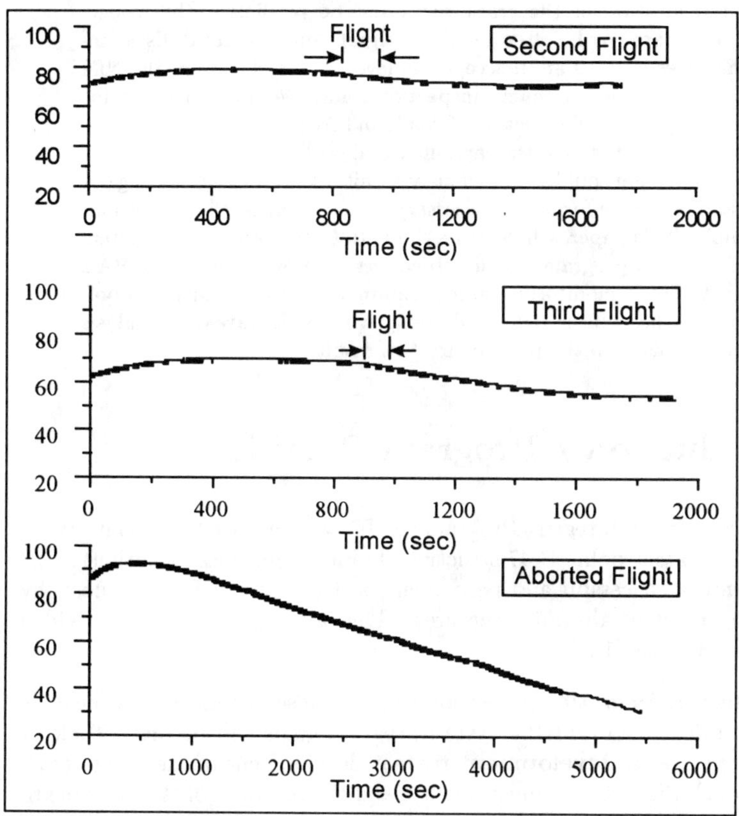

FIGURE 9. Temperature inside AEFIS chassis during DC-XA flight testing.

Damage Detection Experiments and Analysis for the F-16

I. SEARLE*, S. ZIOLA† and S. MAY‡

Abstract

All types of military aircraft suffer from structural cracking. Even though the location of the crack can often be predicted, the inspections required are costly, often requiring significant structural disassembly. Sometimes there is an unacceptable level of uncertainty in the NDI result, leading to more frequent inspections, and even greater costs due to prematurely replaced structure. Broadband Acoustic Emission (BAE) systems have demonstrated the capability to detect crack growth in structure. Such a system could autonomously monitor the structure during operation and provide information necessary to detect crack growth without disassembly. This paper will describe the F-16 Crack Monitoring System Proof of Concept program and its objectives. An overview of the BAE approach to crack monitoring, and preliminary results from the proof-of-concept testing will be presented, along with BAE waveform analysis and comparisons with the preliminary test results.

1 Introduction / Program Overview

The Aircraft Structural Integrity Program (ASIP) was initiated in the late 1950s due to a rash of catastrophic B-47 structural failures. Since its inception, ASIP has evolved into a successful and permanent part of the Air Force philosophy toward management of aircraft structures. The program is currently divided into five primary tasks [1].

Task I (design information): development of those criteria which must be applied during design so that the overall program goals will be met. **Task II (design analysis and development tests):** development of the design environment in which the airframe must operate and the response of the airframe to

*Principal Engineer, Boeing Defense & Space Group

†V.P. Engineering, Digital Wave Corporation

‡Captain, U.S. Air Force, Structures Division, Wright Laboratory, Wright Patterson AFB

the design environment. **Task III (full scale testing):** flight and laboratory tests of the airframe to assist in determining the structural adequacy of the analysis and design. **Task IV (force management data package):** generation of the data required to manage force operations in terms of inspection, maintenance, modifications, and damage assessment when aircraft are flown differently than design. **Task V (force management):** those operations that must be conducted by the Air Force during force operations to ensure damage tolerance and durability throughout the useful life of individual airplanes.

Although all tasks affect the life cycle costs of an aircraft, once an aircraft becomes operational, none affect Operation and Support (O&S) costs more than Task V of ASIP. In fact, a March 1995 Joint Advanced Strike Technology (JAST) Advanced Integrated Diagnostics [2] study found that 27% of overall aircraft costs are related to O&S costs. This figure indicates that there is great potential for savings if personnel, inspection, and maintenance costs can be reduced. It is plausible to believe that a reduction of O&S costs can be realized by developing a means of automating the current inspection and maintenance process. This automated process is the rationale behind the F-16 Crack Monitoring System (CMS) proof of concept effort currently underway as a joint Wright Laboratory, Ogden Air Logistic Center, Boeing Defense and Space Group, and Digital Wave Corporation effort. Ideally, this system could autonomously monitor the structure during aircraft operation and provide information necessary to detect and monitor flaw (crack) growth. Information of this type could then be used to predict inspection intervals, maintenance requirements, and needed structural modifications without conventional and time-consuming aircraft disassembly and inspection techniques. If flaw features (size, rates, location, etc.) can be determined, remaining life and residual strength of the structure can be calculated, leading to more economic and safer aircraft operations.

Simply devising an autonomous flaw monitoring system will not ensure successful technology transition and widespread usage. Two significant obstacles stand in the way of successful deployment: development of a system and processes with widespread technical acceptance, and integration of the system into the existing structural management framework. The first obstacle has been, and continues to be addressed with successful demonstrations of broadband acoustic emission technology. Significant development and numerous demonstrations have been performed by Digital Wave Corporation [3], NASA [4] and Northrop-Grumman [5] on representative aircraft coupons and components. Boeing and Digital Wave have successfully performed crack-growth detection on full-scale aircraft [7], and have been actively developing analysis methods for computer automation of broadband waveform analysis [8, 9]. The Air Force's Smart Metallic Structures (SMS) program [6] has performed additional component testing, and developed a demonstration aircraft health-monitoring system. The F-16 CMS proof-of-concept program adds to the wealth of test data by testing a part that could have significant payoffs if monitored in flight. The second obstacle, integration with the existing ASIP framework, must be addressed in the data analysis and automation efforts currently in-progress at Boeing and Digital Wave Corporation. These efforts are directed toward making data acquisition and processing as easy and transparent to the user as possible. For instance, AE data download should be a part of a standard data download and data processing should be accomplished through the Aircraft Structural Integrity Management Information Systems (ASIMIS). The ASIMIS processing would use software that reduces BAE crack data into flaw feature information required for a crack growth analysis. Periodic Individual Aircraft Tracking (IAT) reports would be based on BAE generated data and not scheduled inspections. This automatic update of ASIMIS data could then circumvent the need for "Hot Spot" or control point inspections until such a time when repair action is required. Thus, a potential for savings has been generated.

Through the previous efforts of the SMS program and the current work on the F-16 Crack Monitoring program, we seek to address the main obstacles facing this technology. Specifically, we will work to develop aircraft operating requirements for an on/off-board system (Task 1), develop preliminary design and pricing details for construction of flight worthy hardware (Task 2), and demonstrate that the technology works on an F-16 Fuselage Station 479 bulkhead (Task 3). A follow-on, Phase II flight test effort to build and test the CMS in-service is currently being planned upon successful completion of this program.

This paper will present an overview of the Task 3 testing. Test setup, and preliminary non-destructive inspection (NDI) test results will be presented, as well as a general discussion about the acoustic emission data system, data, and analysis methods, and results.

2 Concept Testing

There has been a significant amount of previous work [7, 3] done to show that broadband acoustic emission monitoring can be used to detect the presence of a flaw (crack). However, in most of these tests the goal was crack detection. For a crack monitoring system to be useful in an ASIP sense, data that predicts crack geometry as a function of time is important. Thus, the most important goal of this particular test

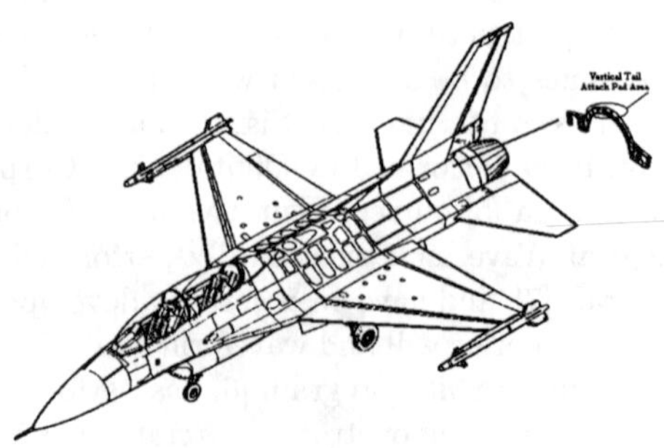

Figure 1: FS479 Bulkhead Detail

was to demonstrate that acoustic emission monitoring and data analysis could predict the crack geometry with sufficient accuracy to provide information beneficial to maintenance and operation of the Air Force F-16 Fleet[1]. The F-16 is particularly well suited for crack monitoring testing and development because of its ongoing problems with bulkhead FS479. Figure 1 shows an F-16, the FS479 bulkhead, and the approximate location of the bulkhead in the aircraft. This bulkhead has two attach-pads for the vertical tail, and carries significant loads, so much so that bulkhead cracking at the intersection of the upper bulkhead flange and the base of the attach-pads has been a chronic problem. The bulkhead is difficult to inspect, and even more difficult to replace. An automated crack-monitoring system that provides crack size and growth-rate information could be extremely useful for inspection and maintenance of this aircraft.

Figure 2 shows a full-scale version of a bulkhead sub-component that was manufactured especially for fatigue testing. The left-hand-side attach pad in the figure is instrumented with acoustic emission sensors. The base of this attach-pad is where fatigue-cracks develop.

The Task-3 testing was performed at Lockheed-Martin Tactical Aircraft System's (LMTAS) facilities in Fort Worth Texas, as an add-on to existing FS479 bulkhead fatigue testing. This particular test was part of a series of fatigue tests titled: *Block 50/52 Fleet Usage Variability Test.*

2.1 Test Setup

The test setup includes the Lockheed-Martin test fixture and test article, and additional equipment supplied from Digital Wave Corporation (DWC) to monitor the test article throughout the test.

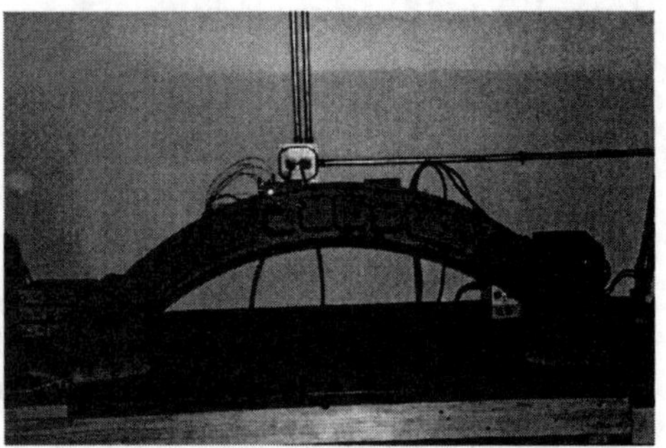

Figure 2: F-16 FS479 Bulkhead Sub-Component

Figure 3 shows a schematic of the test fixture, including the actuator and bulkhead. The test fixture is a simple stand with mounts at the base for the bulkhead. Bolted to the bulkhead attach pads is a stout steel beam that is a dummy tail used to simulate the application of vertical tail loads. At the top of the fixture is an actuator that applies load to the dummy tail. The load is reacted by the bulkhead, which is mounted to the fixture at the two large bolt holes.

[1]And, by extension to other Air Force aircraft.

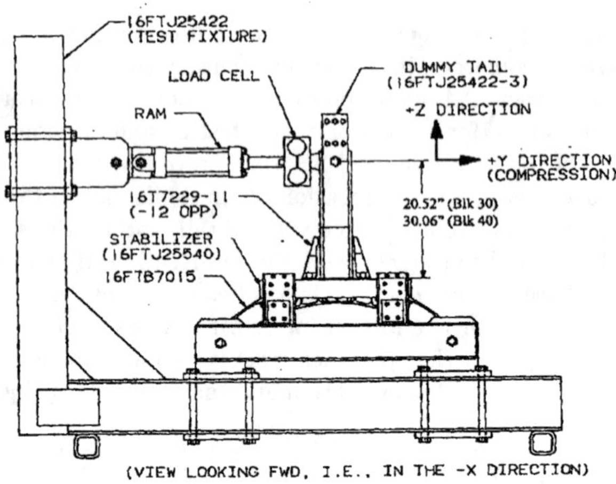

Figure 3: FS479 Bulkhead Fatigue Test Schematic

The test fixture has 0.5 inch thick aluminum plates, with large Teflon pads that serve to stabilize the bulkhead in twist.

Eight BAE sensors were attached to the bulkhead with 5 minute epoxy. Figure 4 shows the configuration of the coordinate system used for locating the sensors, and for performing the acoustic emission source location analyses. Sensor 1 is shown in Figure 4 purely for visual reference.

Eight sensors were used to ensure that accurate source location could be performed. For 3-D source location, a minimum of four sensors must used. Since the method for determining the source location utilizes a non linear least squares approach, the more sensors the better (within reason). Since this bulkhead had never been acoustically monitored prior to this test, eight sensors were chosen as a conservative number. Six sensors would most likely be used in an operational environment.

The signals from each sensor are pre-amplified before entering the Fracture Wave Detector's (FWD) signal conditioning unit. This unit monitors the sensor signals and provides the final signal amplification, filtering, triggering and analog-to-digital conversion functionality. The resulting digitized waveforms are then stored on the computer's hard drive for analysis.

Along with the waveform capture, data from the load and strain gages were also digitized. The strain gage used in this test was an axial gage, located on the upper flange of the bulkhead, near the monitored area. Part of the gage can be seen in Figure 9 at the very bottom center of the picture. A positive strain

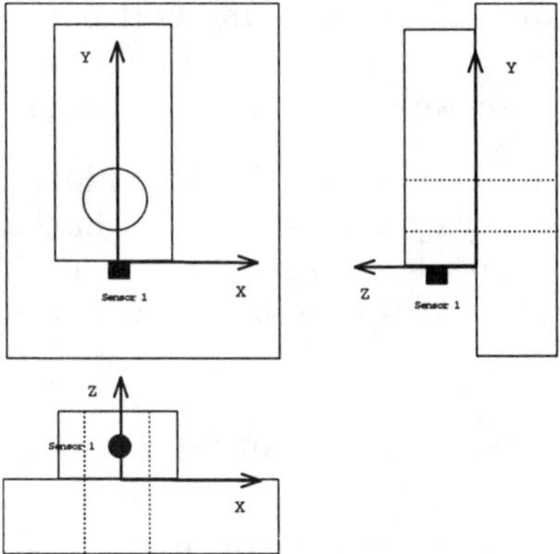

Figure 4: Sensor Test Configuration

indicates tension, or crack opening, in the bulkhead upper flange.

2.2 Wave Propagation

The dimensions and the material of the part under test contribute to the type of wave propagation that will occur. To determine the type of wave expected for fatigue crack growth in the bulkhead, the velocity of propagation needs to be calculated. In aluminum, the bulk compressional mode propagates at a velocity:

$$c = \sqrt{\frac{E(1-\nu)}{\rho(1+\nu)(1-2\nu)}} \tag{1}$$

For aluminum, $E = 70x10^9$ Pa, $\nu = 0.34$, $\rho = 2700$ kg/m^3. If these values[2] are input into the above equation, the velocity is calculated to be 6370 m/sec. The wavelength, l, can be calculated from the velocity and frequency through the following equation:

$$\lambda = \frac{c}{f} \tag{2}$$

where c is the velocity of propagation and f is the frequency of the propagating disturbance. From prior experience, the frequencies excited by crack growth in aluminum can range from roughly 500 kHz to over 2 MHz. This results in

[2]In English units the values are: $E = 10x10^6$ psi, $\nu = 0.34$, $\rho = 2.538x10^{-4}$ slinches/in^3. The velocity is then 250,000 in/sec.

wavelengths from 0.5 to 0.125 inches, much less than the thickness of the attach pad area (approximately 0.75 to 2 inches thick). Because the wavelength is much less than the part thickness, the wave will propagate in a bulk mode [14].

If the part thickness had been thinner (the part thickness much less than the wavelength), then the propagating waves would interact with the upper and lower boundaries of the specimen. This interaction would have converted the bulk modes to plate modes, thus requiring different analysis techniques.

2.3 Acoustic Emission Waveforms

The fatigue test was monitored continuously from initial loading to failure using BAE. A typical fatigue block, consisting of eight hours of spectrum loading (500 hours of flight loads), resulted in approximately 40,000 - 50,000 AE events[3]. This was approximately 300 - 400 MBytes of BAE data per block. The majority of this data consisted of events excited by mechanical sources, such as fretting and rubbing at the bracket/attach pad and bulkhead/restraint interfaces, as opposed to crack growth. Because the force-time function of these noise events are typically much different than from crack growth, they can be easily identified and removed based on simple time-space characteristics [10].

Figures 5 and 6 are examples of typical test signals. Figure 5 shows noise signals generated by mechanical sources, while Figure 6 shows a signal due to crack growth. There are several points to be noted in the three figures. In Figure 6 the crack growth signal has a significant amount of energy in the high frequency range (700 kHz - 1.5 MHz), and has a clearly defined bulk compressional (also known as a primary, or P-wave) mode in all of the waveforms (circled section of the waveforms). Furthermore, all of the initial displacements in the signals are in the same direction, which is consistent with a symmetric source, such as crack growth due to fatigue.

Figure 5 shows a signal that looks similar to the crack signal shown in Figure 6. However, if the signal is carefully analyzed, it can be seen that the initial displacements are not all in the same direction (the arrows show only the initial direction of displacement, and do not reference any particular point on the waveforms), and that the sharp bulk compression mode signal is not observed on all channels. This indicates that the source is not symmetric (possibly a shear source at a mechanical interface), or is not in a direction line of flight to the sensors. Furthermore, source location results show that this waveform originated from an area near the bolt hole between the attach bracket and the attach pad.

[3]50,000 events times 8 channels times 2048 samples per channel adds up to a lot of disk space

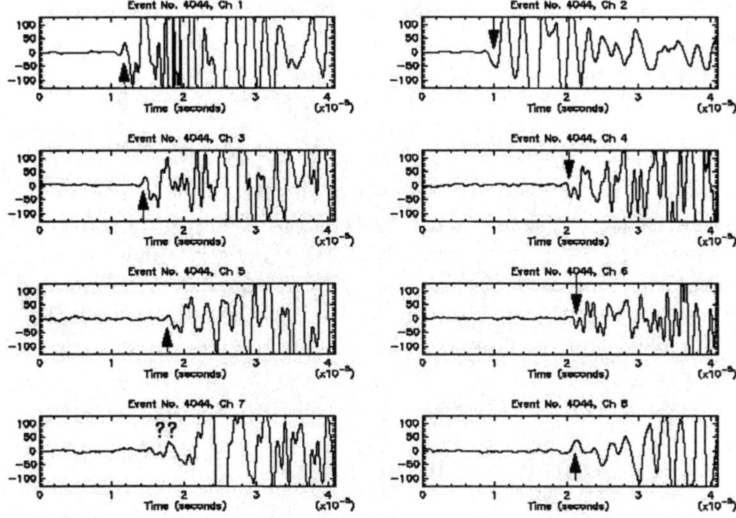

Figure 5: Sample Fretting Event Waveforms

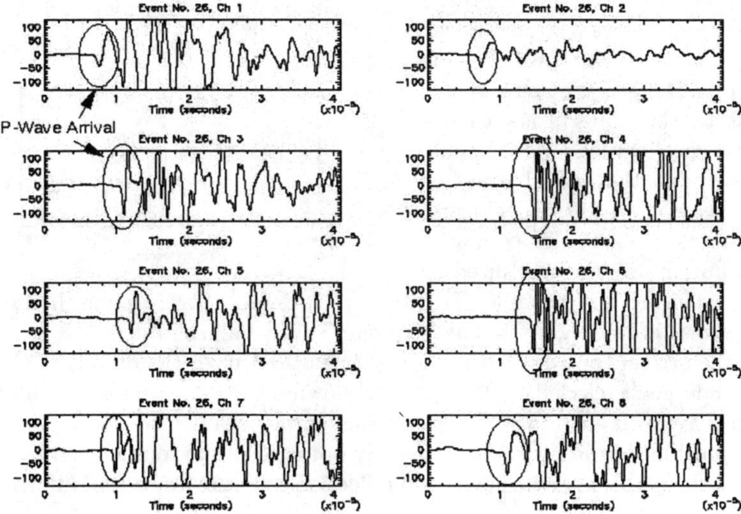

Figure 6: Sample Crack Event Waveforms

2.4 Automated Noise Rejection

Using very conservative sorting routines, approximately 30% of the captured events were eliminated. More stringent routines based on occurrence of the event in the load cycle can be used. Crack growth should only occur during loading which opens the crack (tensile loading). Figure 7 shows a plot of the strain gage time-history and acoustic emission event strain and times (plotted as discrete points). Since the strain gage voltage is positive for crack opening (tension), acoustic emission events with a positive strain, occurring on a positive slope should be retained. All others can safely be discarded. If this information is used along with the frequency content of the event, and mode identification, the amount of data rejected jumps to 90-95%. If source location were to be implemented into the noise rejection routines, this figure would be even higher. Thus, concerns such as data rates and data storage can be eliminated by incorporating simple electronics or software routines into the system. This would allow the rejection of the noise events, reducing the required storage space and analysis time.

2.5 Event Source Location

After the noise signals were removed, 3-D source location was performed on the remaining signals from the tests. The source location routine uses a non linear least squares fit based on the difference of the arrival times of the signals at the transducers to determine the location of the event. To determine the arrival times of the wave at the eight transducers, the arrival time of the compressional bulk mode was measured in each waveform. For example, Figure 6 shows a crack growth waveform; the arrival time is the time index (measured from the origin) of the first bulk mode peak (circled). The signals following the compressional bulk mode in each waveform are a mix of shear modes, surface wave modes and reflections. These portions of the signals are typically not used in the source location analysis since the multiple wave modes and reflections obscure any useful information.

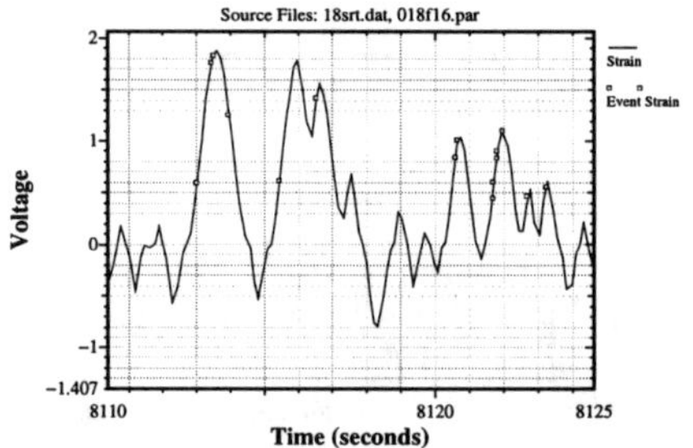

Figure 7: Sample Strain Time History, and Event Strain Values

The acoustic emission event source locations in this report were computed in an automated fashion, as opposed to the labor intensive methods used in the past. Typically, computing event source locations has been difficult, time consuming, and required a highly skilled engineer. Boeing and Digital Wave Corporation have been jointly developing analysis methods, suitable for computer automation, that can identify AE events from cracks, identify the time-of-arrival for each waveform in an event, and compute the 3-dimensional source location of that event. The data from the FS479 testing occupied approximately 300 megabytes of disk space. The AE analysis program spent approximately 10 hours operating on the data to produce the 1302 AE crack-event sources displayed in the figures of this report. At present, the programs that perform this analysis are coded in RLaB, a very high level matrix programming language. Since RLaB programs are interpreted it is not unreasonable to expect the run-times to drop substantially once the program has been translated to a language that can be compiled directly to machine-code (C-language). A total run-time less than an hour, for this particular data set, is not out of the question.

When the waveform data for an event is recorded, the time of the event is also stored. This allows the event source location results to be correlated in time. Figure 8(a) and 8(b)) show the coordinates for the complete set (all 1302) of identified events. These events were extracted, by computer, from the recorded AE data. Figure 8(c) and 8(d) show subsets of the identified events. The subset shown in Figure 8(c) represents acoustic emission events that occurred prior to block 5, while the subset shown in Figure 8(d) represents acoustic emission events that occurred prior to block 9. The reason for breaking out these subsets of acoustic emission source locations is to compare these results with die penetrant observations performed during the test.

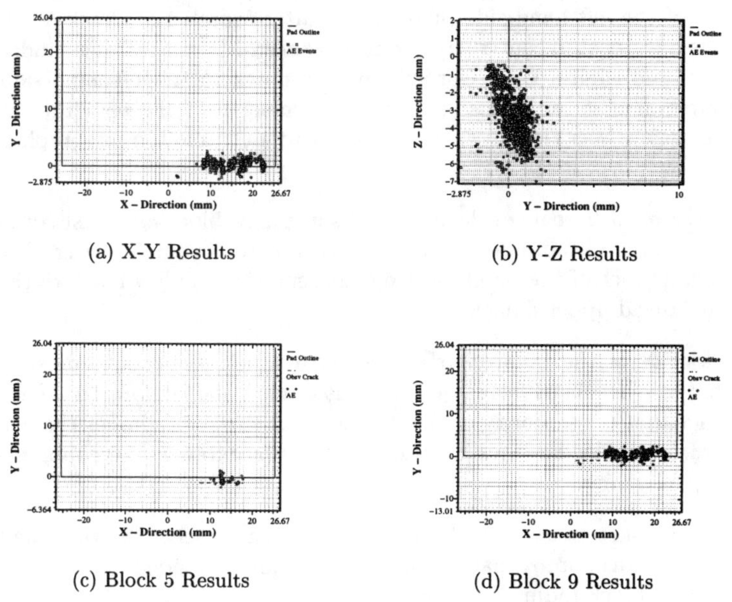

(a) X-Y Results

(b) Y-Z Results

(c) Block 5 Results

(d) Block 9 Results

Figure 8: Acoustic Emission Results

The automated method of determining sorting, selecting, and analyzing the acoustic emission data has yielded some promising results. The present method is a prototype and can certainly stand some improvement, mostly in two areas: improving the robustness of the analysis, and improving the user-interface to the analysis. The addition of other types of time-of-arrival computations, and the ability to select the best method would most likely increase the quantity of identified events. Possibly filling in some of the gaps in the data. A graphical user interface would certainly be simpler than the present command line interface but, would not improve the results any.

2.6 Test and Analysis Results

The fatigue test began on March 12, 1997, and ended March 18, 1997 when the bulkhead failed after the second set of 100% loads, also known as "rogue-loads". The test is performed in "blocks". Each block of loads simulates 500 equivalent flight hours. Each block takes approximately 8 hours, or 1 working shift, to complete. The 100% rogue-loads are applied after every four blocks of fatigue loads.

Lockheed-Martin engineers performed die penetrant inspections of the bulkhead after each block of testing. These inspections detected cracks in the radius at the base of the left-hand-side pillow block after March 12. Crack lengths and positions have been reconstructed from Lockheed-Martin test logs and Boeing, Air Force, and Digital Wave Corporation test notes for the left-hand-side pillow block after block-5, and block-9. At the conclusion of the test, the bulkhead fracture surface was visually inspected prior to a complete fractographic examination.

Figure 9 shows a close up of the left-hand-side pillow block after test completion. The radius area of the pillow block, or attach pad, is completely cracked, as is the upper surface of the bulkhead from corners of the pillow block to the edges of the bulkhead upper flange.

The majority of the failure surface is *not* a result of fatigue cracking. The area that is a result of fatigue cracking is circled and annotated. The reader will notice that the majority of the fatigue crack surface is darker than the final failure surface. The darkness is a result of crack face fretting, or rubbing, during the fatigue loading.

Post test measurements of the fatigue crack surface determined that the fatigue crack surface was approximately 4-6 mm deep, and extended across the aft half of the pillow block radius.

The crack geometry predicted via acoustic emission compares very favorably with the preliminary test results. Acoustic emission predicts (derived from

Z
X
Fatigue Crack
Strain Gage

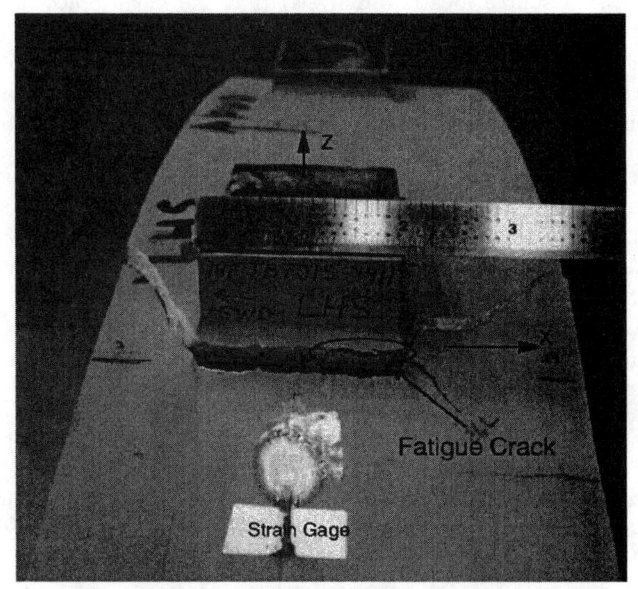

Figure 9: Post-Test Bulkhead

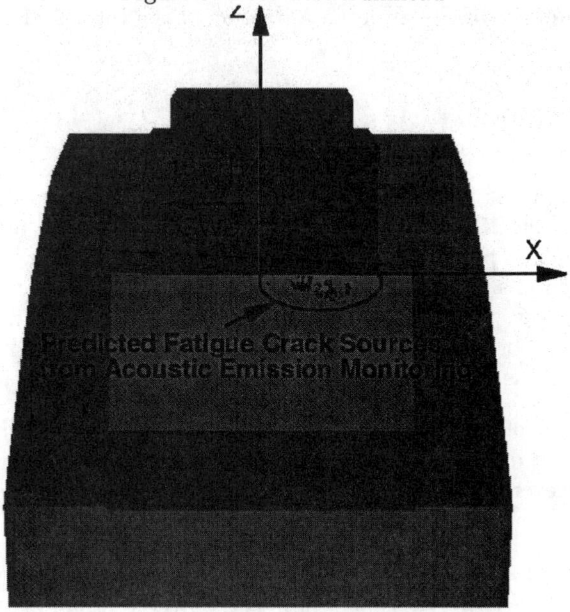

Z
X
Predicted Fatigue Crack Sources
from Acoustic Emission Monitoring

Figure 10: Acoustic Emission Results

acoustic emission source location analysis) that the mean crack depth before failure is approximately 4.8mm, and the crack width is the entire aft half of the pillow block. Figures 8 show the acoustic emission source location results in enough detail to read dimensional data. Figure 10 displays the data shown in Figures 8 in a more realistic[4] fashion.

In addition to the "end-point" test and analysis data correlation, there is good "mid-point" test and analysis correlation as well. Based upon Lockheed-Martin die penetrant inspections, and Air Force, DWC, and Boeing test notes, observed crack position and length are compared with acoustic emission analysis predictions for crack position and length at the end of block 5 and block 9 testing. These data are shown in Figure 8(c) and 8(d). Since the observations are based upon visual inspection, no depth information is present. However, the X-Y direction (surface) data compare well.

It is worth noting that the data shown in Figure 8, and 10 have been automatically generated. The raw data, acquired during fatigue test monitoring, were analyzed in three stages on a computer; without human intervention. The first two stages remove acoustic emission events that are not related to damage growth. The last stage identifies those waveforms that are a result of damage growth, and computes the 3-dimensional source location of the event. This is a great improvement over past methods which required a highly skilled engineer and many hours of work to get even a fraction of the total analysis results.

3 Conclusions

The test data (acoustic emission waveforms), and data analysis compare very favorably with the NDI results. The data analysis produced results that are accurate enough to predict crack geometry and be useful in crack growth and residual life analysis. Furthermore, the analysis results presented herein are the result of computer automation. The software and methods are mature enough that the once labor and skill intensive task of analyzing broadband acoustic emission waveforms can now be automated.

The automation of acoustic emission data does need more work so that the results can more directly support crack growth analysis and residual life predictions. However, there are no foreseeable obstacles left in the automation process.

There is cause for some concern about electronic interference and/or jet engine noise polluting the acoustic emission data. However, there is sufficient data to suggest that this will not be a significant problem. Flight testing of simple

[4]Although the data is displayed in a 3D manner, the scale is *not* identical to the photograph in Figure 9.

test-articles on F-18 aircraft [13] shows that the "background" noise levels are tolerable. Not as closely related, but probably more severe is the flight testing performed on the DC-XA [11] which demonstrated that acoustic emission waveforms could be acquired with sensors and acquisition equipment in very close proximity to a rocket engine. Furthermore, test and numerical experiments [9] have shown that electronic and mechanical noise can be adaptively removed from the waveforms.

At this point there is no practical or technical reason not to pursue a flight demonstration. There are no significant hardware, software, or technical obstacles to developing and producing a flight-capable crack monitoring system that can be used in an ASIP environment.

The remainder of this program will define requirements for an operational crack monitoring system, and scope the cost of developing and manufacturing such a system by performing a preliminary design.

4 Acknowledgements

We gratefully acknowledge the support of Neal Phelps, Lead ASIP engineer for the F-16 ASIP Program. We are also grateful to Lockheed-Martin Engineers Nick Strapko, Rich Luke for providing us with valuable and timely assistance throughout the testing. We would like to thank Mark Defazio, ASC/YP, for affording us the opportunity to monitor the FS479 testing. We would also like to thank Dr. W. Prosser, NASA Langley, and Dr. Roy Ikegami at Boeing Structures Research & Technology for funding to support AE analysis automation efforts.

References

[1] MIL-HDBK-1530, Military Handbook, *Aircraft Structural Integrity Program, General Guidelines for*, November 1996.

[2] "Joint Advanced Strike Technology (JAST) Advanced Integrated Diagnostics Study, Final Report", March 1995, TRW and University of Dayton Research Institute.

[3] M. R. Gorman, "Determining Fatigue Crack Growth in Aircraft by Monitoring Acoustic Emission", Naval Research News, v.XLIV, n. 1, 1992, pp. 24-27.

[4] William H. Prosser, "The Propagation Characteristics of the Plate Modes of Acoustic Emission Waves in Thin Aluminum Plates and Thin

Graphite/Epoxy Composite Plates and Tubes", Baltimore, Maryland, PhD Thesis; Sponsored by NASA Langley Research Center, 1991, (Also Available as NASA-TM-104187).

[5] Martin, C. A., Van Way, C. B., Lockyer, A. J., Kudva, J. N., Ziola, S. M., "Acoustic Emission Testing on an F/A-18 E/F Titanium Bulkhead", Proceedings of Smart Structures and Materials Conference, 1995, v. 244, pp. 204-211.

[6] Van Way, C. B., et al., "Development of an Automated Aircraft Structural Integrity Monitoring System - Overview of the Air Force/Navy Smart Metallic Structures Program", 1996 SPIE.

[7] Searle, I., Ziola, S., Seidel, B. "Crack Detection of a Full Scale Aircraft Fatigue Test", SPIE Smart Structures and Materials Conference Proceedings, 3042-34, San Diego, California, March 1997.

[8] Steve Ziola, "Source Location in Thin Plates" Ph.D Thesis, Naval Post Graduate School, 1990

[9] Ian R. Searle, Steven Ziola, "Identification of Broadband Acoustic Emission Signals", Proceedings of Smart Structures and Materials 1996 Conference.

[10] Ian R. Searle, Steven Ziola, and Paul Rutherford, "Crack Detection in Lap-Joints Using Acoustic Emission", Proceedings of Smart Structures and Materials 1995 Conference, Volume 2444, pp 212-223

[11] D. Pointer, G. Nissen and Q. Huang, "In-flight Acoustic Emission Monitoring of the DC-XA Rocket", Presentation and paper presented at the ASNT 1997 Spring Conference, Houston, Texas, March 1997.

[12] S. L. McBride and Pollard and J. D. MacPhail and P. S. Bowman and D. T. Peters, "Acoustic Emission Detection of Crack Presence and Crack Advance During Flight", pp: 1819-1825, Plenum Press, Edited by Donald O. Thompson and Dale E. Chimenti, volume: 8B, 1989

[13] Dianne M. Granata, Paul Kulowitch, and Willion R. Scott, "In-Flight Acoustic Emission Monitoring of Crack Growth", Review of Progress in Quantitative Nondestructive Evaluation, Volume 13, 1994.

[14] Karl Graff, *Wave Motion In Elastic Solids*, Dover Publications, New York, 1975, ISBN: 0-486-66745-6

Determination of the Fault Position in Rotors for the Example of a Transverse Crack

H. BACH and R. MARKERT

ABSTRACT

In the present paper a model based method is proposed which allows the online identification of malfunctions in rotor systems. The rotor vibrations during operation can be measured at only a few rotor positions. Therefore the full state is reconstructed using an observer and the undamaged linear rotor model. Simultaneously a set of equivalent loads is calculated representing the malfunction. By comparing the estimated equivalent loads to pre-calculated equivalent loads from fault models the type, the position and the extent of the current fault can be detected. For determining the fault position both the observed system state and the determined set of equivalent forces are checked for certain symptoms. The symptoms are used to reduce significantly the number of elements which have to be examined in detail for identifying the fault properties. Finally the fault extent is determined in an iterative process during which also the exact fault position is estimated by different tests resulting in indicators. The suggested identification method was tested on simulated vibration data caused by a transverse crack, but it is also open for the identification of many other faults.

Keywords: Rotating machines, identification, model based monitoring, equivalent load.

1 INTRODUCTION

Rotating machinery installed in power plants such as turbogenerators and feed pumps are normally equipped with vibration sensors, [1]. The information in the vibration data continuously collected during operation are not fully utilized by currently installed signal based monitoring systems, [2]. Therefore, during the last years efforts have been made to develop model based monitoring systems for different technical applications [3] that may be used with or alternatively to conventional signal based monitoring systems. One can expect

Hartmut Bach and Richard Markert, Mechanik II, Darmstadt University of Technology, Hochschulstr. 1, D-64289 Darmstadt, Germany

that a model based monitoring system gives more exact and faster information about the system state than conventional signal based systems, since a-priori information about the rotor system is systematically included in the identification process. The type of fault, its position and its extent can therefore be estimated with more reliability. Catastrophic failures resulting in great monetary loss may be prevented since detailed information about the malfunction is already available at an early stage. Since the malfunction properties are identified by the model based system before stopping, the time devoted for repairing can also be shorter.

Identification methods can be applied in the time or the frequency domain. Which of these is more advantageous depends on the malfunction type and the operating state for which the vibration data can be measured.

An arising fault changes the characteristic vibrations in comparison to the vibration data of the undamaged rotor system. To determine the fault properties in a model based monitoring system using suitable mathematical models for the rotor system and for all potential faults the measured vibration data are compared with synthetic model based vibration data. If the measured and the synthetic vibration data do not match the assumed malfunction properties have to be changed for the calculation.

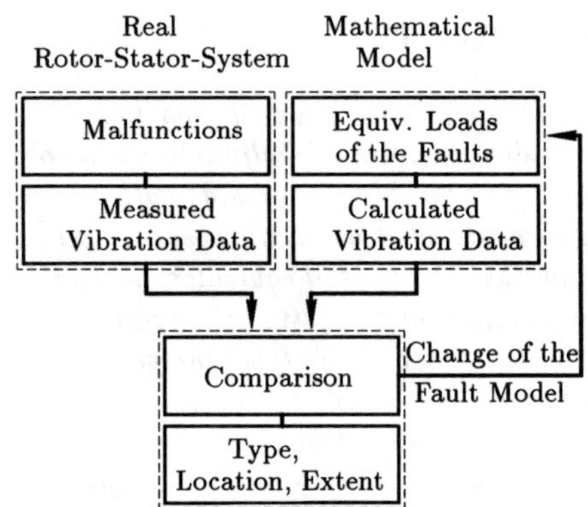

Figure 1: Schematic sketch of a model based monitoring system

In the presented identification process the fault induced system change is taken into account by using equivalent loads in the mathematical model. The equivalent loads are forces and moments acting on the linear undamaged system model to generate a dynamic behavior identical to that one of the damaged real system. By using this approximation the model stays linear and thus the identification process can be carried out on-line.

However the exact modeling of faults would lead in many cases to non-linear equations of motion although the undamaged system is linear. The non-linear equations would require time integration algorithms to calculate the dynamic behavior of the damaged model. An identification process would be extremely time consuming especially since the model of the rotor system has usually many degrees of freedom and the calculations have to be carried out for every possible fault type, every fault position and every fault extent.

The method of equivalent loads ensures that the originally linear system model remains linear also when a non-linear fault is present. The vibrations $\mathbf{q}_0(t)$ of the undamaged rotor system due to the load $\mathbf{F}_0(t)$ during normal operation (e.g. residual unbalance, weight) are described by linear equations of motion,

$$\mathbf{M}_0\ddot{\mathbf{q}}_0(t) + \mathbf{B}_0\dot{\mathbf{q}}_0(t) + \mathbf{K}_0\mathbf{q}_0(t) = \mathbf{F}_0(t). \tag{1}$$

$\mathbf{M}_0$ and $\mathbf{K}_0$ are the mass and the stiffness matrix of the undamaged system including the rotor, the bearings and the foundation. The matrix $\mathbf{B}_0$ includes in addition to the damping also the gyroscopic terms.

The fault induced system change depends on the malfunction type and the malfunction parameters contained in the vector β. One can imagine that the fault induced change in the vibrational behavior could also be caused by additional loads $\Delta\mathbf{F}(t, \mathbf{q}, \dot{\mathbf{q}}, \beta)$ acting on the undamaged system,

$$\mathbf{M}_0\ddot{\mathbf{q}} + \mathbf{B}_0\dot{\mathbf{q}} + \mathbf{K}_0\mathbf{q} = \mathbf{F}_0(t) + \Delta\mathbf{F}(t, \mathbf{q}, \dot{\mathbf{q}}, \beta). \tag{2}$$

Comparing the measured vibrations $\mathbf{q}$ of the damaged system to previously measured reference vibrations $\mathbf{q}_0$ of the system during normal operating conditions, the residual vibrations $\Delta\mathbf{q}$ caused by the fault can be calculated. For the residual vibrations the equation of motion can be given as

$$\mathbf{M}_0\Delta\ddot{\mathbf{q}} + \mathbf{B}_0\Delta\dot{\mathbf{q}} + \mathbf{K}_0\Delta\mathbf{q} = \Delta\mathbf{F}(t, \mathbf{q}, \dot{\mathbf{q}}, \beta). \tag{3}$$

The reference vibrations $\mathbf{q}_0$ are recorded for different operating states after putting the machine into operation for the first time assuming that the new system is undamaged. But these vibrations are not synchronous to the later measured vibrations of the damaged rotor. In order to avoid phase errors during the identification process the reference vibrations are decomposed into their harmonic parts and only the complex Fourier coefficients are stored for using in the later identification process. For the fault identification both the vibrations of the damaged system and the exact angular position φ of the rotor are measured simultaneously to reconstruct the vibrations $\mathbf{q}_0$ by combining the previously stored complex Fourier coefficients with harmonic functions of the current angular position.

Since only a few degrees of freedom of the full system state $\mathbf{q}$ are accessible for measurements an extrapolation process is necessary to obtain the complete system state. But the complete system state and the equivalent loads depend non-linearly on each other and therefore they can be estimated finally only in an iterative process. For calculating the equivalent load in each iteration step the results of the previous step are used.

2 REPRESENTATION OF MALFUNCTIONS BY EQUIVALENT LOADS

In the following some malfunctions of rotor systems and their representation in terms of equivalent loads are described.

An *additional unbalance* (for example due to blade loss) is represented by an equivalent load acting only on one node. For constant rotor speed Ω the equivalent force $\Delta\mathbf{F}(t,\beta)$ depends harmonically on time t,

$$\Delta\mathbf{F}(t,\beta) = [0\ldots|U|\ldots0]^T e^{i(\Omega t + \delta)}. \tag{4}$$

The parameters of the fault model are the extent $|U|$ and the angular position δ of the unbalance $U = |U|e^{i\delta}$.

Rotor to stator rub can occur at seals, in bearings or between blades and housing due to large rotor deflections. Numerical and analytical solutions for the equivalent loads are available for full annular and for partial rub, [4]. The parameter vector β of the fault model contains the location of the rub, the local contact parameters and in case of partial rub also the stator offset. The equivalent loads consisting of the normal and tangential contact forces are periodic with time. A bow due to unisotropic thermal expansion resulting from rub can be taken into account by additional moments acting at the rub position.

Loosening of a bearing bush is characterized by a clearance between the bearing and the pedestal. For small rotor deflection this malfunction can be approximated by a decrease of the bearing stiffness,

$$\Delta\mathbf{F} = -\Delta\mathbf{K}\,\mathbf{q}. \tag{5}$$

In the case of large rotor deflections when the contact behavior becomes dominant the malfunction has to be modeled similar to rotor to stator rub.

A *transverse crack* approximately leads to a time dependent decrease of the rotor stiffness under the assumption of weight dominant rotor deflections. The equivalent load is therefore given by

$$\Delta\mathbf{F} = -\Delta\mathbf{K}(t)\mathbf{q}(t). \tag{6}$$

The matrix $\Delta\mathbf{K}$ depending also on the crack depth a has non-zero coefficients only on the nodes of the crack element. It can be derived by different approximations of the crack, see for example [5] and [6]. In most practical cases the crack induced vibrations $\Delta\mathbf{q}(t)$ are small so that the product $\Delta\mathbf{K}(t)\Delta\mathbf{q}(t)$ can be neglected leading to the approximation

$$\Delta\mathbf{F}(t) = -\Delta\mathbf{K}(t)\mathbf{q}_0(t). \tag{7}$$

3 IDENTIFICATION FROM MEASUREMENT DATA FOR DIFFERENT ROTOR SPEEDS

If measurement data are available for different rotor speeds the identification method proposed by Bachschmid and Dellupi [7] can be applied for detecting faults with equivalent loads being synchronous to the rotor speed,

$$\Delta\mathbf{F} = \Delta\hat{\mathbf{F}}e^{i\Omega t}. \tag{8}$$

Beside the normal loads $\mathbf{F}_0$ also the bearing forces and the equivalent load $\Delta\mathbf{F}$ act on the rotor. In steady state conditions all these forces are periodic and can be represented by Fourier series. Referring to the first order harmonic components the equation of motion leads to the algebraic equation

$$(-\Omega^2\mathbf{M}_{R0} + i\Omega\mathbf{B}_{R0} + \mathbf{K}_{R0})\Delta\hat{\mathbf{q}} = \Delta\hat{\mathbf{F}}(\Omega) + \Delta\hat{\mathbf{F}}_B(\Omega, \hat{\mathbf{q}}) \quad (9)$$

for the amplitudes, where the index $R0$ characterizes the matrices of the undamaged free-free rotor. This methodology allows to take into account the changes of the non-linear bearing forces $\Delta\hat{\mathbf{F}}_B(\Omega, \hat{\mathbf{q}})$ due to the change of the rotor deflection $\Delta\mathbf{q}$. The bearing forces can be calculated from the measured rotor position in the bearings using a suitable non-linear model of the oil film.

The dynamic compliance matrix

$$\mathbf{A}(\Omega) = (-\Omega^2\mathbf{M}_{R0} + i\Omega\mathbf{B}_{R0} + \mathbf{K}_{R0})^{-1} \quad (10)$$

being the inverse of the dynamic stiffness matrix of the rotor can be partitioned so that the measurable deflections are separated from the not measurable ones. By applying a least square approach the corresponding part of the equation for the amplitudes measured at different rotor speeds is finally used to determine the few unknown coefficients in the equivalent load vector $\Delta\hat{\mathbf{F}}(\Omega)$.

4 IDENTIFICATION FROM MEASUREMENTS AT ONLY ONE ROTOR SPEED

4.1 Overview

A complete overview of the identification method is shown in figure 2. From the complete vibration state $\mathbf{q}$ only a few locations are accessible by sensors. The measurable part $\mathbf{q}_e$ and the angular position φ are simultaneously measured and fed into the identification process. At first the residual vibration $\Delta\mathbf{q}_e$ at the sensor locations are calculated. The full state $\tilde{\mathbf{q}}$ is estimated with an observer by using the matrices of the undamaged system.

Symptoms are calculated to carry out a preselection of the fault position. The fault parameters are determined in an iterative process only for the few preselected elements. Finally the fault position is fixed by three indicators calculated during the iteration process.

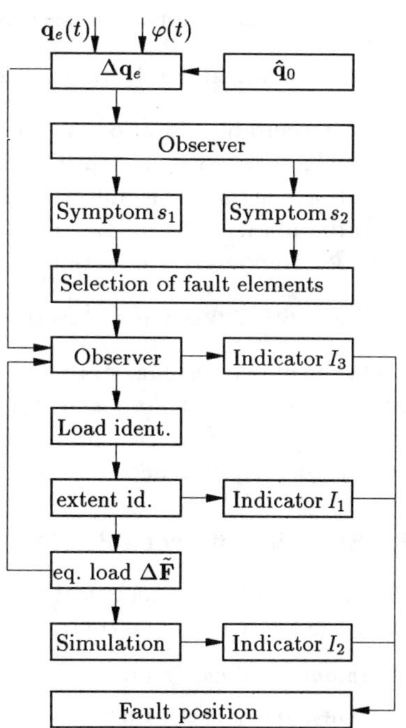

Figure 2: Overview about the identification method

In the following the components of the identification process are explained in detail for a test rotor with a transverse crack shown in figure 3. The test rotor is modeled by 8 finite beam elements having four degrees of freedom at each node. The crack is located in element 4. Only the translational degrees of freedom at the nodes 3, 5 and 7 are assumed to be accessible for measurements which is indicated by symbols for the sensors.

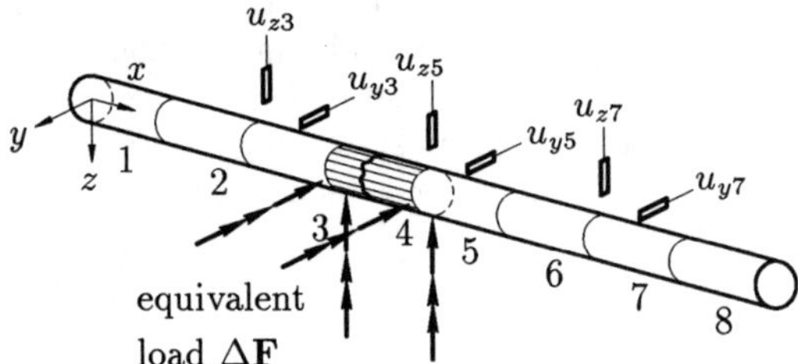

Figure 3: Test rotor

4.2 Observer

The full residual vibration state $\Delta\tilde{\mathbf{z}}$ containing the displacements $\Delta\tilde{\mathbf{q}}$ and the velocities $\Delta\dot{\tilde{\mathbf{q}}}$ of all degrees of freedom is reconstructed by an observer. The well known equation for the observer in the state space

$$\Delta\dot{\tilde{\mathbf{z}}}(t) = \mathbf{A}\Delta\tilde{\mathbf{z}}(t) + \mathbf{B}\Delta\mathbf{F}(t) + \mathbf{H}\left(\Delta\mathbf{q}_e - \mathbf{C}\Delta\tilde{\mathbf{z}}(t)\right) \tag{11}$$

uses the measurable vibration $\Delta\mathbf{q}_e$ in combination with the mathematical model of the undamaged system, [8]. The estimation is better carried out in a modal formulation yielding a significant reduction of the degrees of freedom. The transformation into the modal space is done by using the pre-calculated modal matrix $\mathbf{\Phi}$. The observer equation in the modal space

$$\Delta\dot{\tilde{\mathbf{u}}} = \mathbf{\Phi}^{-1}\mathbf{A}\mathbf{\Phi}\Delta\tilde{\mathbf{u}} + \mathbf{\Phi}^{-1}\Delta\mathbf{B}\mathbf{F} + \mathbf{\Phi}^{-1}\mathbf{H}\left(\Delta\mathbf{q}_e - \mathbf{C}\mathbf{\Phi}\Delta\tilde{\mathbf{u}}\right) \tag{12}$$

leads to the modal residual state $\Delta\tilde{\mathbf{u}}$ from which the full residual vibration $\Delta\tilde{\mathbf{z}} = \mathbf{\Phi}\Delta\tilde{\mathbf{u}}$ and also an estimation of the equivalent forces $\Delta\tilde{\mathbf{F}}$ can be calculated. The system matrix $\mathbf{A}$, the input matrix $\mathbf{B}$ and output matrix $\mathbf{C}$ contain the properties of the undamaged system.

4.3 Selection of probable elements with the fault by symptoms

In analogy to other technical applications [9] *symptoms* are used for determining the likely elements containing the fault.

• *Symptom s_1: Quasi static deflection*

The non zero mean values $\Delta\tilde{\mathbf{q}}_{DC}$ in the components of the residual vibration vector $\Delta\mathbf{q}$ are estimated by averaging time samples of the reconstructed rotor deflections $\Delta\tilde{\mathbf{q}}(t)$.

On the other hand rotor deflections $\Delta \mathbf{q}_{DC,k}$ are calculated for all possible fault positions with only one reasonable extent. By comparing the shapes of $\Delta \tilde{\mathbf{q}}_{DC}$ and $\Delta \mathbf{q}_{DC,k}$ an error can be defined,

$$E_k = \left| \frac{\Delta \tilde{\mathbf{q}}_{DC}}{|\Delta \tilde{\mathbf{q}}_{DC}|} - \frac{\mathbf{q}_{DC,k}}{|\mathbf{q}_{DC,k}|} \right|^2, \quad (13)$$

which is minimal if for the calculations the fault is assumed at the real fault position. Therefore the position number with the smallest error E_k is taken as the symptom s_1 (figure 4).

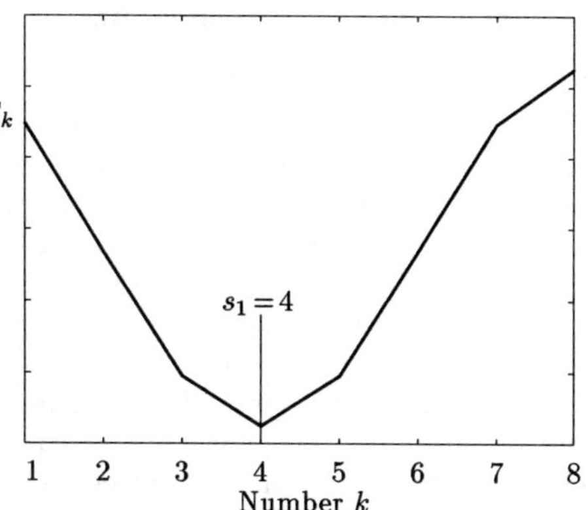

Figure 4: Error E_k versus the number of the element with the fault

- *Symptom s_2: Identified equivalent load*

During the first observation procedure the observer generates a correcting load $\Delta \tilde{\mathbf{F}}$ which excites the mathematical model in a similar way as the fault excites the real system. Therefore the components of the equivalent load are large at those nodes or elements where the real system is damaged (figure 5). Thus the components of the equivalent loads are examined concerning their absolute values. The second symptom s_2 is defined as the position where the identified equivalent load components are maximal. If the single loads lead to different locations the average value is taken.

The mean value of the symptoms s_1 and s_2 is the most probable number of the damaged element. In addition to this probably damaged element its two neighbor elements should be also taken into consideration as possible fault position. Only for these few positions fault extent identifications are carried out.

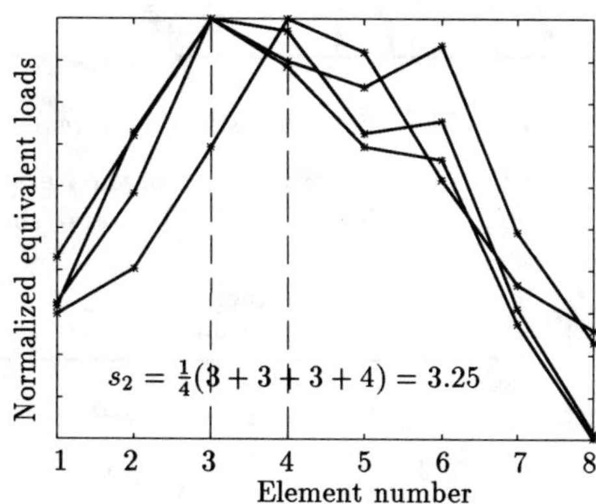

Figure 5: Normalized equivalent loads versus element number

4.4 Fault extent identification

- *Overview about the iteration process*

The extent of the fault is identified in an iterative process shown in figure 6 as part of the complete identification process of figure 2. The measured residual vibration $\Delta\mathbf{q}_e$ is fed into the observer to estimate the full system state. Afterwards a first estimation for the equivalent load $\Delta\tilde{\mathbf{F}}_r$ is made acting only on the fault element. Finally the fault extent is estimated from which a new equivalent load $\Delta\mathbf{F}_a$ can be calculated feeding back into the observer. The identified value for the fault extent converges during the iteration if the identification was carried out for the correct fault position.

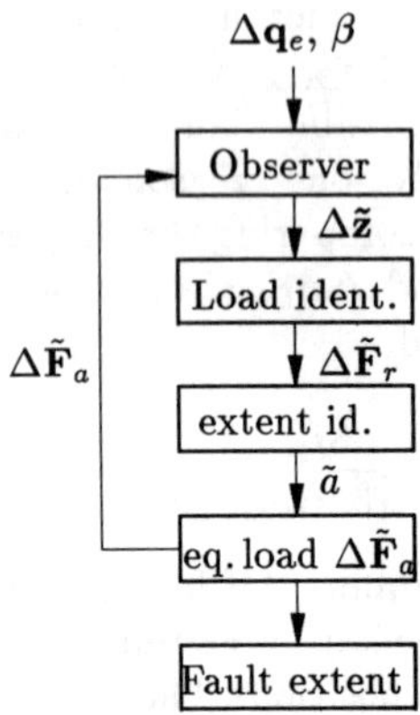

Figure 6: Overview about the iteration process

- *Reconstruction of the equivalent load system $\Delta\tilde{\mathbf{F}}_r$*

Since during the iteration only an approximation of $\Delta\mathbf{F}$ can be fed into the observer the identification process yields equivalent loads $\Delta\tilde{\mathbf{F}}$ acting not only on the fault position but also on all other positions. Taking into account the fact that the equivalent load acts in reality only on the fault element the other components of the equivalent load $\Delta\tilde{\mathbf{F}}$ are neglected as illustrated in figure 7.

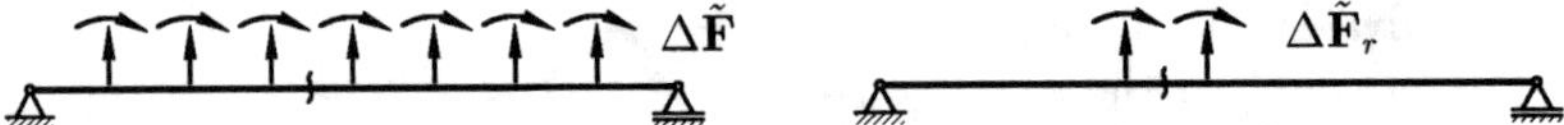

Figure 7: Identified and reduced equivalent load system

To illustrate the convergence of the iteration the maximal values of the identified equivalent loads for the first and the sixth iteration step are shown in figure 8.

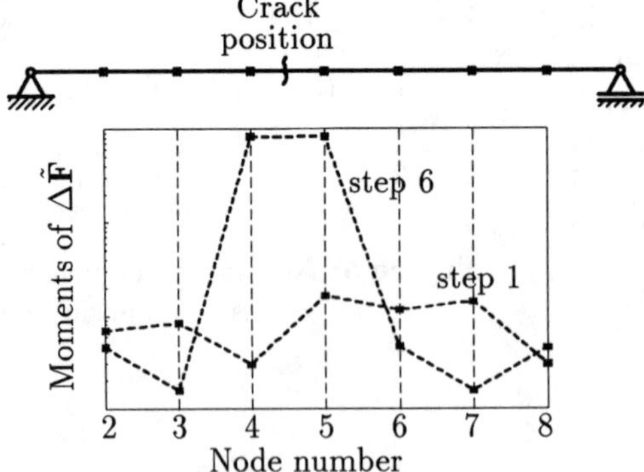

Figure 8: Identified equivalent moments

• *Identification of the fault extent*

The identified equivalent load $\Delta\tilde{\mathbf{F}}_r$ normally is periodic for a constant rotor speed Ω and can be represented by a Fourier series. Based on an appropriate model the fault extent can be calculated from the Fourier coefficients. Finally from the identified fault properties a new approximation of the equivalent load $\Delta\tilde{\mathbf{F}}_a$ is calculated for the next iteration step.

In the case of a transverse crack harmonic signals up to the third order are significant,

$$\Delta\tilde{\mathbf{F}}_r = \sum_{m=0}^{3} \mathbf{c}_m \cos(m\Omega t) + \mathbf{s}_m \sin(m\Omega t). \tag{14}$$

According to [6], in which the stiffness decrease of a finite beam element due to a transverse crack is derived by using the energy release rate, the Fourier coefficients can be stated as third order polynomials of the crack depth a,

$$\mathbf{c}_m = \sum_{k=0}^{3} \mathbf{c}_{mk} \left(\frac{a}{R}\right)^k \quad \text{and} \quad \mathbf{s}_m = \sum_{k=0}^{3} \mathbf{s}_{mk} \left(\frac{a}{R}\right)^k, \tag{15}$$

where the coefficients $\mathbf{c}_{mk}$ and $\mathbf{s}_{mk}$ depend only on the properties of the undamaged system. Finally the crack depth $\tilde{a}$ is estimated by inverting the polynomials (15).

4.5 Fault position confirmation by indicators

• *Indicator I_1: Convergence behavior during the iteration*

The first indicator is extracted from the convergence behavior of the identified value of the fault extent in the iteration. As shown in figure 10 for a cracked rotor the estimated fault extent approaches exponentially with the iteration number a certain value, if the fault element is assumed at the real fault position, whereas a wrongly assumed crack position does not lead to an exponential like approach.

The error of the exponential approximation of the convergence behavior is taken as indicator I_1.

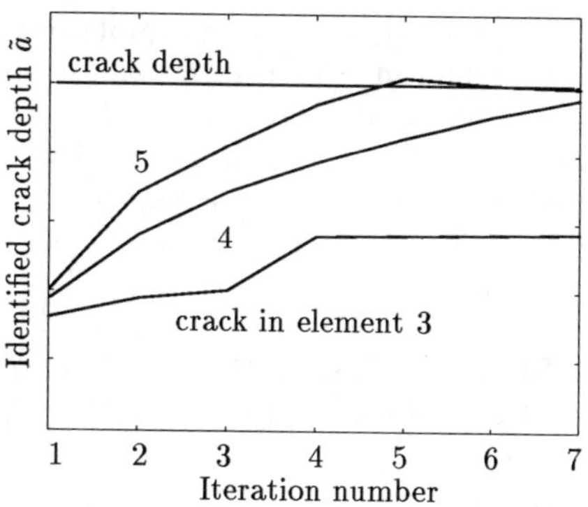

Figure 9: Crack depth versus iteration number

- *Indicator I_2: Simulation error*

The difference between the measured vibration $\Delta \mathbf{q}_e$ and the calculated vibration using the identified fault extent is taken as indicator

$$I_2 = |\Delta \mathbf{q}_{e,s} - \Delta \mathbf{q}_e|^2 , \tag{16}$$

which is minimal for the correctly assumed fault position.

- *Indicator I_3: Observation error covariance*

The observation error defined as the difference between the measured vibrations and the vibrations estimated by the observer in the last iteration step would be zero for correctly assumed fault properties. Otherwise the observer has to compensate the wrong input by a large observer error.

For calculating a third indicator I_3 the correlations between the observation error and sinusoidal test signals with multiples of the rotor speed Ω are calculated. The covariance is large if the reference signal is inherent to the observation error. Therefore, the sum of the absolute values of the correlation between the reference signals and the corresponding observation error is taken as indicator I_3. I_3 is small if the fault is assumed at the correct position because the observation error is in this case uncorrelated to the periodic reference signals.

The fault is assumed to be at the position where the sum of the normalized indicators is minimal.

5 CONCLUSIONS

The presented method working at constant rotor speed allows an on-line detection of the position and the extent of a fault without examining every single element. By evaluating symptoms the range of possible fault positions is reduced significantly so that for only a small group of possible fault locations the time consuming fault extent identification process has to be carried out. The position of the malfunction is finally confirmed by indicators which are calculated during the extend identification process.

References:

[1] BRUEL & KJAER: *Machine-Condition Monitoring Using Vibration Analysis. A Case Study from a Petrochemical Plant*, Application notes BO 0163-11 (1986).

[2] BRUEL & KJAER: *Permanent Vibration-Monitoring*, Application notes BG 0250-11 (1986).

[3] ISERMANN, ROLF: *Modellgestützte Überwachung und Fehlerdiagnose technischer System (Teil 1)* Automatisierungstechnische Praxis 38, *R. Oldenbourg Verlag* (1996) 5.

[4] BLACK, H. F.:*Interaction of a Whirling Rotor With a Vibrating Stator Across a Clearance Annulus.* Journal of Mechanical Engineering Science *10,* 1 (1968) p. 1-12.

[5] GASCH, R.: *Kleiner Beitrag zur Behandlung des dynamischen Verhaltens einer rotierenden Welle mit angerissenem Querschnitt.* ILR-Bericht 8, TU Berlin (1975).

[6] THEIS, W.: *Längs- und Torsionsschwingungen bei quer angerissenen Rotoren – Untersuchungen auf der Grundlage eines Rißmodells mit 6 Balkenfreiheitsgraden.* Fortschr.-Ber. VDI, Reihe 11, Nr. 131, VDI-Verlag, Düsseldorf (1990).

[7] BACHSCHMID, N., DELLUPI, R.: *Malfunction Identification in Rotor Systems from Bearing Measurements Using Partial Models of the System,* to be published in Revista Brasileira de Ciencias Mecanicas, (1997).

[8] MÜLLER, P. C.: *Indirect Measurements of Nonlinear Effects by State Observers.* IUTAM-Symp. Nonl. Dyn. Eng. Syst., University of Stuttgart, Springer-Verlag, (1990), pp. 205 – 215.

[9] BISCHOF, T.: *Ein Beitrag zur Detektion und Bewertung von kugelförmigen Einschlüssen mit holografischer Interferometrie.* Fortschr.-Ber. VDI, Reihe 8, Nr. 398, VDI-Verlag, Düsseldorf (1994).

Acknowledgement:

This project is funded by the European Union (BRITE/EURAM Modiarot BE95-2015 / Contract nr. BRPR-CT95-0022).

KEYNOTE SPEAKERS

K. P. Chong
C. Farrar
A. Kiremidjian

Health Monitoring of Civil Infrastructures

K. P. CHONG

ABSTRACT

Health monitoring of civil infrastructure systems is cost-effective and necessary since these systems are generally the most expensive investments/assets in any country. In the U.S. these assets are estimated at $20 trillion. In addition, these systems have long service life compared with any other kinds of commercial product, and are rarely replaceable once they are erected. Yet the feedback and controls on the "state of health" of constructed systems are practically non-existent. Nondestructive evaluation (NDE), sensing and smart structures are essential parts of this feedback and monitoring system for infrastructures. The NSF NDE initiative as well as workshops and related awards/projects are described in this paper.

INTRODUCTION

The National Science Foundation (NSF) funds research and education in most fields of science and engineering, through grants, contracts and cooperative agreements to colleges, universities, and other institutions. Civil infrastructure systems (CIS) research [1] is one of the Foundation-wide initiatives. The CIS initiative emphasizes systems integration at all levels and specifically addresses basic research in four key areas [2].

- DETERIORATION SCIENCE examines how materials and structures break down and wear out, improving our understanding of deterioration and design, and how to build, and maintain structures that are more durable, safer and more environmentally sound.

- ASSESSMENT TECHNOLOGIES determines durability, safety, and environmental conditions of structures and facilities. Research can lead to nondestructive evaluation techniques, improved sensor technologies, and self-correcting materials. This paper concentrates on the NDE and sensing of civil infrastructures as well as smart materials/structures.

Ken P. Chong, National Science Foundation, 4201 Wilson Blvd., Arlington, VA 22101.

- RENEWAL ENGINEERING extends and enhance the life of civil infrastructure systems and components that would otherwise continue to deteriorate.
- INSTITUTIONAL EFFECTIVENESS AND PRODUCTIVITY recognizes the importance of those factors affecting the decision processes underlying the provision and management of civil infrastructure on the economic and social productivity of society, leading to better decisions that maximize the impact of civil infrastructure investments on the productivity and on the economic and social well-being of the public.

The civil infrastructure ages and deteriorates with time. The deterioration is mostly a result of aging of the materials, excessive use, overloading, climatic conditions, lack of sufficient maintenance, and difficulties encountered in proper inspection methods. All of these factors contribute to the obsolescence of the constructed systems as a whole. As a result, repair, retrofit, rehabilitation, and replacement become necessary to insure the safety of the public. Besides NDE, **accelerated and destructive tests** are also needed to model life-cycle performances based on short term test data. In order to evaluate the safety of a constructed system, it is essential to determine the remaining strength of the total system by a performance analysis based upon the strength and inspection data of the materials and the structural system. A visual inspection process will provide data with respect to the external physical condition of the material such as the degree of corrosion and general deterioration. By past experience and judgment an engineer will evaluate the remaining life of the material and the contribution to the safety of the structure.

Strength determinations may be made by sampling the material and testing it in a laboratory. However, this method destroys some of the structure which subsequently must then be repaired. The repair is usually noticeable and disruptive, as a result, many engineers would prefer a nondestructive technique. Although there are many techniques for nondestructive evaluation currently available for metals which are used in defense and manufactured products, there are only a few which are available for the construction materials such as geomaterials, wood, concrete, masonry, and composites. These current methods for construction materials are not very reliable and in some instances they can only be used on a comparative qualitative basis. Instruments specifically designed for construction materials and with greater accuracy than those currently available are required in order to quantitatively evaluate the safety of the structure and to estimate its remaining life adequately.

RELATED RESEARCH ISSUES

In the past, engineers and material scientists have been involved extensively with the characterization of given materials. With the availability of advance computing, and the new developments in material sciences, researchers can now characterize process, design and manufacture materials with desirable performance and properties. One of the challenges is to model short term micro-scale material behavior, through meso-scale and, macro-scale behavior [3] into long term structural systems performance (Fig.1). Accelerated tests to simulate various environmental forces and impacts are needed. Supercomputers used in parallel are useful tools to solve this scaling problem by taking into account the large number of variables and unknowns to project nano/micro-behavior[4] into infrastructure systems performance, and to extrapolate short term accelerated test results to model the long term lifecycle behavior.

In recent years, researchers from diverse disciplines have been drawn into vigorous efforts to develop intelligent materials or structures that can monitor their own condition, detect impending failure, control damage, and adapt to changing environments. The potential applications of such smart materials/systems are abundant--ranging from design of smart aircraft skin embedded with fiber optic sensors to detect structural flaws; bridges with sensing/actuating elements to counter violent vibrations; and stealth submarine vehicles with swimming muscles made of special polymers. Such a multidisciplinary civil infrastructural systems [2] research front, represented by material scientists, physicists, chemists, biologists, and engineers of diverse fields--mechanical, electrical, civil, control, computer, aeronautical, etc.--has collectively created a new entity defined by the interface of these research elements. Smart materials are generally created through synthesis by combining sensing, processing, and actuating elements integrated with conventional structural materials such as steel, concrete, or composites. Some of these materials currently being researched or in use are [5]:

- piezoelectric composites, which convert electric current to (or from) mechanical forces;
- shape memory alloys, which can generate force through changing the temperature across a transition state;
- electro-rheological (ER) fluids, which can change from liquid to solid (or the reverse) in an electric field, altering basic material properties dramatically.

CONSTRUCTION MAT'LS		**STRUCTURES**	**INFRASTRUCTURE**
micro-level ~	**meso-level ~**	**macro-level ~**	**systems integration**
Atomic Scale	***Microns***	***Meters***	***Up to km Scale***
• micro-mechanics	**• meso-mechanics**	**• beams**	**• bridge systems**
• micro-structures	**• interfacial-structures**	**• columns**	**• lifelines**
• nanofabrication	**• smart materials**	**• plates**	**• cities**

Fig. 1. Materials and structural systems

Current research activities aim at understanding, synthesizing, and processing material systems which behave like biological systems. Smart materials basically possess their own sensors (nervous system), processor (brain system), and actuators (muscular systems)--thus mimicking biological systems. Sensors used in smart materials include optical fibers, corrosion sensors, and other environmental sensors and sensing particles. Examples of actuators include shape memory alloys that would return to their original shape when heated, hydraulic systems, and piezoelectric ceramic polymer composites. The processor or control aspects of smart materials are based on microchip, computer software and hardware systems. Fig. 2 is a sketch of a futuristic bridge system, integrating intelligent sensors, data processing and control, as well as de-icing mechanisms and high performance advanced composites with protective coatings. An artist rendition of this bridge appeared in *USA Today*, 3/3/97.

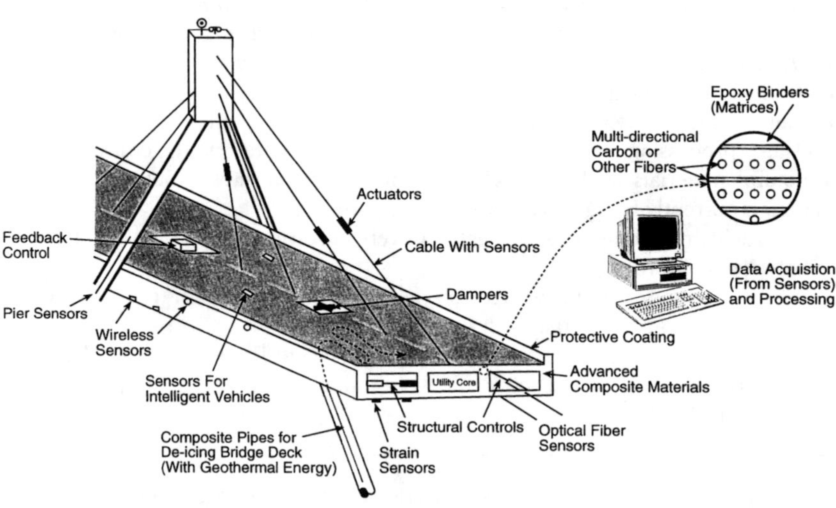

Fig. 2. An intelligent and integrated bridge system

Examples of Smart Materials

Smart materials may heal themselves when cracked. NSF grantees, Carolyn Dry of Illinois and Victor Li of Michigan have been developing self-healing concrete. One idea is to place hollow fibers filled with crack-sealing material into concrete, which if cracked would break the fiber releasing the sealant, according to Surendra Shah, Director of the NSF Science and Technology (S&T) Center at Northwestern. Lehigh researchers at ATLSS, an NSF Engineering Research Center, developed smart paints, which release red dye (contained in capsules) when cracked. Optical fibers which change in light transmission due to stress are useful non-discrete sensors. They can be embedded in concrete or attached to existing structures. NSF-supported researchers at Rutgers University have been studying optical fiber sensor systems for on-line and real-time monitoring of critical components of structural systems (such as bridges) for detection and warning of imminent structural systems failure. NSF grantees at Brown University and the University of Rhode Island investigated the fundamentals and dynamics of embedded optical fibers in concrete. Japanese researchers recently developed glass and carbon fiber reinforced concrete which provides the stress data by measuring the changes in electrical resistance in carbon fibers. Under an NSF Small Grants for Exploratory Research Programs, researchers at the University of California-Berkeley recently completed a study of the application of electro-rheological (ER) fluids for the vibration control of structures. ER fluids stiffen up very rapidly (changing elastic and damping properties) in an electric field. Other NSF supported researchers have been studying shape memory alloys (University of Texas, Virginia Tech, and MIT); surface super-elastic microalloying as sensors and microactuators (Michigan State); and magnetostrictive active vibration control (Iowa State, Virginia Tech); etc. Photoelastic experiments at Virginia Tech demonstrated that NiTiNOL shape memory alloy wires could be used to decrease the stress intensity factor by generating a compressive force at a crack tip. Other examples include:

- Intelligent Structures and Materials, C.A. Rogers, Univ. of S. Carolina. Specifically, this Presidential Young Investigator (PYI) project include: distributed sensing and health monitoring of civil engineering structures, active buckling control structures and active damage control of composites. Micron size magnetostrictive particles are mixed with the composite for health monitoring. Proof-of concept experiments have demonstrated that delaminations could be located using active magnetostrictive tagging. Low velocity impact damage resistance of composites can be improved by hybridizing them with shape memory alloy material (SMA). The graphite epoxy specimen has significantly more fiber breakage compared to the hybrid specimen and is completely perforated at the impact site.
- Research conducted by Prof. M. Taya (Univ. of Washington, Seattle) explores the use of TiNi shape memory alloy fibers and reinforcements in aluminum and epoxy matrix composites. The use of a shape memory alloy for a fiber reinforcement can result in the production of compressive residual stresses in the matrix phase of the composite. This can result in increases in flow stress and toughness of the composite compared to these properties in the same composites when the TiNi was not treated to produce a shape memory effect in the composite. The induced compressive stress in the matrix as a result of the shape memory alloy treatment of TiNi is responsible for the enhancement of the tensile properties and toughness of the composite. The concept of enhancing the mechanical properties of composites by inducing desirable stress states via shape memory reinforcements is innovative.

• Profs. Sharpe and Hemker (Johns Hopkins Univ) developed new test methods to measure the Poisson's ratio and the fracture toughness of Microelectromechanical systems (MEMS). While engineers today are able to design MEMS and predict their overall response, they cannot yet optimize the design to predict the allowable load and life of a component because the mechanical properties of the material are not available. This research is attempting to measure and provide such data. Further, they are performing comprehensive microstructural studies of MEMS materials which, together with the mechanical testing, will enable a fundamental understanding of the micromechanical response of these materials. There is currently a trend toward thicker materials (of the order of hundreds of micrometers) in MEMS because a larger aspect ratio is needed for a mechanical device to be able to transmit usable forces and torques. The mechanical testing techniques are being extended for such applications also.

ACTIVITIES IN NDE

The NSF Division of Civil and Mechanical Systems is encouraging basic research to investigate the requirements to develop suitable instruments and sensors (including MEMS) for nondestructive evaluation of strength properties and other physical conditions for construction materials as well as multi-disciplinary activities involving electrical, mechanical, chemical, computer, electronic, structural, geotechnical engineers and others, as suitable for the project to perform group or team efforts in the research projects. NSF is also encouraging close cooperation and active participation by the industries involved in the manufacture of nondestructive equipment and/or instrumentation. A number of companies are participating in the NDE research in the NSF Small Business Innovative Research (SBIR) programs. By joint efforts viable products can result which will be of great help to engineers responsible for the safety and longevity evaluation of structural systems.

As a source of information on the type of questions which the engineering community would like to have answered by the use of nondestructive evaluation techniques, NSF supported several workshops to outline specific needs of the profession. One was held in February, 1988, at the University of Southern California, Los Angeles, and the Proceedings are titled, "Nondestructive Evaluation for Performance of Civil Structures." The contact person is Dr. Sami F. Masri. The other workshop was held in June, 1989, at the National Institute of Standards and Technology, at Gaithersburg, Maryland, titled "International Workshop on Sensors and Measurement Techniques for Assessing Structural Performance," organized by Dr. Richard D. Marshall and his associates. Some recent workshops are listed at the end of this paper.

In addition to NDE research done at MIT, Illinois, Cornell [7], West Virginia, Virginia Tech, New Jersey Institute of Technology, Georgia Tech, Lehigh, etc., the centers at Johns Hopkins University, NASA Langley NDE Center, NIST Building and Fire Research Laboratory, and Iowa State University concentrate on various aspects of NDE.

NSF NDE INITIATIVE

Believing that "an ounce of prevention is better than a pound of cure," and that NSF should play a leading role in addition to acting as a catalyst in the

"Quantitative Nondestructive Evaluation for Large Structural Systems" in civil engineering, NSF initiatives were issued in April 1990 and January 1992.

Research efforts are to be directed toward increasing fundamental knowledge of quantitative nondestructive evaluation of construction materials and their application to large structural systems of wood, concrete, geomaterials, masonry and steel. The data developed for equipment may be for the material properties alone or for the fabricated forms as total systems. Some data may be used to determine how each system or material deteriorates with time and excessive loads. The data and prototype resulting from the research should be of such a nature that the resulting equipment may be useful for practitioners, and can be produced as a viable portable and economical commercial product.

Some of the recently awarded or completed projects are briefly described below. The projects are mostly in basic areas of interest to the Structures, Geomechanics, and Building Systems programs of the National Science Foundation.

- **Dual Purpose Etched Polarimetric Optical Fiber Sensor for Concrete;** Dr. Farhad Ansari - New Jersey Institute of Technology

In-service performance of concrete structures depends on consistent monitoring or condition for early scheduling of repair and retrofit operations. Proper quality control and condition monitoring procedures are either nonexistent or not readily applicable to large concrete structures. Early detection of large displacements and cracks in concrete elements will result in increased safety and considerable savings in rehabilitation costs. This research will examine the feasibility of using a novel class of fiber optic sensors for real-time condition monitoring of concrete structures. The new embedded sensor will serve a dual purpose by providing in-place strength estimations at early ages during construction, and crack detection capabilities there after.

- **Innovative Optical Fiber Sensors for Civil Infrastructure Systems;** Dr. M.Q. Feng - University of California-Irvine

Experimental research of new optical fiber sensors for monitoring civil infrastructure systems is planned. Optical sensors employ a vibrating wire whose tension can be modulated by external force, strain, or vibration and is translated into the change in the wire vibration frequency. The wire vibration frequency is detected by light sent to and reflected from the wire through an optical fiber cable. Compared to other existing optical fiber sensors which tend to suffer from the lack of reliability and robustness, the sensors have two significant advantages: one is that the sensing head is a vibrating wire (rather than an optical fiber), which can sense a specific physical quantity without interference from miscellaneous effects; the other is that the wire vibration is a well understood physical phenomenon. In fact, with a high level of reliability, its frequency is optically measured and transmitted to recording and other devices through the optical fiber without attenuation or distortion. These advantages should make the sensor system simple, reliable and robust, and hence more readily deployable in civil infrastructure applications.

- **Structural Control for Phase-Related Inputs;** Dr. R. L. Sack, et al - University of Oklahoma

Humans can produce significant dynamic loads, both while remaining in place and moving. Most building codes prescribe uniformly distributed static design live loads for offices, corridors, and public assembly areas such as shopping malls, airport terminals, gymnasium and dance floors. Researchers' previous studies have defined forces caused by in situ activities. The objectives of this research are: (a) to define the dynamic loads generated by activities involving motion such as walking and running; and (b) to establish the benefits and efficiency of low power semiactive control technologies to mitigate excessive structural response due to phase related occupant loads. A special force platform, with a flat transfer function in the frequency range of the anticipated loads, will be constructed, instrumented and used to measure forces generated by one, two and groups of subjects while performing prescribed activities. Loads due to individuals and groups will be characterized using analytical functions, multivariate regression analysis, and random simulation. Simulated results will be compared with measured data for small groups and design loads will be suggested both in time and frequency domains. A lightweight test floor system will be constructed in the second phase of the project. The dynamic characteristics of the floor system will be determined experimentally and analytically. The structure will be instrumented with a variable damper device. The variation in damping will be accomplished by using variable area in the damper mechanisms. The test floor will be loaded by both in situ and moving occupants and the response will be controlled using a closed loop system.

- **Fiber Optic Sensors for Strength Evaluation and Early Warning of Impending Failure in Structural Components;** Drs. Edward G. Nawy, Mohamad H. Maher - Rutgers University, The State University of NJ

The objective of this research is to develop and verify by experimentation the optical fiber strain and stress sensor systems for on-line, accurate, and real time monitoring of critical components of structural systems under service load conditions. The application of these sensor systems will permit easier inspection of structures for impending failures. A series of experiments are to be undertaken on scale models of reinforced and prestressed concrete structural beams, rolled structural steel sections, and beam to column connections.

- **Nondestructive Methods of Assessing Damage in Reinforced Concrete Structures;** Drs. Kenneth H. Stokoe, Jose M. Roesset and James O. Jirsa - University of Texas-Austin

The objective of this research is to study nondestructive methods based on the propagation and attenuation of stress waves for evaluating (1) the degree of damage to structural and foundation elements, (2) the location and extent of that damage, and (3) the extent and quality of repair. One of the key thrusts of this research will be the integrated study of both body and surface stress waves. The techniques to be used will be selected in terms of the type of structural elements and their accessibility for use of nondestructive test equipment. The rapidly-developing technique based on surface wave measurements will be used where there is access to an exposed surface of the structural element which has dimensions similar in magnitude to the depth of the element being investigated.

The Spectral Analysis of-Surface-Waves (SASW) method will be used.

- **Micromagnetic Surface Studies of Materials for Nondestructive Evaluation;** Dr. David Jiles - Iowa State University, Dr. W. A. Theiner - University of Saarland, Federal Republic of Germany

  This project is a joint research effort into the use of magnetic surface studies for nondestructive evaluation between the Center for NDE at Iowa State University and the Fraunhofer Institute for Zerstorungfreie Prufverfahren (NDT) at the University of Saarland in West Germany. The objective of the research is to identify alternating current magnetic measurements which can be made on the surfaces of constructional steel components to evaluate their mechanical condition. Such methods, in which the penetration of the magnetic field is restricted to the surface, will be useful for determining the mechanical state of the surface as a result of shot peening, grinding, surface hardening and residual surface stress. It is also known that in fatigue failure, the fatigue crack initiates at the surface and then propagates into the bulk of the material. Monitoring of the surface condition is therefore critical in identifying the onset of failure in the material as a whole.

- **Fundamentals of Embedded Optical Fibers In Structural Concrete;** Drs. K. S. Kim, T. F. Morse - Brown University

  The thrust of this project will be to combine the knowledge and expertise of faculty members from solid mechanics, materials science, civil engineering, and optical sensing to obtain a more quantitative description of the response of structural concrete to external loads. By embedding optical fibers in a concrete structure during the fabrication process, it is possible to sense, through a variety of techniques, the local internal state of the structure. This research will aid in the prediction of the behavior of such materials, will point toward directions for the improvement of structural concrete, and can be applied to the retrofitting of existing structures with novel sensors.

- **Electronic Shearography for Structural Testing;** Drs. A.K. Maji and M. L. Wang - Universities of New Mexico and Illinois - Chicago

  Shearography, a laser interferometry technique, enables direct observation and measurement of strain concentrations and variations of the displacement field. Shearography performed with the aid of electronic image acquisition and processing can eliminate the practical problems that have precluded the use of other laser interferometry techniques for nondestructive inspection under field conditions. This method, Electronic Shearography (ES), is proposed as an inspection technique for detecting cracking and other sources of stress concentration in structures. The technique offers full-field inspection capability, can be performed in real time, and is easily suited to automated computer aided data analysis. A newly developed geometric filtering technique will be applied to reduce speckle noise. Additional telescopic lenses will be used to facilitate remote measurements. The combination of the hardware and software tools will result in excellent image quality that will make this ES technique acceptable for field inspection of civil engineering structures. The proposed technique will then be evaluated during field inspection of bridges, in conjunction with other state of the art inspection techniques. Theoretical analysis using fracture mechanics and structural dynamics will be used to interpret the results of the ES inspection, and to address the need for repair or retrofitting.

- **Quantitative Nondestructive Evaluation for Concrete Constructed Facilities;** Drs. I.L. Al-Qadi, J.C. Duke, S.M. Riad, R.E. Weyers and G.S. Brown, Virginia Tech.

The objective of the research is to correlate the structural condition of portland cement concrete facilities with the test results using electromagnetic waves, eddy current, and ultrasonic techniques. Electromagnetic and ultrasonic methods have previously shown promise for achieving this objective and it is proposed the use of eddy currents as complementary technique. However, before the objective can be addressed, it is important to clearly establish a scientific/engineering data base from which the electromagnetic, electrical, and ultrasonic properties of portland cement structures can be obtained. This data base will provide insight and understanding of the problems arising from the interaction of these electrical/acoustic methods with concrete structures and will be made available to other researchers considering similar NDE methods. To the proposers' knowledge, this data base represents the first attempt to set up and carry out, under controlled conditions, a comprehensive set of measurements to characterize a select set of concrete specimens using basic electrical and ultrasonic techniques that have only been estimated in previous research.

- **Design and Construction of a Portable Instrument to Measure Stresses in Structures;** Dr. Caesar A. Sciammarella, Illinois Institute of Technology, Chicago, IL

This project will develop a portable field instrument to determine the condition of steel and concrete large structures in order to make an analysis of the safety of the structure. Other uses are in the analysis of cracks in steel and concrete structures, resulting from fatigue, or fabrication, measure residual stresses, determine mechanical properties of materials in-situ, and measure stresses in structures needing repair.

- **Condition Evaluation of Concrete Structures Using High Frequency Ultrasound;** Dr. Sandor Popovics, Drexel University, Philadelphia, PA

An interdisciplinary team of experts in nondestructive evaluation (NDE) from Civil, Electrical and Mechanical Engineering is developing practical ultrasonic techniques for generating high-frequency ultrasound that can penetrate concrete sufficiently to improve NDE of concrete structures, including the estimation of the residual strength of concrete. This will be achieved by transmission of high energy-high frequency (i.e., MHz range) ultrasound using piezo-electric shock waves and pseudo-random codes with large time-band width products; suppression of thermal noise using temporal averaging or pseudo-random code correlation techniques, and reduction of structural noise and internal reverberations by frequency and spatial diversity; nondestructive estimation of the various parameters needed for surveying the condition of concrete structures, including the estimation of residual concrete strength, detection of flaws (voids, cracks, delamination, etc.), and detection of corrosion in reinforcements embedded in concrete structures will be developed by supplementing the usual measurement of velocity, attenuation and scattering of longitudinal waves, with measurements on multiple mode ultrasound propagation, including shear, surface and Lamb waves, as well as higher order plate waves. These advanced NDE methods have successfully been applied by the team members to materials other than concrete. The preliminary tests indicate that the advanced NDE methods can be readily modified for use in concrete.

- **Quantitative Nondestructive Evaluation of Safety and Performance of Geostructures Using Seismic Characterization of Properties and Discontinuous Deformation Analysis;** Drs. Richard E. Goodman and Neville G. Cook; University of California, Berkeley

  Many major constructed civil infrastructure systems in the U.S. were constructed before the development of modern tools and procedures for evaluating the properties of foundation rocks. In addition to problems with foundations, old concrete structures, such as dams, retaining walls, building foundations, pipes, and water tanks, may contain serious cracks hidden from view that can compromise structural safety. Now we face the responsibility of assessing the continued performance and safety of such structures. Recent advances in the understanding of the effects of discontinuities on seismic wave transmission make it feasible to locate and characterize macroscopic discontinuities in rock and concrete structures, as well as to determine their strength and deformability parameters. In this project, recent advances to develop the experimental and interpretive means to characterize fractures in rock and concrete will be used. These characterizations will then be used in a Discontinuous Deformation Analysis (DDA) model to build a numerical simulation that evaluates the overall properties of the defective zone within the structure or foundation. The methods will be developed in the laboratory, tested in bench-scale experiments, and applied in a series of preliminary field trials.

- **An Automated Inversion Procedure for the SASW Test;** Dr. N. Gucunski; Rutgers University

  The objective of this research is to develop the inversion routine for the Spectral-Analysis-of-Surface-Waves (SASW) method. The SASW method is a seismic method for nondestructive testing evaluation of elastic moduli and thicknesses of layered systems, like soils and pavements. One of the major objectives of research conducted in past ten years was to develop a fully automated SASW testing procedure, which would enable "instantaneous" definition of the profile in the field. The biggest obstacle to achieve this represents the inversion process. This problem is especially pronounced in cases of irregular stratification of a layered system, i.e. when the shear wave velocity does not increase with depth. This research suggests three approaches to the solution of this problem. The first approach is based on the comparison of the experimental dispersion curve and the curve from the numerical simulation of the SASW test for an assumed profile. The other two approaches represent simplifications of the first one, that might significantly reduce computational effort and at the same time preserve the desired level of accuracy. Once the problem of the inversion is solved, it is expected that the SASW method will be widely used method for the nondestructive testing of layered systems.

NSF SUPPORTED WORKSHOPS/CONFERENCES ON NDE

The following are some of the recent workshops/conferences funded by NSF:

- "NDE for performance of Civil Structures," Sam Masri (USC), 2/88.
- "NSF-ATLSS Workshop on Sensor and Signal Processing for Structural

Control", 2/91.

- "NDE of Civil Structures and Materials" , Bruce Suprenant, et al (University of Colorado - Boulder) 10/90, 5/92.
- "Recent Advances in Adaptive and Sensory Materials and their Applications", C. A. Rogers and R. C. Rogers (Virginia Tech) 4/92.
- "NDT of Concrete in the Infrastructure," S.P. Shah, et al, SEM Conference, Dearborn, MI, 6/93.
- "NDE of Materials", Robert Green, et al, Boulder, CO, 6/97.

SUMMARY AND CONCLUSION

An overview of NDE at NSF and a discussion on the initiative and projects on the "Quantitative Nondestructive Evaluation for Large Structural Systems" as well as smart structures/materials are presented. The author hopes that this paper will act as a catalyst, sparking interest and further research in the health monitoring of civil infrastructure systems. This paper reflects the personal views of the author, not necessarily those of the National Science Foundation.

ACKNOWLEDGMENTS

Input by his NSF colleagues, including Drs. S. C. Liu, C.S. Hartley, S. Saigal, J. B. Scalzi and O.W. Dillon, are gratefully acknowledged.

REFERENCES

1. NSF 93-5, "*Civil Infrastructures Systems Research: Strategic Issues*," K.P.Chong (Chm), National Science Foundation, Arlington, VA, 1/93.

2. NSF 95-52, "*Civil Infrastructure Systems*" Program Announcement and Guidelines, National Science Foundation, Arlington, VA 3/95.

3. Boresi, A.P. and Chong, K. P., (1987), *Elasticity in Engineering Mechanics*, Elsevier, New York; Chinese edition published by Science Publishers (1995).

4. Kirk, W. P., and Reed, M. A., eds., (1992), *Nanostructures and Mesoscopic Systems,* Academic Press.

5. Chong, K. P., O.W. Dillon, J.B. Scalzi and W. A. Spitzig, Engineering Research in Composite and Smart Structures, *Composites Engineering*, 4(8)(1994), 829-852.

6. Rogers, C. A., and Rogers R.C., eds., (1992), *Recent Advances in adaptive and Sensory Materials and Their applications,* Technomic Publishing, Lancaster, PA.

7. Cheng, C. and Sansalone, M., (1993), "The Impact-Echo Response of Plates Containing Delaminations: Numerical, Experimental, and Field Studies," *Materials and Structures*, RILEM, 26(159), pp. 274-285.

Lessons Learned from Applications of Vibration-Based Damage Identification Methods to a Large Bridge Structure

C. R. FARRAR and S. W. DOEBLING

ABSTRACT

Over the past 30 years detecting damage in a structure from changes in dynamic parameters has received considerable attention from the aerospace, civil, and mechanical engineering communities. The general idea is that changes in the structure's physical properties (i.e., stiffness, mass, and/or damping) will, in turn, alter the dynamic characteristics (i.e., resonant frequencies, modal damping, and mode shapes) of the structure. Properties such as the flexibility matrix, stiffness matrix, and mode shape curvature, which are obtained from modal parameters, have shown promise for locating structural damage. However, the application of these techniques to large civil engineering structures is limited because of the inability to find structures that the owners will allow to be damaged. Also, the cost associated with testing these structures can be prohibitive. In this paper, the authors' experiences with performing modal tests on a large highway bridge, in its undamaged and damaged state, for the purpose of damage identification will be summarized. Particular emphasis will be made on the lessons learned form this experience and the lessons learned from recent tests on another bridge structure.

INTRODUCTION

The interest in the ability to monitor a structure and detect damage at the earliest possible stage is pervasive throughout the civil, mechanical, and aerospace engineering communities. Current damage detection methods are either visual or localized experimental methods such as acoustic or ultrasonic methods, magnetic field methods, radiography, eddy-current methods and thermal field methods (Doherty, 1987) All these experimental methods require that the vicinity of the

---

Charles R. Farrar, Materials Behavior Team Leader, farrar@lanl.gov
Scott W. Doebling, Technical Staff Member, doebling@lanl.gov
Los Alamos National Laboratory, MS P946, Los Alamos, NM 87545

damage is known *a priori* and that the portion of the structure being inspected is readily accessible. Subjected to these limitations, these experimental methods can detect damage on or near the surface of the structure. The need for more global damage detection methods that can be applied to complex structures has led to the development of methods that examine changes in the vibration characteristics of the structure.

Global damage or fault detection, as determined by changes in the dynamic properties or response of structures, is a subject that has received considerable attention in the literature beginning approximately 30 years ago. Based on the amount of information provided regarding the damage state, these methods can be classified as providing four levels of damage detection. The four levels are (Rytter, 1993):

1. Identify that damage has occurred;
2. Identify that damage has occurred and determine the location of damage;
3. Identify that damage has occurred, locate the damage, and estimate its severity; and
4. Identify that damage has occurred, locate the damage, estimate its severity, and determine the remaining useful life of the structure.

The basic premise of the global damage detection methods that examine changes in the dynamic properties is that modal parameters, notably resonant frequencies, mode shapes, and modal damping, are a function of the physical properties of the structure (mass, damping, stiffness, and boundary conditions). Therefore, changes in physical properties of the structure, such as its stiffness or flexibility, will cause changes in the modal properties. A detailed summary of vibration-based damage detection methods can be found in (Doebling et al., 1996) where the general limitations and successes of these methods are discussed. Also discussed in this reference are specific examples of the application of vibration-based damage identification algorithms to bridge structures.

It is the intent of this paper to provide a brief summary of one such study that was made of the I-40 Bridge over the Rio Grande in Albuquerque, NM. This project was performed jointly with researchers from Sandia National Laboratory and New Mexico State University. A detailed discussion of the bridge geometry, testing procedures, data acquisition, and data reduction can be found in (Farrar et al., 1994) In particular, the lessons learned from this study will be emphasized. The authors hope that this discussion will help other investigators avoid some of the pitfalls that were experienced on this test. Additional lessons learned have come from a subsequent test on Alamosa Canyon Bridge in southern New Mexico. Finally, it should be pointed out that many of the lessons have been learned through interactions with other engineers including those at Los Alamos and Sandia National Laboratories and faculty and staff at the University of Cincinnati and Drexel University.

TEST STRUCTURE GEOMETRY

The I-40 Bridges over the Rio Grande in Albuquerque, NM, razed in 1993, formerly consisted of twin spans (there are separate bridges for each traffic direction) made up of a concrete deck supported by two welded-steel plate girders and three steel stringers. Loads from the stringers were transferred to the plate girders by floor beams located at 6.1 m (20 ft) intervals. Cross-bracing was provided between the floor beams. Fig. 1 shows an elevation view of the portion of the bridge that was tested. The cross-section geometry of each bridge is shown in Fig. 2. Each bridge was made up of three identical sections. Each section had three spans; the end spans were of equal length, approximately 39.9 m (131 ft), and the center span was approximately 49.4 m (163 ft) long. All subsequent discussions of the I-40 bridge will refer to the bridge carrying eastbound traffic, particularly the three eastern spans, which were the only ones tested.

DAMAGE SCENARIOS

The damage that was introduced was intended to simulate fatigue cracking that has been observed in plate-girder bridges. This type of cracking results from out-of-plane bending of the plate girder web and usually begins at welded attachments to the web such as the seats supporting the floor beams. Four levels of damage were introduced to the middle span of the north plate girder close to the seat supporting the floor beam at mid-span. Damage was introduced by making various torch cuts in the web and flange of the girder.

The first level of damage, designated E-1, consisted of a two-foot-long, 10-mm-wide (3/8-in-wide) cut through the web centered at mid-height of the web. Next, this cut was continued to the bottom of the web to produce a second level of damage designated E-2. For the third level of damage, E-3, the flange was then cut halfway in from either side directly below the cut in the web. Finally, the flange was cut completely through for damage case E-4 leaving the top 4 ft of the web and the top flange to carry the load at this location. The various levels of damage are shown in Fig. 3. Photographs of the damage can be found in (Farrar et al., 1994).

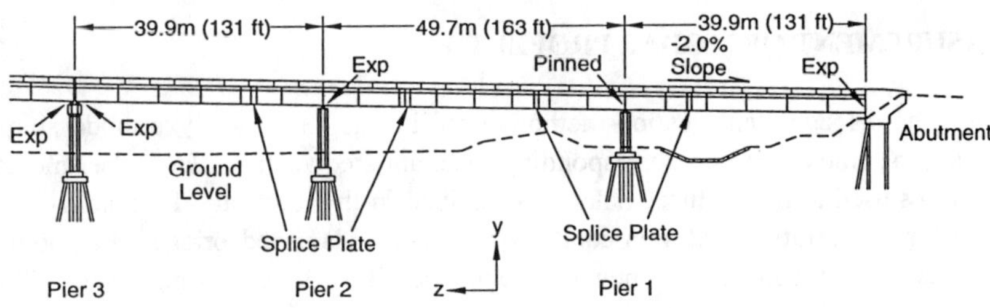

Fig. 1. Elevation view of the portion of the I-40 Bridge that was tested.

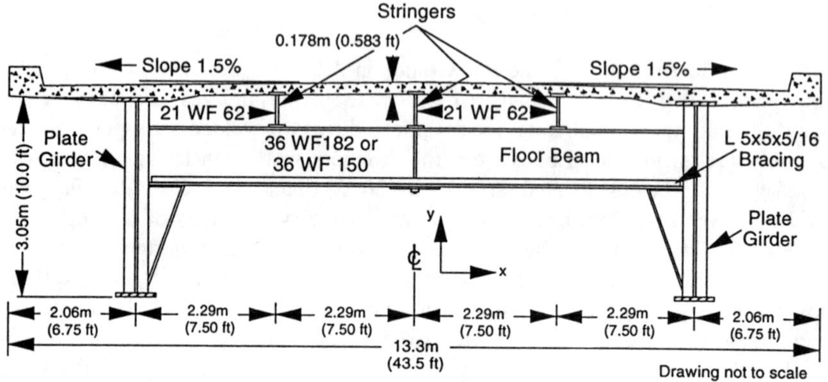

Fig. 2. Corss-section geometry of the I-40 Bridge.

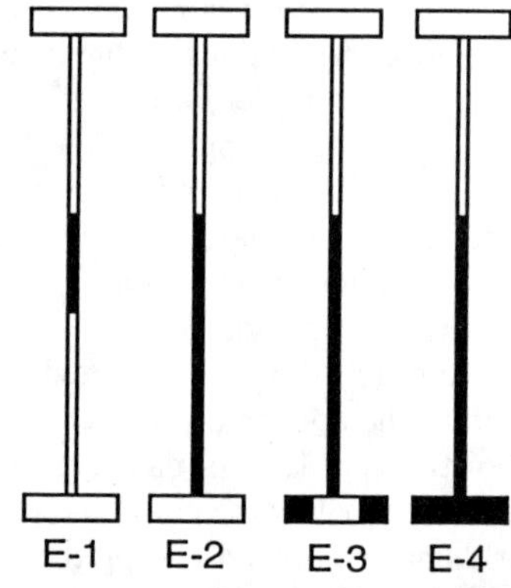

Fig. 3 Damage Scenarios

MEASUREMENT OF MODAL PROPERTIES

The damage identification methods used in this study analyze mode shape data and, in some cases, the corresponding resonant frequencies. The experimental procedures used to obtain these data are described in this section. To obtain these data, a forced vibration test was conducted on the undamaged bridge. Eastbound traffic had been transferred to a new bridge just south of the one being tested. The westbound traffic continued on the original westbound bridge. Excitation from traffic on the adjacent bridges could be felt on the bridge being tested. Next, the four different levels of damage were introduced into the middle span of the north plate girder. Forced vibration tests similar to those done on the undamaged structure were repeated after each level of damage had been introduced. A detailed summary of the experimental procedures can be found in (Farrar et al., 1994).

Excitation

Engineers from Sandia National Laboratory (SNL) provided a hydraulic shaker that generated the measured force input. The SNL shaker consists of a 96.5 kN (21,700-lb) reaction mass supported by three air springs resting on top of drums filled with sand. A 9.79 kN (2200-lb) hydraulic actuator bolted under the center of the mass and anchored to the top of the bridge deck provided the input force to the bridge. A random-signal generator was used to produce a 2000-lb peak-force uniform random signal over the frequency range of 2 to 12 Hz. An accelerometer mounted on the reaction mass was used to measure the force input to the bridge. The shaker was located on the eastern-most span directly above the south plate girder and midway between the abutment and first pier. A more detailed summary of the Sandia shaker can be found in (Mayes and Nusser, 1994).

Data Acquisition

The data acquisition system used in these tests consisted of a computer workstation that controlled 29 input modules and a signal processing module. The workstation was also the platform for a commercial data-acquisition/signal-analysis/modal-analysis software package. The input modules provided power to the accelerometers and performed analog-to-digital conversion of the accelerometer voltage-time histories. The signal-processing module performed the needed fast Fourier transform calculations. A 3500-watt AC generator was used to power this system in the field.

Two sets of integrated-circuit, piezoelectric accelerometers were used for the vibration measurements. A coarse set of measurements (SET1) was first made. These accelerometers were mounted in the vertical direction, on the inside web of the plate girder, at mid-height and at the axial locations shown in Fig. 4. A more refined set of measurements (SET2) was made near the damage location. Eleven accelerometers were placed in the global Y direction at a nominal spacing of 4.88 m (16 ft) along the mid-span of the north plated girder. All accelerometers were located at mid-height of the girder. The spacing of these accelerometers relative to the damage is shown in Fig. 5.

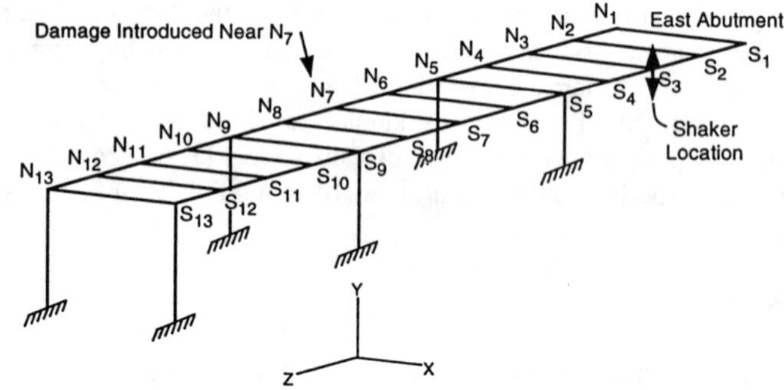

Fig. 4. Set1 (coarse) acceleromater locations.

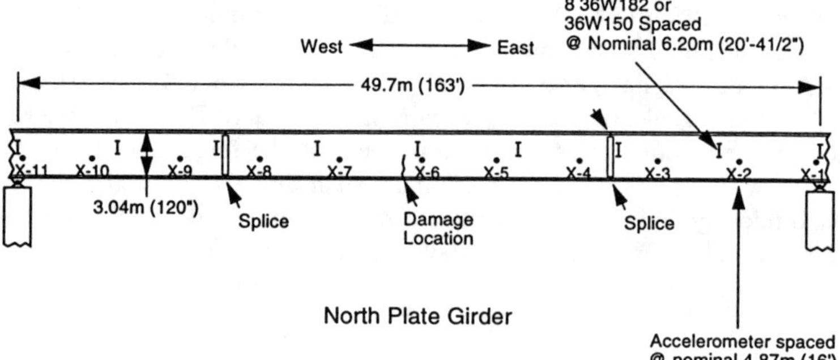

Fig. 5. Set2 (refined) accelerometer locations.

Modal Parameter Identification

Standard experimental modal analysis procedures were applied to data obtained from the SET1 accelerometers during the forced vibration tests to identify the modal parameters of the bridge in its damaged and undamaged condition. In this context experimental modal analysis refers to the procedure whereby a measured excitation (random, sine, or impact force) is applied to a structure, and the structure's response (acceleration, velocity, or displacement) is measured at discrete locations that are representative of the structure's motion. Both the excitation and the response time histories are transformed into the frequency domain in the form of frequency response functions (FRFs). Modal parameters (resonant frequencies, mode shapes, modal damping) can be determined by curve-fitting a Laplace domain representation of the equations of motion to the measured frequency domain data (Ewins, 1985). A rational-fraction polynomial, global, curve-fitting algorithm in a commercial modal analysis software package was used for these fits. Figure 6 shows the first three modes of the undamaged bridge identified from these data. By

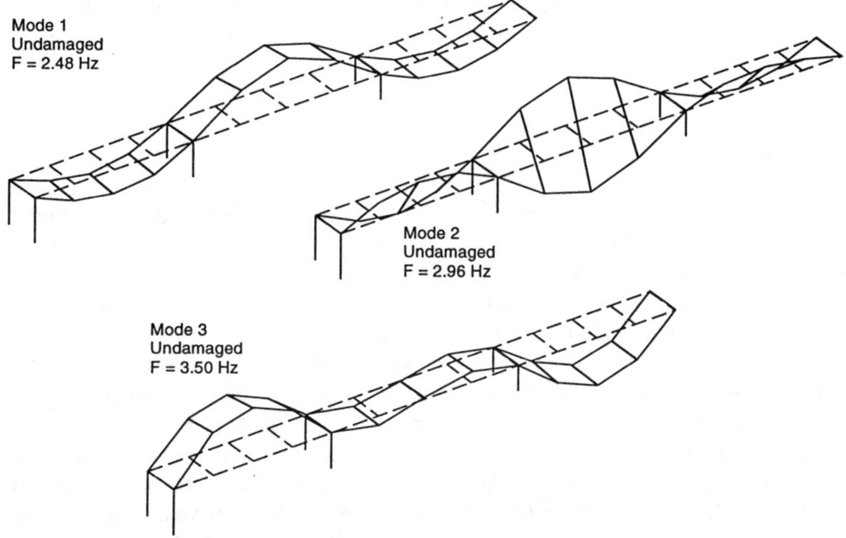

Fig. 6. First three modes measured on the undamaged structure.

measuring the input force and the corresponding driving point acceleration, these mode shapes can be unit-mass normalized.

Immediately after the forced vibration tests with the SET1 accelerometers were complete, the random excitation tests were repeated using the refined SET2 accelerometers. For these tests the input was not monitored. Operating shapes were determined from amplitude and phase information contained in the cross-power spectra (CPS) of the various accelerometer readings relative to the accelerometer X-3 shown in Fig. 5. Determining operating shapes in this manner, as discussed by (Bendat and Piersol, 1980), simulates the methods that would have to be employed when the responses to ambient excitations are measured. For modes that are well spaced in frequency these operating shapes will closely approximate the mode shapes of the structure. However, without a measure of the input force these modes cannot be mass normalized.

INFLUENCE OF DAMAGE ON CONVENTIONAL MODAL PROPERTIES

Changes in Resonant Frequencies and Modal Damping

Table I summarizes the resonant frequency and modal damping data obtained during each modal test of the undamaged and damaged bridge. No significant changes in the dynamic properties can be observed until the final level of damage is introduced. At the final level of damage the resonant frequencies for the first two modes have dropped to values 7.6 and 4.4 percent less, respectively, than those measured during the undamaged tests. For modes where the damage was introduced near a node for that mode (modes 3 and 5) no significant changes in resonant frequencies can be observed.

Changes in Mode Shapes

A modal assurance criterion (MAC), sometimes referred to as a modal correlation coefficient (Ewins, 1985), was calculated to quantify the correlation between mode shapes measured during different tests.

Table II shows the MAC values that are calculated when mode shapes from tests t17tr (damage level E-1), t18tr (damage level E-2), t19tr (damage level E-3), and t22tr (damage level E-4) are compared to the modes measured on the undamaged forced vibration test, t16tr. The MAC values show no change in the mode shapes for the first three stages of damage. When the final level of damage is introduced, significant drops in the MAC values for modes 1 and 2 are noticed. These two modes are shown in Fig. 7 and can be compared to similar modes identified for the undamaged bridge in Fig 6. When the modes have a node near the damage location (modes 3 and 5), no significant reductions in the MAC values are observed, even for the final stage of damage. This result corresponds to the observed similarity in mode 3 shown in Figs. 6 and 7.

From the observed changes in modal parameters it is clear that damage can only be definitively identified after the final cut was made in the bridge. Prior to the final cut, one could not say that the changes observed were caused by damage or were within the repeatability of the tests. In two tests at increasing levels of damage (t17tr and t18tr) the resonant frequencies were actually found to increase slightly from that of the undamaged case. These slight increases in frequency were measured independently by other researchers studying this bridge at the same time (Farrar, et al. (1994)) and are assumed to be caused by changing test conditions. The examination of changes in the basic modal properties (resonant frequencies and mode shapes) demonstrates the need for more sophisticated methods to examine modal data for indications of damage. Also, the need for statistical analysis of the data and quantification of the environmental effects on the measured modal properties is evident when one considers the small changes that are being examined. The topic of variability in modal properties and statistical analysis is discussed in more detail in (Farrar et al., 1997) and (Doebling et al., 1997).

DAMAGE IDENTIFICATION STUDIES

For comparative purposes, five linear modal-based damage identification algorithms were applied to data obtained from the I-40 bridge in its undamaged and damaged

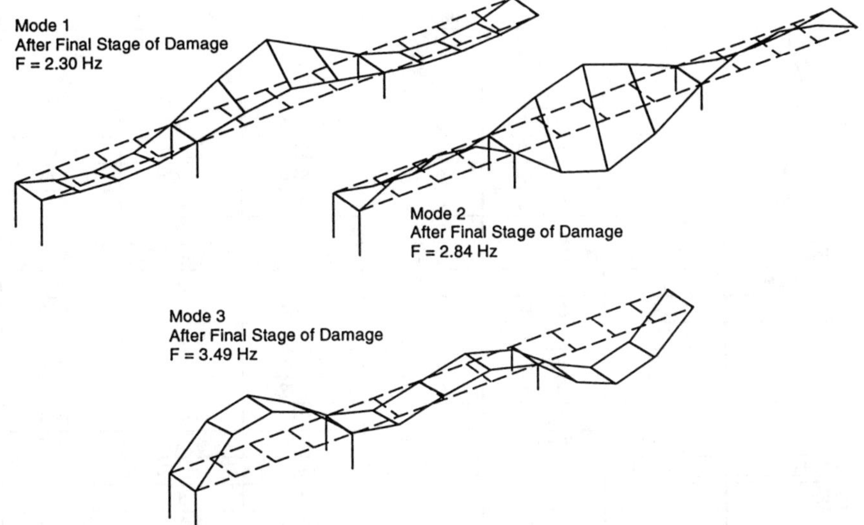

Fig. 7. First three modes measured after the final level of damage.

TABLE I

Resonant Frequencies and Modal Damping Values Identified from Undamaged and Damaged Forced Vibration Tests Using SET1 Accelerometers

		Mode 1	Mode 2	Mode 3	Mode 4	Mode 5	Mode 6
Test Designation	Damage Case	Freq. (Hz)/ Damp. (%)	Freq. (Hz)/ Damp. (%)	Freq. (Hz)/ Damp. (%)	Freq. (Hz)/ Damp. (%)	Freq. (Hz)/ Damp. (%)	Freq. (Hz)/ Damp. (%)
t16tr	Undamaged	2.48/ 1.06	2.96/ 1.29	3.50/ 1.52	4.08/ 1.10	4.17/ 0.86	4.63/ 0.92
t17tr	E-1 cut at center of web	2.52/ 1.20	3.00/ 0.80	3.57/ 0.87	4.12/ 1.00	4.21/ 1.04	4.69/ 0.90
t18tr	E-2 cut extended to bottom flange	2.52/ 1.33	2.99/ 0.82	3.52/ 0.95	4.09/ 0.85	4.19/ 0.65	4.66/ 0.84
t19tr	E-3 bottom flange cut half way	2.46/ 0.82	2.95/ 0.89	3.48/ 0.92	4.04/ 0.81	4.14/ 0.62	4.58/ 1.06
t22tr	E-4 bottom flange cut completely	2.30/ 1.60	2.84/ 0.66	3.49/ 0.80	3.99/ 0.80	4.15/ 0.71	4.52/ 1.06

TABLE II
Modal Assurance Criteria:
Undamaged and Damaged Forced Vibration Tests

Modal Assurance Criteria	Undamaged (test t16tr) X First level of damage, E-1 (test t17tr)					
Mode	1	2	3	4	5	6
1	0.996	0.006	0.000	0.003	0.001	0.003
2	0.000	0.997	0.000	0.005	0.004	0.003
3	0.000	0.000	0.997	0.003	0.008	0.001
4	0.004	0.003	0.006	0.984	0.026	0.011
5	0.001	0.008	0.003	0.048	0.991	0.001
6	0.001	0.006	0.000	0.005	0.005	0.996
Modal Assurance Criteria	Undamaged (test t16tr) X Second level of damage, E-2, (test t18tr)					
Mode	1	2	3	4	5	6
1	0.995	0.004	0.000	0.004	0.001	0.002
2	0.000	0.996	0.000	0.003	0.002	0.002
3	0.000	0.000	0.999	0.006	0.004	0.000
4	0.003	0.006	0.005	0.992	0.032	0.011
5	0.001	0.006	0.008	0.061	0.997	0.004
6	0.002	0.004	0.000	0.005	0.005	0.997
Modal Assurance Criteria	Undamaged (test t16tr) X Third level of damage, E-3 (test t19tr)					
Mode	1	2	3	4	5	6
1	0.997	0.002	0.000	0.005	0.001	0.001
2	0.000	0.996	0.001	0.003	0.002	0.002
3	0.000	0.000	0.999	0.006	0.006	0.000
4	0.003	0.005	0.004	0.981	0.032	0.011
5	0.001	0.006	0.004	0.064	0.995	0.003
6	0.002	0.002	0.000	0.004	0.009	0.995
Modal Assurance Criteria	Undamaged (test t16tr) X Fourth level of damage, E-4 (test t22tr)					
Mode	1	2	3	4	5	6
1	0.821	0.168	0.002	0.001	0.000	0.001
2	0.083	0.884	0.001	0.004	0.001	0.002
3	0.000	0.000	0.997	0.005	0.007	0.001
4	0.011	0.022	0.006	0.917	0.010	0.048
5	0.001	0.006	0.003	0.046	0.988	0.002
6	0.005	0.005	0.000	0.004	0.009	0.965

condition. These algorithms require mode shape data (in some cases unit-mass normalized mode shape data) and resonant frequencies. A summary of these algorithms and their implementation for the study reported herein can be found in (Farrar and Jauregui, 1996). All five methods are based on the observation that in the vicinity of damage there will be a local increase in the structure's flexibility. This increase will alter the mode shapes of the structure in the damage vicinity, hence a comparison of mode shape data processed by these five different methods, before and after damage, should reveal the location of the damage. These methods provide Level 2 damage indication as discussed in the introduction.

Tables III and IV summarize the results from applying the five damage detection algorithms to the experimental modal data from the SET1 and SET2

TABLE III

Summary of Damage Detection Results using
Experimental Modal Data from Coarse Set of Accelerometers (SET1)

Damage Case	E-1	E-2	E-3	E-4
Damage Index Method	**	**	**	*
Mode Shape Curvature Method	**	*	**	*
Change in Flexibility Method	○	○	**	*
Change in Uniform Load Surface Curvature Method	○	○	○	*
Change in Stiffness Method	**	**	**	*

* Damage located, ** Damage located using only 2 modes; ○ Damage not located

TABLE IV

Summary of Damage Detection Results using
Experimental Modal Data from Refined Set of Accelerometers (SET2)

Damage Case	E-1	E-2	E-3	E-4
Damage Index Method	●	●	●	●
Mode Shape Curvature Method	●●●	●●	●	●
Change in Flexibility Method	○	○	○	●
Change in Uniform Load Surface Curvature Method	○	●●●	●	●
Change in Stiffness Method	○	○	○	●

● Damage located, ●● Damage narrowed down to two locations, ●●● Damage narrowed down to three locations, ○ Damage not located

instruments, respectively. In this study, the Damage Index Method was found to have performed the best. All methods were able to definitively locate the damage for the final case, E-4. For the intermediate damage cases mixed results were obtained.

IN HIND SIGHT, THINGS WE SHOULD HAVE DONE

Based on subsequent analysis and observations related to the I-40 bridge tests (Doebling and Farrar, 1997), subsequent tests on another bridge (Doebling et al., 1997 and Farrar et al., 1997), interactions with other researchers in the field (particularly those at the Univ. of Cincinnati and Drexel Univ.), and review of the technical literature related to bridge testing, there are several things that should have been done during these tests (and have been done on subsequent tests) to improve the confidence in the damage ID results. These improvements are listed below:

Perform A More Thorough Pre-Test Visual Inspection.

Visual inspection is used to ascertain the initial condition of the structure. Pay particular attention to boundary conditions and changes to the neighboring vicinity of the test structure. During the tests visual inspection revealed that the east end of the top portion of the south girder was no longer in contact with the concrete at the top of the abutment. During earlier tests observations of the end condition were not made.

Perform Linearity Check.

The system identification portion of the experimental modal analysis procedure typically relies on the assumption that the structure is linear. Linearity can be checked, to some degree, by exciting the structure at different levels and overlaying the measured FRFs for a particular point. Ideally, with the thought of an on-line monitoring system in mind, these different excitation levels would span the range of loading observed during ambient traffic vibration measurements. On subsequent bridge tests it has been observed linearity was only observable over a portion of the spectrum as shown in Fig. 9. Also, a change in the linearity properties can in itself be an indication of damage.

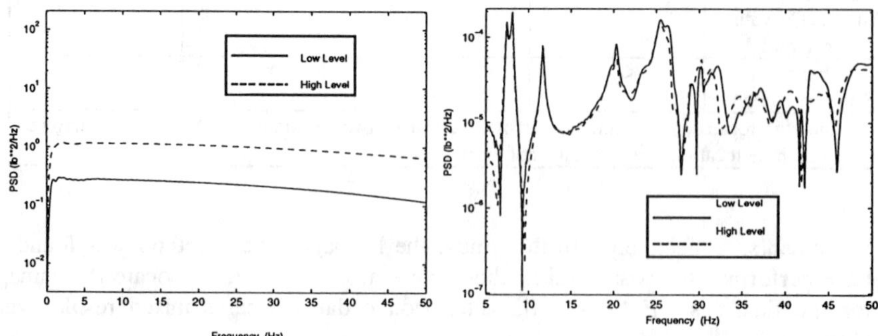

Fig. 9. PSDs of impact excitations used in the linearity check. Corresponding FRF magnitudes are shown on the right.

Perform Reciprocity Checks.

In addition to the assumption of linearity, the system identification portion of the experimental modal analysis typically relies on the assumption that the structure will exhibit reciprocity. Performing a reciprocity check is much more involved when a large shaker is being used because of the setup time involved in relocating the shaker. Also, to check the reciprocity of the structure alone, one must relocate the accelerometers and cables as well as the shaker. Without moving the instrumentation, the reciprocity check will involve reciprocity of the electronics as well as that of the structure. The difference in reciprocity of the structure itself and reciprocity of the structure and the electronics can be seen in Fig. 10 where the check was performed on the Alamosa Canyon Bridge (Farrar, et al., 1997). Because, in general, the electronics will not be moved once the test has started, the latter test is more representative of the reciprocity of the system. Figure 10 shows that reciprocity was only observed over a portion of the spectrum.

Perform As Many Environmental and Testing Procedure Sensitivity Studies As Possible.

Sensitivity of modal test results to environmental conditions and test procedures such as changes in temperature, traffic loading, wind, excitation method, etc. should be quantified to the extent possible. Subsequent tests (after the potential damage has occurred) should be performed under similar environmental conditions using similar test procedures, if possible. Also, a baseline noise measurement should be made for the data acquisition system.

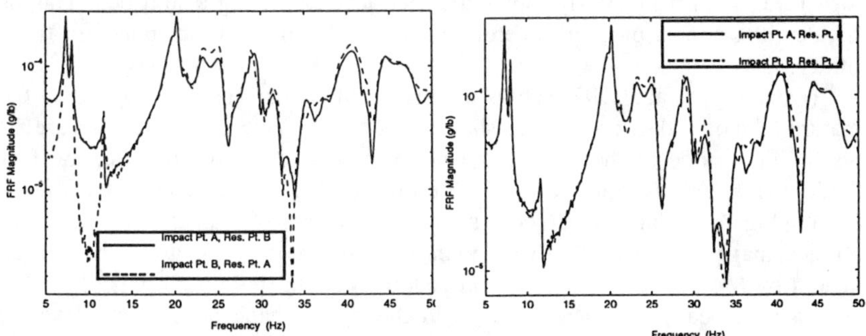

Fig. 10. FRF reciprocity check of the structure and electronics. FRF reciprocity check of the structure only is shown on the right.

Perform False-Positive Studies

As a means of quantifying the effects observed in the sensitivity studies, sets of data from the undamaged structure should be analyzed with the damage id algorithm to demonstrate that the algorithm will not falsely predict damage when in fact none has occurred.

Perform Statistical Analysis

Closely coupled to the false-positive study is the need to perform statistical analysis of the measured modal properties. Performing statistical analysis of the identified modal parameters as discussed in (Doebling, et al., 1997) is imperative if one is to establish that changes brought about by damage are greater than the test-to-test repeatability.

Although the number of papers reporting experimental modal analyses results from bridge structures has greatly increased in recent years, very few of the articles examine the variability in the modal properties that can arise from changes in environmental conditions or from random and systematic errors inherent in the data acquisition/data reduction process. A thorough study of the variability of the modal parameters must be conducted before modal-based damage identification algorithms can be applied with any confidence.

Figure 11 shows the first mode frequencies for the Alamosa Canyon Bridge in southern New Mexico along with their 95% confidence limits plotted as a function of the measurement completion time. Also plotted on Fig. 12 is the change in temperature between the two thermometer readings made on the concrete deck (east - west). This figure clearly shows that the change in modal frequencies are related to the temperature differentials across the deck. The first mode frequency varies approximately 5% during this 24 Hr time period. Similar variations and correlation with deck temperature differentials were observed for the other modes of the structure. Figure 11 clearly shows the need for performing a statistical analysis of the identified modal properties before a damage ID algorithm is applied to these quantities.

(Doebling, et al., 1997) shows the results of similar analyses applied to the estimation of modal damping, mode shapes and mode shape curvature. Figure 12 shows the first mode of the I-40 bridge in its undamaged state and after the first level of damage. In this figure points 1-13 correspond to accelerometer locations S-1 - S-13 in Fig. 4. Points 14 - 26 correspond to accelerometer locations N-1 - N-13 with the damage at point 20. Error bounds for the modal amplitudes have been calculated by Monte Carlo statistical procedures. From this figure it is clear that there is a statistically significant change in the mode shape at this first level of damage. However, the results shown in this plot can not be used to definitively state that the change in the mode shape resulted from damage as opposed to changing environmental conditions.

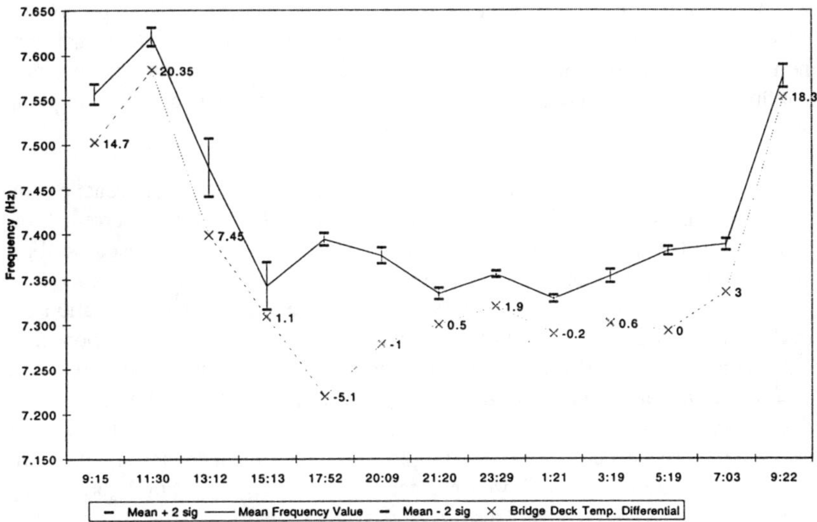

Fig. 11. Change in the first mode frequency during a 24 hr time period.

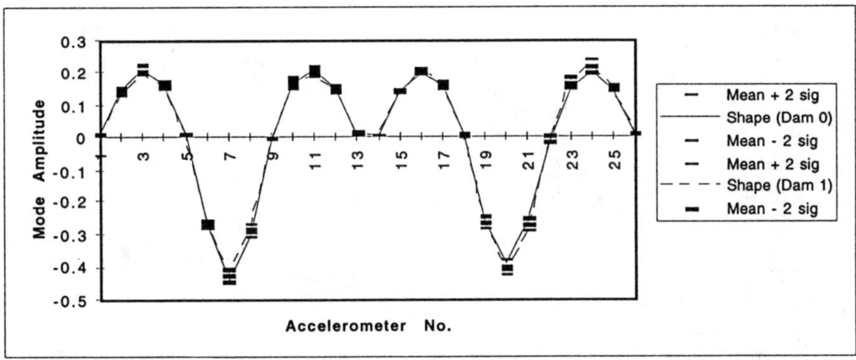

Fig. 12. First mode shape amplitudes and their corresponding 95% confidence limits for the undamaged structure compared to similar quantities measured after the first damage case.

THINGS THAT WERE BEYOND OUR CONTROL

For the I-40 bridge test several conditions occurred that were beyond our control and that could have significantly influenced the experimental modal analyses

have unavoidable conditions arise that are beyond the control of the experimentalist and that can potentially influence the outcome the study. The only thing that can be done is to note the condition and perform additional tests in an attempt to quantify the influence of the changing condition. Examples of some of these unplanned changing conditions on the I-40 bridge are listed below.

1. The load cell located between the Sandia actuator and reaction mass showed that the vibration from traffic on the adjacent bridges, transferred through the ground to the piers and abutment of the bridge being tested, caused the bridge deck to put a peak force of 150 lb. into the reaction mass.

Coherence functions can be used to determine if sources of excitation other than the Sandia shaker are significantly contributing to the measured response. For an ideal linear system the coherence function will yield a value of one. If the response is completely unrelated to the input, this function will yield a value of zero. Values between zero and one result when there is extraneous noise in the measurements, the structure is responding in a nonlinear manner, or sources of input other than the one being monitored are causing the response. For lightly damped structures, low coherence can also occur around resonances when the system response is calculated from a series of time windows as was done in these tests. The response in a particular window is strongly dependent on energy input

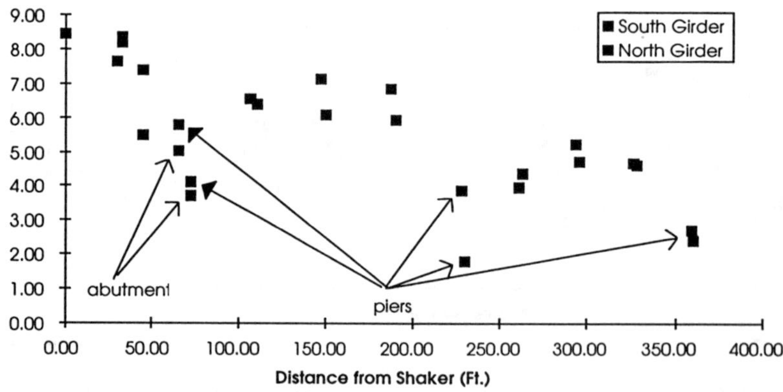

Fig. 11. Area under the coherence function over the frequency range of 2 - 11 Hz plotted as a function of the sensors distance from the shaker.

during the previous window, particularly at resonance, and this response will be uncorrelated with input measured during the current window. A plot of the area under the coherence function for the various measurement locations as a function of their distance from the shaker is shown in Fig. 11. The reduction in coherence that can be observed in this plot is caused by the inputs that result from extraneous sources of noise (traffic on the adjacent spans) causing a greater portion of the measured response at locations further from the excitation source. With the exception of measurement points directly above the support locations where the signal-to-noise ratio is inherently low, there is a distinct trend of poorer coherence as a function of distance from the shaker. The effects of the extraneous inputs are minimized by the averaging process used to calculate the FRFs.

2. Demolition of the concrete deck at the west end of the bridge was started before the forced vibration tests and continued while they were underway proceeding to the third span in from the west end. Portions of the foundation around the north side of the east abutment were removed to build an access ramp for construction work. Both the demolition and the construction of the access ramp can be viewed as changing the boundary conditions of the test structure. Forced vibration measurements taken before and after the access ramp was constructed showed no changes in the resonant frequencies of the structure. Because forced vibration measurements were not made before the demolition of the west end began, the extent of this change on the measured modal properties could not be easily quantified.

WHAT CAN BE DONE PRIOR TO DAMAGE USING INITIAL MEASUREMENTS AND FEM?

Once a finite element analysis has been benchmarked or correlated against the measure modal properties simulated damage scenarios can be introduced into the model and either an eignevalue analysis can be performed or, to better simulate an actual modal test, a time-history analysis can be performed. Mode shape data can then be obtained from either type of analysis and the various damage ID methods can be applied to the observed changes in the modal properties. If a statistical analysis has been applied to the measured modal parameters of the baseline or undamaged structure, then it can be established that the changes in the monitored modal properties such as mode shape curvature resulting from the simulated damage are greater than the variations that can be attributed to experimental repeatability. In addition, the statistical variations calculated for the measured modal properties on the undamaged structure can be assumed to apply to the numerical results from the damaged structure. The use of statistical variations measured on the undamaged structure and assumed for the numerical simulation of the damaged structure can then be used to establish the threshold damaged level that can be reliably detected.

IN RETROSPECT, WHAT WOULD WE DO DIFFERENT

Given infinite resources we would have liked to perform the damage ID process using finite element model updating techniques to obtain a direct comparison of these methods relative to methods that only examine measured modal properties. (Simmermacher et al., 1995) have recently performed a damage assessment of the I-40 bridge using model updating techniques. Ideally, we would have liked to investigate the ability of the various damage ID methods to detect the damage using ambient traffic excitation, but this type of test poses many safety concerns. Also, we would have liked to introduce multiple damage scenarios into the structure.

SUMMARY AND CONCLUSIONS

The application of five linear damage identification methods using experimental modal data gathered from the I-40 Bridge over the Rio Grande in Albuquerque, NM has been summarized. In this study linear damage identification implies that linear dynamic models were used to model the structure both before and after damage. The nature of the damage applied to the I-40 bridge is such that the linear damage models are applicable to these damage scenarios. This study provides a direct comparison of various damage identification algorithms when applied to a standard problem. The authors are not aware of other such comparisons that have been reported in the technical literature.

Examination of results from the experimental modal analyses verify other investigators findings that standard modal properties such as resonant frequencies and mode shapes are poor indicators of damage. The more sophisticated damage detection methods investigated herein showed improved abilities to detect and locate the damage. In general, all methods investigated in this study identified the damage location correctly for the most severe damage case; a cut through more than half the web and completely through the bottom flange. However, for several of these methods, if they had been applied blindly, it would be difficult to tell if damage had not also occurred at locations other than the actual one. The methods were inconsistent and did not clearly identify the damage location when they were applied to the less severe damage cases. Results of this study show that the Damage Index Method performed the best when the entire set of tests are considered. This performance is attributed to the methods of normalizing changes in the damage parameters relative to the undamaged case. Also, the Damage Index Method works with mode shape data that do not have to be unit-mass normalized. This feature is desirable when an on-line monitoring system that uses ambient traffic excitation is being considered.

Another observation from this study, which the authors feel is important, is that the Damage Index Method is the only method tested that has a specific criterion for determining if damage has occurred at a particular location. The other methods only look for the largest change in a particular parameter and it is ambiguous at times to determine if these changes indicate damage at more than one location.

Based on further analysis and observations related to the I-40 bridge tests, subsequent tests on another bridge, interactions with other researchers performing similar tests, and review of the technical literature related to bridge testing, there are several things that should have been done during these tests to improve the confidence in the damage ID results. These improvements include detailed visual inspection of the bridge, performing linearity checks, performing reciprocity checks, performing false-positive studies, performing test condition sensitivity studies, and performing statistical analyses of the measured modal properties.

The lack of application of vibration-based damage detection methods to large civil engineering structures is due, in part, to the limited application and refinement of this technology to *in situ* structures. The authors hope that future researchers in this field will benefit from the lessons learned through trial and error as discussed in this paper.

ACKNOWLEDGMENTS

The authors would like to acknowledge the cooperation and teamwork that was exhibited by all parties involved in these tests including engineers from Sandia National Laboratory; faculty, technicians and students from New Mexico State University; numerous people at the New Mexico State Highway and Transportation Department; and the staff at the Alliance for Transportation Research.

REFERENCES

1. Bendat, J. S. and A. G. Piersol (1980) *Engineering Applications of Correlation and Spectral Analysis*, John Wiley, New York.

2. Doebling, S. W., C. R. Farrar, M. B. Prime, and D. W. Shevitz, (1996), "Damage Identification and Health Monitoring of Structural and Mechanical Systems From Changes in their Vibration Characteristics: A Literature Review," Los Alamos National Laboratory report LA-13070-MS.

3. Doebling, S. W., C. R. Farrar and R. Goodman (1997) "Effects of Measurement Statistics on the Detection of Damage in the Alamosa Canyon Bridge," *Proceedings 15th International Modal Analysis Conference*, Orlando, FL.

4. Doebling, S. W, and Farrar, C. R. (1997) "Using Statistical Analysis to Enhance Modal-Based Damage Identification " in Proceedings, *DAMAS 97 Conference*, Sheffield, UK.

5. Ewins, D. J. (1985) *Modal Testing: Theory and Practice,* John Wiley, New York.

6. Farrar, C. R. W. E. Baker, T. M. Bell, K. M. Cone, T. W. Darling, T. W. Duffey, A. Eklund, and A. Migliori, (1994), "Dynamic Characterization and Damage Detection in the I-40 Bridge over the Rio Grande," Los Alamos National Laboratory report LA-12767-MS.

7. Farrar, C. R., S. W. Doebling, P. J. Cornwell, and E. G. Straser, (1997) "Variability of Modal Parameters Measured on the Alamosa Canyon Bridge," *Proceedings 15th International Modal Analysis Conference, Orlando*, FL.

8. Farrar, C. R. and D. Jauregui (1996) "Damage Detection Algorithms Applied to Experimental and Numerical Modal Data From the I-40 Bridge," Los Alamos National Laboratory report LA-13074-MS.

9. Farrar, C. R., T. A. Duffey, P. A. Goldman, D. V. Jauregui, and J. S. Vigil, (1996), "Finite Element Analysis of the I-40 Bridge Over the Rio Grande," Los Alamos National Laboratory report LA-12979-MS.

10. Mayes, R. L. and M. A. Nusser (1994) "The Interstate-40 Bridge Shaker Project," Sandia National Laboratory report SAND94-0228.

11. Rytter, A. (1993) "Vibration Based Inspection of Civil Engineering Structures," Doctoral Dissertation, Department of Building Technology and Structural Engineering, University of Aalborg, Aalborg, Denmark.

12. Simmermacher, T., D. C. Zimmerman, R. L. Mayes, G. M. Reese, and G. H. James (1995) "The Effects of Finite Element Grid Density on Model Correlation and Damage Detection of a Bridge," *Proc. of the 36th AIAA/ASME/ASCE/AHS/ASC Structures, Structural Dynamics, and Materials Conference.*

13. Doherty, J. E., "Nondestructive Evaluation," 1987, in *Handbook on Experimental Mechanics*, A. S. Kobayashi Edt., Society for Experimental Mechanics, Chapter 12.

Structural Damage Monitoring for Civil Structures

A. S. KIREMIDJIAN,[1] E. G. STRASER,[2] T. MENG,[3]
K. LAW[4] and H. SOON[5]

ABSTRACT

Large, devastating earthquakes and hurricanes of the early 1990's amply demonstrated the need for rapid assessment of critical structures. In parallel, aging infrastructure has focused our attention on preventative maintenance efforts for bridges, airports, ports, and lifelines. Whether long-term deterioration or near-real time post disaster response, the monitoring system employed invariably utilizes sensing, data acquisition, communication, and computation components. Most of the research has focused on damage localization and detection methods, largely ignoring the necessary hardware and other systems to provide input data. This paper presents an overview for structural health monitoring of civil structures from two points: 1) a summary of current research efforts in damage detection and localization using low frequency vibration data, and 2) a systems framework that leverages recent technological advances toward a modular monitoring platform suitable for monitoring civil structures. Fundamental issues of observation of damage in vibration data, quality and stationarity of the measured data, general classes of algorithms, and degree and placement of instrumentation are discussed. A modular hardware platform developed at Stanford University is described. Future trends and possible research directions are presented.

INTRODUCTION

Inspection of existing buildings and bridges after major catastrophic events, such as earthquakes and hurricanes, as well as under normal operating conditions, is often

[1] *The John A. Blume Earthquake Engineering Center, Department of Civil Engineering, Stanford University, Stanford, CA 94305, USA*
[2] *The John A. Blume Earthquake Engineering Center, Department of Civil Engineering, Stanford University, Stanford, CA 94305, USA*
[3] *Department of Electrical Engineering, Stanford University, Stanford, CA 94305, USA*
[4] *The John A. Blume Earthquake Engineering Center, Department of Civil Engineering, Stanford University, Stanford, CA 94305, USA*
[5] *The John A. Blume Earthquake Engineering Center, Department of Civil Engineering, Stanford University, Stanford, CA 94305, USA*

very time consuming and costly because critical members and connections are concealed under cladding and other architectural surface covers. For critical structures, such as hospitals, fire stations, military control/surveillance centers, major bridges, power stations, and water treatment plants, it is imperative that their health be assessed immediately after a major catastrophic event. Similarly, dissemination of information to emergency response officials on major collapses of structures within minutes after the occurrence of a natural or manmade disaster can result in saved lives and prudent resource allocation. Often such information is delayed due to weather conditions, lack of daylight, appropriate survey equipment, or inaccessibility to the site due to terrain obstacles. In many instances, impending collapse of a structure may not be visible from the exterior of the structure. During the January 17, 1994 Northridge, California earthquake several structures that were weakened (but undetected) by the main shock collapsed when a major aftershock occurred. Thus, identification of critically damaged structures will enable timely evacuation of occupants.

While sensing and health monitoring technology has been widely developed and used in the aerospace, automotive and defense industry, it is only recently that attention has focused on civil structures. The deterioration of our infrastructure has pointed to the need for health monitoring of structures under everyday loads. During the last decade considerable theoretical and experimental advances have been made in structural control. In parallel, attempts have been made to design general earthquake damage monitoring systems. For example, conceptual models have been developed for the sensor location, signal transmission, and central processing of information for simple structural systems (e.g., Wu at al., 1990). Laboratory and field experimentation with frame structures and bridges have shown promise for identification of system behavior and critical parameter benchmarking (e.g. Doebling and Farrar, 1997; and Pirner and Fischer, 1997).

This paper summarizes current research efforts in the field and provides an overall framework developed at Stanford for structural damage monitoring systems. Several general classes of damage detection algorithms have emerged and a brief discussion of their merits and drawbacks is warranted. Monitoring systems invariably employ sensing, data acquisition, communication, and computation components. Current advances in wireless communication, MEMS sensors, global positioning systems (GPS), and increased computational power provide the tools for a potentially new solution to the civil monitoring problem. Fundamental issues of data collection, data quality, and damage detection methods are discussed.

STRUCTURAL SYSTEM AND COMPONENT DAMAGE EVALUATION METHODS

Diagnosis of damage in structural systems requires the identification of the location and type of damage and quantification of the degree of damage. Most damage detection methods currently in use rely on visual inspection or on localized measurements. Measurement based methods are still very much at the experimental or research stage with little if any practical deployment. Doebling et al. (1996)

present a comprehensive literature review of damage identification and health monitoring methods for structural and mechanical systems. Their review focuses on methods based on vibration measurements and detection based on changes in vibration characteristics.

A robust health monitoring system requires the following four stages of damage identification (Rytter, 1993):

(1) Determination that damage is present in the structure;

(2) Determination of the geometric location of the damage;

(3) Quantification of the severity of the damage;

(4) Prediction of the remaining service life of the structure.

These stages of damage monitoring may lead to different types of sensor requirements or multi-sensor systems. Similarly, damage algorithms will vary depending on the type of monitoring desired. For example, if stage one information is needed, then sparse vibration measurements may be sufficient to ascertain the existence of damage somewhere in the structure. Identification of the location of damage may require considerably richer sensor network and perhaps sensors that provide more robust local information. In order to determine the degree of damage, in addition to the sensor selection, efficient and robust damage algorithms are needed. Prediction of the remaining service life is typically based on fatigue and fracture measurements and may require different sensors and mathematical tools leading to the estimation of the remaining life or assessment of compliance with design specification. The challenge is in developing systems that can respond to all four stages of damage monitoring after major catastrophic events, such as earthquakes and hurricanes, and under normal operating conditions. This motivates the need to develop two-fold monitoring systems: extreme event and long term monitoring systems.

Earthquakes, hurricanes and tornadoes impose random and extreme loads on structures, thus the response of the structure is also random and usually nonlinear. Damage detection for nonlinear structural systems under extreme events has been studied only to a limited extent (e.g., Loh and Tou, 1995; Masri et al., 1987). There are considerable difficulties with structural parameter identification, damage detection, location detection and damage assessment of such systems. These issues are only now beginning to be addressed by researchers in the field.

Different damage algorithms are typically required for long term health monitoring and for damage detection after a catastrophic event. Among the many methods for long term health monitoring are the methods that estimate changes in natural frequencies, mode shape, mode shape curvatures or strain mode shape changes, matrix update methods which measure changes in mass, stiffness and damping matrices, and hybrid matrix update methods. Doebling et al. (1996) provide extensive discussion on the advantages and disadvantages of these methods. One very difficult aspect of long term monitoring is the variation of the structure system due to environmental changes such as loading, boundary conditions, temperature and humidity.

Damage detection methods also can be categorized into global and local methods. First, global damage detection methods should identify damaged regions and then local detection methods can be employed to conduct more detailed investigation at those regions. Current local detection methods rely on visual inspection or on localized measurements such as acoustic, ultrasonic, magnetic field change, radiograph, eddy current and thermal field change measurements. Most global methods are based on vibration characteristics of the structure to infer damage. These methods are still very much at the experimental or research state with little if any practical deployment.

STRUCTURAL SYSTEM DAMAGE METHODS

The dynamic behavior of structural systems is governed by the properties of the structural members (beams, columns, braces) and their connections (rigid, semi-rigid, etc.). Damage to structural members and joints has a direct effect on the dynamic properties of the overall system. Damage detection methods that have been proposed include the "classical" approach (West, 1988; Lieven, 1988), the eigenstructure assignment approach (Minas and Inman, 1988; Zimmerman and Widengren, 1990), the optimal update method (Baruch, 1984; Kabe, 1985), the design sensitivity method (Hemez, 1993; Flanigan, 1987; Ojalvo et al., 1989), the rank perturbation method (Zimmerman and Smith, 1992), the statistical parameter identification method (Beck et al., 1994, Sohn and Law, 1997) and the neural network applications (Elkordy et al., 1994; Wu et al., 1992). Analytical methods for structural dynamic property identification are typically divided into time domain and frequency domain techniques. System identification methods are further divided into linear and nonlinear methods. The goal of these techniques is to evaluate the dynamic structural characteristics, such as stiffness, damping, structural period, and mode shapes, and monitor changes in their values or signatures under extreme dynamic loads or under normal operation conditions. Changes in modal parameters alone, such as natural frequencies and eigen-mode shapes have shown not be robust estimators of structural damage (Loh and Tou, 1995). Story drifts, large rotations, and shear force, and strain distributions can be benchmarked to establish performance criteria and show promise as reliable indicators of structural damage.

LOCAL DAMAGE DIAGNOSIS

In order to determine local damage, it is first necessary to identify the critical damage modes in structural members (e.g., local flange or web buckling of steel columns, beams and braces; fracture at welded or bolted steel beam to column connections and brace to joint connections; and yielding of beam or column sections). The assessment of the different damage modes requires different types of sensors and sensor locations. For example, yielding and local buckling is typically associated with large strains and thus can be diagnosed by monitoring strains at critical locations. Evaluation of fractures at critical location in members in existing structures poses a more difficult problem. Provided that appropriate sensors can be placed at critical location of members, algorithms that determine the level of damage can be developed for a component based damage index or for benchmark values for specific damage parameters.

It has become apparent with recent extreme natural events, that fracture and cracking at welded and bolted joints is one of the most pervasive modes of failure in steel structures. Shear cracks and failure from under-strength of concrete members also are leading causes of damage and failure of concrete structures. Sensors, such as those based on acoustic or fiber-optic (used with concrete structures) measurements are impractical for steel structures since these sensors, if affixed prior to welding, will be damaged by the welding process. Eddy current inspection methods, however, have been shown to be effective in the detection of cracks, their positions and, through calibration, their dimensions. They have been used in the aerospace industry for material flaw detection and component inspections.

OPTIMAL SENSOR PLACEMENT

A robust, efficient, and economical damage detection system depends on the information extracted from the sensors. Due to economic constraints, it is not feasible to completely instrument civil structures for damage monitoring. Therefore, algorithms and methodologies that address the issue of limited instrumentation and its effect on resolution and accuracy in damage diagnosis are paramount to the application of damage monitoring systems. There are numerous studies regarding optimum sensor placement. Most methods are based on maximizing the trace or determinant (or minimizing the condition number) of the Fisher Information matrix which is expressed as a function of selected parameters corresponding to the objective function. One example is the optimum sensor location (OSL) algorithm proposed by Udwadia (1994) that minimizes the covariance error between the structural parameters that are to be identified and their estimate from the limited measurements. An alternative approach is the effective independence method (Efi) presented by Kammer (1992) which determines the final sensor configuration by selecting sensor location by iteratively removing sensor locations that do not contribute significantly to the linear independence of the mathematical mode shapes. The final sensor location distribution is such that the covariance matrix between the displacement vector in modal coordinates and the modal displacements are minimized. A combination of these methods has been considered by Hemez (1993) and an energy matrix rank optimization method (EMRO) are proposed by Lim (1991). The EMRO method selects the optimal sensor location by maximizing the measured strain energy stored at the sensor locations.

There are many outstanding issues in the structural damage evaluation methods. Currently available structural system damage, local damage and sensor location algorithms pertain to linearly-elastic space trusses. Structures subjected to extreme events, such as earthquakes, hurricanes and tornadoes, necessarily undergo nonlinear deformations. Thus, there is a need for the development of robust damage algorithms for extreme event damage detection for nonlinear multi-degree of freedom frame systems. Nonstationarity in vibration measurements pose difficulties with long term health monitoring. Such nonstationarities arise due to changes in temperature, humidity, loading conditions and structure boundary conditions. The systematic error due to these changes can be as large as 10% (e.g., Doebling and Farrar, 1997). Vibration measurements, in addition, may not be sufficient to identify certain types of damage such as cracks in members and joints. These challenges are primarily analytical. In addition, there are serious difficulties with the implementation and deployment of health monitoring systems. These difficulties stem from the extensive wiring, single data acquisition and processing stations that may be vulnerable to extreme events, availability of continuous power supply and long term maintenance of all hardware and software. Some of these issues are addressed next.

CONCEPTUAL DESIGN OF A DAMAGE MONITORING SYSTEM

Structural damage monitoring systems consist of sensors, communication hardware, and data acquisition and processing components that measure and assess the integrity of a structure. Two types of structural monitoring systems can be identified: (1) systems that measure peak response quantities, such as strain, at specified locations and then correlate these quantities to long-term structural "health", and (2) systems that employ system identification procedures to estimate the changes in various parameters of a structure for damage determinations. Current structural monitoring systems consider either local or global damage parameters.

The conceptual design of the civil structural damage monitoring system that we designed is based on a simple hierarchical scheme consisting of threecomponents:

(1) the sensor unit,

(2) the site master processor,

(3) the central monitoring facility.

Figure 1 shows the schematic configuration of the proposed system. The components of the monitoring system are describedin the following sections

In recent years, the nascent technologies of wireless communication, micromachined sensors, and embedded systems have started to provide the commercial components necessary to design and develop a new class of civil monitoring system based on a modular sensor unit. The sensor unit is comprised of 4 key components: 1) a sensor(s) suited to the desired sensing modality with accompanying analog to digital converters, 2) a wireless modem that allows for untethered data communication, 3) an embedded microprocessor that manages the sensing and communication systems, and 4) batteries to power the unit in the field. The sensor units are managed in a client-server relationship by the site master processor, a PC located at the structure. This architecture creates a very modular and powerful monitoring system that scales to more units easier and at a lower cost than today's wired, single A/D systems.

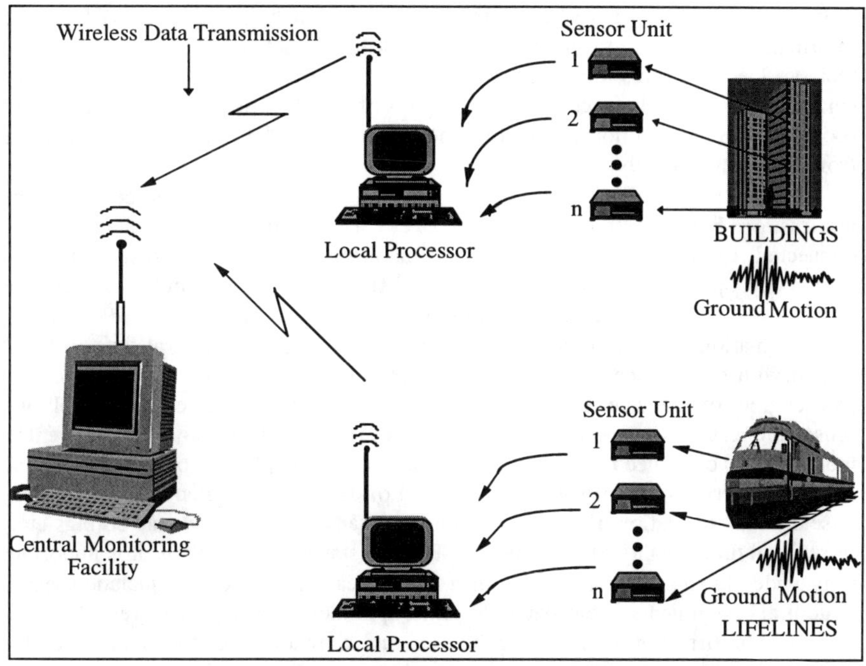

Figure 1. Conceptual Structural Monitoring System.

Data from the sensor unit is transmitted to a data processor. Currently, most sensors provide raw data that has not been processed at the structure site. Site master units typically serve as data collectors but not necessarily as processors. More recently, sensors have been developed that provide partial on site processing and storage of data. Often data processing is performed at locations some distance away from the site. Data transmittal to such locations presently is achieved through telephone communications links.

A well designed sensor system has a site master processor that provides the primary computational engine and is the manager of the structural monitoring system. Functions performed by the site master processor include: coordination and collection of transmitted sensor data, manipulation and analysis of structure specific information, evaluation and determination of structural damage, and transmission of desired structural damage quantities to a central facility. The data from the individual sensors can either be queued or queried by such a site master processor. Currently available digital data acquisition systems and wireless communication capabilities can facilitate rapid data transmission from the sensors to the site master processor without the need of intrusive and vulnerable wires running through the structure. Several analysis tools can be coded within the site master processor, each enabling determination of properties and performance of the overall structure. Examples include system identification, nonlinear time history analysis, frequency domain analysis, and correlation of measured structure quantities to threshold system parameters. If necessary, the site master processor may query any sensor for more information or to perform some simple local analysis and transmit back the information. Damage thresholds can be established to activate alarms at the central facility. Decision tools for selecting appropriate information for transmission to the central facility can be incorporated as part of the functions of the site master processor. Such functions, however, should be able to be overridden by requests from the central facility.

The central monitoring facility is intended to receive and process damage information from all structures in the system. For example, all hospitals may be connected to a central command post monitored by agencies responsible for emergency operations. Following a disaster, the information from the site master processor is relayed to the central facility for further processing.

Information housed at the central facility may include structural data in CAD format with sensor locations identified. Graphical interfaces may show the location and degree of damage throughout the structure. The type of structural and sensor/processor data and the analysis and decision tools to be stored in the central monitoring facility need to be clearly defined. For example, for bridges, it may be sufficient to transmit information on the level of damage (e.g., amount of separation at seat joints, formation of hinge in a column, or amount of settlement footings etc.) or residual functionality of the structure. For fire stations or hospitals, it is important to provide global as well as local information. Local information can include specific structural components, their materials, exits, sprinkler location, intensive care areas, etc. Such information may be preprocessed and warehoused at the central monitor to be retrieved upon request. Retrieval may be automatic or initiated by an operator. Additional analysis and computational tools can be coded in the central monitoring unit. Furthermore, a warning system can be designed to signal the occurrence of a catastrophic failure which may result in possible deaths and injuries. Such a system should have the capability to be turned on by an operator as needed or desired to perform interim system testing or structural integrity identification.

Key issues to be resolved include (a) design and development of reliable hardware for two-way transmission of multitudes of signals over large distances; (b) development of algorithms for damage, loss, and casualty assessment to be hard-coded at the site master processor, (c) development of damage visualization algorithms, (d) development of decision analysis tools as required by key emergency response personnel. Currently, there are no central monitoring systems in existence for commercial use. Utility companies and emergency response organizations in the United States are presently considering the design and implementation of such systems.

SENSOR UNIT COMPONENTS

Improvements in both sensor and communication technology make it possible to create small self-contained units that are able to sense their environment and transmit important observations to a remote site using wireless communication links. Distributing a large number of these units on a structure and allowing them to communicate with each other can create a powerful, and unattended distributed monitoring system that has a variety of applications. Before such systems can be built, there are many technical issues to be addressed in the areas of advanced sensor technology, power-efficient radio systems, low-power computing, and packaging. Several of these issues are discussed as follows.

SENSOR TECHNOLOGY

Through the increasing availability of high-performance micromachined and conventional miniaturized sensors, it is now possible to construct and efficiently distribute sensor systems with multi-modal capabilities, low power, small size and low cost. Typically, sensors for strain, tilt, corrosion, and seismic phenomena are required. Issues that ultimately may determine the actual sensor suite for a given design include: sensor modalities, sensors with appropriate robustness for the application environment, and power, volume/weight, cost and data rate constraints.

The sensors described in this paper are representative examples of "off-the-shelf" (primarily commercially, but in some instances from proven academic research projects) sensors that can be selected for each sensing modality. There is clear potential to utilize silicon micromachined or "MEMS" sensor technology for ultimately scaling down the sizes of the modulesreducing power consumption.

The most widely used sensors for structural monitoring are accelerometers. Popular types include the piezoelectric, force balanced, and capacitive. Accelerometers can also be used for large angle tilt measurement Finer tilt measurements can be made with silicon micromachined electrolytic inclinometers. Such measurements are particularly useful in construction sites, for decoupling displacements from rotations, and for large permanent rotations resulting from seismic response.

Currently, micromachined accelerometers are capable of 0.1 to 1 milli-g levels and should be employed for monitoring signals expected to be well above this level. Therefore, the off-the-shelf silicon micromachined accelerometers are applicable to monitoring forced excitation or seismic response but will most likely not suite ambient monitoring needs. Representative accelerometers include the Analog Devices ADXL, the Kistler 8352A2, and the IC Sensors 3145 families. Another significant advantage of the silicon accelerometers is their healthy economies of scale for industrial quantities. Some of the high-fidelity accelerometers used for years in inertial guidance are beginning to appear repackaged for other commercial applications. A good example would be the Allied Signal QLC400 which provides resolution greater than 10 micro-g for approximately $400.

Crack monitoring, strain measurements, and corrosion determination can be achieved through local diagnostic sensors. Methods to measure and quantify cracks include acoustic emission, ultrasonic detection, magnetic resonance, and a variety of optical and visual techniques. While these methods have their strong points, each requires either constant monitoring (acoustic and ultrasonic), substantial and expensive hardware (magnetic resonance and optical), or physical access to the location of interest (visual inspection).

For local strain monitoring large area strain gages (e.g., metal foil or doped silicon strain gages) can be used for member strain measurement and can undergo cyclic strains compatible with structural steel vibration behavior. Such strain gages, however, need to be placed at critical locations requiring that these locations be identified prior to placement. The strain measurements can provide benchmarks for a strain based mode shape algorithms or can serve as input to jointhysteretic analysis.

The Motorola 6811 microprocessor was chosen based on its wide industrial use, availability of high-level programming environments, wealth of peripherals, and its performance per unit cost. The microprocessor is integrated into a single board computer. The microprocessor operates in two power saving modes making it ideal for our monitoring strategy. Future generations will likely use embedded 486 or Pentium class microprocessors in a PC104 format.

Commercially available radio modems are used for wireless networking. The radio modems are a direct sequence spread spectrum product that operates in the Industrial, Scientific, and Medical (ISM) band between 902 and 928 MHz. Therefore, they require no FCC license to operate, have decent signal penetration through civil engineering materials, and are greatly immune to interference. Range testing has indicated that for indoor situations characterized by concrete walls, steel plates, and concrete slabs, we can expect communication distances of approximately 50 meters. For line-of-sight settings and most outdoor configurations, communication distances of 300 meters are feasible.

The objective of the sensor protocol is to provide intelligent and autonomous sensor polling with a pre-programmed background routine and capability for any sensor to be interrogated if necessary. The bulk of the operating time of a sensor module should be spent in a low power "background mode." Background mode entails scanning some sensors at a low duty cycle appropriate for each. Readings, such as temperature, have typical time constants of minutes to hours, while accelerometers need to be monitored more frequently. For sensors such as the latter case, higher data bandwidths are required, but only if an event of interest occurs. For this type of "bursty" sensor, low-power polling can be obtained using hardware "trigger" circuits that would "wake up" sampling circuits when a programmable signal threshold is exceeded. For this purpose, simple circuits such as a clocked peak-hold circuit and comparator per analog sensor can be employed, generating a hardware interrupt and a burst of analog-to-digital conversion to a buffer memory when activated. The data thus captured can be relayed out through the network and if deemed necessary, the sending module could be reprogrammed to focus on, and continuously transmit from, the triggered sensor(s).

POWER MANAGEMENT OF DATA TRANSFERS

In recent years wireless technology has revolutionized digital communications. In wireless RF transmission, for a minimum usable receiver power at -110 dB and a distance of 100 feet, the transmit power at a carrier frequency of a few 100 MHz (chosen as a compromise between penetration capability and noise immunity) can be as low as 1 mW. With current efficient transmitter electronics, the active power of the transmitter can be a few mW. This low level of transmit power is possible only if strictly local communication is used to maintain network connectivity.

There are two kinds of signals transmitted within such wireless network. One is the synchronization signal for status report at a constant rate. The other is the bursty data signal which may require orders of magnitude higher bandwidth in reaction to detected signals. Thus, the transceiver needs to support both types of signals.

Since the receiver for the high-bandwidth signals is only turned on when the unit is told to listen, which is assumed to be an infrequent event, the dominant power drain is the power needed for the status signals. The difference in duty cycle between the transmitter and receiver means that it may be beneficial to increase the transmit power if that enables a reduction in the receiver power. This component of the system requires considerable further exploration and research.

CONCLUSIONS

Several broad classes of damage detection algorithms have emerged. Most of these are based on vibration measurements and modal analysis of the structural system. Although these developments represent major improvements in our ability to detect damage in certain types of structures there are still significant drawbacks with these methods. All of these methods have shown to work with particular scenarios, but none have shown the ability to identify the damage location. In addition, difficulties with sensor placement and optimal number of sensors are still unresolved in a satisfactory manner. A major effort is also needed to develop damage detection algorithms for nonlinear structures.

In order to evaluate the analytical damage detection methods it is important to focus on testing of real structures. Then, existing techniques should be applied to common data sets to evaluate the merits each technique. Issues of systematic errors due to environmental causes are also best addressed through such an approach.

The modular damage detection system currently under development at Stanford University provides flexibility and can potential solve several of the problems currently faced by monitoring systems. Similarly, problems of installation and maintenance are significantly reduced with the wireless communication system. Local processing by individual sensors can provide a more direct and independent assessment at the sensor location. The system addresses several of the issues raised throughout the paper, however, many problems still remain and will be the subject of future studies.

ACKNOWLEDMENTS

The authors would like to express their gratitude to the National Science Foundation for their support through Grant CMS - 9526102.

REFERENCES

1. Baruch, M., 1994. "Method of Reference Basis for Identification of Linear Dynamic Systems", *AIAA Journal*, **22**, 561-564.
2. Beck, J. L. and Vanik, M. W. and Katafygiotis, L. S., 1994. "Determination of Stiffness Changes From Modal Parameter Changes For Structural Health Monitoring", *The Proc. of The First World Conf. Struct. Control,* Pasadena, CA.
3. Doebling, S. and Farrar, C., 1997. "Using Statistical Analysis to Enhance Modal-Based Damage Identification", in *Structural Damage Assessment Using Advanced Signal Processing*

Procedures, (Dulieu-Smith, J., Staszewski, W., and Worden, K., Eds)., Sheffield Academic Press, Sheffield, England, 199-212.

4. Doebling, S., Farrar, C., Prime, M., and Shevits, D., 1996. "Damage Identification and Health Monitoring of Structural and Mechanical Systems from Changes in Their Vibration Characteristics: A Literature Review", Los Alamos National Laboratory Report No. LA-13070-MS, Los Alamos, New Mexico.
5. Elkordy, M.F. et al., 1994. "Application of Neural Networks in Vibrational Signature Analysis", Journal of Engineering Mechanics, **120**(2).
6. Flanigan, C. C., 1987. "Test/Analysis Correlation Using Design Sensitivity and Optimization", *Aerospace Technology Conference and Exposition*, SAE Paper No. 871743, Long Beach, CA.
7. Hemez, F. M., 1993. "Theoretical and Experimental Correlation between Finite Element Models and Modal Tests in the Context of large Flexible Space Structures", Ph. D. Dissertation, University of Colorado.
8. Kabe, A. M., 1985. "Stiffness Matrix Adjustment Using Mode Data", *AIAA*, **23**(9), 1431-1436.
9. Kammer, D. C., 1992. "Sensor Placement for On Orbit Modal Identification and Correlation of Large Space Structures", *AIAA Journal of Guidance, Control and Dynamics*, **14**(2),251-259.
10. Lieven, N.A.J., and Ewing, D.J., 1988. "Spatial Correlation Mode Shapes, The Coordinated Modal Assurance Criterion (COMAC)", *Proc. 6th IMAC*, 690-695.
11. Lim, T. W., 1991. "Structural Damage Detection Using Modal Test Data", *AIAA Journal*, **29**(12), 2271-2274.
12. Loh, C.H. and Tou, I.C. (1995). "A System Identification Approach to the Detection of Changes in Both Linear and Nonlinear Structural Parameters, *Inter. J. Earthq. Eng. Struct. Dyn.*, **24**, 85-97.
13. Masri, S., et al., 1987. "Identification of Nonlinear Vibrating Structures, Part 1 & 2", *J. App. Mech.*, **54**.
14. Minas, C. and Inman, C. , 1988. "Correcting Finite Element Models with Measured Modal Results Using Eigenstructure Assignment Methods", *Proc. 6th IMAC*, 583-587.
15. Oljavo, I.U. et al., 1989. "PAREDYN-A Parameter Refinement Computer Code for Structural Dynmic Models", *UJMA*, 5(1), 43-49.
16. Peeters,B., and De Roeck, G, 1997. "The Performance of Time Domain System Identification Methods Applied to Operational Data", in *Structural Damage Assessment Using Advanced Signal Processing Procedures,* Dulieu-Smith, J.M., Staszewski, W. J., and Worden, K., Eds., Sheffield Academic Press, Sheffield, England, 377-386.
17. Pirner, M. and Fischer, O., 1997. Monitoring Stress in the GRP Extension of the Prague TV Tower", in *Structural Damage Assessment Using Advanced Signal Processing Procedures,* Dulieu-Smith, J.M., Staszewski, W. J., and Worden, K., Eds., Sheffield Academic Press, Sheffield, England, 451-460.
18. Rytter, A., 1993. "Vibration Based Inspection of Civil Engineering Structures", Ph. D. Dissertation, Department of Building Technology and Structural Engineering, Aalborg University, Denmark.
19. Sohn, H. and Law, K. H., "Bayesian Probabilistic Approach For Structure Damage Detection," accepted for the publication ,*Int. J. Earthq. Eng. and Struct. Dyn.*, 1997.
20. Udwadia, F. E., 1994. "Methodology for Optimum Sensor Locations for Parameteric Identification in Dynamic Systems", *J. Eng. Mech.*, ASCE **120**(2).
21. West, W., 1988. "Structural Fault Detection of a Light Aircraft Structure Using Modal Technology", *JSC Loads and Structural Dynamics Branch Report*.
22. Wu, X. Ghaboussi, J. and Garrett, J.H., 1992. "Use of Neural Networks in Detection of Structural Damage", *Computers and Structures*, **42**(4), 649-659.
23. Zimmerman, D. C. ad Widengren, M. 1990. "Correcting Finite Element Models Using a Symmetric Eigenstructure Assignment Technique", *AIAA Journal*, **28**(9), 1670-1676.
24. Zimmerman, D. C. and Smith, S. W. 1992. "Model Refinement and damage Detection For Intelligent Structures", in *Intelligent Structural Systems* (Tzou, H. S., ed.), Kluwer Academic Publishers.

SESSION 7

CIVIL INFRASTRUCTURES II

Structural Health Monitoring Activities at National Laboratories

C. R. FARRAR, S. W. DOEBLING, G. H. JAMES and T. SIMMERMACHER

ABSTRACT

Sandia National Laboratories and Los Alamos National Laboratory have ongoing programs to assess damage in structures and mechanical systems from changes in their dynamic characteristics. This paper provides a summary of how both institutes became involved with this technology, their experience in this field and the directions that their research in this area will be taking in the future.

INTRODUCTION

At Sandia National Laboratories (SNL) in Albuquerque, New Mexico, and Los Alamos National Laboratory (LANL) in Los Alamos, New Mexico, teams of engineers with structural dynamics backgrounds have been studying the applicability of vibration-based damage identification (ID) methods to a variety of structural and mechanical systems. These engineers, whose experience is primarily in experimental modal analysis, were successful in competing for internal research funds (referred to as Laboratory Directed Research and Development (LDRD) funds) to begin investigations in this field. The vibration-based structural health monitoring efforts of both institutes that are described in this paper have their beginnings as LDRD projects. Both LDRD projects have been successful and follow-on funding has become available such that these efforts will be continuing at both institutes in the future.

SNL and LANL have many complimentary groups that provide an excellent base for the development of health monitoring technologies. At SNL, the joint work performed in structural system ID by the Modal Group of the Experimental Structural Dynamics Department and the Analysis Group of the Structural Dynamics Department have provided the framework for the health monitoring research that has occurred at SNL over the last three years. Similarly, work done by solid-state physi-

Charles R. Farrar, Materials Behavior Team Leader, farrar@lanl.gov
Scott W. Doebling, Technical Staff Member, doebling@lanl.gov
Los Alamos National Laboratory, MS P946, Los Alamos, NM 87545
George H. James, III, Technical Staff Member, GHJames3@aol.com
Todd Simmermacher, Postdoctoral Research Associate, twsimme@sandia.gov
Sandia National Laboratory, Mail Stop 0557, Albuquerque, NM 87185-0557

cists in LANL's Condensed Matter and Thermal Physics Group has been coupled with work from the Engineering Analysis Group and the Measurement Technology Group to enhance the technology of vibration-based damage detection. In addition, the close proximity (organizationally) of SNL's Aging Aircraft Project Department and the Non-Destructive Evaluation Department (NDE group) as well as SNL's Modal group's long term support of the Wind Energy Technology Department (Wind Energy Group) have also contributed to the work in damage detection. SNL's and LANL's contributions have been further strengthened by the close technological relationship that exists between the two institutes.

SUMMARY OF VIBRATION BASED HEALTH MONITORING ACTIVITIES AT SANDIA NATIONAL LABORATORIES

At SNL initial exposure with the health monitoring field was made by a joint delegation from the Analysis and Modal groups. They attended the NASA/Air Force System ID and Health Monitoring Workshop in 1990 from which the first health monitoring proposal was developed. Building upon the initial interest in the field, contact was made with NASA and the American Association of Railroads (AAR). Discussions with the AAR brought to light the problems with the many aging bridges in this country. Discussions were also initiated with the offshore oil industry.

As SNL became more involved in health monitoring, it became apparent that collaboration with universities could foster beneficial relationships. The first university interaction resulted in a joint proposal to the FAA from SNL and Prof. Norris Stubbs at Texas A&M University. Also, early interactions and support from Prof. Ken White's group at New Mexico State University allowed SNL to participate in the ground-breaking Rio Grande Interstate 40 (I-40) Bridge test.

Contact with Virginia Polytechnic Institute (VPI) through Prof. Robert West has provided SNL with state-of-the-art Laser Doppler Vibrometer (LDV) algorithms. The LDV has been used by SNL in a variety of applications including aircraft panels, composite plates, and, most recently, an outdoor test of a wind turbine blade. Collaboration with University of Texas at El Paso (UTEP) through Carlos Ferregut and Roberto Osegueda has further strengthened the understanding of LDV applications to health monitoring problems. This collaboration has produced a significant amount of the nation's experimental activity in this area. Collaborative efforts with Lee Peterson and K.C. Park of the University of Colorado at Boulder (CU) have developed new approaches to analyze and interpret data sets with both high spatial density and high modal density, such as those produced by the LDV. Also, this work produced new damage detection approaches based on structural flexibility matrix estimation and disassembly.

Collaboration with David Zimmerman at the University of Houston was initiated by the Analysis group covering system ID topics. This relationship was later expanded to provide SNL with expertise in the area of Model-Based Damage Detection. Model-based damage detection uses a reference or healthy model of the monitored system which is compared to a set of current measurements. If the model does not simulate the current measurements sufficiently, the system has changed and is assumed to be damaged. And most recently, the use of transmittance functions to detect small damage on relatively large structures has been investigated by Professors Schulz and Pai at North Carolina A&T in collaboration with SNL.

Collaboration was also established with other Government organizations. SNL's relationship with the National Renewable Energy Laboratory (NREL) has provided much motivation and research in the area of monitoring of wind turbine blades. A significant advance in SNL's involvement in Health Monitoring occurred in 1991 when the Federal Aviation Administration (FAA) established the Airworthiness Assurance and NDI Validation Center (AANC). To support this center, an aircraft hangar has been set up with a series of hardware specimens including complete transport and commuter aircraft. The facility replicates a working maintenance environment by incorporating both the challenges of physical inspection as well as the environmental factors which influence inspection reliability. Informal interactions with NASA Johnson Space Center have also been developed through the SNL Special Leave of Absence Program which has allowed one of the authors to work with NASA through a University of Houston program covering model correlation and structural health monitoring topics.

SUMMARY OF VIBRATION BASED HEALTH MONITORING ACTIVITIES AT LOS ALAMOS NATIONAL LABORATORY

Vibration-based damage detection work at LANL had its beginnings almost 15 years ago when engineers in the Advanced Engineering Technology Group attempted to identify the onset of seismically-induced buckling in scale model nuclear reactor containment structures from changes in their measured vibration response. This work was followed by attempts to infer damage in seismically loaded scale-model reinforced concrete shear wall structures from changes in their vibration response.

As a result of an LDRD-funded investigation, physicist in LANL's Condensed Matter and Thermal Physics Group developed and patented a damage ID system referred to as Resonant Ultrasound Spectroscopy (RUS) in the early 1990s. [1] This system combined sine-sweep vibration testing with a homodyne detection system to make very precise measurements of the resonant frequency of small test specimens. For objects of very regular geometry, such as ball bearings, this test system can provide very accurate indications of material or geometric anomalies, such as out-of-roundness of a ball bearing. Subsequent applications of RUS include the detection of salmonella poisoning in eggs from changes in their vibration characteristics, the screening of captured Gulf-War ammunition to determine if artillery shells contain conventional or chemical warheads, and the detection of cracks in machined parts.

Through collaboration with these physicists, engineers from the Advanced Engineering Technology Group (recently re-named the Engineering Analysis Group) were asked to be a primary participant in the damage ID tests on the I-40 Bridge over the Rio Grande.[2] These tests were performed in conjunction with engineers from SNL, faculty and students from New Mexico State University, and the New Mexico State Highway and Transportation Department. The engineers from LANL performed the experimental modal analyses of the bridge in its undamaged and damaged conditions while engineers from SNL ran a hydraulic shaker that provided the input for these vibration tests. Subsequently, data from these tests has been made available to many universities. The physicists contributed to these tests by demonstrating a non-contact vibration measurement system based on a microwave interferometer designed and constructed at LANL.

Participation on this project lead to two additional ongoing LDRD-funded projects in LANL's Engineering Analysis Group related to vibration-based damage detection. Products resulting from these projects include an extensive review of the literature of vibration-based damage ID [3], a workshop held at LANL in September 1995 [4], and a MATLAB-based computer code known as DIAMOND [5] for statistically enhanced modal analysis, damage detection, and finite element model refinement.

OUTCOME OF SANDIA NATIONAL LABORATORIES' LDRD (SUCCESSES AND FOLLOW-ON WORK)

Most of SNL's technological and collaborative efforts have been funded by the LDRD. The one million dollar project spread over two years was entitled "Development of Structural Health Monitoring Techniques Using Dynamic Testing." This project was leveraged with funds from other LDRD projects covering System ID, Virtual Testing, Rapid Prototyping, as well as some LANL projects. Leveraging both these efforts and strong collaborations with private companies such as Holographics, Inc. and Flowind, Inc., a significant number of experimental and analytical sub-projects were performed. In fact, this LDRD supported the efforts of 25 individuals working on over 20 sub-projects and interacting with over 30 internal and external collaborators. Reference [6] contains a complete description of the LDRD efforts. Follow-on projects have been funded on traditional DOE structures, Air Force owned radar antennas, architectural surety, glass response to blast environments, and resonant fatigue testing of adhesive bonds. A field test of wind turbine health monitoring technologies has also been conducted.

APPLICATIONS

In the context of health monitoring, many structures have been studied by SNL. The majority of the work has focused on three main areas: aging aircraft, bridges, and wind turbines. In addition to modal analysis, many different technologies have been used to monitor the health of these structures. The LDV has proven to be valuable as a non-contacting measurement device. The LDV is capable of collecting data at more measurement points than traditional instruments such as accelerometers. The LDV is also proving itself in the field where a modal test can be performed without the usual mounting time of accelerometers. Among the other technologies used is the Natural Excitation Technique (NExT) [7] data reduction technique and neural networks as well as ultrasound and acoustic emission testing. Sub-projects devoted to understanding the utility of Electronic Speckle Pattern Interferometry (ESPI) were also performed.

Aging Aircraft

The aging aircraft facility at SNL has provided many opportunities to conduct health monitoring studies. Initially, work was done on a plate that simulated an aircraft skin. Initial studies on the plate utilized traditional modal techniques and simulated damage. Follow-on work consisted of the use of the LDV on the simulated skins to perform health monitoring studies.

A joint project was also initiated between SNL and a small company called Holographics, Inc. This collaboration was to extend the LDV technology from the initial studies. Frequency Response Functions (FRFs) were gathered from a McDonnell-Douglas DC9. The number of locations where data was gathered was 2233, which is an order of magnitude increase over traditional modal tests. Damage was induced by cutting a stringer, and a modal test was performed initially and after each cut. The data was also sent to CU for additional analysis. The size of the database prompted researchers at SNL and CU to develop procedures to condense the information into a usable set of important parameters.

Some work has also been performed on typical repairs seen on aircraft. A simple bolted repair joint was applied to an induced damaged stringer and the skin was scanned with the LDV. During this series of tests, five composite plates that simulated control surfaces and typical flaws seen in aircraft were also tested with the LDV. Using flexibility matrices and a technique known as flexibility disassembly [8], the flaws in the composite plates were located using the LDV data.

I-40 Bridge Work

SNL became involved with the I-40 bridge test through New Mexico State University, which was the lead institution for the experiment. They were interested in engaging SNL's expertise in modal analysis to develop a shaker excitation system for the bridge. SNL also provided accelerometers and information on the NExT procedure for this test. The NExT technique utilizes a background vibration source (such as automotive traffic or wind) as excitation for the test article. In the case of the I-40 bridge, traffic loading was used as the excitation source. From the response measurements resulting from this type of excitation, natural frequencies, damping, and mode shapes were extracted. SNL also acquired internal funds to support further involvement with the test.

The bridge work, in addition to strengthening the relationship between LANL and SNL, also produced significant advances in SNL's health monitoring technologies. A technique used for localizing errors in a finite element model, STRECH [9], was successfully adapted to be used in a health monitoring capacity. Also, a pseudo-model based damage detection algorithm (MAXCON) [10] was developed and proved useful in analyzing the I-40 bridge data.

Wind Energy

Much of SNL's efforts in health monitoring have been focused at wind energy applications. Structurally, the wind turbine is a fairly simple system. However, catastrophic failure of a working blade can mean damage to other blades, the tower, internal mechanical systems, other wind turbines, or to workers. Also, wind turbines see a tremendous number of fatigue cycles during a typical design lifetime. In conjunction with SNL's active Wind Energy group, many studies relating to failure of a wind turbine blade have been performed.

One of the first studies performed was a quasi-static fatigue test over a five month period. The structure was a Horizontal Axis Wind Turbine (HAWT) and the test was performed at the National Renewable Energy Laboratory (NREL). Modal tests were performed at various stages of the test. The test data included some unexpected phenomena. Following an initial drastic drop in all modal frequencies, most of the modal frequencies stayed constant until failure. At failure, most of the frequencies increased. Typically, frequencies will drop as a structure fails. Set-up dynamics and environmental conditions are believed to have caused the frequency shifts.

Two resonant fatigue tests were performed on Vertical Axis Wind Turbine (VAWT) blade sections. A resonant fatigue test exercises a structure at a natural frequency to decrease the time to failure. These tests also included periodic modal tests to establish a modal history of the fatiguing blades. In one of these tests the blade developed a span-wise crack that caused a shift in the torsional mode frequency while not affecting the frequency of the bending modes. This mode showed a monotonic decrease in frequency as a function of number of fatigue cycles.

A recent project involved the test of a wind turbine blade in the field with a LDV. The work included verifying that the LDV could take reliable data in the field, determining if NExT could be used with LDV data, and determining if simulated damage could be detected in NExT data. The analysis has not currently been completed.

Other Applications

Some other structures that have been studied from the perspective of health monitoring include nuclear power plants, tower guy anchors, and air compressor blades. These projects have been small efforts with little follow on work to date, but have greatly increased the national laboratories' experience base with real operating structures.

One of SNL's main mission is the surety of the nation's aging weapons systems. One weapon system application was an aerospace housing component [11]. Damage was simulated in the housing through a series of cuts in high strain regions. All damage cases were detected through the use of probabilistic neural networks. This highly successful project utilized information condensation techniques developed for use with the LDV.

Work is also being performed on the bearings of a large ground-based satellite dish. These systems are challenging due to the large size of the bearings and the slow, almost quasi-static rotation of the dishes. This work is currently in progress.

OUTCOME OF LOS ALAMOS NATIONAL LABORATORY'S LDRD (SUCCESSES AND FOLLOW-ON WORK)

The investigators at LANL can point to three primary successes that have resulted from the LDRD-funded investigations of vibration-base damage detection. First, the literature review that was recently published is, in the authors' opinion, the most comprehensive summary of the literature in this field to date.[3] Second, the computer code DIAMOND is to the authors' knowledge the only code that assembles many of the recent advances in vibration-based damage detection algorithms into one graphical user interface code.[5] When made available to the public, this code will allow researchers in this field to apply various linear and nonlinear algorithms to modal data to study the relative merits of the various damage ID methods. The code will allow researchers to easily modify existing algorithms and to add new algorithms as they are developed. Current plans are for the code to be distributed with sample data sets such as those from the I-40 bridge test so that researchers can apply their own algorithms to some standard data sets for comparative purposes. The final success of these projects relates to the fact that funding has been made available to continue this work with applications to our nuclear weapons systems. References [3,12-20] provide a summary of much of the work to date that has come from these LDRD projects.

APPLICATIONS

Most of the work conducted at LANL in the area of Structural Health Monitoring has focused on applications to highway bridges. Analysis of data sets from modal tests of bridges has demonstrated the importance of quantifying the variability of the measured modal parameters resulting from environmental conditions. Statistical analysis techniques such as Monte Carlo simulation and Bootstrap analysis have played an important role in the quantification of such variability effects, as well as the incorporation of these effects into various damage ID algorithms. Ongoing work is focused on the testing of idealized structures under laboratory conditions for the purposes of comparing the effectiveness and limitations of various damage ID techniques. Investigators at LANL have also been collaborating with Prof. Emin Aktan and Prof. Arthur Helmicki of the University of Cincinnati Infrastructure Institute to investigate modal parameter ID techniques in the context of structural damage ID [21]. Additionally, investigators from LANL are leading the compilation of a report on the State of the Art in Structural Identification of Constructed Facilities for the American Society of Civil Engineers.

I-40 Bridge

To date, field verification of damage detection algorithms applied to large civil engineering structures are scarce as few full size structures are made available for such destructive testing. Because the I-40 bridges over the Rio Grande in Albuquerque, New Mexico were to be demolished and replaced, the investigators were able to introduce simulated cracks into the structure, perform vibration test before and after each level of damage had been introduced, and then use the test data to validate various damage ID methods. Staff from LANL and SNL performed experimental modal analyses on the bridge in its undamaged and damaged conditions. Researchers from Texas A&M University subsequently applied a damage detection algorithm to these data [22]. The same damage detection algorithm was independently applied by the LANL staff to these data and to numerical data from finite element simulations of the I-40 bridge where other damage scenarios were investigated. The data required by the damage ID algorithm are mode shapes and resonant frequencies for the damaged and undamaged bridge. Results from these investigations are some of the first comparative studies of various damage ID algorithms that have been reported in the technical literature [19].

Alamosa Canyon Bridge

The Alamosa Canyon Bridge in southern New Mexico has been designated as a bridge test facility by the New Mexico State Highway and Transportation Department. Numerous modal tests have been performed on this structure for the purposes of damage detection. With only limited abilities to introduce damage into this structure, recent tests have focused on quantifying the statistical variations in modal properties that result from changing environmental conditions. [15,16] It is imperative that these changes be quantified and that changes resulting from damage are shown to be either greater than or different from those resulting from the test-to-test variations. Recent tests have been performed with the intent of comparing different statistical analysis procedures.

Eight Degree of Freedom Test System

When reviewing the literature on vibration-based damage ID, most studies seem to examine either a beam or a very complex structure such as an offshore oil platform. In an attempt to provide data from some structures of intermediate complexity and at the same time trying to keep the cost of fabrication reasonable, an eight degree of freedom, lumped-mass system was designed such that small quantifiable changes in stiffness or mass can be easily incorporated into the system. In addition, nonlinearities such as a crack opening and closing, variable frictional damping mechanisms, and loose parts rattling around can also be simulated with this system. This system will be tested in a variety of configurations and results of these tests will be used for comparative studies of damage ID algorithms and finite element model updating methods.

Comparative Test Specimens

To provide data sets for comparative studies of various damage ID algorithms, students and engineers at LANL are conducting a series of tests on simple structures. The idea of these tests is to investigate a wide variety of structure types while keeping the cost of specimen fabrication low. These structures include an aluminum I-beam, an aluminum plate, a three-story "unistrut" frame structure, and a fifty-five gallon drum that is intended to simulate a shell structure. If time and funds permit, a composite plate and a reinforced concrete bridge pier will also be tested. Linear and nonlinear damage is being introduced incrementally in these structures and vibration data (both time-histories and frequency domain data) are being measured for the undamaged structure and after each damage state. Various damage ID methods will be applied to these data sets for comparative studies. It is the intent of this study that these data will be made available to other investigators for further investigation.

Weapons Systems

The primary follow-on work that has been generated by LANL's LDRD projects is the application of this technology to nuclear weapons systems. This application will form the mainstay of LANL's future vibration-based damage ID work. The complexity of these weapon systems will require the development of new data analysis techniques that rely more heavily on the identification of changes in nonlinear system response. It is anticipated that technological advancements made through the investigation of these systems will be directly applicable to civilian structures and mechanical systems as well.

WHERE ARE WE GOING FROM HERE?

The national laboratories are repositories of a great deal of the nation's expertise in applying new technologies to real operating structures, both military and civilian. The next steps in the laboratories' development of damage ID technology includes the formation and/or strengthening of collaborations involving technology creators at universities and users with real structures. The national laboratories play key roles in locating appropriate new technologies, developing them for field applications, and brokering these applications. The charter of the national labs also drives the need to attack problems of a national scale which are critical to the reliability, safety, and security of the American people. The decay of the infrastructure including civil, aerospace, industrial, communication, energy, and defense areas represents a problem requiring the attention of the national laboratories.

ACKNOWLEDGMENTS

The authors of this summary would like to thank the following people whose work is summarized in this paper: William Baker, Thomas Carne, Philip Cornwell, Tim Darling, Thomas Duffey, Anthony Gomez, Scott Klenke, Norman Hunter, John Hurtado, Jim Lauffer, Randy Mayes, Albert Migliori, Thomas Paez, and Thomas Rice. The managers, support personnel, university students, and collaborators also deserve more credit than can be offered herein.

Sandia is a multiprogram laboratory operated by Sandia Corporation, a Lockheed Martin Company, for the United States Department of Energy under Contract DE-ACO4-94AL85000. Los Alamos National Laboratory is operated by the University of California for the United States Department of Energy.

REFERENCES

[1] Migliori, A., T. M. Bell, R. D. Dixon, R. Strong, 1993, "Resonant Ultrasound Non-Destructive Inspection," Los Alamos National Laboratory report LA-UR-93-225.

[2] Farrar, C. R. W. E. Baker, T. M. Bell, K. M. Cone, T. W. Darling, T. W. Duffey, A. Eklund, and A. Migliori, June 1994, "Dynamic Characterization and Damage Detection in the I-40 Bridge over the Rio Grande," Los Alamos National Laboratory report LA-12767-MS.

[3] Doebling, S. W., C. R. Farrar, M. B. Prime, and D. W. Shevitz, April 1996, "Damage Identification and Health Monitoring of Structural and Mechanical Systems From Changes in their Vibration Characteristics: A Literature Review," Los Alamos National Laboratory report LA-13070-MS.

[4] Workshop on Damage Identification and Health Monitoring of Structures, Los Alamos National Laboratory Colloquium, September 13-15, 1995.

[5] Doebling, S. W., C. R. Farrar, and P. J. Cornwell, 1997, "DIAMOND: A Graphical Interface Toolbox for Comparative Modal Analysis and Damage Identification," to appear in *Proceedings. of the Sixth International Conference on Recent Advances in Structural Dynamics*, Southampton, UK, July 14-17.

[6] James, G. H., 1996, "Development of Structural Health Monitoring Techniques Using Dynamics Testing," Sandia National Laboratories Report SAND96-0810, Albuquerque, NM.

[7] James, G. H., T. G. Carne, and J. P. Lauffer, 1993, "The Natural Excitation Technique (NExT) for Modal Parameter Extraction from Operating Wind Turbines," Sandia National Laboratories Report SAND92-1666, Albuquerque, NM.

[8] Peterson, L. D., K. F. Alvin, and S. W. Doebling, 1995, "Experimental Determination of Local Structural Stiffness by Disassembly of Measured Stiffness Matrices," AIAA-95-1090-CP, *Proceedings of the 36th Structures, Structural Dynamics, and Materials Conference*, New Orleans, LA.. To appear in ASME *Journal of Vibration and Acoustics*.

[9] Mayes, R.L., 1995, " An Experimental Algorithm for Detecting Damage Applied to the I-40 Bridge over the Rio Grande," *Proceedings of the 13th SEM International Modal Analysis Conference*, Nashville, TN.

[10] James, G., T. Carne, B. Hansche, R. Mayes, G. Reese, and T. Simmermacher, 1995, "Health Monitoring of Operational Structures - Initial Results," *Proceedings of the 1995 AIAA Adaptive Structures Forum*, New Orleans, LA, April 10-13.

[11] Klenke, S. E. and T. L. Paez, 1996, "Damage Identification with Probabilistic Neural Networks," *Proceedings of the 14th International Modal Analysis Conference*, Dearborn, MI.

[12] Farrar, C. R., T. Darling, and A. Migliori, "Microwave Interferometers for Remote Noncontact Vibration Measurements," submitted for publication in *Mechanical Systems and Signal Processing*

[13] Farrar, C. R. and T. A. Duffey, "Simplified Analysis of Bridge Dynamics," accepted for publication in ASCE *J. of Bridge Engineering.*

[14] Farrar, C. R., and G. H. James, "System Identification from Ambient Vibration Measurements on Bridges," accepted for publication in *J. of Sound and Vibration.*

[15] Doebling, S. W., C.R. Farrar, and R. Goodman, February 1997, "Effects of Measurement Statistics on the Detection of Damage in the Alamosa Canyon Bridge," *Proceedings 15th International Modal Analysis Conference*, Orlando, FL.

[16] Farrar, C. R., S. W. Doebling, P. J. Cornwell,. and E. G. Straser, February, 1997, "Variability of Modal Parameters Measured on the Alamosa Canyon Bridge," *Proceedings 15th International Modal Analysis Conference*, Orlando, FL.

[17] Cornwell, P., S. W. Doebling, and C. R. Farrar, February, 1997, "Application of the Strain Energy Damage Detection Method to Plate-Like Structures," *Proceedings 15th International Modal Analysis Conference*, Orlando, FL.

[18] Doebling, S. W. and C. R. Farrar, 1996, "Computation of Structural Flexibility for Bridge Health Monitoring Using Ambient Modal Data," *Proceedings of the 11th ASCE Engineering Mechanics Conference*, Ft. Lauderdale, FL.

[19] Farrar, C. R. and D. Jauregui, March 1996, "Damage Detection Algorithms Applied to Experimental and Numerical Modal Data From the I-40 Bridge," Los Alamos National Laboratory report LA-13074-MS.

[20] Farrar, C. R., T. A. Duffey, P. A. Goldman, D. V. Jauregui, and J. S. Vigil, February 1996, "Finite Element Analysis of the I-40 Bridge Over the Rio Grande," Los Alamos National Laboratory report LA-12979-MS.

[21] Catbas, F. N., M. Lenett, D. L. Brown, S. W. Doebling, C. R. Farrar, and A. Turer, 1997, "Modal Analysis of Multi-Reference Impact Test Data for Steel Stringer Bridges," *Proceedings of the 15th International Modal Analysis Conference*, Orlando, FL, February 3-6, pp. 381-391.

[22] Stubbs, N., J.-T. Kim, and C. R. Farrar, 1995, "Field Verification of a Nondestructive Damage Localization and Severity Estimation Algorithm," *Proceedings of the 13th International Modal Analysis Conference*, Nashville, TN, February, pp. 210–218.

Information Technology in Civil Infrastructure System Monitoring: Opportunities and Issues

A. J. HELMICKI, A. E. AKTAN, L. K. COMFORT and M. KAM

ABSTRACT

The University of Cincinnati Infrastructure Institute has recently joined forces with Drexel University's newly formed Intelligent Infrastructure Institute. Together, the authors representing both of these institutions form a multidisciplinary team with more than 50 years total experience in studying both the technical and non-technical aspects of both Civil Infrastructure Systems (CIS) design, construction, maintenance, and renewal as well as the development and application of Information Technology concepts. This paper shares some of the authors' experiences in trying to exploit the potential of information technologies in the CIS area. It highlights some of the significant capabilities and opportunities presented by information technology as well as indicating some of the issues and further developments which are needed before widespread application of information technology concepts can be successfully implemented in the CIS arena.

INTRODUCTION

The national investment in CIS is estimated at $20 trillion (NSF, 1993). The US construction industry comes second in size to health-care, with about 10% of our GDP estimated as related to construction. In the US, the service performance of many CIS components fall short of their anticipated designs. Well-publicized inadequacies of the transportation system may be reducing the annual growth rate in GDP by as much as 1% (Chong, 1995), equal to insurance industry estimates of the cost of a major East Coast hurricane or a major West Coast earthquake (Sandia, 1997). Thus, the annually compounding economic impact of non-optimum operation

Arthur J. Helmicki, ECECS Dept., UCII, University of Cincinnati, OH, 45221-0030, (513)556-6069.

A. Emin Aktan, DIII, Drexel University, Philadelphia, PA, 19104, (513)556-3689.

Louise K. Comfort, Grad. Sch. of Public and Int'l Affairs, University of Pittsburgh, Pittsburgh, PA, 15260, (412)648-7606.

Moshe Kam, ECE Dept., Drexel University, Philadelphia, PA, 19104.

and maintenance of our CIS may reach significantly higher levels than that of a major disaster, given the probable occurrence of catastrophic events per year. Since surety and the safety and service performance of CIS are interrelated, a lack of optimum CIS performance adversely affects surety, triggering adverse social consequences as well.

This situation is aggravated by the fact that while the CIS and its users are intertwined, financing, legislation, regulation, engineering, operation, maintenance and management are typically undertaken by a very large number of fragmented public and private agencies. For example, the maintenance of the access roadways, the structural system and the lighting facilities of a bridge may be under the jurisdiction of different states, townships and citizens groups, respectively. A highway bridge may carry electric power, water, sewer and telecommunication lines and therefore serve as an intersection node for many CIS, and an accident may bring a dozen agencies together revealing how the vulnerability of one CIS component may affect many other CIS as well as social and economic systems (Comfort, 1994).

Consequently, research and administration of CIS, which have traditionally been the domain of disciplines such as Civil Engineering, Mechanical Engineering and Electric-Power Engineering, has shifted focus towards CIS integration, a key concept in any strategy for advancement and for prudently managing the CIS for sustainability. The US Congress and administration, through the ISTEA (1992) and the NEXTEA (1997) legislation, maintain the need for an integrative, multi-modal approach, and assert strong linkage between sustainable development, the environment and CIS. This, in turn, begs the need for a careful blending of specialists, representing multiple disciplines and backgrounds in order to set a nationwide CIS research and administration policy and strategies which can successfully face the challenges discussed here.

Against this backdrop the US National Science Foundation (NSF) identified the built environment as an important research area in the 1980's, and in 1993 announced CIS as a strategic research field critical for the national interest. In addition to our building stock, the NSF identified electric power, gas and liquid fuels, telecommunication, transportation, and, water and sewer (environmental infrastructure) as some of the critical CIS. The CIS research agenda formulated by NSF (NSF, 1993), and reasserted at a recent workshop (NSF, 1996), which envisioned advances in condition assessment, deterioration science, damage mechanics, rationalization of heuristics, decision and management sciences all brought together in an intelligent systems framework in order to accomplish the scientific, engineering and educational advancements which are needed for advancing and sustaining CIS.

To this list the authors would add the concept of Structural Identification (St-Id) which permits effective integration of experimental research, analysis, information and decision sciences. This concept enables researchers to overcome the single most important technical barrier limiting innovation in CIS engineering, which is the empirical, heuristic foundation of civil engineering practice. Both the education and practice of civil engineering disguise the strong dependence on empirical art and heuristic knowledge-bases, and many practicing engineers and academics may not fully appreciate how detached the design, construction, inspection, maintenance and renewal processes are from the reality of actual loading environments and behavior of structures. Structural identification facilitates the development of a complete, conceptual and objective understanding of real-life (Aktan et al, 1997).

THE ROLE OF INFORMATION AND INFORMATION TECHNOLOGIES

Information and its effective interpretation and communication bind many of the components discussed above together. Information technologies, which can loosely be defined as *the scientific, technological and engineering disciplines and techniques used in information handling and processing* (Stokes, 1983), provides a key enabling technology for improving CIS performance.

In its fullest sense, information technology includes consideration of:

- information gathering/collecting, sensors, sensing, and data acquisition; associated signal processing and signal analysis;
- information storage, archival, retrieval, navigation, and security;
- information communication, transfer, sharing, presentation, and dissemination; and
- information data-basing, mining, analysis, synthesis, and interpretation.

As such, it provides the mechanisms that allow for the integration of essential information into more efficient and cost effective strategies such as: centralized management, intelligent systems approaches, real-time, on-line, remote monitoring, sustainability, life-cycle-based analysis, and the like. In short, information technologies, properly applied, offers the potential for contributing significantly to a renaissance in the design, construction, management, and renewal of CIS.

To date, the main impact of information technology on CIS has been mainly limited to two areas: (i) off-line use of generic design tools (e.g., CAD packages, computer-aided drafting, etc.) and (ii) simple data-basing applications used for inventorying and inventory management. However, the use of information technology for on-line activities is gradually growing and is likely to provide new capabilities and features on a large scale. Among these are on-line monitoring of structures and systems from remote locations, wider use of telemetry, implementation of active status and alarm checks for structures and devices, queries about system status, remote testing, and even active remote control. Direct communication between vehicular traffic-control centers and cars is almost here already, as it has become cost-effective to monitor and direct traffic through arrays of sensors and electronic billboards. In the next two decades, engineers and technicians are likely to employ a much wider array of remotely-operated diagnostic tools, sensors, and controllers.

Despite this promise, much work remains to be done before its full potential can be realized. This paper shares some of the authors' experiences in trying to exploit the potential of information technologies in the CIS area. It highlights some of the significant capabilities and opportunities presented by information technology as well as indicating some of the issues and further developments which are needed before widespread application of information technology concepts can be implemented in the CIS arena.

EXAMPLE #1: THE NATIONAL BRIDGE INVENTORY

Key issues in the application of information management and information technologies in the infrastructure arena can be illustrated by using the example of bridges as an infrastructure component. Bridges serve as a valuable example in representing both critical components of the surface transportation system and also a generic component of model CIS. Moreover, "smart bridges" are being designed and introduced which make use of new sensing devices and robots, and are linked to computer technology, computer networks and telecommunication technology. As a result, such bridges represent not only a permanent critical element of our CIS but also a test-bed to the improvements and hazards involved in linking CIS with new emerging information technologies.

STATE OF THE PRACTICE IN DATA-BASING

The National Bridge Inventory database, maintained and utilized by the Federal Highway Administration (FHWA) for planning and management purposes, represents the current state of the practice in the use of information technology for inventory management. The entries in this data base are obtained from National Bridge Inspection Program which, since the since the early 1970's, has mandated that all bridges in the national inventory undergo bi-annual visual inspections. These inspections are usually performed by non-technical individuals under the guidelines set forth in the ASSHTO Manual for Maintenance and Inspection of Bridges and the AASHTO Bridge Inspector's Training Manual.

However, the National Bridge Inventory Recording and Coding Guide states that the use of these guidelines is optional, and each state may use its own coding scheme. As a result, there is fragmentation and a general lack of standardization and education in both the data collected and the way in which it is collected. Compounding this condition is the fact that the data entries are extremely limited. All entries take on the form of alphanumeric codes or qualitative ratings such as "on a scale from 0 to 9." Thus, the value and reliability of using such data in the decision-making process is questionable at best.

This last statement can be highlighted in two ways: First, in spite of this program, in-service bridge failures precipitated by design errors, and/or construction or maintenance defects still occur today. Thus, the technical challenges in sustaining the service performance of the infrastructure cost-effectively still need to be resolved. Second, in 1996 the US DOT, based on data obtained from the inventory database, announced a backlog of nearly $100 billion for bridge repair, a figure which is increasing continuously with time. The question of how accurate these cost estimates are, how significant are the threats posed by the existing deteriorated inventory awaiting renewal, or how to best prioritized the limited resources available at any time for replacement, retrofit, and renewal cannot be answered due to subjective and limited nature of the database entries.

NEED FOR NEW KNOWLEDGE

The 1995 AASHTO-LRFD Bridge Design Guidelines expect a new-bridge service life of 75 years with "zero or minimal" maintenance. However, many bridges require extensive rehabilitation due to deterioration and damage within just a decade, and evidence points to the lack of a knowledge-base for designing and constructing effective maintenance or renewal. Evidence also points to the lack of a knowledge-base for designing and constructing bridges for durability and optimum life-cycle cost or any rational set of performance measures.

INFORMATION TECHNOLOGY-ORIENTED APPROACH

Reliability-based Evaluation

A more rational approach to rating and management might be based on the concept of reliability evaluation, which is described as the process of establishing the probability of acceptable performance at the service and failure limit-states over a given time period, given its initial condition and maintenance schedule. A reasonably accurate "projection of future performance" is necessary for rational and optimum infrastructure maintenance management (Yao, J.T.P. and Yao, T.H-J.,1997). However, even those bridge owners that have adopted a maintenance management scheme currently follow only empirical rules which do not allow advantages from cost-savings which innovative infrastructure management strategies may provide. Reliability evaluation requires assessing the demands from, and the corresponding capacities of, a facility with some confidence and evaluating the probability of demands exceeding the capacities for critical limit-states. To accomplish this, the relationships between demands, capacities and performance have to be established with reasonable confidence for each critical limit-state all of which could be obtained from properly studied, instrumented, and monitored CIS.

Intelligent Transportation Systems

Constructing new roads and bridges, particularly in densely populated urban areas, has become prohibitively expensive. Improving the performance of existing "congested" infrastructure by innovative policy, strategy and technologies is considered as the only feasible way to meet the demands. Intelligent transportation systems (ITS) offer great promise if we could overcome the barriers in developing, demonstrating, standardizing and commercializing the appropriate technologies.

Life-Cycle Cost-based Analysis

Today, most CIS designs are selected based on initial construction cost and not life-cycle cost. If we could develop the information-base for reliably evaluating the life-cycle cost of various designs, and sought optimal designs based on minimum life-cycle cost, there would be significant cost-savings and improved performance.

EXAMPLE #2: LESSONS FROM CURRENT BRIDGE RESEARCH

In research currently underway, a multidisciplinary team of civil, electrical, and mechanical engineers at the University of Cincinnati Infrastructure Institute (UCII) in conjunction with colleagues at Drexel University's newly formed Drexel Intelligent Infrastructure Institute (DIII) is addressing some of the issues outlined above by conducting exploratory research on actual operating highway bridges. The research team currently operates three highway bridge test-sites consisting of three-span, steel-stringer overpasses located in Cincinnati. Each represents a different phase in the lifecycle of a bridge: birth, middle age, and old age. Soon a fourth bridge will be brought on-line in the Philadelphia area to serve as a test-bed for long span bridge monitoring activities. Using these resources, the team has initiated research in the use of information technology in data acquisition, archival, transmission/real-time interfacing, and dissemination.

INSTRUMENTED MONITORING AND INTERFACE DESIGN

Two of these bridges serve as focal points for the study of instrumented monitoring (see Figure 1). The instrumentation of packages, totaling over 500 sensors, allow for the measurement of ambient conditions, as well as low frequency and high frequency components of structurally related signals (i.e., strain, tilt, displacement, etc.). In addition, each bridge is equipped with a weigh-in-motion (WIM) scale which monitors the weight and speed of all vehicles crossing the bridge. All measurements are recorded by a PC-based data acquisition system. Both bridges are tele-remotely operable from laboratories at UCII via standard telephone/modem connections and custom designed software forming a user friendly data monitor. Such a system is able to augment the existing qualitative, subjective, visual inspection information with quantitative, objective, multidimensional data that can be reliably factored into decisions about management, renewal and sustainability. These data have been used, in turn, for such purposes as the development of rigorous structural capacity ratings, the detection and isolation of structurally significant damage, defects and deterioration, the discovery and characterization of fundamental unknowns/factors affecting structural behavior of the system, etc.

Information technologies, appropriately designed and implemented, are essential to bring the data and corresponding information from this "high tech" approach to full-scale, real-world infrastructure management. However, without proper design, the use of information technologies can also hinder discovery and management activities. The monitors described above can collect as much as 1Gb of data per day. Thus the data representing even just one year's long term monitoring would consist of over 350Gb of data. Without proper pre- and post-processing and effective display of the information, the ability to assimilate and interpret it would greatly be reduced.

EFFECTIVE INFORMATION DISSEMINATION

The third bridge operated by UCII is a decommissioned steel stringer bridge test-bed site which is serving as the test specimen for researching and demonstrating condition assessment and structural damage detection concepts and tools. The research is designed around the iterative application of various damage scenarios followed by structural identification in order to determine the viability and sensitivity of various experimental approaches (including instrumented monitoring and modal testing) for use in objective condition assessment.

To date, the research team has collected and assembled a database of several volumes of documentation on the project covering such aspects of the research as: 1) the original plans and initial condition of the test specimen; 2) how it was prepared for use as a field laboratory; 3) the results of preliminary analyses of the structure and tests conducted in preparation for development of analytical models, instrumentation, and modal testing design; 4) documentation of all testing activities and procedures; and 5) post-processing , evaluation, interpretation, and synthesis of test results. Coupled with this documentation are raw data files obtained from over 100 separate sets of experiments conducted at the bridge for various damage states (several 10's of Gbs).

An international panel of experts in mechanical, electrical, aerospace, civil engineering, and engineering mechanics, specializing in bridge engineering practice, structural reliability, structural identification, fault-detection, modal analysis and information technology, will participate in interpreting the data obtained. In addition, ready access to the data collected would provide a rare opportunity to many other researchers as well as to practicing engineers, policy makers, and the entire bridge engineering community at large.

Providing access to users from a range of disciplines and infrastructure contexts poses the nontrivial issue of how to design the content/layout/organization, and flow of the information database necessary to best convey a full understanding of this research project. The goal is to design a method/medium to communicate effectively to database users all information about the project necessary to allow them access to the experimental data and results obtained in an intelligent and informed manner. Relevant aspects of information technology including development of a CD-ROM, multimedia database documentation of the research would provide ease of access and use as well as maximize information content. This documentation format was chosen because, in addition to the standard one-dimensional textual format of a written report, it will allow the incorporation of information in audio, video, photographic, diagrammatic/CADD, graphical, tabular, and raw numerical data forms.

The total amount of information collected on the damage detection project could easily fill several CD-ROMs. Great care must be taken in the development of user interfaces so that data can be effectively navigated, filtered, and massaged in order to obtain meaningful information, realize crucial insights, and manage the process effectively.

SOME OBSERVATIONS: CAPABILITIES AND LIMITATIONS

One of the most critical research areas in infrastructure development is the emerging use of telecommunication, information technology and robotics in infrastructure monitoring, maintenance, and control. The hardware and software used for telecommunication networks and for computing networks have already become part of the national infrastructure. The growing use of information technology provides great advantages, such as instant access to data bases, improved communication, faster computing, remote control, remote monitoring and improved decision-assist devices. However, this development also has the potential to affect functionality adversely during computer and communication down-time, reduce stability, and introduce security hazards.

HARDWARE AND DEVICES

We need to analyze and understand how new structures should be built, equipped, and connected so that effective and reliable monitoring and control will be realized. This effort involves active sensors, embedded sensors, interface hardware, telecommunication and data-transfer protocols, transmitters, receivers, and antennae or wire/cable interfaces.

SOFTWARE AND ALGORITHMS

The increasing complexity of interacting systems in metropolitan communities has led to the emerging field of study, complex, adaptive systems. Viewed from this theoretical perspective, civil infrastructure, designed to support the economic, social and technical functions of urban areas with concentrations of population at different levels of income and education, has different impacts upon these populations. Information technologies, such as GIS, and the prototype Interactive, Intelligent, Spatial Information System currently under development at the University of Pittsburgh, (Comfort et al, 1997) allow more accurate assessment of the performance of civil infrastructure on the diverse populations of a metropolitan community. Infrastructure both facilitates and constrains the operation of economic functions and social development. New developments in information technology are leading to improved monitoring, analysis and transmission of information to practicing managers in near real-time. These developments create a major shift for practicing managers in their capacity to manage not only the performance of single systems, but also to adjust the performance among the set of systems, achieving a more optimal balance in equity in access and efficiency of operations for the system as a whole.

In addition, emerging technologies such as Decision Fusion and Sensor Fusion (e.g., Kam et al. 1991, Kam et al. 1997) are anticipated to play a key role. These disciplines, originally fostered by DoD research in the early 80's, refer to algorithms and decision tools that integrate information (sensor readings, algorithm decisions, expert opinions) in order to provide decision makers with decisions, estimates and recommendations that are at least as reliable as the most reliable information source. It is necessary to examine the potential of Decision Fusion as the growing use of information technology in infrastructure-related activities leads inexorably to information overload. The challenge of infrastructure management will no longer be lack of information or the inability to transport information to the site where it is needed. Rather, it is likely to be excess of information and the "drowning" of automatic systems as well as human decision makers in a sea of details. The use of information aggregation and data compression algorithms becomes critical.

NETWORKS AND COMMUNICATIONS

The advantages and disadvantages of coupling between stand-alone systems through computer networks need extensive study before they are practical and reliable. We need to interface the infrastructure to entities such as the Global Positioning System, the Internet, cellular telephone networks and communication satellites in a seamless and universally-interpretable and expendable manner. As telecommunications and computers become part of the infrastructure, issues of controllability, authentication, encryption and security need to be addressed.

Figure 2 illustrates how infrastructure components could be connected together via the Internet. It includes details of all the internal connections to emphasize the various Internet workings and bandwidth limitations (Wiggins, 1996, Zgodzinski, 1996). Table 1 gives some basic statistics on the data size and bandwidths for a variety of data types (Cole, 1993). By comparing the entries in Figure 2 with those of Table 1 it is clear that the level of maturity of some components of information technology is not yet at the point where commercially available solutions exist to handle all the required networking and communications outlined in the intelligent bridge monitor example of Figure 1. In particular, it is evident that the incorporation of real-time, on-line monitoring using both transmission of raw sensor data and full-motion video of traffic at the standard 30 frames/sec would exceed the bandwidth limitations currently available on either standard phone or ISDN connections.

STANDARDS AND PROTOCOLS

To realize the benefits of information technology in infrastructure control, we shall need system-wide design, emphasizing robustness, modularity, standardization, expandability and compatibility. The danger is that we create an unwieldy array of implementation methods, design techniques, hardware modules, software standards, communication methods, technologies and protocols, which would become obsolete almost as soon as they are installed, be incapable of communicating with each other, and be inherently incompatible with each other.

In addition to standards development for the purposes of information storage, retrieval, and communication, the work is needed in the development of standards for documentation, as well as data acquisition. These standards will cover hardware and software configuration and interfacing and will extend to cover policies and practices as to how experiments are run, data is collected, processed and documented prior to ultimate storage for external use. Such issues are critical in the effective communications of results and ideas among the various parties in the CIS community.

CONCLUSIONS

Information and the associated infrastructure to handle it are key components necessary for success in addressing the challenges currently facing us from our CIS. Information technologies provide new tools for monitoring systems in real-time and allow for the transmission and communication of virtually instantaneous information to a wide variety of users from engineers to policy makers, to practicing managers, and even directly to the public users. These tools allow us to modulate not only the performance of single systems, such as traffic on roads, but also the interaction between a set of interdependent systems, such as roads, bridges, bus, and rail lines. Such an approach will provide the means for integrated design, construction, and management of CIS thereby allowing us to realize cost savings, performance enhancements, and more effective use of existing and planned CIS.

However, we are not are completely there yet. Much research in the most effective application of information technologies in the CIS arena needs to be carried out. Full scale field-based experimentation is necessary to understand actual, realistic implementation issues and limitations. More mature technologies, standards, an community of industries and commercially available off-the-shelf products, and turn-key systems are needed before these concepts can be put into widespread implementation. Appropriate legislation and research support must be provided to motivate these developments.

As a result, information technologies offer significant promise in developing truly integrative CIS research, development, and management policies. However, in order for it to reach its full potential, the application of information technologies in the CIS arena must be implemented and championed by members of all sectors: government, academe, and industry.

ACKNOWLEGDEMENTS

The authors wish to gratefully acknowledge the support of the Ohio Department of Transportation,, the Federal Highway Administration, and the National Science Foundation under whose auspices this work was performed. In addition, the authors wish to acknowledge the participation of several graduate students and colleagues in the various portions of the research activities described herein. These include: R. Barrish, D. Brown, N. Catbas, Y. Gao, K. Grimmelsman, V. Hunt, S. Iyer, M. Lenett, A. Levi, and A. Wilson.

REFERENCES

1. Aktan, A.E., Farhey, D.L., Helmicki, A.J., Brown, D.L., Hunt, V.J., Lee, K.L., Levi, A. "Structural Identification for Condition Assessment: Experimental Arts." To appear in the Journal of Structural Engineering, ASCE, 1997.
2. Chong K. (1995) "NSF Infrastructure Research Initiative and Potential Synergies with FHWA's Research and Development Programs," North-American Workshop on Instrumentation and Vibration Analysis of Highway Bridges, held at the University of Cincinnati, Cincinnati, OH, July 18-20 (quoted from David Aschauer, Chief Economist, Federal Reserve Bank - Chicago).
3. Cole, B. (1993) "Special Multimedia Report: The Technology Framework," IEEE Spectrum, March, pp.32-39.
4. Comfort, L.K. (1997). "Designing Resilient Communities: Self Organizing Processes in Disaster Management" in Horie, Fukashi and Masaru Nishio, eds. 1997. Future Challenges of Local Autonomy in Japan, Korea, and the United States: Shared Responsibilities between National and Sub-national Governments. Tokyo, Japan: Simul International, Inc.:314-353..
5. Comfort, L.K. 1994. "Self Organization in Complex Systems." Journal of Public Administration Research and Theory, Vol. 4, No. 3 (July):393-410.
6. _____ "Intermodal Surface Transportation Efficiency Act (ISTEA) of 1991," A Summary, U.S. Department of Transportation, FHWA-PL-92-008.
7. Kam, X. Zhu, and P, Kalata (1997) "Sensor Fusion for Mobile Robot Navigation," Proceedings of the IEEE, Vol. 87, No. 1, January, pp. 108-119.
8. Kam, Q. Zhu, and W. Chang (1991) "Hardware Complexity of Binary Distributed Detection Systems with Isolated Local Bayesian Detectors," IEEE Transactions on Systems, Man and Cybernetics, Vol. 21, No. 3, pp. 565-571, May.
9. NSF (1993), Civil Infrastructure Systems Research: Strategic Issues, A Report of the Civil Engineering Systems Task Group, NSF 93-5.
10. NSF (1996), Workshop on Integrated Research for Civil Infrastructure, July 15-27, Washington, DC.
11. Sandia (1997) Proceedings, Conference on Architectural Surety, held at Los Alamos, MN, May.
12. Stokes, A. (1983), Concise Encylopedia of Information Technology, Prentice-Hall.
13. Wiggins, R., (1996) "How the Internet Works," Internet World, October, pp. 54-60.
14. Yao, J.T.P. and Yao, T., H-J. (1997) "Optimal Performance of Civil Systems Using Symptom Based Realiability and Health Monitoring," Paper presented and published in the Proceedings of the NSF Workshop on Optimal Performance of Civil Infrastructure Systems, Portland, OR, April.
15. Zgodzinski, D. (1996) "Enter ASDL," Internet World, October, pp. 72-75.

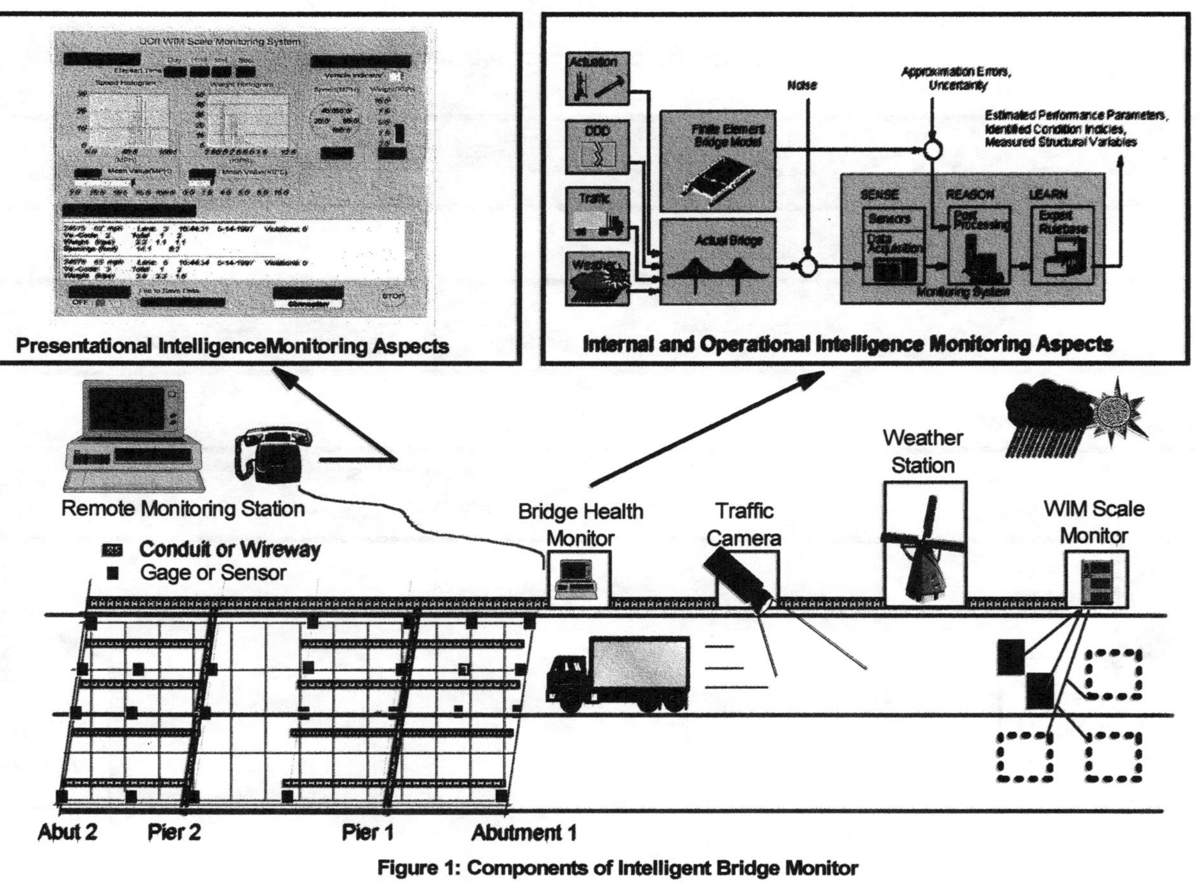

Figure 1: Components of Intelligent Bridge Monitor

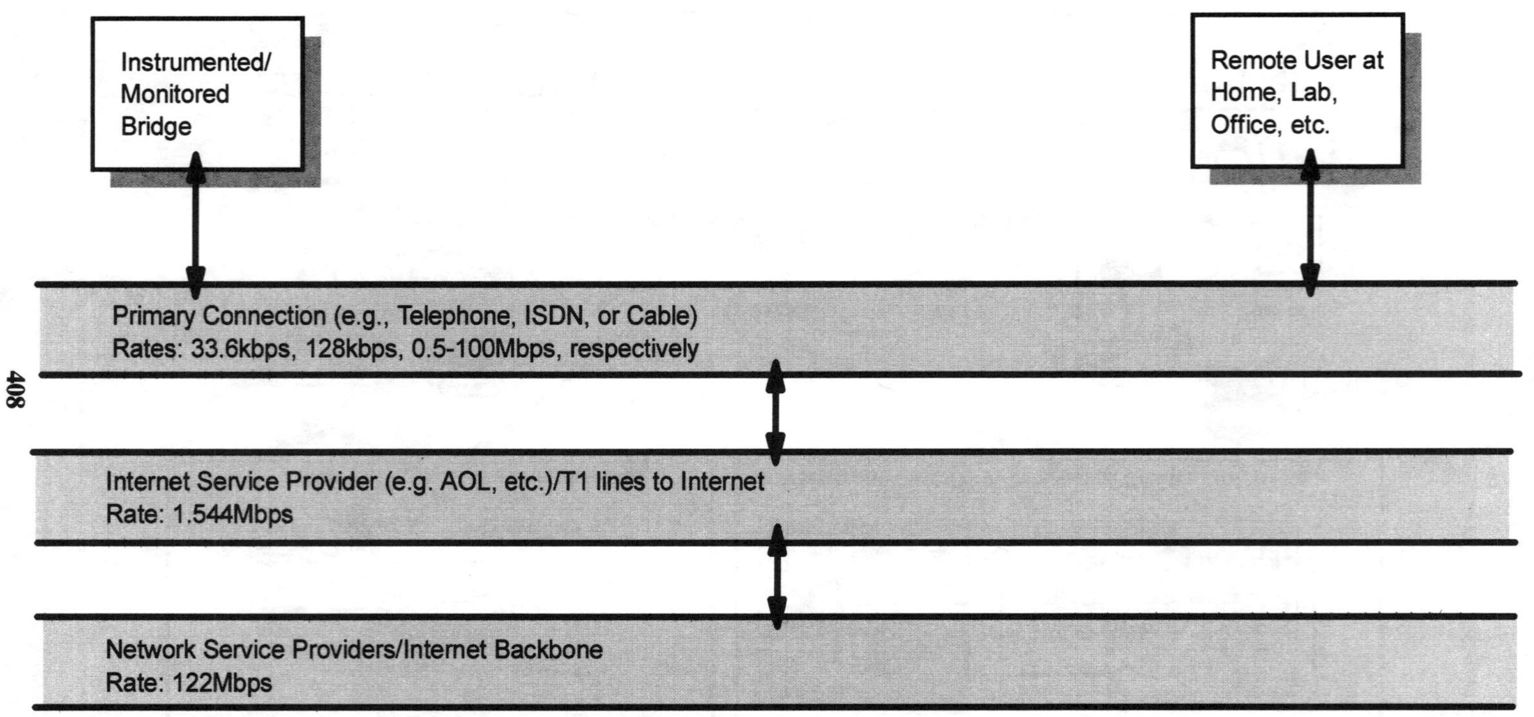

Figure 2: Internet-based Monitoring Configuration

	Text	Image	Audio	Video
Object Type	ASCII	bit-mapped, photo	digitized audio	TV, analog, or digitized stream
Size and Bandwidth	2kb/page	64-7,500kb/ image depending on resolution	6-44kbps	27.7Mbps for 640x480x24 pixels/frame, color

Table 1: Size and bandwidth requirements for various multimedia data/objects

Two Years' Experience Using OASIS Real-Time Remote Condition Monitoring System on Two Large Bridges

R. L. NIGBOR and J. G. DIEHL

ABSTRACT

The multiple needs of long-term health monitoring, extreme event recording, and real-time alerting for civil structures have proven a significant challenge to instrumentation engineers. On the one hand, real-time alerting communications following an extreme event requires very little data and a simple alarm. Such systems have been in use in nuclear power facilities for decades. However, the quality and amount of data needed for proper health monitoring is immense, and real-time communications and alerting using a system adequate for health monitoring yields vast amounts of data that must be managed. Furthermore, the reduction of complex information to terms that are easier for control center operators to evaluate and respond to (i.e. green/yellow/red-light) is an important factor in bridge management. This paper reports on the performance experience of a new monitoring system which addresses all of these goals. The system has been installed on two bridges of different types and in different countries.

OASIS "On-Line Alerting of Structural Integrity and Safety"

In June of 1995 Agbabian Associates introduced a new real-time, on-line monitoring system for continuous evaluation of the structural integrity of critical structures. The system is called OASIS: On-Line Alerting of Structural Integrity and Safety. OASIS is designed to improve management and safety decisions because:

- Response time is reduced through fast communications, intelligent processing

Robert L. Nigbor and John G. Diehl, Agbabian Associates, Inc., 1111 South Arroyo Pkwy, Suite 470, Pasadena, CA 91105

- On-line analysis offers better information about structural status
- Visual displays create better understanding of critical problems

OASIS performs four vital functions simultaneously, in real time.: (1) a remote, real time alerting system using visual, on-screen imaging and audible alarms; (2) an event-triggered, high dynamic range, high speed accelerograph which operates in the background; (3) remote control and display of system functions through direct feedback, "hands-on", Windows graphical environment and visual icons; and (4) a health monitoring system.

The standard acquisition system utilizes a Kinemetrics Mt. Whitney with a 19-bit resolution (better than 1 part in 500,000!), high dynamic range digital event recorder with true multitasking (the Mt. Whitney can monitor/record and communicate simultaneously), remote interrogation and programmability.

Standard OASIS system monitoring software includes the following:

- Visual display of subject structure including sensor icons
- Real-time, dual-level alerting, including location
- Sensor icon, click-on waveform display and statistics from any channel
- Remote command/control of acquisition system

RAMA IX BRIDGE, BANKOK, THAILAND

In 1995 OASIS was installed on the Rama IX bridge in Bangkok, Thailand. The purpose of this system is to continuously monitor structural motion, stresses, and related structural/environmental data and to remotely alert the bridge operators if any parameters exceed specified thresholds. The system also measures long-term changes in structural parameters.

The Rama IX Bridge is one of the world's largest cable-stayed bridges. Spanning the Chao Praya River between Bangkok and Thonburi at Wat Sai, it measures 782 meters in total length including the 450 meter main span and two 166 meter side or back spans. The reinforced concrete approaches extend the overall length to three kilometers.

The monitoring system includes sixteen channels of data including (see Figure 1):

- 14 channels of acceleration monitoring
- 3 channels of wind
- 1 channel temperature

This relatively simple monitoring system utilizes the symmetry of the structure by monitoring only one side of the bridge.

Data is monitored using a Kinemetrics' Mt. Whitney. This is an 18 channel, 24-bit digital recorder with 19-bits of resolution. The Mt. Whitney is located at the base of the West tower. Event triggered data is recorded using a PCMCIA recording module, in this case two 20Mbyte Sundisk.s.

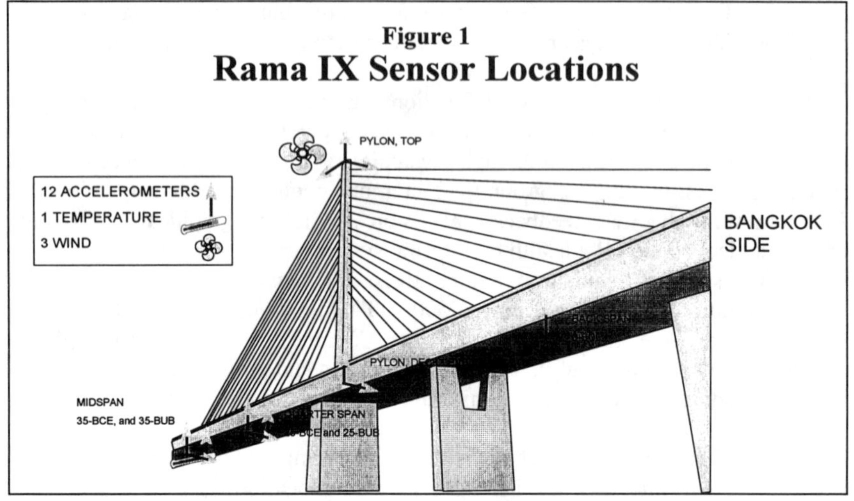

Figure 1
Rama IX Sensor Locations

Typically the Mt. Whitney functions as an event recorder using programmable threshold trigger and weighting algorithms. However, this Mt. Whitney's firmware was modified to Agbabian Associates' specifications to provide a continuous, real-time, serial data stream output of all channels. This stream provides continuous data at 10 samples per second for all sixteen channels.

The data is received at the management operations center about 5 kilometers distant via short haul modem and twisted pair cable. The data is continuously processed and displayed using Agbabian Associates' OASIS software.

NAMHAE GREAT BRIDGE, CHINJU, REPUBLIC OF KOREA

In 1996 a more extensive and modernized version of the same monitoring system was installed on the Namhae Great Bridge in the south coast of the Republic of Korea. The Namhae Great Bridge is a suspension bridge with main span of 400 meters, and side spans of 125 meters. The towers rise 60 meters above the mean water level. The deck and towers are all structural steel.

The Namhae monitoring system provides 36 channels of data including (see Figure 2):

- 16 channels of acceleration monitoring
- 6 channels of wind
- 10 channels of strain
- 4 channels of displacement

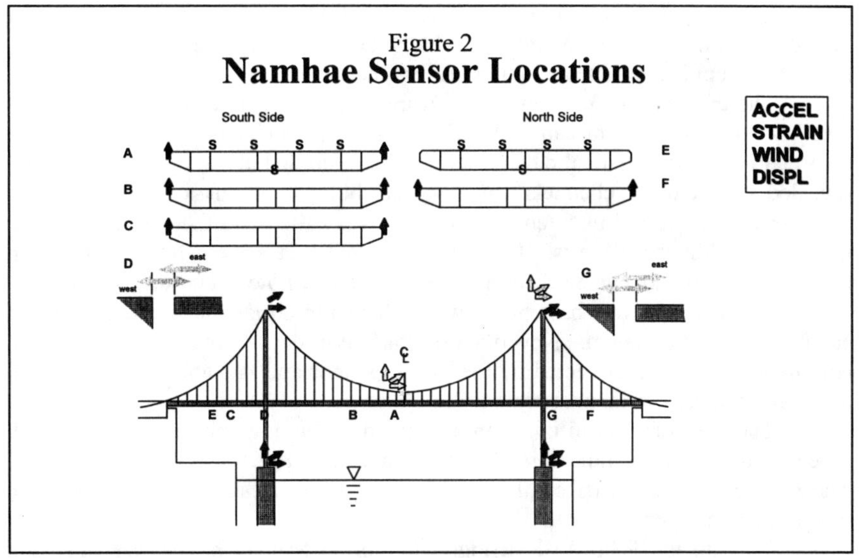

Figure 2
Namhae Sensor Locations

Data is monitored using two Kinemetrics' Mt. Whitneys, one at the base of each tower. Event data is recorded using a PCMCIA recording module. Real time data is transmitted via fiber-optic cable to the operations center located at the North side of the bridge. A desk top computer with a large color monitor accepts both data streams, processes the data using Agbabian Associates OASIS software and displays the on-line data on a single screen. The display looks something like Figure 2. As with the RAMA IX system, real-time waveforms can be accessed by double-clicking any sensor icon.

COMPARISON OF PERFORMANCE AND EXPERIENCE

RAMA IX

The RAMA IX Bridge Monitoring System has functioned continuously since June of 1995 with few disruptions. One loss of function was due to worn bearings in the wind anemometers. This failure, which occurred in October of 1996, was predicted but occurred sooner than expected. It affected only the horizontal wind channels. The bearings were replaced from parts in stock. The second was due to construction on the bridge which temporarily closed down the communication lines between the Mt. Whitney and the remote operations center. The Mt. Whitney continued to provide all other functions including automatic threshold triggering and event recording. Only the real-time display was affected.

This disruption serves to illustrate one of the principle advantages of the OASIS approach. There is a useful redundancy inherent in utilizing a multi-tasking system like the Mt. Whitney. In the foreground, data is being acquired, digitized, and filtered, and forwarded in real-time to the OASIS on-line display. In the background, the Mt. Whitney performs the traditional functions of an event recorder, continuously evaluating data. In the event that the threshold is exceeded, the entire event is recorded including prevent memory.

The third disruption occurred in April of 1997. This was characterized by "red" or secondary alarms caused by high offsets on all channels. Analysis of the recorded logs and statistical data (automatically recorded at the operations center) revealed a pulse on all channels which was attributed to an unknown power event. The local support and maintenance operator was called and the system restarted normally within three hours. It should be noted that system reboot, and even updating of firmware can be accomplished from the remote operations center.

The system owner has until now not elected to utilize the health monitoring capabilities of the system. Nevertheless, long term data (more than 30 minutes, continuous) has been collected by the local operator and Agbabian Associates on an approximately semiannual basis for future study.

The wisdom of utilizing an easily understood graphical interface with green, yellow, and red indicators of system status has proven itself again and again. The reliance of the owner on the monitoring system became evident during the recent spurious event in April.

To date the RAMA IX Bridge Monitoring System has successfully and reliably met every requirement by the owner. It provides and excellent base for future expansion.

Namhae Great Bridge

Basic elements of the Namhae Great Bridge Monitoring System have been installed since December 1996 including event recording and most of the sensor channels. The system was finally completed in May of 1997.

The system is more complex than the RAMA IX system on several levels. First, the basic number of channels are more than doubled. Second, for the first time fiber optic communications were used to transmit the data from the bridge to the operations center. This greatly improved the quality of the communications. Third, the type of data being recorded is more complex, including wind at two locations, displacement between the deck and the towers, and dynamic strain in the steel deck.

Of these the strain data is perhaps the most important for future evaluation of the bridge. One of the purposes of the system was to utilize static and dynamic strain gages to collect data on the fatigue capacity of the bridge. The static data was recorded at low data rates using a separate recording system which could be accessed off-line using OASIS communications. The dynamic strain data recorded by the system was needed for evaluating the long term performance of the orthotropic steel deck for several reasons:

- Need confirmation of actual strain levels at critical locations to permit:
  - calibration of structural models
  - predict strain at unmeasured locations
  - develop stress range spectra for fatigue studies
- Continuous strain data is needed for statistical analysis

- Need to develop baseline data for future fatigue studies

One of the results of having more channels was a more complicated display. The graphical approach still proved more useful for presenting information.

NEW ISSUES FOR MODERN SYSTEMS

The principal improvements in modern instrumentation systems in the last five years have been in areas of dynamic range, the multitasking capabilities of acquisition systems, the on-line processing capabilities of acquisition systems which provide on-line evaluation and presentation of data, and the speed and variety of communications for off-site display. OASIS is one system that attempts to utilize these improvements.

Performance Visualization: Graphical or Textual, Interactive, On-Site or Remote?

Modern communications, instrumentation, and software offer a wide variety of data display and visualization, as well as command access for system maintenance, control, and data retrieval. Graphical displays are highly recommended for non-technical operators, and can be had for very little extra cost. Graphical displays also enable rapid evaluation of complex data for more rapid decision making. OASIS offers an interactive graphical display, allowing multifunction, menu driven actions. Figure 4 shows the RAMA IX display using the new 32-bit Windows 95 version of the software.

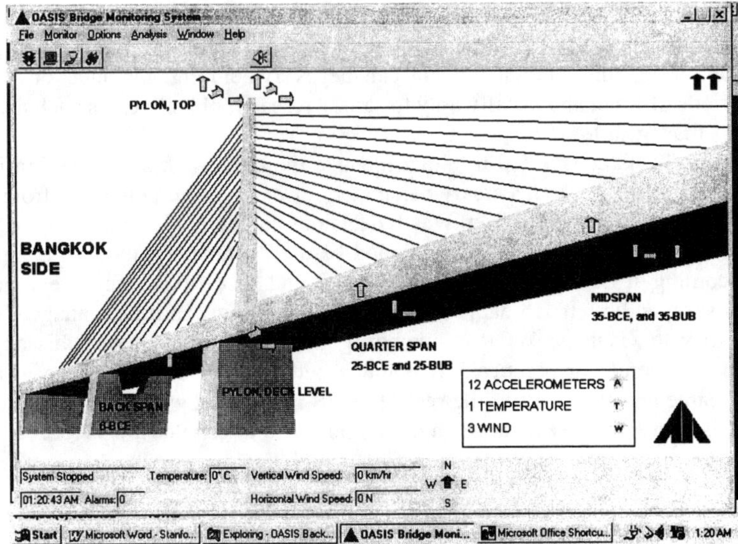

Figure 4 **OASIS GRAPHICAL DISPLAY**

Finally, displays can be local or on-site, or remote, or both. Complex data need not be transmitted off-site. Modern computers can quickly evaluate data and reduce it to simple parameters more useful for management and safety functions. Complete data sets can always be downloaded for further analysis.

Alarms? Visual, Audible, Automatic, Interactive, Remote?

Alarms or warning signals are an important part of the management and safety functions of a monitoring system. Modern computer systems and displays offer a wide variety of signals, including lights, horns, relay outputs, and graphical displays. These signals can be displayed locally on-site, or remotely to multiple locations.

Some alarms need not be continuous. A brief audible and visual signal followed by a printed record is fully adequate to respond to the event. Other alarms are serious and require manual intervention to reset.

The philosophy and logic of these controls is well understood. What is new is the opportunity to combine these systems with complex instrumentation systems in local and wide area networks. Given the enhanced processing capabilities of modern instruments, alarms can now be based on sophisticated on-line analysis for better decision making.

The version of OASIS used for both RAMA IX and Namhae Bridge offers dual-level alerting with icons temporarily changing to Yellow for first stage alerts, and Red for second stage alerts. Red alerts must be automatically cleared. All alarms are printed automatically. All channels have independent thresholds.

Data Management

A large, modern, high speed digital monitoring system can generate gigabytes of data in a short amount of time. Depending on the objectives for monitoring, this amount of data can be overwhelming. In fact, only for short periods of time, and usually only to satisfy research objectives, is such a volume of data manageable.

Tools are needed to either manage the data, such as processing it in real-time, or restrict it to necessary parameters. Data must be converted from numbers into useful information from which decisions can be made. This is especially true for safety applications. As mentioned before, fortunately, modern computers are becoming less expensive, more reliable, and more powerful every day. The question is, how much data should be kept, and how can it be managed? While it is easy to visualize WORM jukebox gigabyte storage systems, this approach is neither useful nor practical. It is much better to decide in advance what data is valuable and throw away the rest. OASIS records only statistics on a continuous basis, and extreme events when preprogrammed thresholds are exceeded.

CONCLUSION

The new monitoring system will have been in successful operation for more than two years by the time this paper is published. The remote real-time data analysis and display provides green/yellow/red-type indication of bridge condition. Recorded high-resolution data are currently being analyzed for initial dynamic fatigue-based condition monitoring. Future augmentations include realizing the long-term health monitoring capabilities for which the system was originally designed.

New Directions in Seismic Monitoring of Multi-Story Buildings

E. ŞAFAK

ABSTRACT

Seismic response of multi-story buildings has customarily been formulated as a vibration poblem in terms of mode shapes, modal frequencies, damping ratios, and participation factors. The same problem can also be formulated as a wave propagation problem in terms of wave travel times, and reflection and transmission coefficients. When compared to the vibration approach, the wave propagation approach is simpler, more accurate, and can account for the energy absorbtion by the soil layers under the foundation. It also provides better tools for identification and damage detection from vibration records, and allows utilization of digital simulation software packages. The standard instrumentation schemes currently used to monitor the response of multi-story buildings to earthquakes do not provide the data necessary to study wave propagation in buildings. The U.S. Geological Survey is currently instrumenting two buildings in Southern California to investigate this problem. The buildings are being instrumented with multiple sensors at every floor, and will be supplemented with synchronized free field and downhole sensors in the near future. Records from these buildings will provide the crucial data to investigate wave propagation in buildings.

INTRODUCTION

Seismic monitoring of structures is an important component of research on earthquake engineering. Monitoring involves installing permanent sensors (usually accelerometers) and recorders in selected structures in seismically active urban areas. The two main structural instrumentation programs in the United States are those run by the CDMG (California Division of Mines and Geology) and the USGS (United States Geological Survey).

Research Structural Engineer, U.S. Geological Survey, 1711 Illinois Street, Golden, CO 80401

The standard practice followed by both the CDMG and the USGS when instrumenting multi-story buildings has been to install 10 to 20 sensors, depending on building size, to capture the main vibration modes. The sensors are typically installed at the foundation level, roof level, and a few intermediate levels. A large number of records have been obtained from the instrumented buildings during the recent California earthquakes. These records have provided critical data needed for research and improvements in seismic design codes. However, it is becoming clear that the standard instrumentation schemes currently used in multi-story buildings are not sufficient to investigate some of the important factors that influence building response and damage, such as the interstory drift, soil-structure interaction, nonproportional damping, flexibility of floor slabs, and the effects of nonstructural elements.

Realizing the inadequacy of standard instrumentation, the USGS has recently started designing new structural instrumentation schemes to investigate specific seismic response problems. The two buildings that are currently being instrumented in Southern California are examples of this new approach. The objective in these two instrumentation is to investigate the propagation of seismic waves in the buildings. As will be discussed below, the wave propagation approach provides a more complete assessment of the seismic response of buildings than the classical vibration approach.

WAVE PROPAGATION FORMULATION OF SEISMIC RESPONSE

The standard approach to calculate seismic response of multi-story buildings has been to formulate it as a vibration problem in terms of mass, damping, and stiffness matrices, and then solve the resulting equations by using modal analyses or time integration techniques. Although the vibrations of a structure during an earthquake are a direct consequence of the propagation of seismic waves in the structure, the wave propagation approach has not been widely used to calculate seismic response, except for structures that can be approximated as a continuous medium (e.g., Clough and Penzien, 1975; Uzgider and Aydoğan, 1986; Todorovska and Lee, 1989; Ivan, 1997).

To show the propagation of seismic waves in a multi-story building, we plot in Figure 1 the accelerations recorded in a 52-story building during the Northridge earthquake (Shakal, et al., 1994). The accelerations on each floor are fully synchronized. If we follow a characteristic peak in the time histories, such as the one marked by 'o' in the figure, we clearly see the time delay of the peak's location as the seismic wave moves up the building. The time shifts represent the travel times of the waves between the floors. There is a 1.2 second time shift from the basement to the top of the building, corresponding to the total travel time of a seismic wave in the building.

For multi-story buildings founded on layered soil media and subjected to vertically propagating seismic shear waves, we can formulate the response as a one-dimensional wave propagation problem. We consider the building as an extension of the layered media, and model each story of the building as another layer in the wave propagation path. The only difference between the soil and the building layers (including the foundation) is that the layer interfaces in the building have concentrated masses because of the floor and the foundation slabs. For simplicity, we assume that the building does not slide or rock, and the wave transmission from ground to the building can be approximated by a one-dimensional model. Figure 2 shows a multi-story building resting on a multi-layered soil medium. The seismic waves propagating vertically through the soil layers and the building are composed of the upgoing and downgoing waves as shown in the figure. Since all the records are given at discrete times, it would be appropriate to derive the equations of wave propagation in the discrete-time domain. To do this, we specify two auxiliary variables at each layer, the discrete amplitude of the upgoing wave at the top of the layer, $u(t)$, and the discrete amplitude of the downgoing wave at the bottom of the layer, $d(t)$. They are shown in Figure 3 for three successive layers, layers $j-1$, j, and $j+1$. The upgoing wave $u_j(t)$ in layer j is composed of the reflected portion of the downgoing wave from the bottom of the same layer, plus the transmitted portion of the upgoing wave from the layer below. With the notation given in Figure 3, we can express this relationship by the following equation:

$$u_j(t) = R_{d,j-1} \cdot d_j(t-\tau_j) + T_{u,j-1} \cdot u_{j-1}(t-\tau_j) \tag{1.a}$$

where $R_{d,j-1}$ and $T_{u,j-1}$ are the reflection coefficient for the downgoing waves and the transmission coefficient for the upgoing waves, respectively, at interface $j-1$. τ_j denotes the one-way travel time of the waves in layer j. Similarly, we can express the downgoing wave $d_j(t)$ in layer j as the sum of the reflected portion of the upgoing wave from the top of the same layer, plus the transmitted portion of the downgoing wave from the layer above. This corresponds to the following equation:

$$d_j(t) = R_{u,j} \cdot u_j(t-\tau_j) + T_{d,j} \cdot d_{j+1}(t-\tau_j) \tag{1.b}$$

where $R_{u,j}$ and $T_{d,j}$ denote the reflection coefficient for the upgoing waves and the transmission coefficient for the downgoing waves, respectively, at interface j. Equations $1.a$ and b are valid for all intermediate soil and building layers. For the first and the top layers, we modify the equations slightly to incorporate the boundary conditions. It can be shown that the equations for the first layer (i.e., the layer overlying bedrock) are:

$$u_1(t) = R_{d,0} \cdot d_1(t-\tau_1) + T_{u,0} \cdot x_0(t-\tau_1) \tag{2.a}$$

$$d_1(t) = R_{u,1} \cdot u_1(t-\tau_1) + T_{d,1} \cdot d_2(t-\tau_1) \tag{2.b}$$

where $x_0(t)$ is the incident seismic wave at the bedrock-soil interface. The equations for the top layer (layer $m+N$) are:

$$u_{m+N}(t) = R_{d,m+N-1} \cdot d_{m+N}(t - \tau_{m+N}) + T_{u,m+N-1} \cdot u_{m+N-1}(t - \tau_{m+N}) \tag{3.a}$$

$$d_{m+N}(t) = R_{u,m+N} \cdot u_{m+N}(t - \tau_{m+N}) \tag{3.b}$$

The wave amplitudes at layer interfaces are calculated as the sum of the upgoing and downgoing waves at the bottom of the layer above the interface (or at the top of the layer below the interface). For example, the amplitudes at the top, $y_{j,top}$, and the bottom, $y_{j,bot}$, of interface j are:

$$y_{j,top}(t) = u_{j+1}(t - \tau_{j+1}) + d_{j+1}(t) \tag{4.a}$$

$$y_{j,bot}(t) = u_j(t) + d_j(t - \tau_j) \tag{4.b}$$

The continuity of motions at the interface requires that $y_{j,top}(t) = y_{j,bot}(t)$. The wave amplitude at the bottom of the top floor slab can be calculated from Eq. $4.b$ by putting $j = m + N$. The wave amplitude at the top of the top floor slab is:

$$y_{m+N,top}(t) = T_{u,m+N} \cdot u_{m+N}(t) \tag{5}$$

If the sampling interval of the input bedrock motion $x_0(t)$ is small enough such that the travel times τ_j in the layers are integer multiples of the sampling interval, the above equations all become finite-difference equations. For a given incident wave, reflection and transmission coefficients, and the travel times the equations can be solved recursively starting from the lowest layer and continuing upward. Therefore, to describe the soil-building system, all we need are the wave travel times at each layer, and the reflection and transmission coefficients at each interface. The damping in the soil and the building can be incorporated in the calculations either in the form of time- or frequency-domain filters, or by making the velocities in the layers complex-valued quantities (Şafak 1995, 1997*a*).

The wave travel times in the layers are calculated by simply dividing the layer thickness with the wave velocity. The wave velocities in building layers are related to the story masses and stiffnesses. If we choose to incorporate damping by making the velocities complex, the wave travel times in the layers also become complex. To derive the equations for the reflection and transmission coefficients at layer interfaces, we subject the interface to an upgoing (or a downgoing) incident wave of a specified frequency and unit amplitude. We calculate its reflected and transmitted components by using the conditions that the wave amplitudes above and below the interface are equal, and the shear forces are in equilibrium. Since the interfaces at the foundation and building floors have concentrated masses (the foundation and floor slabs), the inertia force acting on the mass should be included in the shear equilibrium equations of these interfaces. The effect of a concentrated mass at an interface is similar to that of a low-pass filter; it transmits the waves unchanged at low frequencies while strongly attenuating the high frequencies. Explicit forms of the reflection and transmission coefficients for different types of interfaces can be found in Şafak (1997a).

DAMAGE DETECTION USING WAVE PROPAGATION APPROACH

As stated earlier, the seismic response of multi-story buildings is ordinarily formulated as a vibration problem and the response is expressed in terms of the natural frequencies and the modes of the building. Consequently, the majority of the methods suggested for identification and damage detection are based on the identification and detection of changes in modal characteristics (e.g., mode shapes, modal frequencies, damping ratios, and participation factors). It is well known that the modal parameters are not good indicators for damage, because they are not sensitive to small and local changes in the building's characteristics. In the wave propagation formulation, the seismic response is expressed in terms of the wave travel times between the floors, story dampings, and the wave reflection and transmission coefficients at floor levels. For a given set of records, we can identify these parameters by using the formulation given above.

Let $z_{j,top}(t)$ and $z_{j,bot}(t)$ denote the recorded motions at the top and the bottom of layer j, respectively. Note that $z_{j,top}(t)$ is the motion at the bottom of interface j, and $z_{j,bot}(t)$ is the motion at the top of interface $j-1$. Therefore, using Eq. 4, we can write the following:

$$z_{j,top}(t) = y_{j,bot}(t) = u_j(t) + q^{-\tau_j} d_j(t) \qquad (6.a)$$

$$z_{j,bot}(t) = y_{j-1,top}(t) = q^{-\tau_j} u_j(t) + d_j(t) \qquad (6.b)$$

where q^{-j} denotes the backward-shift operator and is defined as $q^{-j}x(t) = x(t-j)$. We can extract the upgoing and downgoing waves from Eq. 6 as

$$u_j(t) = \frac{z_{j,bot}(t) - q^{-\tau_j} z_{j,top}(t)}{1 - q^{-2\tau_j}} \qquad (7.a)$$

$$d_j(t) = \frac{z_{j,top}(t) - q^{-\tau_j} z_{j,bot}(t)}{1 - q^{-2\tau_j}} \qquad (7.b)$$

To calculate the travel times τ_j in the layers, we determine the travel time of a characteristic phase through each floor. It should be noted here that in order to have a precise estimate of travel times we need to record the motions at a high sampling rate. Also, in order to determine the travel time in a layer we need to have records both at the top and the bottom of the layer. Once the travel times are determined, we can extract the upgoing and downgoing waves by using Eq. 7. For noisy data, $u_j(t)$

and $d_j(t)$ should not be calculated directly from Eq. 7. Instead, we can estimate them in the least-squares sense from Eq. 6. Knowing u_j and d_j in every layer, we can estimate the reflection and transmission coefficients by using Eq. 1. The detail of these calculations can be found in Şafak (1997*b*).

For damage detection, we look for changes in the propagation parameters, namely the wave travel times, and the reflection and transmission coefficients. Experiments have shown that the wave propagation parameters are far superior to vibration parameters in detecting damage, because they are more robust and more sensitive to damage. (Şafak, 1997*b*).

BUILDINGS INSTRUMENTED TO STUDY WAVE PROPAGATION

To study the propagation of seismic waves in buildings, the USGS is currently instrumenting two multi-story buildings in Southern California. The first building is the 17-story, steel-frame Health Sciences Building at UCLA in Los Angeles. The building is being instrumented with 72 accelerometers, four horizontal at every floor plus four vertical at the basement level (Figure 4). The second building is the 10-story, reinforced-concrete, frame/shear-wall Millikan Library at Caltech in Pasadena. The building is being instrumented with 36 accelerometers, three horizontal at every floor plus three vertical at the basement level (Figure 5). Both instrumentation projects will be completed by the end of 1997. We plan to enhance the instrumentation in both buildings by adding synchronized ground and downhole sensors in the near future, so that we can track not only the propagation of waves in the buildings but also the transmission of waves from ground to the buildings.

SUMMARY AND CONCLUSIONS

The wave propagation approach provides a superior alternative to the commonly used vibration approach when studying the seismic response of multi-story buildings. The wave propagation approach is: (1) simpler to implement, (2) more accurate at high frequencies, (3) can incorporate damping more accurately, (4) can account for the energy absorbtion by the soil layers under the foundation, (5) provides better tools for identification and damage detection from vibration records, and (6) allows utilization of digital simulation software packages.

The standard instrumentation schemes currently used for multi-story buildings do not provide the data necessary to study the propagation of seismic waves in buildings. The USGS is currently instrumenting two multi-story buildings in Southern California to investigate this problem. The buildings are being instrumented with multiple sensors at every floor, and will be suplemented with syncronized free field and downhole sensors in the near future. Records from these buildings are expected to provide the data necessary to investigate wave propagation in buildings.

REFERENCES

Clough, R.W. and Penzien, J., 1975, "*Dynamics of Structures*, McGraw-Hill Book Co., New York, NY.

Ivan, W.D., 1997, "Drift spectrum: Measure of demand for earthquake ground motions",*Journal of Structural Engineering*, ASCE, Vol.123, No.4, April, p.397.

Şafak, E., 1995, "Discrete-time analysis of seismic site amplification",*Journal of Engineering Mechanics*, ASCE, Vol.121, No.7, July, p.801.

Şafak, E., 1997*a*, "Wave-propagation formulation of seismic response of multi-story buildings", submitted to the ASCE *Journal of Structural Engineering*.

Şafak, E., 1997*b*, "Identification and damage detection in multi-story buildings using wave propagation approach", in preparation, will be submitted to the ASCE *Journal of Engineering Mechanics*.

Shakal, A., Huang, M., Darragh, R., Cao, T., Sherburne, R., Malhotra, P., Cramer, C., Sydnor, R., Grazier, V., Maldonado, G., Petersen, C., and Wampole, J., 1994, CSMIP strong-motion records from the Northridge, California earthquake of January 17, 1994, *Report No. OSMS 94-07*, California Department of Conservation, Division of Mines and Geology, Office of Strong Motion Studies, Sacramento, California.

Todorovska, M.I. and Lee, V.W., 1989, "Seismic waves in buildings with shear walls or central core", *Journal of Engineering Mechanics*, ASCE, Vol.115, No.12, December, p.2669.

Uzgider, E., and Aydoğan, M., 1986, "Simple and efficient method for the dynamic response of 2D frames subject to ground motions", *proceedings*, 8th European Conference on Earthquake Engineering, September 7-12, 1986, Lisbon, Portugal, Vol.3, p.6.2/17.

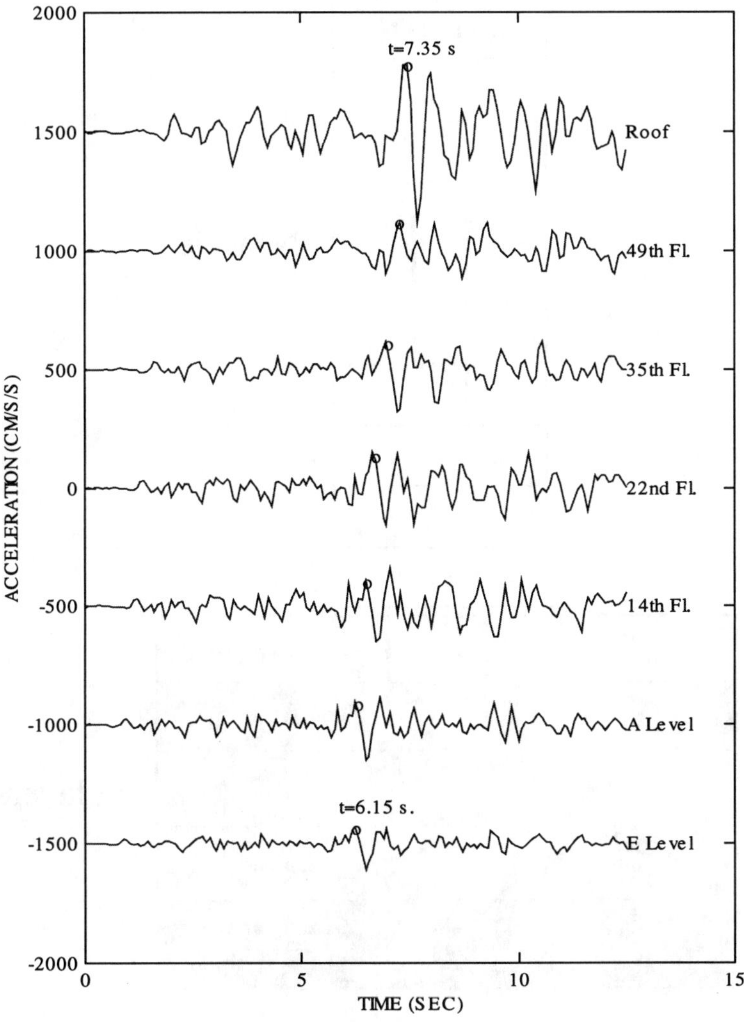

Figure 1 - Recorded Northridge accelerations of a 52-story high-rise building.

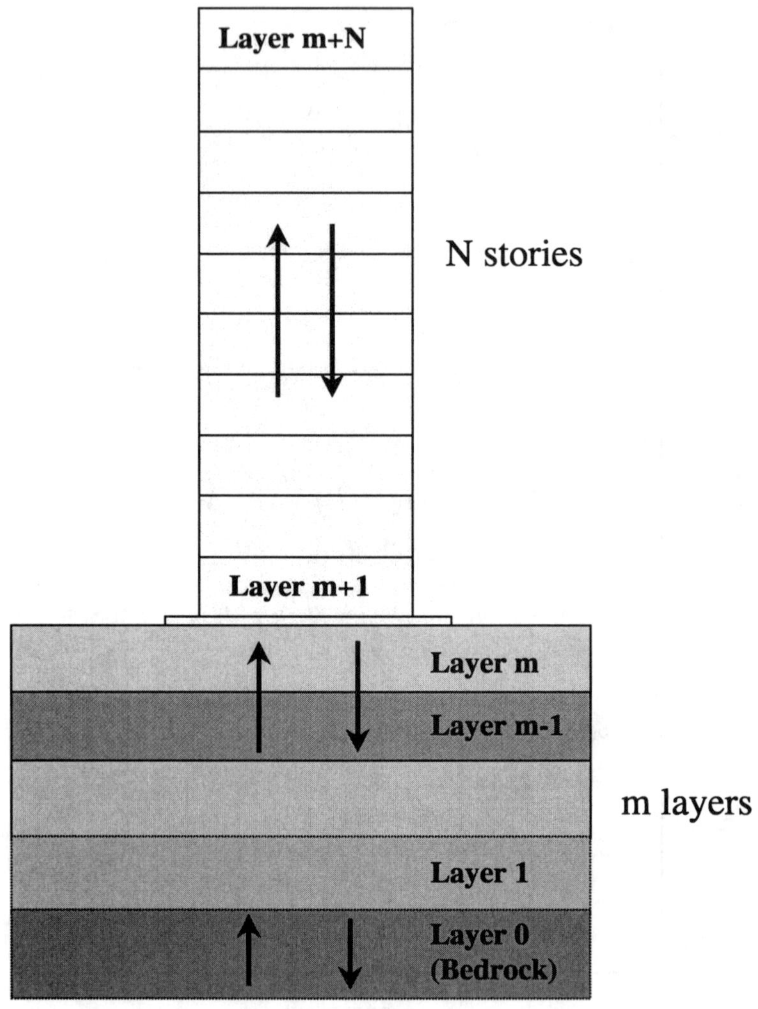

Figure 2 - A multi-story building on a layered soil medium with upgoing and downgoing seismic waves.

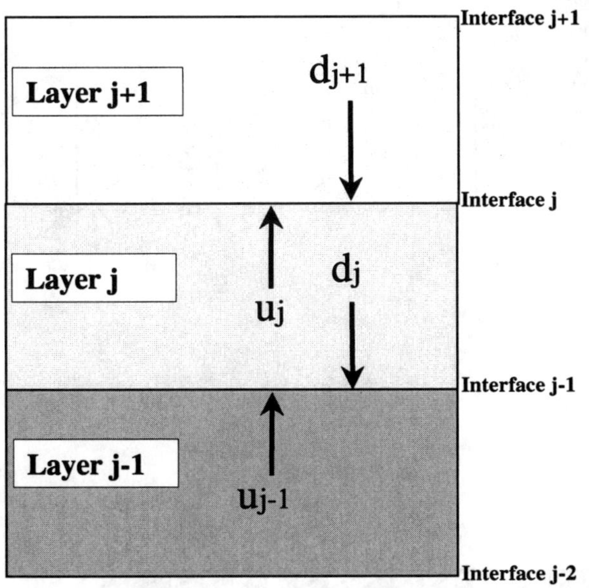

Thickness of the layer	=	h_j
Velocity in the layer	=	V_j
Travel time in the layer	=	$\tau_j = h_j/V_j$

Figure 3 - Three successive layers with upgoing and downgoing waves.

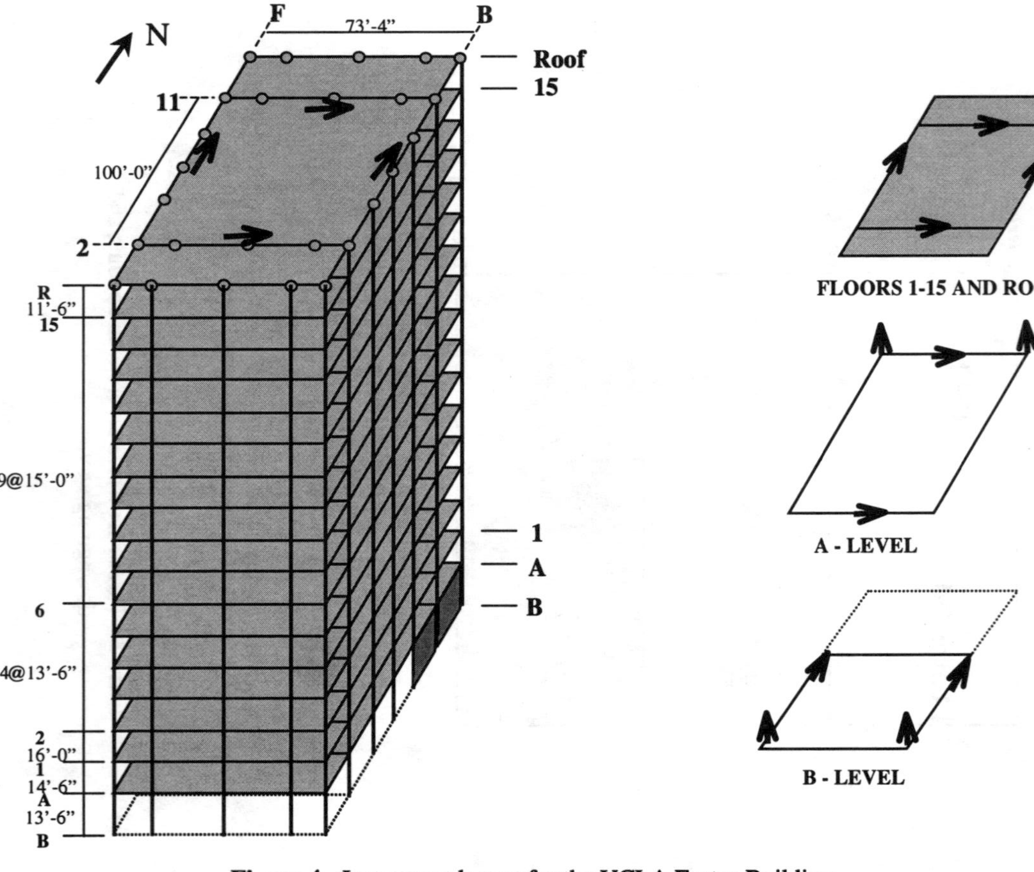

Figure 4 - Instrument layout for the UCLA Factor Building

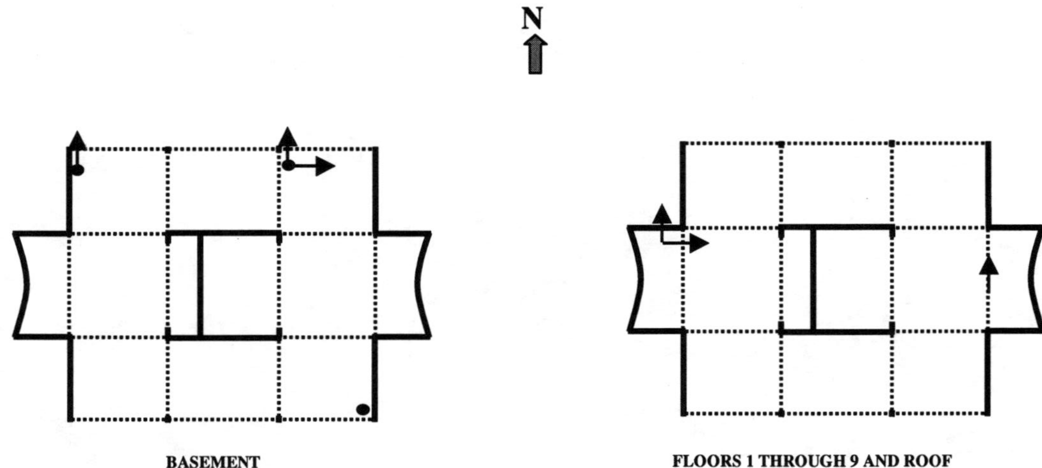

Figure 5 - Instrumentation layout for the Caltech's Millikan Library.

SESSION 8

MODELING AND DIAGNOSTIC METHODS III

Electro-Mechanical (E/M) Impedance Method for Structural Health Monitoring and Non-Destructive Evaluation

V. GIURGIUTIU[1] and C. A. ROGERS[2]

ABSTRACT

This paper presents the electro-mechanical (E/M) impedance method for structural health monitoring and non-destructive evaluation. The precursors to this method have been developed over a period of several years by Professor Craig A. Rogers and a number of his Ph.D. students and research collaborators, but a unified approach has only recently been established. Section 1 of the paper gives a brief listing of the precursor work and the related research. Section 2 presents a description of the E/M technique, indicating the basic equipment, the conceptual system diagram and typical impedance response and damage index results. Section 3 deals with the main physical ingredients of the E/M technique, i.e., the direct and converse piezoelectric effects, and the elastic wave propagation through an elastic medium. Formulae for the electro-mechanical impedance and for the damage index are briefly presented. Section 4 presents the advantages and special characteristics of the E/M impedance method, and compares it with other NDE techniques such as vibrations, ultrasonics, eddy currents, E/C impedance, etc. Section 5 shows a number of experimental proof-of-concept demonstrations that illustrate the applicability of the E/M technique to a large number of engineering areas. The paper finishes with conclusions and a comprehensive list of references.

1. INTRODUCTION

The structural-integrity health-monitoring for detection of incipient damage in aging structures remains a major preoccupation of the engineering community. During scheduled maintenance and overhaul of a variety engineering structures ranging from bridges, to aircraft, Navy ships and nuclear reactors, a battery of damage detection methods are applied to assess the structural health. However, none of these NDE techniques can perform continuous monitoring of the structure using non-intrusive small transducers that could be remotely interrogated in an automated mode. The electro-mechanical impedance technique aims to do just this.

The precursors to the electro-mechanical (E/M) impedance technique for structural health monitoring and non-destructive evaluation were developed over a

[1]Associate Professor, Department of Mechanical Engineering, University of South Carolina, Columbia, SC 29208.

[2]Professor and Dean, College of Engineering, University of South Carolina, Columbia, SC 29208.

period of several years by Professor Craig A. Rogers with the assistance of a number of Ph.D. students and research collaborators. Liang, Sun, and Rogers (1993), and Giurgiutiu, Chaudhry and Rogers (1994) described the coupling between the electrical and mechanical impedance of a piezoelectric transducer affixed to an elastic structure. Chaudhry, Sun, and Rogers (1994) and Sun, Chaudhry, Liang and Rogers (1995) use impedance measurements on a model space-truss to identify damage. Sun, Chaudhry, Rogers, and Majmundar (1995) used switching electronics and real-time computer data processing to interrogate several transducers placed on the space-truss model. Chaudhry, Joseph, Sun and Rogers (1995) studied local-area health monitoring on a tail-fuselage aircraft junction. Chaudhry, Lalande, Ganino, Rogers, and Chung (1995) measure the structural integrity of a composite patch repair specimen. Ayres, Rogers, and Chaudhry (1996) investigated the health monitoring of a ¼-scale steel bridge junction. Childs, Lalande, Rogers, and Chaudhry (1996) studied the NDE of complex precision parts. Estaban, Lalande, Rogers, and Chaudhry (1996) attempted the modeling of wave-propagation in bolt-jointed 1-D bars.

Other investigators used different approaches for structural health monitoring. Keilers and Chang (1995) identified delamination in composite beams using an array of PZT wafers, some acting as actuators, others as sensors. Lakshmanan and Pines (1997) used wave propagation models to detect transverse cracks in a rotating composite beam. Saravanos, Birman, and Hopkins (1996) proposed the use of surface mounted piezoelectric transducers to detect changes in local curvature of a composite beam under sub-resonance excitation. Moetakef, Joshi, and Lawrence (1996) considered the generation of elastic waves through piezoceramic patches.

2. DESCRIPTION OF THE ELECTRO-MECHANICAL IMPEDANCE TECHNIQUE

The electro-mechanical impedance technique utilizes the direct and the converse electro-mechanical properties of piezoelectric (PZT) materials, allowing for the simultaneous actuation and sensing of the structural response. The variation of the electro-mechanical impedance of a PZT sensor-actuator interrogator intimately bonded to the structure is monitored over a large frequency spectrum in the high kHz frequency band (Figure 1). The basic ingredient for the method are: an

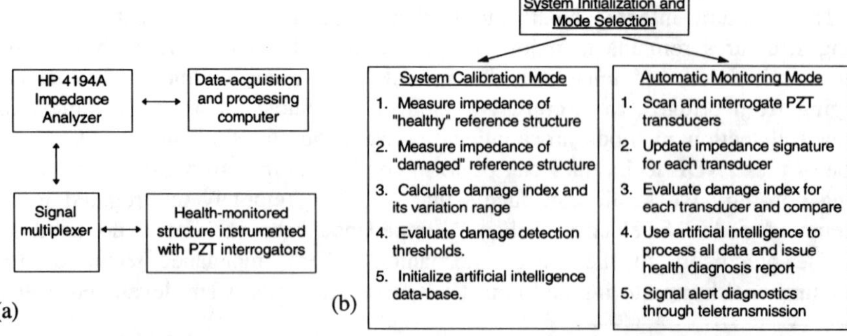

Figure 1 Principles of the electro-mechanical impedance technique: (a) Schematic diagram of the experimental set-up; (b) conceptual diagram of the automated monitoring system.

array of piezo-electric PZT sensor-actuators applied to the monitored structure, and a high-precision impedance analyzer coupled to a data-acquisition computer. The size of the PZT sensor-actuators is typically small (less than 0.5 sq. in., 0.01 in. thick), allowing for non-intrusive installation in the monitored structure. The PZT sensor-actuator is excited electrically by high-frequency low-power voltage, and its complex electro-mechanical impedance is measured over a given frequency range. Figure 2a shows the frequency response diagrams obtained during a typical laboratory demonstration of the electromechanical impedance technique. Figure 2b presents a typical damage index diagram that distinguishes between damaged and undamaged locations during the health-monitoring process.

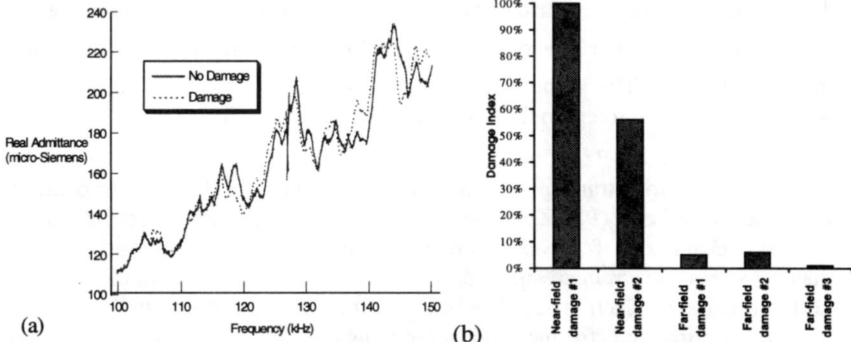

Figure 2 Typical damage detection results: (a) impedance signature pattern; (b) damage index comparison (after Chaudhry, Joseph, Sun and Rogers, 1995).

The high resolution of this technique is ensured through the intimate electro-mechanical coupling between the electrical impedance response of the piezo-electric sensor-actuator affixed to the structure and the local mechanical impedance of the adjacent material present in the structure and in the structural joints. Localization of the sensing area ensures sensitivity of the impedance signature only to damage and/or structural changes in the near-field area of the PZT sensor-actuator. The insensitivity of the methods to far-field changes ensures rejection of the unwanted far-field information and prevents the method from giving "false alarms" due to changes that are part of the normal structural usage (changes in boundary conditions and mass distribution, service loads, etc.).

The electro-mechanical impedance technique utilizes well-developed equipment that is currently available for high-frequency accurate measurements of electronic and electro-chemical impedance. This aspect is a significant advantage for quickly bringing this NDE technique to wide-spread practical implementation.

3. PHYSICAL MECHANISM OF THE ELECTRO-MECHANICAL IMPEDANCE TECHNIQUE

The electro-mechanical impedance technique relies on two main physical principles: (a) the piezo-electric material coupling between mechanical and electrical fields that permits the interrogation of the structural health through electric means; and (b) the elastic wave propagation that allows the interrogation signal to travel into the structure. A brief review of these aspects is given next.

3.1 PIEZO-ELECTRIC TRANSDUCERS

3.1.1 Basic Equations of Linear Piezo-Electric Material Behavior

The general constitutive equations of linear piezo-electric material behavior, given by ANSI/IEEE Standard 176-1987, describe a tensorial relation between mechanical and electrical variables (mechanical strain S_{ij}, mechanical stress T_{ij}, electrical field E_i, and electrical displacement D_i) in the form:

$$\begin{aligned} S_{ij} &= s^E_{ijkl} T_{kl} + d_{kij} E_k \\ D_j &= d_{jkl} T_{kl} + \varepsilon^T_{jk} E_k \ , \end{aligned} \qquad (1.)$$

where s^E_{ijkl} is the mechanical compliance of the material measured at zero electric field ($E=0$), ε^T_{jk} is the dielectric permittivity measured at zero mechanical stress ($T=0$), and d_{kij} is the piezo-electric coupling between the electrical and mechanical variables. The second equation reflects the *direct piezo-electric effect*, while the first equation refers to the *converse piezo-electric effect.*

3.1.2 Surface -Mounted PZT Wafer Transducers

The piezo-electric transducers used in the electro-mechanical impedance technique are thin Lead (Pb) Zirconate Titanate (PZT) ceramic wafers intimately bonded to the surface of the host structure (Figure 3a). In this configuration, mechanical stress and strain are applied in the 1 and 2 directions, i.e. in the plane of the surface, while the electric field acts in the 3 direction, i.e., normal to the surface. Hence, the significant electro-mechanical couplings for this type of analysis are the 31 and 32 effects. The application of an electric field, E_3, induces surface strains, S_{11} and S_{22}, and vice-versa.

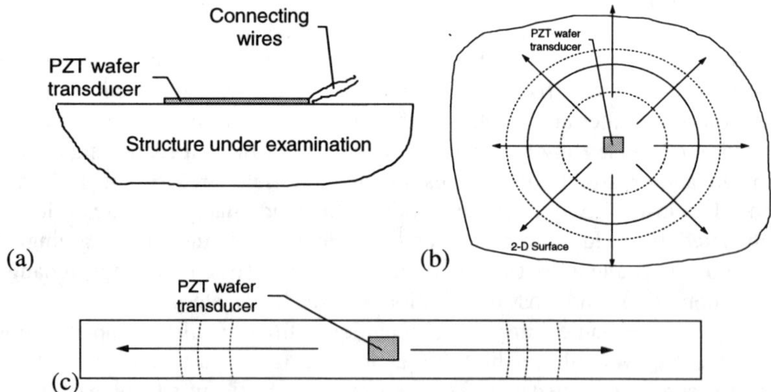

Figure 3 Details of the E/M impedance technique: (a) PZT wafer transducer affixed to the host structure; (b) PZT wafer transducer acting on a 2-D surface;(c) PZT wafer transducer acting on a 1-D structure.

For a PZT transducer affixed to 1-D member, e.g., a beam along the 1-direction (Figure 3c), the analysis is mainly one-dimensional. In this case, the dominant electro-mechanical coupling constant is d_{31}. If the transducer is placed on a 2-D the surface, the analysis is, in principle, two-dimensional (Figure 3b). Since the electro-mechanical coupling constants, d_{31} and d_{32}, have essentially same value, radial symmetry can be applied, and the analysis can be reduced to a one-dimensional case in the radial coordinate, *r*.

3.1.3 Driving-Point Impedance

The effect of the piezo-electric transducer bonded to the structure surface is to apply a local strain parallel to the surface that creates strain waves in the structure. In response, the structure presents to the transducer a local drive-point mechanical impedance, $Z_{str}(\omega) = i\omega m_e(\omega) + c_e(\omega) - ik_e(\omega)/\omega$. Through the coupling between the mechanical and electrical effects taking place in the transducer, the structural impedance modifies the effective electrical impedance of the piezo-electric transducer. The electro-mechanical impedance technique for health monitoring and NDE utilizes the changes that take place in the driving point impedance to identify incipient damage in the structure.

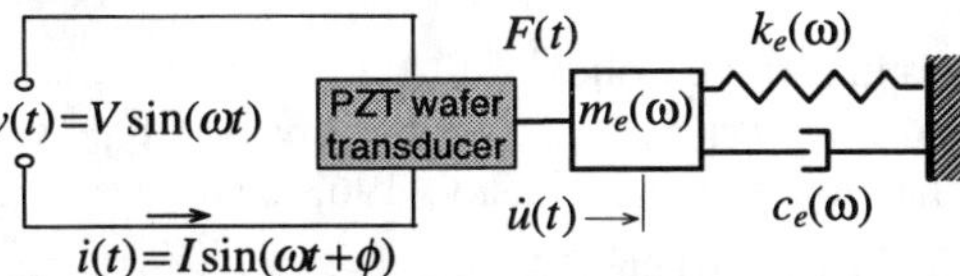

Figure 4 Electro-mechanical coupling between the PZT transducer and the structure.

The driving point impedance change is sensed electrically through the electro-mechanical coupling of the piezo-electric transducer,

$$Y(\omega) = i\omega C\left(1 - \kappa_{31}^2 \frac{Z_{str}(\omega)}{Z_{PZT}(\omega) + Z_{str}(\omega)}\right), \quad (2.)$$

where $Y(\omega) = 1/Z(\omega)$ is the equivalent electro-mechanical admittance, and κ_{31} is the electro-mechanical cross-coupling coefficient, i.e., $\kappa_{31} = d_{13}/\sqrt{\bar{s}_{11}\bar{\varepsilon}_{33}}$.

3.1.4 Complex Impedance Spectrum and Scalar Damage Index

The complete application of the electro-mechanical impedance method is performed by scanning a predetermined frequency range in the hundreds of kHz range and recording the complex impedance plots. By comparing these plots for at various intervals during the service of a structure, meaningful interpretation can be extracted pertinent to the modifications taking place in the structure and the appearance of defects. The frequency range has to be high enough for the wavelength to be significantly smaller than the defect size.

A qualitative estimation of the structural health can be rapidly achieved through the *damage index*. The damage index is a scalar quantity that serves as a metric of the damage taking place in the structure. A convenient damage index is:

$$M = \sum_N \left[\mathrm{Re}(Y_i^1) - \mathrm{Re}(Y_i^0)\right]^2, \quad (3.)$$

where *N* is the number of sample points in the spectrum, and the upperscripts 0 and 1 signify the initial and the present state of the structure.

3.2 WAVE PROPAGATION IN ELASTIC MEDIA

A wave propagating along the *x*-axis follows the differential equation

$$\partial^2 u / \partial t^2 = c^2 \partial^2 u / \partial x^2, \quad (4.)$$

where u is a generic disturbance, and c is the wave speed. The general solution of Equation (1) is given in the form (Graff, 1991):

$$u(x,t) = Ae^{i(\gamma x - \omega t)} + Be^{i(\gamma x + \omega t)}, \quad (5.)$$

where ω is the circular frequency and γ is the wave number. The first term in Equation (2) signifies an advancing wave, while the second term represents a retreating wave. The constants A and B are determined from the initial conditions. If the medium is bounded, boundary conditions are also imposed, and wave reflection and refraction takes place at the boundary. Equation (2) can also be expresses as

$$u(x,t) = Ae^{i2\pi(\frac{x}{\lambda} - \frac{t}{T})} + Be^{i2\pi(\frac{x}{\lambda} + \frac{t}{T})}, \quad (6.)$$

where T is the period of oscillation ($T = 1/f$) and λ is the wavelength ($\lambda = cT$).

3.2.1 Elastic Waves in Infinite Solids

In an unbounded 3-D elastic solid, two basic waves type can exist: dilatational and rotational. Dilatation waves, (Kolsky, 1963), are described by the equation $\rho \partial^2 \Delta / \partial t^2 = (\lambda + 2\mu) \nabla^2 \Delta$, where $\Delta = \varepsilon_{xx} + \varepsilon_{yy} + \varepsilon_{zz}$ is the dilatation of the medium, λ is Lame's constant, μ is the shear modulus (a.k.a. G), and ρ is the mass density. (The constants λ and μ can be related to the Young's modulus, E, and Poisson's ratio, ν, by $\lambda = \nu E / [(1-2\nu)(1+\nu)]$ and $\mu = E / [2(1+\nu)]$. The propagation speed of dilation waves is $c_\Delta = \sqrt{(\lambda + 2\mu)/\rho}$ (Table 1). In seismic studies, dilatational wave are called P-waves, where "P" stands for "principal" or "pressure".

Table 1 Propagation velocities c_Δ, c_ω, and c_R for various materials (adapted after Graff, 1991, pp.278)

Material	Dilatational wave speed c_Δ		Rotational wave speed c_ω		Rayleigh wave speed c_R	
	m/sec	ft/sec	m/sec	ft/sec	m/sec	ft/sec
Aluminum	6,150	20,177	3,100	10,171	5,705	18,718
Gold	3,140	10,302	1,170	3,839	2,913	9,557
Steel	5,710	18,734	3,160	10,367	5,289	17,351
Concrete	2,425	7,956	1,617	5,305	2,165	7,103

Rotational waves are described by $\rho \partial^2 \vec{\omega} / \partial t^2 = 2\mu \nabla^2 \vec{\omega}$,, where $\vec{\omega}$ is the rotation vector ($\omega_1 = \partial u_3 / \partial u_2 - \partial u_2 / \partial u_3$, etc.). The propagation speed of rotational waves is $c_\Delta = \sqrt{2\mu / \rho}$ (Table 1). Rotational waves corresponds to incompressible distortion of the solid, e.g., shear, and are often referred to as distortion waves or shear waves. Distortional waves can be polarized in a preferred plane. In seismic studies, distortion waves are called as S-waves, where "S" stands for "secondary" or "shear". SH-waves signify shear motion taking place in the horizontal plane, while SV-waves refer to the vertical plane.

3.2.2 Elastic Waves in Bounded Solids

When the solid medium is not infinite, two additional aspects have to be considered in connection to the boundary effects: (a) the effect of wave reflection and refraction on the boundary; and (b) the existence of additional wave type closely related to the boundary effects. The main effect of wave reflection at the boundary is that the presence of only one type of wave cannot be maintained. E.g., when a pure pressure wave traveling at an oblique angle hits a free boundary, both pressure and shear waves are generated in the reflection process.

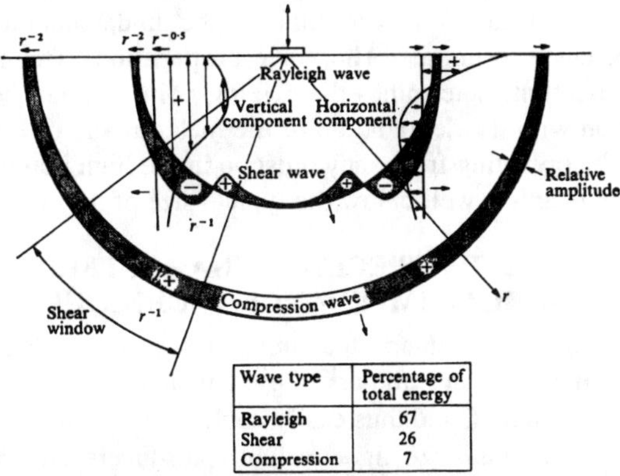

Wave type	Percentage of total energy
Rayleigh	67
Shear	26
Compression	7

Figure 5 Distribution of displacement and energy in dilatational, shear, and Rayleigh waves generated in an elastic half-space by a harmonic normal load (after Woods, 1968).

A free boundary gives rise to new wave types connected with the movement close to the solid surface: Rayleigh waves and Lamb waves. Rayleigh wave amplitude decreases rapidly with depth, and becomes almost zero at a depth of approximately 1.6λ. At the surface, the energy contained in the Rayleigh waves greatly exceeds the energy contained in the shear and pressure waves, as illustrated in Figure 5. This aspect is essential not only in seismology, but also in wave-based NDE, for the detection of defects must be inferred from surface interrogation of the material. The propagation speed of Rayleigh waves is close to that of shear waves, i.e. $c_R / c_2 \approx (0.87 + 1.12\nu)/(1+\nu)$ (Table 1). Love wave are confined to a superficial layer existing on top of a homogeneous solid. Their propagation speed depends on frequency and approaches the shear wave speed in the substrate material for $f \to 0$.

3.2.3 Elastic Waves in Rods, Plates and Shells

The elastic waves propagating in rods can be axial, torsional, flexural and shear. The longitudinal waves involve uniform axial pressure, while torsional waves involve torsional shear. The propagation of axial waves takes place with the velocity $c_0 = \sqrt{E/\rho}$, while the torsional waves propagate with $c_S = \sqrt{G/\rho}$. The flexural waves involve normal stress varying linearly across the cross-section. At relatively low frequencies (e.g., $f < 100$ kHz in a 1-in diameter steel bar), flexure waves are described by the classical Euler-Bernoulli theory of bending, and their wave speed varies linearly with the wave number ($c_F = \gamma\sqrt{EI/\rho A}$). At higher frequencies, Timoshenko's theory of flexural vibrations predicts the simultaneous presence of two types of waves, one associated with bending, the other with shear. At very high frequencies and wave numbers ($\gamma \to \infty$), their propagation speeds become $c_0 = \sqrt{E / \rho}$ and $c_{S'} = \sqrt{\kappa G / \rho}$, respectively, where κ is a cross-section related numerical factor. Similar effects are also present in plates and shell, though the mathematical description of the phenomenon becomes more elaborate. Of special interest are also the Lamb waves which appear in plates just as Rayleigh waves appear in solids.

Elastic wave propagation is essential in the understanding of the electro-mechanical impedance technique. The wave propagation dynamics (transmission, refraction and reflection) determines the driving-point impedance of the structure, and it modification with the degradation of the material and of the structural joints. For this reason, the operating frequency must be in the high hundreds of kHz range, such that the wavelength is well below the typical size of the structural defects.

4. ADVANTAGES AND SPECIAL CHARACTERISTICS OF THE ELECTRO-MECHANICAL IMPEDANCE TECHNIQUE

The electro-mechanical impedance technique has several advantages over other health monitoring methods. In summary, these advantages are:

- Not based on any model, and thus easily applied to complex structures.
- Non-intrusive and small-size light weight transducers that add no significant mass to common structures and can be placed in inaccessible locations.
- Insensitive to unwanted disturbances associated with changes in boundary conditions, loading, or normal operational vibrations.
- Capable of on-line health monitoring, and suitable for continuous monitoring that can replace scheduled depot-based inspections.

Several special characteristics differentiate the electro-mechanical impedance technique from other NDE techniques such as modal analysis, frequency spectrum, ultrasonics, eddy currents, etc., as detailed next.

4.1 VIBRATIONS-BASED TECHNIQUES

The vibration based NDE techniques bear resembles with the present method inasmuch that it uses vibrations to identify damage. Most vibrations-based NDE techniques rely on performing a system vibrations identification of the structure before and after damage. The presence of damage is inferred from subtle modification appearing either in the structural frequencies, or in the modal, stiffness, damping and mass, or in the structural modeshapes. The main drawback of the vibrations approach is that it relies on the measurement of global properties to identify localized changes. Global parameters do not change significantly when small local damage takes place. At low frequencies, small local cracks cannot significantly affect the modal frequency to permit effective detection. And the long wavelength of the low frequency modes across the local crack or damage without sensing it. The presence of small flaws and localized or incipient damage in the structure though it can be critical to the structural safety, may only induce minute changes in the global vibration response of the structure. In contrast, the E/M impedance technique performs damage detection on a local level and can identify minute changes that would pass unobserved by a global vibrations based method. The E/M impedance technique can detect incipient localized damage and is ideal as an early-warning technique.

4.2 ULTRASONICS

Ultrasonic techniques have gained wide acceptance and popularity as NDE methods due to their capability of detecting local damage (Krautkramer, 1990). They rely on elastic waves propagation and reflection within the material, and identifies the field inhomogeneities due to local damage and flaws. The ultrasonic techniques use large transducers, that transmit pulses of elastic waves and interpret the reflection and/or the transmission of these to extract information about the state of the structure. Ultrasonic detection is best suited for through-the-thickness excitation. Since ultrasonic waves cannot be practically induced at right angles to the structural surface, localized surface flaws, and transverse cracks cannot be readily detected. In contrast, the E/M impedance technique uses small inexpensive transducers, permanently bonded to the structural surface. It can produce waves parallel to the surface, and thus detect damage that would escape an ultrasonic method.

4.3 EDDY CURRENTS

Eddy current methods can detect well surface damage and cracks perpendicular to the surface, but it is restricted to conductive material. The eddy current techniques resemble the E/M impedance method inasmuch that they perform stead-state harmonic interrogation of the structure, but they rely on electric and magnetic fields and hence can only be applied to conductive materials. In contrast, the E/M impedance technique is based entirely on elastic waves propagation and can be applied to all common engineering structural materials.

4.4 THE ELECTRO-CHEMICAL (E/C) IMPEDANCE TECHNIQUE

The electro-chemical (E/C) impedance technique bears a phonetic resemblance to the E/M impedance method, but it is based on different principles and has a different domain of application. The E/C impedance method measures the electro-chemical impedance of pure substances and compounds at small and very small frequencies. In contrast, the E/M impedance method measures the electro-mechanical impedance of the structure at high and very frequencies.

5. EXPLORATORY DEMONSTRATIONS OF THE ELECTRO-MECHANICAL IMPEDANCE TECHNIQUE

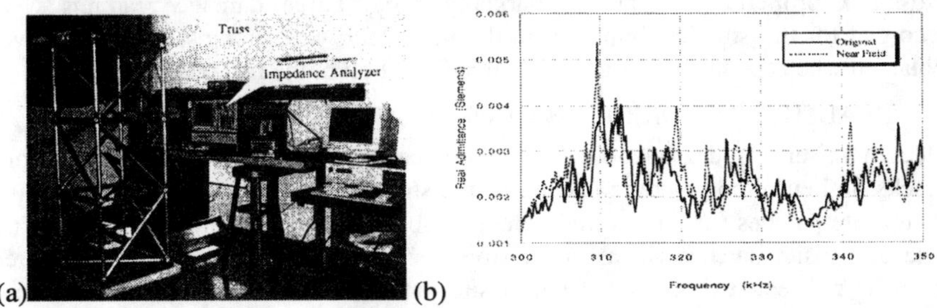

Figure 6 E/M impedance health monitoring of a 3-bay space truss (a) the experimental set-up; (b) E/M impedance frequency spectrum before and after application of near-field damage (Chaudhry, Sun, and Rogers, 1994)

High-frequency electro-mechanical (E/M) impedance measurements offer two clear benefits over vibration-based NDE methods: (a) ensures high resolution to incipient material and structural damage; (b) ensures localization of the actuation/sensing area. A few exploratory experiments are presented next.

5.1 THREE-BAY ALUMINUM TRUSS

Chaudhry, Sun, and Rogers (1994) described the use of the E/M impedance technique for health monitoring a three-bay aluminum truss (Figure 6). The purpose of this experimentation was the on-line implementation of a system using PZT actuator/sensors at multiple critical locations. Small PZT's (approximately 8 x 8 x 0.2 mm) were bonded to the eight nodes of the middle-bay. Damage was simulated by loosening one of the member's connection with the nodal Derlin-ball.

5.2 TAIL-FUSELAGE AIRCRAFT JUNCTION

Chaudhry, Joseph, Sun, and Rogers (1995) used this technique to perform exploratory testing of a typical aircraft joint structure using (Figure 7a). The spectral difference between the electro-mechanical impedance curves can be quantified in a scalar value using a least-squares approach, called damage index.

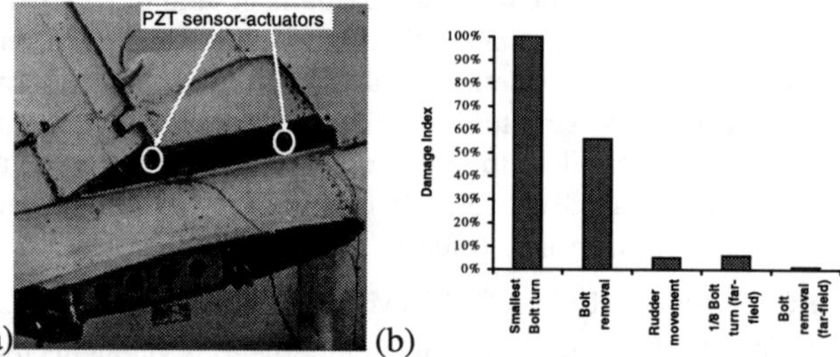

Figure 7 The tail-fuselage bolted junction of a Piper Model 601P airplane: (a) location of PZT interrogators; (b) damage index bar-chart (after Chaudhry, Joseph, Sun, and Rogers, 1995).

Figure 7b shows the damage index measured in the exploratory demonstration of the E/M technique performed on the Piper Model 601P aircraft fin/fuselage bolted junction. It is important to notice that the method is highly sensitive to actual damage, while it is relatively insensitive to other types of changes taking place during the normal operation of the aircraft. The sensing localization and sensor cross-talk properties are also remarkably good. Large damage readings were recorded for the smallest bolt turn in the near-field, while almost no reading was obtained when the same change was applied to a bolt in the far field.

5.3 HEALTH MONITORING OF A COMPOSITE PATCH REPAIR

Due to in-service degradation, a structure can become damaged and develops cracks. In lieu of complete replacement, the structure can be locally repaired using composite patches (advanced high-strength fibers embedded in a polymeric resin). The composite patches act as crack stoppers: the load field around the repaired crack is redirected through the composite patch, and the crack is arrested. In a typical application, cracks developed in an aircraft metallic skin are repaired with carbon/epoxy patches applied with a wet lay-up technique. However, its effectiveness depends to a very large extend on the adhesive bond between the substrate and the composite patch repair. NDE methods for monitoring and early detection of cracks and delaminations in the adhesive interface are essential for the acceptances of this repairs technique. To test the applicability of the E/M impedance technique to the repair health monitoring, an experiment was performed using a 700 mm × 126 mm dog-bone specimen repaired with a 200 mm × 70 mm carbon/epoxy patch interface (Chaudhry, Lalande, Ganino, Rogers, and Chung, 1995). A 10 mm × 10 mm PZT interrogator was used to monitor the crack growth under fatigue loading and to detect disbonds in the patch/substrate (Figure 8).

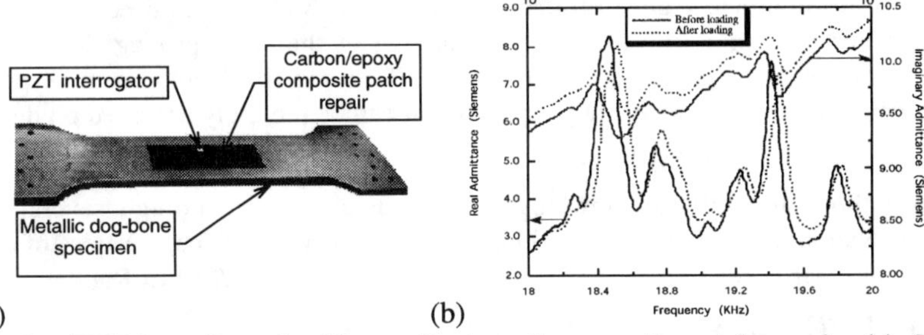

Figure 8 E/M impedance health monitoring of composite patch repair: (a) dog-bone specimen; (b) electro-mechanical impedance plots (after Chaudhry, Lalande, Ganino, Rogers, and Chung, 1995).

5.4 CONCEPTS FOR SPOT-WELDED AND WELD-BONDED STRUCTURES

The newly developed electro-mechanical (E/M) impedance technique is particularly suited for NDE and health monitoring of spot-welded and weld-bonded structures. This novel technique utilizes inexpensive piezo-electric probes that can be permanently affixed to the specimen and interrogated at various time intervals. We propose to use this technique, in conjunction with ultrasonics, for the evaluation of spot-welded and weld-bonded joints(Figure 9).

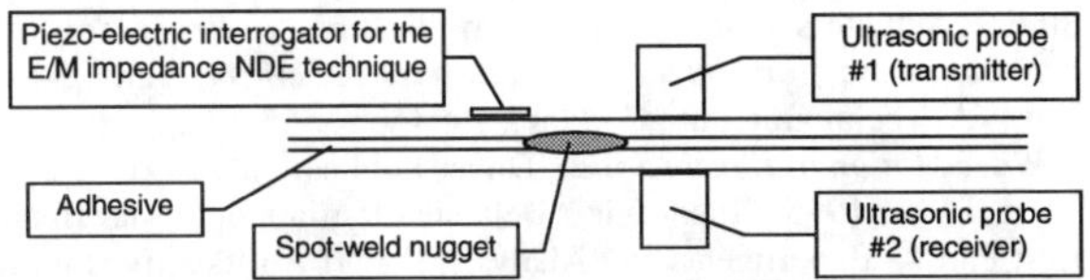

Figure 9 Schematic representation of the proposed NDE technique using the E/M impedance technique and its comparison with conventional ultrasonics.

6. CONCLUSION

The electro-mechanical (E/M) impedance technique for high-frequency structural health monitoring has been presented. The method utilizes the changes in the apparent electrical impedance of a surface mounted piezo-electric transducer (PZT wafer) intimately coupled with the drive point impedance of the host structure. High frequency impedance spectra are measured with an impedance analyzer, while data collection and interpretation is performed with a microcomputer. Because of its high frequency, the technique is very sensitive to incipient damage in the host structure, while the interrogation area is restricted to the transducer vicinity. This sensing/actuation localization provides the capability of monitoring specific critical areas while being insensitive to far field changes. Laboratory tests covered applications from aerospace to civil engineering, composites and fine machinery. Further developments and refinements of the method are required in two main directions: (a) a more profound understanding of the wave-propagation and flaw detection issues on material and structural NDE; and (b) development of portable and deployable impedance analyzer units that can be remotely accessed on demand.

7. ACKNOWLEDGMENTS

The authors gratefully acknowledge the financial support through several grants from National Science Foundation, Army Research Office, Office of Naval Research, Air Force Office of Scientific Research, and NSF/EPSCoR.

8. REFERENCES

Ayres, T., Z. Chaudhry, and C. A. Rogers, 1996. "Localized Health Monitoring of Civil Infrastructure via Piezoelectric Actuator/Sensor Patches," *Proceedings, SPIE's 1996 Symposium on Smart Structures and Integrated Systems*, SPIE Vol. 2719, pp. 123-131.

Chaudhry, Z., F. Lalande, A. Ganino, C. A. Rogers, and J. Chung, 1995. "Monitoring the Integrity of Composite Patch Structural Repair via Piezoelectric Actuators/Sensors," *36th AIAA/ASME/ASCE/AHS/ASC SDM Conference*, New Orleans, LA, pp. 2243-2248.

Chaudhry, Z., F. P. Sun, and C. A. Rogers, 1994. "Health Monitoring of Space Structures Using Impedance Measurements," *Fifth International Conference on Adaptive Structures*, Sendai, Japan, 5-7 December, 1994; pp. 584-591.

Chaudhry, Z., T. Joseph, F. Sun, and C. Rogers, 1995. "Local-Area Health Monitoring of Aircraft via Piezoelectric Actuator/Sensor Patches," *Proceedings, SPIE North American Conference on Smart Structures and Materials*, SPIE Vol. 2443, pp. 268-276.

Childs, B., F. Lalande, Z. Chaudhry, and C. A. Rogers, 1996. "High-Frequency Impedance Analysis for NDE of Complex Precision Parts," *Proceedings, SPIE's 1996 Symposium on Smart Structures and Integrated Systems*, SPIE Vol. 2717, pp. 237-243.

Esteban, J., F. Lalande, and C. A. Rogers, 1996. "Theoretical Modeling of Wave Localization due to Material Damping," Proceedings, *SPIE's 1996 Symposium on Smart Structures and Integrated Systems*, San Diego, CA, 25-29 February, 1996.

Giurgiutiu, V., Chaudhry, Z., Rogers, C.A., 1994, "The Analysis of Power Delivery Capability of Induced Strain Actuators for Dynamic Applications", *Proceedings of the Second International Conference on Intelligent Materials, ICIM'94*, June 5-8, 1994, Colonial Williamsburg, VA, Technomic Pub. Co., Inc., 1994, p.p. 565-576.

Graff, K. F., 1991. *Wave Motion in Elastic Solids*, Dover Publications, 1991.

Keilers, C. H., Chang, F.-K., 1995, "Identifying Delamination in Composite Beams Using Built-in Piezoelectrics: Part I - Experiments and Analysis; Part II An Identification Method", *Journal of Intelligent Material Systems and Structures*, Vol. 6, pp. 649-672, September, 1995.

Kolsky, H. 1963. Stress Waves in Solids, Dover Publications, 1963.

Krautkramer, J. and Krautkramer, H., 1990, Ultrasonic Testing of Materials, Springer Verlag, 1990.

Lakshmanan, K. A. and Pines, D. J., 1997, "Modeling Damage in Composite Rotorcraft Flexbeams Using Wave Mechanics", *Journal of Smart Materials and Structures* (in press)

Liang, C., F. P. Sun, and C. A. Rogers, 1993, "An Impedance Method for Dynamic Analysis of Active Material Systems, *Proceedings, 34th AIAA/ASME/ASCE/AHS/ASC SDM Conference*, La Jolla, CA, 19-21 April 1993; pp. 3587-3599.

Mansfeld, F., 1993, "Analysis and Interpretation of Electrochemical Impedance Spectroscopy (EIS) Data for Metal and Alloys", #12606010, Schlumberger Technologies, UK, 1993.

Moetakef, M. A., Joshi, S. P., Lawrence, K. L., 1996, "Elastic Wave Generation by Piezoceramic Patches", *AIAA Journal*, Vol. 34, No. 1966, pp. 2110-2117.

Saravanos, D. A., Birman, V., and Hopkins, D. A., 1996, "Detection of Delamination in Composite Beams Using Piezoelectric Sensors", Report # LEW-16308, NASA LeRC, , OH 44135

Woods, R. D., 1968. "Screening of Surface Waves in Soils", J. Soil Mech. Founds. Div. Am. Soc. Civil Engrs., Vol. 94, July 1968, pp. 951-979.

Embedded Self-Sensing Piezoelectric for Damage Detection

C. PARDO DE VERA and J. A. GÜEMES

ABSTRACT

In this paper a method for damage detection based on self-sensing piezoelectrics is presented. When a piezoelectric actuator is electronically excited it causes stress in the structure to which it is attached. Since the structure deformation is influenced by the presence of damage, so the response of the sensor will be. The goal of the method is to use the same element as both actuator and sensor, analyzing the response of the system to a given excitation in the frequency domain. In such way the damage and undamaged structure can be distinguish and the amplitude of damage determined. The distinction between the exciting and response signals is achieved by an electronic bridge circuit, the same kind as the one previously used by Yellin and Shen to feedback an active damping system. A comparison between this self-sensing method and the most common cross-talk method is done, showing the similarity and validity of the technique. A method to distinguish between the spectrum changes due to real damage and temperature shift is also presented. As the temperature rises the density decreases and PZT characteristics changes and so do the resonance frequencies of the spectrum. The acquired signal has then to be manipulated in order to compensate the temperature changes.

C. Pardo : Aeronautical Engineer ; e-mail: cpardo@dmpa.upm.es
J. A. Güemes: Professor; e-mail: aguemes@dmpa.upm.es

Escuela Tecnica Superior de Ingenieros Aeronauticos
Universidad Politecnica de Madrid
Pza. Cardenal Cisneros 3
28040 MADRID
SPAIN

INTRODUCTION

Monitoring techniques are commonly based on the comparison of certain properties like eigenfrequencies, modal shapes, damping, etc. between the structure to be monitored and a reference pattern obtained by means of experiments in an undamaged structure or by analytic ways like finite elements models, etc. The first way is the more widely used due to the difficulty of getting a correct model of a complex structure or situation. The use of piezoelectrics in damage detection is founded on the stress generation caused in the material by a piezoelectric actuator. This stress is transmitted through the structure to a piezoelectric sensor that converts the experimented strain into electric energy, allowing us to relate the obtained voltage to the damage condition of the structure, since the way the material is deformed is a function of its internal condition. This paper shows the results obtained by means of a selfsensor, that is, a single piezoelectric that acts both as actuator and sensor, reducing the number of PZT's necessary to be attached for the monitoring of the whole structure. The selfsensor make use of an electronic bridge circuit that allows us to distinguish between the exciting and sensing signals. Exciting the system at different frequencies and analyzing its response we get the structure signature, which will be subsequently compared with the pattern, verifying the structural integrity. However, there are other situations in addition to damage presence, in which the structure signature is modified, like temperature changes, which modify the resonant frequencies of the structure and PZT characteristics. So, it is important to take into account this fact when we are trying to determine whether or not there is damage in the structure. In this paper a temperature compensation method is proposed, showing experimental results in the application of such method.

ELECTRONIC BRIDGE CIRCUIT DESCRIPTION

To use a piezoelectric as both actuator and sensor simultaneously we have to resort to some type of electronic device that allows us to excite the system and acquire the response that is generated when the PZT is deformed in the structure. Some authors, like Z. Chaudhry and F. Sun have utilized a circuit to measure the electric power consumption or the electronic impedanze of the circuit, getting the PZT-structure electromechanical impedanze. In this way, by means of a frequency sweep the structure signature at the present condition is generated. The method used in our research consists in the usage of an electronic bridge circuit that enables us to obtain the sensor signal directly. With this bridge, similar to the one previously used by Yellin and Shen to feedback a PZT in an active damping system, we avoid the use of high cost and high precision systems like those used for the electromechanical impedanze determination. The bridge circuit utilized is presented in figure 1.

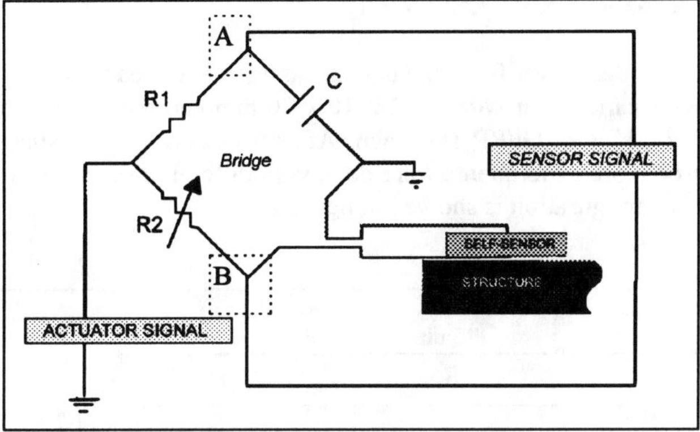

Figure 1. Electronic bridge circuit used for selfsensing.

Electronically, the PZT could be considered as a capacitor in series with a voltage source (sensor signal), so the voltage between A and B points when an exciting signal is generated responds to the following expression:

$$V_{AB} = V_E \left[\frac{1}{R_2 C_{PZT} \omega j} - \frac{1}{R_1 C \omega j} \right] - V_S \left[\frac{R_2 C_{PZT} \omega j}{R_2 C_{PZT} \omega j + 1} \right] \tag{1}$$

where:

V_E : Exciting signal.
V_S : Sensing signal
ω : Exciting frequency
C_{PZT} : PZT capacity.

If the bridge is correctly balanced ($R_2C_{PZT} = R_1C$), the voltage between A y B only has information about the response of the structure to the excitation. The exciting signal being considerably greater than the response signal, it is extremely important to get a precise equilibrium of the bridge to prevent any amount of the exciting signal that might filtrate to the output from masking the sensing results.

DAMAGE ASSESSMENT RESULTS

With the idea of verifying the damage assessment capability of the method an experiment was carried out with a 0.2 x 10 x 20 mm ceramic piezoelectric attached to a 4 x 30 x 115 mm GFRP specimen. After acquiring the undamaged structure response, three more experiments were done with different hole sizes at 35 mm from the PZT. This configuration is showed in figure 2.

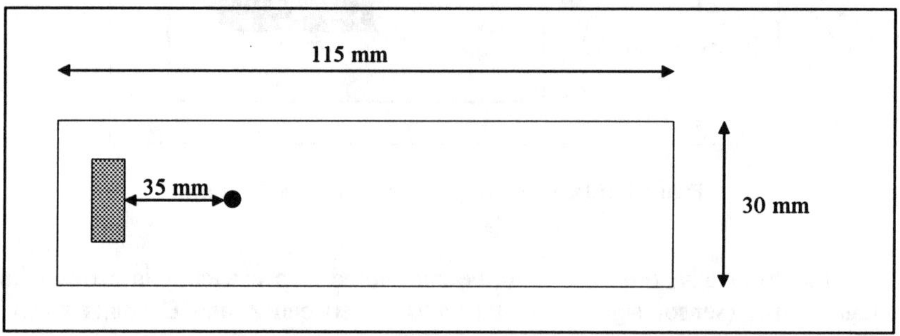

Figure 2. Specimen utilized in damage sensibility tests

The hole was made using subsequently a 3, 4, and 6 mm diameter bit, exciting the PZT with a high acquisition rate board, controlled via software with Labview and acquiring the sensor signal simultaneously with the same acquisition board. The transfer function obtained for the 3, 4 and 6 mm holes is presented in figures 3, 4 y 5 respectively together with the undamaged one.

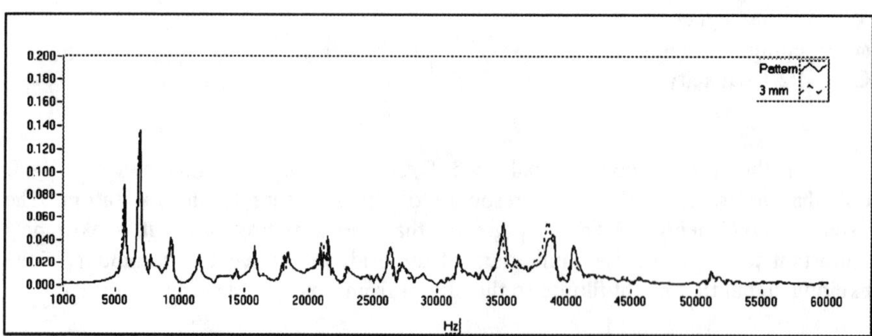

Figure 3. Transfer function with a 3 mm diameter drill

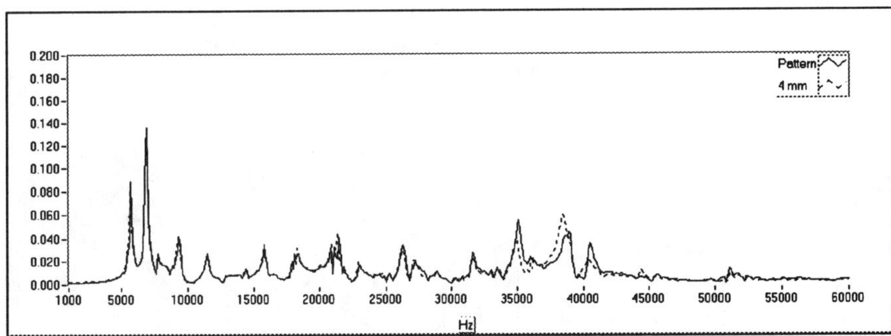

Figure 4. Transfer function with a 4 mm diameter drill.

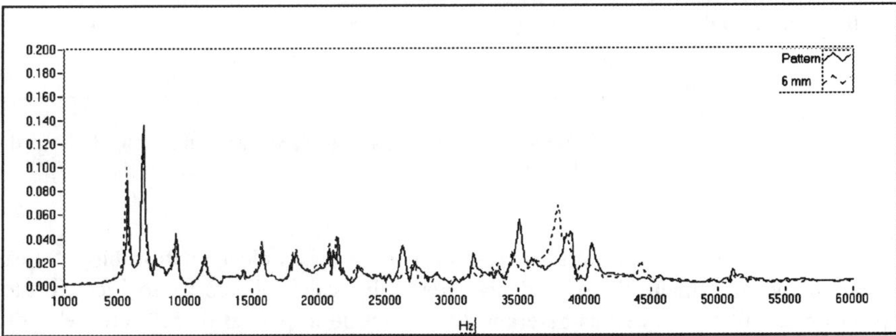

Figure 5. Transfer function with a 6 mm diameter drill

The most important effect when the hole grows appears around 38 KHz, increasing the amplitude of the corresponding resonant frequency. We can define a damage factor based on the difference between the transfer function of the damage and undamaged structure in the following manner:

$$D = \sum_{1}^{N} |TF_i| - |TF_i^o| \tag{2}$$

where $|TF_i|$ is the absolute value of the transfer function at a certain frequency, N is the number of points in which the TF is divided in and 0 denotes the reference pattern.

Applying this damage factor to the acquired TF the values of table 1 are obtained, as shown in figure 6.

Taladro	DF
3mm	1.42
4 mm	1.73
6 mm	2.94

Table 1. Damage factor as a function of the drill diameter

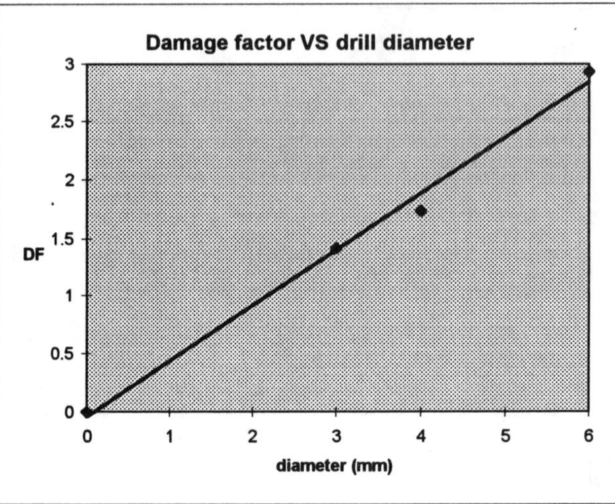

Figure 6. Table 1. Damage factor as a function of the drill diameter.

We can conclude from these results that there is a clear relationship, almost linear, between damage factor and diameter drill, which allows us to identify the structure condition, as well as determine if it must be inspected more accurately by establishing a threshold.

CROSS-TALK COMPARISON

To compare the self-sensing method with the more widely used cross-talk technique a test was carried out in which the TF were obtained using these two methods, both in the case of an undamaged and a simulated damage specimens. The TF obtained are showed in figures 7 and 8.

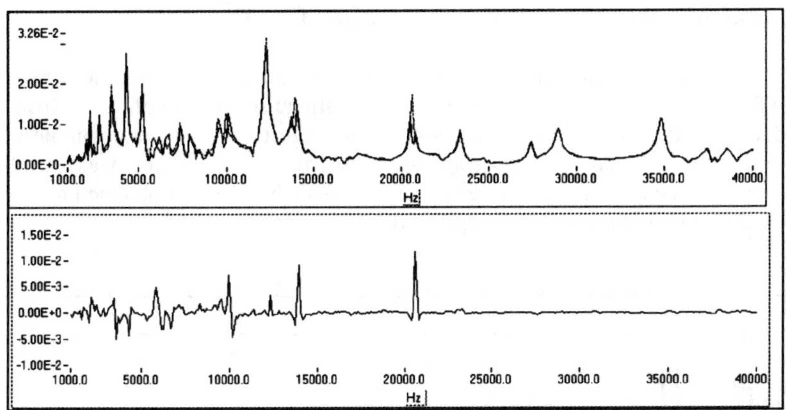

Figure 7. Transfer function after and before the damage and difference between them using the selfsensing technique.

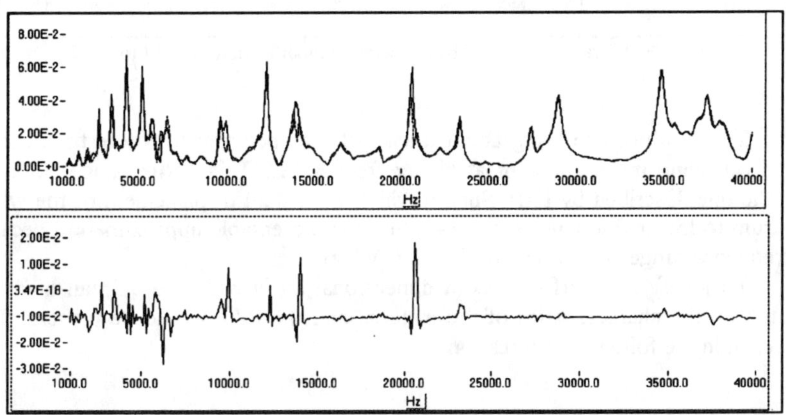

Figure 8. Transfer function after and before the damage and difference between them using the cross-talk technique.

In figures 7 and 8 is shown the similarity between the TF obtained by means of both techniques, as well as in the differences between those corresponding to the structure with and without damage, proving that the presence of damage causes modification of certain frequencies of the spectrum and certifying the validity of this method against the cross-talk technique.

TEMPERATURE INFLUENCE AND CORRECTION

Variation in environmental condition, like temperature causes a variation in the TF which has to be known to be able to distinguish this modification from that produced by the presence of real structural damage that we want to monitor. To make a study of the influence of temperature over the TF a specimen was introduced in an oven, raising the temperature from 34 to 47°C. The results obtained are shown in figure 9, where the curves are separated 1°C.

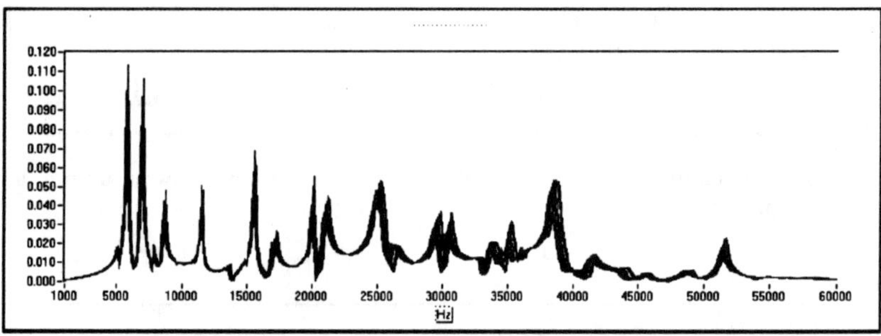

Figure 9. TF variation caused by temperature modification from 34 to 47°C

Temperature rise bring about a shift of the spectrum to lower frequencies, this effect being more apparent at higher frequencies. This performance coincides with the one described by F. P. Sun, in which a constant displacement of the whole spectrum to lower frequencies occurs. This is an acceptable approximation because his frequency range is very narrow (0.5-0.6 MHz)

To justify this performance a dimensional analysis of the TF can made, in which both the characteristics of the material and the PZT would be present. This will result in the following expression:

$$\frac{V_S}{V_A} = f\left(\frac{\varepsilon d_{31}^2 V_A^4}{L^5 \rho \omega^2}, \frac{E}{L^2 \rho \omega^2}, \frac{E_{PZT}}{L^2 \rho \omega^2}, \frac{\rho_{PZT}}{\rho}, \ldots, Geom\right) \tag{3}$$

Temperature changes have influence over the expansion of the PZT and the structure ($4\ 10^{-6}$), as well as over the dielectric and piezoelectric properties $2.5\ 10^{-3}$ and $5\ 10^{-4}$ respectively.

In view of the expression (3) we can assert that temperature variations cause a modification in the spectrum that responds to the following law.

$$\frac{\omega}{\omega_0} = \frac{d_{31}}{d_{310}} \sqrt{\frac{\varepsilon \rho_0}{\rho \varepsilon_0}} \tag{4}$$

where the density changes are negligible against the other two terms. Thus the spectrum variation could be related to temperature as follows:

$$\frac{\omega}{\omega_0} \approx 1 - \left(a - \frac{b}{2}\right)\Delta T \qquad (5)$$

where a and b are the temperature variation coefficient of d_{31} and ε respectively.

In figure 10 are represented the experimental and theoretical frequency scale factors. The experimental ones are those which provides a lesser difference with the reference pattern.

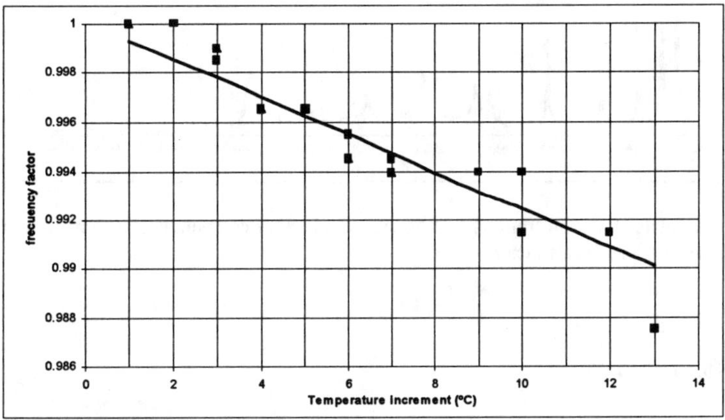

Figure 10. Experimental and theoretical frequency scale factor

In figure 10 is shown the good approximation of the theoretical results with the experimental ones, that enables us to compensate in temperature the obtained TF and identify those changes due to another causes like the presence of structural damage. In figure 11 are represented the transfer functions after and before be compensate.

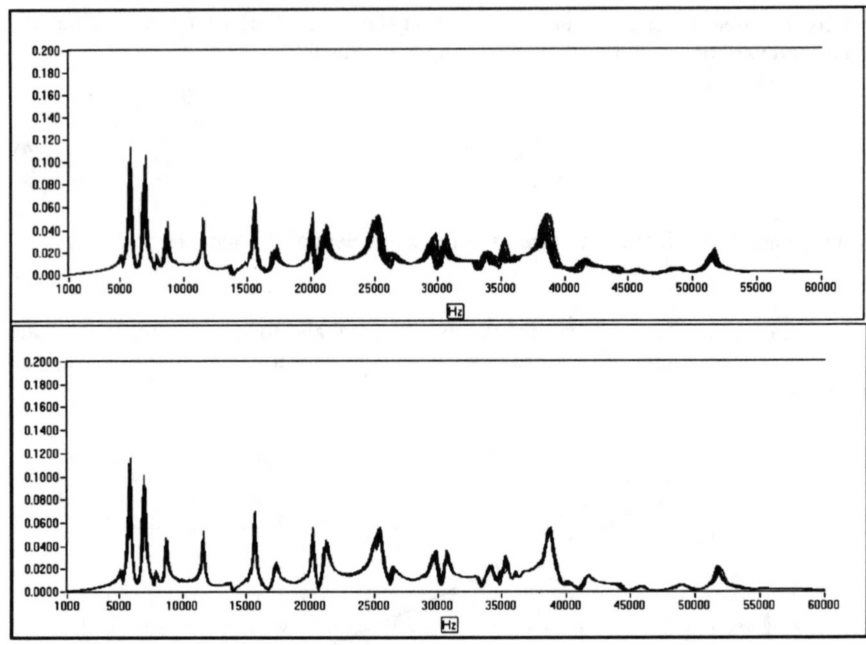

Figure 11. Transfer function from 34 to 47°C before and after compensate the temperature shift by means of the frequency scale factor.

6. CONCLUSIONS

In this paper has been proved the capability of the selfsensor method to detect damage, separating the sensing and exciting signal of PZT by means of an electronic bridge circuit, obtaining the transfer function of the system in the present condition, being subsequently compared with a reference pattern. The results have been compared with those obtained by means of separated actuator and sensor (cross-talk-method), showing the selfsensor technique validity. In order to avoid false alarms due to the effects of temperature changes on the transfer function the causes of that change have been analyzed and compensated.

7. REFERENCES

Bao, X., Varadan, V. V., and Varadan, V. K. "Crack detection in fastener holes using surface acoustic wave", SPIE vol 2443 (1995)

Basu, A. K., and Matthews, L. K. "Assessment of the current condition of structures using non destructive, modeling and parameter stimation techniques", SPIE vol 2446 (1995)

Betti, R., and Testa, R. B. "Vibration and damage control for long span bridges", Smart materials and structures vol 4 num 1A (1995)

Butler, R. K., and Rao, V. S. "Identification and control of two-dimensional smart structures using distributed sensors", SPIE vol 2442 (1995)

Chaudhry, Z. A., Joseph, T., Sun, F. P., and Rogers, C. A. "Local-area health monitoring of aircraft via piezoelectric actuator/sensor patches", SPIE vol 2443 (1995)

Gallego-Juárez, J. A. "Piezoelectric ceramics and ultrasonic transducers", CSIC (1989)

Garcia, G. V., and Stubbs, N. "Effect of damage size and location on the stiffness of a rectangular beam", SPIE vol 2446 (1995)

Im, S., and Atluri, S. N. "Effects of a piezoactuator on a finitely deformed beam subjected to general loading", AIAA JOURNAL vol 27 num 12 (1989)

Just, F. A., and Hendriks, S. L. "Crack detection on a cantilever beam using the nearest approximation method", SPIE vol 2446 (1995)

Lalande, F., Chaudhry, Z. A., and Rogers, C.- A. "Impedance-based modeling of actuators bonded to shell structures", SPIE vol 2443 (1995)

Liang, C., Sun, F., and Rogers, C. A. "Electromechanical impedance modeling of active material systems", Smart materials and structures vol 5 num 2 (1996)

Okafor, A. C., Chandrashekhara, K., and Jiang, Y. P. "Delamination prediction in composite beams with built-in piezoelectric devices using modal analysis and neural network ", Smart materials and structures vol 5 num 3 (1996)

Okafor, A. C., Chandrashekhara, K., and Jiang, Y. P. "Damage detection in composite laminates with built in piezoelectric devices using modal analysis and neural network", SPIE vol 2444 (1995)

Resch, M., Berg, H., and Elpass, W. J. "System identification and vibration control for composite structures with embedded actuators", SPIE vol 2443 (1995)

Sheng, M. H. M. "Analysis and vibration control of beams containing piezoelectric sensors and actuators", SPIE vol 2442 (1995)

Sun, F. P., Chaudhry, Z. A., Rogers, C. A., Majmundar, M., and Liang, C. "Automated real-time structure health monitoring via signature pattern recognition", SPIE vol 2443 (1995)

Yellin, J. M., and Shen, I. Y. "A self-sensing active constrained layer damping treatment for a Euler Bernoulli beam", Smart materials and structures vol 5 num 5 (1996)

Application of Neural Networks to Damage Detection in Turbine Blades

G. KAWIECKI

ABSTRACT

This paper shows the feasibility of using a very simple feedforward backpropagation neural network for fast and accurate estimation of the location and size of a crack in a cantilever beam. The presented network is trained and tested using data generated by a closed-form, one-dimensional theoretical model of the cracked beam. It is shown that the neural network is a very attractive alternative to presently used methods.

INTRODUCTION

Evaluation of structural dynamic response has been shown to be an effective method of nondestructive structural damage detection (Stubbs and Kim, 1996; Lyon, 1995). The presence of a defect affects dynamic characteristics of a structure, such as natural frequencies and mode shapes. Changes in those parameters can be used to define the location of a damage and in many cases also its size, particularly for simple structures. However, the interpretation of vibration testing results is a very tedious and error-prone process. The task of experimental data analysis can be greatly facilitated by computer expert systems. Such systems should be capable of properly evaluating the complex relationships among causes, i.e., the damages, and effects, i.e., the anomalies in structural response. Neural networks are one of the very few available tools that are capable of learning to associate the changes in dynamic response with particular structural damages.

The feasibility of using neural networks for structural damage detection has been recently an objective of a significant research effort (Ganguli, Chopra and Haas, 1996; Napolitano and Kosmatka, 1996; Rhim and Lee, 1994; Povich and Lim, 1994; Zgonc and Achenbach, 1996). The emerging consensus is that neural networks perform best when applied to relatively simple, lightly damped structural systems (Napolitano and Kosmatka, 1996). Good examples of such structures are turbine or aircraft propeller blades. They can be modeled as cantilever beams, and the associated damping levels are very low. However, very little attention has been devoted to such a simple, yet important, problem as detecting damages in cantilever beams. Chukwujekwu, Chandrasekhara and Jiang, 1996, demonstrated an application of a feedforward backpropagation neural network to estimate the size of a

Department of Mechanical and Aerospace Engineering and Engineering Science, The University of Tennessee, Knoxville, TN 37996-2210

delamination in a composite beam. The problem of finding damage location has not been considered. Thomas and Flockton, 1994, have shown an application of auto-associative and classifier networks to damage detection in cantilever beams. Neither location, nor size of the damage were considered. Also, the results were somewhat inconclusive.

The purpose of the present paper is to show the feasibility of applying a simple neural network to estimate the location *and* size of a crack in a blade modeled as a cantilever beam. A comprehensive solution of this problem has an immediate application in energy generation and aeronautical industries. At least one serious aircraft accident has been caused by a failure of a cracked propeller blade (Phillips, 1995).

FORMULATION

The goal of this study is to design a simple neural network capable of predicting the size and location of a crack in a cantilever beam using information about the first natural bending frequency and the amplitude of steady-state displacements under a known steady state excitation. The data necessary to train the neural network have been produced using a simple model of a cracked cantilever beam based on previously published studies (Rizos, Aspragathos and Dimarogonas, 1990; Ostachowicz and Krawczuk, 1991). A one-sided crack is modeled as a rotational spring, as shown in Fig. 1. The displacement functions for both parts of the beam (from the root to the crack and from the crack to the tip) can be expressed as

$$y_1(z) = a_1 \cosh(kz) + a_2 \sinh(kz) + a_3 \cos(kz) + a_4 \sin(kz) \tag{1}$$

$$y_2(z) = b_1 \cosh(kz) + b_2 \sinh(kz) + b_3 \cos(kz) + b_4 \sin(kz) \tag{2}$$

where

$$k = L\left(\frac{\omega_n^2 \rho A}{EI}\right)^{\frac{1}{4}} \tag{3}$$

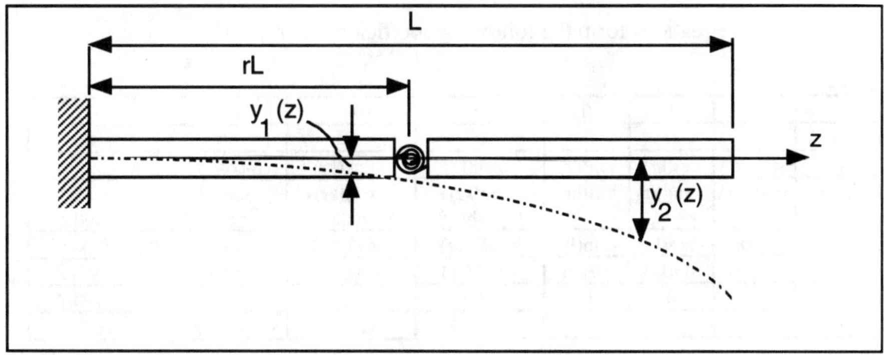

Figure 1 Model of a cracked beam.

The following boundary conditions describe the investigated beam

$$y_1(0) = 0 \tag{4}$$

$$\frac{dy_1(0)}{dz} = 0 \tag{5}$$

$$y_1(r) = y_2(r) \tag{6}$$

$$\frac{dy_1(r)}{dz} = \frac{dy_2(r)}{dz} - q\frac{d^2y_2(r)}{dz^2} \tag{7}$$

$$\frac{d^2y_1(r)}{dz^2} = \frac{d^2y_2(r)}{dz^2} \tag{8}$$

$$\frac{d^3y_1(r)}{dz^3} = \frac{d^3y_2(r)}{dz^3} \tag{9}$$

$$\frac{d^2y_2(1)}{dz^2} = 0 \tag{10}$$

$$\frac{d^3y_2(1)}{dz^3} = 0 \tag{11}$$

where (Ostachowicz and Krawczuk, 1991)

$$q = 6H/L(\gamma^2(72.21\text{-}117.1\gamma + 420.7\gamma^2\text{-}585.5\gamma^3 + 854.2\gamma^4\text{-}829.3\gamma^5 + 281.7\gamma^6)) \tag{12}$$

and

$$\gamma = \frac{d}{H} \tag{13}$$

The above equations form the following coefficient matrix

[cm]=

1	0	1	0	0	0	0	0
0	1	0	1	0	0	0	0
cosh(kr)	sinh(kr)	cos(kr)	sin(kr)	-cosh(kr)	-sinh(kr)	-cos(kr)	-sin(kr)
sinh(kr)	cosh(kr)	- sin(kr)	cos(kr)	-sinh(kr)+ kqcosh(kr)	- cosh(kr)+ kq sinh(kr)	sin(kr)- kqcos(kr)	- cos(kr)- kqsin(kr)
cosh(kr)	sinh(kr)	- cos(kr)	- sin(kr)	- cosh(kr)	- sinh(kr)	cos(kr)	sin(kr)
sinh(kr)	cosh(kr)	sin(kr)	-cos(kr)	- sinh(kr)	-cosh(kr)	-sin(kr)	cos(kr)
0	0	0	0	cosh(k)	sinh(k)	-cos(k)	- sin(k)
0	0	0	0	sinh(k)	cosh(kr)	sin(k)	-cos(k)

(14)

Eqs. 4 - 11 are used to find the fundamental frequencies and tip displacements corresponding to a given combination of a nondimensional crack location r and size d, for a given excitation. The fundamental frequencies are found by equating the determinant of matrix cm to zero and solving the resulting characteristic equation for k. Tip displacements are obtained as follows. Eq. 11 is modified to account for the tip driving force of magnitude F

$$\frac{d^3 y_2(1)}{dz^3} = \frac{FL^3}{EI} \tag{15}$$

and Eq. 3 is modified to account for the excitation frequency ω

$$k = L\left(\frac{\omega^2 \rho A}{EI}\right)^{\frac{1}{4}} \tag{16}$$

Then, the system of Eqs. 4 - 11 is solved for coefficients a and b (Eqs. 1 and 2). Tip displacements are determined by computing the value of y_2 at z=L.

Eight nondimensional crack depths and twelve crack locations are considered. Crack depths range from 0.01H to 0.51H at equal increments and crack locations span the length of the beam from 0.01L to 0.61L at equal intervals. The cracked beam problem has been solved using *Mathematica* algebraical manipulation package. The following beam parameters have been assumed: elasticity modulus $E = 2.06 \cdot 10^{11}$ Pa, beam length L = 0.3 m, cross section height H = 0.02 m, cross section width B = 0.02 m, and specific density $\rho = 7.86 \cdot 10^3$ kg/m^3.

Natural frequencies and tip displacements associated with these crack size/location combinations are used to train the neural network. That network has been designed as a function approximator. The problem of crack detection in a cantilever beam lends itself well to a solution using that type of a network. Detailed justification is given in the Results portion of this paper. The chosen network has two inputs, one for feeding the measured tip displacements and the other for natural frequencies, one hidden layer with log-sigmoidal neurons, and an output layer with two linear neurons producing estimates of crack depth and location. The input to the network has been scaled between 0.1 and 0.9 to avoid saturation of log-sigmoidal activation functions. The network was trained using a Levenberg-Marquardt algorithm. A schematic of the network is shown in Fig. 2.

The performance of the network has been tested using an additionally generated set of natural frequency/tip displacements. These data have been produced for crack depths d = 0.003 m, 0.005 m, and 0.007 m, and crack locations r = 0.045 m, 0.075 m, 0.1050 m, 0.1350 m, and 0.1650 m. Therefore, crack parameters in the test data set were different from those in the training set. The neural network portion of this study has been done using the Matlab Neural Network Toolbox.

RESULTS

The natural frequencies and tip displacements corresponding to the 96 combinations of crack size and location have been presented in Figs. 3 (a) and (b). Fig. 3 (a) shows the tip displacement y as a function of crack parameters d and r. Fig. 3 (b) presents the relationship between the first natural bending frequency f and crack parameters. Fig. 3 helps also in visualizing the nature of the problem. We can note that a given tip displacement y_1 and the associated fundamental frequency f_1 correspond to an infinite number of crack size/location combinations, as shown by curves $y_1(d, r)$ and $f_1(d, r)$ in Figs. 3 (a) and 3 (b), respectively. However, the projections of these curves on the *dr* plane, namely $y'_1(d, r)$ and $f'_1(d, r)$, will intersect at a location with coordinates (d_c, r_c), corresponding to the parameters of the crack. Therefore, the role of the neural network will be to find the coordinates of the intersection point (d_c, r_c), i.e., crack characteristics corresponding to a given natural frequency/steady-state tip displacement combination. Experiments have shown that as few as sixteen neurons in the hidden layer were enough for an adequate performance of the network.

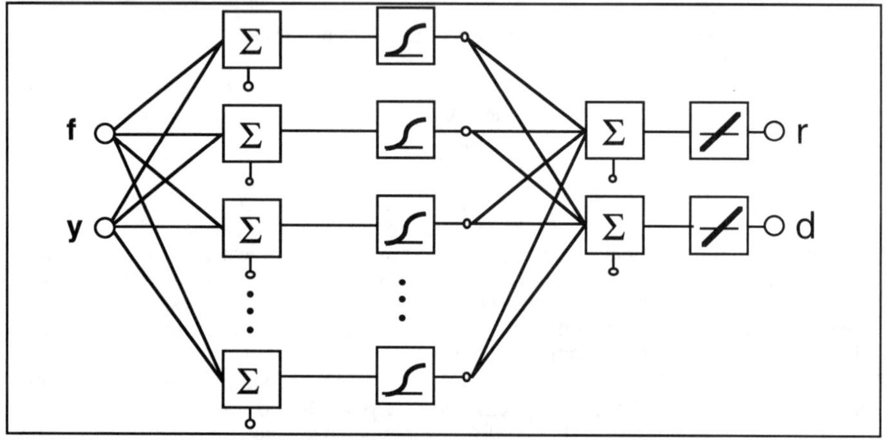

Figure 2 Neural network architecture.

Fig. 4 shows the performance of the network with 16 neurons in the hidden layer for the 15-element test data set. Exact solutions obtained using the theoretical model are compared with the predictions generated by a network trained using the 96 element data set. The array of plots shows exact and predicted relationships between crack size and tip displacement, crack location and natural frequency, crack depth and natural frequency and crack location and tip displacement. The overall error of neural network predictions is shown in Fig. 5, where the solid line indicates the size error, and the dashed line shows the location error. The training of the presented network was discontinued after 1000 epochs when the sum of squared errors reached 0.00082.

For comparison, the performance of a network with 20 neurons in the hidden layer was evaluated. Exact solutions are compared with modified network predictions in Figs. 6 and 7. For this network, training was discontinued after 7800 epochs when the sum of squared errors reached $3 \cdot 10^{-7}$. The error did not exceed 2 %, as shown in Fig. 7. The performance of this network was checked also for noise-contaminated data. Random numbers with magnitudes falling within ± 0.1 % of signal amplitude were added to natural frequency and tip displacement amplitude vectors. In this case, random signal amplitudes corresponded to about 0.2 Hz and $5 \cdot 10^{-7}$ m, what matched the resolution of modern equipment. The performance of this network for noisy data is shown in Figs. 8 and 9 where, again, the solid line indicates the size error, and the dashed line shows the location error. We can note that the network missed only one data point out of fifteen. In the remaining cases, the overall error of neural network prediction was less than 15 %. This result could be significantly improved if a larger training data set was used.

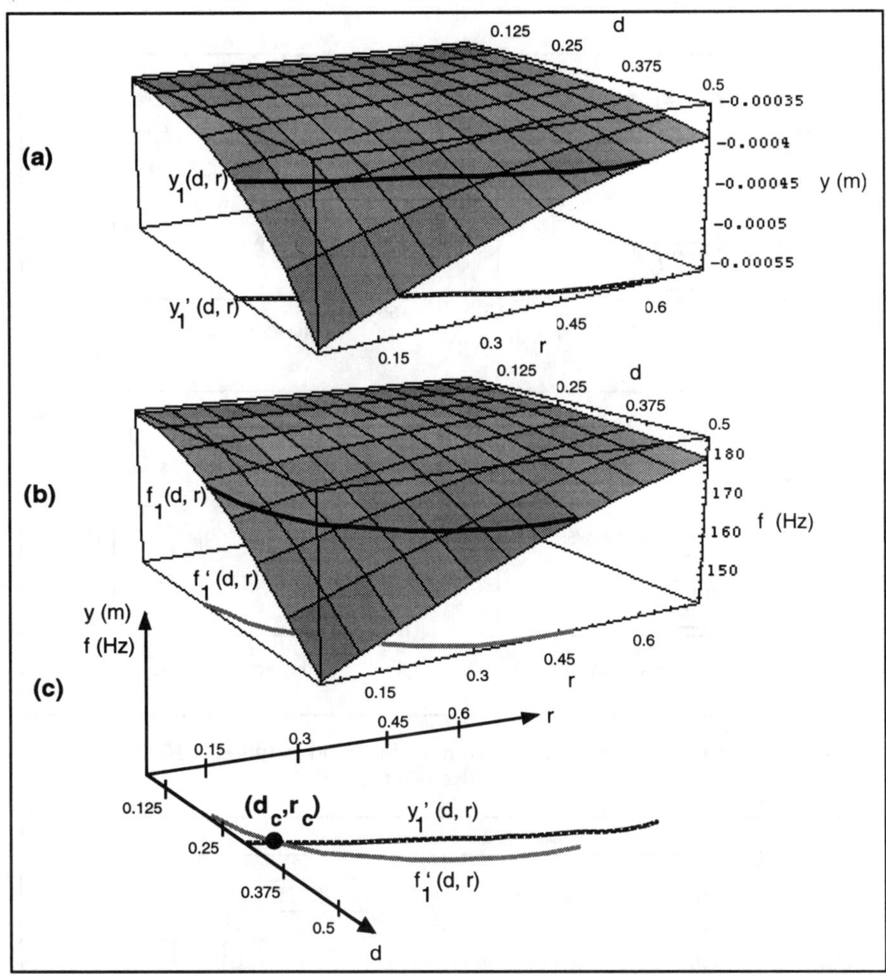

Figure 3 Dynamic response characteristics vs. crack size and location.

CONCLUSIONS

This paper shows the feasibility of applying a very simple neural network to predict effectively the location and size of a crack in a cantilever beam. The results of this work can be used in inspection of structural elements that can be modeled as cantilever beams, e.g., turbine blades or aircraft propeller blades. The proposed method solves efficiently and accurately an inverse problem of estimating damage size and location from dynamic response characteristics. It is shown that the neural network performs adequately for data contaminated by measurement errors.

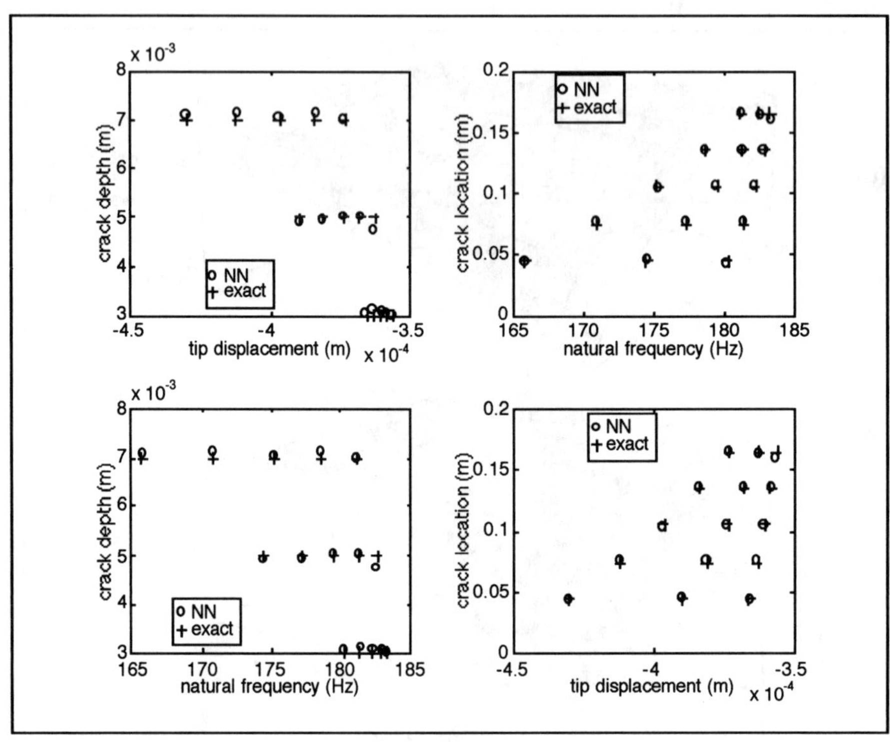

Figure 4 Comparison between exact and neural network solutions - 16 neurons in the hidden layer.

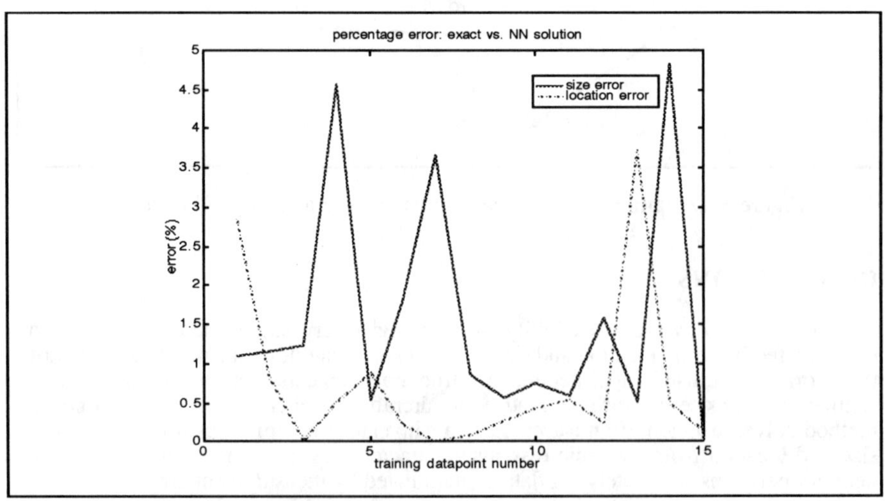

Figure 5 Percentage error for network solution - 16 neurons in the hidden layer.

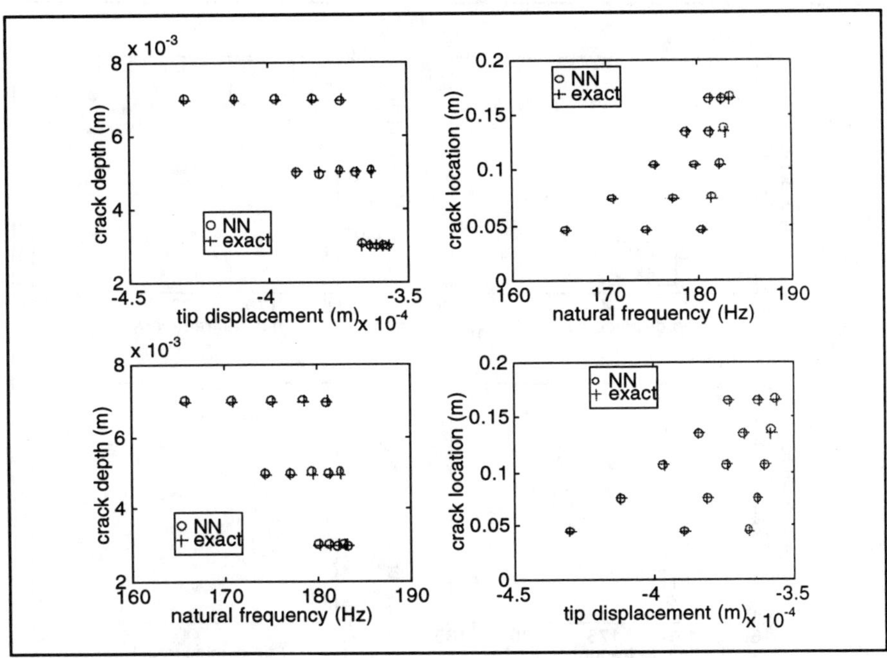

Figure 6 Comparison between exact and neural network solutions - 20 neurons in the hidden layer.

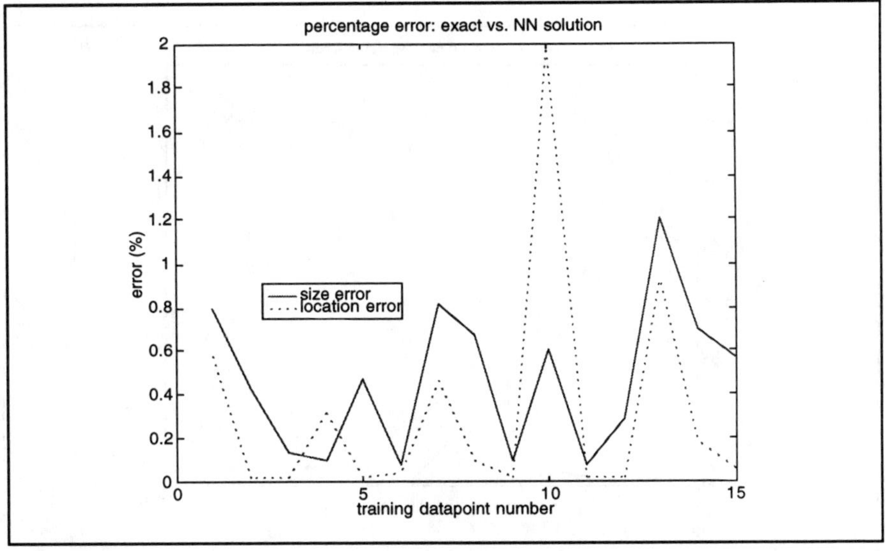

Figure 7 Percentage error for network solution - 20 neurons in the hidden layer.

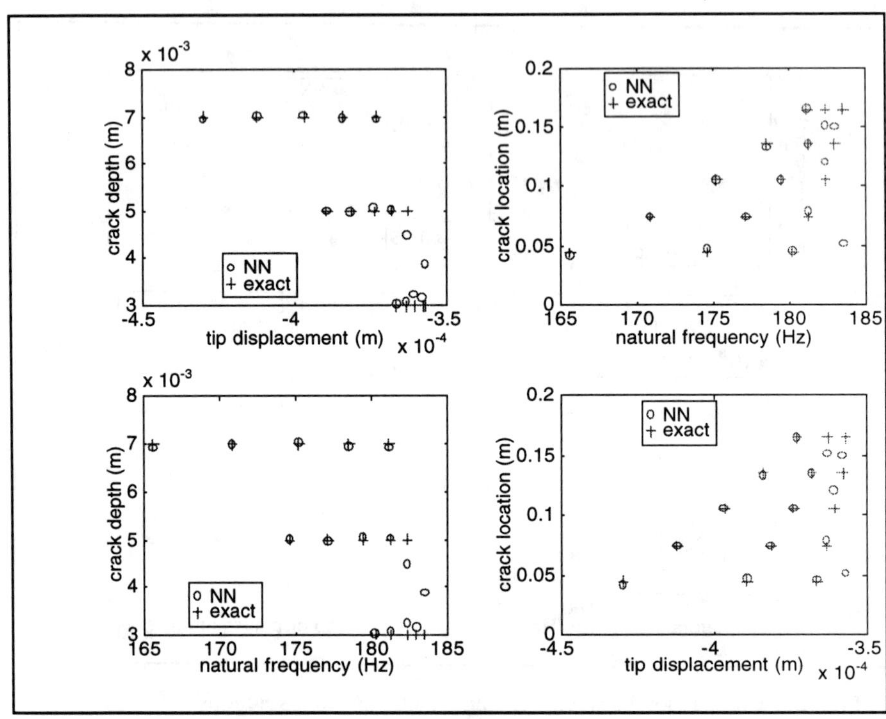

Figure 8 Network performance for noisy data - 20 neurons in the hidden layer.

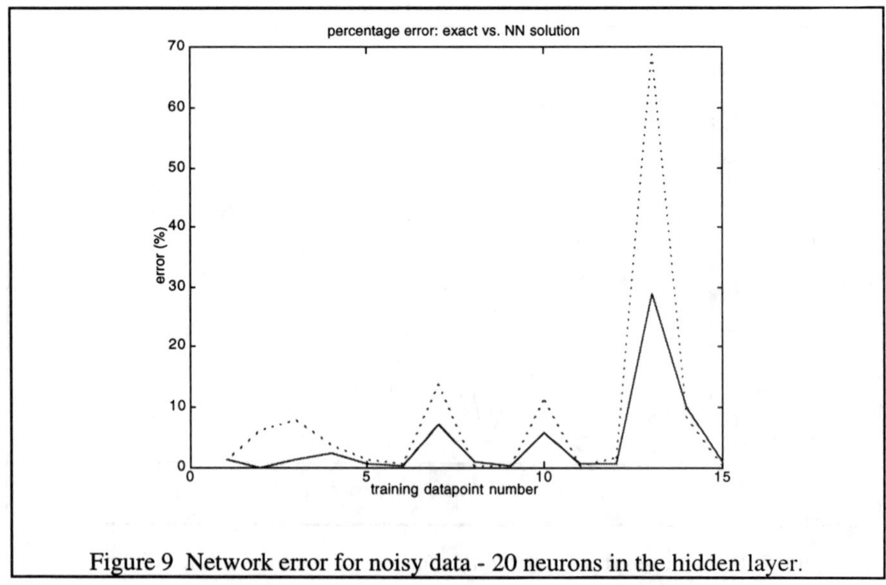

Figure 9 Network error for noisy data - 20 neurons in the hidden layer.

ACKNOWLEDGMENTS

The support provided by an NSF grant CMS9402802 is gratefully acknowledged. Special thanks go to Dr. Wesley Hines for sharing his knowledge of neural networks.

REFERENCES

Chukwujekwu Okafor, A., Chandrashekhara, K., and Jiang, Y. P., June 1996, "Delamination prediction in composite beams with built-in piezoelectric devices using modal analysis and neural network," Smart Materials & Structures, Vol. 5, Number 3, pp. 338-347.

Ganguli, R., Chopra, I., and Haas, D. J., April 15-17, 1996, "Detection of Helicopter Rotor System Simulated Faults Using Neural Networks," AIAA-96-1646-CP, Proceedings of the 37th Structures, Structural Dynamics and Materials Conference, Salt Lake City, Utah, pp. 1246-1263.

Lyon, R. H., January 1995, "Structural Diagnostics Using Vibration Transfer Functions," Sound & Vibration, pp. 28-31.

Napolitano, K. L., and Kosmatka, J. B., April 15-17, 1996, "Damage Detection of a Damped Structure with Neural Networks," AIAA-96-1225-CP, Proceedings of the 37th Structures, Structural Dynamics and Materials Conference, Salt Lake City, Utah, pp. 209-213.

Ostachowicz, W. M. and Krawczuk, M., 1991, "Analysis of the Effect of Cracks on the Natural Frequencies of a Cantilever Beam," Journal of Sound and Vibration, Vol. 150, Nr. 2, pp. 191-201.

Phillips, E. H., "Blade Failure Focus of NTSB Crash Probe, August 28, 1995, "Aviation Week and Space Technology, p. 31.

Povich, C. R., and Lim T., W., April 21-22, 1994, "An Artificial Neural Network Approach to Structural Damage Detection Using Frequency Response," AIAA/ASME Adaptive Structures Forum, Hilon Head, SC., pp. 151-159.

Rhim, J., and Lee, S. W., April 21-22, 1994, "A Neural Network Approach for Damage Detection and Identification of Structures," AIAA/ASME Adaptive Structures Forum, Hilton Head, SC., pp. 173-180.

Rizos, P. F., Aspragathos, N. and Dimarogonas, A. D., 1990, "Identification of Crack Location and Magnitude in a Cantilever Beam from the Vibration Modes," Journal of Sound and Vibration, Vol. 138, No. 3, pp. 381-388.

Stubbs, N. and Kim, J.-T., August 1996, "Damage Localization in Structures Without Baseline Modal Parameters," AIAA Journal, Vol. 34, No. 8, pp. 1645-1649.

Thomas, M. D., and Flockton, S. J., 12-14 October 1994, "Fault Detection in a Mechanical System by Neural Networks," Second European Conference on Smart Structures and Materials, Glasgow, Scotland, pp. 55-58.

Zgonc, K., and Achenbach, J. D., 1996, "A neural network for crack sizing trained by finite element calculations," NDT&E International, Vol. 29, No. 3, pp. 147-155.

Search Strategies for Identifying a Composite Plate Delamination Using Built-In Transducers

C. H. KEILERS, JR.

ABSTRACT

Search strategies are compared that can identify single, rectangular, mid-plane delaminations in composite plates using the signals from surface-mounted piezoceramics. The strategies are demonstrated by repeatedly assuming a single delamination and then running a delaminated plate model that predicts transducer frequency responses. An objective function compares the responses to those for a hypothetical plate with unknown damage. When the objective function is minimized, the responses agree, and the assumed delamination is the best-estimate. The chosen objective function is well-behaved but has several local minima; therefore, random search strategies are used to find the global minimum. The most efficient strategy starts with assumed delaminations selected from a uniform random distribution at discrete points in the parameter space (i.e., a combinatorial search). Some of these points will be close to local minima. The strategy continues by using a random walk with shrinking step size across the parameter space. This latter search is made more robust by using probabilistic hill climbing acceptance tests, also known as simulated annealing. The overall technique systematically investigates several possible local minima and predicts the delamination to a preset accuracy.

INTRODUCTION

Composites are susceptible to nearly invisible damage following an impact that induces delaminations between plies. Several investigators are pursuing research relevant to identifying structural damage using built-in health monitoring systems. This paper will functionally describe a baseline, built-in system for delamination detection, as well as compare search strategies for improving the best-estimate of delamination size and location. For demonstration purposes, this work is based on a dynamic model of the behavior of a simply supported, orthotropic plate with a single delamination and with surface-mounted piezoelectric transducers.

Senior Technical Specialist
Technical Staff, Defense Nuclear Facilities Safety Board
625 Indiana Ave., NW, Suite 700
Washington, DC 20004

The author previously demonstrated an iterative approach that identified the size and location of mid-plane delaminations in composite beams by using surface-mounted piezoceramics (Keilers and Chang, 1995). This was a two-parameter identification problem. Some of the challenges were that the results were sensitive to instrument noise and that the single-input/single-output configuration predicted two equally likely delamination locations. More recent work has focused on using the signals from multiple sensors and actuators to identify the best-estimate size and location of a single rectangular mid-plane delamination in a simply-supported plate (Keilers, 1997). This is a four parameter identification problem.

Using some of the concepts discussed, a "baseline" damage identification system might be developed. Such a system might consist of a distributed sensor array operating initially in a passive or "listening" mode. Following an impact, the location of potential damage can be estimated based on high-amplitude transients from nearby sensors (Choi, et al, 1994). At that time, the transducer array could switch to an "active" mode, with actuators as well as sensors. Frequency responses from each sensor-actuator combination can be compared to baseline, "healthy-structure" frequency responses to improve the damage estimate. The estimate can be further improved by repeatedly running a model with different assumed delaminations and comparing the calculated responses to those just measured. By appropriately selecting the assumed delaminations, the number of model runs to achieve convergence can be minimized.

This paper will describe the first indications of possible damage provided by built-in transducers. The paper then compares several random search strategies for selecting candidate delaminations to model and, thereby, minimizing runtime. For many minimization problems like this one, random strategies have advantages over gradient or Hessian-based methods. For example, approximating a gradient by finite difference requires many function evaluations in a local area. This area may be far from the global minimum or, even worse, close to a local minimum, resulting in false convergence. For the same effort (i.e., the same number of function evaluations), a random strategy provides more global information about the parameter space, but penalizes rate of convergence (Otten and van Ginnekin, 1989, Hajela, 1997).

DEMONSTRATION EXAMPLE

The simply supported plate in Figure 1 is used here to test different search strategies. The plate substrate is assumed to be T300/976 in a $[0_4/90_4]_s$ layup. A single square mid-plane delamination is located centered at a quarter-span point and covers 0.8 percent of the plate area.

The objective of the search strategies is to use shifts in amplitude responses from attached transducers to find the best-estimate mid-plane delamination, without any apriori knowledge. For transducers, thin piezoceramic patches are assumed to be mounted on opposing sides of the plate, as shown, covering 1.8 percent of the plate. Each of these can be used as an actuator at one time, exciting the plate in bending, and as a bending strain sensor at another time. A "sensor-actuator pair" then consists of two of the nine piezoceramics shown in the figure. Since switching the sensor for the actuator does not change the frequency response, there are 8 factorial possible combinations. For simplicity here, only adjacent piezoelectrics are considered for use as sensor-actuator pairs (e.g., no. 1 and 4, 1 and 5, and 1 and 2 in the figure).

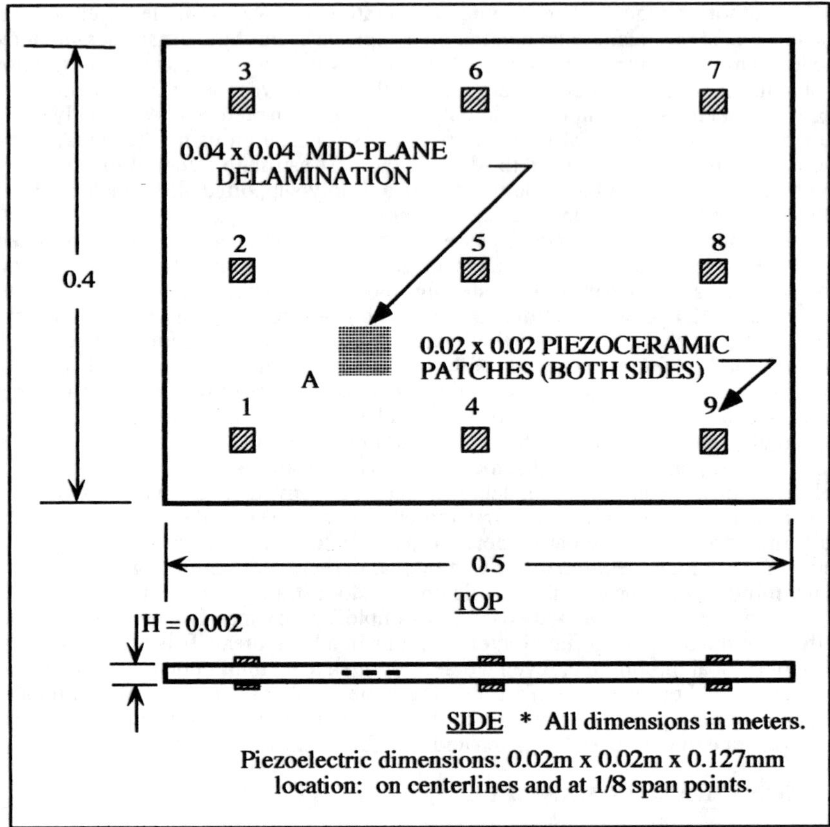

Figure 1: The example - finding a delamination of less than 1 percent of plate area.

The analytical model used was previously developed for predicting the frequency response from piezoelectrics attached to a delaminated composite plate (Keilers, 1997). The principal assumptions are thin, specially orthotropic material; simply supported boundary conditions; normals remain straight and normal to the mid-plane; negligible in-plane and rotary inertia; negligible transverse normal stress; negligible damping; no net in-plane force; piezoelectrics are thin, isotropic in the plane, and have negligible effect on plate stiffness. The model is based on a variational principle and uses the lowest 86 mode shapes for the undelaminated plate. To include the effect of a single delamination, the delamination contact surfaces are assumed to slide frictionlessly across each other without the delamination opening up. This leads to a linear model that predicts the stiffness loss.

This approach to dynamically modeling a delamination has been successfully used before for small delaminations (e.g., Mujumdar and Suryanarayan, 1988). Performance degrades for larger flaws since it neglects the delamination buckling effect. Work by the author has focused on embedding the model in a search loop and essentially running the model repeatedly until a scalar objective function is minimized. This minimum corresponds to a best-estimate mid-plane delamination. A weakness of this approach is that it returns a delamination description matching the frequency response, even if some other type of damage has occurred. This is a topic for further study.

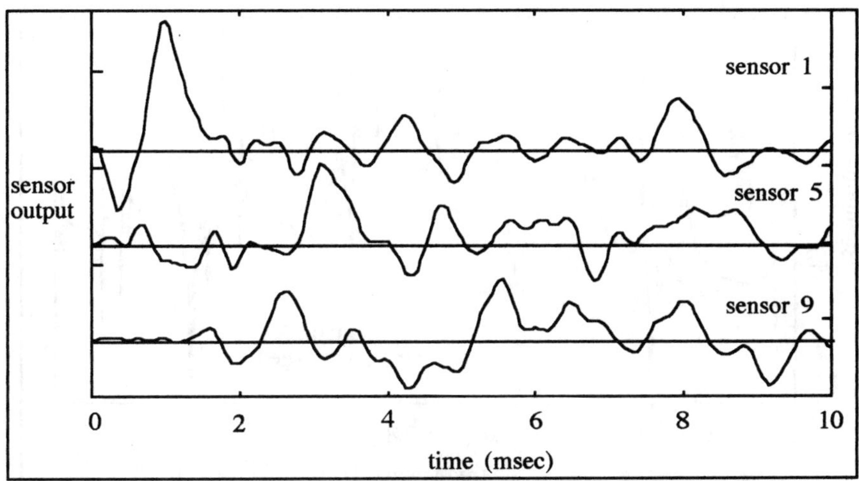

Figure 2: Transient sensor outputs for an impact at point A of Figure 1. A triangular input loading, peaking at 0.5 msec, was assumed. The signals were calculated using the model discussed but with 1 percent damping.

INITIAL DAMAGE INDICATORS

Identifying a delamination is easier if there is a good initial guess. For example, if all the transducers are initially in a listening mode, the first indication of impact damage is likely to be high amplitude transients such as those shown in Figure 2. Comparing these transients, the damage location would appear to be near sensor 1. Several authors have discussed using such transient signals to identify the impact point and peak loading (Choi, et al, 1994).

After a potentially damaging event, structural integrity might be assessed by switching to an active mode - in other words, using one or more piezoceramics as an actuator while the others are used as sensors. Figure 3 illustrates the subtle shift in a predicted amplitude response due to this small delamination. This comparison is even more difficult in practice since the shifts may be obscured by noise or may be caused by effects other than damage, such as changes in loading or temperature. These effects are not discussed here.

A useful way to compare two amplitude responses is to use a scalar quadratic objective function or figure of merit, *fom* (Keilers and Chang, 1995). Based on previous studies, one choice of *fom* is as follows:

$$fom = \text{MAX}\left(fom_j\right) \tag{1}$$

where:

$$fom_j = \sum_{i=1}^{nfreq} W_i \left(\frac{\log\ y_i^c}{\log\ y_i^m} - 1 \right) \tag{2}$$

and y_i^c and y_i^m are the amplitudes at the ith frequency for the two responses being compared. W_i is a weighting factor that can be chosen to reduce noise sensitivity (assigned to 1 here). fom_j is a single scalar that compares the responses from a single sensor-actuator pair, pair *j*. The *fom* of (1) is appropriate in this example with responses from multiple sensor-actuators.

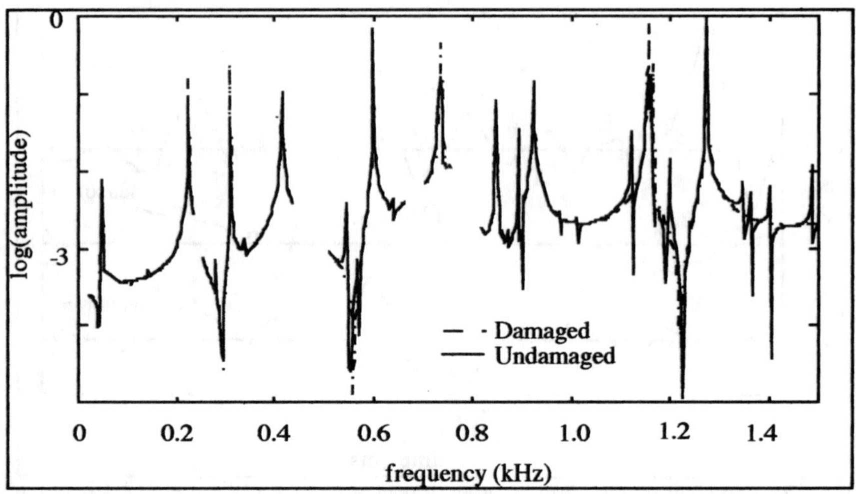

Figure 3: Amplitude responses for sensor-actuator pair 1/5, with and without the Figure 1 delamination. Only frequencies within +5/-25 Hz of resonances are included.

Previous studies have shown that this gives a generally well-behaved objective function over the parameter space. However, the topology still has local minima, which discourages the use of gradient-based methods. This paper uses this objective function, together with random search techniques discussed in the next section, to find this subtle delamination.

SEARCH STRATEGIES

The first strategy considered is the uniform random (UR) search, which consists of running the model repeatedly with different feasible delaminations.

```
while not done,
        dl   = uniform(constraints);
        fom  = model(dl);
        test if done;
end;                                            (UR)
```

Within the function "uniform," potential delamination sizes and locations are randomly selected from uniform distributions and then checked against feasibility constraints. For all the demonstrations here, the constraints applied are (1) the delamination must fit entirely within the plate area; (2) the delamination length and width are less than 25 percent of the corresponding plate dimensions; and (3) the delamination aspect ratio is less than 3. The second constraint ensures that delaminations are still small enough that delamination buckling is limited. The third constraint restricts the delamination to practical length-to-width ratios.

The main advantages of the UR technique are its simplicity and robustness against converging to a local minima. However, the number of model runs until convergence is high. The top of Figure 4 shows, on a relative scale, that the errors in the predicted delamination area and center location vary widely over the 150 runs made. The bottom shows the 20 candidate delaminations with the smallest *fom* found at the end of this process. Even after 150 runs, the actual delamination is not evident.

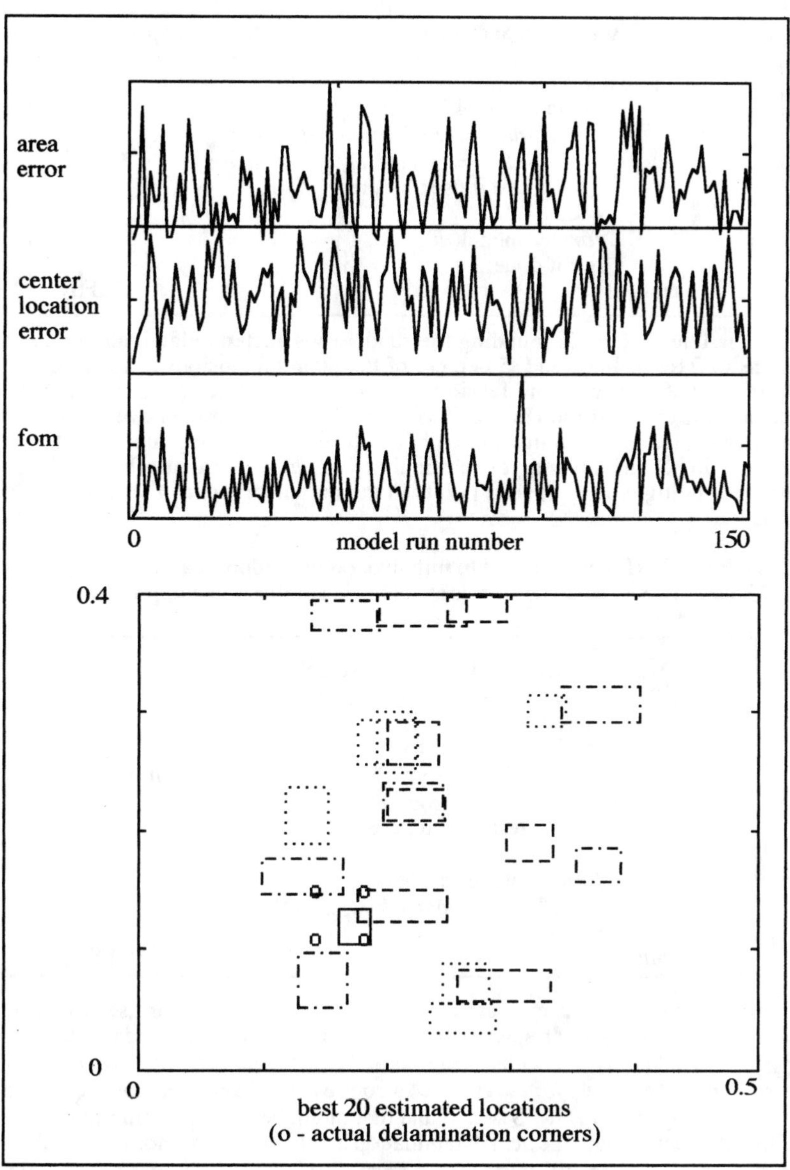

Figure 4: Results from applying a uniform random (UR) search to identify the Figure 1 delamination. The top shows that errors in the estimated delamination size and location, as well as *fom*, vary widely throughout the search. The bottom shows that the 20 delaminations found with lowest *fom* are dispersed in size and location.

The second strategy is the uniform random fixed step (URFS) search. This is essentially still a UR search but at discrete points in the parameter space.

```
while not done,
        while not used before,
                dl_1  = uniform(constraints);
                dl    = round(dl_1 , step size);
                test if used before;
        end;
        fom = model(dl);
        test if done;
end;                                                    (URFS)
```

The new feature here is rounding the randomly selected delamination sizes and locations to fixed values (e.g., 5 percent of the plate dimensions in each direction) while still satisfying the same feasibility constraints as UR. Since the interval is fixed, an additional test loop is added to avoid running the model over again with a previously evaluated delamination. While still requiring a large number of runs to converge, URFS has advantages; particularly, a guarantee that effort will not be wasted evaluating closely spaced points in the parameter space that add little new information.

The UR and URFS can be used to initialize other random search techniques, such as the Gaussian fixed step (GFS) search. This is similar to the approach in (Keilers and Chang, 1995).

```
DL_0 = truncated UR or URFS search
[dl_0 dl_std] = update(fom, dl, DL_0 );
while not done,
        while not used before,
                dl_1  = normal(dl_0 , dl_std , constraints);
                dl    = round(dl_1 , step size);
                test if used before;
        end;
        fom = model(dl);
        [dl_0 dl_std] = update(fom, dl, DL_0 );
        test if done;
end;                                                    (GFS)
```

Initially, the output of a truncated UR or URFS search (DL_0) is used to estimate properties of the parameter space. In the "update" function, standard deviations (dl_{std}) for the delamination center locations and side lengths are found for a subset of these runs with the lowest *fom*. "Update" also provides a delamination, dl_0, that had a low *fom* during the UR/URFS search and that might be close to a minimum.

Within the search loop, the next delamination is selected from a normal distribution using dl_0 and dl_{std} while still meeting the feasibility constraints. After the model is run, if the new *fom* is low, the corresponding delamination dl is used to reset dl_{std}. Also, the next center dl_0 for use with "normal" is selected by cycling through delaminations with low *fom* from the initial uniform search (DL_0).

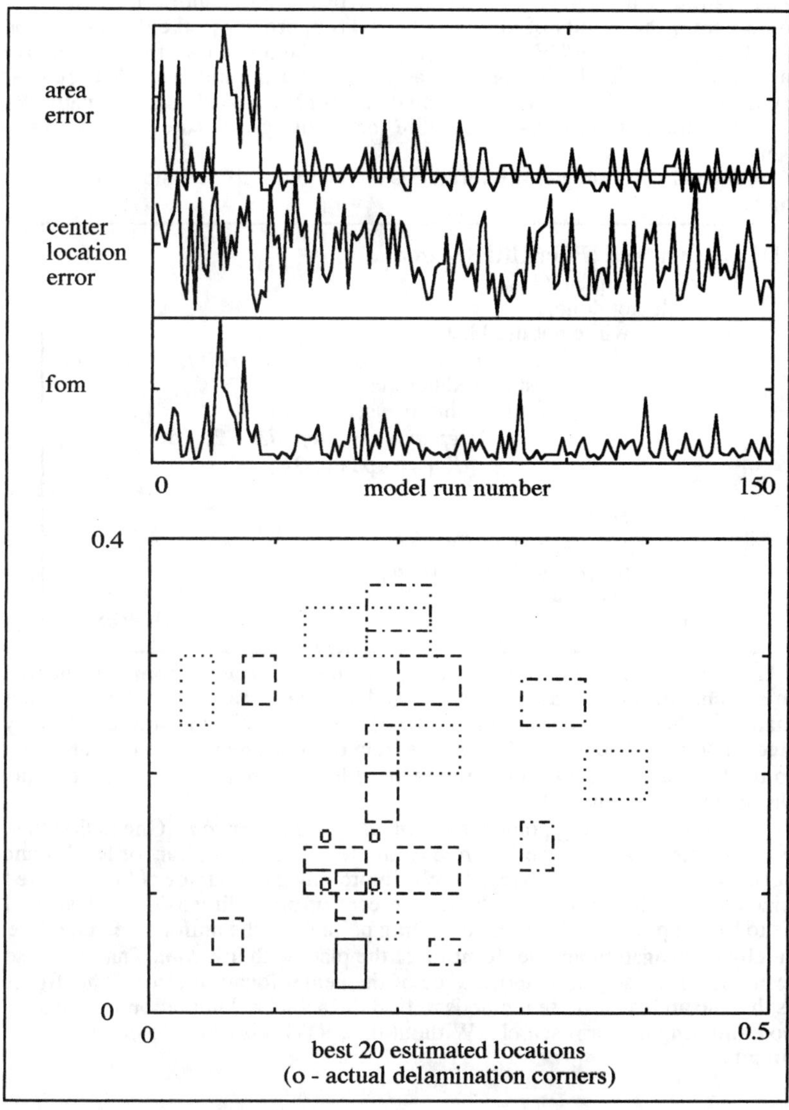

Figure 5: Results from applying a Gaussian fixed step (GFS) search to identify the Figure 1 delamination. Delaminations for the first 25 runs are selected using URFS. The top shows that errors and *fom*, drop somewhat throughout the search. The bottom shows that the 20 delaminations found with lowest *fom* are not as dispersed in size and location as with the UR strategy.

In this manner, the GFS strategy systematically searches for the global minimum by cycling through several potential minima indicated by the initial uniform search.

Figure 5 shows the results of applying the GFS approach to the demonstration problem. Compared to the UR strategy (Figure 4), the top shows that the relative errors are decreasing slowly during the search. The bottom indicates that progress may be occurring toward identifying the actual delamination. Although still painfully slow, GFS is quicker than UR/URFS while maintaining robustness against false convergence.

The fourth and fifth strategies are variations on random walk techniques with fixed step length.

```
DL0 = truncated UR or URFS search
      [dl0] = update(DL0);
      while not done,
            while not used before,
                  dl  = random walk(dl0, constraints, step size);
                  test if used before;
                  if all neighbors used before,
                        step size = step size / 2;   or
                        [dl0] = update(DL0);
                  end;
            end;
            fom = model(dl);
            if (fom < fom_min) dl0 = dl;
            test if done;
      end;                                          (RWFS)
```

Like GFS, a candidate delamination, dl_0, is initially found from a truncated uniform search. In the "random walk" procedure, one parameter describing that delamination is chosen and perturbed by a fixed step size. Which parameter to vary is selected randomly. The search then moves from one discrete point in the parameter space to the first neighboring point found with a lower *fom* than the current value, hence the name "random walk."

Eventually, no neighboring point can be found with a lower *fom*. One option then is to divide the step size. In the Figure 6 example (which is similar for RWFS and RWPH), the step size is divided twice, thereby improving the accuracy of the identified flaw's size and location from 5 to about 1 percent of plate dimensions. A second option is to "give up" and pick a second starting point from the uniform search phase, DL_0, thereby interrogating another location of the plate with low *fom*. That was also done, resulting in the step-like appearance of the center location error. This figure clarifies that the ability of these methods to find the actual delamination is enhanced by a good initiating uniform search. Without this, RWFS is likely to converge to a false minimum.

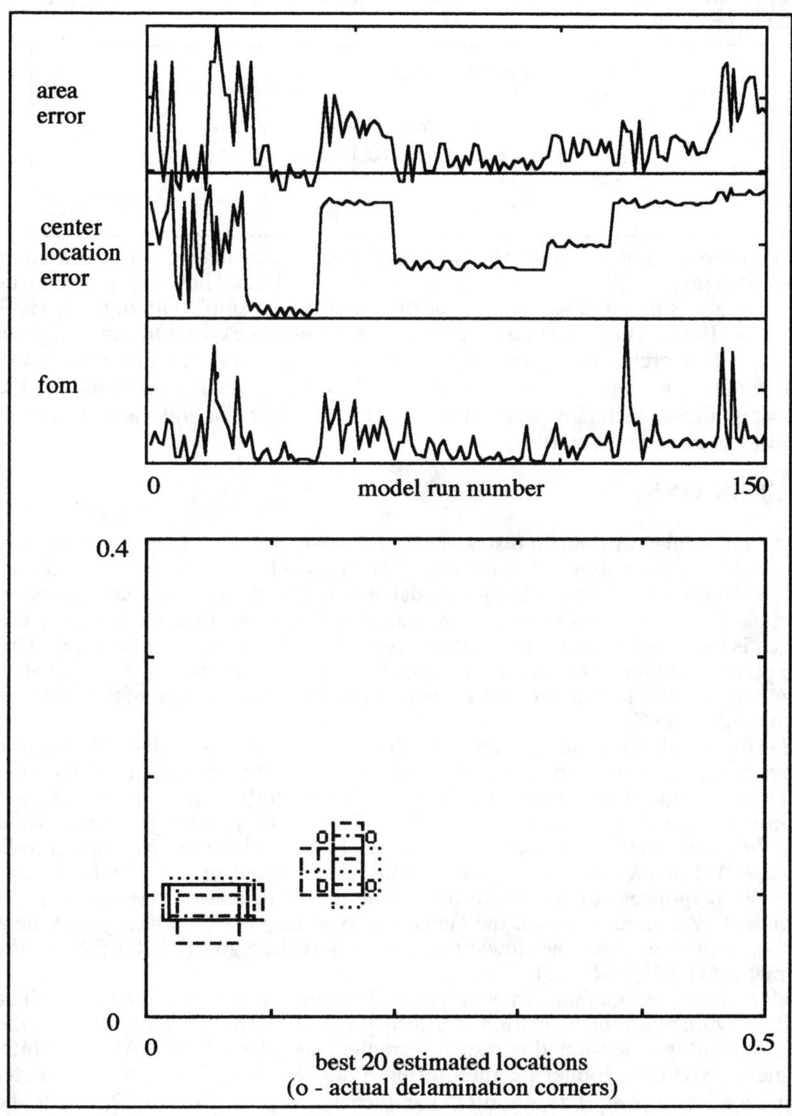

Figure 6: Results from applying a random walk search with probabilistic hill climbing (RWPH), also known as simulated annealing. This search was initiated by the same 25 URFS chosen delaminations used for GFS. By run no. 50, the actual delamination was identified. RWPH then interrogated other possible minima, causing the step-like appearance in the center location error.

The robustness of RWFS can be improved by not automatically rejecting neighboring points that are found to have higher fom. A possible alternative is to replace the acceptance test with:

```
if  (fom < fom_min)  dl_0 = dl ;
else
        r = exp( (fom_min - fom) / tfom);
        if ( r > uniform(0,1) ) dl_0 = dl ;
end;
                                                  (RWPH)
```

In other words, the neighbor is accepted if r is greater than a uniform random number in the range of [0,1]. In the literature this alternate test is known as "probabilistic hill climbing", "simulated annealing," or the "Metropolis step" (Otten, et al, 1989, Kirkpatrick, 1983). As the search progresses, the acceptance criterion can be tighten by reducing the search control parameter, *tfom*. The approach here is to automatically assign *tfom* to the standard deviation of the 7 lowest values of *fom* found for the particular minimum being investigated. *tfom* is reset to its initial high value whenever the algorithm picks a new starting point from DL_0.

CONCLUSIONS

Time and frequency domain responses from built-in transducers may provided an early warning of a possible delamination. The approach here consists of assuming candidate delaminations and running a model until the predicted amplitude responses align closely with those measured, as indicated by low *fom*. One weakness of this approach is that it predicts a delamination, even if the damage is another type. This may be partially compensated by recent studies that indicate that knowing the equivalent damage may be sufficient to estimate a structure's residual buckling strength (Salunkhe and Mujumdar, 1997).

The effort to identify the best-estimate delamination can be reduced if random search strategies are used together appropriately. For example, if the differences between the damaged and undamaged responses are subtle (e.g., Figure 3), then a long uniform search (UR/URFS) is the best first step. The longer the uniform search, the more likely the global minimum will be found later by the random walk methods (RWFS/RWPH). This process can be expedited by precalculating and storing the responses for the uniformly selected delaminations, since they are independent. As an alternative, the Gaussian fixed step (GFS) strategy may be a good compromise between the slower but more robust strategies (UR/URFS) and the faster methods (RWFS/RWPH).

Finally, fixed-step methods may be preferable over random step methods, since the former avoid wasteful evaluation of nearby points in the parameter space (i.e., the same disadvantage mentioned earlier for gradient-based methods). After a while, these methods will no longer be able to find a discrete neighbor in the parameter space with a lower *fom*. The algorithm can then either (1) terminate; (2) divide the step size, thereby increasing the result's accuracy; or (3) start all over at another promising point in the delamination space. This allows achieving a specific tolerance on the result (e.g., 1 percent of each plate direction) and systematically interrogating local minima around the plate until the global minimum is found.

The opinions expressed in this paper are those of the author and do not necessarily reflect those of any Government agency, including the Defense Nuclear Facility Safety Board.

REFERENCES

Choi, K., C. H. Keilers, and F. K. Chang, "Impact Damage Detection in Composite Structures Using Distributed Piezoceramics," 35th AIAA/ASME/ASCE/AHS/ASC Structures, Structural Dynamics, and Materials Conference, Apr 18-20, 1994, Hilton Head, SC.

Hajela, P., "Non-Gradient Methods in MDO Status and Future Directions," AIAA 97-1570, 38th AIAA/ASME/ASCE/AHS/ASC Structures, Structural Dynamics, and Materials Conference, Apr 7-10, 1997, Kissimmee, FL.

Keilers, C. H. and F. K. Chang,"Identifying Delamination in Composite Beams Using Built-In Piezoelectrics: Part I - Experiments and Analysis," Journal of Intelligent Material Systems and Structures, Vol 6, No. 5, Sep 1995, pp 649-663.

Keilers, C. H. and F. K. Chang,"Identifying Delamination in Composite Beams Using Built-In Piezoelectrics: Part II - An Identification Method," Journal of Intelligent Material Systems and Structures, Vol 6, No. 5, Sep 1995, pp 664-672.

Keilers, C. H., "Identifying Composite Plate Delaminations Using Built-In Sensors and Actuators," AIAA 97-1343, AIAA/ASME/AHS Adaptive Structures Forum, Apr 7-10, 1997, Kissimmee, FL.

Kirkpatrick, S., C. D. Gelatt, M. P. Vecchi, "Optimization by Simulated Annealing, *Science*, Vol. 220, No. 4598, May 13, 1983, pp 671-680.

Mujumdar, P. M., and S. Suryanarayan, "Flexural Vibrations of Beams with Delaminations," *Journal of Sound and Vibration*, Vol. 125, No. 3, 1988, pp 441-461.

Salunkhe, A. and P. Mujumdar, "Estimation of Buckling Loads of Laminated Composite Plates with Flaws Using a System Identification Approach," AIAA 97-1160, 38th AIAA/ASME/ASCE/AHS/ASC Structures, Structural Dynamics, and Materials Conference, Apr 7-10, 1997, Kissimmee, FL.

Otten, R. H. J. M, and L. P. P. P. van Ginneken, *The Annealing Algorithm*, Kluwer Academic Publishers, 1989.

SESSION 9

SENSING TECHNOLOGY DEVELOPMENT II

Dry Contact Ultrasonic Sensors for Structural Monitoring

B. T. KHURI-YAKUB, F. L. DEGERTEKIN and J. PEI

ABSTRACT

New applications in in-situ monitoring for defects and failure detection impose significant challenges for conventional ultrasonic techniques which involve wetting or large area contacts. Hertzian contact Lamb wave sensors solve some of these problems for structural monitoring and material characterization. These sensors minimize the contact size and provide accurate, in-situ information under harsh conditions. A normal mode analysis of the Lamb wave excitation mechanism shows the high mode selectivity of the Hertzian contact transducers which are confirmed by experiments. These transducers can be effectively used for structural monitoring and tomographic imaging. As an example, we present results obtained from erosion/corrosion monitoring experiments in water carrying pipes in nuclear reactors.

INTRODUCTION

As natural modes of propagation in plate-like structures, Lamb waves emerge as a valuable tool for ultrasonic sensing techniques (Victorov, 1967). Conventionally, bulk mode conversion and interdigital transducers are used to excite and detect Lamb waves for defect detection and material characterization (Nayfeh and Chimenti, 1988; White 1987). These techniques rely on immersion, coupling fluid or permanent contact with the structure to be monitored. In many applications this is undesirable and/or not practical. Therefore, ultrasonic transducers, which can excite and detect Lamb waves using dry, temporary or permanent contacts can be useful for certain applications in high temperature, clean environments (Degertekin and Khuri-Yakub 1994; Pei et al. 1995). Insensitivity to surface conditions and high mode selectivity are other desired features for many other applications (Alleyne and Cawley 1992).

In this paper, we summarize the development and applications of the Hertzian contact Lamb wave transducers in structural characterization and defect detection. These transducers use a piezoelectric transducer to excite extensional waves in a buffer rod which in turn couples energy to the Lamb waves in the plate at a point contact between the buffer rod and the plate (Fig. 8).

B.T. Khuri-Yakub, Stanford University, E.L. Ginzton Laboratory, Stanford, CA 94305

Using the normal mode theory, we show that these transducers can be used to excite the lowest order Lamb waves selectively and with high signal to noise ratio (SNR) and present experimental results. Finally, we implement Lamb wave sensors for materialcharacterization and structural monitoring and present results for erosion/corrosion monitoring in pipes and tomographic imaging of defects.

LAMB WAVE EXCITATION BY SURFACE SOURCES

THEORY

The 2-dimensional normal mode theory is widely used to analyze waveguide problems (Auld 1990). The approach is generally applicable to layered, anisotropic plates since Lamb waves are the normal modes of these structures. To derive the Lamb wave excitation efficiency by surface sources we start with the complex acoustic reciprocity relation for a non-piezoelectric plate,

$$\frac{\partial}{\partial x}\{-\mathbf{v}_2^* \cdot \mathbf{T}_1 - \mathbf{v}_1^* \cdot \mathbf{T}_2\} \cdot \hat{\mathbf{x}} + \frac{\partial}{\partial z}\{-\mathbf{v}_2^* \cdot \mathbf{T}_1 - \mathbf{v}_1^* \cdot \mathbf{T}_2\} \cdot \hat{\mathbf{z}} = \mathbf{v}_2^* \cdot \mathbf{F}_1 + \mathbf{v}_1^* \cdot \mathbf{F}_2 \,. \tag{1}$$

In this expression, $\mathbf{F}_1$ and $\mathbf{F}_2$ are the volume forces driving the corresponding stress and particle velocity fields with the same subscripts and * denotes complex conjugation. The modes in the wave guide can also be excited by traction forces $\mathbf{T} \cdot \hat{\mathbf{z}}$ and the velocity sources on the waveguide surfaces, as in most practical applications, such as wedge transducers, surface-bonded transducers etc.

In the far-field, where the evanescent fields can be neglected, arbitrary velocity and stress fields with $e^{j\omega t}$ time variation can be expressed as a sum of orthogonal modes, i.e. the Lamb wave modes in a free plate, as

$$\begin{aligned} \mathbf{v}(x,z) &= \sum_n a_n(x)\mathbf{v}_\mathrm{n}(z) \\ \mathbf{T}(x,z)\cdot\hat{\mathrm{x}} &= \sum_n a_n(x)\mathbf{T}_\mathrm{n}(z)\cdot\hat{\mathrm{x}} \,. \end{aligned} \tag{2}$$

By choosing the field $\mathbf{v}_1$ as an arbitrary field distribution as in (2) and $\mathbf{v}_2$ being a normal mode of propagation in $+x$ - direction in the free plate we have,

$$\mathbf{v}_2 = e^{-j\beta_n x}\mathbf{v}_\mathrm{n}(z), \tag{3}$$

where β_n is a real propagation constant. Setting $\mathbf{F}_1 = \mathbf{F}_2 = 0$ eliminates the volume sources, and leaves the surface traction and velocity fields as the only sources. Integrating (1) along the waveguide and invoking mode orthogonality, we obtain the equation for the mode amplitude for the $\mathrm{n^{th}}$ mode excited by surface sources as

$$4P_{nn}\left(\frac{\partial}{\partial x} + j\beta_n\right)a_n(x) = f_{sn}(x) \tag{4}$$

whose solution has the form

$$a_n(x) = \frac{1}{4P_{nn}} e^{-j\beta_n x} \int_0^x f_{sn}(\zeta) e^{j\beta_n \zeta} d\zeta \tag{5}$$

assuming that $a_n(0) = 0$. P_{nn} is the acoustic power flow associated with the nth mode,

$$P_{nn} = \frac{1}{4} \int_{\substack{cross \\ section}} (-\mathbf{v}_n^* \cdot \mathbf{T}_n - \mathbf{v}_n \cdot \mathbf{T}_n^*) \cdot \hat{\mathbf{x}}. \tag{6}$$

Finally, we write the forcing function $f_{sn}(x)$ in (4) explicitly in terms of the applied tractions and the velocity components of the propagating mode at the plate surfaces as

$$\begin{aligned} f_{sn}(x) &= \mathrm{v}_{nx}^*(b)\mathrm{T}_{xz}(x,b) + \mathrm{v}_{nz}^*(b)\mathrm{T}_{zz}(x,b) \\ &\quad - \mathrm{v}_{nx}^*(0)\mathrm{T}_{xz}(x,0) - \mathrm{v}_{nz}^*(0)\mathrm{T}_{zz}(x,0) \end{aligned}. \tag{7}$$

Since the plate is stress free, the tractions at the surface due to the normal modes do not appear in the forcing function. Knowing the particle velocity distribution along the thickness of the plate and the power carried by the normal mode, one can evaluate the mode amplitude due to different surface sources. In figure 1, we plot the two-way relative excitation efficiency of the A_0 and S_0 modes in an aluminum plate of 1 mm thickness as a function of frequency. In this case, a 2 dimensional normal traction source (T_{zz}) with 20 μm lateral dimension is used. The efficiency is normalized to the power in the A_0 mode for frequency-thickness product (fd) of 0.05 MHz*mm. The results indicate that the A_0 mode is selectively excited by 55 dB relative to the S_0 mode for fd < 0.2 Mhz*mm using a normal traction source. Noting that in this fd range only these two Lamb wave modes exist, single mode operation with the A_0 mode should be possible. In figure 2, the variation of efficiency at 500 kHz is given as a function of aluminum thickness.

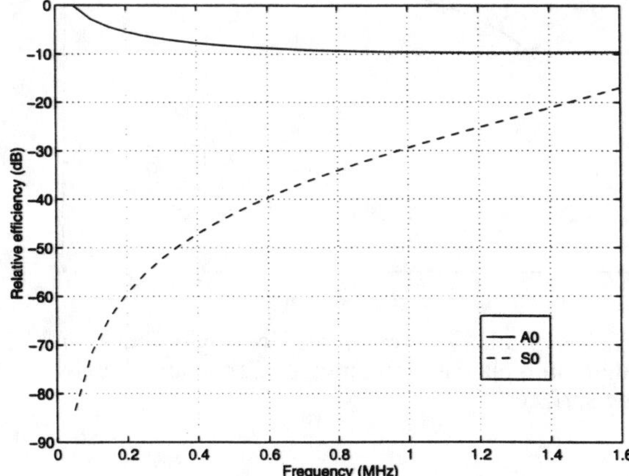

Figure 1. Variation of relative excitation efficiency of the A_0 and S_0 modes in a 1 mm thick aluminum plate with frequency.

When a lateral traction source is considered (T_{xz}), similar mode selectivity figures can not be achieved in this fd range as depicted by figures 3 and 4. In these figures the excitation efficiency is normalized to the S_0 mode excitation efficiency at fd=0.05 MHz*mm Applying lateral traction at one surface of the plate generates both modes with comparable amplitudes. However, equation (7) suggests that if same lateral stress in applied to the top and bottom surfaces of the plate, the S_0 mode can be excited due to the symmetric nature of the source.

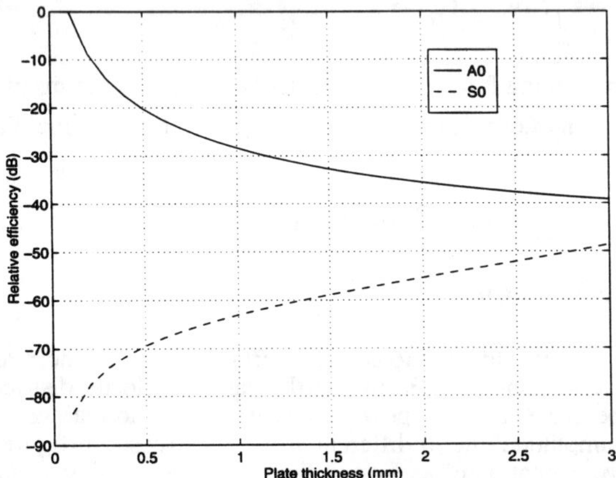

Figure 2. Variation of relative excitation efficiency of the A_0 and S_0 modes in aluminum plates at 500 kHz with plate thickness.

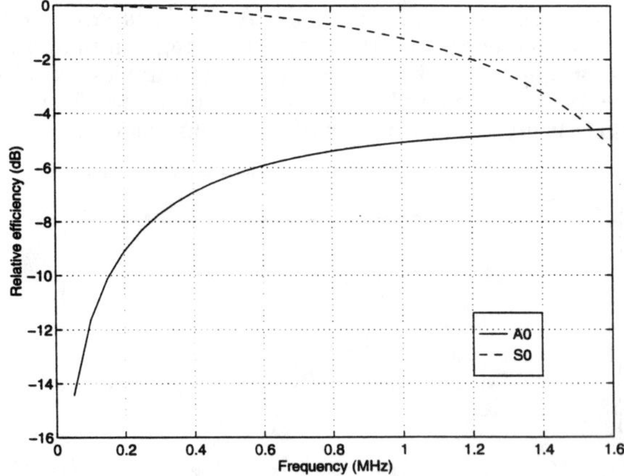

Figure 3. Variation of relative excitation efficiency of the A_0 and S_0 modes in 1 mm thick aluminum plate as a function of frequency. Lateral traction is applied on one surface.

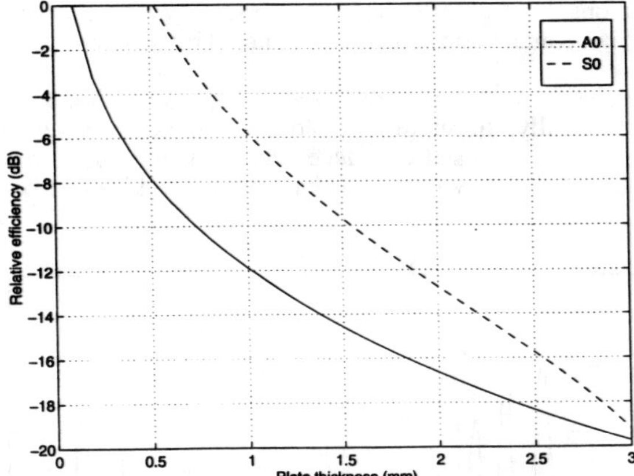

Figure 4. Variation of relative excitation efficiency of the A_0 and S_0 modes in aluminum plates at 500 kHz with plate thickness. Lateral traction is applied on one surface.

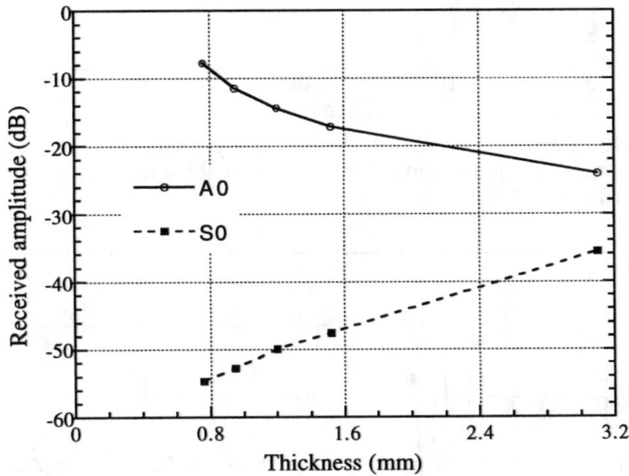

Figure 5. Received amplitude of the A_0 and S_0 modes in aluminum plates of varying thickness at 500 kHz.

EXPERIMENTAL RESULTS

The theoretical results of the previous section are tested using Hertzian contact transducers in different combinations. In figure 5, we plot the received amplitude of the A_0 and S_0 modes as a function of the aluminum plate thickness at 500 kHz when the transducer is contacted to the top surface of the plate as in figure 8. In this case, normal traction is applied on a small area, resembling the assumptions for the theoretical calculations for figure 2. The experiments verify the single mode A_0 operation by Hertzian contact transducers as predicted by theory.

The excitation efficiency of different dry contact lateral traction sources are also investigated experimentally. In figure 7 we plot the digitized received signal when a Hertzian contact transducer is used to induce lateral traction on the top surface of a 0.77 mm thick aluminum plate with a transducer with 500 kHz center frequency.

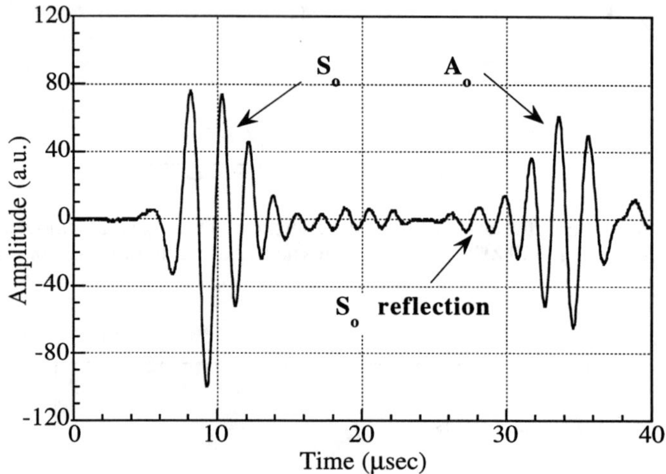

Figure 6. Received signal at the receiving dry contact transducer terminals when lateral traction is applied on a 0.77 mm thick aluminum plate on one of the surfaces.

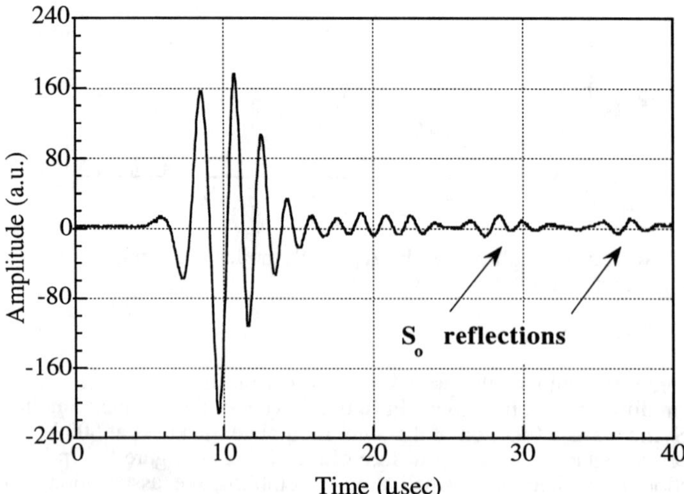

Figure 7. Received signal at the receiving dry contact transducer terminals when lateral traction is applied on a 0.77 mm thick aluminum plate on both of its surfaces for single mode S_0 operation.

The S_0 and A_0 mode arrivals are easily distinguished since their group velocities are different. This data confirms the prediction shown in figure 4 that lateral traction applied at one surface of the plate will generate both modes with comparable amplitudes. This method is useful when dual mode operation is desired for differential measurements.

When the transducers are used to induce lateral traction on both surfaces of the plate, matching the particle velocity profile of the symmetric mode (S_0), the excitation of the antisymmetric mode (A_0) should be suppressed. This can also be achieved by the dry Hertzian contact transducers, as shown in figure 7. In this case, all the parameters are the same as the experiment in figure 6, except the contact locations. It is clearly seen that the S_0 mode is selectively excited and the SNR is around 55 dB.

EROSION AND CORROSION DETECTION

An application of the dry contact transducers is realized by monitoring the erosion and corrosion in fluid carrying steel pipes. Figure 8 shows the experimental setup. Steel is used as the transducer buffer pin material to obtain the best acoustic impedance match between the pin and the steel plate or pipe wall under inspection. The tip of the pin is rounded to have a radius of curvature of 100 μm. The spherical tip gives a dry point contact to a test plate and the transducer/steel rod assembly is spring loaded to ensure stable contacts every time the steel pin is pushed against plate surface.

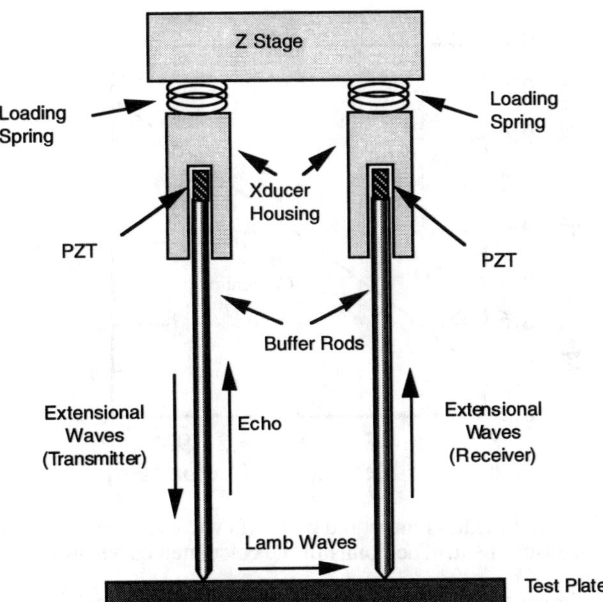

Figure 8. Experimental setup of plate thickness measurement. Two PZT-5H transducer and steel buffer pin set are used as the Lamb wave transmitter and receiver.

An identical transducer/steel pin set is used as a receiver to detect the Lamb wave transmitted through the test plate. It is worth noting that different types of pins with different radii of curvature have been used depending on the application and materials involved.The PZT-5H is machined to a cylindrical shape with a diameter of 6.35 mm and a height of 12.7 mm. The resonant frequency of the transducer when bonded to steel is 70 kHz with 50% fractional bandwidth ensuring that only the extensional mode is generated in the buffer rod.

Lamb wave phase velocity is obtained using a time of flight (TOF) measurement technique. A short electrical pulse of 500 volts is applied to the transmitter which generates the extensional wave in the steel pin. At the contact interface between pin tip and test structure, part of the extensional mode energy is reflected back to the transmitter and generates an echo electrical pulse, the other part of the energy is coupled into the plate as the A_0 Lamb wave. After propagation through the plate, the Lamb wave is converted back to an extensional mode at the receiver tip and then, to an electrical signal in the receiving transducer. The pin-to-pin time of flight is measured by monitoring the time interval between the transmitter echo and the received signal. The first zero crossing of the echo signal triggers the start of the time delay counter and the first zero crossing of the transmitted signal stops the counter. The time delay measured is the time that takes the Lamb wave to travel through the pipe wall from the transmitter tip to the receiver tip. The effect of the steel pins is eliminated due to the subtraction of the time delay in the pins from both transmitter and receiver.

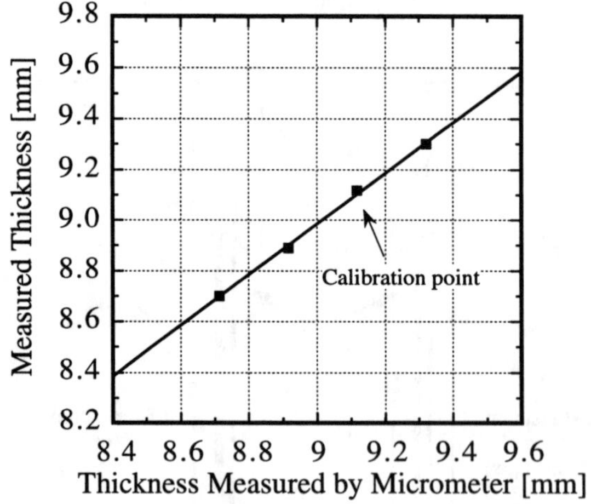

Figure 9. Plate thickness measured by Lamb waves as compared to the external measurement. The points fit the calculated curve within 1% error.

Because the electronic time delays introduced by the amplifiers and filters are also included in the TOF data, a proper calibration is necessary. The calibration is performed on a steel step wedge plate with its thickness varying from 8.25 mm to 9.25 mm. With known distance between the transmitter and receiver pin tips, a TOF data point is taken in a region of the steel plate whose thickness is measured using a micrometer caliper. This data point is fitted to the theoretical dispersion curve and hence the fitting parameters corresponding to the electronic delays are obtained. With the same fitting parameters, the TOF data of other regions are converted to plate thickness accordingly. Figure 4 shows the measurement of regions with different thickness vs. the measurement done with a micrometer. Note that the fit of the data is better than 1%.

PIPE WALL THICKNESS MEASUREMENT

Similar measurements are then performed on a steel pipe elbow that was removed from service. The diameter of the pipe is 35 cm, and the wall thickness is about 1cm. A calibration run was also done on the steel pipe at a location where the thickness could be independently measured with a micrometer caliper. Wall thickness variations were measured to range from 7.8 mm to 10.1 mm which indicated the presence of extensive corrosion at some locations inside the pipe. This thickness variation was further verified with a traditional pulse echo measurement using a longitudinal wave transducer operating at a frequency of 10 MHz. Figure 10 shows a comparison of the pipe thickness as measured with both methods. Overall, there is excellent agreement between the two measurements.

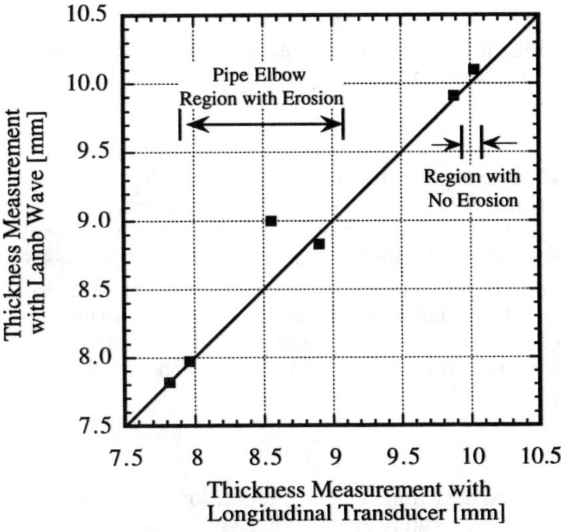

Figure 10. Thickness measurement performed on a section of steel pipe.

TOMOGRAPHIC IMAGING SYSTEM

Figure 11(a) shows the experimental setup of a scan imaging system used to detect a depression defect. The thickness of the aluminum plate under inspection is around 1 mm. The transducers, therefore, are chosen according to this thickness to have their operating frequency around 200 kHz. Only the A_0 mode Lamb wave can be generated in the plate at this frequency. The PZT-5H piezoelectric material is machined to a cylindrical shape with a diameter of 3 mm and a height of 5 mm to obtain the 200 kHz resonance. Quartz rods are used as the buffer pins with their tips sharpened to have a radius of curvature of 100μm. The transducers are spring loaded and pressed on the plate surface to form a dry contact.

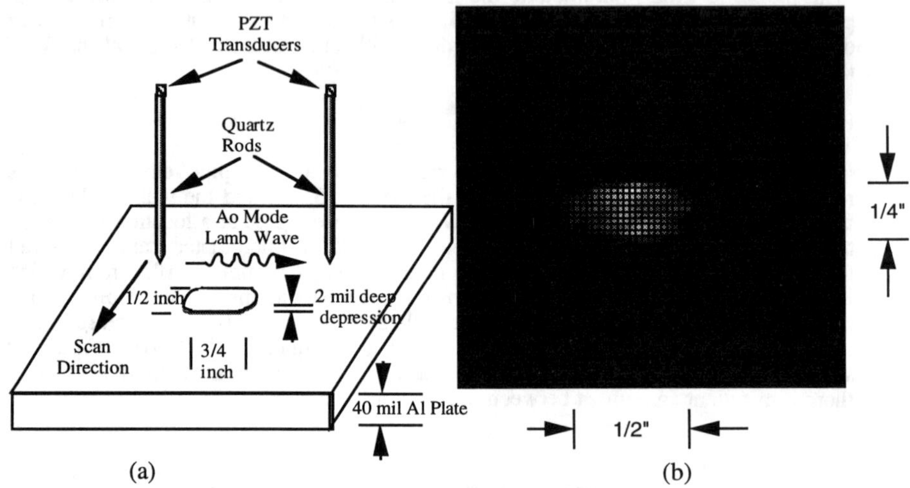

Figure 11. (a) Imaging system for aircraft skin. (b) Reconstructed time of flight image of the depression.

A 1/2" by 3/4" and roughly 0.5 mm (2 mil) deep depression is milled into the aluminum plate to simulate a defect. The plate is mounted on an X-Y and rotation stage and the transducers scan over the area with defect. Due to the thinning of the defect area, the time of flight increase as the pins scan across the depression. Filtered back projection[8] tomographic inversion method is used to reconstruct the defect image. 64 points are taken along each scan direction and a total of 60 scans are taken at an interval of every 3°. The reconstructed time of flight image is shown in figure 11(b) with a resolution of 64×64 pixels. The depth resolution with the obtained SNR is around 2.7μm.

CONCLUSION

The dry Hertzian contact transducers enable ultrasonic sensing applications in different environments. By carefully choosing the frequency of operation and the nature of the contact to the specimen, these transducers can selectively excite and detect Lamb waves with high SNR. The normal mode theory provides a guide for selecting the critical parameters. When applied to erosion/corrosion monitoring, dry contact ultrasonic transducers emerge as a sensitive defect sensors which do not require any surface preparation. It is also possible to obtain tomographic images by implementing sensor arrays that are spring loaded to conform with contoured structures. Some other applications of these sensors for material characterization and defect detection can be found elsewhere (Degertekin and Khuri-Yakub, 1997).

REFERENCES

Alleyne, D.N., and Cawley, P. 1992. "The interaction of Lamb waves with defects." *IEEE Trans. on UFFC,* 39:381-97.

Auld, B.A. 1990. *Acoustic fields and waves in solids.* Krieger Publishing Co. Vol. 2.

Degertekin, F.L., and Khuri-Yakub B.T. 1994. "In-situ acoustic temperature tomography of semiconductor wafers.", *Appl. Phys. Lett.*, 64(11):1338-40.

Degertekin, F.L., and Khuri-Yakub B.T. 1997. "Lamb wave excitation by Hertzian contacts with applications in NDE." *IEEE Trans. on UFFC*, 44(4).

Nayfeh, A., and Chimenti, D. 1988. "Propagatoin of guided waves in fluid-coupled plates of fiber-reinforced composite." *J. of Acoust. Soc. Am.*, 83:1738-43.

Pei, J., Degertekin, F.L., Khuri-Yakub, B.T., and Saraswat, K.C. 1995. "In situ thin film thickness measurement with acoustic Lamb waves.", *Appl. Phys. Lett.*, 66(17):2177-79

Victorov, I.A. 1967. *Rayleigh and Lamb waves*: Plenum Press.

White, R.M., Wicher, P.J., Wenzel, S.W., and Zelles, E. 1987. "Plate-mode ultrasonic oscillator sensors.", *IEEE Trans. on UFFC*, 34(2):162-70.

An Integrated Circuit to Operate a Transponder with Embeddable MEMS Microsensors for Structural Health Monitoring

P. NEUZIL, F. M. SERRY, O. KRENEK and G. J. MACLAY

ABSTRACT

We present the design and the results of tests performed on a CMOS IC which interfaces with MEMS capacitive sensors and with an antenna coil for embedding inside physical structures to monitor local stress and strains. Using inductive coupling, the IC powers itself from a pilot signal at 130kHz, which it receives from an interrogating device called the Reader. The IC then makes measurements of the sensor's output and converts the measurement to a binary code, which is then transmitted to the Reader by differential phase-shift-key (DPSK) modulation of a 65kHz signal across the antenna coil. The IC has roughly 3000 transistors and was fabricated using 2-micron double poly-silicon, double metal technology at a commercial foundry. The IC uses a $\Delta\Sigma$ A/D converter for the measurement of the capacitance of a sensing capacitor as compared with that of a reference capacitor.

This project was funded by the US Naval Research Laboratory (NRL), contract number N00014-94-C-2231, with Mr. Lee Gause as COTR., MTS Systems Corporation as the prime contractor and system integrator.

INTRODUCTION

Capacitive sensors in conjunction with sigma-delta type A/D converters offer the advantages of low power consumption, high resolution, and good linearity with relatively simple circuitry. This combination is well-suited for telemetry applications in health monitoring of structures. Such applications rely on wireless transmission of the information to and from a transmitter-responder (transponder) attached to or embedded in the structure. In many applications, it is imperative that the size of the transponder remains as small as possible. This is the case, for instance, when the transponder is embedded inside an object, the structural integrity of which is monitored using on-board measurements made by and transmitted by the transponder to the outside. An example of such an embedding object is the wing of an aircraft or the hull of a submarine. Smaller transponders are better because they alter the structure of the embedding object less than large transponders.

The bulk of the power available to the transponder is consumed for transmission, and therefore little power is left for sensor operation, data acquisition and conditioning since large storage capacitors are to be eliminated from the transponder. Therefore, low power consumption is essential for the IC which runs the telemetry cycle at the transponder.

The IC is a part of an integrated device (the transponder), which also includes one each off-chip reference and sensing capacitors, and the antenna coil, which is also off-chip. The sensing and reference capacitors are identical in design and construct except that the plates of the reference capacitor are fixed in place and relative to each other, whereas the relative position of the plates of the sensing capacitor is allowed to change in response to local changes in stress and strain in the embedding object. The acquisition circuit interfaces directly with these capacitors, and measurements are made to compare the capacitance of the sensing capacitor to that of the reference capacitor. This is to help compensate for effects such as temperature variation.

The measurement is made using a switched capacitor technique, and the measured analog signal is digitized and then encoded. The code is then transmitted out using the antenna coil. In the following sections of this article, we describe the design of an IC which addresses many of the problems associated with telemetry involving microsensors embedded inside structures. Then we describe the results of some tests we have performed on fabricated IC's based on our design.

Figure 1 shows the block diagram of the IC, interfaced with sensing and reference capacitors (not shown in Fig. 1), zener diode to protect circuit against overvoltage, capacitor for power supply filtering and a coil for data transmission. The key parts of the circuit are further described in details.

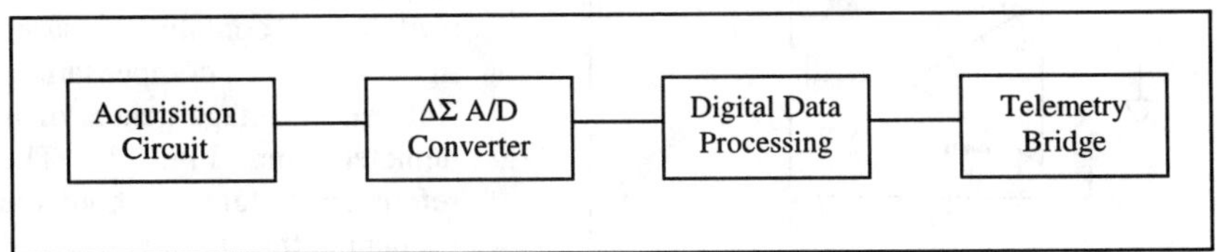

Figure 1. Block diagram of the IC.

I. ACQUISITION CIRCUIT

The acquisition circuit (Fig.2) works as an offset-free sample and hold circuit. During phase I the external capacitor CX (pressure sensing capacitor) is charged through the switch S1 up to the charging voltage, Vref. During phase II the internal path is discharged (S3, S9 closed) and the holding capacitor CQX is charged up to the offset voltage of the OTA (S6 closed). Then during phase III the charge from the capacitor CX is transferred on to the holding capacitor CQX (S3,

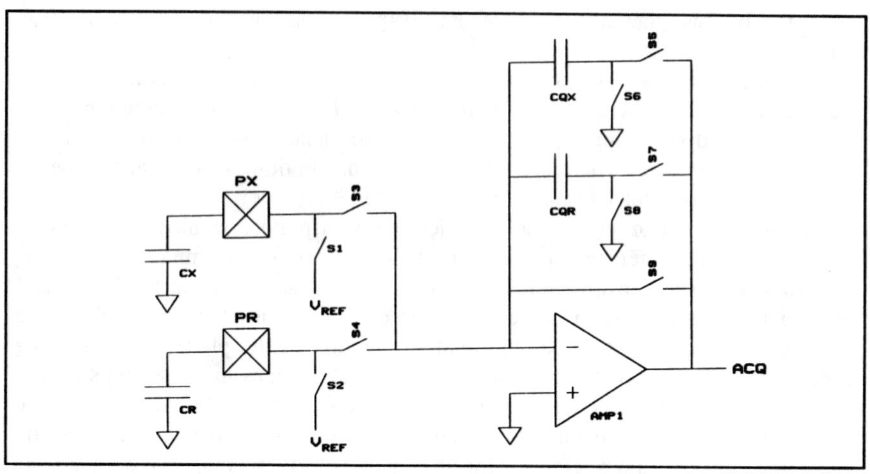

Figure 2. Acquisition Circuit

S5 closed) and the signal on the node ACQ is further processed in the next block. In the same time the external reference capacitor CR is charged up to the reference voltage, Vref. During phase IV the internal path is again discharged (S4, S9 closed) and the second holding capacitor CQR is charged up to the offset voltage (S8 closed). Finally during phase V the charge from capacitor CR is transferred on to the holding capacitor CQR (S4, S7 closed). The signal on the node ACQ is further processed, the signal on the node ACQ is hold during phase I as well. All switches contain charge injection compensation. Reference voltage generator is depicted on Fig. 3. The reference voltage is about the P-channel threshold voltage.

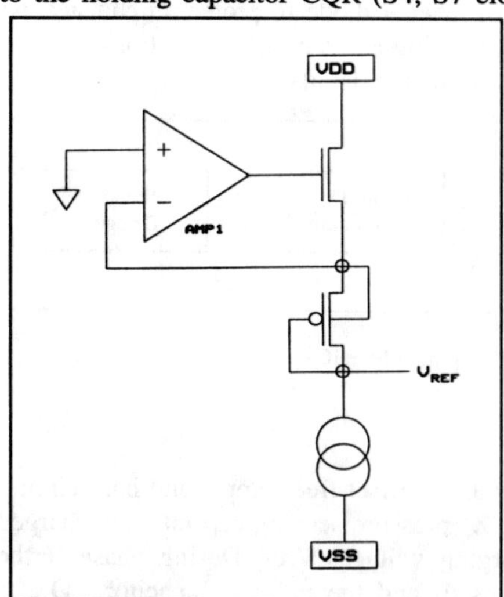

Figure 3. Reference Voltage Generator

The output of the acquisition circuit as described above is the input to the ΔΣ A/D modulator.

II. DELTA-SIGMA MODULATOR

First-order delta-sigma modulator consists of an OTA as an balancing integrator and the comparator (Fig.4). Before the conversion starts, the integrator capacitor C2 is discharged (S5 closed). The output from the acquisition circuit serves as an input to the delta-sigma modulator. During phase I capacitor C1 is grounded (S2, S3 closed), then during phase II the charge from capacitor C1 is transferred onto the integrator capacitor C2 (S1, S4 closed). The signal on the balancing node (BAL) controls the polarity of the reference voltage. If BAL<0 then the reference voltage have to be subtracted and two phases are necessary to complete this task. As long as the reference signal from acquisition circuit is available during phases IV and V, then during phase IV the capacitor C1 is charged (S1, S3 closed) and during phase V the charge from that capacitor is

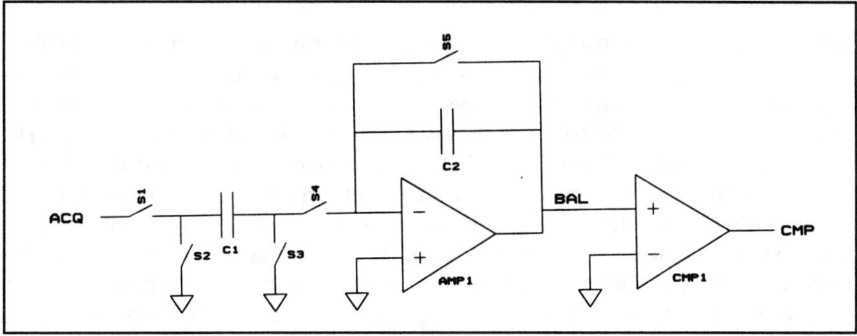

Figure 4. Delta-Sigma Modulator

transferred (subtracted) to the integrator capacitor C2 (S2, S4 closed). If BAL>0 then for charge transfer one phase is enough - phase V (S1, S4 closed), but during phase IV the capacitor C1 was grounded (S2, S3 closed). Then phase I occurs again to start new conversion cycle. All switches contain charge injection compensation.

III. DIGITAL PROCESSING

The signal CMP from delta-sigma modulator gives the information about relation between the measured and the reference signals. For 8-bit-resolution conversion 512 conversion cycles of delta-sigma modulator are necessary. Digital signal CMP from delta-sigma modulator is fed into 9-bit presettable counter, which is clocked at the end of phase III (after measured signal is transferred on the integrator capacitor C2). At the beginning of the conversion the counter is preset to 256 (hex), after the conversion is completed the 8 LSBs represent the conversion value called "Neg". The formula for extracting converted value of C_x is:

$$C_x = \frac{2 * Neg - 256}{256} C_r, \quad (1)$$

where C_r is capacitance of the reference capacitor and 256 assumes that 8 bit resolution is used.

The 8-bit conversion value (Neg) and the other necessary information (overflow, polarity and parity) is then parallel loaded into the shift register, which serves during the transmission period. The information consisting of the header, conversion value, sign, overflow and parity is then eight-times serially transmitted.

IV. DIFFERENTIAL PHASE-SHIFT KEYING TRANSMISSION

Digital communication of data for health monitoring is inherently more robust, and when augmented with error-detection/correction code, add reliability to the operation of health monitoring system based on wireless transmission of information. The IC described here was interfaces with an inductive coil, across which a sinusoidal signal at 130kHz, transmitted by a commercial Reader, coexists with a 65kHz signal, which is derived from the 130kHz signal by the IC. The phase 65kHz signal, which is the outgoing signal (from transponder to Reader) is modulated by circuitry on the IC. The phase shifts are always 180 degrees, and are designed to transmit binary data. Every time the binary "0" is to be transmitted, the phase of the outgoing signal is shifted by 180 degrees. Sample of the encoded signal is in Fig. 5 and 6.

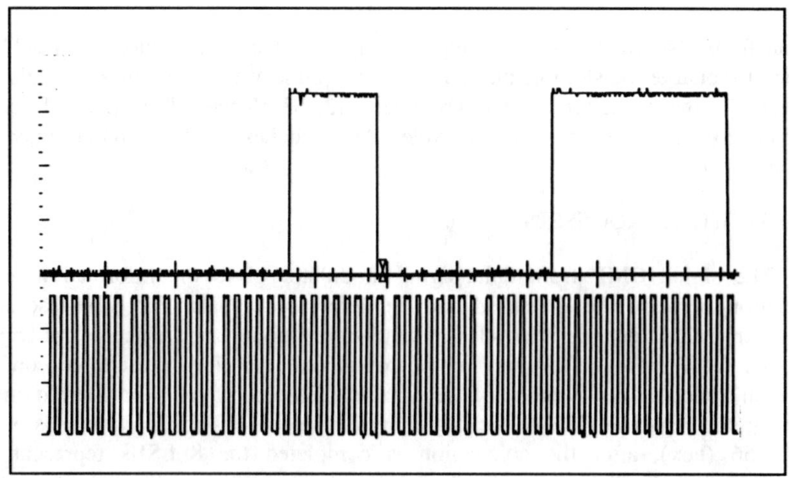

Figure 5. Output Data Signal (top) and DPSK Signal (bottom) versus time.

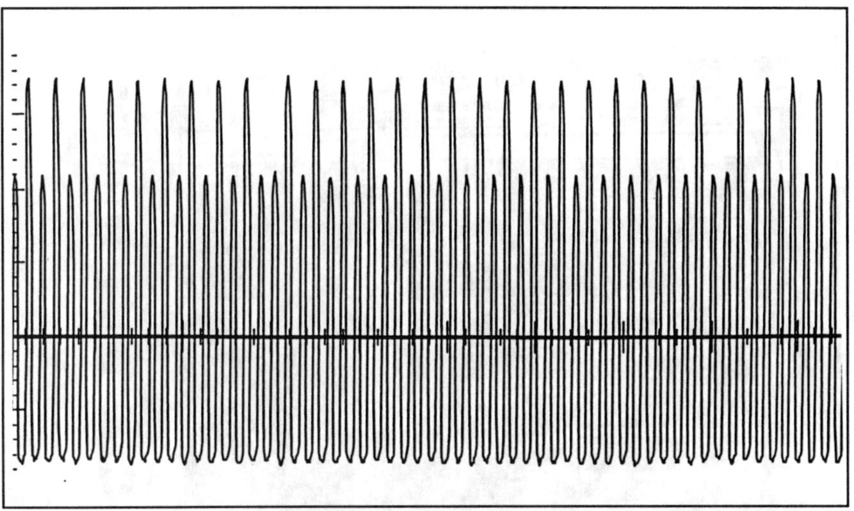

Figure 6. DPSK Signal Versus Time Taken from the Output Coil

V. CLOCK, POWER AND SIZE REQUIREMENT

As mentioned earlier, the size requirements on the sensing device are quite stringent in many health monitoring applications. This is, for example, true of those applications where structural integrity of the embedding object may be compromised by the embedded sensor if the latter is not small enough. A prime example of this is components of aircraft such as the wing. On the other hand wireless communication with the embedded sensors are preferred due to obvious advantages. A wireless sensor and transponder need to extract the power for their operation from an external source. In some telemetry applications, where size is not an issue, SMD capacitors on the transponder are used to store the extracted power and once the necessary amount of energy is stored, measurement and transmission follow. These commercially available SMD capacitors are however currently too thick for the type of thickness-sensitive applications for which we designed our IC. The IC described here extracts the power for the operation of the transponder from the 130kHz interrogation signal from the Reader, and uses no storage capacitor. However there is a capacitor between power supply and ground to remove the ripple of power supply. The IC also extracts its clock signal from the 130kHz signal from the reader.

IC chip was fabricated in commercial CMOS foundry (Orbit Semiconductor) using N-well 2 μm double polysilicon, double metal process. Layout of the chip is shown in Fig. 7.

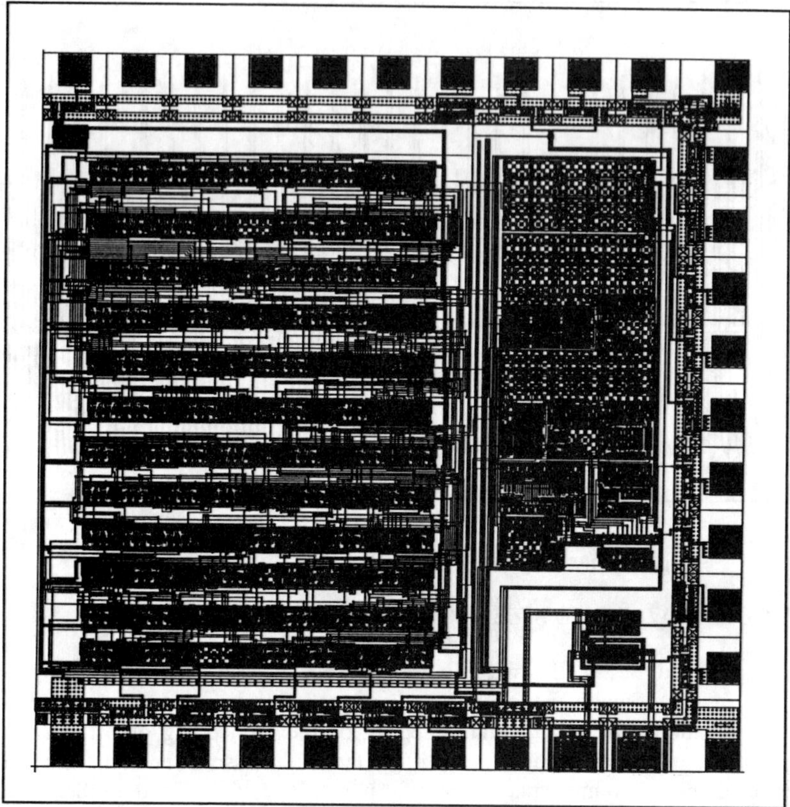

Figure 7. Chip Layout. On the left side are the digital blocks, on the top right are the analog blocks. The telemetry bridge is on the bottom right.

VI. MEASUREMENTS ON THE IC

The IC has an array of small (1.7pF) on-chip capacitors, which are used for testing the linearity of response of the acquisition and A/D conversion circuitry and for calibration of the transmitted data from the transponder to the Reader. Figure 8 shows a typical measurement which shows the linearity of the circuit. The offset needs to be improved in future designs, however, the linearity is reasonable.

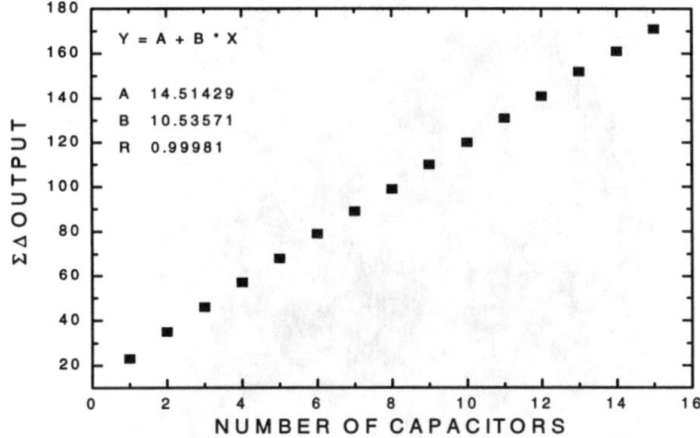

Figure 8. Linearity of measurements made by the IC. The output number of ΔΣ is the value "Neg" in formula (1)

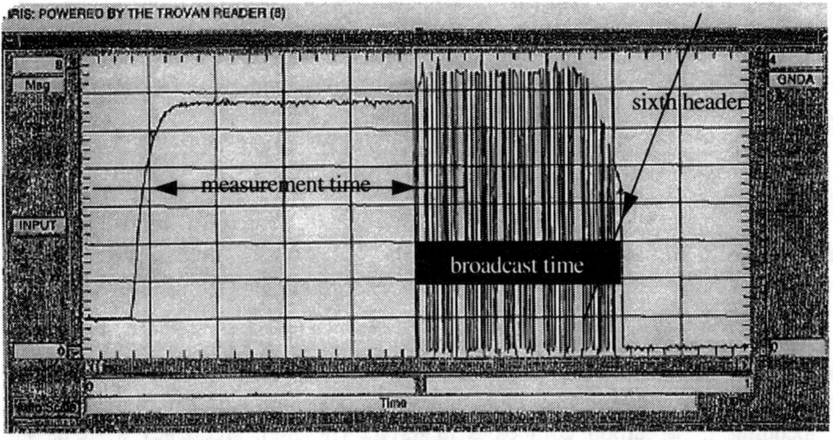

Figure 9. Signal transmitted by the transponder and measured at the Reader Antenna

In Figure 9, we have the measurement from the Reader's receiving antenna of the signal transmitted by the transponder. The first portion of the signal corresponds to measurement time, during which data is transmitted from the transponder. The second portion includes the encoded data.

Figure 10. Complete circuit mounted on PC board with coil prior embedding. Photo courtesy of Donald Krantz, MTS Systems.

The chip was also assembled at small PC board with all devices and embedded into carbon composite. Final device prior embedding in show in Fig. 10. This part of work was done by Donald Krantz, MTS.

CONCLUSION

We have designed and tested an IC which interfaces with capacitive microsensors, makes measurements of the microsensor output, digitizes the readout, encodes the digitized data, and transmits this data out to a receiver (the Reader).

There are currently many commercial telemetry systems available on the market. These systems can be divided into two classes: low frequency (10's of kHz to 2 MHz), and high frequency (900 MHz and higher). Our IC is designed for embedding in a carbon composite material. The high frequency systems are unattractive for telemetry to an embedded sensor in a carbon composite for several reasons. High frequency systems are typically active systems which means they require some type of energy storage device, such as a battery for the operation of the transponder. Such systems also have limited lifetime unless the battery is re-charged periodically. Also, the storage device would have to withstand high temperature curing processes associated with fabrication of some of the commonly used composite materials. Lower frequencies transmit relatively easily through carbon composites, whereas high frequency signals do not.

We have designed our IC to interface with an existing low frequency telemetry Reader (TROVAN). These Prior to choosing this platform, we investigated several of these commercial telemetry systems, including the popular TIRIS system from Texas Instruments. TIRIS required a resonance circuit with a large inductive coil with a high Q, which is quite difficult to make. Also, the resonance frequency of the transponder shifts significantly when the transponder coil is embedded in or near a conductive object, and this shift is not easy to track; this became a major liability because TIRIS uses Frequency Shift Keying for digital data communication.

The TROVAN system has a low-Q coil and a simple method for implementing DPSK for digital wireless transmission of the measurement data. Our chip can also work with piezoresistive sensors with few small modifications.

This project was funded by the US Naval Research Laboratory (NRL), contract number N00014-94-C-2231, with Mr. Lee Gause as COTR., MTS Systems Corporation as the prime contractor and system integrator.

The authors would like to acknowledge the help of John Belk, McDonnel-Douglas, Donald Krantz, MTS and Phil Troyk, IIT.

Effectiveness and Limits of Self-Sensing Piezoelectric Actuators

J. TANI, G. G. CHENG and J. QIU

ABSTRACT

A self-sensing piezoelectric actuator is a single piezoelectric element simultaneously used for both sensing and actuation functions. The motion sensing signal from the element is separated from the applied control signal by a special electric circuit, and can then be amplified and fed back to the same element to induce control forces. In this paper, the equations of motion for a piezoelectrically coupled electromechanical system are presented and used as the basis for an analysis of the simultaneous sensor/actuator problem. The electrically equivalent model and transfer function of a self-sensing actuator are investigated. It is demonstrated that factors such as the equivalent resistance and capacitance change of the piezoelectric element have significant effects on the measured transfer function. Finally, several suggestions are proposed to enable the successful realization of the self-sensing actuator.

INTRODUCTION

Piezoelectric elements have been used extensively as actuators or sensors in active structural control experiments. In most of these, the piezoelectric element performs a single transducer function: either sensing or actuation. Recent advances in the application of piezoelectric materials provide a method for combining both functions in a single device and producing a "self-sensing actuator." The self-sensing actuator described in 1992 [1,2] has attracted much interest, and many papers have been published on this topic [4-8]. Dosch et al [1] demonstrated the concept of using a single piece of piezoceramic material to simultaneously sense the vibrations of a cantilever beam and apply a control moment to it. This is possible because there is a duality in the piezoelectric material between the effect which relates mechanical stress and electric charge displacement and the effect which relates mechanical strain and electric field. This is advantageous in the control of flexible structures since the

Junji Tani, Tohoku University, Katahira 2-1-1, Aoba-ku, Sendai 980-77, Japan
Gen Guo Cheng, Tohoku University, Katahira 2-1-1, Aoba-ku, Sendai 980-77, Japan
Jinhao Qiu, Tohoku University, Katahira 2-1-1, Aoba-ku, Sendai 980-77, Japan

resulting sensor and actuator are perfectly collocated with each other; as several authors have pointed out [2,3], in the absence of actuator and sensor dynamics, a structure controlled with collocated velocity feedback is unconditionally stable. In addition, combining a sensor/actuator pair in a single package cuts instrumentation in half and reduces the number of system components.

The key problem in self-sensing is to separate the sensor signal caused by mechanical stress of a piezoelectric element from the control signal applied to the same element. Since the amplitude of the control signal is usually much larger than that of the sensor signal, it is very difficult to separate the sensor signal from the mixed signals. Many researchers use a bridge circuit to achieve this objective. This paper begins with a discussion of the balance of the bridge circuit, which is the most essential part of self-sensing, and investigates some important factors that often are ignored in the previous work; in fact, these factors are usually present and play a critical role in determining the success of the implementation of a self-sensing actuator.

Finally, several suggestions are proposed to enable the realization of the self-sensing actuator.

PIEZOELECTRIC TRANSDUCER ELECTRIC CIRCUIT ANALOG

Several papers have presented the equations of motion for a piezoelectrically coupled electromechanical system [9-10]. These equations will be called the actuator and sensor equations of the system. In the case of a single piezoelectric element (for example, the application to a cantilevered beam), the charge, voltage and capacitance are scalars, and the equations are expressed as follows:

Actuator equation

$$(M_s + M_p)\ddot{\Gamma} + (K_s + K_p)\Gamma - \Theta V = F \tag{1}$$

Sensor equation

$$\Theta^T \Gamma + C_p V = q \tag{2}$$

where M_s, M_p, K_s, and K_p are the mass and stiffness matrices of the structure and the piezoelectric element, and C_p is the capacitance of the piezoelectric element. The vector Γ represents the generalized displacement coordinates, F is an external force, and Θ is the piezoelectric coupling matrix. V is the voltage applied to the piezoelectric element and q is the charge. The $\Theta^T\Gamma$ term is proportional to mechanical strain. The appearance of Θ in both equations implies duality in collocated control when the equations are written for a single piezoelectric element.

Now let

$$y = \dot{\Gamma} \quad \text{and} \quad I = \dot{q} \tag{3}$$

so that y is generalized velocity and I is current. Then, taking the Laplace transform

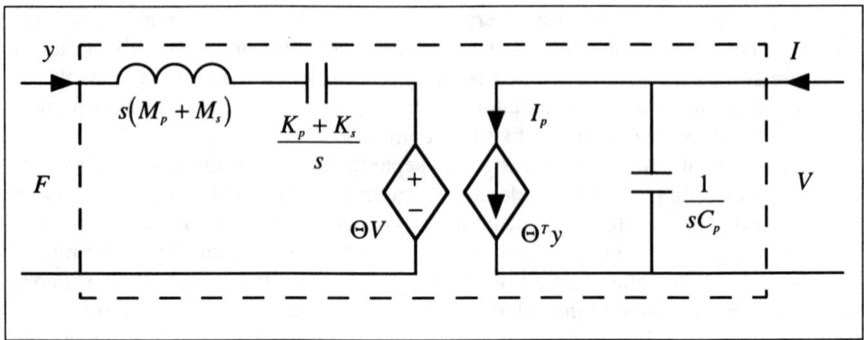

Fig. 1 Electrical analog relating voltage to velocity and velocity to current

of Eqs. (1) and (2), we obtain

Actuator equation

$$s(M_s + M_p)y + \frac{K_p + K_s}{s} y = F + \Theta V \tag{4}$$

Sensor equation

$$\Theta^T y + sC_p V = I \tag{5}$$

where s is the Laplace variable.

We can represent Eqs. (4) and (5) schematically as shown in Fig. 1. This figure illustrates the coupling between electrical and mechanical variables. The mass M_s+M_p corresponds to an inductor and the stiffness K_p+K_s corresponds to a capacitor. The voltage V (or current I) can be used to control the mechanical character of the piezoelectric element through the coupling quantity ΘV, while the velocity y (or exciting force F) affects the sensor voltage through the coupling quantity $\Theta^T y$.

Applying electrical circuit laws to Fig.1, we have

$$M_s + M_p \Theta^T y = I - C_p \dot{V} \tag{6}$$

If we let

$$I_p = \Theta^T y \tag{7}$$

then

$$I = I_p + C_p \dot{V} \tag{8}$$

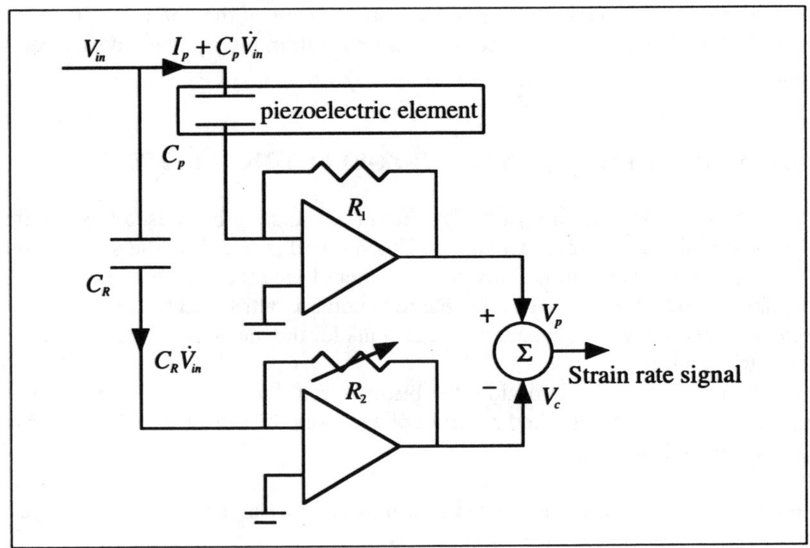

Fig.2 Circuit for a self-sensing actuator

In this expression, I_p is the sensor signal we want to obtain, while I and V are the control current and voltage, respectively.

Some researchers [2] have used the circuit shown in Fig. 2 to extract the sensor signal I_p from the total current I. The control voltage V_{in} is applied to both the piezoelectric element and a reference capacitor C_R. The current I flows through the piezoelectric element, while $C_R\dot{V}_{in}$ flows through C_R. The op-amp/resistor circuits convert these currents to voltages V_p and V_c, which are then differenced to yield a voltage proportional to strain rate. The two voltages are

$$V_p = -R_1 I = R_1(I_p + C_p\dot{V}_{in}) \tag{9}$$

and

$$V_C = -R_2 C_R \dot{V}_{in} \tag{10}$$

If we choose R_1, R_2, and C_R so that $R_1C_p = R_2C_R$, the difference of the two voltages is

$$V_p - V_C = -R_1 I_p = -R_1\Theta^T\dot{\Gamma} \tag{11}$$

which is proportional to the strain rate and independent of the control voltage, as desired. With this, we have achieved the desired result in theory; the physical realization, however, is actually very difficult.

ELECTRICAL MODEL OF THE PIEZOELECTRIC ELEMENT

In research on self-sensing, a bridge circuit has usually been used to separate the sensor signal from the control signal. This method is based on the assumption that the piezoelectric element is equivalent to either of the models in Fig. 3, where V_p and Q_p are the equivalent voltage and charge generators attributed to the piezoelectric effect, respectively. The capacitor C_p accounts for the dielectric properties of the piezoelectric material [1].

On the basis of this assumption, the bridge circuit shown in Fig. 4 is used to eliminate the effect of the applied control voltage from the sensor signal. Here, Z is an arbitrary impedance.

From Fig. 4, using the Laplace transform, we can write the sensor signal V_s as

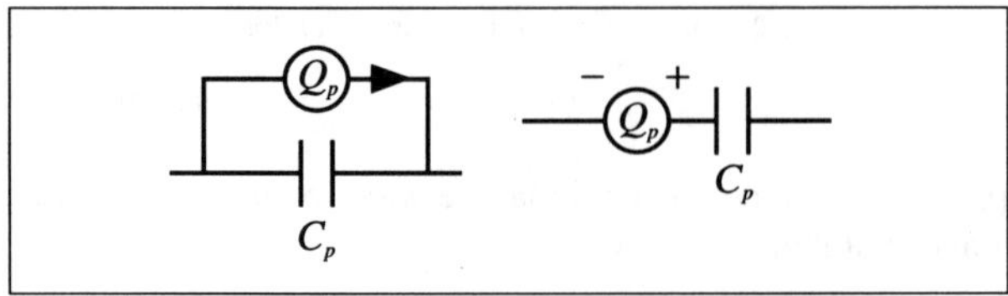

Fig.3 Equivalent models of the piezoelectric element

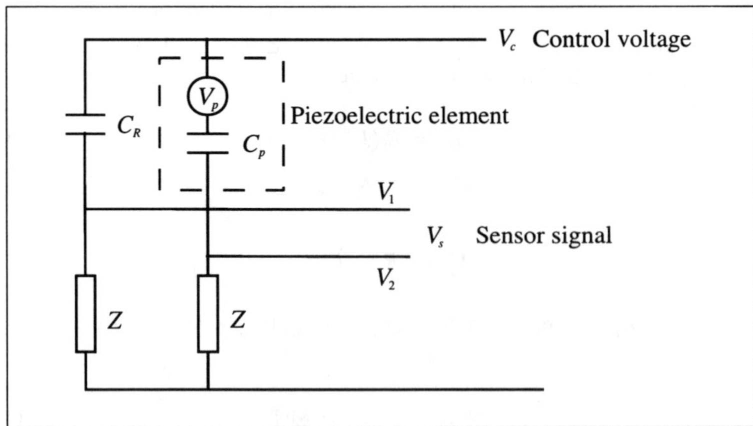

Fig.4 Bridge circuit for a self-sensing actuator

$$V_s(s) = V_1 - V_2 = \frac{ZC_p s}{1+ZC_p s} V_p(s) + [\frac{ZC_R s}{1+ZC_R s} - \frac{ZC_p s}{1+ZC_p s}] V_c(s) \tag{12}$$

where $V_s(s)$, $V_p(s)$, and $V_c(s)$ are the Laplace transform of V_s, V_p, and V_c, respectively.

If we choose the reference capacitor so that

$$C_R = C_p \tag{13}$$

the bridge is said to be balanced, and Eq. (12) becomes

$$V_s(s) = \frac{ZC_p s}{1+ZC_p s} V_p(s) \tag{14}$$

Here the sensor output voltage V_s is a function of only the piezoelectric voltage V_p due to mechanical deformation. In this case, self-sensing is theoretically achieved. But in practice, the piezoelectric element may not match the ideal model. More realistic equivalent circuits for the piezoelectric element are shown in Fig. 5, where R is the resistance of the element and ΔC represents a variation in its capacitance. We let ΔZ be the impedance of the parallel combination of R and ΔC; in the frequency domain

$$\Delta Z(j\omega) = \frac{R}{1+j\omega\Delta CR} \tag{15}$$

so the corresponding admittance is

$$\Delta Y(j\omega) = \frac{1}{\Delta Z(j\omega)} \tag{16}$$

and the admittance of the parallel combination of C_p and ΔY is

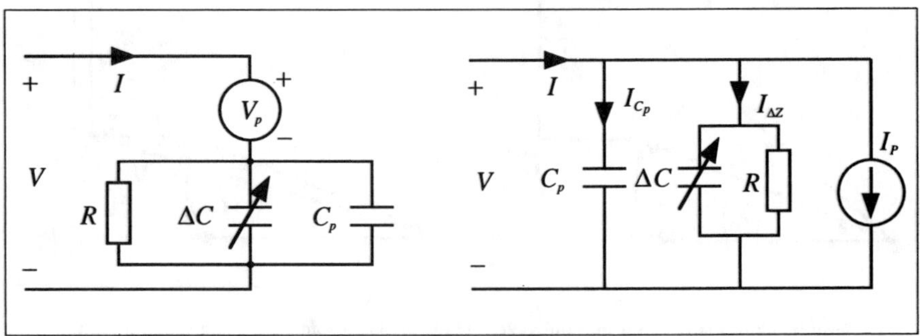

Fig.5 Modified models of the piezoelectric element

$$Y(j\omega) = j\omega C_p + \Delta Y(j\omega) = |Y|\angle\alpha \tag{17}$$

If all variables are assumed to be sinusoidal with frequency ω, a vector analysis approach may be used in which each variable is a complex number represented by a capital letter with an overbar. Then the total current flowing into piezoelectric element may be written as

$$\bar{I} = \bar{V}Y(j\omega) + \bar{I}_p = \bar{I}_{C_p} + \bar{I}_{\Delta Z} + \bar{I}_p \tag{18}$$

To simplify the graphical depiction, we assume for convenience that the sensor signal current $\bar{I}_p$ and the current $\bar{I}_R$ drawn by the resistance are in phase; then the currents in each branch may be represented as vectors as shown in Fig.6.

If the reference capacitor is again chosen so that C_R=C_p, then

$$\bar{I}_{C_p} = \bar{I}_{CR} \tag{19}$$

where $\bar{I}_{CR}$ is the current flowing into the reference capacitor, and the difference $\bar{I}_d$ of the currents flowing in the piezoelectric element and reference capacitor is

$$\bar{I}_d = \bar{I}_{C_p} + \bar{I}_{\Delta Z} + \bar{I}_p - \bar{I}_{CR} = \bar{I}_{\Delta Z} + \bar{I}_p \tag{20}$$

The vector components in this expression are also shown in Fig. 6. In this case, the bridge circuit is as shown in Fig. 7. Since the parameters ΔC and R may change with time, it is very difficult to keep the circuit balanced. In addition, the control

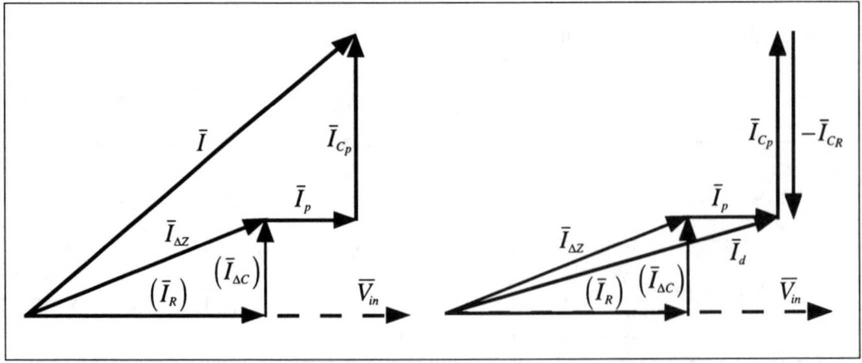

Fig.6 Currents flowing into piezoelectric element (left) and the result of subtracting the reference capacitor current (right)

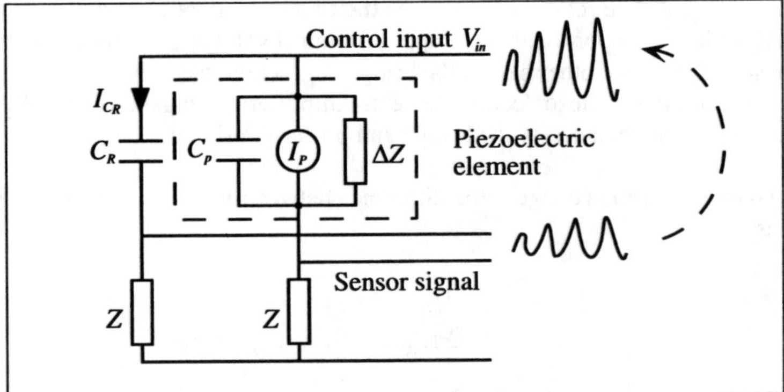

Fig.7 Bridge circuit with feedback path

voltage V_{in} is usually much larger than the sensor voltage. If the bridge does not stay balanced, a part of the control voltage appears in the sensor output, and undesirable positive feedback may occur through the controller as shown in Fig. 7. This phenomenon has been observed in many experiments.

SELF-SENSING ACTUATOR CIRCUIT TRANSFER FUNCTION

In this section we discuss the transfer function of a self-sensing actuator circuit

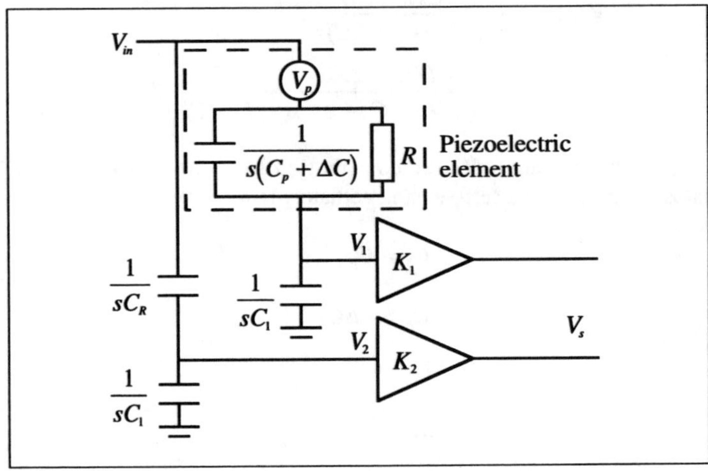

Fig.8 Bridge circuit for a self-sensing actuator

as shown in Fig. 8. The reference path uses the two capacitors C_R and C_1 to yield a voltage V_2 which is proportional to the applied control voltage V_{in}. This is subtracted from the signal from the other half of the bridge, V_1, to produce a signal proportional only to the strain in the piezoelectric element. Amplifiers with gains K_1 and K_2 may be used to adjust the balance of the bridge and buffer the signal.

The output (sensor) voltage is the difference between V_1 and V_2, where V_1 can be written as

$$\begin{aligned} V_1(s) &= \frac{\frac{1}{sC_1}}{\frac{1}{sC_1} + \frac{R}{1 + R(C_P + \Delta C)s}}(V_{in}(s) + V_P(s)) \\ &= \frac{1 + R(C_P + \Delta C)s}{1 + R(C_1 + C_P + \Delta C)s}(V_{in}(s) + V_P(s)) \end{aligned} \tag{21}$$

and V_2 is

$$V_2(s) = \frac{\frac{1}{C_1 s}}{\frac{1}{C_1 s} + \frac{1}{C_R s}} V_{in}(s) = \frac{C_R}{C_1 + C_R} V_{in}(s) \tag{22}$$

Then the output signal is

$$\begin{aligned} V_s(s) = &\frac{1 + R(C_P + \Delta C)s}{1 + R(C_1 + C_P + \Delta C)s} V_P \\ &+ \left(\frac{1 + R(C_P + \Delta C)s}{1 + R(C_1 + C_P + \Delta C)s} - \frac{C_R}{C_1 + C_R} \right) V_{in} \end{aligned} \tag{23}$$

The second term is an undesired contribution from the control voltage V_{in}; it must equal zero to achieve a self-sensing actuator. If we have

$$\begin{aligned} C_R &= C_p \\ C_P &>> \Delta C_P \end{aligned} \tag{24}$$

and

$$\left.\begin{aligned} RC_1\omega \\ RC_p\omega \end{aligned}\right\} >> 1 \tag{25}$$

in the frequency range of interest, we have

$$V_S \cong \frac{C_P}{C_1 + C_P} V_P \tag{26}$$

which is the desired result. To achieve this goal, we should select a piezoelectric material and a bridge capacitor C_1 in such a way that the products RC_p and RC_1 are as large as possible, and ΔC_p is as small as possible.

The voltage applied to the piezoelectric element V_a is the difference between V_{in} and V_1:

$$\begin{aligned} V_a(s) &= V_{in}(s) - V_1(s) \\ &= V_{in}(s) - \frac{1 + R(C_P + \Delta C_P)s}{1 + R(C_1 + C_P + \Delta C_P)s}(V_P(s) + V_{in}(s)) \\ &= \frac{RC_1 s}{1 + R(C_1 + C_P + \Delta C_P)s} V_{in}(s) - \frac{1 + R(C_P + \Delta C_P)s}{1 + R(C_1 + C_P + \Delta C_P)s} V_P(s) \end{aligned} \tag{27}$$

Solving Eq. (23) for V_p and substituting into Eq. (27), the actuator voltage can be written as

$$\begin{aligned} V_a(s) &= \frac{RC_1 s}{1 + R(C_1 + C_P + \Delta C_P)s} V_{in} \\ &\quad + \left(\frac{1 + R(C_P + \Delta C_P)s}{1 + R(C_1 + C_P + \Delta C_P)s} - \frac{C_R}{C_1 + C_R} \right) V_{in} - V_s \\ &= \frac{C_1}{C_1 + C_R} V_{in} - V_s \end{aligned} \tag{28}$$

It is convenient to define coefficients β_1, β_2, and β_3 as

$$\beta_1 \equiv \frac{C_1}{C_1 + C_R} \tag{29}$$

$$\beta_2(s) \equiv \frac{1 + R(C_P + \Delta C_P)s}{1 + R(C_1 + C_P + \Delta C_P)s} \tag{30}$$

$$\beta_3(s) \equiv \frac{1 + R(C_P + \Delta C_P)s}{1 + R(C_1 + C_P + \Delta C_P)s} - \frac{C_R}{C_1 + C_R} \tag{31}$$

If the conditions expressed in Eqs. (24) and (25) are satisfied, then

$$\beta_3 \approx 0 \tag{32}$$

When a piezoelectric element is bonded to a cantilevered beam assumed to be in pure bending [1], the expression for the moment applied by the piezoelectric element is given by:

$$M(x,t) = K_a V_a(t)[h(x - x_1) - h(x - x_2)] \tag{33}$$

where h(·) is the Heaviside step function, x_1 and x_2 are the locations of the ends of the piezoelectric element, and the constant K_a depends on the geometric and material properties of the beam and the piezoelectric element:

$$K_a = bd_{31}E_P(t_P + t_s) \tag{34}$$

where t_p and t_s are the thickness of the piezoelectric element and the beam, respectively, b is the width of both the beam and the piezoelectric element, and E_p is Young's modulus of the piezoelectric material. The piezoelectric strain constant is denoted by d_{31}.

The expression for the sensor voltage generated by the piezoelectric effect is

$$V_P = K_s\left(\frac{\partial w(x_2,t)}{\partial x} - \frac{\partial w(x_1,t)}{\partial x}\right) \tag{35}$$

where

$$K_s = \frac{E_P d_{31} b h_c}{C_P} \tag{36}$$

and w is the deflection of the beam and h_c is the distance from neutral axis to the midpoint of the piezoelectric element.

The open-loop block diagram of the system is shown in Fig. 9, where the mechanical transfer function of the beam from the applied moment to the displacement is represented by G. From this block diagram we see that the transfer function from the applied control voltage to the sensor output is

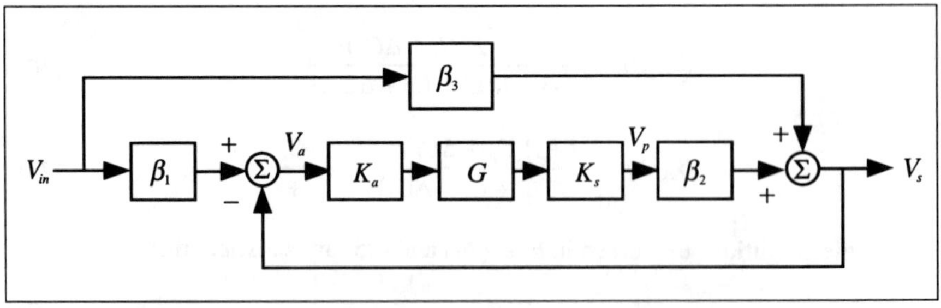

Fig.9 Open-loop block diagram of beam with self-sensing actuator

$$\frac{V_s(s)}{V_{in}(s)} = \frac{\beta_3 + \beta_2 K_s G K_a \beta_1}{1 + \beta_2 K_s G K_a} \tag{37}$$

The output V_s is a combination of both desired the piezoelectric voltage and an undesired input feedthrough term. If there is a finite parallel leakage resistance R or a non-zero piezoelectric capacitance change ΔC, then the feedthrough term β_3 will not be zero and the output voltage of sensor will not correspond to the mechanical strain. In that case, the self-sensing actuator will not be achieved.

CONCLUSION

Recently the concept of a self-sensing actuator has attracted much interest, due to the advantages of collocation and system component reduction. But it is not sufficient to model the piezoelectric element as a lossless capacitor in parallel with a charge generator. In practice, the equivalent leakage resistance of the piezoelectric element and the change in piezoelectric capacitance over time or with applied voltage may cause unacceptable errors.

To successfully implement a self-sensing actuator, the following points should be considered:
(1) A circuit which acts as a negative resistor could be used to offset the leakage resistance of the piezoelectric element.
(2) The piezoelectric material should be chosen so that the capacitance of the piezoelectric element changes very little due to environmental conditions or applied electric field.

REFERENCES

1. J. Dosch, D. J. Inman and E. Garcia, 1992, "A Self-Sensing Piezoelectric Actuator for Collocated Control." *Journal of Intelligent Material Systems and Structures*, Vol. 3, pp. 116-185.

2. E. H. Anderson, N. W. Hagood and J. M. Goodliffe, 1992, "Self-Sensing Piezoelectric Actuator: Analysis and Application to Collocated Structures." *Proceedings 33rd Struct. Dynam. Mater. Conf.*, pp. 2141-2155.

3. D. J. Leo and D. J. Inman, 1993, "Modeling and Control Simulation of a Slewing Frame Containing Active Members." *Journal of Smart Materials and Structures,* Vol. 2, pp. 82-95.

4. H. S. Tzou and J. J. Hollkamp, 1994, "Collocated Independent Modal Control with Self-Sensing Orthogonal Piezoelectric Actuators (Theory and Experiment)." *Journal of Smart Materials and Structures*, Vol. 3, pp. 277-284.

5. N. W. Hagood and E. H. Anderson, 1991, "Simultaneous Sensing and Actuation Using Piezoelectric Materials." *Active and Adaptive Optical Components, SPIE* Vol. 1543, pp. 409-421.

6. S. E. Miller and H. Abramovich, 1995, "A Self-Sensing Piezolaminated Actuator Model for Shells Using a First Shear Deformation Theory." *Journal of Intelligent Material Systems and Structures*, Vol. 6, pp. 624-638.

7. J. Pratt and A. B. Flatau, 1995, "Development and Analysis of a Self-Sensing Magnetostrictive Actuator Design." *Journal of Intelligent Material Systems and Structures*, Vol. 6, pp. 639-648.

8. J. H. Lang, 1988, "Electromechanical Devices as Simultaneous Motion Actuators and Sensors." *MIT Symposium on Intelligent Components in Aerospace.*

9. C. K. Lee and F. C. Moon, 1990, "Modal Sensors/Actuators." *Trans. of the ASME*, Vol. 57, pp.434-441.

10. N. W. Hagood, W. H. Chung and A. von Flotow, 1990, "Modeling of Piezoelectric Actuator Dynamics for Structural Control." *Journal of Intelligent Material Systems and Structures*, Vol. 1, pp. 327-354.

A Peak Strain Sensor for Damage Assessment and Health Monitoring

B. WESTERMO* and L. D. THOMPSON*

ABSTRACT

The paper discusses the development and testing of a passive, peak strain sensor technology for use in damage assessment and health monitoring of bridges and buildings. The technology is based on the irreversible magnetic property changes that occur in a class of steel alloys when strained. This feature provides for sensors which can passively monitor the peak applied strain (or deflection) at a specific location within a structure without the need for constant power or data acquisition equipment. A test system, which has been installed into a private residence near the Hayward fault in Berkeley, CA and has been funded by an NSF program, will be discussed. Sensors have also been developed for monitoring the peak strains within concrete column reinforcing bar. These are currently being tested in conjunction with Caltrans and an NSF grant.

THE TECHNOLOGY

The authors have been involved in the development of structural health monitoring systems based on a passive technological approach over the course of the last several years (Thompson and Westermo, 1993; Westermo and Thompson, 1994; Dunning and Thompson, 1995; Thompson and Westermo, 1996; Westermo, 1996). The deformation sensors utilized are composed of metastable alloys which gradually transform, via a strain-induced martensitic phase transformation, from a paramagnetic, austenitic parent phase to a thermodynamically stable, ferromagnetic, martensitic product phase. The ferromagnetic response (or change in same) of the sensors acts as a means to uniquely determine the peak structural strain in the locales where the sensors are physically attached to the structure. The peak structural strains are useful in assessing the structural health since the degradation

* Strain Monitor Systems, Inc., 1425 Russ Blvd., San Diego, CA 92101 & College of Engineering, San Diego State Univ., San Diego, CA 92182

mechanisms operative over long-term service, such as corrosion, stress corrosion cracking, fatigue, etc., result in a gradual loss of structural stiffness, i.e., a gradual increase in structural compliance. By monitoring the high-stress locations in structures it is then possible to identify the early stages of structural deterioration and to address the remedies to repair or replace specific components as necessary. The principal advantage to this monitoring approach lies in the fact that the sensors can function completely without power or data acquisition systems. The retention of the peak strain information is inherently contained in the amount of martensite produced during structural deformation. Since the martensitic transformation is irreversible, the extent of transformation, indicated by the sensor ferromagnetic response, is directly related to the peak strain which in turn can be correlated with the damage condition.

Figure 1 shows the typical output voltage from an externally attached sensor that has been developed for monitoring peak tensile strains in structures. The gauge operates by applying tensile strains to a sensor element of the types shown in fig. 2. The ferromagnetic content is measured by an electrical coil which encases the necked portion of the sensor element. An inductance bridge circuit within the sensor housing translates the amount of ferromagnetic material within the coil into a voltage output. Elements with different gauge section lengths (the length of the necked down region) produce different ranges of measurable deflection or strain. Figure 3 shows a photograph of such a sensor (the model shown here is about 11.5 cm. long). The sensor is attached by way of the threaded rods at each end.

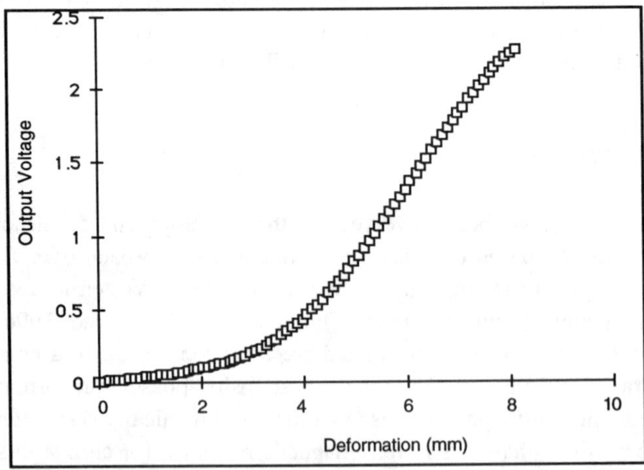

Figure 1. Sensor Output vs. Peak Displacement

Figure 4 shows a comparison of the true strain vs. the measured peak strain from such a sensor. The design earthquake response of the California Dept. of Transportation (Caltrans) I-5 Gavin Canyon replacement bridge was calculated and used as input motion for testing of the sensor. The dashed line indicates the longitudinal, normal strain history experienced at the top of one of the columns and the solid line shows the measured sensor output. Only the tensile (or positive) strains were measured here. It is also possible to measure peak compressive strains through the mechanical yoking of the housing so as to convert compressive strains on the structure to tensile strains on the sensor element. Generally, structural integrity is most readily assessed by examining the level of tensile strains displayed by the most highly stressed structural components although in some applications it may be advantageous to measure compressive strains as well. As can be seen in this figure, the output increases with each new cyclical peak; if one were to monitor the output only after the earthquake, the final value (indicative of the maximum tensile strain occurring at 19 secs.) would be recorded.

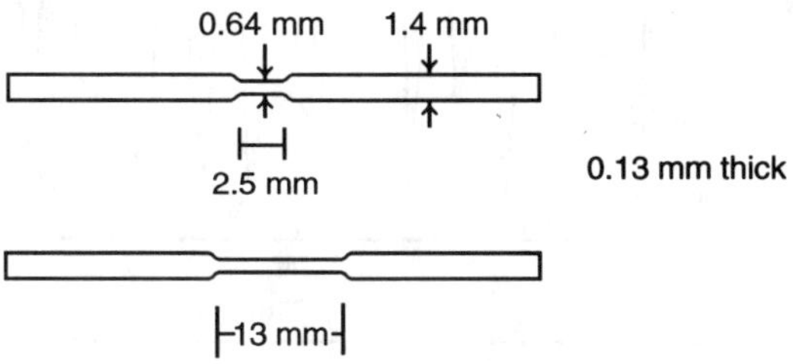

Figure 2. Sensor Element Profiles.

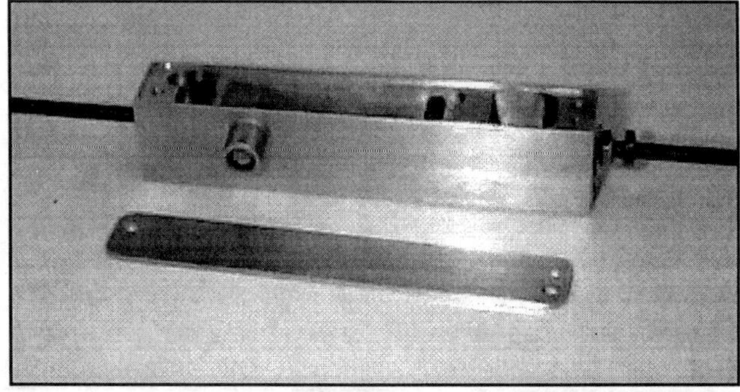

Figure 3. Peak Strain Sensor Housing (with cover removed).

Figure 5 shows the results of a beam bending test where a peak sensor and a bonded, resistive strain gauge were placed on the bottom flange of a steel beam in the maximum moment region. The beam was loaded to failure (buckling of the top flange). The plot shows the output voltage vs. the measured strains at the gauge location. In this instance, the sensitivity of the gauge provides for a measurement threshold of less than 10% of the yield value for the beam. The upswing of the curve at high strain levels is due to the severe geometry change brought on by the initiation of buckling in the compression flange.

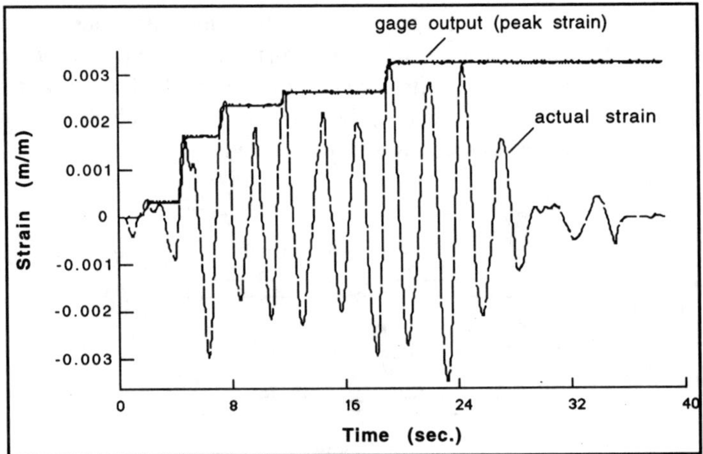

Figure 4. Comparison of Applied Strains vs. the Peak Sensor Output for an Earthquake Excitation.

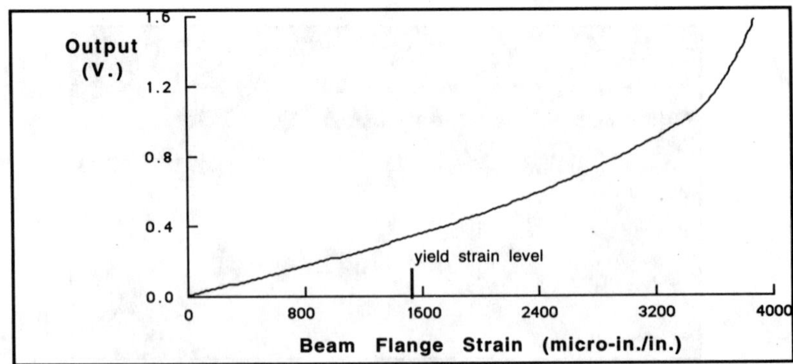

Figure 5. Sensor Output vs. the Peak, Normal Strain on the Bottom Flange of a Steel, Box Beam

APPLICATIONS

Earthquake Damage Monitoring; Berkeley, CA Installation

A demonstration system of the sensor applications for residential damage monitoring was installed into a large, woodframe house in January of this year. The structure is a three-floor, wood frame building with approximately 300 sq. m area and a concrete pile and grade beam foundation. The three year old structure is of typical woodframe and drywall construction which was built to current code specifications for seismic construction. The house sits on a 30 degree slope near the top of the Berkeley Hills overlooking the San Francisco Bay. The property is within 100 m of the Hayward fault system.

An analyses of the structure and foundation indicated that the most vulnerable structural mechanism was the in-plane shear resistance of the framed walls. The foundation consists of 18 concrete piles extending down to bedrock. These are tied into the concrete retaining wall and grade beams with continuous reinforcing bar. The sections of the first and second floor which sit on the foundation are securely anchored with bolts and straps. The center of the house is open to the top two stories with little internal wall or framing for support. The upper floor deadload and lateral load are entirely supported by the first level framed walls. These walls have no intentional diagonal bracing; their shear rigidity comes only from the wall board. In this situation, the lateral loads are most likely to "rack" the walls in a damage scenario. The equivalent shear strength of the walls was roughly the same for the east-west and north-south directions of motion. Based on this, it was decided to monitor the wall racking, or rotation, as an indicator of incurred damage. A sensor mounting was configured which anchors the gauge diagonally between a stud and the cover plate or top beam. Figure 6 shows a picture of this configuration.

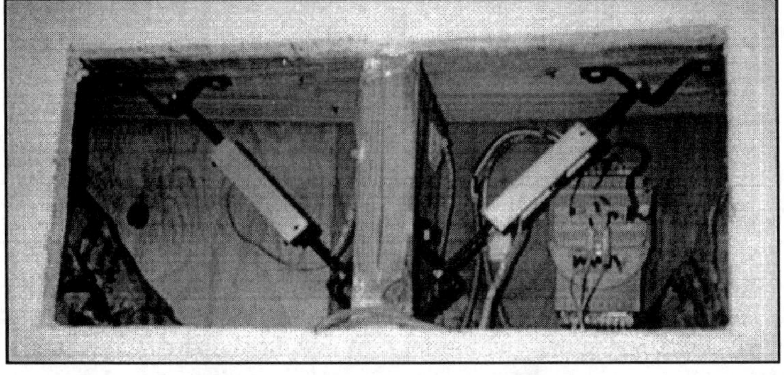

Figure 6. Typical Gauge Installation Site.

The sensor system is shown schematically in figure 7. Three sites in the frame were chosen for monitoring. Each site consisted of a pair of diagonal gauges for primarily monitoring the wall racking movement. The three gauge junctions were networked with a digital RS-485 twisted wire pair and the 12V DC power cable. An embedded microprocessor controls the digital network, interrogates the junctions, and stores pertinent data. It has a watchdog timer with automatic restart to eliminate any program or power failure problems from halting the system's operation. The processor is connected to a cellular modem with fax and pager communication capability. An accelerometer (monitoring the horizontal foundation acceleration) was also connected into the system. The system was intended to routinely (continuously in this case) interrogate all of the sensors and store pertinent data or changes on each cycle. It also monitors the accelerations to prompt an automatic data capture (assuming an earthquake was occurring) regardless of whether the gauges sense any strain changes or not. The power is provided by a 7 A-Hr, 12V battery. The battery is continuously trickle charged from house current. In the event of a power failure, the battery will run the system for at least 12 hours, more than sufficient time for the processor to initiate an outcall on the cellular modem and download its status to the host PC and monitoring station.

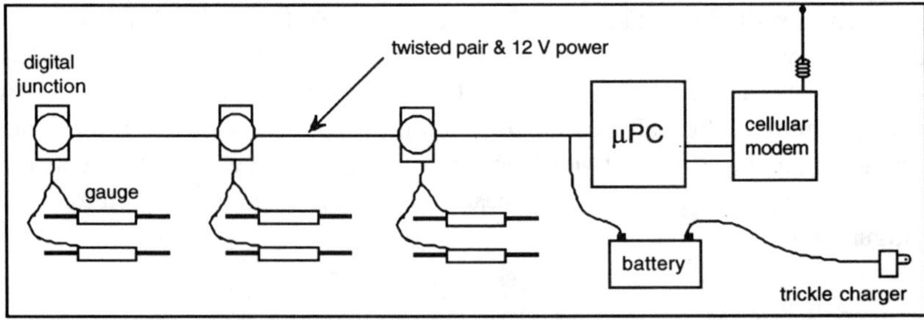

Figure 7. Monitoring System Schematic

The sensor is mounted at a 45 degree angle with 12.7 cm long legs. The sensor configuration provided for a maximum measurement range of about 1.2 degrees or a differential displacement of about 6.35 cm. on a wall with a 3 m. height. Two sensors were mounted at each stud so that the peak rotations in both directions could be monitored. The sensors have an accuracy of roughly 2% of full scale (including long-term drift of the output voltage). In this configuration, the sensor is strained by rotations of the stud (or column) in either horizontal direction. By the geometry, the straining is 40% more sensitive for rotations the plane of the wall, however the out-of-plane rotations do contribute. For this structure, the out-of-plane rotations are resisted by the in-plane shear strength in the connected perpendicular walls. Therefore, the out-of-plane straining is indicative of shear racking in these walls.

The program continuously monitors each of the peak strain sensors, a temperature sensor, an accelerometer, and the battery supply voltage. The accelerometer is monitored at about 50 Hz to insure capturing close-to-peak accelerations from a strong motion event and the gauges, voltages, and temperatures are monitored about once every 5 seconds or immediately if the processor thinks an earthquake is occurring. Data is stored to RAM whenever a sensor output is exceeded or when the accelerometer detects an exceedence of 5% g. For deflections or accelerations which exceed a preset limit, the processor initiates a fax transmission and an alpha-numeric pager alert. The cellular modem is set to receive incoming calls from a PC for data downloading or reprogramming. A password is required for telephone access to the system. The system was installed in late January of 1997 and is currently being monitored and modified to decide on the best formats for data collection, alarms, and program operation. Thus far, there has been no activity, either real or indicated by the system.

Reinforcing Bar Sensor Development and Testing

A program to develop peak strain sensors for use on the reinforcing bar in concrete structures is currently in progress. The envisioned sensor would be mounted onto the rebar before the concrete pour or would be retrofit by exposing the bar, installing the sensor and then grouting the cavity. The initial step in this product development was to design a tensile gauge that would be robust enough to survive the concrete pour and cure. A pair of gauges were installed onto the longitudinal reinforcing bar of a 0.4 scale (0.6 m. diameter) column that was tested to failure at the Powell Structural Labs of the University of California, San Diego.

The sensors consisted of a tensile gauge, of the type shown in fig. 3, which was anchored to the rebar at the rod ends with steel brackets. These brackets were epoxied to the rebar and spaced 13 cm. apart. With this configuration the gauges had a measurement range of 200 to 10,000 micro-m/m (or roughly 10% to 500% of yield). At levels lower than this range the damage would typically be nonexistent or cosmetic; at levels higher it would be visibly obvious and there would probably be no question as to the repair or replacement of the column. The entire gauge assembly was then covered with a plastic casing and sealed with caulk to protect it. Figure 8 shows a photograph of the installed gauge taken before the pour. The ferromagnetic content of the element was measured with a Hall sensor and driver magnet combination. The column employed a staggered design for the foundation bar ends, intended to focus the plastic region about 0.6 m. above the base, and had continuous spiral reinforcement with a pitch of about 18 cm. per turn. The longitudinal bars were #7, grade 60 steel.

Figure 8. Installed and Sealed Gauge

The two gauges (denoted as north and south) were installed on the longitudinal bars on both sides of the column and in the plane of the load actuator. The north gauge was located about 0.76 m. up from the column base (about 2.5 cm above the end of the foundation bar). The south gauge was mounted onto the first bar to the right of the main bar on the south side of the column about 0.81 m. up from the base.

The testing consisted of a series of slow, push-pull sequences of single or triple cycles at increasing levels of top displacement. The test started with displacements at one-quarter of yield and concluded at a displacement ductility level of 8. A dead load of 200 kips was applied to the column during the test. A positive displacement at the top corresponds to a southern movement (i.e. tension on the north bar) and vv. Figure 9 shows a photograph of the south face of the column taken at the conclusion of the test (ductility=8). The white outlined area shows the location of the sensor on that side. The spiral reinforcing had broken in four levels across the plastic zone and the rebars on both the south and north face had buckled out by as much as 70 mm.

Figure 10 shows the measured peak strain results plotted along with the scaled record of the top displacement. The displacements are quantified in terms of the yield level (y) or the ductility (μ) which are listed next to each peak. The yield displacement at the top was calculated and designated as 15 mm. and a ductility of one was calculated at 19.8 mm. The yield strain in the rebar was 2,000 micro-m/m. The peak displacement at each cycle is labeled by the top displacement in yield or ductility. The horizontal axes are related to time but not linear due to the frequent pausing of the test for crack inspection. As seen from the data, the south gauge is measuring the tensile strains induced by negative (or northward) top deflections and the north gauge is measuring the peaks in southward displacement. The south gauge began detecting peak strains at the first negative cycle to one-quarter of yield and gave a peak strain of about 170 micro-m/m. The north gauge indicated a peak strain of around 150 micro-m/m at the positive one-quarter yield level. The gauges proceeded to indicate the peak strains incurred during the loading up to a peak ductility of 2.

Figure 9. South Face of Plastic Region at Later Stage of Testing (dashed zone outlines location of gauge)

The small peaks, or bumps, that occur in the gauge outputs at higher strain levels were caused by the use of a DC Hall sensor to detect the ferromagnetic content. Ferromagnetic steels exhibit the behavior of changing their magnetic susceptibility in proportion to the applied stress. Common steels increase their susceptibility in tension. These local peaks represent the portion of the response where the new peak is straining the element beyond its last peak and inducing slightly higher susceptibility values due to the loading stress. Once the gauge displacement is reduced enough to relieve the elastic straining (about 0.005 mm.) the gauge output settles down to its static peak output level. This effect becomes more noticeable at higher strain levels because more martensite is present, and it helped to indicate that the gauges were still bonded and operating at ductility levels up to 6 although their static value was not increasing.

The data shown here is for displacements up to a ductility of 2. The test continued up to its failure (dictated by a large reduction in stiffness) at a ductility of 8. Beyond a ductility of 2 the gauges did not indicate any further significant increase in the peak strain at their location. But, as mentioned, they did exhibit the small rise and drop in output at peak positive or negative displacement cycles up to a ductility of 6. This indicated that the gauges were still intact (being strained at top dead center of the peak) and operating up to that level. The south gauge output indicated a peak strain of about 3,500 micro-m/m and the north gauge estimated strains of about 3,400 micro-m/m. Resistive foil gauges mounted onto the rebar had failed by a ductility of 2. The bending of the bars at ductility=6 detached the peak strain gauges and they stopped functioning at this point.

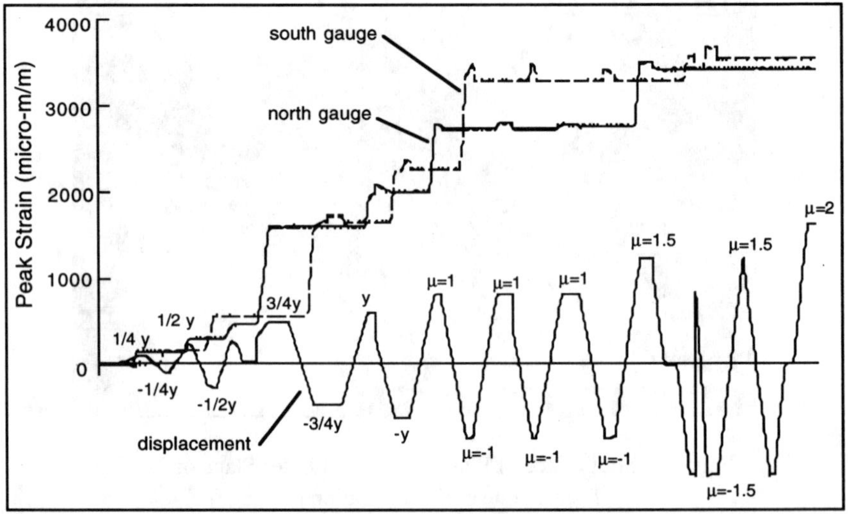

Figure 10. Peak Strain Response (both gauges) and Top Displacement History

Composite Column Retrofits

Another approach to utilize these phase transforming sensor materials is to embed them within a structural material and then monitor their ferromagnetic content with an external probe, this will work for any non-ferromagnetic material. The fact that the magnetic susceptibility of the sensor material changes so appreciably (roughly four orders of magnitude) as a consequence of the straining makes it applicable to embedment. The usable size and embedment depth of the elements depends on the sensitivity of the magnetometer. A SQUID (superconducting quantum interference device) can detect 30 micron diameter sensor wire at distances up to 0.5 m. Inductance coils and Hall sensors were used in these tests, they are capable of detecting a sensor within a distance of 5 mm. As part of a development project with XXsys Technologies, Inc. the sensors were used to monitor the peak hoop strains incurred in the carbon fiber wrap placed around bridge columns as a retrofit.

Sensor elements (similar to the shapes shown in fig. 2) were placed within the carbon wrapping and monitored during their destructive tensile testing. The sensors were placed horizontally in the jacket to measure the peak tensile hoop strains incurred during the loading. The center portion of the elements were sandwiched in plastic sleeves to isolate it and cause the element to bond with the matrix only at its ends, thus producing an amplification in the measured average strains. In this configuration the sensors could measure peak strains of 15,000 micro-m/m with a resolution of about 200 micro-m/m.

The column was a 0.4 scale circular (0.8 m diameter) reinforced concrete column which was fixed at the base and loaded horizontally at the top (3 m tall). All of the tests were quasi-static, push-pull type with three repeated cycles at each ductility level for ductilities of one and above. The column was pre-damaged (with a loading series up to the yield value) before the wrap application. The sensor strips were placed 3 plies under the surface on opposite sides of the column in a plane perpendicular to the load direction plane. Both sensors were embedded about 10 cm up from the base of the column. A Hall sensor was placed over the long strip sensor located on the north side of the column and a pickup coil sensor was placed over the short sensor located on the south side. The pickup coil used an inductance bridge circuit to measure the ferromagnetic content in the sensor. The sensors were monitored actively during the testing, in the true filed application the magnetometer probe would be manually scanned over the surface after an earthquake to obtain a reading.

The reduced data is shown in fig. 12. The test began with a series of single cycles at displacement levels below a ductility of one, after that the test consisted of three full cycles at each ductility level up to the test conclusion. Each point indicates the peak reading for that half cycle of displacement. Data previous to a ductility of 1.5 was not significantly different from zero and were not plotted. At a ductility of 4, the south sensor reached a peak value of about 1,100 micro-in/in which did not change significantly with higher ductilities. It seems unlikely that the true strains did not increase beyond that.

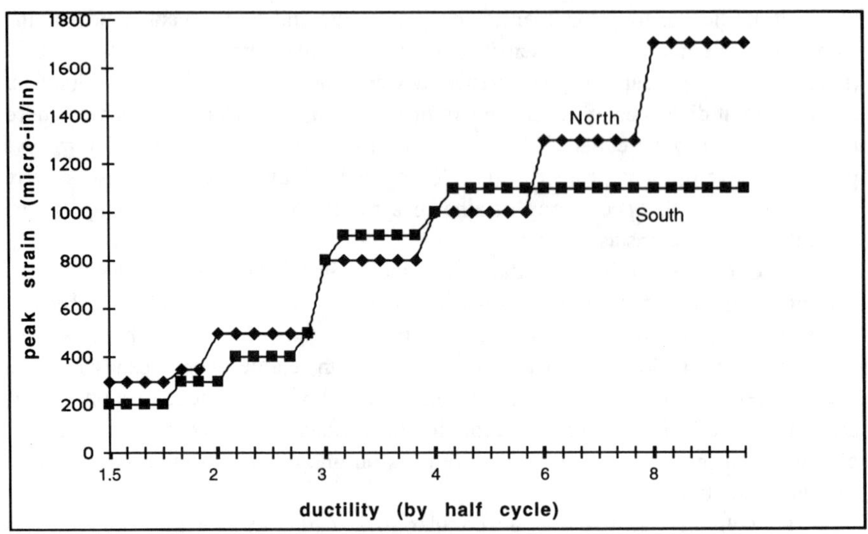

Figure 12. Measured Peak Hoop Strain Within the Carbon Jacket

REFERENCES

1. Westermo, B.D. and L.D. Thompson, 1993. "A New Testing and Evaluation Technology for Damage Assessment and Residual Life Estimation in Aircraft Structures." Proceedings of the 14th Aerospace Testing Seminar, Inst. of Environmental Sciences and the Aerospace Corporation, Manhattan Beach, CA, 5-10.

2. Westermo, B.D. and L.D. Thompson, 1994. "Smart Structural Monitoring: A New Technology." SENSORS, Nov:15-18.

3. Dunning, J.S. and L.D. Thompson, 1995. "Development of Smart Solid-State Structural Damage Assessment Systems for Underground Facilities." Proceedings 1995 North American Conf. on Smart Structures and Materials, SD, CA.

4. Thompson, L.D. and B.D. Westermo, 1996. "Development of Smart Structural Attachment Fixtures." Proceedings 1996 North American Conf. on Smart Structures and Materials, San Diego, CA.

5. Westermo, B.D., 1996. "A Passive Structural Health Monitoring System for Bridges." Proceedings of 1996 CalTrans-FHWA NDT Conference, San Diego, CA.

KEYNOTE SPEAKERS

P. Cawley
C. Boller

Quick Inspection of Large Structures Using Low Frequency Ultrasound

P. CAWLEY

ABSTRACT

Lamb wave testing for the quick inspection of large structures is compared to the use of natural frequency and other modal property measurements. The location of the source of a change in the response is more straightforward with Lamb waves than with vibration measurements so it is easier to determine, for example, whether a change in the measurements is due to a change in the boundary conditions or to damage in the structure itself. Lamb wave testing is more complicated than low frequency modal measurements since it is necessary to control the signal input to the structure very carefully in order to avoid problems with multiple modes and dispersion. However, Lamb wave inspection is pratctical and examples of its use to detect corrosion under insulation in chemical plants and delaminations in composite materials are given.

INTRODUCTION

VIBRATION TESTING

The quick, reliable inspection of large structures has been a major research goal for at least three decades, the work being driven by the difficulty and cost of assuring the integrity of structures such as steam turbines in power stations, oil rigs, highway bridges, pressure vessels and airframes. Most inspection techniques, such as conventional ultrasonics or eddy currents, require a transducer to be scanned over each part of the structure which is to be inspected. This is at best slow, and at worst impossible due to access or other constraints.

The critical defect size in many big structures is relatively large so there is the potential to trade a reduction in test sensitivity for a reduction in the testing time. This realisation has led to a great deal of research on the use of measurements of the vibration characteristics of the structure to infer its integrity. The vibration measurements require access to only a limited number of points on the structure, but can be used to assess the integrity of the whole structure. Two excellent, comprehensive reviews of this field have recently been published [1,2] so no attempt will be made here to cover the wealth of published material.

The most attractive parameters to measure are the natural frequencies and damping of the lower structural modes since these can be measured at a single

Peter Cawley, Dept Mechanical Engineering, Imperial College, London SW7 2BX, UK.

point, yet are characteristics of the whole structure. Unfortunately, while damping is very sensitive to the presence of some types of damage, it is also strongly affected by many other factors such as friction at joints and so it is difficult to obtain sufficiently reproducible measurements. The reproducibility of natural frequency measurements is generally much better, so they are more commonly used, even though their sensitivity to damage is often lower [3]. If strain or displacement measurements are made at a series of points, it is possible to define the mode shapes of the structure and these may sometimes be more sensitive to localised damage. Having obtained the natural frequencies and/or mode shapes, they can be compared with baseline measurements either (ideally) on the same structure or on another, nominally identical, structure which is known to be good, changes in the characteristics indicating the presence of damage.

It is also possible to use the measurements in conjunction with theoretically or experimentally derived models of the structure to predict the likely location and severity of the damage. This is particularly important in large structures since the vibration method can then be used as a screening tool, any areas which are identified as potentially defective subsequently being inspected using conventional non-destructive testing techniques. The slow, conventional methods are therefore only used in the regions which are identified as suspect. The author was involved in early work in this area [4,5]. Since then, a wide range of possible parameters obtained from the basic experimental data have been used to locate damage. These include the flexibility matrix, the stiffness error matrix, changes in the measured stiffness matrix, a variety of model updating methods and neural networks [1].

The major difficulty with the vibration methods is that they are not very sensitive to localised damage, and they tend to be much more sensitive to other likely changes in the structure or its surroundings. This is illustrated in Fig 1 which shows the natural frequency changes in the first flexural mode of vibration due to full-width cracks at the root of a cantilever beam as a function of crack depth. The diagram also shows the changes due to increases in the length of the beam. The frequency reduction due to a crack 2% of the beam thickness deep is 40 times smaller than that due to a 2% increase in the length of the beam. This means that the frequency change produced by a crack through 2% of the beam depth could also be produced by a length change of only 0.05%. Hence, if defects are to be detected by comparisons with baseline measurements made on a nominally identical, but not the same, structure, it is essential that the structures be produced to very good dimensional tolerances [6]. The problem is less severe when the baseline measurements are made on the same structure, but shifts in, for example, the boundary conditions or temperature can still produce frequency changes that swamp the changes due to damage. The frequency changes due to damage tend to be relatively small because the damage is usually localised in the structure, whereas dimension or temperature changes, for example, affect the whole structure.

LAMB WAVE TESTING

The sensitivity of vibration response measurements to structural damage arises from the effect of the damage on the propagation of the waves which travel through the structure, combining at the natural frequencies to produce characteristic standing wave patterns. The sensitivity of the vibration characteristics to the dimensions of the structure is caused by changes in the wave transit time across the structure due to dimensional changes, which results in changes in the natural frequencies. Similarly, sensitivity to changes in the boundary conditions is a result of variations in the the amplitude and/or phase of the reflection coefficient of the waves incident on the boundary. The sensitivity to these factors could potentially be reduced if, instead of measuring the vibration of the structure after many wave interactions with the boundaries, the measurements were limited to the short period between the generation of the wave and the onset of multiple reflections.

For example, consider a 0.5 m long cantilevered steel beam with a 10 mm square cross section. From simple analysis described by, for example, Timoshenko et al [7], the first flexural natural frequency of the beam is about 33 Hz and the wavelength in the first mode is approximately four times the length of the beam. The velocity of the flexural waves is given by the product of the frequency and the wavelength and so is about 66 m/s. Now suppose that the beam has a Q factor of 50 and its natural frequencies are to be measured by an impact test. It may readily be shown that the time taken for the amplitude of the beam in its first mode to decay to 1% of its initial amplitude is approximately 2.2 seconds, which for a wavespeed of 66 m/s corresponds to about 150 wave transits from the source position to one end of the beam, reflection back to the other end of the beam and then reflection back to the source.

An alternative testing technique would be to send a wave selectively in one direction along the beam and to monitor the returning echoes until the first boundary reflection was received. Any reflections arriving before the boundary reflection would indicate the presence of a defect; a change in the nature of the boundary reflection would indicate a change in the boundary condition. Variations in the time of arrival of the boundary reflection would suggest either a change in the dimensions of the beam or a change in the wavespeed; both of these effects could be caused by temperature variations. In order to test the other end of the beam, the direction of wave propagation from the transducer would be reversed. This idea is contrasted with modal measurements involving many boundary reflections in Fig 2.

The waves propagating in beam and plate-like structures which produce resonant vibration are usually either flexural or extensional waves. (Torsional waves are also sometimes important, particularly in shafts.) These are part of a class of waves known as Lamb waves. Viktorov has published extensively on the use of Rayleigh and Lamb waves in nondestructive testing and monitoring applications, and his book [8] has become a standard text in the field. However, it was probably Worlton [9] who first recognised the advantages of using Lamb waves to inspect plates. Since then there has been a great deal of interest in these and other guided waves in NDT applications and Doyle [10] has reviewed the early work on the subject. A selection of references to more recent work may be found in [11,12].

Fig 3 shows the dispersion curves for Lamb waves propagating in a steel plate. The diagram shows the group velocity (the velocity at which energy propagates along the plate) as a function of the product of the frequency and the thickness of the plate. At frequency-thickness products below about 1.6 MHz-mm, two modes, a_0 and s_0, can propagate. At very low frequency-thicknesses, these correspond to flexural and extensional modes respectively. As the frequency-thickness increases above 1.6 MHz-mm, more modes are possible and the interpretation of signals tends to become more complicated. In practical testing, it is generally desirable to excite a single mode. However, even if this is achieved, mode conversion will in general occur at boundaries and at any other discontinuities such as defects, so multi-mode signals must be interpreted. Indeed, if a single mode is succesfully launched, one means of detecting defects is to look for the presence of other modes in the signal received at a remote point. The excitation of single Lamb modes is discussed by Alleyne and Cawley [13,14]

Fig 3 also shows that the group velocity is a function of the frequency-

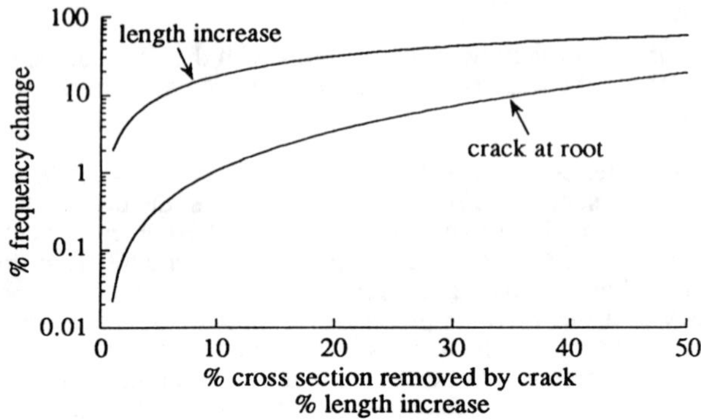

Figure 1. Comaprison of frequency change in first flexural mode of cantilever beam with depth of crack at root and beam length.

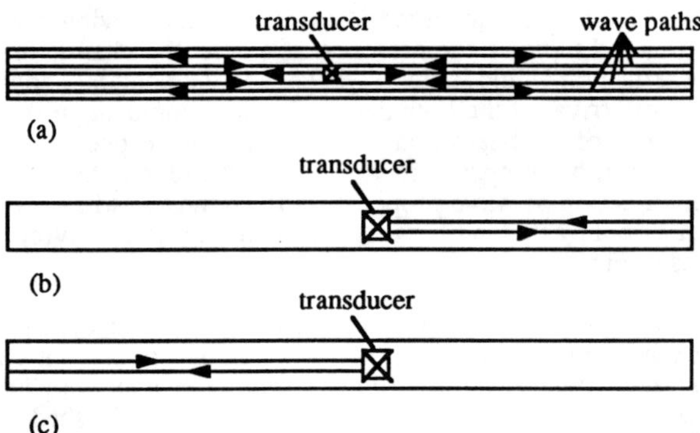

Figure 2. Wave paths in beam (a) resonance testing; (b) Lamb wave testing to right of transducer; (c) Lamb wave testing to left of transducer.

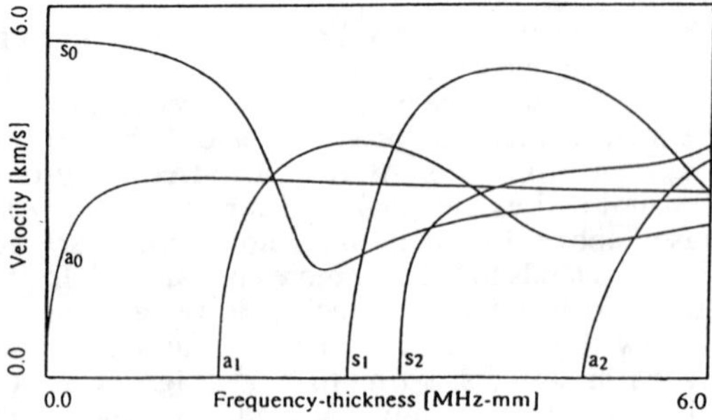

Figure 3. Group velocity dispersion curves for steel.

thickness product. Therefore if a particular mode is excited by a broadband pulse, the different frequency components of the wave will travel at different speeds and the pulse shape will change as it propagates along the plate. This makes long-range inspection very difficult so attempts must be made to limit the bandwidth of the excitation to a range over which there is little dispersion (i.e. the group velocity does not change significantly with frequency). It would not be possible to do the Lamb wave test of Figs 2b and 2c on the 0.5 m long beam of the above example using the flexural wave at the frequency of the first natural frequency (33 Hz). At this point, the frequency-thickness product is only 3.3 x 10^{-4} MHz-mm and Fig 3 shows that the a_0 (flexural) mode is extremely dispersive in this region.

The effect of dispersion is illustrated in Fig 4 which shows numerical predictions using the method described by Alleyne and Cawley [11] of the propagation of the s_0 mode along a 3 mm thick steel plate. The excitation was a 5 cycle toneburst enclosed in a Hanning window, the frequency of the toneburst being 0.67 MHz giving a centre frequency-thickness of 2.0 MHz-mm. Fig 4a shows the surface displacement in the z direction (normal to the plate surface) at x=0, the excitation position, while Figs 4b to 4d show the corresponding waveforms 100, 200 and 500 mm along the plate. It is clear that in a practical test, the signal to noise ratio will reduce rapidly with distance along the plate. The spreading of the signal also greatly reduces the spatial resolution of the test (i.e. the ability to distinguish between closely spaced reflectors and hence, for example, to see the reflection from a defect close to a boundary).

In contrast, Fig 5 shows the corresponding signal at x = 500 mm when the centre frequency-thickness was 1.35 MHz-mm. Here, the pulse shape is maintained and the signal to noise ratio and spatial resolution would be satisfactory over a significant distance in a practical test. The maximum distance which can be covered depends on the attenuation in the material, the power which can be input by the transmitting transducer, the quality of the receiver and associated amplifiers and other factors such as the uniformity of the structure. However, propagation distances of several metres can be obtained, particularly in essentially one dimensional structures such as pipes where lateral beam spreading effects are minimised.

Hence, in a long range test, it is advantageous to excite the required mode in a non-dispersive region. This is frequently most conveniently achieved by testing at frequency-thickness products close to the maxima in the group velocity dispersion curves shown in Fig 3. Testing in these regions is particularly advantageous with the symmetric modes as their maximum propagation velocities are significantly higher than the velocities of the other modes which can propagate at the same frequencies, which facilitates the use of time domain gating to separate out the mode of interest from other modes which may be present in the received signal. For a given velocity of the wave, the spatial resolution obtained in the test increases with frequency. Consider, for example, a 10 cycle toneburst of a Lamb wave propagating at a phase velocity of 5 km/s and a frequency of 1 MHz. The wavelength is 5 mm so ten cycles occupy a length of 50 mm parallel to the direction of propagation. This means that the echoes from two reflectors 25 mm apart will just be separable. If the frequency is reduced to 100 kHz, the spatial resolution is reduced to 250 mm. Lamb wave testing is therefore usually carried out at frequencies much higher than those of the lower structural natural frequencies.

Before choosing a particular Lamb mode, it is obviously essential to check that it is sensitive to the defect types of interest. The stress mode shapes give a good indication of the likely sensitivity to defects at different positions through the plate or shell thickness and the precise sensitivity may be determined either experimentally or by finite element predictions. This is discussed in more detail in [11]. In general the sensitivity of Lamb wave inspection is lower than that of more localised ultrasonic testing which is generally carried out at higher frequencies. However, it is usually better than that obtained with natural frequency and mode shape measurements largely because, as discussed above, it is usually done at higher frequencies in order to improve the spatial resolution of the test and to avoid the problems of dispersion.

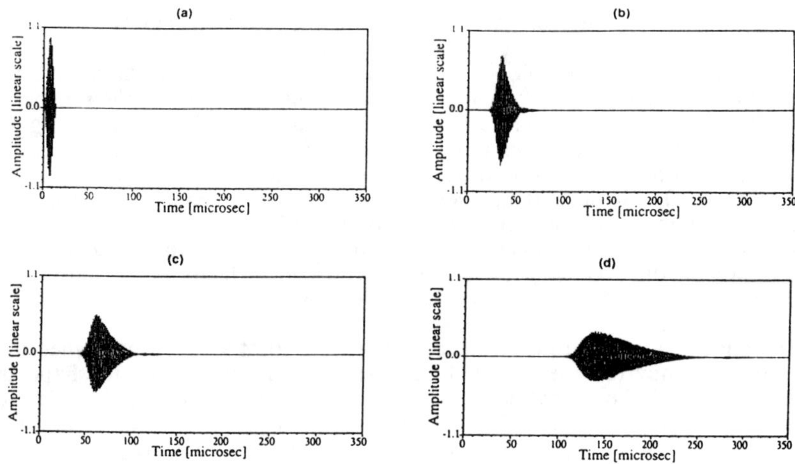

Figure 4. Predicted time histories of propagation of a pure s0 mode in a 3 mm plate excited by a 10 cycle, 0.67 MHz Hanning windowed toneburst. (a) at source (x=0); (b) x=100 mm; (c) x=200 mm; (d) x=500 mm.

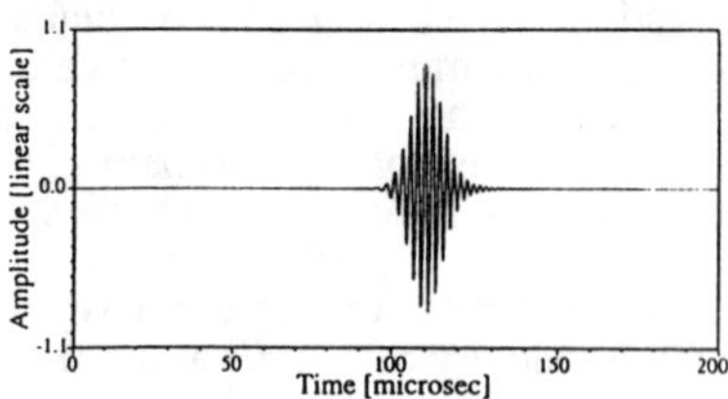

Figure 5. As Figure 5d (x=500 mm) but frequency 0.45 MHz.

THE APPLICATION OF LAMB WAVE TESTING

Space constraints mean that only two examples will be given here, the first being a laboratory test on the detection of delaminations in a composite plate and the second a field trial of the detection of corrosion under insulation in chemical plant pipework.

DETECTION OF DELAMINATIONS IN COMPOSITE PLATES

Figs 6a and 6b show tests on 600 mm long composite laminates using the s_0 mode, the transmitting and receiving transducers being placed at one end of the laminate. The transmitting transducer was a conventional ultrasonic transducer mounted on a perspex wedge so that it was oriented at the appropriate angle to the plate surface to generate the s_0 mode, while the receiver was a similar transducer oriented normal to the plate surface. The excitation signal was a 10 cycle, 0.5 MHz toneburst enclosed in a Hanning window. The first part of the response from a good laminate shown in Fig 6a is a well defined wavelet, T_{S0}, which is the s_0 mode generated by the transmitter and received by the adjacent receiver; this is followed by a series of small signals, labelled T_{PEX}, which were caused by reverberation in the wedge of the transmitter; at about 190 μs, there is another wavelet, $R_{S0\text{-}R}$, which is the s_0 reflection from the far end of the laminate. The time delay between the transmitted and reflected signals is about 160 μs, which corresponds to a propagation distance of about 1000 mm at the s_0 mode velocity in this composite material of 6.35 mm/μs. This corresponds to twice the distance between the receiver and the far end since the waves have done a 'round trip'. Both the T_{S0} and R_{S0} signals are similar in form to the input 10 cycle windowed toneburst, showing that the wave has propagated with minimal dispersion. Fig 6b shows the corresponding result on a laminate with a 20 mm long delamination about 100 mm from the far end. In comparison with Fig 6a there is an extra signal, labelled $R_{S0\text{-}D}$, ahead of the reflected signal from the end of the laminate. Further details of this work are given in [15].

THE DETECTION OF CORROSION UNDER INSULATION IN PIPES

Corrosion in pipework is a major problem, particularly in the oil, gas, chemical and petro-chemical industries. Since a high proportion of industrial pipelines are insulated, this means that even external corrosion cannot readily be detected without the removal of the insulation, which in most cases is prohibitively expensive. There is therefore an urgent need for the development of a quick, reliable method for the detection of corrosion under insulation (CUI).

The author is part of a team working on a project whose ultimate aim is to develop a guided wave testing technique for the inspection of pipework in chemical plants, the target being to detect any areas of corrosion larger than D/2 x D/2 in area and T/2 deep where T is the pipe wall thickness and D is the pipe diameter. The technique is to work on insulated pipe in the 2 -12 inch nominal bore diameter range and an inspection range allowing successive transducer positions to be at least 15m apart is required. (While most of the test work has been on pipes in the 2-12 inch diameter range, the technique is not limited to these sizes.)

The L(0,2) mode (the terminology used to describe cylindrical Lamb modes is discussed in [16]) in the frequency range around 70 kHz is very attractive to use for long range testing since it is practically non-dispersive and is also the fastest mode, which means that it will be the first signal to arrive at the receiver and so can readily be separated by time domain gating. Its mode shape is similar to that

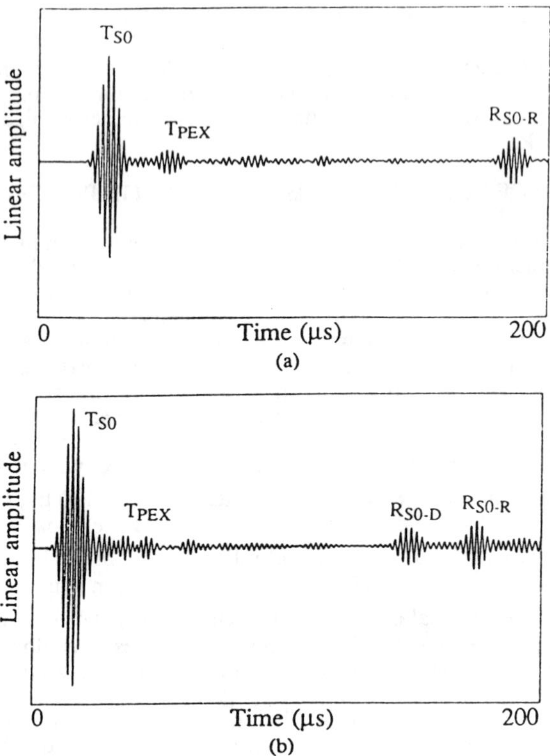

Figure 6. Lamb wave measurements in 600 mm long composite laminates. (a) no defects; (b) 20 mm long delamination 100 mm from right hand end.

of the s_0 mode in plates at low frequency-thickness products, the particle motion being predominantly axial and the strain being roughly uniform through the pipe wall. It is therefore well suited to the detection of corrosion which may initiate at either surface of the pipe.

One of the operating companies supporting this work was planning to replace insulation along an 8 inch schedule 30 (7 mm wall thickness) atmospheric residue line in a pipe track from a refinery. The insulation comprised 50 mm thick mineral wool loosely wrapped round the pipe and covered with galvanised steel sheeting. Problem areas associated with loose fitting insulation at some points along the lines in this pipe track had been identified in the past, with repairs being made to the pipeline. The company had decided to remove all the insulation and to inspect the line visually prior to re-commissioning. This provided an ideal opportunity to test the Lamb wave technique in the field and also to gain invaluable experience on real corrosion defects. The Lamb wave tests could be conducted before the insulation was removed and then the results obtained correlated with the visual inspection.

The overall length of the line inspected using the Lamb wave technique was 550m and contractors removed 1m lengths of insulation at intervals of at least

20m along the line in order to provide access for transducer placement. The pipe was supported at regular intervals of about 8m and these pipe supports provided useful reference points. The majority of the supports were essentially steel 'Tee' brackets welded to the pipe. Each bracket was about 450mm long, 70mm high and 10mm thick and was welded to the pipe along its length on both sides so that it was normal to the pipe surface and in-line with the pipe axis. A few pipe supports were of a 'U' shape yoke design. The dimensions were similar to the 'Tee' design, but the support was welded over about half of the pipe circumference.

In all the tests the excitation signal was a 5 cycle tone burst modified by the application of a Hanning window function, most tests being carried out at a centre frequency of 64 kHz. The transducer system used was similar to that described in [14] and was positioned at a series of locations along the pipe in turn. At each transducer location tests were carried out with the L(0,2) mode being excited in the forward and backward directions in turn [12]. Figs 7a and 7b show the signals obtained by sending waves in the forward and backward directions respectively at one test location. The time axis has been converted to a distance scale using the velocity of the L(0,2) mode obtained from the theoretical dispersion curves. There is a 'dead zone' of about 1m at the start of the traces due to the operation of the rudimentary diode bridge circuit which was connected between the ring and the function generator and capture unit to isolate the receiver amplifier from the large amplitude excitation signals. The length of the 'dead zone' could be reduced by improving the circuit design.

The first large signal marked Y on Fig 7a is due to a welded yoke pipe support (support number PS64) and this is closely followed by a reflection (W) from a butt weld. The three remaining significant echoes correspond to two further welded yoke supports (Y) with a weld (W) roughly midway between them. The weld signals were identified at the time of the test and this was confirmed when the insulation was removed. Propagation in the backward direction (Fig 7b) revealed three welds; in this case the reflections from the pipe supports were smaller than those in the forward direction. This is because they were welded 'Tee' rather than yoke supports.

Fig 8 shows the results of a test at the location where the most severe corrosion was encountered. The three signals marked C1, C2 and C3 at the start of the trace were identified as likely corrosion sites. The later visual inspection revealed that these locations corresponded to deep pits up to 4 mm deep (57% of the wall thickness) in a generally corroded area about 2m long extending over 30% of the pipe circumference. The full scale amplitude in Fig 8 has been reduced to 3 mV compared with 11 mV in Fig 7. This suggests that there was also a considerable amount of attenuation produced by scattering from the rough corroded surfaces. The location marked C4 was not originally identified as a corrosion site since it was thought that the signal could be caused by reverberation between the previously identified features. However, the visual inspection revealed a further area of corrosion covering about 25% of the pipe circumference with some pits extending to between 4 and 5 mm deep. This indicates that the reliable detection of defects downstream of large defective areas is likely to be difficult. This would not be a problem in practice since once the first region had been identified, the insulation would be removed from regions on either side in order to check the extent of the problem.

Tests were carried out at 13 locations on the pipe and the defect detection results are summarised in Fig 9. The depth data must be treated with some caution since it was obtained from rough measurements on site; there was no opportunity to produce replica mouldings. The horizontal and vertical lines on the plot

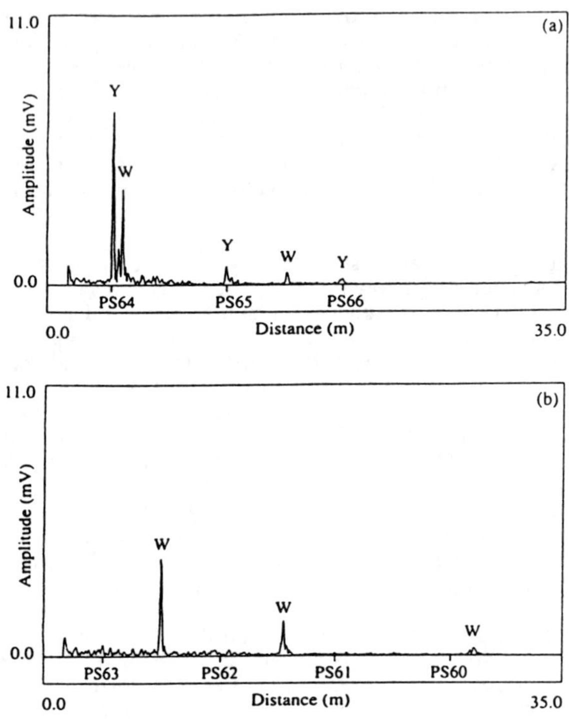

Figure 7. Site tests on 8 inch pipe at good location (a) forward direction; (b) backward direction.

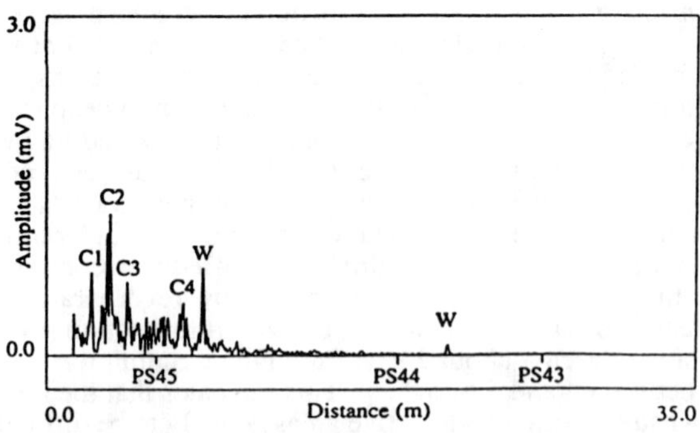

Figure 8. Result at location with most severe corrosion.

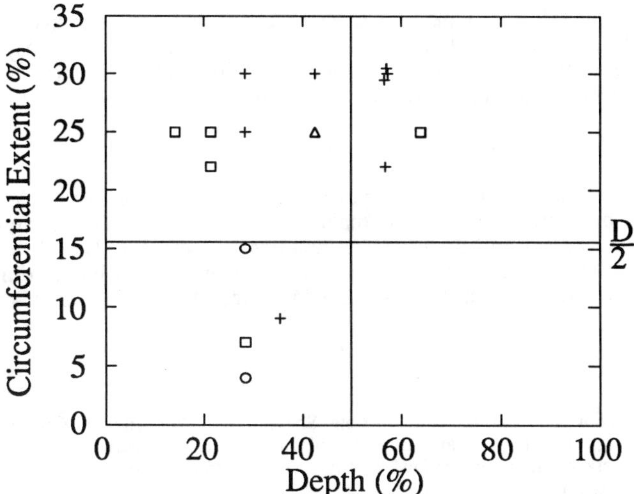

Figure 9. Summary of defect detection results on 8 inch pipe. (+ corrosion identified correctly at time of ultrasonic tests; □ corrosion not identified at time of test, but subsequently correlated with distinct echo; o no correlation with ultrasonic echo; Δ no correlation with ultrasonic echo, but defect 17.5m from nearest transducer position.

correspond to the detection target of 16% (D/2) circumferential extent and 50% depth. All but one of the defects shown in Fig 9 whose circumferential extent exceeded 16% corresponded to clear echoes in the ultrasonic data, the exception being a defect 43% of the wall thickness deep extending over 25% of the circumference which was located 17.5 m from the nearest transducer position. The results suggest that on this pipe the reliable range is around 12-15m in either direction from a transducer location so successive test positions can be up to about 25-30m apart. The only defect whose depth was greater than 50% of the wall thickness which was not identified at the time of the test was corrosion site C4 of Fig 8 which was downstream of three other severe corrosion sites, C1-C3. The visual inspection also identified a further 18 corrosion sites which were not measured sufficiently accurately to be plotted on Fig 9. However, none of them had a depth greater than 50% of the wall thickness and most of them corresponded to minor wastage. Further details of these tests can be found in [17].

CONCLUSIONS

Lamb wave testing for the quick inspection of large structures has been compared to the use of natural frequency and other modal property measurements. Lamb wave inspection has the advantage that the measurements are made before the waves have undergone multiple boundary interactions. The location of the source of a change in the response is more straightforward with Lamb waves than with vibration measurements so it is easier to determine, for example, whether a change in the measurements is due to a change in the boundary conditions or to damage in the structure itself. Lamb wave testing is

control the signal input to the structure very carefully in order to avoid problems with multiple modes and dispersion. However, Lamb wave inspection is pratctical and examples of its use to detect corrosion under insulation in chemical plants and delaminations in composite materials have been given.

ACKNOWLEDGEMENTS

The pipe inspection work was carried out under the European Commission THERMIE programme in a project managed by TWI with financial support from a consortium of oil and chemical companies and the UK Health and Safety Executive. The results summarised here were obtained in collaboration with Dr David Alleyne of Imperial College and Mr Peter Mudge and Mr Alan Lank of TWI.

REFERENCES

1. Doebling, SW, Farrar, CR, Prime, MB and Shevitz, DW 'Damage identification and health monitoring of structural and mechanical systems from changes in their vibration characteristics: a literature review', Los Alamaos national Laboratory, Report LA-13070-MS, UC-900, May 1996.
2. Dimarogonas, AD 'Vibration of cracked structures: a state of the art review', Engineering Fracture Mechanics, Vol 55, pp831-857, 1996.
3. Adams, R.D. and Cawley, P. 'Vibration techniques in non-destructive testing', Research Techniques in NDT, R.S. Sharpe (ed), Vol 8, pp303-360, 1985.
4. Adams, R.D., Cawley, P., Pye, C.J. and Stone, B.J. 'A vibration technique for non-destructively assessing the integrity of structures', J. Mechanical Engineering Science, Vol 20, pp93-100, 1978.
5. Cawley, P. and Adams, R.D. 'The location of defects in structures from measurements of natural frequencies', J. Strain Analysis, Vol 14, pp49-57, 1979.
6. Cawley, P. 'Non-destructive testing of mass produced components by natural frequency measurements', Proc I Mech E, Vol 199, Part B, pp161-168, 1985.
7. Timoshenko, S.P., Young, D.H. and Weaver, W. 'Vibration problems in engineering', 4th edition, John Wiley, New York, 1974.
8. Viktorov, IA 'Rayleigh and Lamb waves', Plenum Press (New York), 1970.
9. Worlton, D.C. 'Ultrasonic testing with Lamb waves', Non-Destructive Testing, Vol 15, pp218-222, 1957.
10. Doyle, PA 'Crack depth measurement by ultrasonics: a review', Ultrasonics, Vol 16, pp164-170, 1978.
11. Alleyne, D.N. and Cawley, P. 'The interaction of Lamb waves with defects', IEEE Trans Ultrasonics, Ferroelectrics and Frequency Control, Vol 39, pp381-397, 1992.
12. Alleyne, D.N. and Cawley, P. 'Long range propagation of Lamb waves in chemical plant pipework', Materials Evaluation, Vol 55, pp504-508, 1997.
13. Alleyne, D.N. and Cawley, P. 'Optimisation of Lamb wave inspection techniques', NDT&E International, Vol 25, pp11-22, 1992.
14. Alleyne, D.N. and Cawley, P. 'The excitation of Lamb waves in pipes using dry coupled piezoelectric transducers', J NDE, Vol 15, pp11-20, 1996.
15. Guo, N. and Cawley, P. "Lamb wave reflection for the quick NDE of large composite laminates', Materials Evaluation, Vol 52, pp404-411, 1994.
16. Silk, M. G., and Bainton, K. F. "The propagation in metal tubing of ultrasonic wave modes equivalent to Lamb waves," Ultrasonics, Vol 17, pp11-19, 1979.
17. Alleyne, D.N., Cawley, P., Lank, A.M. and Mudge, P.J. 'The Lamb Wave Inspection of Chemical Plant Pipework', Review of Progress in Quantitative NDE, Vol 16, DO Thompson and DE Chimenti (eds), Plenum Press, New York, 1997 (in press).

Structural Health Monitoring in Aircraft–State-of-the-Art, Perspectives and Benefits

C. BOLLER and C. BIEMANS

ABSTRACT

An overview of actual activities related to monitoring aircraft structures is presented. This includes the wide range from existing loads monitoring systems in aircraft to future integration of sensors and thus NDT-techniques into metallic and polymerbased composite structures. Potential methods, techniques and parameters especially with regard to piezoelectric sensing are discussed. Methods for validating structual health monitoring will be presented.

INTRODUCTION

Aircraft structures are designed to accomplish their operational requirements at minimum cost. The ratio of life-cycle-cost versus operational capability has been significantly reduced over the past decades. This has been possible through the introduction of fracture mechanics analysis, better fatigue life evaluation concepts and improved non-destructive testing, thus allowing to switch from safe-life to fail-safe design.

Fail-safe conditions require a substantial amount of monitoring. This amount significantly depends on the failure criticality to be monitored. Generally speaking a function can be established saying: The amount (cost) of monitoring increases (possibly exponentially) with increased criticality of the failure to be monitored when the level of available monitoring technology is kept constant. Since sensing and data processing technology, reliability and cost have however very much improved over the past years, the above mentioned function has been quantitatively shifted towards a reduction in monitoring cost.

The percentage of aircraft that are being operated beyond their design lives is ever increasing. As of 1993, approximately 51% of the aircraft in the U.S. Air Force inventory were over 15-years and 44% over 20-years old. Yet some aircraft models that have already served NATO for 30 years or more may need to be retained for

Christian Boller*) and Christian Biemans, Daimler-Benz AG, Research and Technology Exchange Group, D-70546 Stuttgart/Germany, *)on leave from: Daimler-Benz Aerospace AG, Military Aircraft D-81663 München/Germany

another two decades. One of the problems with aging aircraft is the rise in time needed for inspection and repair. An example has been given by Sampath (1996), that for the EF-111A aircraft, on an average, the man-hours required for scheduled inspection and repair of each aircraft in the depot has risen from about 2200 hours in 1985 to about 8000 hours today. Automated inspection could therefore be of great benefit to an aging aircraft and fleet.

Monitoring of aircraft structures is generally performed on ground. It is based on the knowledge of structural criticality resulting from design. Inspection techniques are either visual or using non-destructive testing techniques in areas of poor visibility. Although knowing that this is not the case, damage accumulation during operation has traditionally been assumed to be constant by taking the worst flight condition as a basis. Over the past years however conventional strain gauges have been either implemented into the aircraft structures or sensors being already available in the aircraft have been used for monitoring operational load sequences [AGARD 1991].

Besides strains and thus loads other physical parameters can be monitored such as temperature, pressure or vibrations. This is for example done with Engine Condition Monitoring (ECM) systems, where a network of sensors informs if critical engine parameters are within the operational range or not. ECM has received FAA and JAA approval and is used within various operators' fleets [Spragg et al. 1989; Haberding 1985]. Integrated health and usage monitoring systems (IHUMS) are another example, which have become quite popular for helicopters and commercially available [Bristow, 1992]. These systems have been developed for gears and are based on monitoring acoustic signals being generated from the various rotating parts of the gear.

It is expected that benefits can be achieved in eliminating the manpower effort associated with aircraft inspections [Kudva et al., 1993]. Numbers being reported from estimated values are in excess of 35 million US $ per year for a F-18 (assuming 33 hours of flight per aircraft per month and 1000 aircraft fleet) and more than 9 million US $ per year for a T-38 (based on 420 flight hours per aircraft per year, 720 fleet aircraft). The automation of just one logistics function could result in an approximate savings of 100,000 US $ per year in manpower and equipment.

The following paragraphs will therefore deal with monitoring techniques being available today and will describe the attempts being done regarding enhanced structural monitoring.

STRUCTURAL HEALTH MONITORING VALIDATION

The major question driving the aircraft operator regarding structural health monitoring is: *How is structural health monitoring able to reduce the life-cycle-cost of my aircraft?*

A first step for answering similar typed questions in other fields has been the application of Quality Function Deployment (QFD) Analysis by establishing a 'house of quality' being schematically shown in Figure 1. This diagram relates the customer needs to the available monitoring technologies. A rating system allows to determine which of the monitoring technologies is best suited for the specific customer needs.

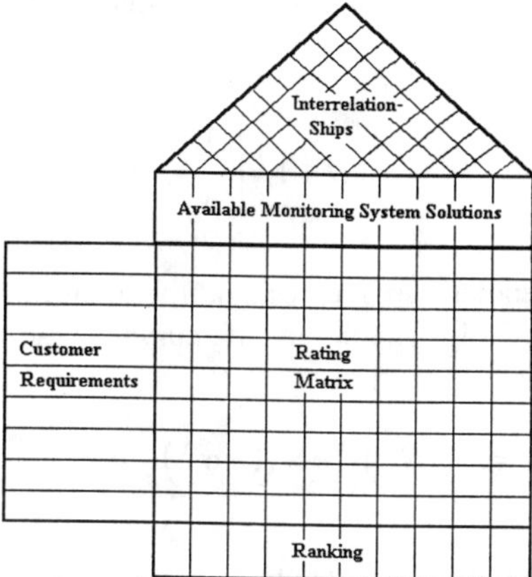

Figure 1 Principle of QFD-analysis for structural health monitoring

The next step then requires to quantitatively determine the amount of manpower related inspection effort and cost that can be replaced by a structural health monitoring system (SHMS) as well as the cost for purchasing, operating, maintaining and disposing of that system. Only if the difference between these two cost terms turns out to be positive and operational availability of the aircraft is not negatively influenced, the SHMS turns out to be beneficial. It may be useful also mentioning under that context that it is worthwhile analyzing the damage criticality of the various locations of an aircraft. This can easily lead to a 20/80-solution saying, when monitoring the top 20% of the damage critical locations alreday 80% of the required monitoring effort may be achieved.

Another point strongly to be considered is the reliability of sensor information. Any false information can again significantly increase the inspection cost and even negatively influence operational availability of the aircraft. Operators have therefore been cautious in adding further sensors to their aircraft fearing additional sensors to reduce their overall aircraft reliability. The SHMS being implemented therefore requires to be at least as reliable as manpower related inspection.

The major question driving the engineering technologist regarding all work being done in structural health monitoring is: *What is the relevant parameter for monitoring damage accumulation?*

The first assumption being mainly used today is considering stress to be that parameter. Stress can be determined ‘relatively easy’ from loads and strains which is even true for complex notch geometries when using analytical methods such as finite element analysis. To make these analyses manageable in a reasonable time, a lot of linearities have to be assumed. This is however not true when taking full advantage of the structural material’s true constitutive and damage behaviour. So far, materials have been desired to be designed as linear as possible in their behaviour. The introduction of structural health monitoring however proves that this desire can be turned the other way around which can be easily explained by the following example.

Assume a material or structure has to withstand a major impact damage occuring once during its operational life, where damage accumulation of that material/structure follows the relationships given in Figure 2. Damage accumulation is assumed to follow the relationship

$$D = 1 - \left(\frac{N_r}{N_f}\right)^a \tag{1}$$

where *D*, N_r, N_f and *a* denotes damage, residual life, total life and a constant describing non-linearity respectively. The fatigue life curve is described as follows:

$$\log\left(\frac{N_i}{N_{ref}}\right) = \log(N^*) = k \times (1 - \sigma^*) = k \times \left(1 - \frac{\sigma_i}{\sigma_{ref}}\right) \tag{2}$$

with *N*, *k* and σ being the number of loading cycles, slope of the fatigue life curve and stress and *i* and *ref* denoting the actual and reference value respectively.
The problem to be solved by the designer can therefore be expressed by the following question: *How much has the allowable stress* σ^* *to be reduced, that a sufficient portion of life* N_p *can be provided, that allows to account for the impact damage to be induced?*
Conventional design, where *no SHMS* has been implemented requires to consider the most critical case to be such that the structure is hit by the major impact at the very beginning of its life. Following Eq. (1) the residual life $N_{r,i}$ after the impact can therefore be determined to be

$$\frac{N_{r,i}}{N_{ref}} = (1 - D_{im})^{1/a} \tag{3}$$

The life $N_{p,i}$ that has to be provided is then

$$\frac{N_{p,i}}{N_{ref}} = 1 - \frac{N_{r,i}}{N_{ref}} = 1 - (1 - D_{im})^{1/a} \tag{4}$$

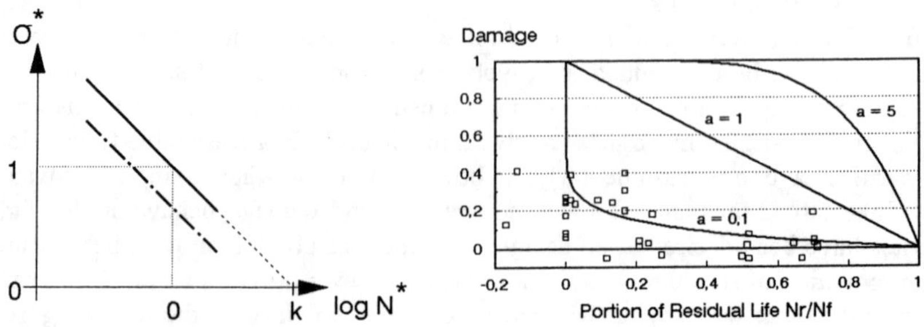

Figure 2 Example of damage accumulation relationships

where D_{im} is the damage induced by the impact. To still achieve a last mission (e.g. a flight) safely a further portion of life has to be provided which is N_m/N_{ref}. This results in the total life N_p that has to be provided to be

$$\frac{N_p}{N_{ref}} = \frac{N_{p,i}}{N_{ref}} + \frac{N_m}{N_{ref}} = 1 - (1 - D_m)^{1/a} + \frac{N_m}{N_{ref}} \tag{5}$$

When converting Eq. (2) to

$$\sigma^* = b - \frac{1}{k} \times \log N^* \tag{6}$$

the residual life for $\sigma^* = 1$ can be determined which is

$$N^* = 1 - \frac{N_p}{N_{ref}} = (1 - D_m)^{1/a} - \frac{N_m}{N_{ref}} \tag{7}$$

Constant b in Eq. (6) can now be determined which finally leads to the new reduced life curve being

$$\sigma^* = 1 + \frac{1}{k} \times \left[\log\left\{ (1 - D_{im})^{1/a} - \frac{N_m}{N_{ref}} \right\} - \log N^* \right] \tag{8}$$

Considering a *SHMS to be implemented* in the structure leads to a similar calculation process. However the most critical case is here that the major impact damage may now be initiated at the beginning of the last mission (flight), because any earlier damage can be assumed to be repaired and thus removed. Damage $D_{a,i}$ after that impact is therefore only allowed to be

$$D_{a,i} = 1 - \left(\frac{N_m}{N_{ref}} \right)^a \tag{9}$$

Including the impact damage D_{im} leads to the allowable damage before the impact $D_{b,i}$ which is

$$D_{b,i} = D_{a,i} - D_{im} = 1 - \left(\frac{N_m}{N_{ref}} \right)^a - D_{im} = 1 - \left(\frac{N_p}{N_{ref}} \right)^a \tag{10}$$

Thus the life N_p/N_{ref} to be provided results in:

$$\frac{N_p}{N_{ref}} = \left[\left(\frac{N_m}{N_{ref}} \right)^a + D_{im} \right]^{1/a} \tag{11}$$

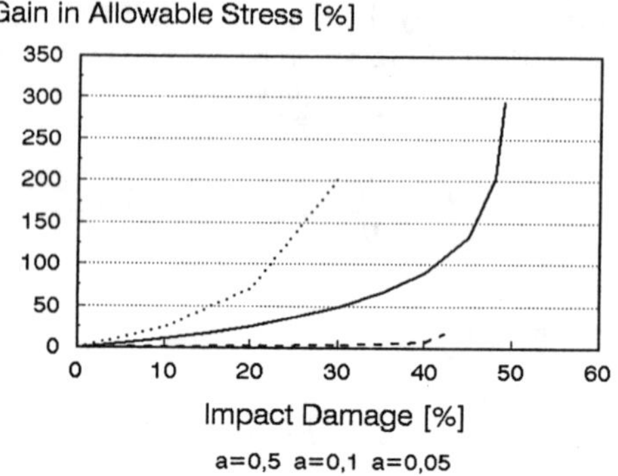

Figure 3 Influence of damage size and nonlinearity in damage accumulation

Constant b from Eq. (6) can now be again determined, which finally leads in the reduced life curve to be

$$\sigma^*_{SHM} = 1 + \frac{1}{k} \times \left[\log\left\{ 1 - \left[\left(\frac{N_m}{N_{ref}} \right)^a + D_{im} \right]^{1/a} \right\} - \log N^* \right] \tag{12}$$

A comparison between Eq.s (8) and (12) leads to

$$\frac{\sigma^*_{SHM}}{\sigma^*} = \frac{k + \log\left\{ 1 - \left[\left(\frac{N_m}{N_{ref}} \right)^a + D_{im} \right]^{1/a} \right\}}{k + \log\left\{ (1 - D_{im})^{1/a} - \frac{N_m}{N_{ref}} \right\}} \tag{13}$$

which finally allows to determine the benefits from using a SHMS and the parameters influencing that benefit. For a realistic example the gain in allowable stress σ^*_{SHM}/σ^* versus the amount of impact damage D_{im} is shown in Figure 3 for different nonlinearities in damage accumulation described by the parameter a.

Generally the gain in allowable stress increases with an increase in allowed induced damage and reaches an infinite value when the induced damage reaches its maximum allowable value. However it also has to be kept in mind, that an increase in induced damage requires an increased effort for repair, which thus may again absorb the benefit gained so far. The right optimum therefore needs to be determined. Furthermore, the way how damage accumulates has a strong influence on the gain in allowable stress. This information can be quite interesting for material designers and may convert designers in their desire for a linearity in damage accumulation.

Generally the result given in Figure 3 shows that it is worth pursuing the development in SHMS and to more specifically determine their benefits in experimental work.
Beside stress and strain a lot of other physical parameters can be used for monitoring which will be described with its appropriate sensors in the subsequent.

AVAILABLE SHMS

The traditional way of loads/stress monitoring is by bonding strain gauges to well selected areas in the aircraft and measuring strain sequences. These sequences are then converted to stresses and/or loads thus allowing numerical fatigue life evaluation. Various examples are described in [AGARD 1991].
Although strain gauges are widely used, this technique might also be performed with other types of sensors measuring strains such as made of optical fibres or piezoelectric materials.
An alternative to strain measurement being applied on the Panavia Tornado military aircraft is the use of flight parameter sensors such as for speed, altitude, air data, pressure, fuel quantity, flaps position, etc. and converting their information into a load sequence [Bauer, 1987]. Although this solution does not increase aircraft complexity and reliability, stress and fatigue analysts still claim more precision in the load sequences being provided.
Both systems are actually used for differing between special events (hard landings and limit load exceedances) and the general load spectrum.
Visual inspection is sometimes one of the quickest monitoring techniques and can be psychologically quite important in critical situations. To explore the benefits of a system allowing visual monitoring of the aircraft's outside during flight, the CAA has therefore installed three external video cameras on a Boeing 747. A camera on the tail fin views the starboard wing, one on the port tailplane views the port-wing trailing edge, and a camera with a panning capability is mounted on the forward underbelly. The major events considered to be monitored are ground operations, ice/snow accreation on aircraft, incorrectly configured landing gear or flaps/slats, engine fire/failure, or during emergency evacuation.

STRUCTURAL NDT TECHNOLOGY INTEGRATION

The forementioned available SHMS suffer from the fact that damage can only be predicted but not really determined itself. This can only be done when applying NDT and carefully inspecting the critical area itself. However the significant progress achieved in sensor performance, miniaturisation and cost has made the idea more realistic of NDT becoming an integral part of a material and structure. When considering this idea, a first step is to determine which of the NDT technniques being commonly used are suitable for being integrated into a material and structure which is given with respect to aircraft structures in Table I.
Apart from visual inspection, use of ultrasonics and eddy current are the procedures generally applied during on-ground maintenance. The other techniques have recent-

Table I TECHNOLOGY SELECTION FOR STRUCTURAL HEALTH MONITORING SYSTEMS

Monitoring Technology / Monitoring Qualities	In-Flight Applicability	Degree of Development	-Loads Monitoring	Damage Identification	Damage Propagation	Fatigue Crack	Impact (BVID)	Delamination	Corrosion
Visual/Borescope	No	high				Yes			Yes
Strain	Yes	high	Yes			(Yes)			
Flight Parameters	Yes	high	Yes			(Yes)			
Magnetic Particle	No	high		Yes	Yes	Yes			
Eddy Current	No	high		Yes	Yes	Yes	(Yes)	(Yes)	
Penetrant	No	high		Yes	Yes	Yes		(Yes)	
Paintings	No	low		Yes			Yes		
Chem. Sensing	Yes	low		Yes					Yes
Radiography	No	high		Yes	Yes	Yes	Yes	Yes	Yes
Modal Analysis	Yes	high	Yes	Yes	Yes	Yes	Yes	Yes	?
Acoustic Emission	Yes	high	Yes	Yes	Yes	Yes	Yes	Yes	(Yes)
Holography	?	low		?	?	?	?	?	?
Ultrasonic	Yes	high		Yes	Yes	Yes	Yes	Yes	?
Acousto-Ultrasonics	Yes	high		Yes	Yes	Yes	Yes	Yes	?
Lamb Waves	Yes	high		Yes	Yes	Yes	Yes	Yes	?
Shearography	?	low		?	?	?	?	?	?
Thermography	No	low		Yes	Yes	Yes	Yes	Yes	
Barkhausen Noise	?	low	?	?		?			?
Magneto. Opt. Eddy Curr.	No	low		?	?	?			?
Comp. Aid. Tomography	No	low		Yes	Yes	Yes	?	?	?

ly gained more attraction because of improved availability of these techniques. Major applicability and experience gained with these techniques has been with metals. A limited number of them is however also applicable to composite materials. Fatigue fuses and crack propagation gauges are other simple means for monitoring of crack initiation and propagation.

The techniques that can be considered to be suitable for structural integration are strain, flight parameters, chemical sensing, modal analysis, lamb waves, acoustic emission, and acousto ultrasonics.

Compared to the well established NDT procedures chemical sensing might still be excluded from the techniques mentioned before. In-situ monitoring of damage is a further major criterion for an advanced aircraft health monitoring system which reduces the candidate NDT procedures to modal analysis, lamb-waves, acoustic emission and acousto-ultrasonics. These latter procedures are also able to monitor various kinds of damage such as fatigue cracks, BVID, delamination or even corrosion. An overview of the different techniques is also given in [Boller 1996].

SENSORS AND DATA PROCESSING

Beside the strain gauges having been mentioned before, there is a variety of other sensors available such as those made of piezoelectric materials or using optical fibres. Both can be used for either monitoring strain and acoustical signals. Measures (1992) performed a variety of laboratory tests in integrating optical fibres into the leading edge of a composite aircraft wing for damage detection. Blaha and McBride (1992) and Fürstenau et al. (1992) have both shown fibre optic sensors to show good performance even under flight conditions. The equipment required for signal data processing is actually still comparatively large and costly, but progress in data processing and electronic units leads to the expectation, that this can be significantly reduced over the next years.

For handling the data generated by a large number of sensors, *neural networks* have become a viable technique. So far a lot of numerical work has been done for dama-

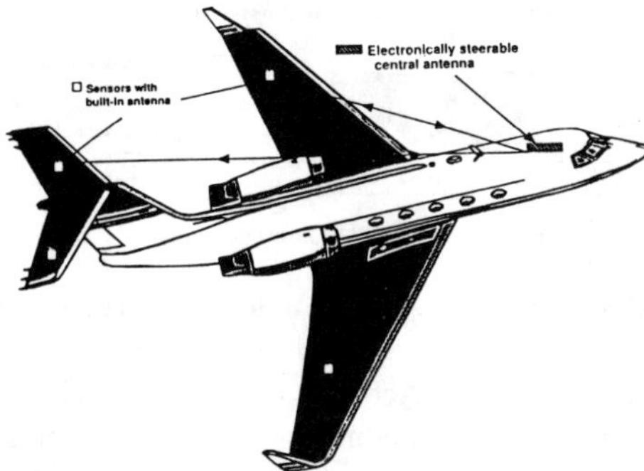

Figure 4 Wireless telemetry for health monitoring of aircraft [Varadan and Va radan, 1996]

ge detection (e.g. Tomlinson, 1996) which shows to be promising with regard to future experimental work.
Antennae used today in aerospace systems are mainly applied for communication between the aerospace vehicle and the ground, between different vehicles or for determining any other environmental conditions. Whenever considering the integration of various sensors into an aerospace vehicle, antennae could very much help to collect the information generated by these sensors and thus avoiding a possibly large amount of wiring. Varadan and Varadan (1996) have made suggestions on how to remotely transmit the sensor signal information via electronically steerable antennae (Figure 4).

SHMS PROOF OF CONCEPT

Modal analysis, which is based on the fact that vibration modes change due to a change in structural integrity, is a further NDT procedure which is widely applied for monitoring space structures. Hickman et al. (1991) have shown on a demonstrator aircraft, that modal analysis could well be used for monitoring damage such as lost rivets or bolts. Analytical and experimental results published in various papers (Tracy and Pardoen 1989) show, that the delamination's size must be at least 10% of the component's surface to be reliably detected by that technique.
Acoustic Emission (AE) has been successfully used for monitoring discontinuities, fatigue failures, materials behaviour, welds including welding processes or stress corrosion cracking in pressure vessels, aerospace and engineering structures. Acoustic emission is the elastic energy being suddenly released when materials undergo deformation. The F-111 fighter/bomber aircraft has been tested in a chamber where the aircraft was periodically chilled to -40°C and stressed between +7,3 g and -3,0 g and an AE system was used to locate sources of structural failure [Carlyle 1989]. Techniques were developed to eliminate loading noise and a strategy was established to identify locations where sensors should be placed to obtain optimized signals. These developments have become feasible since handling, processing and interpretation of data has been improved through better computer technology and

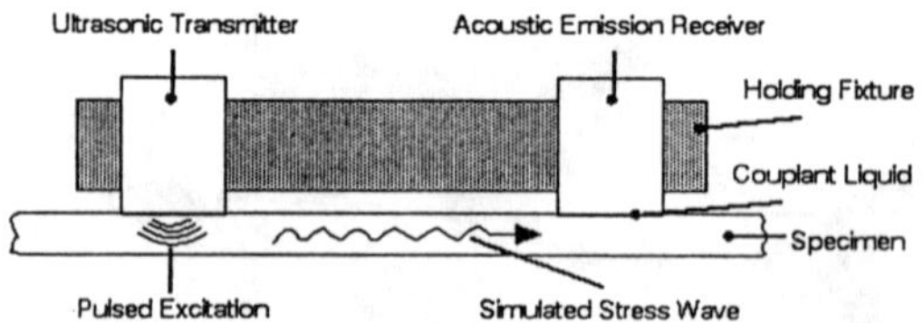

Figure 5 Schematic Diagram of the Acousto-Ultrasonic Technique

new attempts [McBride et al. 1991]. Acoustic emission can also be well applied for determining damage in polymerbased composites. However the high ratio of acoustic signal damping in these materials must be kept in mind, which can lead a sensor to be required every 10 cm, depending on the frequency to be monitored.

Acousto-ultrasonics is a technique which has been proved to be even more sensible than acoustic emission [Vary and Lark 1979]. Figure 5 illustrates the principle of this technique. It requires two probes, one of which is used to introduce ultrasonic stress waves into the structure (actuator) and the other to pick up these stress waves at another position (sensor), where sensor and actuator can both be piezoelectric elements. As soon as the damaged area lies between the two probes the shape of the received acoustic signal changes because of change in material damping characteristics due to the damage (crack, delamination) occured.

Initial tests having been performed by Keilers and Chang (1995) on composite plates joints have shown that the method works and is worth to be pursued.

Lamb waves is another method being also related to the actuator/sensor principle which is based on Lamb wave theory (e.g. Victorov 1967; Kaczmarek et al. 1996). In this case a longitudinal and transverse plate wave is emitted into the structure. The ability of these waves to propagate over long distances is highly advantageous. However much care is required to find the right angle for inducing the lamb waves, especially when the structure to be monitored is of a geometric shape significantly different to that of a plate.

ONGOING ACTIVITIES AND EMERGING TECHNOLOGIES

For over 20 years the Aircraft Structural Integrity Program (ASIP) [MIL-STD-1530A 1975] as well as others is now going on at the United States Air Force (USAF). ASIP itself requires that an airframe be capable of withstanding the growth of an assumed initial flaw under normal operational usage over a prescribed time interval. This can be well done by use of a monitoring system. To this extent a SHMS has been designed for the USAF on a modular basis [Kudva et al., 1996] (Figure 6). The system includes sensors, local preprocessors, a central processor, and software capable of making aircraft maintenance and logistics decisions. Individual sensors track strain, acceleration, temperature, corrosive environment, and structural damage. Due to the modular architecture of the SHMS with physically

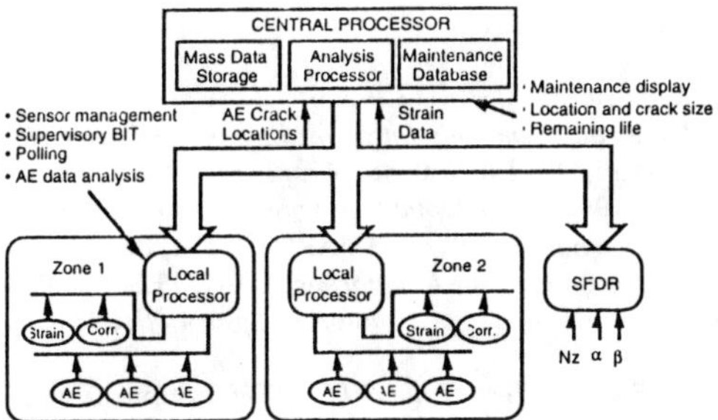

Figure 6 SHMS Architecture [Kudva et al., 1996]

distributed units being logically centralized, excellent flexibility is achieved, allowing for system growth and ease of replacement without impacting the baseline design.

In Europe a major EU-funded programme is going on being entitled Monitoring ON-line Integrated Technologies for Operational Reliability (MONITOR), which includes major aerospace manufacturers, suppliers, research establishments and universities. The objective is to reduce operating cost of aerospace structures by implementing eiter a structure-integrated loads or damage monitoring system. MONITOR aims to (1) understand transport operators requirements and translate these into a health and usage monitoring system specification, (2) develope a monitoring system that will allow structural usage to be effectively monitored and assessed, (3) develop a damage detection system that will allow structural health and integrity to be effectively monitored and assessed, (4) confirm the performance of prototype health and usage monitoring systems in ground and in-flight evaluation, and (5) provide guidelines to best practise in the design, manufacture and qualification of structurally integrated sensing systems.

The extensive activities in developing smart technologies and structures has structural health monitoring to become one of the main topics. The objective of sensors becoming an integral part of structural materials and thus of the aircraft structure itself has opened new perspectives regarding reduced inspection cost, better confidence in what is called advanced materials and even more light weight design. The structural health monitoring initiative has furthermore obtained a significant push due to what is to be developed as smart skins. These skins are designed for control of aerospace structures and may be used to reduce acoustic noise and vibration, drag and skin friction using advanced polymeric smart materials, MEMS (Microelectromechanical Systems) and built-in antennas (Varadan and Varadan, 1993). Applications include smart helicopter rotorblades with microstrip patch antennas and detection and discrimination of hostile threats resulting from laser, radio-frequency and x-rays such as having been performed in the satellite attack warning and assessment flight experiment (SAWAFE) for actively filtered transparencies and conformal antennae (Obal et al., 1992a). Wireless remote and continuous telemetry for application to rotorcraft and smart skin aerospace structures are further areas discussed by Varadan and Varadan (1994).

REFERENCES

AGARD, 1991: *Fatigue Management*; AGARD-CP-506.

Bauer W., 1987: *Event Monitoring Functions Introduced by the Onboard Life Monitoring System (OLMOS) into a German Aircraft*; DFVLR-Mitt. 88-04, pp. 131 - 154.

Blaha F.A. and S.L. McBride, 1992: *Fiber-Optic Sensor Systems for Measuring Strain and the Detection of Acoustic Emissions in Smart Structures*; AGARD CP-531, Paper 21.

Boller Chr., 1996: *Fundamentals on Damage Monitoring*; AGARD LS-205, Paper 4.

Bristow, 1992: *Integrated Health and Usage Monitoring System (IHUMS)*; Aircraft Engineering, February 1992, pp. 12-13.

Carlyle J.M., 1989: *Acoustic Emission Testing the F-111*; NDT-Internat., Vol. 22, No. 2, pp. 67 - 73.

Fürstenau N., D.D. Jantzen and W. Schmidt, 1992: *Fiber-Optic Interferometric Strain Gauge for Smart Structures Applications: First Flight Tests*; AGARD CP-531, Paper 24.

Haberding R., 1985: *Schwingungsüberwachung an Turbo-flugtriebwerken am Beispiel des Airbus A310*; VDI-Berichte 568, pp. 167 - 181.

Hickman G.A., J.J. Gerardi and Y. Feng, 1991: *Application of Smart Structures to Aircraft Health Monitoring*; J. of Intell. Mat. Syst. and Struct., Vol. 2, pp. 411 - 430.

Kaczmarek H., C. Simon and C. Delebarre, 1996: *Evaluation of Lamb Wave Performances for the Health Monitoring of Composites Using Bonded Piezoelectric Transducers*; Proc. of ICIM'96 and ECSSM 96, SPIE Vol. 2779, pp. 130 - 135.

Keilers C.H. and F.-K. Chang, 1995: *Identifying Delamination in Composite Beams Using Built-In Piezoelectrics*; J. of. Int. Mat. Syst. & Struct., Vol. 6, pp. 649-672

Kudva J.N. et al., 1993: *Smart Structures Concepts for Aircraft Structural Health Monitoring*; SPIE Smart Structures Conference, Albuquerque/NM.

Kudva J.N., A.J. Lockyer and C.B. Van Way, 1996: *Structural Health Monitoring of Aircraft Components*; Paper 9; AGARD LS-205.

McBride S., M. Viner and M. Pollard, 1991: *Acoustic Emission Monitoring of a Ground Durability and Damage Tolerance Test*; in: D.O. Thompson and D.E. Chimenti: Review of Progress in Quantitative Nondestructive Evaluation; Vol. 10B, Plenum Press, pp. 1913 - 1919.

Measures R., 1992: *Fibre Optic Sensing for Composite Smart Structures*; AGARD CP-531, Paper 11

MIL-STD-1530A, 1975: *Military Standard, Aircraft Structural Integrity Program, Airplane Requirements*

Obal M. and J.M. Sater, 1992: *Adaptive Structures Programs for the Strategic Defense Initiative Organization*; Proc. of the 33rd Structures, Structural Dynamics, and Materials Conf., Dallas/TX.

CP.Sampath S.G., 1996: *Aging Combat Aircraft Fleets - Long Term Applications*; AGARD LS-206.

Spragg D., U. Ganguli, R. Thamburaj, R. Hillel and R.W. Cue, 1989: *The Role of Inflight Engine Condition Monitoring on Life Cycle Management of CF-18/F404 Engine Components*; AGARD R-770; Paper 4.

Tomlinson G.R., 1996: *Use of Neural Networks/Genetic Algorithms for Fault Detection and Sensor Location*; AGARD LS-205.

Tracy J.J. and G.C. Pardoen, 1989: *Effect of Delamination on the natural Frequencies of Composite Laminates*; J. of Composite Material, Vol. 23, pp. 1200 - 1215.

Varadan V.K. and V.V. Varadan, 1994: *Smart Materials, MEMS and Electronics Integration for Aerospace Applications*; Proc. of the Internat. Aerospace Symp. (IAS), Nagoya/Japan, pp. 115-123.

Varadan V.K. and V.V. Varadan, 1996: *Smart Structures, MEMS and Smart Electronics for Aircraft*; Paper 8; AGARD LS-205.Vary A. and R.F. Lark, 1979: *Correlations of Fiber Composite Tensile Strength with the Ultrasonic Stress Wave Factor*; J. of Testing & Evaluation; pp. 185 - 191.

Victorov I.A., 1967: *Rayleigh and Lamb Wave*; Plenium, New York.

SESSION 10

CIVIL INFRASTRUCTURES III

Optical Fiber Monitoring System of Bridges in Korea

K.-S. KIM and S.-H. PAIK

ABSTRACT

Recently, the interest in safety assessment of civil infrastructures is increasing in Korea. Especially, as bridge structures become large-scale, it is necessary to monitor and maintain the safety state of bridges, which requires the monitoring system that can make a long-term measurement during the service time of bridge.

In Korea, civil engineers have been applied monitoring system to several bridges, as New Haeng Ju Bridge and Riverside Urban Highway Bridge, etc., but these applications have some problems such as sensors for long-term measurement, setup techniques of bridge monitoring system, and the assessment of measured data.

In the present study, optical fiber sensor system was tested and confirmed by laboratory test of concrete member. Attaching or embedding optical fiber sensors to structural members in Sung San Bridge, the response of the bridge by load test was measured.

These monitoring system will be applied to the mock-up test of a bridge, and the applicability of the system and the load capacity of the bridge will be assessed.

1. INTRODUCTION

Optical fiber sensors having advantages of no electromagnetic interference, good resolution, good durability, good efficiency of remote

Ki-Soo Kim and Sang-Hyun Paik, Ssangyong Research Center, SSangyong Cement Industrial Co., Ltd., YUSEONG, TAEJEON, 305-345, Republic of Korea.

signal transmission, they can be a useful measure for displacements of bridges.

In Korea, there has been no example of the application of optical fiber sensors to civil structures. Therefore, we developed the optical fiber sensor and its signal processing system. To assess the applicability of optical fiber sensors to bridge monitoring system, we have performed laboratory test of reinforced concrete member, and applied these system to the existing real bridges.

2. SENSOR CONSTRUCTION

An intrinsic Fabry-Perot fiber optic sensor with circular cross-section is, mainly, composed of two parallel, partially reflecting mirrors spliced into the optical fiber at Lo distance apart. After cleaving the fiber, the end of fiber is coated by TiO_2 film with the thickness of 500-1000Å, and then the TiO_2 film become partially reflecting mirror. The two mirrors in the fiber made by previously explained manner compose the Fabry-Perot interferometric system.

A coherent laser beam of light passed along the optical fiber, the light intensity reflected from the first and second mirror is measured.

[Fig. 1] shows the diagram of Fabry-Perot fiber sensor developed in this study.

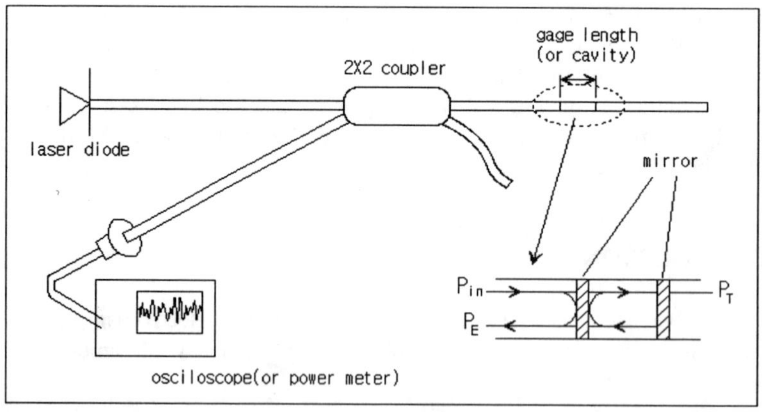

[FIGURE 1] DIAGRAM OF FABRY-PEROT FIBER SENSOR

3. LABORATORY BEAM TEST

3.1 TEST PROCEDURE

Laboratory tests were performed to compare the data from fiber optic sensor with those from conventional strain gauges and to verify the effectiveness of optical fiber sensor.

The cavity length Lo between two mirrors in optical fiber sensor is about 1cm. The sensors were embedded in 20cmX20cmX150cm cement concrete specimens. We attached an intrinsic Fabry-Perot sensor on a reinforcing steel bar in specimens and strain gauges on the other steel bar near the Fabry-Perot sensor[Fig. 2]. We placed these steel bars in the wooden mold and poured the concrete mixtures. After curing this specimens for one month in the air, we put it on the UTM(universal Testing Machine) and carried out 4 point bending test. Data from Fabry-Perot sensor were measured with peak counting method using eye detection and with X-Y Plotter.

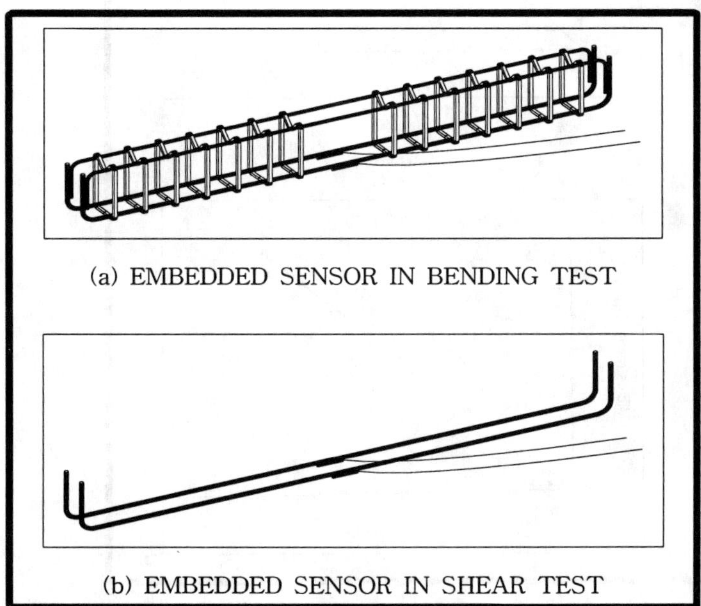

(a) EMBEDDED SENSOR IN BENDING TEST

(b) EMBEDDED SENSOR IN SHEAR TEST

[FIGURE 2] DIAGRAM OF TEST SPECIMEN

3.2 EXPERIMENTAL RESULTS

Two types of tests were carried out. First, load test was performed to induce the failure of specimen by bending force. The specimen had several steel stirrups across the tensile steel bars to prevent the shear failure[Fig. 2(a)]. We measured data from Fabry-Perot sensors and strain gauges. From the analysis of data, they showed good linearity as shown in [Fig. 3]

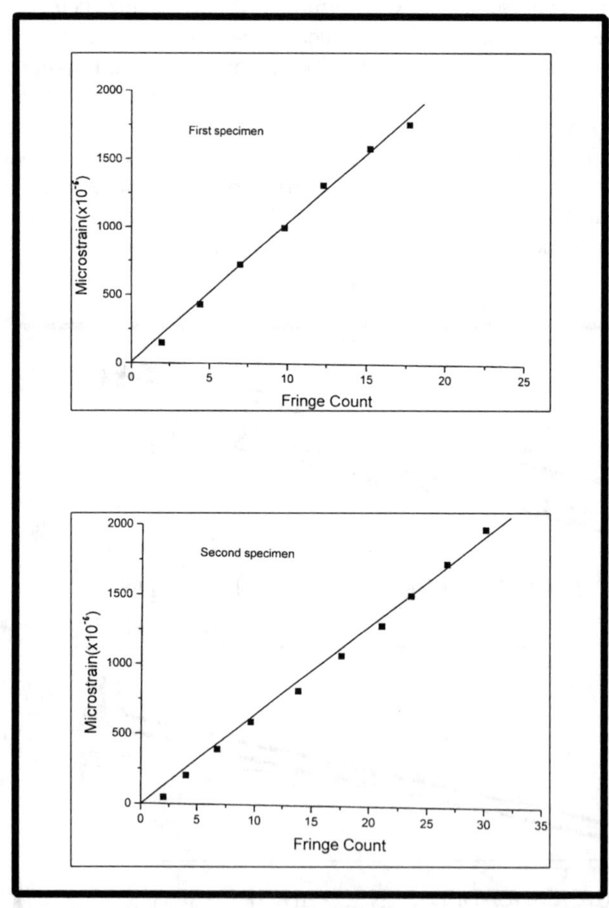

[FIGURE 3] CORRELATION BETWEEN OPTICAL FIBER SENSOR AND STRAIN GAUGE DATA

Second, to induce the shear failure of concrete specimen, the other specimen had only two tensile steel bars without any stirrups and the sensors were attached to the tensile steel bars[Fig. 2(b)]. The specimens were failed by shear force, and data from strain gauges and Fabry-Perot sensors were obtained as shown in [Fig. 4] and [Fig. 5]. Data from Fabry-Perot sensors were recorded by X-Y plotter, and they showed small fluctuations due to the crack propagation and the amplitude change according to increasing load. The phenomenon due to shear strain in the cross-sectional plane of the sensor occurs, and it was predicted by K. Kim, et'.al. The conventional strain gauges cannot measure any strain after the failure of the specimen, but the Fabry-Perot sensors showed good response even after failure.

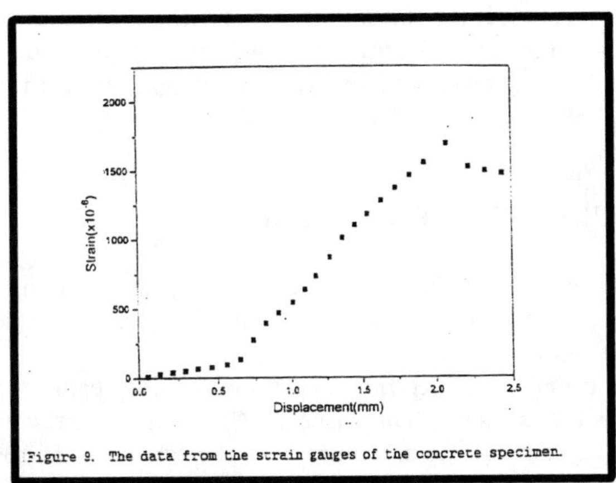

[FIGURE 4] DATA FROM CONVENTIONAL STRAIN GAUGES

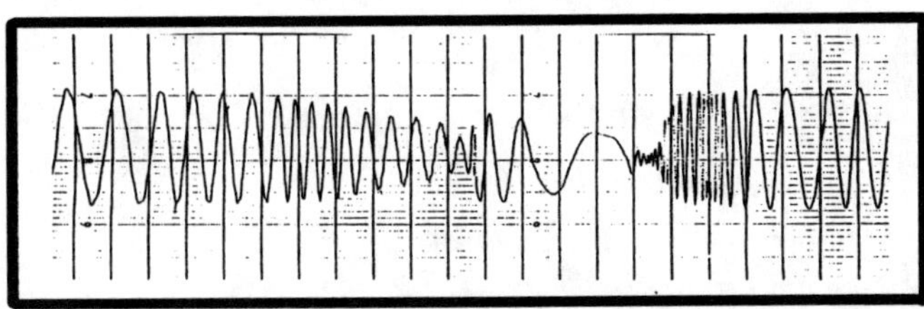

[FIGURE 5] DATA FROM FABRY-PEROT SENSOR (BY X-Y PLOTTER)

4. APPLICATION TO EXISTING REAL BRIDGE

4.1 TEST PROCEDURE

From the results of laboratory tests, outputs from Fabry-Perot sensors showed good linearity to output from conventional strain gauges, so we can apply Fabry-Perot sensors to monitor the behaviour of a real bridge.

Therefore, we chose Sungsan Bridge, which is one of the longest bridges in Korea, to apply these system to real bridge, and attached intrinsic Fabry-Perot sensors to steel girders of this bridge. We applied numbers of strain gauges, acceleration sensors and deflection sensors as well as optical fiber sensors. In this study, we will discuss the application of Fabry-Perot sensors. After the execution of visual inspection, three spans of the bridge were chosen, then optical fiber sensors as well as conventional strain gauges were attached to a steel girder at midspan. Static and dynamic loads were applied to the bridge by 30ton trucks.

4.2 STRUCTURE OF THE SYSTEM

[Fig. 6] shows the applied optical fiber sensor system, which consists of sensor, laser source, detector, data processor, and control and display subsystem.

Laser diode transforms electric signal into laser beam, then the laser transfers to the sensors. The laser in the sensor part varies by the changes in strains and temperatures of the material surrounding the sensor. The laser changed into electric signals by detector, and then the Processor changes the signals to readable data to the computer. Obtained data are analyzed by data processor, and finally the structural safety of bridge is assessed.

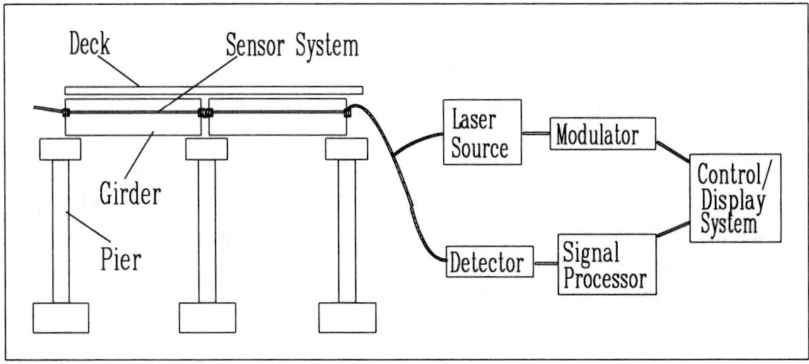

[FIGURE 6] BRIDGE MONITORING SYSTEM OF OPTICAL FIBER SENSOR

4.3 STATIC LOAD TEST

[Fig. 7] shows the results of static load test. The strain before test truck starts, and the strain when the truck is located at the sensor point, are measured. They showed stable response, and the sensor maintains constant output intensity. Also, it looks like almost no noise except natural vibration. The amount of length change of the sensor can be calcualted approximately by Equation(1).

$$\Delta l = \frac{\Delta I}{I_{max} - I_{min}} \times \frac{\lambda}{2} \qquad (1)$$

If we suppose that 2π phase change means 1.3μm length change and the intensity change linearly, the strain when the truck is placed to the sensor point is about 12.3μ strain. Supposing that the accuracy of 1/10 can be read, the resolution is approximately 0.12μ strain. Therefore the optical fiber shows very good resolution.

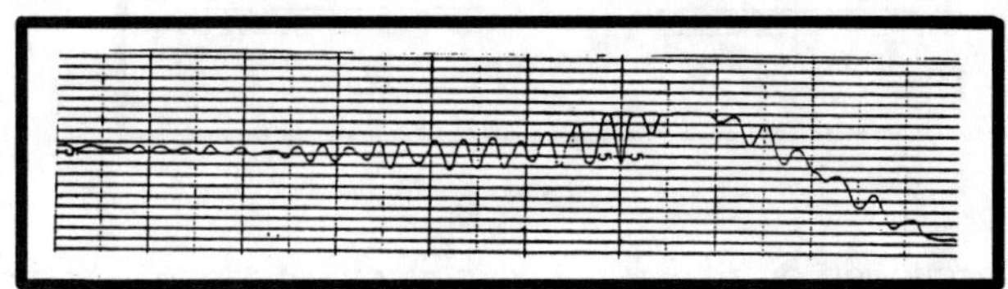

[FIGURE 7] RESPONSE OF FABRY-PEROT SENSOR IN STATIC LOAD TEST

4.4 DYNAMIC LOAD TEST

Strain patterns were measured by Optical Fiber Sensors at each dynamic test step passing by the velocity from 10km/h to 60km/h with 30ton truck.

[Fig. 8] shows the change of measured strains by optical fiber sensors in dynamic load test according to the change of truck speed.

The bridge experiences similar amounts of stress at the truck speed of 20km~40km, and less amounts of stress when the truck speed is below or above the speed.

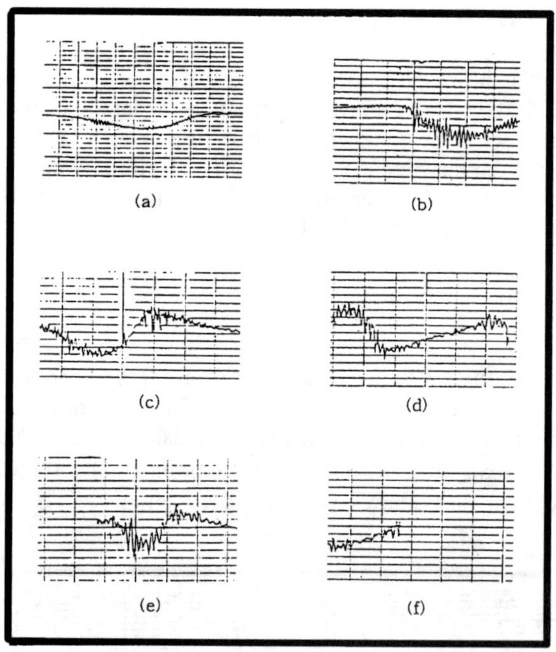

[FIGURE 8] RESPONSE OF FABRY-PEROT SENSOR ACCORDING TO THE CHANGE OF TRUCK VELOCITY(x-axis ; time, y-axis ; strain)
(a) 10 km/h (b) 20 km/h (c) 30 km/h
(d) 40 km/h (e) 50 km/h (f) 60 km/h

5. CONCLUSION

To develop the bridge monitoring system using optical fiber sensor, we developed Fabry-Perot optical fiber sensor, and confirmed this system by laboratory test and the application to Sungsan Bridge.

Output from Fabry-Perot sensor embedded in concrete specimen showed good linearity to output from the conventional strain gauge, and so Fabry-Perot sensors can be applied to maintenance and control system instead of strain gauges.

Measurement by Fabry-Perot optical fiber sensors was performed as a part of safety diagnosis of Sungsan Bridge, and showed good applicability. In the static load test, optical fiber sensor system showed the high resolution of approximately 0.12μ strain. In the dynamic load test, it clearly shows the tendency of strain in the bridge with the change of truck velocity.

REFERENCES

1. Kim, K. S., M. Breslauer and G. S. Springer, "The Effect of Embedded Sensors on the Strength of Composite Laminates" J. of Reinforced Plast. and Comp. Vol. 2, pp 949-958, 1992.
2. Kim, K. S., A. Segall and G. S. Springer, "The Use of Strain Measurements for Detecting Delarminations in Composite Laminates", Composite Structures Vol. 23, pp 75-84, 1993.
3. Kim, K. S., Y. Ismail and G. S. Springer, "Measurement of Strain and Temperature with Embedded Intrinsic Fabry-Perot Optical Fiber Sensors", J. of Composite Materials Vol. 27, pp 1663-1667, 1993.
4. Kim, K. S., L. Kollar and G. S. Springer, "A Model of Embedded Fiber Optic Fabry-Perot Temperature and Strain Sensors" J. of Composite Materials Vol. 27, pp 1618-1662, 1993.

Sustainability of Civil Infrastructure Systems

E. AKTAN, A. HELMICKI and V. HUNT

ABSTRACT

Infrastructure sustainability is a new paradigm focusing on the integration of organizational, financial, informational, managerial, and technical approaches needed to maintain infrastructure systems or change them in a planned way over the course of their lifetimes. Sustainability can be discussed in terms of four topics: life-cycle engineering, technology investment, performance measures, and project management as well as their interrelationships. Integrative, inter-disciplinary and multi-institutional research and technology development, conducted within university-government-industry partnerships, are deemed essential for innovation which is necessary to sustain civil infrastructure systems (CIS). The writers approached highway bridges both as representing a critical component of the transportation system, and also a generic component within a CIS. Actual operating bridges were used as field-test specimens, in collaboration with federal and state DOT engineers, consultants, contractors and suppliers. When applied in the context of structural identification, the experimental tools of modal analysis and instrumented monitoring lead to an objective evaluation of flexibility, capacity, damage, and to a detailed analytical characterization of a bridge in terms of a field-calibrated finite-element model. The organizational and other non-technical barriers obstructing the implementation of advanced technologies have been understood as well. These observations and a synthesis of the data led to an insight on the life-cycle behavior of the most populous bridge type, and a type-specific maintenance management strategy has been formulated.

INTRODUCTION

The national investment in civil infrastructure systems (CIS) is estimated at $20 trillion (NSF, 1992). The US construction industry comes second in size only to health-care, and about 10% of our GDP is estimated to be related to construction. In the US, the service performance of some CIS components fall short of what has been anticipated in design. For example, well-publicized inadequacies in the service performance of the highway transportation system causes unnecessary wear and tear to vehicles, delays in moving people and goods, and accidents which may be altogether reducing the annual growth rate in GDP by as much as 1% (David Aschauer,

---

Emin Aktan, Arthur Helmicki, and Victor Hunt, Cincinnati Infrastructure Institute, University of Cincinnati, OH, 45221-0071, Ph: 513-556-3689; e-mail: aaktan@boss.cee.uc.edu

Chief Economist, Federal Reserve Bank - Chicago). Damage to highways (mainly bridges) due to natural hazards, such as the recent Loma Prieta and Northridge Earthquakes in California, have an equivalent impact. Hence, the annually compounding economic impact of non-optimum operation and maintenance of CIS may reach significantly higher levels than that of major disaster, given the probability of occurrence for catastrophic events per year. We note that population growth, demographic changes, increased expectations for service from deteriorated systems, and new CIS required for communication and information systems have led to expanded CIS that are increasingly complex and difficult to manage intelligently (Institute, 1997).

The US National Science Foundation (NSF) has identified the built environment as an important research area since the 1980's, and announced CIS as a strategic research area critical for the national interest in 1993. Building Stock, Electric Power, Gas and Liquid Fuels, Telecommunication, Transportation, and, Water and Sewer (Environmental Infrastructure) were identified as some of the critical CIS. While the CIS and their users are intertwined, their financing, legislation, regulation, engineering, operation and maintenance management are typically undertaken by a very large number of fragmented public and private agencies. A highway bridge may carry electric power, water, sewer and telecommunication lines and therefore serve as an intersection node for many CIS, and an accident may bring a dozen agencies together revealing how the vulnerability of one CIS component may affect many other CIS.

The US Congress and administration, through the ISTEA (1992) and the NEXTEA (1997) legislations, maintain the need for an integrative, multi-modal approach, and asserts the strong linkage between sustainable development, the environment and the CIS. Infrastructure sustainability is a new paradigm focusing on the integration of organizational, financial, informational, managerial, and technical approaches needed to maintain infrastructure systems or change them in a planned way over the course of their lifetimes. Sustainability can be discussed in terms of four topics: life-cycle engineering, technology investment, performance measures, and project management as well as their interrelationships.

INTEGRATIVE, INTER-DISCIPLINARY RESEARCH AND DEVELOPMENT

A recent workshop (Report, 1966) asserted that new knowledge is needed to provide the intellectual support for infrastructure decisions, and that such knowledge can only be initiated through research which is interdisciplinary. A complete conceptualization of the problem requires closely-coordinated research and demonstrations which can only be conceived, financed, designed and executed by committed CIS champions. These champions must integrate many different technical and non-technical expertise areas in sciences and engineering, and should come from academe, industry and government to represent the many viewpoints which sometimes conflict. Many believe that a combination of financial, legal, organizational, political and societal issues pose a much greater challenge than technical limitations. However, the issues which have to be resolved for innovating CIS engineering are so closely intertwined that a demarcation between technical and non-technical barriers is not possible. A renaissance in civil engineering education and practice appears to be essential to satisfy the societal expectations for CIS performance in the next Century.

A number of experiments exploring multi-institutional partnerships, consortia or alliances have been conducted for CIS research and development in the last decade. Such an experiment was conducted at a modest scale by establishing an infrastructure institute at the University of Cincinnati in 1988, forming a partnership with Ohio DOT, the Federal Highway Administration (FHWA) and several industries. Cincinnati researchers approached highway bridges as representing not only a specific and critical component of the transportation system, but also a component within a generic CIS. The research teams included faculty and students from civil, mechanical, electrical, aerospace and materials engineering, DOT engineers, contractors, materials and hardware suppliers and consulting engineers. Research did not only generate new knowledge on bridge behavior and loading environments, but also permitted an in-depth understanding of a number of issues which have to be resolved to make progress towards an integrated and effective approach to CIS engineering practice. Our experience in the coordination of team research is that it must be designed around an actual operating test-bed which will permit the observation of the complex integrated technical and non-technical systems which control the operation and maintenance management of any CIS. The objective of this paper is to provide a synopsis of the writers' decade-long experience, taking advantage of their specific findings through research on actual operating highway bridges for exemplifying some of the policy and strategy, as well as technology needs for innovating CIS engineering.

RESEARCH DESIGN AND STRUCTURAL IDENTIFICATION

SOCIETAL ISSUES

Table 1 outlines how societal issues drive research focal areas and fundamental knowledge needs for advancing the state-of-the-art in CIS engineering. The research tools which are considered most relevant for generating the fundamental knowledge and to serve as a foundation for new technologies, and, which have been further investigated, are listed in Figure 1.

In spite of the National Bridge Inspection Program started in the 1970's, in-service bridge failures precipitated by design errors, and/or construction or maintenance defects still occur today. The probability of such failures may be controlled by developing proper policy and strategy, since research in the last decade nearly eliminated the technical challenges in identifying bridges which are safety concerns and correcting the deficiencies including earthquake vulnerability. We consider the probability of an undesirable CIS component failure within its life-cycle a ***safety concern*** (say, $>10^{-5}$). Undesirable failures are those which pose a risk to human safety, limit any immediate response needs and/or adversely impact economic recovery following a disaster such as an earthquake, flood or accident.

In 1996 the US DOT announced a backlog of nearly $100 Billion for bridge repair, a figure which is continuously increasing with time. The 1995 AASHTO-LRFD Bridge Design Guidelines expect a new-bridge service life of 75 years with "zero or minimal" maintenance. However many bridges require extensive rehabilitation due to deterioration and damage within just a decade, and evidence points to a lack of a knowledge-base for designing and constructing effective maintenance or renewal. Evidence also points to a lack of knowledge-base for designing and constructing for durability and optimum life-cycle cost or any rational set of performance measures. ***Service performance*** requires functional adequacy, stiffness, vibration-control, durability, ease-of-inspection and maintenance, and, life-cycle cost while recognizing the importance of aesthetics.

Today, current financing mechanisms for the bridge inventory offer no federal aid for routine maintenance, while aid is available for rehabilitation or replacement. Consequently, there is no incentive to states for optimizing maintenance or to accurately evaluate bridge conditions, as opposed to simply requesting federal aid for rehabilitation of safety concerns. Moreover, current federal regulations still dictate bids for new construction to be awarded based on lowest initial construction cost and not ***life-cycle cost*** which deters innovation. It is well-accepted that if we could develop the information-base for reliably evaluating the life-cycle cost of various designs, and sought optimal designs based on minimum life-cycle cost, there would be significant cost-savings and improved performance. The irony is that constructing new roads and bridges, particularly in densely populated urban areas, has become prohibitively expensive. Improving the performance of existing "congested" infrastructure by innovative policy, strategy and technologies is considered as the only feasible way to meet the demands. Intelligent transportation systems (ITS), for example, offer great promise if we could overcome the barriers in developing, demonstrating, standardizing and commercializing appropriate technologies.

RESEARCH FOCAL AREAS

The CIS research agenda formulated by NSF envisioned condition assessment, deterioration science, renewal engineering and institutional effectiveness as focal areas where new-knowledge was needed (NSF, 1992). Table 1 adopts these focal areas with some modification.

A reasonably accurate "projection of future performance" is necessary for rational and optimum infrastructure maintenance management (Yao, J.T.P. and Yao, T.H-J..1997). ***Reliability evaluation*** is described as the process of establishing the probability of acceptable performance at the service and failure limit-states over a time window given the as-is condition and a maintenance scenario. Reliability evaluation requires assessing the demands from and the corresponding capacities of a facility with some confidence and evaluating the probability of demands exceeding the capacities for critical limit-states. To accomplish this, the relationships between demands, capacities and performance have to be established with reasonable confidence for each critical limit-state. ***Identification and Integration of CIS*** is therefore necessitated as a focal area since many interrelationships and interactions between different CIS, sub-systems or components are sometimes never clearly established or discovered only by accident.

FUNDAMENTAL KNOWLEDGE

The third column of Table 1 indicates the fundamental knowledge needed in the areas of deterioration science, damage mechanics, structural identification and decision sciences in order to accomplish the scientific, engineering and educational advancements which are needed for sustaining CIS (NSF, 1992).

Structural Identification (St-Id) is a concept which permits effectively integrating experiment, analysis, information and decision sciences. Arguably, the single most important technical barrier limiting innovation in CIS engineering may be the empirical, heuristic foundation of civil engineering practice. Unfortunately both the education and practice of civil engineering disguise the strong dependence on empirical art and heuristic knowledge-bases, and many practicing engineers and academics may not fully appreciate how detached the design, construction, inspection, maintenance and renewal processes are from the reality of actual loading environments and behavior of structures. Structural identification is a critical enabling concept for developing a complete, conceptual and objective understanding of real-life as illustrated in Figure 1 (Aktan et al, 1997).

In order to rationalize civil engineering practice it is important to use St-Id for making progress towards the conceptualization and measurement of fundamental unknowns for constructed facilities, including:

- Quantitative knowledge on the as-built ***state parameters*** (e.g., initial local and global stiffness, stresses, strains, and displacements) and their variation at different limit-states and over time;
- Clear and quantitative definitions for ***performance parameters*** (e.g., safety, life-cycle cost, functionality, serviceability, etc.) and their relationships with state parameters;
- Clear and complete understanding of the ***loading environment (including intrinsic loads) and defects, deterioration and damage mechanisms*** which influence the state-of-force in a structure, lead to changes in state parameters, and/or affect performance;

- Actual capacities of ***load-resisting mechanisms*** and how the capacities and failure mechanisms are affected by various types of defects, deterioration, and damage;
- Observable and easily measurable ***condition-and-damage indices*** sensitive to changes in state properties, and that correlate to the capacity of load resisting mechanisms and performance.

STRUCTURAL IDENTIFICATION TOOLS

The experimental, analytical and information tools which are needed for structural identification are listed in the individual blocks of Figure 1. We define ***physical similitude*** as the art-science of designing experiments in real-life so that critical phenomena are properly represented and the results may be generalized. For example, in physical model testing it is quite common to test an element or a subassembly in 2D without simulating 3D interactions or all the boundary and continuity conditions accurately and completely. ***Measurement technologies*** to document existing global and local geometric details of large constructed facilities and the changes which occur in these with time are envisioned as a means of improving our accurate and complete visualization capabilities so that we can design experiments which would satisfy physical similitude. ***Modal analysis and instrumented monitoring*** are the two experimental technologies which the writers have been exploring for field applications to operating constructed facilities (Aktan et al, 1997). These experimental technologies should be designed and implemented within a structural identification framework, and the synergy between them should be realized, in order to gain maximum benefit (in some cases, any benefit) from their applications.

While a multitude of structural responses, state parameters or load effects may be directly measured or extracted from modal analysis or instrumented monitoring, the writers directed their applications towards measuring ***flexibility*** in terms of various temporal and spatial resolutions. Flexibility has been demonstrated to be a quantitative, comprehensive, and damage-sensitive signature, and a conceptual condition index (Raghavendrachar and Aktan, 1992; Farrar et al, 1997). It is possible to obtain flexibility in terms of a fine spatial resolution by modal testing, enabling condition evaluation at the regional or element-level. Flexibility may be measured practically in terms of sparsely distributed coordinates by contact or non-contact displacement measurement techniques during diagnostic truck-load tests. Non-contact measurements by radar, falling-weight deflectometer applications or conversion of strain, acceleration, and tilt measurements obtained during instrumented monitoring offer promise as well.

STRUCTURE-SPECIFIC VERSUS TYPE-SPECIFIC KNOWLEDGE

The prevailing approach for the design, construction, inspection, evaluation, maintenance and management of CIS is that ***"every constructed facility is a different and unique structure."*** Indeed, we recognize that even two "identical" test specimens fabricated under controlled circumstances in a laboratory exhibit variations in their mechanical characteristics which typically become larger with advanced limit-states. Moreover, significant scale-and-loading rate influence has been observed in the behavior of physical models tested under realistic loading with a concerted effort for physical similitude (Bertero, et al, 1984). On the other hand, structural systems such as plates, diaphragms, plane-grids, etc. constructed with similar materials and subjected to similar loading environments are known to exhibit many similarities in behavior, even in the failure limit-states. An understanding of common threads in behavior has been useful in civil engineering education, particularly in the case of material-specific design subjects such as reinforced-concrete, steel, prestressed concrete, etc.

Type-specific management is a strategy advocated by the writers for capitalizing on the common threads in behavior for a practical evaluation of the nation's bridge inventory. In the current inspection and inventory practice, bridges are broadly classified as reinforced concrete deck-on-steel girders, also known as steel-stringer bridges, concrete slab, steel-truss, etc. Experienced engineers also base their decisions on a heuristic approach which recognizes the typical behavior patterns and problems of each bridge category. The writers' research indicates that accurate and complete structural identification does lead to powerful and reliable simulation capabilities, permitting an understanding of the principal design, construction, and maintenance attributes that influence state and performance characteristics and their interrelationships for a broad class of bridges such as steel-stringer, etc. Therefore, it appears possible to rationally classify (for example, "steel-stringer") bridges into several groups of like behavior, based on the most critical attributes that govern behavior.

For example, some of the ***critical loading and structural parameters*** which govern the serviceability and damage limit-state behavior and performance of steel-stringer bridges through their life-cycle are illustrated in Fig. 2. Once the threads between common design, construction and maintenance defects (many of which are not yet identified and occur due to the empirical nature of civil engineering practice), and the deterioration and damage mechanisms are clearly understood, it should be possible to represent the population of almost 200,000 steel-stringer bridges in the National Bridge Inventory in terms of a large-size statistical sample of say, thirty, for further study. The statistical population would be rigorously tested and studied by St-Id, and practical techniques for type-specific condition assessment, evaluation, and maintenance can be developed for the entire population. In this manner, we can examine, rationalize, and take full advantage of the heuristic knowledge-base we have accumulated on the behavior of different recurring bridge types through the National Bridge Inspection Program since its inception in the early 1970's. We will be learning how to design and maintain for optimum life-cycle performance from a study of both the successful and unsuccessful bridges which are in the Inventory. Naturally, one-of-a-kind, unique bridges, many of which are major, monumental and/or historic structures would have to be managed in a bridge-specific manner by taking advantage of structural identification.

LIFE-CYCLE BEHAVIOR OF STEEL-STRINGER BRIDGES

Steel-stringer (reinforced-concrete deck on steel girders) bridges make up both the most populous bridge type and the largest number of "deficient" bridges in the US National Bridge Inventory (Federal, 1993). Design of steel-stringer bridges evolved since the 1930's. While the early designs generally used simple-spans, the construction of continuous steel-stringer bridges starting from early 40's, and the construction of integral bridges starting from the 70's were major innovations.

Since the 1980's, composite designs and Grade 50 (and more recently, Grade 70) steels have become quite common. These changes, in conjunction with the strength design provisions have led to decreased girder sizes and increased spacing, leading to increased span deflections and vibration. The 1995 AASHTO-LRFD code is expected to lead to even more flexible designs. The relation between operating span deflections and serviceability is well-known, however, the desired range of bridge flexibility which would provide an optimum life-cycle cost has not been established. Furthermore, we still follow empirical guidelines in the new or renewal design decisions regarding girder sizing, spacing and camber, deck thickness and reinforcing details, design of the interface between the deck and the girders, cross-frame sizing, spacing and their girder-connection details, bearings, abutments and sub-structure. We also have considerable difficulty in successfully designing and constructing bridges with curved geometry or large skew.

BRIDGE BEHAVIOR DURING CONSTRUCTION

A new 3-span steel-stringer bridge currently under construction, HAM-126-0881L, has been designated as a test-site for measuring the fabrication and construction-induced intrinsic responses, followed by demonstrations of health-monitoring and intelligent-infrastructure. The measurement and documentation of construction and service effects will permit an evaluation of the complete and absolute state-of-force in the bridge, together with the corresponding causative effects or events. A complete sensor suite of over 300 sensors and associated hardware, including a weigh-in-motion (WIM) roadway scale, is being incorporated in the construction. The data collection during the construction stage has been augmented by conducting modal testing and diagnostic tests under controlled truck-loading following the completion of construction during April 1997. An accurate finite-element computer model is being developed and calibrated to transfer the recorded strain, distortion, inclination, displacement, and temperature data into the corresponding stresses, forces, and reactions.

Data has been collected by appropriate instrumentation of the foundation and substructure components, girders and the cross-braces prior to critical fabrication steps in the shop (including heat-cambering), and by adding instruments at each subsequent step in the construction culminating with the pouring of the deck and parapet concrete and the approach slabs. Data acquired through different construction phases and during service has and will be used to conceptualize less-understood or unknown phenomena that influence bridge performance, and to verify design assumptions and rating models. For example, the net strain offset due to the dead load of the deck within the deep-foundations (drilled-shafts), piers and at the girder flanges have been compared with analytical results obtained from computer simulations. The measured strains due to the weight of the deck concrete differed from the predicted values, particularly in the steel-girder. For example, rather than 7.6 Ksi (52 MPa) as expected, stresses of about 10 Ksi (69 Mpa) were measured due to the dead-load effect of the deck concrete.

SERVICE BEHAVIOR OF A 10-YEAR OLD BRIDGE

Since 1993, the writers conducted a comprehensive survey of commercially available sensors and data acquisition systems, and the most promising sensors have been investigated through a rigorous system-calibration conducted in the laboratory (Levi, et al, 1996). Then using the lessons learned in the first phase, a pilot 64-sensor monitor system was designed and installed on HAM-42-0992, a 10-year-old, 55 m (180 ft), three-span, steel-stringer bridge. The bridge has been subjected to numerous experiments and structural identification since 1991 (Aktan et al, 1993). The monitor system incorporates strain, displacement, rotation, acceleration, and temperature sensors, and an ambient temperature, wind, and humidity monitor. The monitor has successfully operated for over two years and valuable data has been recorded (Fig. 4).

The strains captured by the monitor indicate significant stress accumulation in the girders, particularly adjacent to the acute angles of the abutment. Over the two-year period shown, the bridge witnessed an average temperature differential of amplitude 110^0F. The resulting net micro-strain change in the girder at the bridge abutment was observed to exceed 350, which corresponds to 10.2 Ksi of intrinsic stress. In addition, it was observed that the intrinsic stress due to the average temperature differential was less at the inner girders and decreased with longitudinal distance from the abutments. Meanwhile, the recorded responses under even the highest traffic loads indicated levels of strain substantially less than that required of the bridge live-load design (180 Micro-strain). The temperature-related responses overwhelmed the truck responses such that the critical regions of the bridge depended on the changes in intrinsic forces.

It is clear from these observations that the combined effects of environmental and traffic responses may be "in phase" or "out of phase" depending on the location and time of occurrence. Thus, because of their magnitudes and variability, unless the changes in intrinsic responses of an actual bridge are monitored for several years, it may not be possible to differentiate between the impact of defects/deterioration/damage, environment, and traffic from natural changes in intrinsic responses. By the same token, it is not possible to conduct just one proof-test to rate a bridge, since performance and responses greatly depend on the interactions between the soil, pavement and the bridge which, in turn, are significantly affected by long-term climate changes, seasonal changes and the ambient conditions at the time of the test.

DAMAGE LIMIT-STATE BEHAVIOR OF A 40-YEAR OLD BRIDGE

A decommissioned, forty-year-old three-span steel-stringer bridge, HAM-561-0683, has been serving as the test specimen for evaluating different concepts, experimental approaches, algorithms, and hardware/software tools for detecting various types and levels of existing or artificially induced deterioration and damage (Aktan et al, 1997 (b), Catbas et al, 1997, Lenett et al, 1997 and Levi et al, 1997). Scenarios included long-term deterioration effects such as disintegration of deck concrete due to chemical attack, displacement of abutments, dislocated bearing, frozen bearing, girder fracture, loss of effective area and inertia of girders, breaking of connections between girders and cross-frames, and loss of bond between reinforced-concrete deck and girders providing composite action.

The test-bridge has been subjected to a comprehensive set of structural identification experiments both before and after the application of each deterioration, damage, or retrofit scenario. The two basic experimental approaches that have been explored for condition and damage assessment are instrumented monitoring and dynamic testing. The latter included multiple-reference impact tests, forced-excitation sine-sweep tests, and ambient monitoring. The most important benefit expected from this research is the scientifically collected data on the impact of typical deterioration and damage scenarios on the local, regional, and global state properties of the test bridge, and the effects of environmental conditions on the bridge mechanical characteristics.

The space limitations of this paper do not permit a detailed discussion of the preliminary findings. However, a major finding that is relevant for health-monitoring and control has been the interactions between changes in the: (a) ambient conditions; (b) the boundary conditions of the bridge; (c) mechanical characteristics of the bridge; and, (d) the impact of damage on the mechanical characteristics of the bridge. For example, when boundary conditions are "ambiguous," e.g., when bearings at the abutments are not all in contact or expansion joints are locked, the bridge mechanical characteristics were very sensitive to temperature. While in this condition, the bridge could not be considered stationary during periods of large temperature changes within a day. Similar non-stationary behavior was observed after certain types and levels of damage which disturbed the equilibrium of intrinsic forces in the bridges.

CONCLUSIONS

Research conducted by a multi-disciplinary team of engineers, in the context of an academic-government-industry partnership is summarized. The global objective of the research is to generate fundamental and generic knowledge of the actual loading environments and behaviors of recurring highway bridge types. Once the critical design, material, loading and location parameters which impact bridge behavior, and their interrelationships, are clearly understood and rationalized, it will be possible to select, test and study a statistical sample for a generic characterization of the large population of recurring bridge types. The understanding and insight gained from such research will pave the way to type-specific management of populous bridge types which have common threads in behavior such as the steel-stringer bridges.

The steel-stringer bridge type has proven economical in the US for small-to-medium spans (individual spans of 40-250 feet, or, up to about 75 m.) and continuous-girder bridges have an excellent safety record. While the safety performance of the large majority of steel-stringer bridges have been perfectly satisfactory, the technical challenges in cost-effectively sustaining the service performance of these bridges is a serious concern which remains to be resolved.

The writers have rigorously tested and studied eight steel-stringer bridges. They are currently monitoring three steel-stringer bridges as national test-beds as briefly described in this paper. Each of the three bridge test-beds and the associated research projects described, when viewed in isolation, provides insight towards the questions facing CIS researchers and managers. However, when taken together, the sum total of the results obtained from these test-beds has the added benefit of providing a view of the complete life-cycle of a steel-stringer bridge. For example, the maximum steel girder stresses of short-span integral steel-stringer bridges due to construction and dead load are observed to be in the order of 10 Ksi. The stresses which accumulate at various locations due to seasonal changes in the ambient conditions in the climate zone of the Ohio Valley are of the same order of magnitude. The seasonal changes and the daily variations in temperature affect bridge stresses and critical locations of stress accumulation significantly. In relation to intrinsic stresses, truck stresses are considerably less. The largest truck-induced stress recorded on the operating bridge did not exceed 5 Ksi. The largest live-load induced stress which the writer ever measured on a steel-stringer bridge under permit loading was under 10 Ksi. Obviously, the typical, short-span continuous steel-stringer highway bridge has ample capacity, but life-cycle service performance of many bridges have not been satisfactory. Life-cycle performance and durability of steel-stringer bridges may be improved by rationalizing their design, construction and maintenance, based on a complete structural identification of a statistical sample of these bridges.

The researchers have developed experimental techniques which permit accurate measurement of bridge structural properties, understand global condition and evaluate future performance. The experimental tools permit an objective identification of any damage, and also lead to a detailed analytical characterization of the test bridge in terms of a field-calibrated finite-element model. Finite-element models which accurately simulate the actual 3D geometry, stiffness, boundary and continuity of an actual bridge permit a rational study of how different design, construction, maintenance and renewal details influence bridge flexibility and force-distribution. The challenge is now in the applications to sufficiently large numbers of bridges, both recurring and one-of-a-kind types of bridges to improve our insight.

ACKNOWLEDGEMENTS

The writers wish to acknowledge the participation of many graduate students and colleagues in the various portions of the research activities described herein. These include: D. Brown, N. Catbas, M. Lenett, A. Levi, and A. Wilson. In addition, the authors gratefully acknowledge the support of the FHWA, the NSF, the Ohio Department of Transportation (ODOT), and the Ohio Board of Regents (OBoR) under whose support and auspices this research has been performed.

REFERENCES

1. Aktan, A.E., Chuntavan, C., Toksoy, T. and K.L. Lee. 1993. "Structural Identification of a Steel-Stringer Bridge for Nondestructive Evaluation." *TRR 1393*, National Academy Press, Washington, D.C., pp. 175-185.

2. Aktan, A.E., Dalal, V., Farhey, D., Helmicki, A., Hunt, V., and S. Shelley. Sept., 1996. "Condition Assessment For Bridge Management." *Journal of Infrastructure Systems*, 2 (3), ASCE.

3. Aktan, A.E., Farhey, D.L., Helmicki, A.J., Brown, D.L., Hunt, V.J., Lee, K.L., and A. Levi. 1997. "Structural Identification for Condition Assessment: Experimental Arts." To appear in the *Journal of Structural Engineering*, ASCE.

4. Aktan, A.E., Brown, D., Farrar, C., Helmicki, A., Hunt, V., and J. Yao. 1997. "Objective Global Condition Assessment." Proceedings, *15th International Modal Analysis Conference*, Orlando, FL.

5. Aktan, A.E., Helmicki, A.J. and V.J. Hunt. 1997. "Issues In Health-Monitoring For Intelligent Infrastructure." To appear in *Smart Materials and Structures on Large Civil Structures*, U.S.-Japan Workshop on Smart Structures Technology.

6. Federal Highway Administration Report. 1993. *The Status of the Nation's Highways, Bridges, and Transit: Conditions and Performance*, Publication No. FHWA-PL-93-0.

7. Catbas, F.N., et al. 1997. "Modal Analysis Of Multi-Reference Impact Test Data For Steel Stringer Bridges." Proceedings, *15th International Modal Analysis Conference*, Orlando, FL.

8. Institute for Civil Infrastructure Systems (ICIS). April, 1997. *NSF Initiative Announcement*.

9. Lenett, M., et al. 1997. "Issues In Multi-Reference Impact Testing of Steel Stringer Bridges." Proceedings, *15th International Modal Analysis Conference*, Orlando, FL.

10. Levi, A., et al. 1996. "Instrumentation, Testing and Monitoring of Reinforced Concrete Deck-On-Steel Girder Bridges." *Report UC-CII-96*, UCII, Cincinnati, OH.

11. Levi, A., et al. 1997. "Instrumented Monitoring and Diagnostic Load Testing of Steel Stringer Bridges." Proceedings, *15th International Modal Analysis Conference*, Orlando, FL.

12. NSF Civil Infrastructure Systems Task Group (CISTG). 1992. *NSF Workshop on Civil Infrastructure Systems Research*, Chair: Ken Chong, Washington, D.C.

13. Raghavendrachar, M. and A.E. Aktan. August 1992. "Flexibility by Multireference Impact Testing for Bridge Diagnostics." *Journal of Structural Engineering*, ASCE.

14. *Report of Workshop on Integrative Research for Civil Infrastructure*. July, 1996. Washington, D.C. (New York University: www.nyu.edu/urban).

15. Yao, J.T.P. and T. H-J. Yao. April, 1997. "Optimal Performance of Civil Systems Using Symptom Based Realiability and Health Monitoring." Proceedings, *NSF Workshop on Optimal Performance of Civil Infrastructure Systems*, Portland, OR.

SOCIETAL ISSUES	FOCAL AREAS FOR RESEARCH	FUNDAMENTAL KNOWLEDGE NEEDS
Service Performance Safety Concerns Effectiveness of: Maintenance and Renewal Improved Performance Life-cycle Cost	Systems Id and CIS Integration Objective Condition Assessment Reliability Evaluation For Service & Safety Intelligent Renewal Management: Technical and Organizational	Deterioration Science Loading Environment; Defects, Deterioration, Damage Mechanisms Damage Mechanics Identify Damage and its Impact on Structural Reliability Structural Id Identify and Relate Design, Construction, Maintenance & Renewal Techniques to Structural State and Performance Decision Sciences Heuristics, Information & Data Engineering/Sciences, Uncertainty and Risk Analysis

Table 1: Infrastructure Research Design

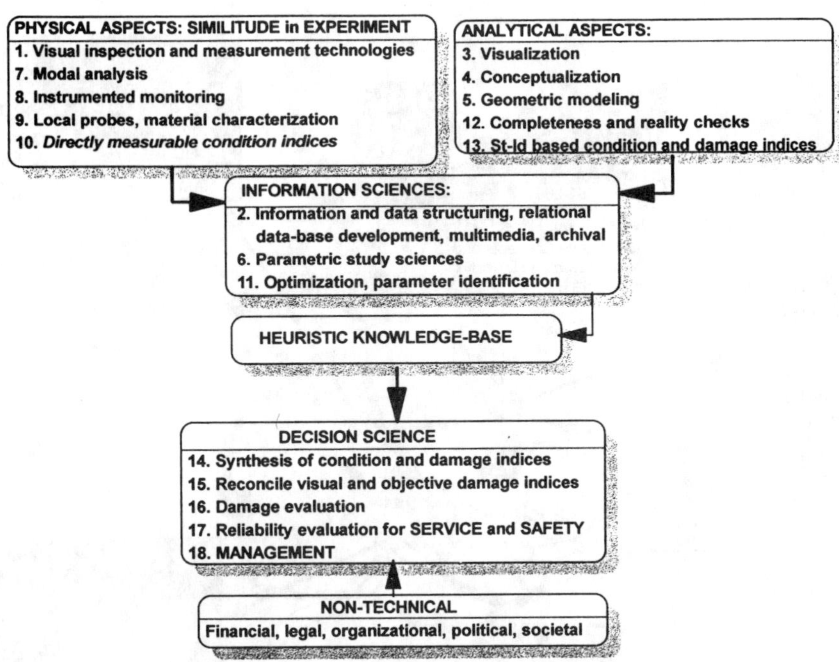

Fig 1. Structural Identification of Infrastructure Management

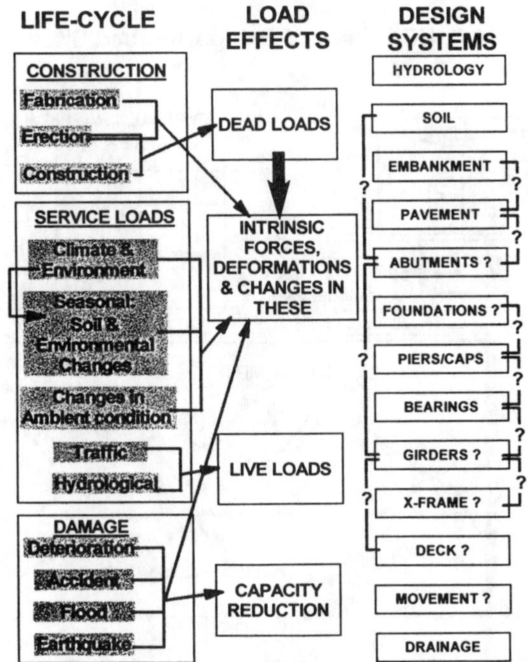

Fig 2. Steel Stringer Bridge, Life-Cycle, Load Events and Design Issues

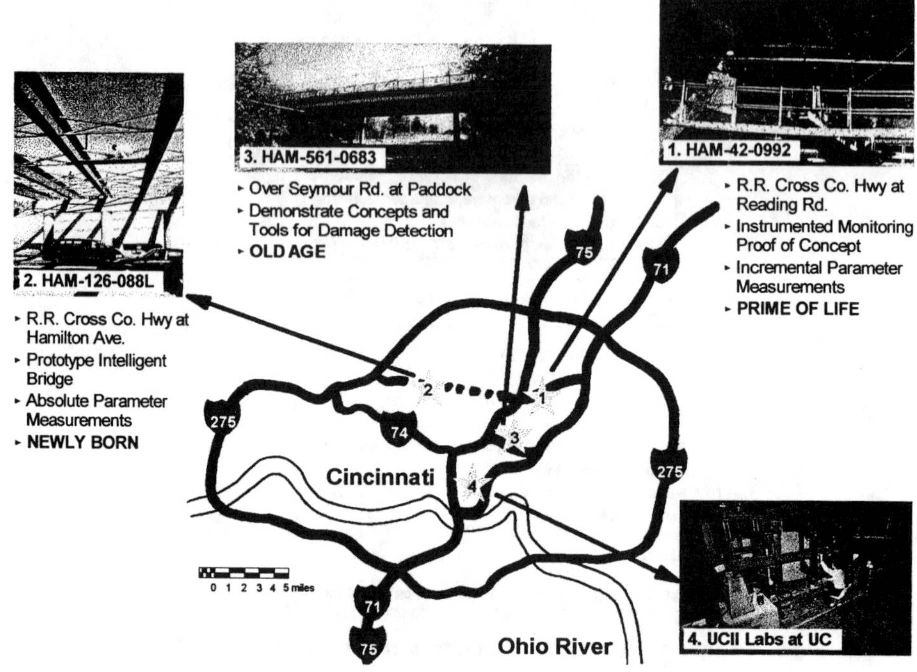

Fig 3. UCII National Bridge Testbed Sites

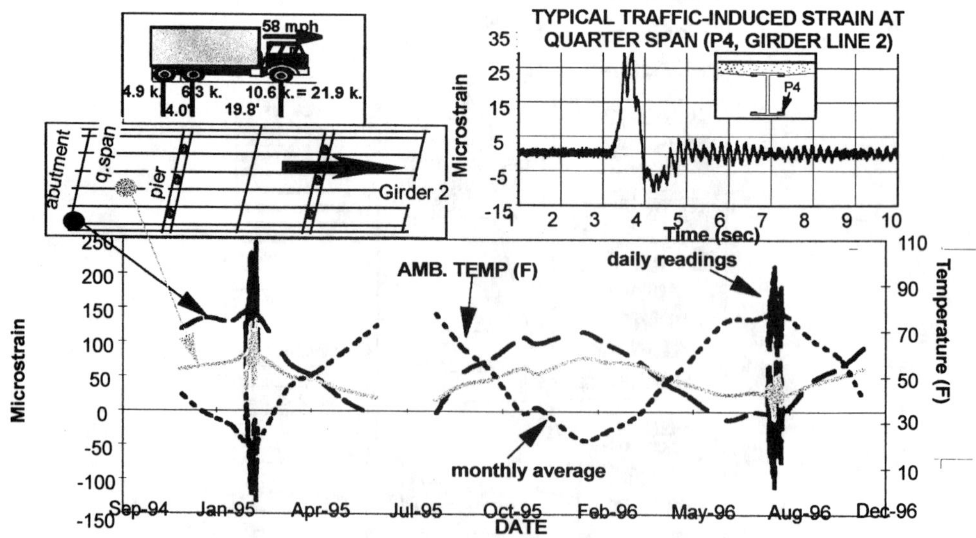

Fig 4. Typical service level load effects and responses

Practical Aspects of Testing Large Bridges for Structural Assessment

A. FELBER

ABSTRACT

This paper will highlight a few of the practical aspects of dynamic tests conducted on long span bridges. The ambient vibration testing procedure will be described and compared to the forced excitation testing method. Common constrains which are particular to long span bridge testing will be presented. In many instances, the test results were used to update finite element models which in turn were utilized to assess the structural health of bridges. The updating of the finite element models based on the experimental results will be discussed briefly. Using the tests of the Port Mann Bridge as an example, the effects of the testing results on the structural retrofit design will be described. Finally a brief outlook on the future testing applications on large bridges will be given.

INTRODUCTION

Experimental dynamic tests of full scale civil engineering structures have been performed by many researchers and practitioners for several decades now. Dynamic testing of civil engineering structures can be performed using either Forced Vibration Testing (FVT) or Ambient Vibration Testing (AVT) techniques. It has been demonstrated in the early seventies that both of these methods lead to similar results if they are applied correctly (Trifunac, 1972). With some exceptions, the FVT approach has historically been used for small to medium size bridges while the AVT method has been used extensively on long span bridges.

FVT requires a controlled excitation of the structure applied at one or more points along the structure. The structure's response to this forced excitation is then recorded at a large number of points. The key aspect of this technique is that the response of the structure is referenced to a controlled and measured excitation. If all the structural responses are due to the controlled excitation, then well established data analysis techniques can be used to determine the natural frequencies and mode

Andreas Felber, EDI Experimental Dynamic Investigations Ltd., 1633 Stephens Street, Vancouver, BC, CANADA, V6K 3V4. Phone: (604) 737-3636, Fax: (604) 737-4284, E-mail: edi@intergate.bc.ca.

shapes, as well as the modal mass, stiffness and damping of the structure. For small structures, the response to forced excitation is generally much greater than any response due to wind or other natural sources, and thus the method has be applied quite successfully for several decades all over the world (for examples see Agardh 1991; Cantieni, Deger and Pietzko 1994, Hudson 1964).

For large and flexible structures, such as long span bridges, it becomes much more difficult and costly to provide controlled excitation at levels which are significantly higher than the excitation provided by wind. This can be readily illustrated by assessing the amount of force which a typical shaker can provide at the frequency of the first vertical mode of a bridge. The frequency of the fundamental vertical mode, fv_1 (in Hertz), of a bridge as a function of span length, L (in meters), can be estimated from:

$$fv_1 = 100/L \tag{1}$$

Forced vibration test methods often apply the known force to the structure either using inertial shakers. The maximum force which can be provided by an inertial shaker is a function of the amount of mass, M, which is moving through a certain stroke, S, at a specific frequency, f. In order to maximize the amount of force exerted on the bridge at a given frequency the mass is moved in a sinusiodal fashion in the direction where the maximum force needs to be applied. The magnitude of the force, $F(f)$, as a function of the excitation frequency provided by an inertial shaker is given by:

$$F(f) = 4\pi^2 f^2 M(S/2) \tag{2}$$

Typical inertial shakers used on bridges have strokes in range of 0.2 m to 0.6 m and a moving mass in the range of 200 kg to 500 kg. Substituting equation 1 into equation 2, the force which can be applied to the structure in the lowest fundamental mode can be expressed as a function of bridge length. This relationship is shown in Figure 1 for the range of shakers mentioned above. From this figure it can clearly be seen that for long bridges, the amount of force that can be produced is very small. It should also be considered that long bridges have a large mass which needs to be excited. For example the steel superstructure of the Port Mann arch bridge which has a 365 m long main span weighs in the order of 11850 tons. Using the largest shaker in the range discussed above will only result in an excitation force in the order of 1 kN. This figure clearly illustrates that for long bridges, inertial shakers can not provide enough force to provide controlled excitation.

For long span bridges, AVT has been applied quite successfully (Buckland et al. 1979; Abdel-Gaffar and Scanlan, 1985; Luz 1987; Law and Ko, 1995). This method only requires the measurement of the structure's responses to ambient excitation which might be caused by wind, micro seismic tremors or vehicle traffic. If it is assumed that the excitation is relatively smoothly distributed in the frequency band of interest, then the natural frequencies and mode shapes of the structure can be identified quite readily. In addition, it is possible to estimate damping values associated with individual modes. The main advantage of this method is the that no artificial excitation has to be applied to the structure. Furthermore, on bridges, vehicle

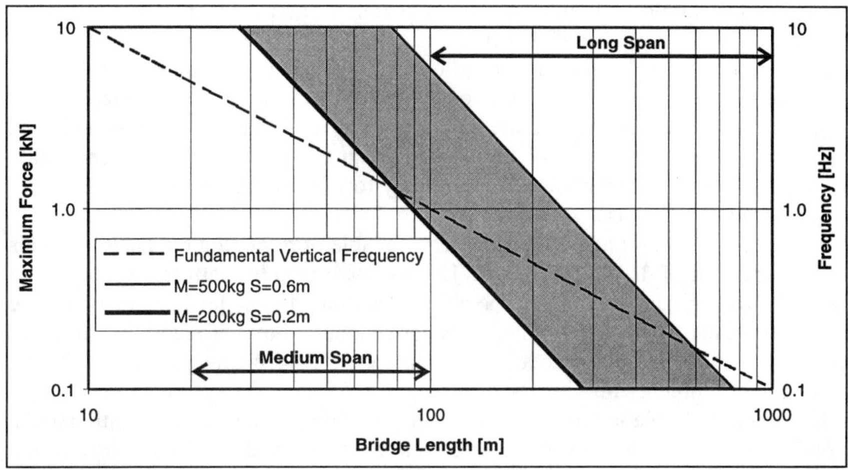

Figure 1: Maximum force output of an inertial shaker at the fundamental vertical mode of a bridge

traffic does not have to be shut down during the testing. Traffic is generally a welcome source of excitation, as it usually is a good source of broad band excitation.

TESTING HARDWARE

The hardware which forms a typical AVT system is shown schematically in Figure 2. A complete description of an actual system is presented elsewhere (Felber and Cantieni, 1996).

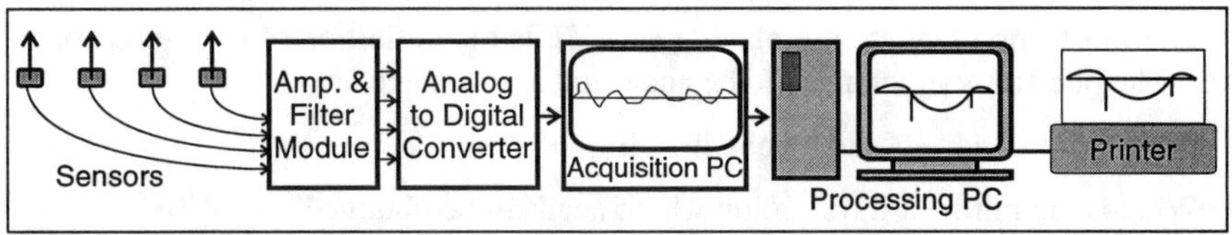

Figure 2: Schematic of a typical ambient vibration testing system

The measurement hardware needs to be specifically selected for ambient vibration testing. The sensors used to measure ambient vibrations are usually Force Balanced Accelerometers (FBA). Typical FBAs have a range of ±0.5 g and a sensitivity of 5V/g. The full scale range of ±0.5 g is chosen because most ambient excitations cause vibrations of less than ±0.5 g. However, traffic induced accelerations larger than ±0.1 g have been measured on steel bridges. The excellent dynamic range (130 dB from 0-50 Hz) of these sensors makes it possible to reliably sense accelerations as small as $1\mu g$. The FBA has a natural frequency of 50 Hz and damping of 67% of critical to ensure an even amplitude response in the region of interest (typically 0 to 10 Hz). At the same time, the sensors also act as low-pass filters to remove unwanted high frequency accelerations.

The cables should be tough, shielded and ensure minimal signal loss and interference over large distances (500 m). The cable weight is also a major factor that should be considered as it has to be handled safely in constricted areas.

Amplifier and filter units should allow large number of amplification steps covering gains up to 1000. If incremental gain increases of a factor two can be selected then the maximum resolution of the digitized signals can be achieved. Selectable low- pass filtering as low as 2 Hz is necessary to remove all unwanted higher frequencies from the signals.

The data acquisition hardware should be capable of digitizing the analog signals with a minimum of 16 bit resolution. The software which controls the Analog to Digital Converter hardware should be able to facilitate the acquisition of very long records (typically 32768 to 524288 points per channel). Great productivity increases can be utilized in the field if data acquisition software specifically designed for the acquisition of ambient vibration data is used.

It is often desirable to have one computer coordinating the data acquisition and a second computer to perform all on site data analysis, mode shape animation and printing.

TEST PLANING AND EXECUTION

This section gives a description of the procedure for the planning and execution of an AVT study. While each bridge will present a unique set of problems, many of the elements described below apply.

DEFINITION OF TEST OBJECTIVES

The most important step in planing an AVT is the definition of clear test objectives. The questions which need to be answered at this stage are:

- What information should be obtained from the test?
- What is the minimal information which needs to be obtained?
- What additional information would be useful?

Only if these questions are answered can the planning of the test begin. The test should be planned so that the absolutely necessary information is obtained at the beginning of the test. Data obtained near the end of a test will then merely add more definition to mode shapes. Using this approach means that unforeseen problems can be accommodated more easily.

PRELIMINARY FIELD VISIT

Once the test objectives have been defined, a field visit is highly recommended as many important details can not always be obtained from plans and photographs. During a field visit the logistics of cable arrangement and sensor mounting should be clarified and all parts of the structure which need to be instrumented should be visited. Particular attention should be given to the hazards, from traffic or other risks, which the test crew will be exposed to during the test. At this stage access restrictions to the structure should also be investigated.

Access to large structures which form part of the life lines that support a metropolitan center is often very difficult to secure since any closures may have a dramatic impact on the traffic flow of the entire city. Therefore it is often impossible to shut down these bridges for testing. Even night closures or partial lane closures may not be authorized in many cases. Under such conditions acceleration records often have to be obtained from locations along the sidewalks. In some case, there may be no side walks and then measurements can be taken on catwalks or travelers. It is not unusual for transportation authorities to restrict access to the side walks during rush hour. This is done to prevent accidents which can occur when motorist watch the testing crew instead of the vehicles around them.

PREPARATION FOR FIELD WORK

AVT involves obtaining a series of sets of acceleration records which cover many locations on the structure. In each set of records some signals are from reference sensors (which remain in the same spot for the entire test) and the rest of the signals are from roving sensors.

Based on the predefined test objectives a measurement point grid should be defined. Once the measurement point grid has been defined, the location of the reference sensors has to be established. Suitable reference station locations for the identification of the fundamental modes of vibration can generally be chosen from experience. Using several reference sensors reduces the probability of having all the references at nodes of a particular mode shape. Furthermore multiple references permit the comparison of mode shapes obtained with respect to the individual references.

After the reference locations have been selected, the roving sensor placements for the individual setups have to be planned. The setup planning should attempt to achieve the following:

- Setups should be arranged so that maximum mode shape definition is obtained early in the test.
- Sensor movements between measurements should be as short as possible.
- Sensor movements should avoid cable tangles.

It has been our experience that all carefully worked out setup plans have to be changed during the field work. Thus it is important to remain flexible so that changes can be made to the setup plan at any time during the testing.

EXECUTION OF FIELD WORK

The following steps are typical for an AVT:

1. Mark measurement locations (sometimes done during preliminary site visit).

2. Install reference sensors.
3. Start data acquisition of reference signals with initial data acquisition settings.
4. Install roving sensors for first setup.
5. Interpret data from reference sensors.
6. Adjust acquisition parameters if required.
7. Acquire signals from all channels.
8. Transfer new data from the acquisition PC to the analysis PC.
9. Move sensors to new locations and simultaneously analyse the new data.
10. Start acquisition for the next setup and continue analysis of previous measurements and animate partial mode shapes.
11. Repeat steps 8 to 10 until all measurement points have been covered.
12. Print out spectra and mode shapes while packing up the equipment.

During step 10, bad data can be identified and corrective measures can be taken. Also, it can be verified that the reference locations are indeed suitable for the test objectives. This is done by comparing the spectra from all locations with those obtained at the reference locations. If the reference location spectra contain all the peaks which are present in all the other spectra, then the reference location is suitable. Otherwise, if important peaks are missing from all reference spectra, the reference sensors should be relocated and the test restarted from step 3.

TIPS AND TRICKS

While there is an endless number of tips and tricks which one acquires with time, here are a few which we have found to be particularly useful.

Color coding of sensors and cables makes connectivity checking and communications much easier. Since sensor locations are generally numbered, no confusion between sensor location and sensor numbers occurs. Also, if the cables are colored along their entire length, cable handling and untangling becomes much easier.

Once the data acquisition parameters have been selected and a series of measurements has been made, it is best to keep these parameters for the remainder of the test, unless a very good reason arises to change parameters. It should be kept in mind that all the data obtained during a test has to be compatible. Thus data acquisition parameter changes could require a restart of the whole test.

Occasionally during the relocation process it will become apparent that an individual sensor placement will take much longer than anticipated. In such cases it is often better not to wait but to start acquiring data from all the other sensors. The problem sensor can then be relocated and the data can be obtained during the next setup. While this causes a change of the setup plan it generally saves a lot of time.

Each member of the testing team should be equipped with a complete set of tools which can be stored conveniently in a surveyor's vest. Having all the necessary equipment on hand avoids the waste of time associated with getting tools and materials from the equipment vehicle.

DATA PROCESSING PROCEDURE

Due to a lack of space, only a brief description of the processing methodology is presented. For an extensive description refer to (Felber and Cantieni, 1996). All the data processing is performed on site using the ambient vibration data specific programs U2, V2 and P2 (EDI, 1997).

DETERMINING NATURAL FREQUENCIES

The peaks in the Power Spectral Density (PSD) of ambient vibration acceleration signals are used to identify the structure's natural frequencies. However, using peaks of individual PSDs to identify natural frequencies has the following disadvantages:

- During an ambient vibration study many acceleration time histories are recorded and the interpretation of all the individual PSDs is a considerable task.
- Some measurements are most likely taken at or near locations corresponding to nodes of certain modes and hence their PSDs will not exhibit significant peaks corresponding to these modes.

To overcome these shortcomings, the Averaged Normalized Power Spectral Density (ANPSD) has been introduced. This function is defined as:

$$ANPSD(f_k) = \frac{1}{l}\sum_{i=1}^{i=l}\frac{PSD_i(f_k)}{\sum_{k=0}^{k=n} PSD_i(f_k)} \tag{3}$$

where: f_k = kth discrete frequency and n = number of discrete frequencies.

It should be noted that in practice, machine vibration and other artificial sources can cause peaks in the ANPSD which do not correspond to natural frequencies of the structure. Such peaks are often quickly discounted using judgement and experience.

An example will be used to illustrate the versatility of the ANPSD function. The data used for the example are the recordings obtained from the three span continuous steel plate girders of the Queensborough bridge (Felber and Ventura, 1996). The data consists of vertical ambient vibration records from 46 locations spaced evenly along each sidewalk of the bridge. To identify all the vertical modes of this bridge, the ANPSD for all the vertical acceleration signals was computed (see Figure 3a). In total, 13 vertical modes were identified in the range of 0 - 10 Hz for this bridge.

The vertical ambient vibration records of the Queensborough bridge, which were obtained from pairs of sensors mounted on opposite sides of the deck, can be added or subtracted to generate new signals. If two signals obtained from opposite sides of a symmetric bridge deck are added, then accelerations corresponding to twisting of

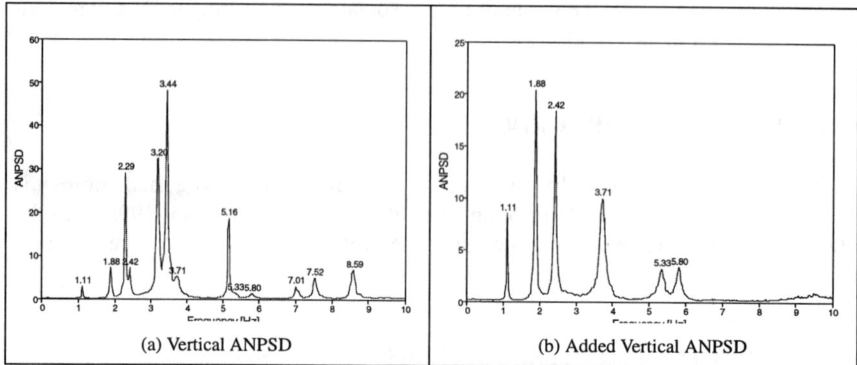

(a) Vertical ANPSD

(b) Added Vertical ANPSD

Figure 3: ANPSD's for the vertical signals of the Queensborough bridge

the deck about its centerline will be cancelled. At the same time portions of the signals corresponding to vertical translation will be enhanced. These added signals can then be used to compute a new ANPSD which contains dominant peaks corresponding to vertical bending modes of the structure. In this 'added' ANPSD the peaks of torsional modes of the deck are suppressed. This can be clearly observed in Figure 3b where the peaks at 1.11, 1.88 and 2.42 Hz stand out more clearly. In particular, the peaks at 3.71, 5.33 and 5.80 Hz are defined much better.

DETERMINING MODAL RATIOS

If two acceleration records, $\ddot{x}_a$ and $\ddot{x}_b$, are measured simultaneously then using their fast Fourier transforms the transfer function, $T_{a,b}(\omega)$, can be computed and used to approximate their modal ratio, $\phi_{a,j}/\phi_{b,j}$, at a frequency ω_j:

$$T_{a,b}(\omega_j) \approx \phi_{a,j}/\phi_{b,j} \tag{4}$$

While the transfer function is defined for all frequencies, the approximation for the mode shape ratios is only valid at frequencies corresponding to natural frequencies. The transfer function information can be utilized using two different approaches: one assumes that the modes are all real while the other assumes more general complex modes. A detailed discussion of the merits of both of these approaches is beyond the scope of this paper but can be found in (Felber and Cantieni, 1996).

ANIMATING MODE SHAPES

After the natural frequencies have been obtained from the ANPSDs and the transfer functions have been computed with respect to the reference stations, the mode shapes can be assembled and animated for visual verification. The importance of visually verifying the modes shapes can not be overstated. This is particularly important while the testing is still in progress and when only partial mode shapes can be assembled (from readings for those setups which are completed). At this stage an experienced operator can often spot a mis-aligned or malfunctioning sensor just by looking at the computed partial mode shapes. Thus errors can be corrected and critical measurements can be repeated immediately.

CASE STUDY

The above described approach will be illustrated using a recent seismic assessment study of the Port Mann Bridge as an example. This study was commissioned by the Ministry of Transportation and Highways of British Columbia in March of 1994. The seismic assessment was carried out by Buckland & Taylor Ltd. and the dynamic testing was performed by EDI Ltd.

DESCRIPTION OF THE PORT MANN BRIDGE

A general overview of the Main Span of the Port Mann Bridge is shown in Figure 4. It is a stiffened tied steel arch which has a centre span length of 365 m over the main navigation channel. The Main Span and the two 110 m anchorage spans weigh 11850 tons. The Main Span is supported by two bents, one at each end (N2 and S2), and two interior piers (N1 and S1). The arch is restrained vertically but allowed to move longitudinally at N2 and S2. At S1 the Main Span is supported on rollers, which permit longitudinal movement of the structure, while at N1 it is fixed longitudinally and transversely.

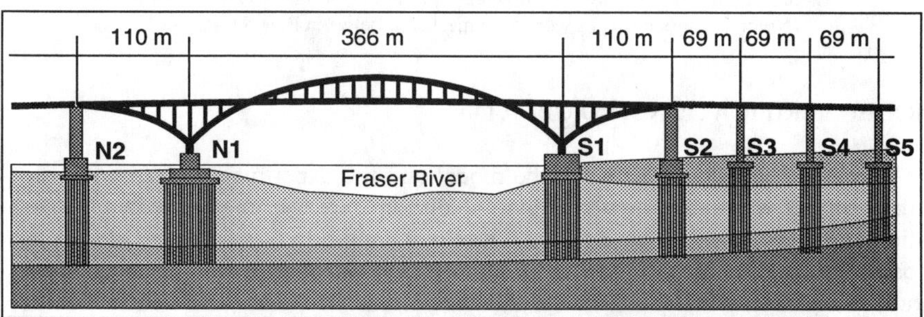

Figure 4: Elevation of the Port Mann Bridge Main Span

DESCRIPTION OF THE COMPUTER MODEL OF THE PORT MANN BRIDGE

The dynamic model of the Port Mann Bridge was created using CAMIL, an inhouse, 3D, dynamic, structural analysis program of Buckland & Taylor Ltd. For this correlation, a model encompassing the stiffened tied arch and three span continuous segments of the approach structure on either side was used. This model consisted of about 1200 3D beam elements for the superstructure and 60 discrete masses and 30 springs for the foundations. The original model was created using the as built drawings and reports.

DYNAMIC TESTING RESULTS

More than 400 acceleration records were obtained and analyzed to determine significant dynamic characteristics of the bridge. Measured acceleration levels ranged from about 0.01 mg for horizontal motions at ground level up to 30 mg for vertical motions of the approach superstructure. The dominant frequencies of the Main Span were determined using ANPSDs. The first four modal frequencies are 0.46, 0.50, 0.68 and, 0.84 Hz. The mode shapes corresponding to these frequencies are shown in Figure 5 using two different views: a plan view of the deck and an elevation view. The solid lines represent the deformed measured portions of the structure and the dashed lines show the corresponding undeformed reference shape.

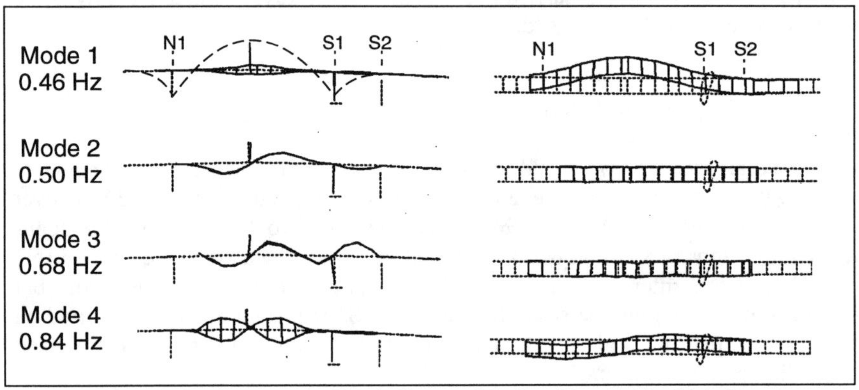

Figure 5: First four measured Mode Shapes of the Port Mann Bridge Main Span (Note: The mode shapes were only measured between Pier N1 and S2)

MODEL VERIFICATION AND UPDATING

The initial agreement between the model and the experimental results was not satisfactory. It was quickly realized that the discrepancy was due to the fact that the roller bearings at pier S1 were seized. Once this condition was accounted for in the model, the agreement improved considerably. At this stage it was not clear if the bearings were only seized under the relatively small dynamic forces due to traffic, which might be insufficient to overcome the friction in the bearing. Thus a detailed optical survey of the piers was conducted which confirmed that the bearings were also seized at force levels induced by daily temperature fluctuations. It was found that the piers N1 and S1 move horizontally relative to each other approximately 25 mm during one thermal cycle. Since the force required to move the massive foundations 25 mm is quite considerable the bearings were considered seized.

Further study of the discrepancies between computed and measured transverse mode shapes suggested that the assumed properties of members in the braced frames directly above N1 and S1 were incorrect. It was subsequently confirmed that these members were indeed different from what was shown on the as-built drawings. This was confirmed from field measurements of these members. It turned out that these members had been resized by the steel contractor to carry large temporary loads.

Once the member seizes were corrected to reflect the actual condition on site, very good agreement between measured and computed mode shapes was reached.

The model updating was carried out using judgement and experience in adjusting key parameters of the model rather than utilizing model updating routines which commonly have difficulty when model boundary conditions are incorrect.

IMPACT ON THE RETROFIT DESIGN

Once the analytical model was calibrated, the seismic demands on the structure could be computed with more confidence. Since the bearings at S1 were seized it was decided to modifiying them so that they can share the longitudinal seismic forces on the superstructure with the bearings at N1. Measures were also recommended to make the members of the braced frames above N1 and S1 more ductile as they provide a significant transverse load path. Both of these retrofit design decisions are a direct result of the AVT and model verification process.

DISCUSSION AND OUTLOOK

AVT provides a quick and economically method of determining the dynamic characteristics of long span bridges. Often AVT reveals that the real bridges behave different from the computer model predictions. If sufficient measurements are taken to properly define the dynamic characteristics of the bridge, then it should be possible to update the computer model of the structure. Once the key elements of the model have been identified and adjusted to result in better agreement between experimental and analytical results, an analytical investigation of the structure can begin.

Alternatively, the calibrated computer model can be utilized to investigate various damage states of the structure. This will reveal the sensitivity of various structural responses to particular types of damage. If a particular damage condition is considered likely to occur and has a significant effect on the overall safety of the bridge, then specific instrumentation can be designed and installed to monitor the key parameters which would indicate a damaged condition.

On the other hand, to instrument a structure permanently with a variety of sensors which are well distributed but not targeted for specific damage scenarios may be less effective. However, the data collected from such a system should be useful for the calibration of analytical models. Recently a number of new long span bridges has be equipped with extensive structural health monitoring systems. Unfortunately most of these systems have not been in operation long enough to tell if they are more effective than traditional visual bridge inspections.

CONCLUSIONS

Ambient vibration testing is a cost effective method for determining the dynamic characteristics of long span bridges which does not interfere with the regular operation of these structures.

AVT provides experimental data which can be used to calibrate computer models. This calibration is essential if the structural assessments based on analytical model predictions are to be carried out with accuracy and confidence.

Since most permanent structural health monitoring systems on long span bridges have not been in operation long enough, it is too early to comment on the effectiveness of such systems compared to traditional visual inspections.

REFERENCES

1 Abdel-Gaffar, A. M., and Scanlan, R. H., 1985. "Ambient Vibration Studies of Golden Gate Bridge: II. Pier-Tower Structure", Journal of Engineering Mechanics, Vol. 111, No. 4, pp. 483-499, April.

2 Agardh, L., 1991. "Modal Analysis of Two Concrete Bridges in Sweden", Structural Engineering International, Journal of the International Association of Bridge and Structural Engineering, Vol. 1, Issue 4, pp 35-39.

3 Buckland P. G., Hooley R., Morgenstern, B. D., Rainer J. H. and van Selst A. M., 1979. "Suspension Bridge Vibrations: Computed and Measured", Journal of the Structural Division ASCE, Vol. 105, pp. 859-874.

4 Cantieni, R., Deger, Y., Pietrzko, S., 1994. "Large Structure Investigation with Dynamic Methods: The Bridge on the River Aare at Aarburg", Prestressed Concrete in Switzerland, Report of the Swiss FIP Group to the 12th FIP Congress, Washington, D.C., pp. 20-26, May 29 to June 2.

5 EDI Experimental Dynamic Investigations Ltd, 1997. "U2, V2 & P2, Manual, Version 2.2", 1633 Stephens Str., Vancouver, B. C., Canada.

6 Felber, A., and Cantieni, R., 1996. "Indtroduction of a new Ambient Vibration Testing System - Description of the Sytem and Seven Bridge Tests", EMPA Report-No. 156'521, Dübendorf, Switzerland.

7 Felber, A., and Ventura, C. E., 1996. "Frequency Domain Analysis of the Ambient Vibrqation Data of the Queensborough Bridge Main Span,. 14th International Modal Analysis Confrerence Proceedings, pp. 459-465, Dearborn Michigan.

8 Felber, A. J., Taylor, P. R., van Selst, A., Ventura, C. E., Prion, H. G. L. 1995. "Seismic Assessment of Existing Bridges One Role for Dynamic Testing", 'Bridges into the 21st Century' Conference Proceedings, pp. 765-722, Hong Kong, October 2-5.

9 Hudson, D. E. 1964. "Resonance Testing of Full-Scale Structures", Journal of the Engineering Mechanics Division ASCE, Vol. 90, pp. 1-19.

10 Law, S. S. and Ko J. M., 1995. "Ambient Vibration Measurements of the Tsing Ma Bridge Towers", 'Bridges into the 21st Century' Conference Proceedings, pp. 585-592, Hong Kong, October 2-5.

11 Luz, E., 1987. "Experimental Modal Analysis of Large Scale Structures", Proceedings of the International Conference on Mechanical Dynamics, pp. 257-262, Senyang, China, August 3-6.

12 Trifunac, M. D., 1972. "Comparison between Ambient and Forced Vibration Experiments", Earthquake Engineering and Structural Dynamics, Vol. 1, pp. 133-150.

Damage Detection of a Model Bridge Using Modal Testing

M. L. WANG, D. SATPATHI and G. HEO

Abstract

The transportation infrastructure in this country is getting overburdened by aging bridges and increasing traffic flows. According to recent Federal Highway Administration estimates, nearly 35% of all bridges are either structurally or functionally deficient. In view of this situation, a lot of emphasis is being placed on the development of newer and more efficient methods of inspection and monitoring. A structural dynamics based method for detecting damages in a bridge superstructure has been proposed in this article. A benchmark experiment has been carried out using a scaled model of a typical steel plate girder bridge.

1. Introduction

Bridge inspections during the late 1980's revealed that out of the 576,000 highway bridges in the United States, about 236,000 were rated deficient by present day standards [Professional Engineering News, 1987]. Repair or replacement of these bridges will cost billions of dollars and involves time. It is however necessary to keep the deficient bridges open to traffic, yet minimize the risk of a catastrophic collapse. In view of this situation it is important to implement a rigorous bridge inspection program for these critical bridges. While a variety of nondestructive techniques using methods such as visual, optical radiography, ultrasonic, acoustic emission, dynamic property, magnetic particle, eddy current, microwave, thermal and so on [Bray 92, White 92, Tang 91 and Topole 95] are available, visual inspection is the oldest and most commonly used inspection method. This type of traditional method for bridge inspection is typically time consuming and is highly subjective, and requires highly experienced inspectors. A significant amount of research effort is being dedicated towards the development of efficient bridge inspection systems. In this article the authors have proposed a structural dynamics based system.

Structural dynamic methods [Kaouk 95, Kashangaki 92, Ken 80, Kim 95] show a lot of promise in global health monitoring, because damage that is significant to the bridge will result in a reduction of its stiffness. This produces changes in certain structural dynamics parameters such as modal frequencies and mode shapes. Although frequencies may not be good indicators of bridge health, some techniques [Yao 92 & 94, Zimmermann 94] utilizing mode shape data have demonstrated the ability to indicate and locate damage.

The modal analysis is a technique whereby a structure's resonant frequencies, mode shapes and damping ratios are determined from vibration measurements. The basic methods of modal analysis are well established. Several hardware elements are required for the vibration measurements such as an exciter used to provide a known input force into the structure and a transducer to convert the motion of the structure into an electrical signal. The analog time-domain signals from the transducer are converted into digital frequency-domain information by an analyzer, which also performs a number of computations on the digitized signal such as calculating the spectral density functions from transducer outputs. From this, the analyzer calculates the Frequency Response Function (FRF). The coherence is a measure of the noise in the signal and is used to indicate the accuracy of the measurement process over a given range of frequencies. In general, the coherence should be equal to one at frequencies near the structure's resonant frequency. After the frequency response has been calculated, various vibration parameters may be extracted. These parameters include the natural frequencies, damping ratios, and modal amplitudes associated with each resonant peak of the measured FRF. From this platform, a specific procedure for evaluating the health of bridges can be formulated. The procedure measures the dynamic response of the candidate bridges, evaluates the response and provides an in-depth condition evaluation that provides critical information for an overall Bridge Management System. Considerable research has been done in the area of using structural dynamics parameters for NDT [Brinker 95, Chou 90, Farrar 94, and Kaouk 94]. A basic premise of damage identification theories will be reviewed and described in the following sections.

2. Background of Damage Detection Methods Using Dynamic Property Measurement

A damage in a structural system alters its dynamic character. The change is characterized by changes in the eigenvalues (frequencies), eigenvectors (mode shapes), damping ratios. A number of damage identification theory of the structures has been developed [Li 94, Lam 95, Kon 94, Ko 94, Lin 93, Maduakolam 95]. One such damage identification method known as the Damage Index Method has been described below.

Damage Index Approach

For solving the effects of modal uncertainty and degree of severity of damage, damage index method was developed by Stubbs and Kim [Stubbs 93, Stubbs 92 and Stubbs 95] to detect, locate, and estimate the severity of damage in structures based on a few of their characteristic mode shapes. It requires a few modes of the structure. For a structure that can be represented as a beam, a damage index, β_{ij}, based on changes in the curvature of the ith modes at location j is defined as

$$\beta_{ij} = \frac{\left(\int_a^b \left[\phi_i^{*''}(x)\right]^2 dx + \int_0^L \left[\phi_i^{*''}(x)\right]^2 dx\right)\int_0^L \left[\phi_i^{''}(x)\right]^2 dx}{\left(\int_a^b \left[\phi_i^{''}(x)\right]^2 dx + \int_0^L \left[\phi_i^{''}(x)\right]^2 dx\right)\int_0^L \left[\phi_i^{*''}(x)\right]^2 dx} \tag{1}$$

where $\phi_i^{''}(x)$ and $\phi_i^{*''}(x)$ are the second derivatives of the ith mode shape corresponding to the undamaged and damaged structures, respectively. L, is the length of the beam, and a and b are the limits of a segment of the beam where damage is being evaluated. Statistical methods are then used to examine changes in this index and associate these changes with possible damage locations [Kim 93]. Assuming that the collection of damage indices, β_{ij}, represents a sample population of a normally distributed random variable, a normalized damage localization indicator is obtained as follows,

$$\zeta_j = \frac{\beta_j - \overline{\beta}_j}{\sigma_j} \tag{2}$$

where $\overline{\beta}_j$ and σ_j represent the mean and standard deviation of the damage indices, respectively. The disadvantage of the method is that it may not show the clear location and magnitude of the damage of the structural system when there are multiple damage locations with the same severity of the damage because σ_j in Eq.(2) increases.

3. The Bridge Model

In order to prove that the concepts described it the above section worked it was necessary to perform experiments on an actual structure. A model of a simply supported long span bridge was built towards this end. In real life, long bridges are characterized by very low closely spaced resonance frequencies. This is primarily due to large span over width ratios, and structural redundancies; which in turn introduce non-linearity in the dynamic behavior of structure. The guiding design principle behind the model bridge was to be able to simulate this real life behavior of actual bridges. The design process was based on a dynamic simulation on I-DEAS (SDRC) where different combinations of geometrical parameters like span, width, beam sizes and configuration, slab thickness and boundary conditions were simulated to obtain acceptable mode shapes and frequencies. In spite of the best of efforts, it was not possible to obtain the low closely spaced resonance frequencies that typify actual complex bridges. This was to maintain constructional simplicity which in turn restricted the degree of redundancy that the structure could develop. The model structure that was finally chosen is shown in Fig. 1.

The bridge superstructure is a scaled down model of a single span plate girder bridge with two main beams transferring the load to the supports. A third middle beam was added to increase the redundancy . The middle beam transfers its load onto four asymmetrically placed floor beams in order to introduce some degree of non-linearity to the system. A two inches thick reinforced concrete slab was used as the deck of the superstructure. The thickness of the slab was based on a FE simulation so that the mode shapes could be measured. During the simulation process, the use of a thinner plate resulted in insufficient stiffness which in turn produced mode shapes which were dominated by the deformation of the plate instead of the entire structure. One of the goals of this research program was to be able to investigate the effects of the boundary conditions on the modal parameters. To incorporate this goal, the supports were specifically designed to introduce the following types of boundary conditions a) fixed-fixed; b) roller-fixed; and c) pinned-pinned.

4. Experimental Modal Testing of the Model Bridge

In order to implement the damage index approach it was necessary to carry out a modal analysis of the bridge before and after the artificial damage was introduced. For the experimental modal analysis the structure was instrumented with Dytran model 3187B1 accelerometers at 24 locations as shown in Fig. 2. The accelerometers were mounted on the steel beams with magnetic bases. The data acquisition system was a 32 channel Zonic PC-7000 system that was software controlled using Zeta a proprietary data acquisition and post processing software. The structure was excited with an impact hammer at a fixed point and the acceleration responses were measured at all the 24 sensor locations. The data was stored in the form of FRFs (Frequency Response Function) which is mathematically defined as

$$H_1(f) = \frac{\overline{G}_{y,x}(f)}{\overline{G}_{x,x}(f)} \tag{3}$$

In Eq.(3) the numerator, $\overline{G}_{y,x}(f)$, represents the frequency domain cross correlation function between the response signal (y) and the reference signal (x); and the denominator, $\overline{G}_{x,x}(f)$, represents the auto correlation function of the reference signal which in this case is the output from the impact hammer.

The FRF data were then imported into ME'Scope (Vibrant Tech. Inc.) for modal parameter extraction and mode shape animation. The experimental FRF data was curve fitted by using a method of residues to obtain the modal frequency, damping ratio and mode shapes. The first six experimental modes that were obtained from the bridge prior to the introduction of damage is shown in Fig. 3.

4. Damage Detection Algorithm Applied to Experimental Data

In order to evaluate the efficiencies of the damage detection algorithms, modal experiments were carried out on the model bridge that has been discussed earlier in section 3. To evaluate the effect of the boundary conditions on the damage, three different boundary condition sets were tested (Fig. 4). Modal test for each boundary condition were run before and after a damage was introduced. The sensor location scheme for these tests is shown in Fig. 2. The damage consisted of a cut through the bottom flange and half of the web at the midpoint of one of the main beams (Fig. 5). During the test procedure FRF data was acquired from the sensors and the data processed as discussed earlier. The results of the damage detection algorithms as applied to these tests is discussed in the following sections.

During modal animation the structure had been represented as a 2-D structure. For the damage analysis the structure was reduced to a 1-D beam elements (Fig. 6). Thus the reduced structure is made up of 3 lines with 8 nodes, where each node represents a measurement point. Subsequently, the internodal elements were subdivided into ten elements each to produce a finer mesh (Fig. 6). From the FRF data the modal vectors, modal frequencies and damping ratios were extracted. Subsequently, the mode shapes were interpolated between nodes using cubic natural spline functions shown in Fig. 7. Subsequent computations were carried out as shown in Eq. (1) & (2).

The results of this analysis for all the three boundary condition sets are shown in Figs. 8-10. The data from the fixed-fixed boundary condition set, the fixed-roller and roller-pin boundary conditions indicate damage around the 45th element which is essentially the center of the beam. This matches the actual location of the introduced damage. Since the integral of the second derivative of the mode shape is a representation of the strain energy, the damage index method in essence is calculating the strain energy contribution of each and every element. Thus by comparing the relative strain energy contributions of each and every element at two different states it is feasible to detect damage from a global response measurement.

5.Conclusions

Based on the results obtained, the damage index method has performed quite well. It is however to be noted that it will be difficult to perform a blind damage identification with any of these methods unless the extent of the damage is severe. In the current experimental study, the introduced damage was quite severe. The reasons for the superior performance of the damage index method lies in the fact that the damage criteria is based on statistics. The damage index β, is considered to be a random variable. During the subsequent calculation of ζ, the statistical variability of the sample population is removed to quantify the actual damage location. Secondly, a process of differentiation is similar to a high pass filtering process. In calculating the relative contribution, some of the noise resulting from the differentiation is smoothed out during the integration process. Nevertheless, this method will have difficulties if multiple damage locations are present because standard deviation used by damaged index approach is increased in the population of all damage indices. Also errors may creep in depending on the interpolation functions used. In the future further experiments in this area will be conducted to study the effect of damage size and occurrence of multiple damage.

6. Acknowledgment

This research was funded under NSF grant # 9622576. Dr. S. C. Liu is the program director.

7. References

1.Bray, D.E., and McBride, D., "Nondestructive Testing Techniques," A Wiley-Interscience Publication, John Wiely & Sons, Inc. 1992.
2.Brincker, R., Kirkegaard, P.H. and Anderson, P., "Damage Detection in an Offshore Structure," Proc. of the 13th International Modal Analysis Conference, 1., pp. 661-667, 1995.
3.Chou, Chaur-Ming, and Wu, Chi-Hsing, "System Identification and Damage Localization of Dynamic Structures," AIAA-90-1203-CP.
4.Farrar, C.R., "Dynamics Characterization and Damage Detection in the I-40 Bridge over the Rio Grande," Los Alamos National Laboratory, LA-12767-MS.
5.Kaouk, Mohamed and Zimmerman, David C., "Reducing the Required Number of Modes for Structural Damage Assessment," AIAA-95-1094-CP, 1995
6.Kashangaki, Tomas A-L, Smith Suzanne w. and Lim, T., "Underlying Modal Data Issues for Detection Damage in Truss Structures," AIAA/ASME/ASCE/AHS/ASC Structures, Structural Dynamics, and Materials Conference, AIAA-92-2264-CP, April 1992.
7.Kenley, R.M. and C.J. Dodds, Structural Monitoring Limited, "West Sole We Platform: Detection Of Damage By Structural Response Measurements," 12th Annual Offshore Technology Conference, 1980.
8.Kim, Jong-Tae, "Assessment of Relative Impact of Model Uncertainty on the Accuracy of Global Nondestructive Damage Detection in Structures," Ph.D. Dissertation, Texas A&M University, Aug. 1993.
9.Kim, J.T. and N. Stubbs, "Damage Detection In Offshore Jacket Structures From Limited Modal Information," International Journal of Offshore and Polar Engineering, Vol.5, No.1, March, 1995.
10.Lam, H.F., J.M. Ko and C.W. Wong, "Detection Of Damage Location Based On Sensitivity Analysis," Proceedings of the 13th International Modal Analysis Conference, 1995.
11.Li, Cuiping and Smith, Suzanne Weaver, "A Hybrid Approach for Damage Detection in Flexible Structures," Journal of Guidance, Control and Dynamics, No.3, pp. 419-425, 1995.

12.Lindner, Douglas K., and Goff, Richard, "Damage Detection, Location, and Estimation for space Structure," Smart Structures and Intelligent Systems, SPIE, 1917, pp. 1028-1093, 1993

13.Maduakolam, Mishael N., and Stubbs, Norris, "Global non-Destructive Detection of mass and structural damage to conventional and floating Bridges," SPIE-The International Society for Optical Engineering, Smart Structures and Materals, 1995, Smart Systems for Bridges, Structures, and Highways.

14.Stubbs, N, Kim, J.T., and Topole, K.G. "An Efficient and Robust Algorithm for Damage Location in Offshore Platforms,"

15.Stubbs, N., Kim, Jeong Tae, and Farrar, Charles R., "Field Verification of a Nondestructive Damage Location and Severity Estimation Algorithm," Proceedings of the 13th International Modal Analysis Conference, Feb. 1995.

16.Stubbs, Norris, and Osegueda, R., "Global Damage Detection in Solids - Experimental Verification," The International Journal of Analytical and Experimental Modal Analysis, 5(2), 81-97, 1990.

17.Tang, J.P. and K.M. Leu, "Vibration Tests and Damage Detection Of P/C Bridges," Journal of the Chinese Institute of Engineers, Vol.14, No.5, 1991.

18.Topole, K.G., Ingenieurbau, G., and Stubbs, N., "Nondestructive Damage Evaluation in Complex Structures from a Minimum of Modal Parameters," Modal Analysis: the International Journal of Analytical and Experimental Modal Analysis, V 10, N 2, pp. 95-103, Apr. 1995.

19.White, Kenneth R., Millor, J., and Derucher, Kenneth N., "Bridge Maintenance Inspection and Evaluation," 2nd Edition, Marcel Dekker, Inc. 1992.

20.Yao, G.C., Chang, and K.C., Lee, G.C., "Damage Diagnosis of Steel Frames using Vibrational Signature Analysis," Journal of Eng. Mechanics, Vol. 118, No. 9, September, 1992.

21.Yao, G. C., and Lien N., "Modal moment Analysis for Damage Diagnosis of Steel Structures," 5th US National Conference on Earthquake Engineering, Chicago, IL, 1994.

22.Zimmerman, D.C., and M. Kaouk, "Structural Damage Detection Using A Minimum Rank Update Theory," Transactions of the ASME, 116, April, 1994.

23.Zimmermann, D.C., S.W. Smith, Kim, H.M., and Bartkowicz, T.J., "An Experimental Study of Structural Damage Detection Using Incomplete Measurements," AIAA-94-1712-CP, 1994.

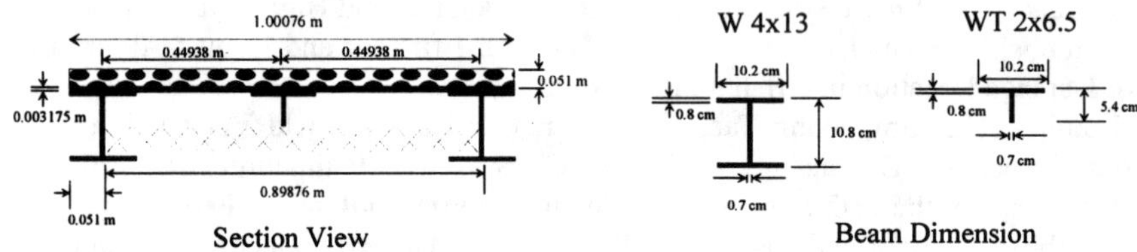

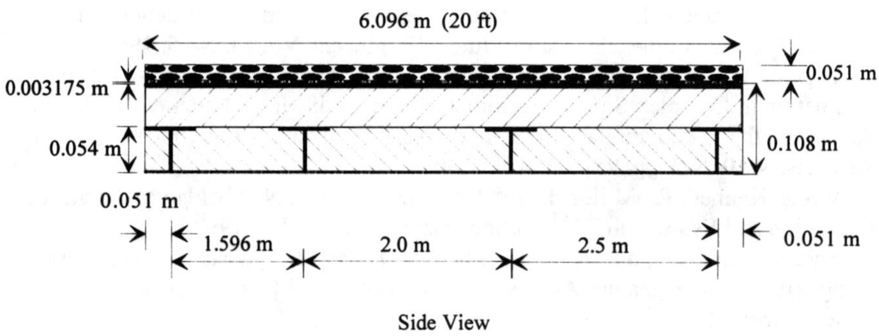

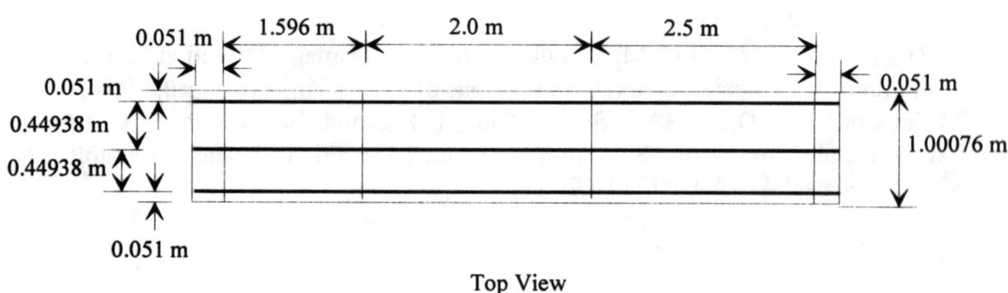

Fig. 1 Details of model bridge

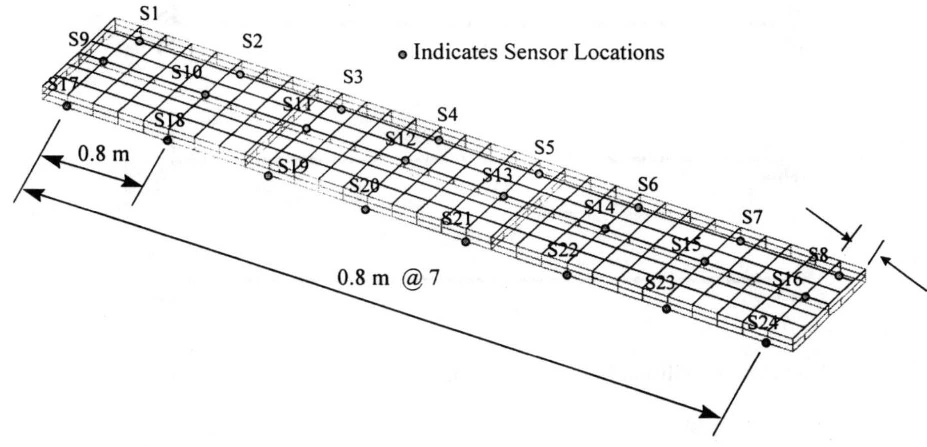

Fig. 2 Accelerometer Locations for Experimental Modal. Test a Numerically Simulated Long Span Bridge

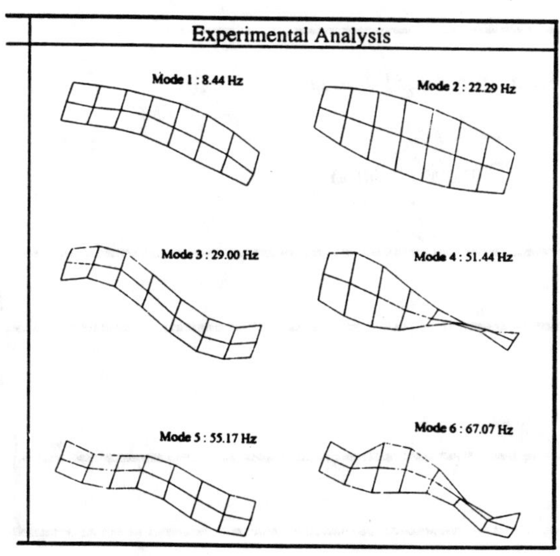

Fig. 3 Experimental mode shapes for the model bridge

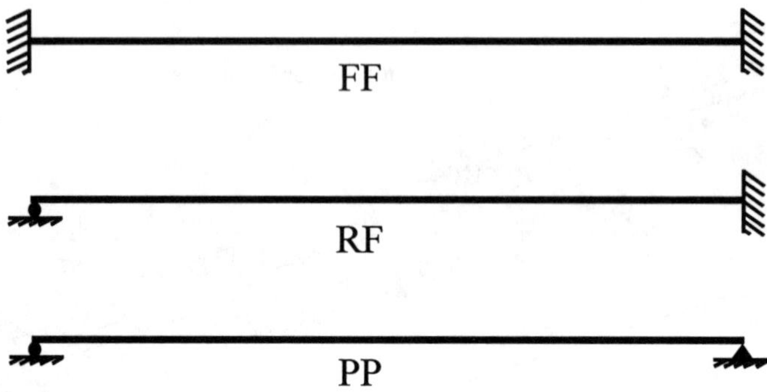

Fig. 4. Boundary Condition Sets; FF (BC 1), RF (BC 2), and RP (BC 3)

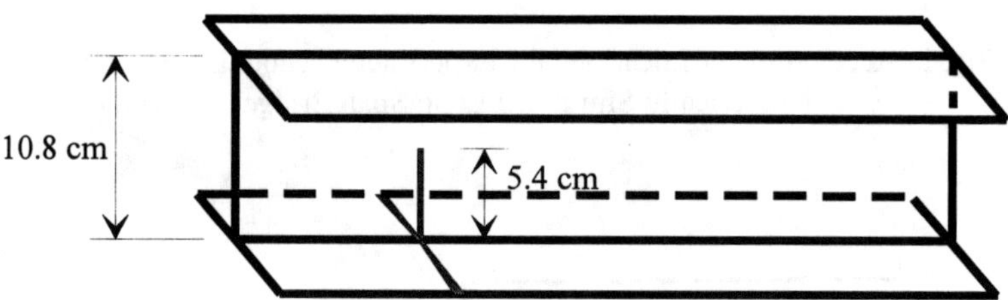

Fig. 5 Details of introduced damage

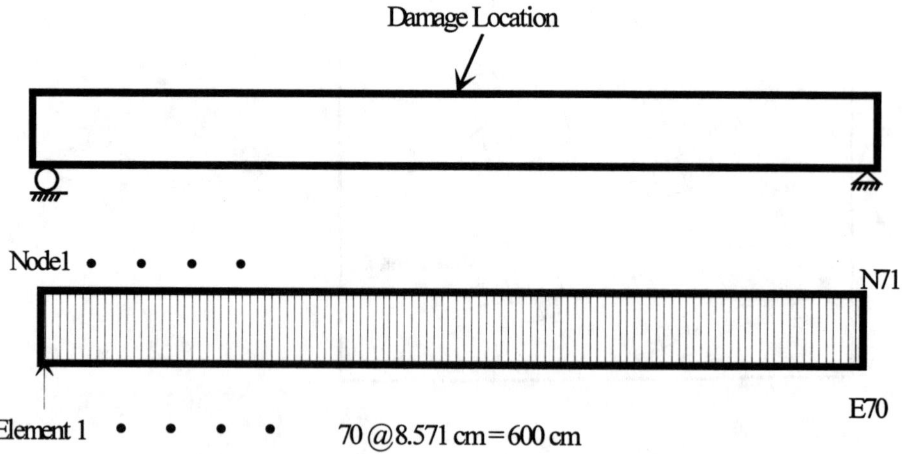

Fig. 6 Schematic Damage Detection Models of the Bridge

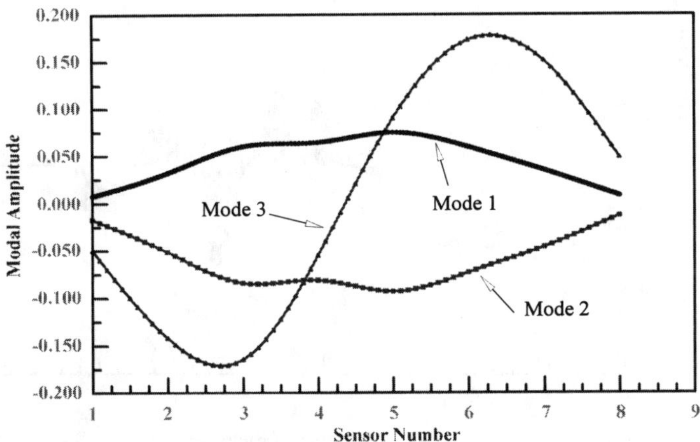

Fig. 7 Interpolated Modal Shapes of the Model Bridge; BC 1 Before Damage

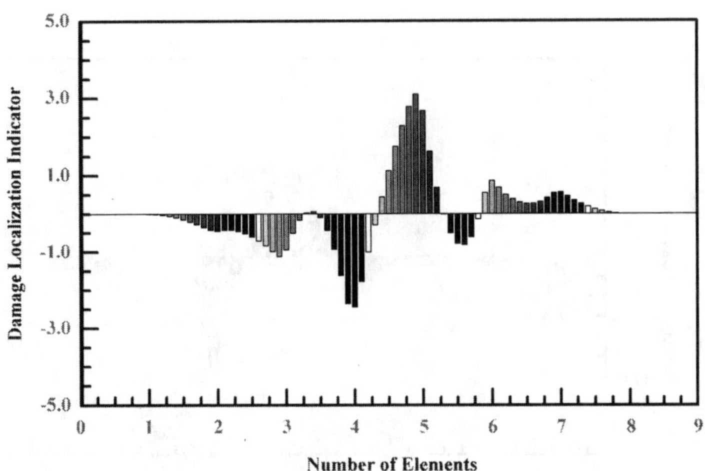

Fig.8 Damage Localization Indicator for BC 1; Mode Shapes are Interpolated with Cubic Spline Function.

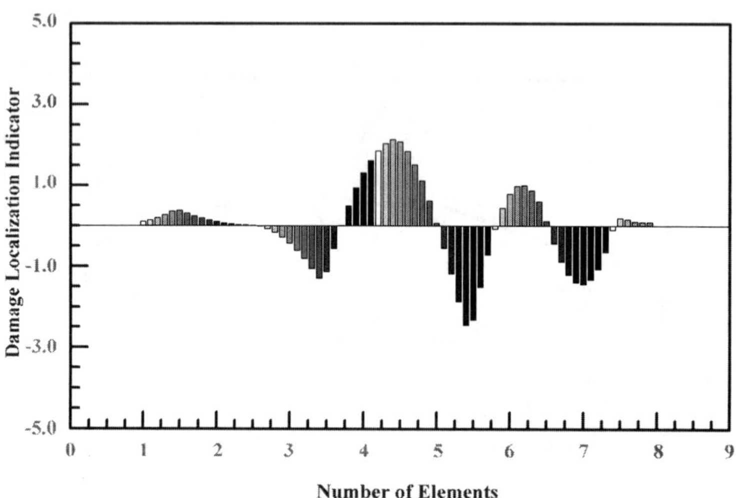

Fig.9 Damage Localization Indicator for BC 2; Mode Shapes are Interpolated with Cubic Spline Function.

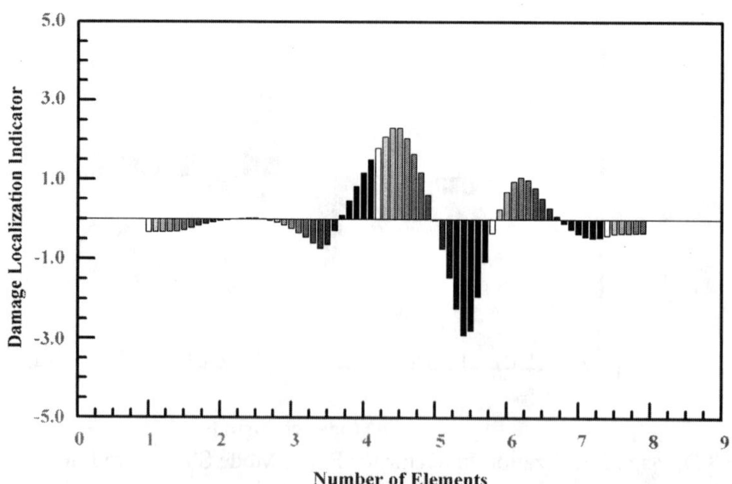

Fig.10 Damage Localization Indicator for BC 3; Mode Shapes are Interpolated with Cubic Spline Function.

SESSION 11

GENERAL APPLICATIONS

A New Concept of Maintenance Inspection

K. EGAWA

ABSTRACT

Maintenance inspection is classified one of nondestructive inspection for condition monitoring in which many nondestructive inspection tools are used. But present status of maintenance inspeciton is criticized as schedule–driven, time–consuming, labor–intensive and expensive. The weakest point of maintenance inspection today is that it has no function for giving us warning when incipient damage occurs.

For constructing a new concept of maintenance inspection without these defections, the author thinks of introducing smart structure's concept into maintenance inspection. In concrete, it means that one attaches many sensors on equipment and structures to be maintained and reforms them sensory ones. For backing up this new concept of maintenance inspection, two cases of bolt inspection are taken as examples, and great advantages coming from this new concept is shown here.

1. INTRODUCTION

The purpose of stress/strain measurements(SSM), in other words experimetnal mechanics, is to check whether designed and manufactured structures have enough strength for the expected loads or not, and also aged structures can endure to the expected maximum load or not, that is, whether we can continue to use it or not. Talking about nondestructive inspection(NDI), we can devide it into manufacturing control and maintenance inspection (condition monitoring). By the former one we can reduce defective products in production line and by the latter one we can check the safety of structures like bridge. In the point of the judgement of the safety of designed and used structures, we can say that stress/strain measurements and nondestructive inspection have the same aim, inspite of their tedchnological difference. But it looks us that

Koichi EGAWA, Niigata Institute of Technology, 1719 Fujihashi, Kashiwazaki, Niigata, 945–11, JAPAN

the people in stress/strain measurements world have not been eager to develop their studies for maintenance of equipment and structures. We guess that deterioration is one of the phenomena having very low speed of changing, so that it was not easy for SSM people to handle it.

As third technological region related with maintenance inspection, there is so–called "diagnosis and maintenance of equipment". Here, this technology is abbreviated as Maintenance. The relation of these three kind fo sciences and technologies may show as Fig.1; NDI and SSM relate each other in judgement of the degree of harm of defects, Maintenance and NDI relate in inspection technology, and Maintenance and SSM in measurement technology.

These three kind of science and technologies have same purpose on keep–ing equipement and struc–tures safe during their opera–tion, in spite of calling their technology in different names, as maintenance inspection in NDI, judgement of safety of aged structures in SSM, diagnosis in Maintenence. Just as these three different names show, this region of science and technology has been thought as three differ–ent ones and has not been aware fo common study region. In Fig,1, the author shows it as region No.4. Then the author tries to develop his essey from the point of finding a concept on what this common technological region should be.

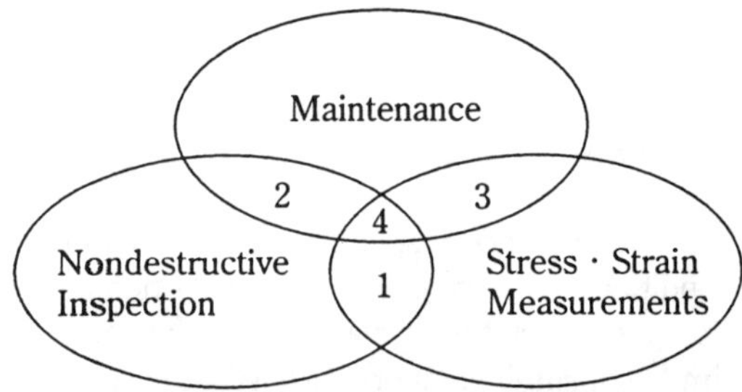

1. Judgement of harm of defect
2. Maintenance inspection
3. Judgement of safety of aged structures
4. New maintenance inspection

Fig. 1 Relations of maintenance technology, nondestructive inspection and stress / strain measurements

Picking up the speciality of two main categories of NDI, those are manufac–turing control and maintenance inspection. The former has speciality of; (1)just one time of inspection in short time, and (2)inspecting many number of, and relatively small size products in production line, and the latter has; (1)inspec–tion many times up to the end of life of equipment or structures with labor–intensive, time–comsuming processes, and (2) inspecting relatively large and sometimes huge size structures, like oil reservoir tanks or long bridges. So, the former is done mainly in a factory , often by using automated inspection devices, and the latter, on the contrary, done in field by hands. For these specialities, maintenance inspection is mainly done periodically, and also it takes long time up to obtain inspection results and becomes expensive.

Maintenance inspection has another important speciality that we would suffer heavy damage, sometimes with loss of human lives, when breaking the structures we inspected may occur. So, concerning about maintenance inspection, we should think it's influence on our daily life.

Talking about progress of maintenance inspection, Maintenance people shows their goal with steps to to reach it, that is, from today's periodical inspection(they call it as time–based inspection) to predictive, then preventive ones as shown in Table 1.

Table 1 Expected steps of progress in maintenance technology

1. Time Based Maintenance
2. Condition Based Maintenance
3. Predictive Maintenance
4. Preventive Maintenance

The author knows that, periodical inspection stands on the hypothesis that great damages would not be happened during former inspection time to coming one, but we know that this hypothesis may not always correct from our experiences of, take for instance, hearing bridge falling in Soeul, Korea and upper fuselage breaking off of Aloha airline in Hawaii. And the worst of all, by periodical inspection, we can not take warning signs or signals in emergency.

On the predictive and preventive maintenance, Maintenance people did not say what those are, so the author suppose them that they are the system by which we can predict damages which lead to breaking down of equipment or structures and one can repair them before breaking down of them by using the signal of warning.

3. CONCEPT OF INTELLIGENT (SMART) MAINTENANCE INSPECTION

Here, let's think about the meaning of participation of stress/strain measurements in the common place of NDI and Maintenance shown as region No.4 in Fig.1. There are many ways of this participation, take for instance, offereing new measuring(or inspection) technologies to NDI and/or Maintenance, but, the author proposes the following ways of joining; by introducing the concept of health monitoring derived from intelligent (or smart) structures concept, we make equipment or structures into sensory ones. Thus, we will be able to do quick, automatic(or minimum human involving) inspection on inspective equipment or structures and to obtain its data quickly by measuring instrumentation. This is the type of new maintenance inspection the author would like to suggest. In other words, the goal of new maintenance inspection is to establish the measuring system by which we can do on–line monitoring which has automaticity and minimum human involvement and self–judging warning system. This new maintenance inspection system would be one of the type of predictive maintenance shown in Table 1 which would be a expected goal in near future by Maintenance people. In this new maintenance inspection system one will be use AE method and stress/strain measurement methods–especially electric strain gage(ERSG) method–, instead of other NDI techniques like ultrasonic and X–ray method, because with AE and ERSG method one can catch the change of condition (stress field) and also follow up this change continuously.

4. INTELLECTUALIZATION OF MAINTENANCE INSPECTION

At first the concept of intelligent structures should be explained here, but the author believes that it is enough to show the follwing figure (Rogers,1993) .

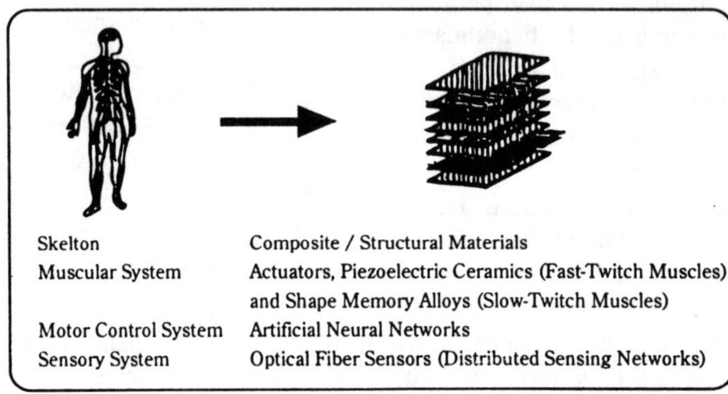

Fig. 2 Prevailing concepts of smart structures

Maintenance inspection today have two main shortcomings, as explained already in former chapter; lack of capability to cope with emergency and high–cost. These weak–points come from periodical inspection method and working system of inspection that demands many human hands. The former one is more important than the latter one, the author thinks, because we have to do maintenance inspection for preventing hazardous damage caused by the accident of equipment or structures. From this point of view, the problem of expensiveness has only secondary importance.

Now, as the capability of coping with emergency, following two characteristics are thought; (1) monitoring inspective object restlessly and sensing change of condition of it, and if necessary, stopping operation of the object before it going to worse condition, and (2) holding initiated damage in some duration not extending into serious one. The ways of (1) is better and easy to establish than that of (2).

4.1 INTELLECTUALLIZATION OF EQUIPMENT AND STRUCTURES

For improving maintenance inspection today by the method of (1) in the chapter before, we should monitor machines, equipment and structures restlessly. Thus we should attach sensors to these inspective objects and build up automatic (or minimum human involving) monitoring and warning system with using these sensors. It is nothing but making an inspective object into a sensory structure defined (Wada,1990) as first step of intellectualization of structures. This is called, in other words, as on–line health monitoring(Rogers, 1995) or predictive maintenance as Maintenance people said.

4.2 SENSORS

The aim of posting sensors is to get warning signals automatically before change of condition of watching point becoming serious. For this purpose, we should decide two threshold; first– one will be used as judging level whether present state of condition of watching point being over its steady or safety state or not, and second– one whether present state in dangerous or not.

4.2.1 Required property of sensor

The sensors which will be attached to equipment and structures for making them sensory structures subject general enviromantal effect as vibration, noise, dust, high– humidity, sun heat radiation, and in some case more severe effects like salt– water splash, flowing liquid, high/low temperature, high pressure, vacuum, strong electro/magnetic field and radiation. So, one should think compensation of these effects and sometimes correction of obtained data. Adding to this, one should think about their protection method.

In general, required property of sensors for the purpose here are; (1)endurance for general environmental injury, (2)long life for at least 5– 10 years, and (3)simple and easy handling and attachment. In Table 2 required property of sensor are shown in 13 items. Also one should care that most of these requirements are extended to the total measuring system including lead wires to instruments.

Table 2 Required properties of sensors

1.	Small weight and volume
2.	Small change occurrence in measuring part with attaching sensor
3.	Good following property for rapid / sudden change of measuring physical quantities
4.	High sensitivity
5.	Remote measurement
6.	Simultaneous (or during very short time) measurement of multi - point
7.	Intensitive for environmental change like temperature and humidity
8.	Insensitive for electric and magnetic disturbance
9.	Toughness for vibration and noise
10.	Little aging deterioration
11.	Capable to connect to computer and / or warning system
12.	Simplicity of sensor attachment and of wiring
13.	Not requiring for tough and heavy wires

4.2.2 Attaching of sensor

In general, sensors are attached outside of inspective equipment and structures for this purpose, because one can not embed sensors into them for most materials, and for existing equipment and structures one should attach sensors only from outside. But one can embed sensors only in concrete, composite laminates and some plastic materials.

There are three attaching method generally used; (1)mechanical attachment, (2)adhering, and (3)welding. Their advantages and disadvantages are shown in Table 3. We choose suitable method under consideration of the material of monitoring structure and its environment.

Next, we should think about protection of these sensors and also, about routing and fixing of lead wire from sensor to monitoring and warning system. In general, light weight, fine wire which is capable to use in long distance and immune to general and hostile environmental effects is the most suitable. In this sense, optical fiber lead wire is dirable.

Table 3 Comparison of attaching methods for sensors

Items		Mechanical Method	Adhering	Welding
1. Attachment to Metal	①General	Possible	Possible	Limited only for iron alloy and titanium
	②Thin M.	Occurrence of reinforcement effect	Possible	Limited only for iron alloy and titanium
2. Attachment to Non-metallic M.	①General	Possible	Possible	Impossible
	②Ductile M.	In some cases impossible	Possible	Impossible
3. Working Efficiency		Inadequate	Good	Poor
4. Working Tools / Facilities		Many	Some	A little many
5. Cost		A little much	A little	Much
6. Conditions after attachment	①Weight	Heavy	Light	A little heavy
	②Volumn	Great	Small	Small
	③Toughness	Considerable	Some	Much
	④Toughness for shaking	Good	Best	Better
	⑤Toughness for environmental attack	Good	Protective coating required	Good
7. Merits		Tough, Applicable to almost all M.	Very small weight, Applicable to almost all M.	Toughest

M. : Materials

4.2.3 Posting of sensor

Considering about posting of sensors for monitoring equipment and structures, problems we should think about are; (1) where, and (2) how many. First, about location of sensors, we should select the place where initiation and progression of damage affect much on structure's total strength and integrity, and where damages are difficult to find. As an example, in airframe structure, bolts and metal fittings for connection of wing root and fuselage, and engine hanging bolts would be selected at first, then members under heavy loads like wing's main beams, bolts and fitting parts and those of tail wing, stabilizer, flaps, elevators, and then main members of fuselage and so on.

Next, let's think the number of sensors to be attached with taking an example of the accident of Aloha airline. It is reported that the front, upper part of fuselage broke off as the result of progession and linkage of crack around rivet holes on central part of the fuselage. For detecting these kind of damage, should we attach sensors all bolts and rivets in an airplane? The answer is "No". It is impossible to attach sensors on all of them, because of their numerous numbers. Thus, we should put grade of importance in the location of monitoring, and then, according to this grade, we selelct locations to putting sensors with considering the number of sensors can be used.

On the contray of this story, we may have to put sensors all of the bolts and fixing parts in a long, large bridge, because the breaking of them will induce heavy damage, and the number of them are not so many to handle.

4.2.4 Limitation of damage detection by sensor

For considering this problem, following two points are chosen, that is, roles and limitation of present NDI and SSM technologies for making equipment and structures to sensory ones.

Point No.1: Many NDI tools are used for checking the quality of joints especially made by welding when large equipments and structures are built. As we think about X–ray method which is the most popular and effective tool of checking the quality of welding, we know present X–ray device is too large to use as the sensor for sensory structures. Luckily our purpose of putting sensors on structures is not checking the quality of joints, that is, checking the numbers and the size of exsiting defects, but detect the change of condition of monitoring location, that is, change of stress/strain field, deformation, position etc. So, even if there are some defects in monitoring location, they have no importance if there will not happen changes in the location during operation. Thus, we need not to think of direct appliction of NDI and SSM tehchnologies for making the sensor for sensory structures, but to think what kind of phenomena or physical quantities we should detect as for condition monitoring of equipment and structures.

Point No.2: It is told that we can detect unbond of composite laminates even if they attached closely, but it is very difficult for us to measure deterioration and its rate of adhesives used by present NDI technologies (Achenbach, 1996). In SSM we have same kind of limitation in crack measurement by ER SG that we can not measure the size of crack under it, and also the location of crack apart from it. That is problem of sensitivity and range of measurement of ERSG for crack detection. As other problem, how we attack the problem of size reduction and deformation by corrosion in which changing speed is very low.

According to these consideration, we get conclusion that we should, at first, think about what kind of quantities should we detect for condition monitoring of given structure, then choose inspecting or measuring method and sensors from market.

4.2.5 Non–sensing area

When we will plan to make equipment and structures as sensory ones, it is impossible to put numerous sensors on them, so there should be non–sensing area even in sensory structure. Here, again, as for considering this problem, an example of leakage of sodium from a pipe of a fast breeder reactor "Monjyu" in Japan is taken. This accident is caused by the breaking of temperature sensing pipe projected in the steam pipe and through this sensing pipe, sodium came out. This was a small accident occured in non–sensing area and did not cause heavy damage occurance in whole system of the fast breeder reactor. There is some possibility, the author guess, that this accident would be leave in some duration, because of no breaking sensors on the location(pipe).

This example shows us a weak point of "point by point" measurement(sensing) method, and we should learn from this accident that even in a sensory structure, we have to have same problem, so we should design that non–sensing area should be limited only in the place in which damage occurance would not induce great accident, and also that the arrangement and posting of sensors should be done for catching some signals with the sensors apart from damaged point when the damage will spread in some extent.

We hope that optical fiber sensor(FOS) will develop enough quickly to be used as line sensor. When it would be realized, we will be able to use it as line, area and cubic sensor by routing it around on monitoring structure. FOS has also another advantage of sensing condition change by its breaking.

4.3 ACTIVE SENSORS

Up to now, the sensors we think about are the devices which detect condition changes in the place where they are attached. Different from this type of sensors, we can imagine so–called active sensor that can, take for instance, hit a bolt to be inspected by hammer like action and detect its vibration responses. As this type of sensor, there is PZT devices. As an example of the application of this type of sensors, an idea was presented (Rogers, 1995) that PZT sensors were put on joints or other suitable points of members of truss, which is considered as a main form of space structures, and were supplied one of them electricity to vibrate a member of the truss and measured other PZT sensor's output voltage, then this process was repeated in order. By comparing vibration output voltages of each PZT sensors during operation with those of healthy condition, one could detect some condition change of truss and its location.

As learning from this example, the use of active sensor has a great benifit for new maintenance inspection proposed here.

4.4 PROMISING SENSORS FOR NEW MAINTENANCE INSPECTION

The sensors which are planned to be used in new maintenance inspection should be suitable for automatic, no human–involving, on–line health monitor–ing. Several sensors are picked up which are thought to be suitable for this purpose and showed in Table 4 with their merits and attaching methods.

Table 4 Promising sensor for new maintenance inspection

Kind	Name	Merits	Attaching Method
(1)	①Optical fiber	Possibility of line sensing of deformation and temperature, Location estimation by breaking	Attachment (A) Embedding (E)
(2)	①Electric resistance strain-gauge	Possible to use in hostile environments	A
	②AE sensor	Sensing only changes in physical property	A, E
	③Fine electric lead wire	Sensing crack and large deformation by breaking Simple, Low-cost	A
	④Magnetic sensor	Sensing crack and large deformation by magnetic leakage soft magnetic piece (like magnetic rubber)	A
(3)	①Carbon fiber string	Line sensing of large deformation, but not location estimation, Substitute of iron bar in concret	E
	②Piezoelectric paint	Vibration and / or crack detection, Easy application for complicated shapes	A
(4)	①Piezoelectric ceramics	Active sensor, vibration detection	A, E
	②Shape memory alloy	Active sensor, deformation detection	A, E

In this table, kind(1) shows line sensor by which one can measure or detect condition change along sensor line, kind(2) point sensor, kind(3) line or area sensor which detect change of condition along line or in area as gross amount, but have no capability to detect the location of the chanage , and kind(4) active sensor. ERSG satisfies most of the requirements for measurement as sensors of this purpose, and ERSG method has many know–hows to measure strains and some other physical quantities under ordinary and hostile environments, and has also sensor protection method and techniques how to deal with given measuring condition, and most of the measuring systemare made easy to con–nect to personal computor system.

Talking about AE method, it is thought as a capable candidate for condition monitoring, because it can detect condition change, and so, it is told that it is already used in civil structures.

5. EXAMPLES OF INTELLECTUALIZATION OF MAINTENANCE INSPECTION

Up to now, the author explains his essey of new maintenance inspection and it necessity. Here he will introduce three examples by which he tries to show his thought of new maintenance inspection; the first one is a simple case of detecting loads working in a bolt, and the other two are good, praftical ones.

5.1 MONITORING ON BOLT LOOSENING

This report (Makabe et al, 1996) shows a new monitoring method of bolt loosening which is simple and capable to continuous monitoring with two ERSG adhered on bolt head along with bolt axis and cross it rectangularly, or cross bolt axis with 45 degrees each others. Two component of a test jig tightened with the bolt is attached on fatigue testing machine and shaked. By dynamic strain waves from the ERSGs under tesion–compression load, one can detect loosening of bolt, the report says.

5.2 PREDICTION OF FATIGUE OF TIE–BAR BOLT OF DIE–CASTING MACHINE

This paper (Oike, 1996) shows an experimental result fo crack detection in tie–bar bolt of 500 tons load die–casting machine (shown in Fig.3) with four ERSGs (adopting four–gage method) on tie–bar bolts. Until that time, NDI inspection have been doing by stopping operation and tightening tie–bar bolt up to the design value, and then trying to detect cracks in the bolts by attach–ing ultrasonic transducer on the cross–section of the bolts. This maintenance inspection have been done periodically once a year and it took about 1200 hours. As a result they found about ten cracks in 460 tie–bar bolts every year.

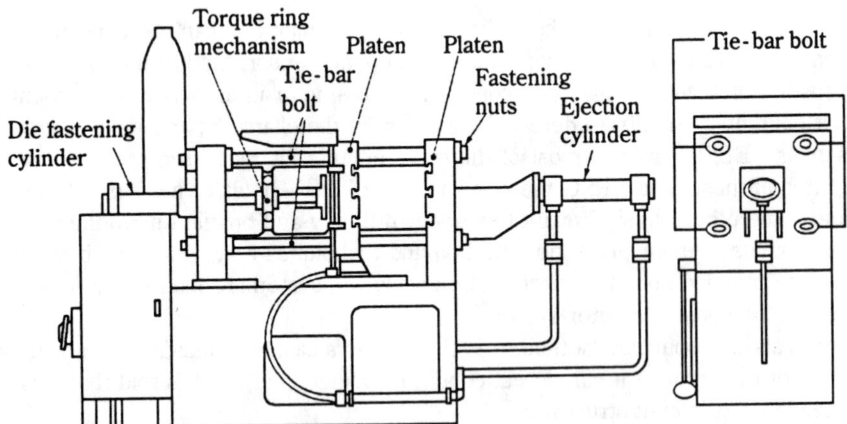

Fig. 3 A die casting machine and its tie-bar bolts monitored for fatigue failure

After attaching ERSG as sensors for making these die–casting machines as sensory ones, they are able to detect crack in tie–bar bolts and measure load variation in the bolts during operation. As the results of this improvement they can reduce their inspection time and work remarkably, and as fringe benefit, they can rise up working rate and reduce replaicng parts cost by the result of reducing trouble during operation. Adding to these benefits, they could find un-balanced load in four tie-bar bolts during operation and this made them possible to more reduction of the machine troubles.

This is a good example of showing advantages of intellectualization of equipments. Also we can presume that the level of maintenance of this paper has already advanced from time-based maintenance to predictive one.

5.3 DIAGNOSES OF EQUIPMENTS IN NUCLEAR POWER PLANT

This article (Takahashi et al, 1988) shows an example of attaching accelerometers on a nuclear power reactor and its steam generator for detecting strange vibration, loosening of bolts in its cooling system. The location and number of attached vibration detectors in primary cooling system are shown in Fig.4. (This primary cooling system is in the second stage of development of BRW nuclear power plant, now it is in the fifth stage-advaced BRW). Vibration detection is one of important measurements for maintenance, and PZT sensors, accelerometers and noise sensors are widely used for this purpose.

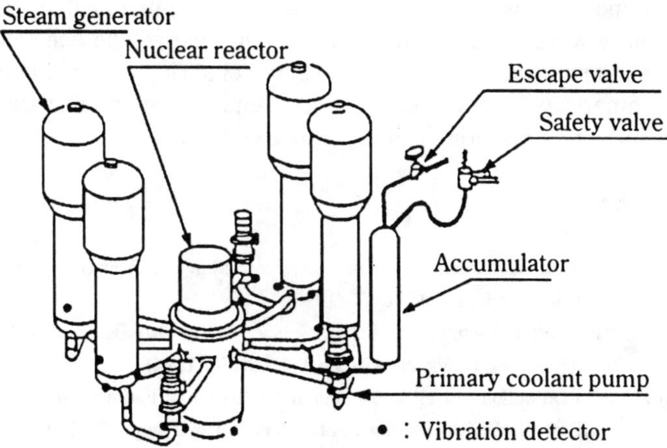

Fig. 4 An example of sensor arrangement on primary cooling system of nuclear power reactor

In this example, dignosis of strange behavior of the system and sending warning signals were done automatically, but moniotring of the system was done by men for taking quick responses in emergency. This example shows us a type of new maintenance inspection of important infrastructures with minimum human involvement.

6. CONCLUDING REMARK

In this essey, the author tries to propose a kind of new maintenance in spection for improving present status of maintenance, which is done periodically with using much of human labor, by introdiucing the concept of intelligent structures(or sensory structures). The most heavy defect of periodical inspection is that it has no capability of sending warning signals in emergency, and the second time–consuming, labor–intensive and expensive. The new maintenance inspection the author proposed here can improve these defects (but on the cost we need some further discussion). In concrete, equipment and structures to be inspected are changed into sensory ones by attaching sensors on them and connecting monitoring/warning system with these sensors. Thus, these inspective equipment and structures can monitor their condition change and send warning signals in emergency by themselves. For keeping these equipment and strucutres in safe during operation, condition changes caused by sudden external load change, crack initiation and progression, and deterioration of materials by age etc. should be monitored restlessly for long period. Thus the sensors and related inspection or measuring systems are demanded to satisfy long time stability in continuous use.

As conclusion, summing up the concept of coming maintenance (inspection) the author proposed, it is shown as "no human–involvement, automatic and prompt". Those words means that restless monitoring will be done automatically with no (minimum) human involvement and monitoring data will be judged promptly whether it will be safe or not, and in emergency warning signals will be sent instantly for preventing coming damage or disaster.

REFERENCES

Achenbach, A.J.: Recent Advances in NDE in USA, Proceedings of First US–Japan Symposium on Advances in NDT, pp.5, (1996).

Oike, I.: Prediction Technique of Fatigur Failure of Tie–bar Bolt by ERSG, Kyowa Engineering News, No.459, pp.3475–3480 (1996).

Makabe, A et al: On Detection of Bolt Loosening and Initiation of Fatigue Cracks, Proceedings (II) of JSME, 73th Spring Conference, Vol.96(1), pp.240–241, (1996).

Rogers, C.A.: One–day Seminar Note by J.Sirkis, JSNDI, (1993).

Rogers, C.A.: On–line Health Monitoring Concept and Challenges, Proceedings of International Symposium on Intelligent Materials & Robots, pp.407–410, (1995).

Takahashi, K. et al: Diagnoses on Nuclear Power Plant, Handbook of Dignoses and Maintenance of Equipments, Vol.2, Fuji–Techno. System, pp.1568, (1988).

In Situ Measurement of Laminated Composite Stiffness

J. D. GENTRY

ABSTRACT

This paper presents a new nondestructive inspection technique which enables accurate reconstruction of the D_{11}, D_{22}, A_{44}, and A_{55} stiffness matrix components of laminated composite plate structures. This innovative technique, which involves measuring the velocity of propagating Lamb waves, is introduced by discussing the theoretical basis for the measurement. Requirements for test fixturing and data acquisition software are then reviewed. Finally, results correlating stiffness values obtained using traditional destructive methods and the new nondestructive approach are presented.

INTRODUCTION

The most common method for measuring the stiffness of a composite laminate is to mechanically load representative specimens and measure strain versus applied stress. Ultrasonic techniques have been demonstrated including a leaky Lamb wave technique (Mal and Bar-Cohen, 1988) and a bulk wave approach (Every and Sachse, 1990). However, none of these methods are practical for measuring the stiffness of composites *in situ*.

Digital Wave Corporation in Englewood, Colorado has developed an instrument that measures the lowest order anti-symmetric Lamb mode or flexural mode velocity to reconstruct the D_{11}, D_{22}, A_{44}, and A_{55} stiffness matrix values. The instrument, called the F-Scan™, is capable of scanning relatively thin composite specimens and is commercially available. The F-Scan™ only requires access to one side of the plate and uses dry coupled sensors that allow measurements to be made in the laboratory, on the production line, or in the field.

The remainder of this paper provides a brief theoretical background for computing stiffness values from measured velocities, describes an experiment to

---

Jeffrey D. Gentry, Digital Wave Corporation, 11234-A East Caley Avenue, Englewood, CO 80111

compare stiffness values measured using the F-Scan™ with theoretical values, and discusses potential applications for in situ measurement of composite stiffness values.

THEORETICAL BACKGROUND

To understand the wave propagation in laminated composite plates, theory which accounts for the variation in material properties must be developed. For the composite laminates, higher order plate theory developed by Tang et. al. (1988) is used for the calculation of the flexural mode velocities. In classical plate theory the shear moduli are taken to be very large so that the transverse shear deformation can be neglected. In composites, this is no longer the case and must be accounted for by including shear deformation and rotary inertia effects. The equations of motion higher order plate theory are

$$\frac{\partial N_x}{\partial x} + \frac{\partial N_{xy}}{\partial y} = \rho \frac{\partial^2 u_0}{\partial t^2} + R \frac{\partial^2 \varphi_x}{\partial t^2} \tag{1}$$

$$\frac{\partial N_{xy}}{\partial x} + \frac{\partial N_y}{\partial y} = \rho \frac{\partial^2 v_0}{\partial t^2} + R \frac{\partial^2 \varphi_y}{\partial t^2} \tag{2}$$

$$\frac{\partial Q_x}{\partial x} + \frac{\partial Q_y}{\partial y} = \rho \frac{\partial^2 w_0}{\partial t^2} \tag{3}$$

$$\frac{\partial M_x}{\partial x} + \frac{\partial M_{xy}}{\partial y} - Q_x = R \frac{\partial^2 u_0}{\partial t^2} + I \frac{\partial^2 \varphi_x}{\partial t^2} \tag{4}$$

$$\frac{\partial M_{xy}}{\partial x} + \frac{\partial M_y}{\partial y} = R \frac{\partial^2 v_0}{\partial t^2} + I \frac{\partial^2 \varphi_y}{\partial t^2} \tag{5}$$

where r is the mass density and

$$(\rho, R, I) = \int_{-h/2}^{h/2} \rho \; (1, z, z^2) \; dz \tag{6}$$

and

$$\begin{bmatrix} N_x \\ N_Y \\ Q_Y \\ Q_x \\ N_{XY} \\ M_x \\ M_Y \\ M_{XY} \end{bmatrix} = \begin{bmatrix} A_{11} & A_{12} & 0 & 0 & A_{16} & B_{11} & B_{12} & B_{16} \\ A_{12} & A_{22} & 0 & 0 & A_{26} & B_{12} & B_{22} & B_{26} \\ 0 & 0 & A_{44} & A_{45} & 0 & 0 & 0 & 0 \\ 0 & 0 & A_{45} & A_{55} & 0 & 0 & 0 & 0 \\ A_{16} & A_{26} & 0 & 0 & A_{66} & B_{16} & B_{26} & B_{66} \\ B_{11} & B_{12} & 0 & 0 & B_{16} & D_{11} & D_{12} & D_{16} \\ B_{12} & B_{22} & 0 & 0 & B_{26} & D_{12} & D_{22} & D_{26} \\ B_{16} & B_{26} & 0 & 0 & B_{66} & D_{16} & D_{26} & D_{66} \end{bmatrix} \begin{bmatrix} \partial u_0/\partial x \\ \partial v_0/\partial y \\ \partial w/\partial y + \varphi_y \\ \partial w/\partial x + \varphi_x \\ \partial u_0/\partial y + \partial v_0/\partial x \\ \partial \varphi_x/\partial x \\ \partial \varphi_x/\partial y \\ \partial \varphi_x/\partial y + \partial \varphi_y/\partial x \end{bmatrix} \tag{7}$$

where

$$\left(A_{ij}, B_{ij}, D_{ij}\right) = \int_{-h/2}^{h/2} \left(Q_{ij}\right)_k \left(1, z, z^2\right) dz \quad \text{where} \quad i,j = 1,2,6 \tag{8}$$

and

$$A_{ij} = \kappa_i \kappa_j \int_{-h/2}^{h/2} \left(Q_{ij}\right)_k dz \quad \text{where} \quad i,j = 4,5 \tag{9}$$

N and Q are the force resultants and M is the moment resultant, A_{ij}, B_{ij}, and D_{ij} are the laminate stiffnesses, calculated using laminated plate theory, $k_i k_j$ are the shear correction factors, u_0, v_0 are mid-plane displacements of the plate, yx and yy are the rotation components along the x and y axes, respectively. Substituting plane wave solution results into the characteristic equation for the flexural mode the characteristic equation for a symmetric laminate is given by

$$\begin{vmatrix} D_{11}k^2l_1^2 + 2D_{16}k^2l_1l_2 + D_{66}k^2l_2^2 + A_{55} - I\omega^2 & D_{16}k^2 + (D_{12}+D_{66})k^2l_1l_2 & -iA_{55}kl_1 \\ D_{16}k^2 + (D_{12}+D_{66})k^2l_1l_2 & D_{66}k^2l_1^2 + 2D_{16}k^2l_1l_2 + D_{22}k^2l_2^2 + A_{44} - I\omega^2 & -iA_{44}kl_2 \\ -iA_{55}kl_1 & -iA_{44}kl_2 & A_{55}k^2l_1^2 + A_{44}k^2l_2^2 - \rho\omega^2 \end{vmatrix} = 0 \tag{10}$$

The characteristic equation above is solved by assuming values for *k* and then solving the determinant for *w* using a numerical root finding technique. From this the phase (*w*/*k*) and group (d*w*/d*k*) velocities can be calculated.

EXPERIMENTAL RESULTS

Using the above equations, the phase velocities for the flexural mode were computed for a uni-directional, 16 ply, AS4/3501 graphite/epoxy composite plate. Figure 1 shows the lamina material properties and the D_{11}, D_{22}, A_{44}, and A_{55} computed using laminated plate theory.

Material AS4/3501			
Lamina Properties		**Computed Stiffness**	
E11	144.8 GPa	D11	145 Nm
E22	9.65 GPa	D22	9.66 Nm
G12	5.97 GPa	A44	6.85 MN/m
G13	0.0 GPa	A55	11.36 MN/m
n12	0.3		
n23	0.34		

Figure 1. Material Properties and Computed Stiffness Constants

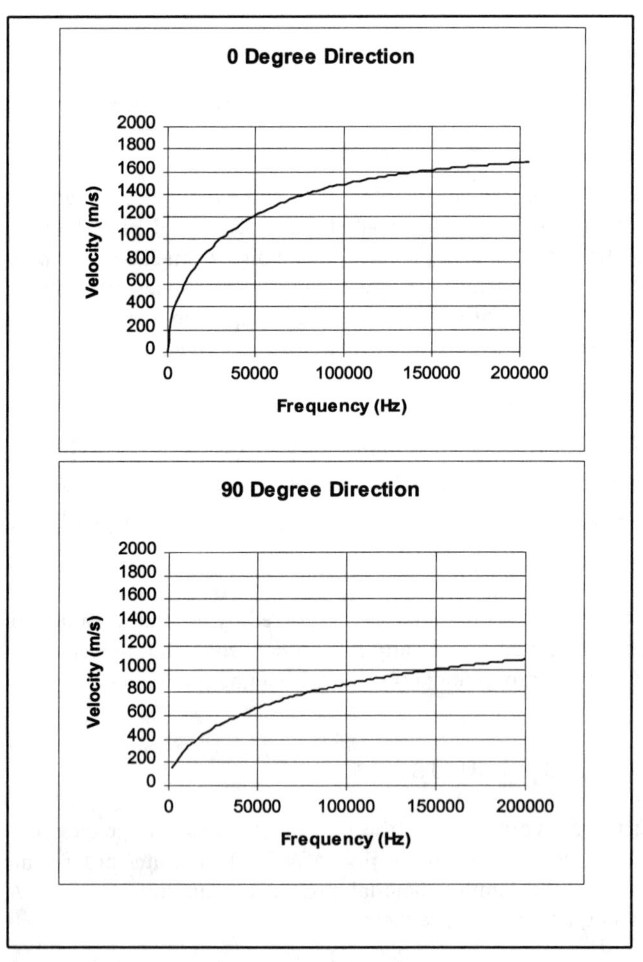

Figure 2. Theoretical Flexural Mode Dispersion Curves

The flexural mode dispersion curves, which are a plot of velocity as a function of frequency are shown in Figure 2. This figure shows the phase velocity dispersion curves in the 0 and 90 degree directions.

It should be noted that the 0 degree velocity is predominately influenced by D_{11}, and A_{55}. This is readily explained by examining the classical plate solutions for the low frequency region of the dispersion curve:

$$V_{phase} = \left(\frac{D_{11}\omega^2}{\rho h} \right)^{1/4} \tag{11}$$

and the limiting solution for the high frequency region:

$$V_{phase} = \sqrt{\frac{A_{55}}{\rho h}} \tag{12}$$

Substituting D_{22}, and A_{44} will result in the limiting equations for the 90 degree velocity.

The F-Scan™ system consists of a arbitrary function generator board, pulse amplifier, transmitting and receiving sensors, 2-D scanning bridge, analog signal conditioning, high speed analog-to-digital converter, and a CPU. A functional block diagram of the F-Scan™ system is shown in Figure 4.

The F-Scan™ system prototype was used to measure the flexural mode phase velocities propagating in the 0 and 90 degree directions of the graphite/epoxy plate. The velocity measurements were made over a range of frequencies from 60 kHz to 200 kHz. A photograph of the prototype 2-D scanning table is shown in Figure 5.

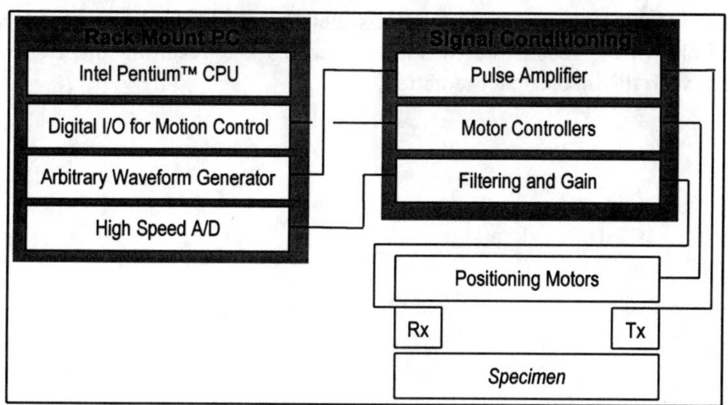

Figure 4 Functional Block Diagram of the F-Scan™ System

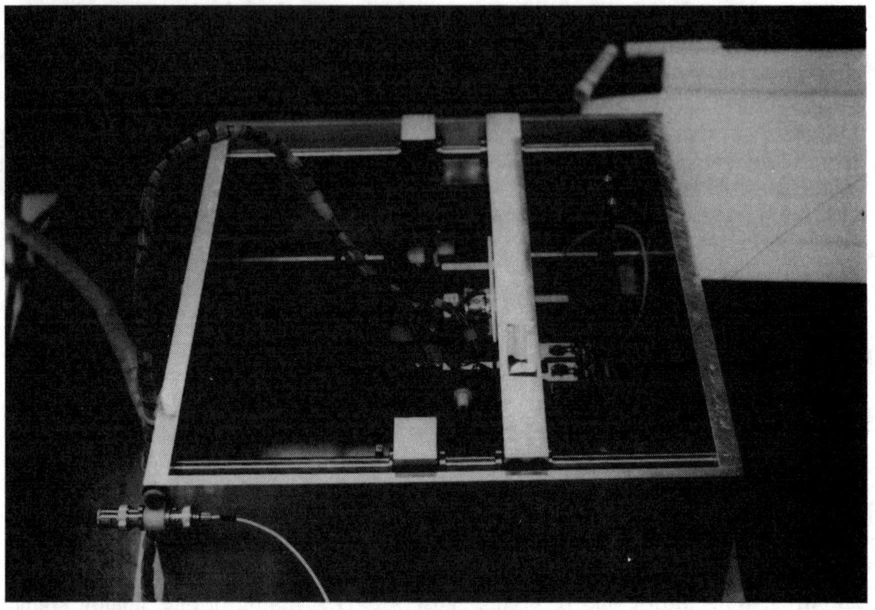

Figure 5. The F-Scan™ System Prototype

The measured phase velocities are shown with the theoretical dispersion curves in Figure 6.

Using a proprietary routine built into the F-Scan™ system software, the measured velocities were used to reconstruct the D_{11}, D_{22}, A_{44}, and A_{55} values. Figure 7 shows the reconstructed stiffness values and resulting dispersion curve compared with the theoretical prediction.

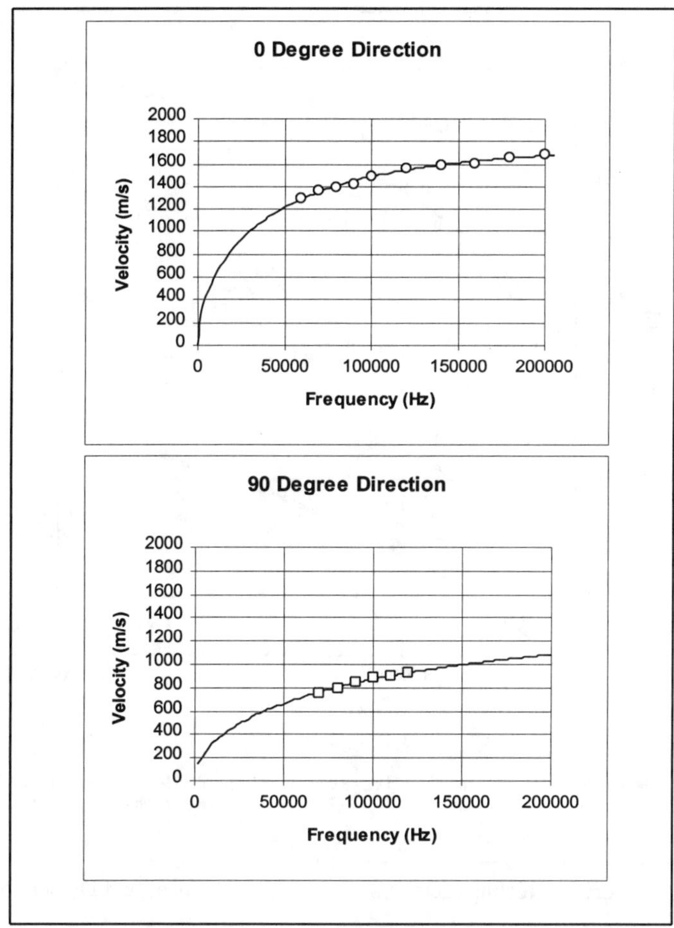

Figure 6. Experimental Vs. Theoretical Phase Velocities

CONCLUSIONS

A nondestructive measurement of the D_{11}, D_{22}, A_{44}, and A_{55} stiffness matrix components of laminated composite plates has been demonstrated. The technique uses highly accurate measurements of propagating anti-symmetric Lamb waves and shows great promise for measuring composite stiffness values *in situ*. A commercial instrument, called the F-Scan™ system, has been developed for characterizing composite materials in research laboratories, providing quality assurance and process control monitoring, as well as offering composite stiffness measurements in the field.

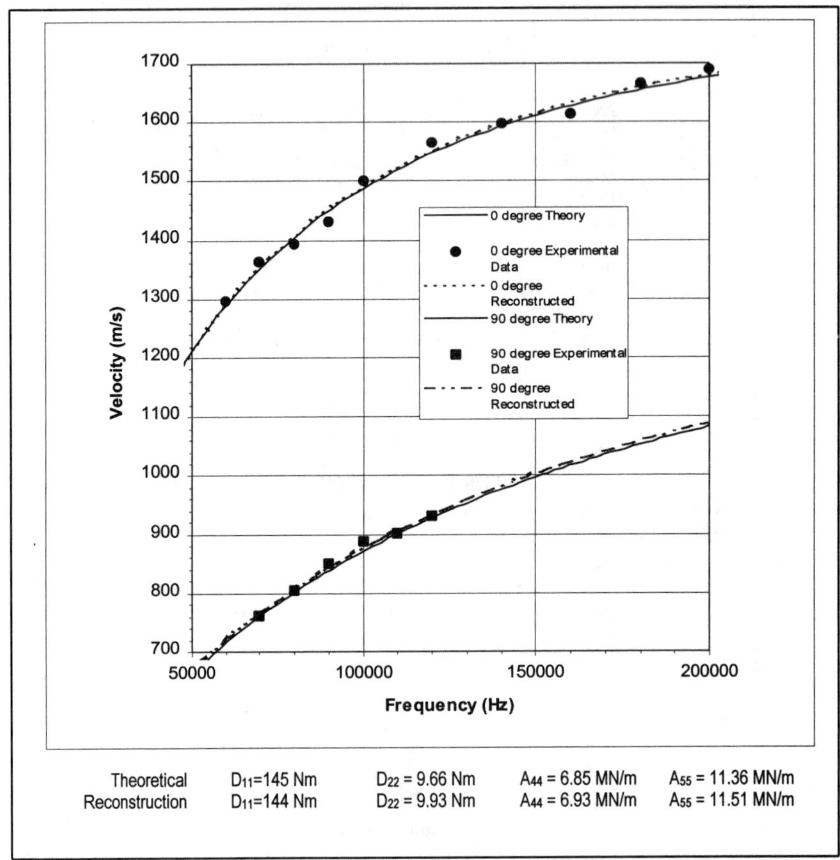

Figure 7. Reconstructed vs. Theoretical Stiffness and Dispersion

Preliminary results obtained with the F-Scan™ system prototype show an extremely good match with the theoretical predictions. Reconstructed D_{11}, D_{22}, A_{44}, and A_{55} stiffness values for a 16 ply, uni-directional AS4/3501 graphite/epoxy plate were within 1% of the material manufacturer's published values. Furthermore, a statistical study showed that the repeatability of the velocity measurement using the prototype fixturing is within plus or minus 3%.

Future work with the F-Scan™ system will include a direct comparison of reconstructed stiffness values with values obtained with mechanical testing for a wide range of materials and lay-ups.

ACKNOWLEDGEMENTS

The author would like to acknowledge Dr. Steve Ziola and Dr. Wei Huang of Digital Wave Corporation who are responsible for the development of the F-Scan™ instrument and provided key contributions in the theoretical background sections of this paper.

REFERENCES

Every, A.G., and Sachse, W., (1990). "Determination of the Elastic Constants of Anisotropic Solids from Acoustic Wave Group Velocity Measurements," *Phys. Rev.* B 42:8196-8205.

Mal, A.K., and Bar-Cohen, Y., *Ultrasonic Characterization of Composite Laminates*, Wave Propagation in Structural Composites, Proceedings of the Joint ASME and SES Meeting, ASME AMD Vol 90, A.K. Mal and T.C.T. Ting, Ed., 1988, pp. 1-16.

Tang, B., Henneke II, E.G., and Stiffler, R.C., 1988. "Low Frequency Flexural Wave Propagation in Laminated Composite Plates," in *Acousto-Ultrasonics Theory and Application*, Plenum Press, pp. 45-65.

Condition Monitoring of Silicon-Wafer Slicer Cutting Crystal Ingots

Z. JIANG,[1] S. CHONAN,[1] K. KAWASHIMA,[2]
K. MUTO[1] and W. ICHIHARA[1]

ABSTRACT

The silicon ingot must be cut to yield wafers with flat and smooth surface. However, the flatness and smoothness are easily affected by the cutting condition of the inner-diameter saw blade. In this paper we present an experimental study on evaluation and real time monitoring of the slicer cutting the silicon ingot. The circumferential, radial and lateral cutting forces are measured by using the piezoelectric 3-axis load sensor fixed onto the mount of the silicon ingot. The collected data is analyzed by wavelet analysis and the correlation of the sensor signals with the condition and the cutting efficiency of the blade is investigated. The results show that the sharpness of the cutting edge and the anisotropic stiffness of the blade are both clearly predicted by analyzing the high-level frequency signal in the data with the wavelet method.

INTRODUCTION

The ID(inner-diameter) saw blade is nowadays mainly used in the crystal wafering. The blade is adhered with diamond whetstones at the inner edge, and tensioned initially at the outer periphery to increase the stiffness and to make the inner edge as circular as perfect. The diamond abrasive should be distributed uniformly along the cutting edge and the in-plane tension of the blade should be made uniformly too. However, these situation is almost impossibly realized because the asymmetric dressing of the abrasive and the anisotropic stiffness in the roll and the cross directions of a blade plate can not be avoided in manufacturing. The imperfection of manufacturing blade causes the degradation of the cutting effect and sometimes leads to the damage of wafers.

As the asymmetric distribution of the abrasive causes the blade to cut into the ingot strayed from the desired pass, it results in an appearance of a variation

[1]Zhongwei Jiang, Seiji Chonan, Koji Muto and Wataru Ichihara, Department of Mechatronics and Precison Engineering, Tohoku University, Sendai, Japan.

[2]Kazuo Kawashima, Asashi Diamonds Company, Chiba, Japan.

of the cutting force. Also the anisotropic stiffness of the blade comes out a variation of the cutting force. It is therefore of technological importance to evaluate or predict the condition of the saw blade by monitoring of its cutting force.

A lot of papers have been published on the vibration characteristics of the OD(outer-diameter) and ID saw blades[1–4], and some of them on the vibration and flatness control of the blade[5]. The feedback signals in control system is usually the displacement or deflection of the blade. However, the deflection at the cutting edge or the warp of a wafer is difficult to be measured directly in real time. But the cutting force transmitted through the ingot can be measured easily by a force sensor mounted on its clamp. If the cutting force can be measured and processed in good correspondence with the cutting condition of a blade and the quality of a wafer, it is miraculous for wafer manufacturing that the cutting condition of the blade and the flatness of the wafer are monitored and controlled in real time.

With this point in mind, this paper presents an experimental study on the measurement and data processing of the force signals obtained in the cutting process of the crystal wafering. Several different types of blades have been tested to cut a 6″-diameter silicon ingot. The circumferential, radial and lateral forces were measured by a piezoelectric 3-axis load sensor fixed onto the mount of the ingot, further the data were first analyzed by the wavelet method and then condensed by means of statistics. Some methods are proposed for monitoring, evaluating and predicting the cutting condition of a blade by using the high-level frequency force signals.

MEASUREMENT AND DATA ANALYSIS

Measurement and Experimental Setup

Figure 1 shows an outline of the experimental setup. As it is fed at speed of 50mm/min in x-direction, the silicon ingot is cut by an ID saw blade rotated with 1480 rpm. The circumferential, radial and lateral cutting forces are measured by a piezoelectric 3-axis load sensor fixed onto the mount of the ingot, and stored in an oscilloscope and then transmitted to a computer for data analysis.

The ID blades used in the experiment were 0.665m and 0.24m in outer and inner diameters, and the silicon ingots were 6 inch in diameter. The cutting signals are recorded at three locations denoted by the depth of cut, 10mm, 70mm and 130mm, as shown in Fig.2. The sampling time is set at 40kHz and the recording time for a serial data is 0.2sec.

The time response of the cutting signals are plotted as an example in Fig. 3.

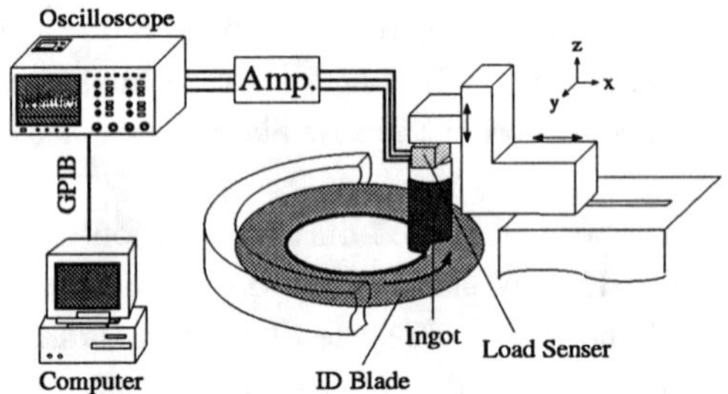

Fig.1. Experimental setup of slicer cutting ingot.

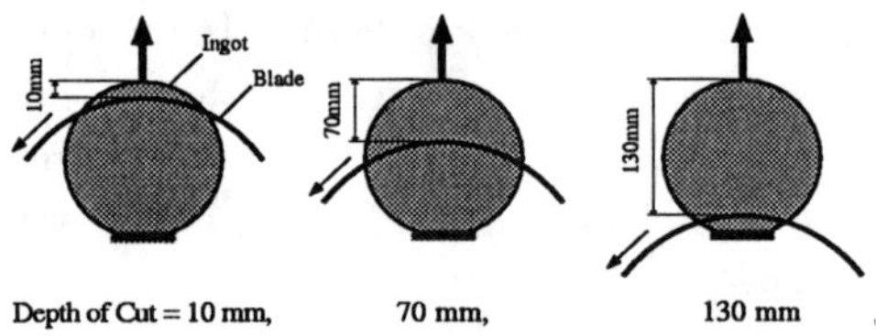

Fig.2. Definition of the depth of cut of wafer.

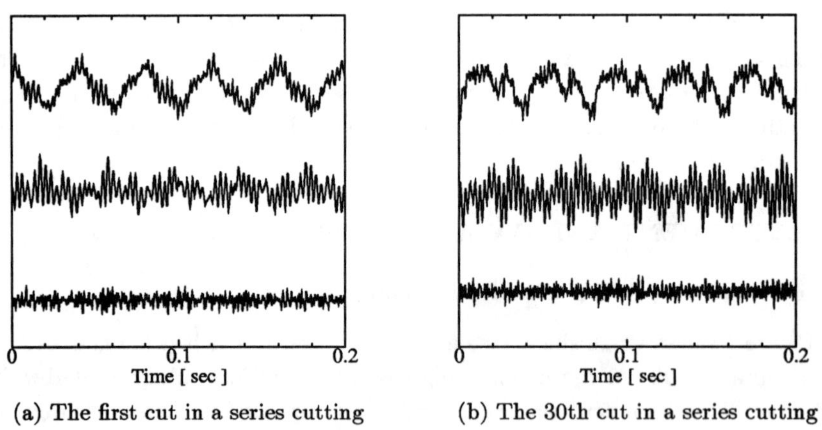

(a) The first cut in a series cutting

(b) The 30th cut in a series cutting

Fig.3. Time response of cutting signals in different situation.

Figure (a) is for the case when the first cut of wafer was made, and Fig.(b) shows that of the 30th cut of wafer. Each curve from top to bottom of the figure represents the signals of the radial(x), circumferential(y) and lateral(z) loads respectively. It was found that the lateral forces are not varied so much even in the different cutting situation. The analysis on the lateral signals is therefore ignored hereinafter. Furthermore, the x and y-direction cutting forces appear a great difference between the first and the 30th cut of the wafer. This means the cutting condition or the situation of the blade can be predicted and evaluated if an apt data processing method is introduced. It is shown later that the wavelet analysis method is one of the efficient way to solve this kind of problems.

Wavelet Analysis Method

The wavelet signal processing method is introduced and applied to the signals obtained in the experiment in order to make the evaluation and prediction more efficient.

Based on the wavelet theory, the measured signals $f(x)$ can be decomposed into a series expansions in form of

$$f(x) \sim \sum_j g_j(x) = \sum_j \sum_k d_k^{(j)} \psi(2^j x - k), \tag{1}$$

where $\psi(x)$ is a mother wavelet and $d_k^{(j)}$ is the expansion coefficient. If the a scaling function $\phi(x)$ is defined by a selected two-scale sequence $\{p_k\}$ as

$$\phi(x) = \sum_k p_k \phi(2x - k), \tag{2}$$

the mother wavelet $\psi(x)$ is then given by

$$\psi(x) = \sum_k q_k \phi(2x - k), \tag{3}$$

where the coefficient sequence $\{q_k\}$ is obtained from the two-scale relation. If the coefficient sequences $\{p_k\}$ and $\{q_k\}$ are known the coefficient $d_k^{(j)}$ of Eq.(1) can be calculated by the decomposition algorithm of

$$d_k^{(j-1)} = \frac{1}{2} \sum_l h_{2k-l} c_l^{(j)}, \quad c_k^{(j-1)} = \frac{1}{2} \sum_l g_{2k-l} c_l^{(j)}. \tag{4}$$

Now, the following shows the detail process to calculate the above coefficients. First, one selects the fourth order B-spline[6] as the scaling function $\phi(x)$, since the fourth order B-spline can be easily created by substituting $m = 4$ into the mth order B-spline such as,

$$\begin{aligned} N_m(x)|_{m=4} &= \frac{1}{(m-1)!} \sum_{k=0}^{m} (-1)^k \binom{m}{k} (x-k)^{m-1}|_{m=4} \\ &= \frac{1}{6} \sum_{k=0}^{4} (-1)^k \binom{4}{k} (x-k)^3, \\ \text{OR}: \qquad N_4(x) &= \sum_{k=0}^{4} p_k N_4(2x-k) \equiv \phi(x), \end{aligned} \tag{5}$$

and it satisfies the tow scale function. The mother wavelet is in this case given in form of

$$\psi_{N_4}(x) = \sum_{k=0}^{3*4-1} q_k N_4(2x-k). \tag{6}$$

The sequences $\{p_k\}$ and $\{q_k\}$ are obtained by applying Fourier Transform method and given by

$$p_k = \frac{1}{2^{4-1}}\binom{4}{k} \quad (k=0,\ldots,4),$$
$$q_k = \frac{(-1)^k}{2^{4-1}}\sum_{l=0}^{4}\binom{4}{l} N_{2*4}(k+1-l), \quad k=0,\ldots,(3*4-2). \tag{7}$$

Next solve the decomposition sequences g_k and h_k from following formulas[6]

$$G(z) = \sum_{k=-2n-7}^{2n+3} \frac{1}{2} g_k z^k = \frac{E_7(z)}{z^3}\Big(\frac{1+z^{-1}}{2}\Big)^4 \sum_{k=-n}^{n} \alpha_k z^{2k} \tag{8}$$
$$H(z) = \sum_{k=-2n-7}^{2n-3} \frac{1}{2} h_k z^k = \frac{7!}{z^7}\Big(\frac{1-z}{2}\Big)^4 \sum_{k=-n}^{n} \alpha_k z^{2k} \tag{9}$$

where

$$E_7(z) = z^6+120z^5+1191z^4+2416z^3+1191z^2+120z+1=0,$$
$$\alpha_k = \sum_{i=1}^{3} C_i a_i^{|k|},$$
$$a_i = r_i - 20 + \sqrt{(r_i-20)^2-1}, \quad i=1,2,3,$$
$$r_i = 2\rho^{1/3}\cos\Big(\frac{\varphi+2\pi i}{3}\Big),$$
$$\rho = 301\sqrt{301}, \quad \tan\varphi = \frac{7\sqrt{11905}}{-5166}, \quad \frac{\pi}{2} < \varphi < \pi,$$
$$C_i = \prod_{k=1}^{4-1}\frac{1}{(a_k - a_i^{-1})}\prod_{j\neq i}\frac{a_j}{(a_j-a_i)}. \tag{10}$$

Solving the coefficients of Eq.(4), one further needs to know the initial value $c_k^{(0)}$ which is given by,

$$c_k^{(0)} = \sum_l \beta_{k+2-l}^{(4)} f(l), \quad \beta_k^{(4)} = \sqrt{3}(\sqrt{3}-2)^{|k|}. \tag{11}$$

So the multiresolution approximation $g_j(x)$ are then obtained in from of

$$g_j(x) = \sum_k d_k^{(j)} \psi(2^j x - k). \tag{12}$$

Intending to make the evaluation and prediction as quantitative as possible, some definitions are introduced as follows.

$$\begin{aligned}
&\textit{Average of } g_j \textit{ for each level } j: && Ga(j) = \frac{1}{N}\sum_{i=1}^{N}|g_j(i)|,\\
&\textit{Maximum of } g_j \textit{ for each level } j: && Gm(j) = \max[|g_j(i)|], \quad i = 1....N,\\
&\textit{Average of Maximum } Gm(j): && Gma(j) = \frac{1}{K}\sum_{k=1}^{K}Gm(j),\\
&\textit{Standard Deviation of } Gm(j): && S_{Gm}(j) = \sqrt{\frac{1}{N}\sum_{k=1}^{N}[Gm_k(j) - Gma(j)]^2}.
\end{aligned} \tag{13}$$

RESULTS AND DISCUSSION

Damage Monitoring

Figure 4 shows a damaged blade with a very small warp on the diamond whetstone layer. Fortuitously some data were obtained as this damaged blade cut several pieces of the wafer. In fact it was very dangerous using this kind warped blade to cut the ingot. Comparing the cutting signals of a normal blade with those of the damaged one, as shown in Fig.5, it is easy to find the big difference between the each other. Here, the upper and middle curves in each figure are the force signals of x and y-direction, while the lower spikes represent the rotation period of the blade.

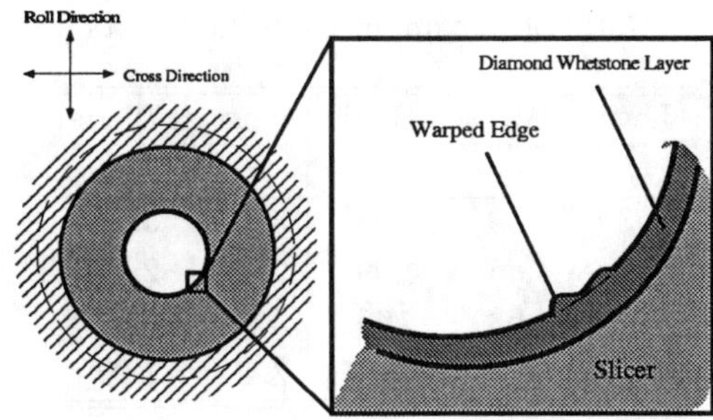

Fig.4. Aspect of warp on the inner edge of blade.

As an example Fig.6 shows the multiresolution approximations $g_j(x)$ obtained by applying the wavelet analysis to those cutting signals. The original cutting signal, which is marked by $j = -0$, has decomposed into eight levels labeled from $j = -1$ to $j = -8$. Comparing Fig.(a) to (b), it is found that several ripples appear at level $j = -3$ to $j = -5$ in case of the warped blade. It is clear that the damage of blade can be evaluated if the ripples is quantified by

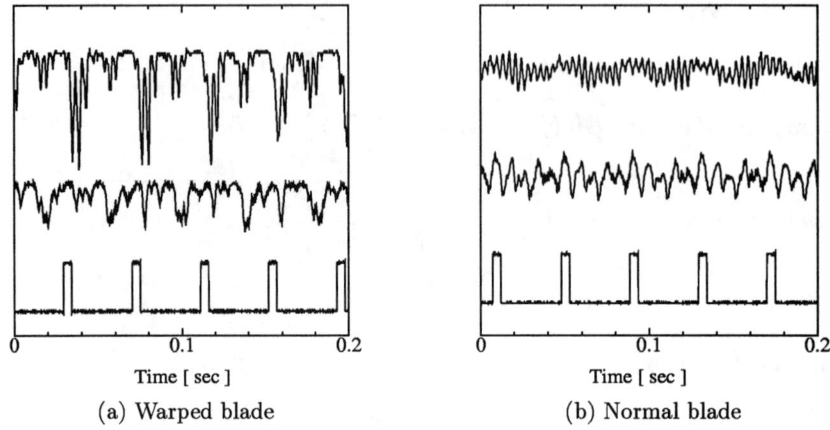

(a) Warped blade (b) Normal blade

Fig.5. Force signals of damaged blade and normal blade.

some numerical methods.

A simple way to quantify the value of the ripples is to calculate the ratio of the maximum amplitude to its average for each level. Table 1 shows the results of level $j = -3$ to $j = -6$, and it is evident that the warped blade has higher values than that of a normal one.

Table 1. Ratio of maximum amplitude to its average.

Level	Warped Blade		Normal Blade	
j	x	y	x	y
-3	5.09	5.95	3.95	3.70
-4	4.86	4.69	2.83	2.81
-5	4.63	3.65	3.56	2.80
-6	2.28	2.15	2.49	1.90

Effect on Ellipticity of Inner Cutting Edge

Usually the ID saw blades need to be tensioned at the outer periphery in order to stiffen the blades and to make their inner edges be a perfect circle. The lack of circularity at the inner edge easily makes the blade damaged and the wafer bastardized. This section shows how the cutting signals are affected by the ellipticity of the inner edge of a blade.

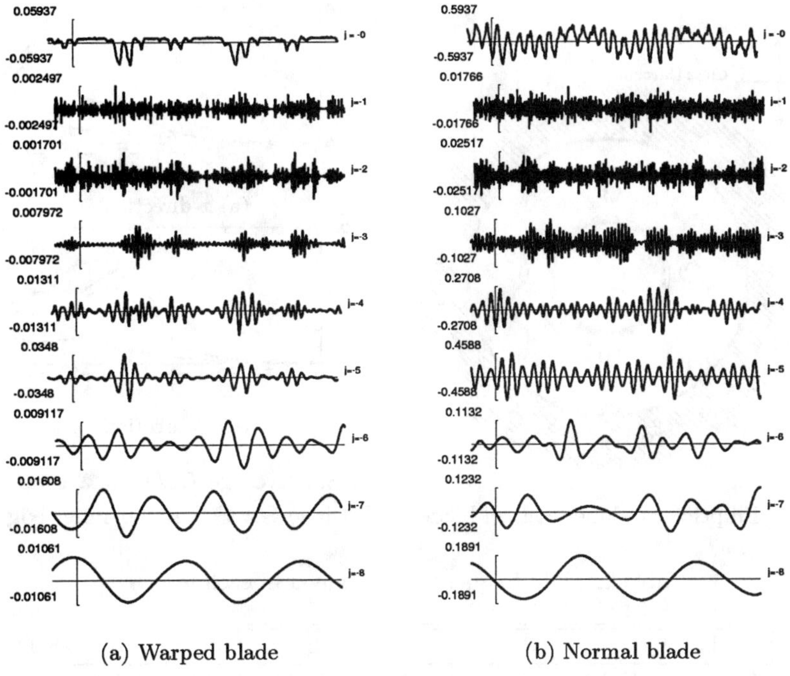

(a) Warped blade (b) Normal blade

Fig.6. Example of wavelet analysis results.

In this experiment, the blade is first tensioned uniformly along the outer periphery till the inner edge becomes circular with 900μm radial extension and then strained more 200μm only in one direction either the roll direction (1–3), which is named "Blade F", or the cross direction (2–4), named "Blade G", as shown in Fig.7.

Figure 8 shows the average values $Ga(j)$ calculated by Eq.13. The solid lines connecting symbols ■ and □ are the results of Blade F and Blade G respectively. After cutting several pieces of wafer Blade F and Blade G were strained again to make their inner edge circular. The dashed line connecting symbol ● is just the case for the re-circled blade.

The level j, which represents the frequency range, is associate with some cutting behaviors corresponding to the ellipticity of the cutting edge, the vibration of the slicer and so on. For instance, $j = -8$ represents the frequency around 20Hz which corresponds to the rotating frequency of the blade. The main frequencies contained in each level are listed in Table 2. It is found that the levels from $j = -6$ to $j = -8$ indicate the behaviors correlated to the rotation speed or the ellipticity of the blade as well as the anisotropic stiffness of the blade,

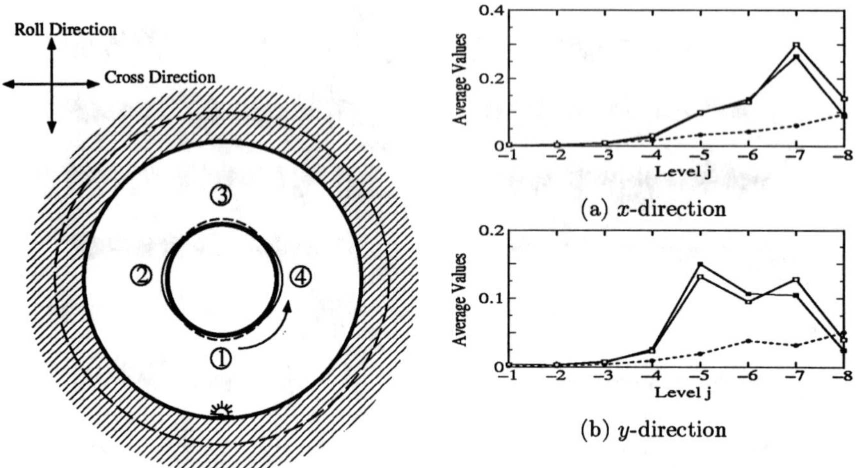

Fig.7. Ellipticity of inner edge of blade.

Fig.8. Average $Ga(j)$; ■:Blade F, □:Blade G, ●:Circular inner edge.

Table 2. Main frequencies contained in each level j.

Level j	-3	-4	-5	-6	-7	-8
Frequency(Hz)	730-850	300-420	170-270	80-130	40-70	25

and the levels from $j = -3$ to $j = -5$ indicate the performance correlated to the vibration of the slicer.

Figure 8 shows that the average amplitude of Blade F and Blade E is higher at levels $j = -5$ to -7 than that of a normal circular blade. This means that the lack of the circularity of the inner edge causes the cutting force varied significantly and periodicly along the cutting edge and leads easily to diminish the durability of the balde.

Evaluation and Cutting Condition Monitoring

Three blades with different properties, as listed in Table 3, have been tested to delve into some methods for evaluation and monitoring of the cutting condition.

More than 40 pieces of wafer were cut for each blade listed in Table 3. Blade L and Blade M have different Young's modulus but same diamond whetstone layer, while Blade M and Blade N have same Young's modulus but different size of the diamond whetstones. The cutting signals both of the x and y-direction were measured and recorded every three pieces of the 40 cuts and all the obtained

Table 3. Properties of the blades

Blade ID	L	M	N
Lode No.	7335	8575	8575
Yang's Module E_{roll}(GN/^{2})	174	191	191
Yang's Module E_{cross}(GN/^{2})	193	217	217
Width of whetstone layer(mm)	2.0	2.0	2.0
Diameter of diamond whetstone(μm)	40∼60	40∼60	50∼70

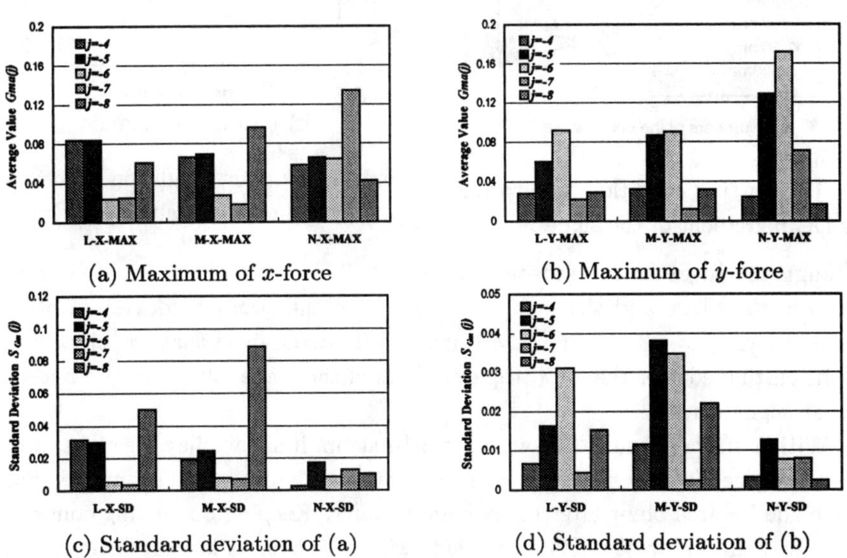

(a) Maximum of x-force (b) Maximum of y-force

(c) Standard deviation of (a) (d) Standard deviation of (b)

Fig.9. Maximum and standard deviation of each level(j).

data were analyzed by the wavelet method.

Figure 9 illustrates the bar graphs of values of $Gma(j)$ and $S_{Gm}(j)$ for level $j = -4$ to $j = -8$. The value $Gma(j)$ expresses the amplitude of the cutting force and $S_{Gm}(j)$ describes the deviation from its average amplitude. Comparing with those figures, it is denoted that the amplitude of the cutting force of Blade N is bigger but the standard deviation is smaller than the other blades. This means Blade N, which is adhered with large diamond whetstones, cuts the ingot more stably and durably. Also the graphical bar patterns of the maximum forces in x-direction are divide into two groups, for instant, Blade L and M have same pattern with higher value at levels $j = -4$, -5 and -8 whlie Blade N has the highest value at level $j = -7$. This indicats that the blades with small diamond whetstones are easily oscillated or the vibration of the slicer are easy transmitted

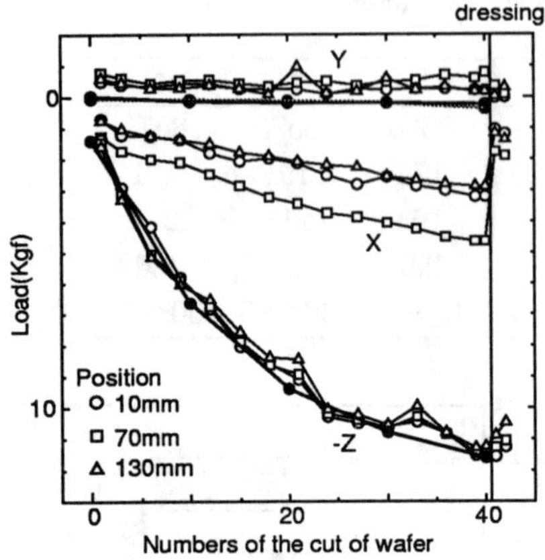

Fig.10. Cutting condition monitoring by DC ingredient of the signal.

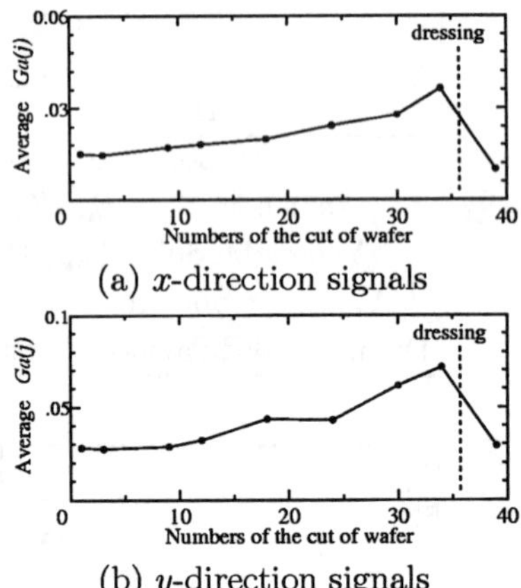

(a) x-direction signals

(b) y-direction signals

Fig.11. Cutting condition monitoring by the average values $Ga(j)$.

through the ingot to the sensor conceivably due to the restricting condition between the blade and the ingot. Furthermore, the standard deviation of x-force at level $j = -8$ is extremely large, and it is conceived that the ellipticity of the cutting edge or the anisotropic stiffness of the blade influences the x-force signals significantly.

With a view of Figure 3(a) with Fig.3(b) again, it shows that the signal amplitude of the 30th cut of the wafer is higher than the first cut. This means that the blade became blunt after many cuts of wafer. Keeping on cutting continuously without any treatment on the blade will make a degradation or even break of wafers. Usually the wafer makers have to stop the machine and sharpen the blade under this kind situation, which is so-called table dressing. To investigate the dressing effect, the table dressing was carried out after cutting continuously 35 pieces of the wafer in the experiment. Figures 10 and 11 are the results showing the dressing effect on the cutting signals. Figure 10 shows the results obtained by extracting the DC ingredient from the original signals. It is denoted that the DC ingredient forces in x-direction are amplified with an increase of the cut numbers of wafer and diminished immediately and significantly after a table dressing. The graph described by x-curve has been used in wafer manufacturing to diagnose if the table dressing is needed or not. However, the y-directon signals have almost no change on the cut numbers of wafer and the z-direction ones are just decreased with the weight of the ingot.

Figure 11 shows the case that both x- and y- direction signals can be used for monitoring of the dressing effect if their high-level frequency signals are operated by means of the wavelet analysis. The curve in the figures is plotted by value of $Ga(j)$ of Eq.(13). It clear that both x- and y-direction cutting signals are getting greater with an increase of the cut numbers of wafer and return back after the table dressing.

CONCLUSIONS

An experimental study on the evaluation and the monitoring of the cutting condition of the wafer slicer has been presented. The cutting force signals, which are measured by a piezoelectric 3-axis load sensor and processed by the wavelet analysis, are available and efficient for the evaluation and the prediction of the cutting condition of slicer. The obtained results are summarized as follows.

1. A warped cutting edge can be detected by the quantification of the ripples from the signals.
2. The vibration of the slicer, the ellipticity of the cutting edge, the anisotropy of the stiffness of the blade and the condition of the diamond whetstones are able to be evaluated by the average, the maximum and the standard deviation defined in Eq.13.
3. The sharpness of the cutting edge and the requirement of table dressing can be monitored directly in real time by the force signals.

REFERENCES

1. Forman, S.E., and W.J.Rhines, 1972. "Vibration Characteristics of Crystal Slicing ID Saw Blades," *Journal of the Electrochemical Society,* 119:686-690.
2. Carlin, J.F., H.C.Bridwell and R.P.Dubois, 1975. "Effects of Tensioning on Buckling and vibration of Circular Saw Blades," *ASME Journal of Engineering for Industry*, 97:37-48.
3. Chonan,S., Z.W.Jiang and Y.Yuki, 1993. "Stress Analysis of a Silicon-Wafer Slicer Cutting the Crystal Ingot", *ASME Journal of Mechanical Design,* 115:711-717.
4. Chonan,S., Z.W.Jiang and Y.Yuki, 1993. "Vibration and Deflection of a Silicon-Wafer Slicer Cutting the Crystal Ingot", *ASME Journal of Vibration and Acoustics,* 115:529-534.
5. Chonan,S., Z.W.Jiang and K.Masui, 1994. "Deflection Control of a Large-Scale Silicon-Wafer Slicer Cutting a Crystal Ingot", *Transactions of the Japan Society of Mechanical Engineers,* 60C(570):498-505.
6. Sakakibara, S., 1995. "Wavelet Beginer's Guide," *Tokyo Denkidaigaku Publishing Group.*

Static Shape Control of Smart Beams with Laminated Sensors and Actuators

C.-C. LIN[1] and C.-Y. HSU[2]

ABSTRACT

The paper presents a novel scheme capable of controlling beam shapes without relying on information of external loads. An adaptive control algorithm is used for a laminated beam integrated with sensors and actuators, and a theoretical simulation is formulated and performed to demonstrate its feasibility. Fourier sine series incorporated with Stokes transformation is employed for the simulation of beam deflections. Clamped-clamped and simply supported beams are used for illustrative purposes. Simulation results indicate that the scheme is effective.

1. INTRODUCTION

Recent technological advancements in microsensing, microactuation, and active changes of shapes of aerofoils and turbine blades have stimulated the progress of research on shape control of structures with piezoelectric actuators. An example of a specific application is the bonding of piezoelectric actuators on an aeroelastic lifting surface discussed by Crawley et al. (1988) for altering the shape of the lifting surface, thereby changing the lift on the vehicle.

Satellite structures demand high precision of desired shapes which may be achieved by using piezoelectric actuators for shape control. Varadajan et al. (1996) applied a finite number of actuator patches to control the shape of structures. Static shape control of structures is a topic of current interest, especially for large space structures and smart structures. Haftka et al. (1985) investigated the static shape control of large space antennas, and Austin et al. (1994) studied adaptive wings. Koconis and Kollar (1994) used an analytical method to find optimal actuator voltages in order to control the structures to remain closest to a desired shape. Agrawal et al. (1994) developed a finite difference method, incorporating Kuhn Tucker optimality conditions, for finding optimal control voltages. The mentioned studies rely on information of loads except for the work of Haftka et al. (1985).

---

1 Professor, author to whom correspondence should be addredded.

2 Graduate student.

Institute of Applied Mathematics National Chung-Hsing University Taichung, Taiwan 402, Republic of China

The objective of this paper is to present a novel scheme for shape control of beam structures without load information. The scheme involving an adaptive control algorithm is used for beam structures integrated with sensors and actuators. The sensors and actuators are laminated into the beam, and their electrodes are trimmed to shapes of sine functions. A controller with a gradient projection algorithm connects each sensor-actuator pair of same shape which controls one specific mode without influencing other modes. The advantage of such control strategy is to assure a desired precision without load information. Numerical results show that the laminated beams with adaptive control algorithm is a feasible scheme to control the deformed shape of beams without load information.

2. FORMULATION

A beam with laminated sensors and actuators trimmed in shapes of sine functions, under static loading is considered. The geometry, coordinate system and some symbols are shown in Fig. 1(a). Same voltages are applied on the upper and lower actuators with opposite signs as shown in Fig. 1(b).

2.1 Sensors

Piezoelectric material produces charges when deformed. The physical phenomenon is called direct piezoelectric effect. The charge q_s produced by the sth sensor lamina resulting from integration of electric displacement D_{es} $(Culombs/m^2)$ of the sth sensor according to Dosch and Inman (1992) becomes

$$q_s = \int_0^L D_{es} w_s R_s(x) dx \ , \qquad D_{es} = \frac{2K_{31}^2}{g_{31}} \frac{[Z_{us} + Z_{ls}]}{2} \frac{d^2W}{dx^2} \tag{1}$$

in which x is the axial coordinate, W is the transverse displacement, L is the length of the beam, g_{31} is the piezoelectric stress constant, W_{31} is the electromechanical coupling factor, w_s is the width, Z_{us} and Z_{ls} are the distances measured from the upper and lower surfaces of the sth sensor, respectively. $R_s(x)$ is the electrode profile function. Shapes for $R_1(x)$ and $R_2(x)$ are shown in Figs. 1(c) and 1(d). The subscript s ranging from 1 to S, denotes the sth sensor, and $2S$ is the total number of sensors laminated in a beam-plate. Since the voltage V_s equals the charge divided by capacitance C_v, hence

$$V_s = \frac{2K_{31}^2}{C_v g_{31}} \int_0^1 \frac{[Z_{us} + Z_{ls}]}{2} \frac{d^2\overline{W}}{L d\overline{x}^2} w_s R_s(x) d\overline{x} \tag{2}$$

where $\overline{W} = W/L$ and $\overline{x} = x/L$.

2.2 Beams and Actuators

The relations of shear force Q, bending moment M and displacement W accounting for the pizeoelectric effect of actuators are

$$M = D\frac{d^2W}{dx^2} + U \ , \qquad Q = -\frac{dM}{dx} = -\frac{d}{dx}(D\frac{d^2W}{dx^2} + U) \tag{3}$$

in which

$$U = \frac{d^2}{dx^2}[\sum_{a=1}^{n} D_a R_a(x) w_a d_{31} \frac{V_a}{t_a}], \qquad D_a = E_a w_a [Z_{ua}^2 - Z_{la}^2] \tag{4}$$

D is the effective bending stiffness; E_a, t_a and w_a are respectively Young's modulus, thickness and width for each lamina of actuators; V_a is the voltage applied on the ath actuator lamina, d_{31} and $R_a(x)$ are the piezoelectric strain constant and electrode profile function of the ath actuator, respectively. Z_{ua} and Z_{la} are the distances measured from the upper and lower surfaces of the ath actuator as shown in Fig. 1(b), respectively. The subscript a ranging from 1 to n, denotes the ath actuator, and $2n$ is the total number of sensors laminated in a beam-plate.

Some detailed discussions on fundamental equations can be found in the study by Wang and Lin (1995). The equilibrium equation of a beam integrated with laminated sensors and actuators becomes

$$\frac{Dd^4\overline{W}}{L^3 d\overline{x}^4} = f(\overline{x}) - \frac{d^2}{L^2 d\overline{x}^2}(\sum_{a=1}^{n} D_a R_a(\overline{x}) w_a d_{31} \frac{V_a}{t_a}) \tag{5}$$

in which $\overline{W} = W/L$ and $\overline{x} = x/L$. The general elastically restrained boundary conditions are

$$\text{at } \overline{x} = 0: \; k_0 L\overline{W} = -D\frac{d^3\overline{W}}{L^2 d\overline{x}^3}, \quad K_0 \frac{d\overline{W}}{d\overline{x}} = \frac{D}{L}B_0 + \sum_{a=1}^{n} D_a R_a(0) w_a d_{31} \frac{V_a}{t_a} \tag{6}$$

$$\text{at } \overline{x} = 1: \; k_1 L\overline{W} = -D\frac{d^3\overline{W}}{L^2 d\overline{x}^3}, \quad K_1 \frac{d\overline{W}}{d\overline{x}} = \frac{D}{L}B_1 + \sum_{a=1}^{n} D_a R_a(1) w_a d_{31} \frac{V_a}{t_a} \tag{7}$$

in which k_0 and k_1 are linear spring constants, K_0 and K_1 are rotational spring constants, and B_0 and B_1 are curvatures at $\overline{x} = 0$ and 1, respectively.

3. METHOD OF ANALYSIS

The basic concept behind the scheme proposed in the study is to control arbitrarily loaded and generally supported beam-plates to desired shapes, through a control algorithm and optimization process. Since any function can be expressed in terms of Fourier series, by trimming the shapes of actuators laminated in the beams in the form of sine functions, one may use such scheme to suppress undesirable Fourier components of deformation. As a result, Fourier sine series is used as the general solution in the analysis which would be adaptable to the proposed control scheme. Knowing that the conventional Fourier sine series solution is only adequate for simply supported beams, Fourier sine series in conjunction with Stokes transformation is used to account for other types of supporting conditions. Series expressions for the deflection function $W(\overline{x})$ and its derivates for the beam analysis are as follows:

$$\overline{W}(\overline{x}) = \sum_{m=1}^{\infty} A_m \sin \alpha_m \overline{x}, \qquad 0 < \overline{x} < 1 \tag{8}$$

where $\alpha_m = m\pi$ with $\overline{W}(x) = \overline{W}_0$ and $\overline{W}_1$ at $\overline{x} = 0$ and 1, respectively. The Fourier series representations of derivatives of $\overline{W}$ are

$$\frac{d\overline{W}}{d\overline{x}} = \overline{W}_1 - \overline{W}_0 + \sum_{m=1}^{\infty}(a_m + \alpha_m A_m)\cos\alpha_m\overline{x}, \qquad 0 \le \overline{x} \le 1 \tag{9}$$

where $a_m = 2[(-1)^m W_1 - W_0]$.

$$\frac{d^2\overline{W}}{d\overline{x}^2} = -\sum_{m=1}^{\infty}\alpha_m(a_m + \alpha_m A_m)\sin\alpha_m\overline{x} \qquad 0 < \overline{x} < 1 \tag{10}$$

with $d^2\overline{W}/d\overline{x}^2 = B_0$ and B_1 at $\overline{x} = 0$ and 1, respectively.

$$\frac{d^3\overline{W}}{d\overline{x}^3} = B_1 - B_0 + \sum_{m=1}^{\infty}\{2[B_1(-1)^m - B_0] - \alpha_m^2(a_m + \alpha_m A_m)\}\cos\alpha_m\overline{x}, \quad 0 \le \overline{x} \le 1 \tag{11}$$

$$\frac{d^4\overline{W}}{d\overline{x}^4} = -\sum_{m=1}^{\infty}\alpha_m\{2[B_1(-1)^m - B_0] - \alpha_m^2(a_m + \alpha_m A_m)\}\sin\alpha_m\overline{x}, \qquad 0 < \overline{x} < 1 \tag{12}$$

Such Fourier series solutions in conjunction with Stokes transformation have been used successfully in the paper by Wang and Lin (1996).

The exact solution of a simple case is presented here. The solution will be used for comparing results in later illustrative examples to verify the computer program made on the basis of the scheme proposed in the study. A simply supported beam bonded symmetrically by one pair of sensor ($s = 1$) and actuator ($a = 1$) having a uniform electrode profile, $R_a(x) = 1$, is considered. For a beam under a uniformly distributed load $f(\overline{x}) = q_0$, equation (5) becomes

$$\frac{Dd^4\overline{W}}{L^3 d\overline{x}^4} = q_0 \tag{13}$$

The general solution of equation (13) is

$$\overline{W}(\overline{x}) = \frac{q_0 L^3}{24D}\overline{x}^4 + \frac{C_3}{6}\overline{x}^3 + \frac{C_2}{2}\overline{x}^2 + C_1\overline{x} + C_0 \tag{14}$$

Using equations (6) and (7) for simply supported boundary conditions, the constants C_0, C_1, C_2 and C_3 are determined by the following equations

$$\begin{aligned} -\frac{D_1^* L}{D}V_1 &= C_2, \qquad & -\frac{D_1^* L}{D}V_1 &= \frac{q_0 L^3}{2D} + C_3 + C_2 \\ 0 &= C_0, \qquad & 0 &= \frac{q_0 L^3}{24D} + \frac{C_3}{6} + \frac{C_2}{2} + C_1 + C_0 \end{aligned} \tag{15}$$

in which V_1 is the voltage applied on the actuator and $D_1^* = 2D_1 w_a d_{31}/t_a$. Substituting the four obtained constants C_0, C_1, C_2 and C_3 into equation (14), the deflection can be written as

$$\overline{W}(\overline{x}) = \hat{W}_0\overline{x}[\overline{x}^3 - 2\overline{x}^2 - 12e^*\overline{x} + (1+12e^*)] \ , \ \hat{W}_0 = \frac{q_0L^3}{24D}, \quad e^* = \frac{D^*V}{q_0L^2} \qquad (16)$$

When $e^* = 0$ the deflection is reduced to the familiar result,

$$\overline{W}(\overline{x}) = \hat{W}_0\overline{x}[\overline{x}^3 - 2\overline{x}^2 + 1] \qquad (17)$$

For the electrode profile function $R_s(x) = 1$, the voltage can be obtained from equation (2) as

$$V_s = C_s\hat{W}_0(-2 - 24e^*)\overline{w}_s, \qquad C_s = \frac{2K_{31}^2}{LC_v g_{31}}\frac{[Z_{us} + Z_{ls}]}{2}w_s \qquad (18)$$

For the sake of convenience in discussion, displacements at both ends of the beam-plate are considered to be zero. The sine series solution for beams having ends with rotational restraints in general can be obtained by the substitution of equations (12) into the governing equation (5) as follows:

$$A_m = \frac{2f_mL^3}{D\alpha_m^4} + \frac{2[B_1(-1)^m - B_0]}{\alpha_m^3} - \frac{D_a^*L}{D\alpha_a^2}\delta_{am}V_a \times IFI \qquad (19)$$

in which $D_a^* = 2D_aw_ad_{31}/t_a$, and

$$IFI = \begin{cases} 1 & \text{if } R_a(x) \text{ of actuator is a sine function.} \\ 0 & \text{if } R_a(x) \text{ of actuator is a constant.} \end{cases}$$

By substituting equation (8) into boundary conditions described in equations (6) and (7), we find

$$B_0 = \frac{LK_0}{D}\sum_{m=1}^{\infty}\alpha_mA_m - \sum_{a=1}^{n}\frac{D_aL}{D}w_ad_{31}\frac{V_a}{t_a} \times BFI \qquad (20)$$

$$B_1 = \frac{LK_1}{D}\sum_{m=1}^{\infty}\alpha_mA_m(-1)^m - \sum_{a=1}^{n}\frac{D_aL}{D}w_ad_{31}\frac{V_a}{t_a} \times BFI \qquad (21)$$

where $BFI = \begin{cases} 0 & \text{if } R_a(x) \text{ of actuator is sine function.} \\ 1 & \text{if } R_a(x) \text{ of actuator is a constant.} \end{cases}$

Substituting equation (19) into (20) and (21), one obtains

$$\begin{bmatrix} BC_{11} & BC_{12} \\ BC_{21} & BC_{22} \end{bmatrix}\begin{Bmatrix} B_0 \\ B_1 \end{Bmatrix} = \begin{bmatrix} BF_1 \\ BF_2 \end{bmatrix} \qquad (22)$$

in which

$$BC_{11} = 1 + \frac{K_0}{D}Q_{11}, \quad BC_{12} = -\frac{K_0}{D}Q_{12}, \quad BF_1 = \frac{K_0}{D}QF_1$$

$$BC_{21} = \frac{K_1}{D}Q_{21}, \quad BC_{22} = 1 - \frac{K_1}{D}Q_{22}, \quad BF_2 = \frac{K_1}{D}QF_2$$

and

$$Q_{11} = \sum_{m=1}^{\infty} \frac{2}{\alpha_m^2}, \quad Q_{12} = \sum_{m=1}^{\infty} (-1)^m \frac{2}{\alpha_m^2}, \quad Q_{21} = \sum_{m=1}^{\infty} (-1)^m \frac{2}{\alpha_m^2}, \quad Q_{22} = \sum_{m=1}^{\infty} (-1)^m \frac{2}{\alpha_m^2}$$

$$QF_1 = \sum_{m=1}^{\infty} \{\frac{2 f_m L^3}{D\alpha_m^3} + \frac{L D_a^* V_a \delta_{ma}}{D\alpha_a} \times IFI\} - \sum_{a=1}^{n} \frac{D_a L}{K_0} w_a d_{31} \frac{V_a}{t_a} \times BFI$$

$$QF_2 = \sum_{m=1}^{\infty} \{(-1)^m \frac{2 f_m L^3}{D\alpha_m^3} + (-1)^m \frac{L D_a^* V_a \delta_{ma}}{D\alpha_a} \times IFI\} - \sum_{a=1}^{n} \frac{D_a L}{K_0} w_a d_{31} \frac{V_a}{t_a} \times BFI$$

Solving equation (27), B_0 and B_1 are obtained as follows:

$$B_0 = \begin{vmatrix} BF_1 & BC_{12} \\ BF_2 & BC_{22} \end{vmatrix} / \det, \quad B_1 = \begin{vmatrix} BC_{11} & BF_1 \\ BC_{21} & BF_2 \end{vmatrix} / \det, \quad \det = \begin{vmatrix} BC_{11} & BC_{12} \\ BC_{21} & BC_{22} \end{vmatrix} \tag{23}$$

If both ends are clamped, $B_0 = NB_0 / QB$, $B_1 = NB_1 / QB$, where $NB_0 = QF_2 Q_{12} - QF_1 Q_{22}$, $NB_1 = QF_2 Q_{11} - QF_1 Q_{21}$, $QB = Q_{12} Q_{21} - Q_{11} Q_{22}$.For the simply supported case, $B_0 = 0, B_1 = 0$.For the simply supported-clamped case, $B_0 = 0$, $B_1 = -QF_2 / Q_{22}$.

By using equations (8) and (19), the displacement $\overline{W}(\overline{x})$ is expressed as

$$\overline{W}(\overline{x}) = \sum_{m=1}^{\infty} \{\frac{2 f_m L^3}{D\alpha_m^4} + \frac{2[B_1(-1)^m - B_0]}{\alpha_m^3} - \frac{D_a^* L}{D\alpha_a^2} \delta_{am} V_a\} \sin(\alpha_m \overline{x}) \tag{24}$$

Substituting equations (10) and (19) into equation (2), one obtains

$$V_s = -2C_S \sum_{m=1}^{\infty} \{\frac{f_m L^3}{D\alpha_m^2} + \frac{[B_1(-1)^m - B_0]}{\alpha_m} - \frac{D_a^* L}{2D} \delta_{am} V_a\} \int_0^1 \sin(m\pi\overline{x}) R_s(\overline{x}) d\overline{x} \tag{25}$$

For $R_s(x) = 1$ the voltage of the sensor is

$$V_S = -4C_S \sum_{m=1,3,5,\ldots}^{\infty} \{\frac{f_m L^3}{D\alpha_m^2} + \frac{[B_1(-1)^m - B_0]}{\alpha_m} - \frac{D_a^* L}{2D} \delta_{am} V_a\} \tag{26}$$

For $R_s(x) = \sin(k\pi\overline{x})$ the voltage of the sensor is

$$V_S = -2C_S \{\frac{f_m L^3}{D\alpha_m^2} + \frac{[B_1(-1)^m - B_0]}{\alpha_m} - \frac{D_a^* L}{2D} \delta_{am} V_a\} \tag{27}$$

4. SHAPE CONTROL

While the design objective of shape control is to achieve the goal of maintaining a structure in a desired shape with high precision under various loading conditions. The conventional scheme is to use sensors and actuators of uniform width along the beam. The effect of such configuration is quite restrictive as the actuation produces a specific form of deformation to compensate deformation of structures under load. As the deflection of beams may be represented by a Fourier sine series, and is often governed by a few dominant components; sensors and actuators trimmed in sinusoidal forms are used in the present study. The piezoelectric material of polyvinylidene fluoride (PVDF) is used for sensors and actuators. A sensor having electrode profile represented by $R_s(\overline{x}) = \sin(s\pi\overline{x})$, called the sth mode sensor, can sense the contribution of the $\sin(s\pi\overline{x})$ term of the deflection. An actuator having electrode profile represented by $R_a(\overline{x}) = \sin(a\pi\overline{x})$, called the ath mode actuator is used to control the contribution of the $\sin(a\pi\bar{x})$ term of the deflection. When $s = a$, the sensor and actuator are of the same shape. They are linked together as one pair. The scheme of the shape control is to use each mode sensor to guide the actuation voltage of actuator of the same pair in order to reduce the mode sensor voltage, thus to reduce deformation. By activating an adequate number of pairs of senors and actuators, the shape of a structure is considered to be well controlled when each sensor has provided an optimum guidance on the voltage for the actuator of the same pair. The present configuration is much more flexible and effective for controlling beam shapes without loading information. Clearly, the main design objective is to obtain optimal voltage of actuators to minimize the difference of the desired and actual deflections.

4.1 Optimization of the Control Voltage

The control procedure is established to bring the deformed structure close to the original shape by applying optimal voltage of the actuator. Each sensor output is used to construct an objective function O as the measure of deformation, and the electric voltage parameter e^* of corresponding actuator is used as a design variable. It can be stated as a mathematical model as

$$\min O(e^*) \qquad \text{subject to} \qquad G_l \le G(e^*) \le G_u \tag{28}$$

The objective function O can be chosen as $|V_s|^2$, and G_l and G_u are the lower and upper limits of the voltage parameter of actuators.

4.2 Optimization Procedure

The step by step procedure for arriving at the optimization of the problem is :

1. Choose the initial design variables.
2. Calculate gradient information by the central difference method.
3. Obtain the search direction vector and step size when a constraint is active, a series of correction steps need to be executed to return the design point into the feasible region.
4. Check the convergence criteria.
5. Repeat steps 2 to 4 until the convergence criteria are satisfied.

In order to search the optimal value of the design variables, the gradient projection method is employed. The method is a feasible direction method in which gradient information are necessary to determine the search direction. Here, the gradients of the objective function $O(e^*)$ and constraint functions $G(e^*)$ are obtained by the central difference method. If the design point is inside the feasible region, then the steepest direction $-\nabla O(e^*)$ is used as the search direction. The central-difference of the derivative of the objective function can be expressed as

$$\frac{dO(e^*)}{de^*} = \lim_{\Delta e^* \to 0} \frac{O(e^* + \Delta e^*) - O(e^* - \Delta e^*)}{2\Delta e^*} \tag{29}$$

While the design point reaches constraint boundaries, the search direction can be computed by using the steepest descent direction to project into the linear manifold of the supporting hyperplanes of the active constraints,

$$\bar{d}_i = -[PR]\nabla O(e^*) \tag{30}$$

where $[PR] = [I] - [\nabla G]^T([\nabla G][\nabla G]^T)^{-1}[\nabla G]$ is a gradient matrix and ∇G is a projection matrix of active constraints. The new design point e_{i+1}^* can be obtained from the last design point e_i^* by the relation,

$$e_{i+1}^* = e_i^* + \alpha \bar{d}_i \tag{31}$$

where the step size α specification is arbitrary if the objective function value decreases.

The new design point which moves along this direction may enter the infeasible region. For the purpose of turning back the design point from infeasible region to feasible region, a series of correction steps need to be executed to change the design point close to the active constraints. Same procedure must be repeated until the convergence criteria are satisfied, and the optimum point is then obtained.

4.3 Computer Program

A computer program which combines the Fourier series analysis and optimal design algorithm for calculating the optimal voltage parameter of actuators is developed. The flow chart is shown in Fig. 1(e). This is a close loop control by using the sensor output as a feedback signal. The gradient projection subroutine calls the Fourier series analysis for deflection calculations at each design step. There is a do-loop controlled by the convergence criteria. The optimal voltage parameters of actuators are obtained when the criteria are satisfied.

5. NUMERICAL RESULTS AND DISCUSSIONS

Some numerical results are obtained for illustrative purposes. The actual beam deflections are obtained by using 100 sine function terms. The loading conditions considered in examples are uniformly distributed, piece-wise uniformly distributed, and concentrated loads. The optimal values of the actuators are obtained by the procedure described in section 4.2.

Case 1: Uniformly Distributed Load

The loading is uniformly distributed from $\bar{x} = 0.0$ to 1 as $f(x) = q_0$. For the purpose of checking the validity of series solution and the computer program for the shape control problem, beam deflections are calculated by the exact solution and the sine series solution. The electrode profiles of sensor and actuator are uniform. The controlled deflections of beam under different actuator voltage parameter e^* are shown in Fig. 2. The objective function varying with respect to the actuator voltage parameter obtained by the two solutions are shown in Fig 3. These results show the validity of the sine series solution to be used for the analysis of the shape control problem.

Four different beams are considered to show their shape control efficiencies. All electrode profiles have the same dimensions with two sensor laminae and two actuator laminae. They are uniform electrodes for sensors and actuators, a sine shape for sensor and a uniform actuator, a uniform sensor and a sine shape for actuator, and a sine shape for both. The objective functions of these four configurations varying with respect to the actuator voltage parameter are shown in Fig. 4. It should be noted that the optimal values of these configuration to minimize the objective function are very close. But the controlled deflections as shown in Fig. 5

are different. The controlled deflection of the last two configurations are closest to the original undeformed shape with $W(\bar{x}) = 0$. The shape control efficiency of the first configuration is the lowest. The second one is in the middle of all. These results are obtained on the basis that only one pair of sine shaped modal senors and actuators are active.

In Fig. 6, it shows that one pair of (1) mode, two pairs of (1,3) and three pairs of (1,3,5) modes of sine shaped sensors and actuators are active. Numbers in the parenthese denote shapes of Fourier components involved. The shape control efficiency is improved by using more pairs of sensors and actuators. From the results using two pairs of (1,3) and three pairs (1,3,5) sine shaped sensors /actuators, it may be concluded that the use of three pairs of sensors/actuators is sufficiently adequate to control the beam shape.

Case 2: Piece-Wise Uniformly Distributed Load

The loading is distributed from $\bar{x} = 0.25$ to 0.75 as , $f(x) = q_0$ and zero for the remaining part of the beam. The objective functions varing with respect to the actuator voltage parameter are shown in Fig. 7. The magnitude of the objective function of a uniform sensor is larger than the sine shaped sensor. In Fig. 8, it shows that the best result for the shape control efficiency is the use of two pairs of (1,3) sine shaped sensors and actuators.

The loading is distributed from $\bar{x} = 1/6$ to $1/3$ as $f(x) = q_0$, and zero for the remaining part of the beam. The objective function is shown in Fig. 9 and the deflection of beam is shown in Fig. 10. The use of two pairs of (1,2) sine shaped sensors and actuators can control the beam shape close to the original shape.

Case 3: Concentrated Load

The loading function $f(\bar{x}) = P_0\delta(\bar{x} - \bar{x}_0)$ for $\bar{x}_0 = 0.5$ and $1/3$ on a simply supported beam are considered. The objective functions varying with respect to the voltage parameter are displayed in Figs. 11 and 13, and the controlled deflections of beam are shown in Figs. 12 and 14. The load $f(\bar{x}) = P_0\delta(\bar{x} - \bar{x}_0)$ for $\bar{x}_0 = 0.5$, and $1/6$ on a clamped-clamped beam are considered next. The controlled deflections of the clamped-clamped beams are shown in Figs. 15 and 16. Fig. 17 shows the deflection of a clamped-clamped beam under two concentrated loads, $f(\bar{x}) = 0.75P_0\delta(\bar{x} - \frac{1}{3}) - P_0\delta(\bar{x} - \frac{2}{3})$. The controlled deflection is nearly zero when four pairs of sensors and actuators are used. It may be concluded that the use of sine shaped sensors and actuators is a suitable choice for shape control of beams with various boundary conditions when concentrated loads are applied at various locations.

6. CONCLUSIONS

A simulation procedure by using piezoelectric sensors and actuators to control the shape of beam has been proposed. The electrodes of these sensors and actuators are trimmed to shapes of various sine functions. The beam deflections are influenced by the loading and boundary conditions. By using several pairs of sensor-actuators, deflections of beams with various supporting conditions can be controlled for various loading conditions. It can be concluded that the sine shaped profile used for the electrode of sensors and actuators are feasible choices for shape feedback control. Four pairs of sine shaped sensors and actuators may be adequate to bring the deformed shape closely back to the original shape of a beam under various loading conditions. The controlled results are excellent for the piece-wise uniformly distributed loads and concentrate loads according to the numerical results. The proposed scheme should have practical applications.

REFERENCES

1. Crawley, E.F., D.J. Warkentin, and K. B. Lazarus. Jan 1988. "Feasibility Analysis of Piezoelectric Devices," Space Systems Laboratory, Massachusetts Institute of Technology, Cambridge, MA, Rept. MIT-SSL No. 5-88.
2. Varadajan, S. and K. Chandrashekara and S. Agarwal. 1996. "Adaptive Shape Control of Laminated Structures," Forum , Salt Lake City, UT, pp. 197-206.
3. Haftka, R. T. and H.M. Adelman. 1985. "An Analytical Investigation of Static Shape Control of Large Space Structures by Applied Temperature," *AIAA Journal*, Vol. 23, pp. 450-457.
4. Austin, F. and M.J. Rossi, W. Van Nostrad, G. Knowles and A. Jameson. 1994. " Static Shape Control of Adaptive Wings " *AIAA Journal*, Vol. 32, pp. 1895-1901.
5. Koconis, D. B. and L.P. Kollar. 1994. "Shape Control of Composite Plates and Shells with Embedded Actuators. II. Desired Shape Specified," *Journal of Composite Materials*, Vol. 28(3), pp. 262-285.
6. Agrawal, S. K. and D. Tong. 1994. "Modeling and Shape Control of Piezoelectric Actuator Embedded Elastic Plates," *Journal of Intelligent Material Systems and Structures*, Vol. 5, pp. 514-521.
7. Dosch, J. J. and D.J. Inman. 1992. "A Self-Sensing Piezoelectric Actuator for Collocated Control, " *Journal of Intelligent Material Systems and Structures*, Vol. 3, pp. 166-184.
8. Wang, J. T. -S. and C.-C. Lin. 1995. "Vibration of Beam-Plates having Multiple Delaminations," AIAA paper no. 95-1502-CP, Proceeding of the AIAA/ASME/ASCE/AHS /ASC 36th SDM Conference and AIAA/ASME Adaptive Structures Forum, New Orleans, LA, April 10-13.
9. Wang, J. T. -S. and C.-C. Lin,. 1996. "Dynamic Analysis of Generally Supported Beams using Fourier Series, " *Journal of Sound and Vibration*, Vol. 196(3), pp. 285-293.

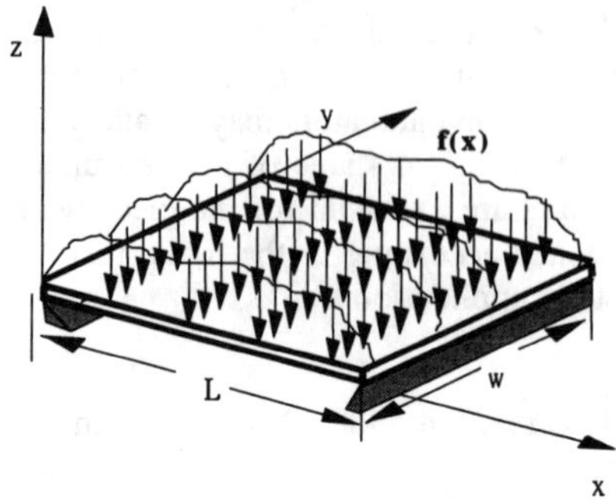

Fig.1(a) Laminated beam-plate and coordinate system.

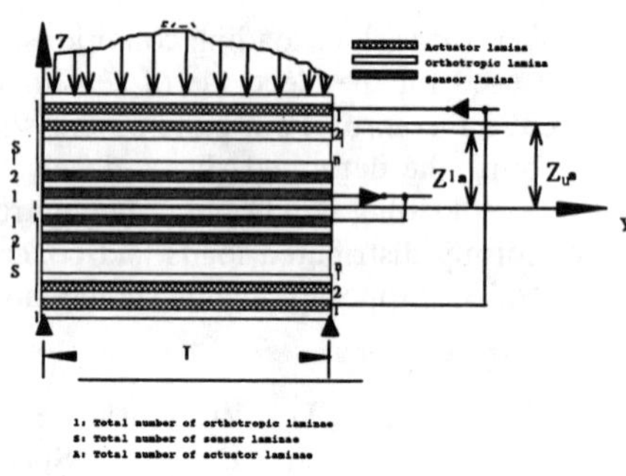

Fig.1(b) The configuraion of a laminated beam-plate with modal sensors and actuators.

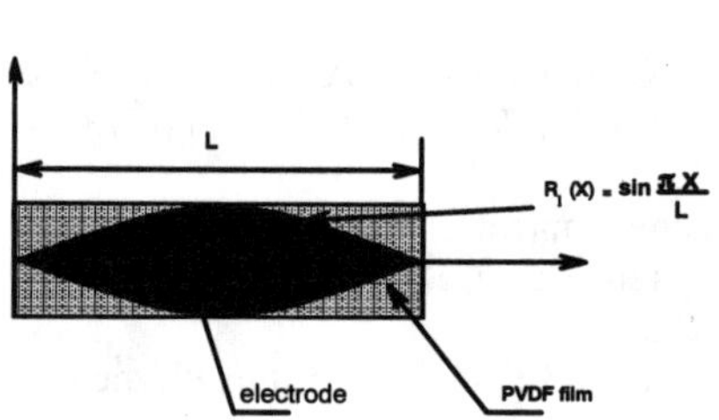

Fig. 1 (c)The electroe of film used as a PVDFde profil sensor/actuator on the simply supported beam to control mode 1.

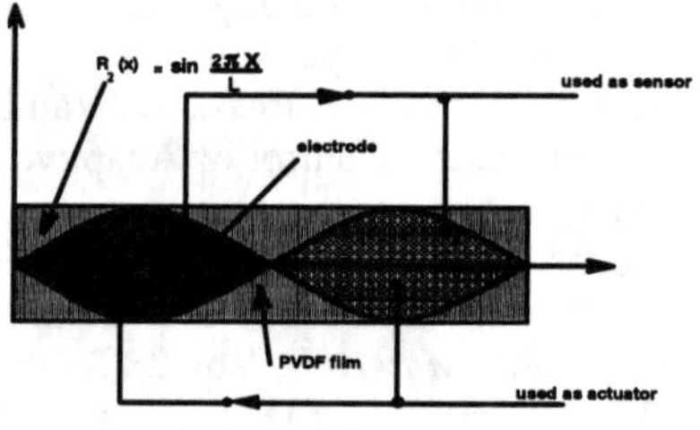

Fig. 1 (d)The electrode profile of PVDF sensor/actuators for the simply supported beam to control mode 2.

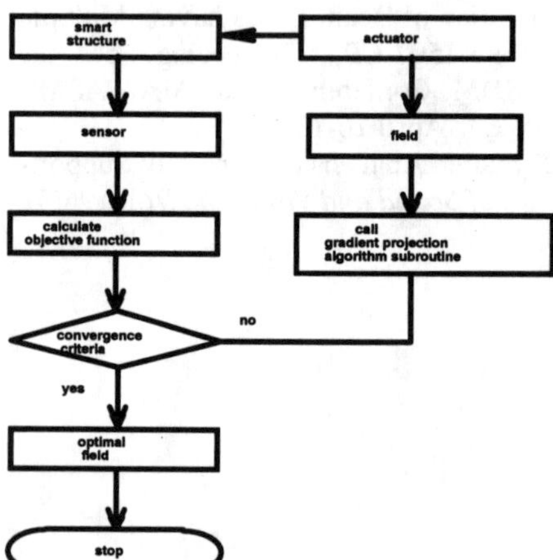

Fig. 1(e) Flowchart for adaptive shape control procedure.

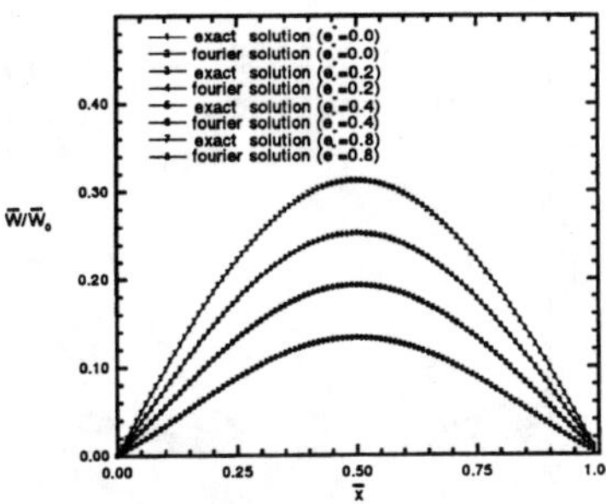

Fig.2 Deflections of the simply supported beam with various actuator voltage parameters.(for case1:uniformly distributed load)

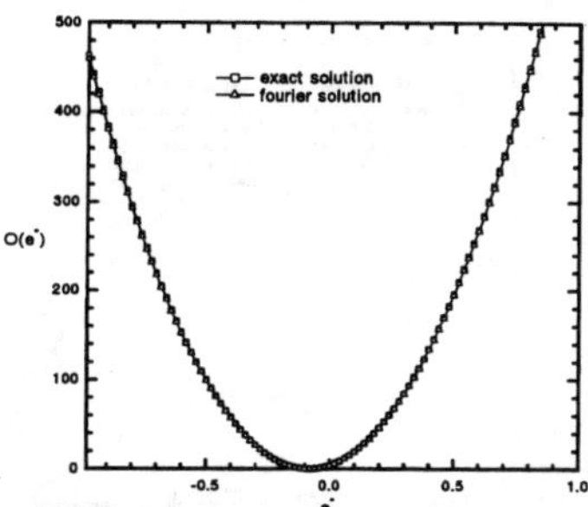

Fig.3 Objective function vs. voltage parameters of uniform actuator. (for case1:uniformly distributed load)

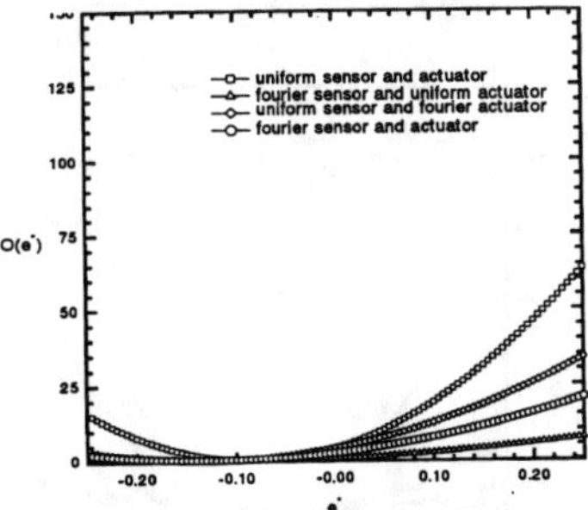

Fig.4 Objective functions vs. voltage parameters of actuator.(for case1 with different shapes of sensors and actuators)

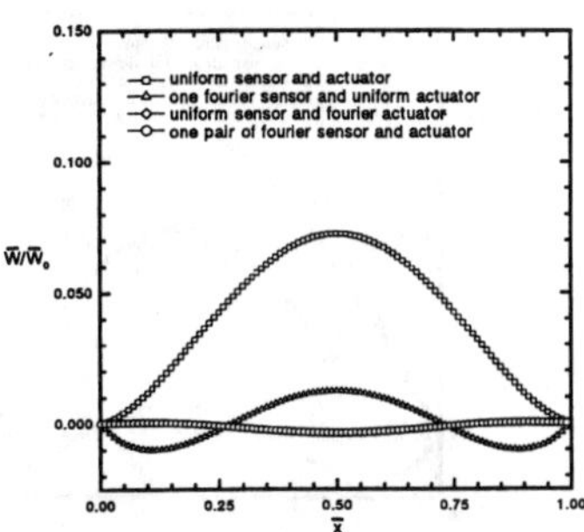

Fig.5 Deflection of the simply supported beam controlled with optimal actuator voltage parameters. (for case1 with different shapes of sensors and actuators)

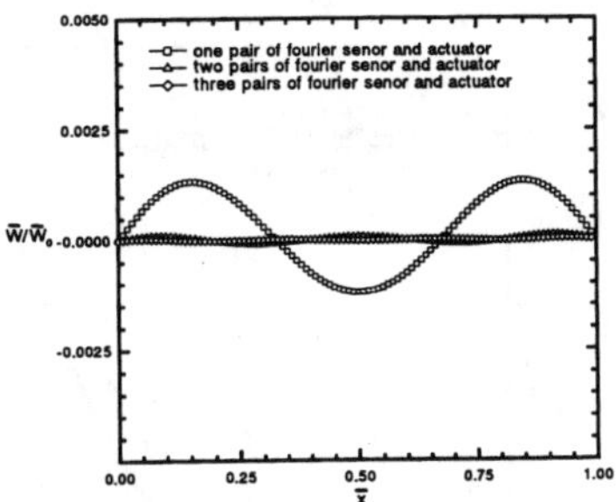

Fig.6 Deflection of the simply supported beam controlled with optimal actuator voltage parameters.(for case1 with different pairs of sensors and actuators)

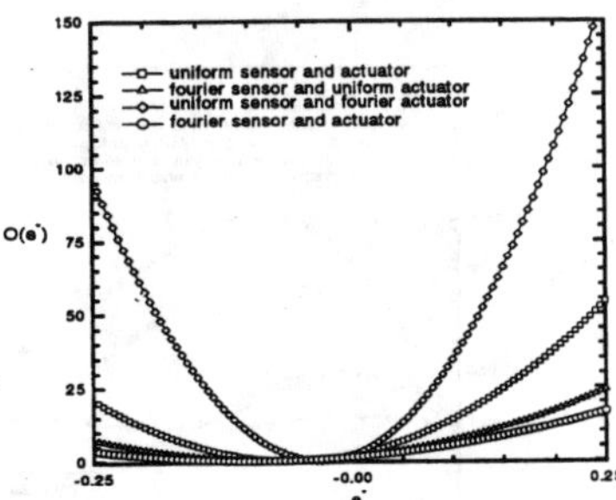

Fig.7 Objective functions vs. voltage parameters of actuator.($f(\bar{x}) = q_0$ for $0.25 < \bar{x} < 0.75$ and zero elsewhere)

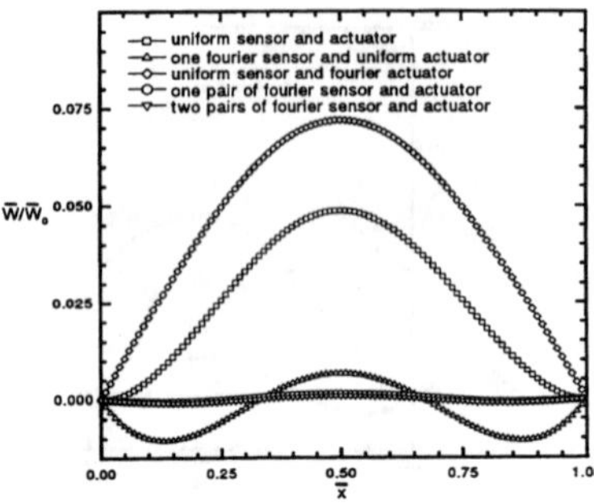

Fig.8 Deflection of the simply supported beam controlled with optimal actuator voltage parameters.
($f(\bar{x}) = q_0$ for $0.25 < \bar{x} < 0.75$ and zero elsewhere)

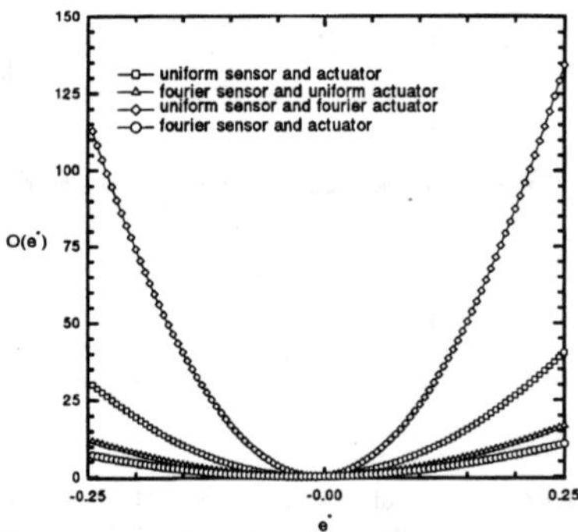

Fig.9 Objective functions vs. voltage parameters of actuator.
($f(\bar{x}) = q_0$ for $1/6 < \bar{x} < 1/3$ and zero elsewhere)

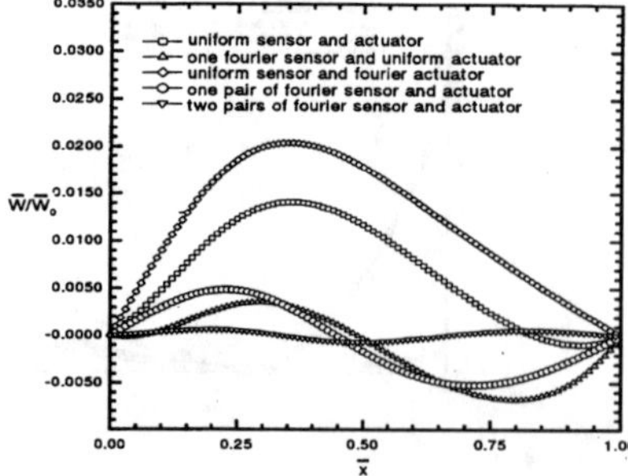

Fig.10 Deflection of the simply supported beam controlled with optimal actuator voltage parameters.
($f(\bar{x}) = q_0$ for $1/6 < \bar{x} < 1/3$ and zero elsewhere)

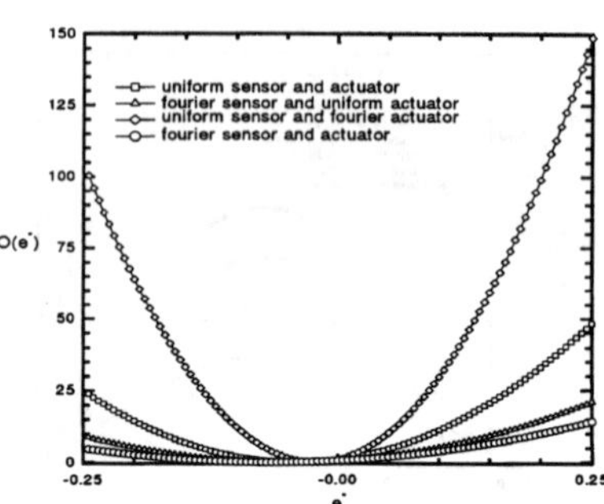

Fig.11 Objective functions vs. voltage parameters of actuator.
($f(\bar{x}) = P_0\delta(\bar{x} - 0.5)$)

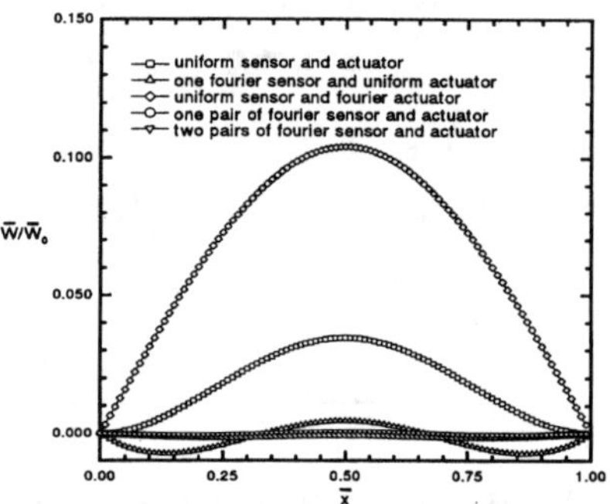

Fig.12 Deflection of the simply supported beam controlled with optimal actuator voltage parameters.
($f(\bar{x}) = P_0\delta(\bar{x} - 0.5)$)

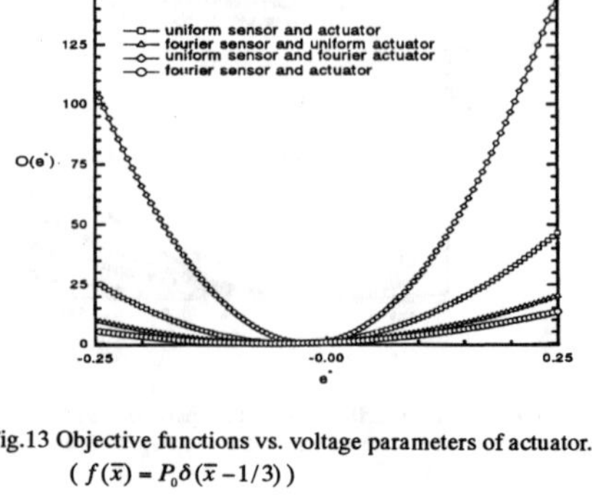

Fig.13 Objective functions vs. voltage parameters of actuator.
($f(\bar{x}) = P_0\delta(\bar{x} - 1/3)$)

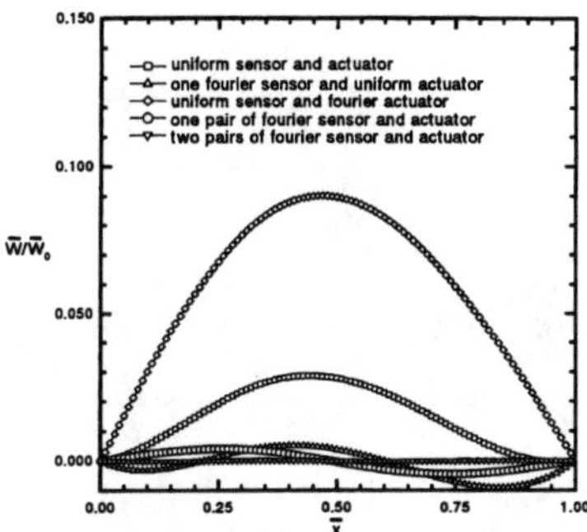

Fig.14 Deflection of the simply supported beam controlled with optimal actuator voltage parameters.
$(f(\bar{x}) = P_0\delta(\bar{x} - 1/3))$

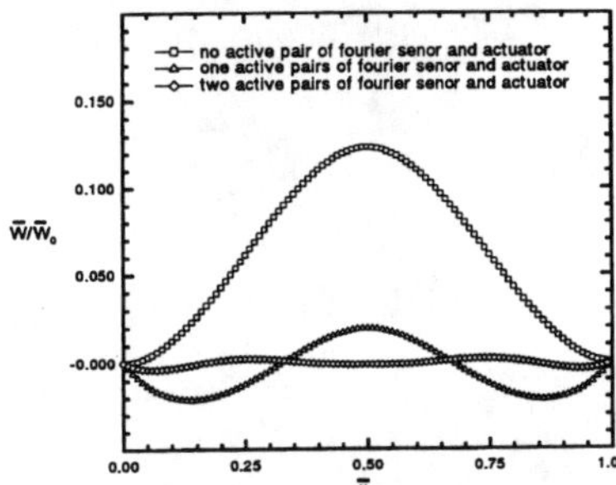

Fig.15 Deflection of the clamped-clamped beam controlled with optimal actuator voltage parameters.
$(f(\bar{x}) = P_0\delta(\bar{x} - 0.5))$

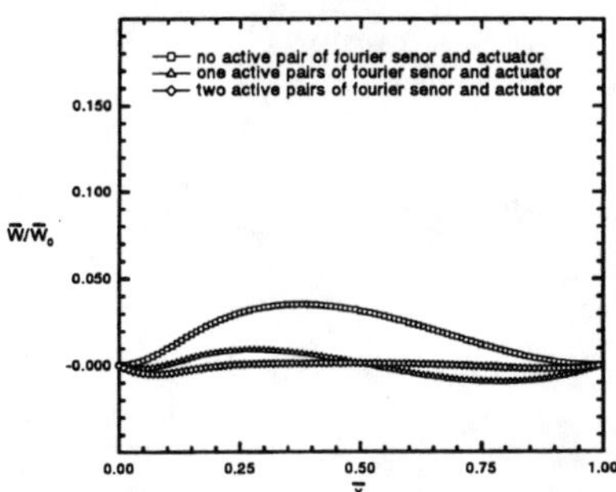

Fig.16 Deflection of the clamped-clamped beam controlled with optimal actuator voltage parameters.
$(f(\bar{x}) = P_0\delta(\bar{x} - 1/6))$

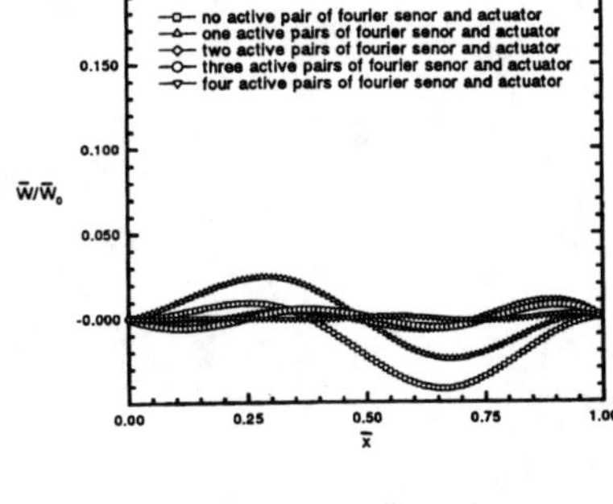

Fig.17 Deflection of the clamped-clamped beam controlled with optimal actuator voltage parameters.
$(f(\bar{x}) = 0.75P_0\delta(\bar{x} - 1/3) - P_0\delta(\bar{x} - 2/3))$

SESSION 12

MODELING AND DIAGNOSTIC METHODS IV

Systems Fault Detection, Localization and Assessment by the Use of Adjusted Models

H. G. NATKE

ABSTRACT

Systems faults can be essential for the systems' life, and for economic serviceability. The application of systems engineering, taking into account the complete life cycles of the systems, is thus important. Systems monitoring is the first step in fault detection. Measured states of the system at various life times inform us about the evolution of the system under monitoring. Resulting mathematical models adjusted to the system states at required life times are the optimum knowledge basis about the system, if the adjusted models are verified, validated and usable. The tool for model adjustment is system identification, which uses the prior information from system analysis combined with measured data.

The adjusted mathematical models will be used for fault detection, localization, and for state condition assessment of the system. Additionally, they serve for cause finding, for studies of remedies, and for trend prediction.

However, in addition to the above-mentioned procedure and the foregoing system analysis, significant modifications in the state condition first have to be detected by monitoring. This is done by the use of symptoms as sensitive quantities with respect to the expected modifications (faults) or with respect to the required performance of the system in operation/service. In the conclusions some new ideas are mentioned in this context.

All these statements are summarized in the formulation of holistic dynamics as a part of systems engineering.

INTRODUCTION

Damage detection, localization, characterization, assessment and trend prediction are parts of health condition investigations to prolong the life span of the systems and to prevent catastrophic failures. Measured data of the existing system contain maximum information about the system under investigation, if

H.G. Natke, Curt-Risch-Institute (CRI), University of Hannover, Appelstraße 9A, D-30167 Hannover, Germany, fax (+49.511)7622236, e.mail: Prof.Natke@mbox.cri.uni-hannover.de

they are suitable - whatever this means - and if the measurement errors are taken into consideration. Additionally, prior knowledge from system analysis with its uncertainty completes the information about the system. If this is taken life time-dependent, which means including the system's evolution, then the resulting model permits us to describe the holistic dynamics of the system (Natke and Cempel, 1995a). Consequently, such a model enables us to detect modifications of the current state from previous states, to locate faults, to investigate the severity of the faults, to find out causes of the damage by sensitivity analysis and simulations, and to assess the damage with decision-making based on trend prediction and limit states, on restricted use or possible actions.

The paper first discusses the methodology of model-supported diagnosis. Then model adjustment during health monitoring of the structural system is discussed. The importance of an adjusted mathematical model is worked out. An example illustrates the procedure discussed. Finally, some recent research results regarding symptom-based investigations are mentioned.

METHODOLOGY OF MODEL-AIDED DIAGNOSIS

Figure 1 from (Natke and Cempel, 1997a) contains the general procedure of model-based diagnosis. Damage/fault here is defined as parameter modifications of the structured mathematical model at the recent state compared with that of the reference model. Weak point analysis and sensitivity studies for finding symptoms for monitoring have to be performed at the stage of theoretical system analysis. Additionally, possible faults are classified including their symptoms, discriminants, features and patterns. The result is a damage catalogue as a theoretical basis for weak point observation and inspection (observation is used as being synonymous with measurement). If an alert threshold is exceeded, the system is monitored by symptoms and/or measured states. The monitoring may be completed by additional detailed studies, or/and NDT etc. Then follow investigations to determine whether the detected modifications are significant. If the detected modifications in the monitored quantities are not significant, the procedure goes back to the weak point observation etc. Otherwise, the previous model $\mathcal{M}_{i-1}$ or reference model $\mathcal{M}_0$ has to be adjusted to the current state: $\mathcal{M}_i$. The adjustment has to take probably changed loadings (environment conditions) into account. It may be necessary for additional measurements to be performed. Significant modifications between $\mathcal{M}_{i-1}$ or $\mathcal{M}_0$ and $\mathcal{M}_i$ leads to diagnosis and assessment (see Section on Information Resulting from the Adjusted Models).

Adjustment is performed by the use of prior knowledge (the previously adjusted models) and by measured data at requested discrete life times $\theta_i, i = 0, 1, .., N$, with θ_N the breakdown time of the system.

MODEL ADJUSTMENT

The mathematical model used in practice is generally spatially discretized:

$$\mathcal{M}_i : \quad M_i\ddot{u}(t) + C_i\dot{u}(t) + K_i u(t) = p(t), \tag{1}$$

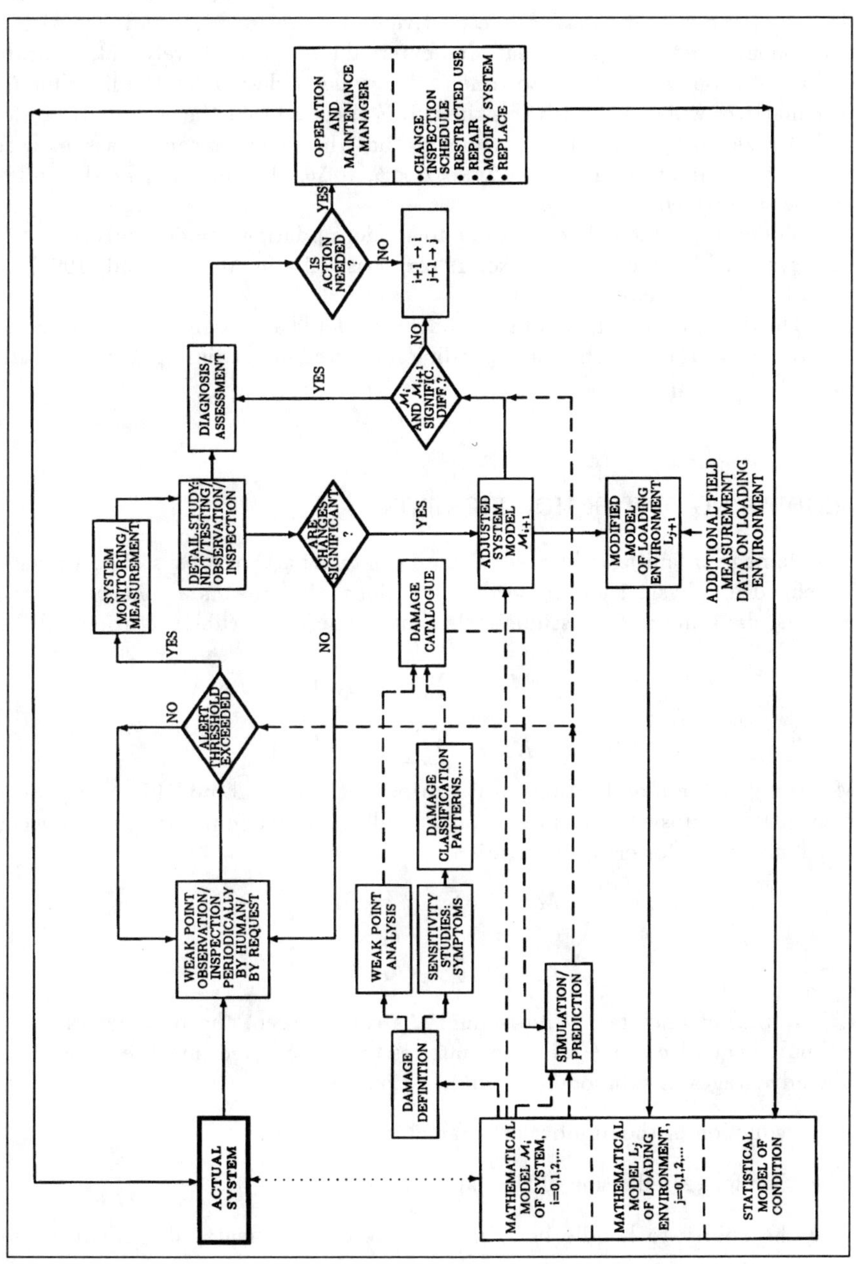

Figure 1: Methodology of th model-supported diagnosis

where M, C, K are quadratic matrices of order n representing the inertia, viscous damping, and stiffness matrices respectively. The vectors $u(t)$ and $p(t)$ with n components are the displacements and external forces respectively, dots indicate differentitation with respect to time t. The index i designates the life time θ_i. The matrices will evolve with this time. It is assumed that the system dynamics will change slowly with the life time, so that the matrices can be assumed as constant within an interval $\theta_i - \Delta\theta_i \leq \theta < \theta_i + \Delta\theta_i$. In the following this index will be suppressed.

Model adjustment is equivalent to model updating, model correction within system identification (Natke, 1992a; Friswell and Mottershead, 1995) as described briefly below.

The dynamic quantities of the model considered are assembled in the vector α_r, where $r = 1(1)M$. The corresponding measured quantities α_r^m together with α_r perform the residuals

$$v_r = \alpha_r - \alpha_r^m \epsilon \mathbb{C}. \tag{2}$$

MODELLING OF THE MODIFICATIONS

Damage or faults are located in components as parts of subsystems. Damage is defined as physical parameter modifications, that means as changes in the inertias, damping and/or stiffness elements. Therefore, substructuring will be applied:

$$\left.\begin{array}{lcl} M & = & \sum_{\sigma=1}^{S} M_\sigma, \\ C & = & \sum_{\rho=1}^{R} C_\rho, \\ K & = & \sum_{\iota=1}^{I} K_\iota. \end{array}\right\} \tag{3}$$

Modelling of the modifications of the prior mathematical model is now performed with the subsystems and the corresponding summand matrices introduced, which also serve for error localization:

$$\left.\begin{array}{lcl} M^c & = & \sum_{\sigma=1}^{S} a_{M_\sigma} M_\sigma, \\ C^c & = & \sum_{\rho=1}^{R} a_{C_\rho} B_\rho, \\ K^c & = & \sum_{\iota=1}^{I} a_{K_\iota} K_\iota. \end{array}\right\} \tag{4}$$

It is a kind of uncertainty modelling. As can be seen, the parameters to be estimated equalized to 1 turn the matrices to be corrected into the prior ones. The advantages of submodel-parametrization are the

- reduction of the number of parameters to be estimated
- therefore a reduction of the expenditure, and
- the ability to handle large models ($\rightarrow$ object-orientated modelling and programming).
- It permits a first localization of the observed modifications.

The choice of the submodels is done using

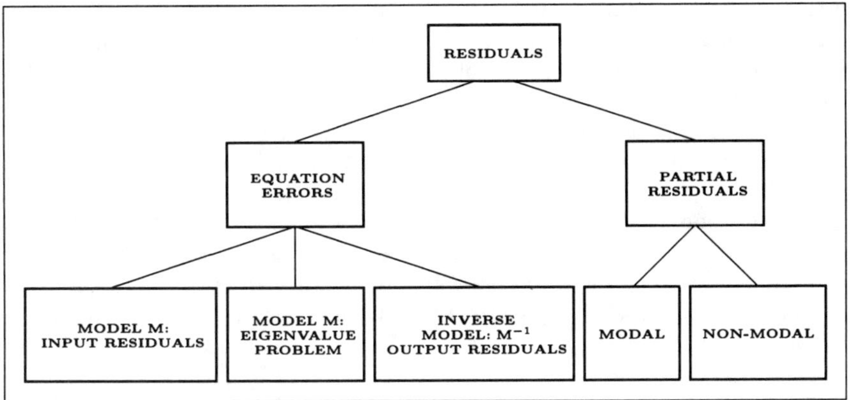

Figure 2: Classification of residuals

- prior knowledge of probable (sub-)model faults,
- prior knowledge from system analysis about the system behaviour (physical interpretation),
- the values of the residuals, and if necessary,
- by recalculations.

The choice of residuals depends on the problem. Possible residuals are shown in Fig. 2.

The residuals depend on the adjustment parameters to be estimated, and on the measuring uncertainties. If systematic errors in the measurements are avoided or computationally corrected, and if irregular measurement errors are modelled stochastically, estimators consequently have to be applied.

MODEL ADJUSTMENT PROCEDURE

Model adjustment is classified as an inverse problem (Natke, 1997) which is generally ill-conditioned for discrete models. Therefore regularization methods have to be applied (Natke, 1992b). The regularized problem takes into account additional information which changes the mathemathematical operator. Here the extended weighted least squares is emphasized. This assumes normal probability distributions for the random variables:

$$J(a) = v^*(a)G_v v(a) + (a-e)^T G_e(a-e). \tag{5}$$

The vector a contains the $J = S + R + I$ adjustment parameters, and e is the corresponding vector of prior knowledge, for example from the previous models. The weighting matrix G_v is the inverse covariance matrix of the measuring errors, and the weighting matrix G_e describes the confidence in the additional information e. As can be seen, the first term in Eq. (5) satisfies the classical weighted least squares (WLS) problem, and the penalty term modifies it; it controls the solution of the WLS.

The estimates $\hat{a}$ put into the parameter-dependent parameter matrices (4) result in the estimated parameter matrices $\hat{M} = M^c(\hat{a}_M), \hat{C} = C^c(\hat{a}_C), \hat{K} = K^c(\hat{a}_K)$.

In addition, of course, the covariances of the estimates have to be estimated.

Conditions and hints for choosing the submodels, for using order-reduced models, and for handling incomplete measurements, can be found by the reader in (Natke and Cempel, 1997a) and the mentioned references on system identification.

INFORMATION RESULTING FROM THE ADJUSTED MODELS

Parameter errors can be described by the Hessian matrix. The diagonal elements of the inverse Hessian matrix serve for the variances of the parameters. But one should take into account that the inverse Hessian matrix is in general only a lower bound for the covariance matrix (Rao-Cramér inequality).

Instead of this, one can estimate the variances of the estimates by taking into account several measurement sets (samples) (Natke and Cempel, 1997a).

Additional, confidence intervals can be determined (Bendat and Piersol, 1986).

On one hand the errors of the estimates have to be known in order to verify and validate the adjusted mathematical model (Natke, 1997), and on the other hand the comparison with the estimates of the reference model (at least their deviations from 1) will explain the significance of the modifications.

Damage detection can be done by subsystem modelling and with the estimates of the adjustment factors. Significant modifications of the adjustment factors estimated for the recent state from the previous ones of the reference model serve for damage detection and will indicate the location(s) of the fault(s).

If **fault localization** with pre-chosen subsystems is too coarse, the repetition of the adjustment procedure with successive halving of the submodels will lead to an enclosure of the damage. If necessary, additional measurements have to be performed, because the parameter domain will be enlarged.

Damage information about type and size can be obtained from models with local properties and by damage model building. Dynamic models are generally global models (energy-based), which can be used for damage detection and localization. However, if additional local information is needed, then, for example, finite element models with many degrees of freedom have to be established for the subsystems involved. These models have local properties if the number of degrees of freedom is chosen sufficiently high. Fault type and fault size can be obtained by additional information (based on prior knowledge by theoretical system analysis or by experience): this means by fault model building.

Assessment is now possible with the various adjusted mathematical models describing the state conditions of the system under investigation at the chosen life times. Consequently, one can use all the advantages of system analysis. Additionally, these mathematical models are verified, validated and (hopefully) usable[1]. This is the best known knowledge base concerning the system.

All the features of system analysis are available. Various loading conditions and various faults with their effects on dynamic and static performances can be simulated. Limit stresses, pre-given acoustic levels, limit displacements etc. lead to an assessment when compared with the simulated corresponding quantities by the use of the adjusted model. As already mentioned, one has to take into account the characteristics of the models used (global, local). However, if the system modification is restricted to local causes, then static subsystem modelling restricted to the submodel with the local modification is sufficient.

It is noted here that eigenquantities are generally seldom suitable for serving as symptoms. Eigenfrequencies are proportional to the energies of the respective degrees of freedom, therefore they are global quantities with some local properties (dependent on the number of degrees of freedom). The mode shapes with their local properties are not very sensitive with respect to local system modifications, and additionally, measured mode shapes are corrupted by noise. However, the second and third spatial derivatives of the modal shapes are very sensitive to local stiffness modifications (Cempel and Natke, 1992). Consequently, they are appropriate in this sense, but one has to avoid numerical differentiation (by taking the equations of motion (Natke, 1989)). However, the advantage of the eigenquantities is their use in model adjustment to obtain a validated mathematical model, which means that the structure of the model can be corrected directly by the known eigenstructure given by the identified modal quantities.

Significant deviations compared with the previous values can lead to the causes of these modifications. The causes can be due to weak points of the system, and to erroneous assumptions of the loadings. The tools for **cause finding** are given by prior knowledge of system analysis (including, of course, the symptoms chosen, sensitivity analysis, weak point analysis etc.), and by simulation, because various assumptions have to be checked.

In many cases the causes of failures are primary causes described by patterns in the symptom domain. In such cases, pattern recognition techniques or neural networks can be used in order to recognize on-line the emerging primary cause of faults.

In this context mention must be made of input identification as a systematic approach when the loading has changed during the system's life time. If it is possible to find a parametric model for the forcing, then the problem is reduced to parameter estimation, and the known methods of estimation can be applied.

Trend Prediction is based on the various adjusted mathematical models at different life times $\theta_i, i = 1(1)N$. These models make it possible to investigate

[1] Verification, validation and usability are defined in (Natke and Cempel, 1995b). In engineering terms verification means finding the correct solution of the given equations; validation means working with the correct equations, and usability requires that the errors included are sufficiently small for the goal of the model.

the fault/damage evolution, and to establish a mathematical model for the life time θ_{N+1} by the application of deterministic (extrapolation) or stochastical methods. The future state condition can also be predicted and assessed with this predicted model (see above).

Many prediction techniques can be used, even some taken from econometry, like Brown's exponential smoothing, which gives satisfactory results in machine diagnostics. However, the best approach is to find a proper evolutionary model of the system under consideration as an energy-transforming system. Weibull and Fréchet distributions are suitable for this purpose. They result in minimum prediction error and are easy to program for computing.

EXAMPLE OF MODEL-SUPPORTED DIAGNOSIS

The Norderelb bridge serves to demonstrate the procedure, without direct application of the adjustment procedure, which can be found elsewhere (Natke and Cempel, 1997a; Natke, 1992a; Friswell and Mottershead, 1995). A finite element (FE) model was built for the Norderelb bridge. Experimental modal testing was then carried out. The results of the theoretical and experimental modal analysis are compared. They agree very well in the interesting modal quantities, so that an adjustment was unnecessary. This model was taken as a reference model. Then damage was **simulated** by model modification, and it is shown that the comparison of the eigenfrequencies and modal shapes of the model for the damaged state with those of the reference model permit the detection and localization of the simulated damage.

THE SYSTEM AND ITS FE-MODEL

The Norderelb motorway bridge is a cable-stayed steel bridge consisting of five fields of 31 m - 64 m - 171 m - 64 m - 80 m with a total span of 410 m. Fig. 3 is an overview drawing, and Fig. 4 shows the cross-section.

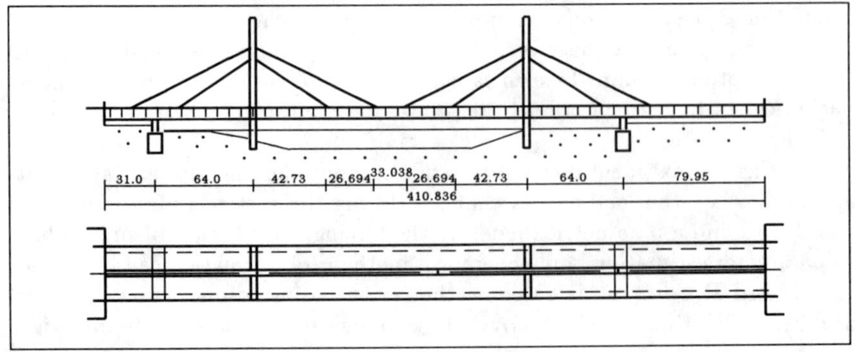

Figure 3: Overview drawing

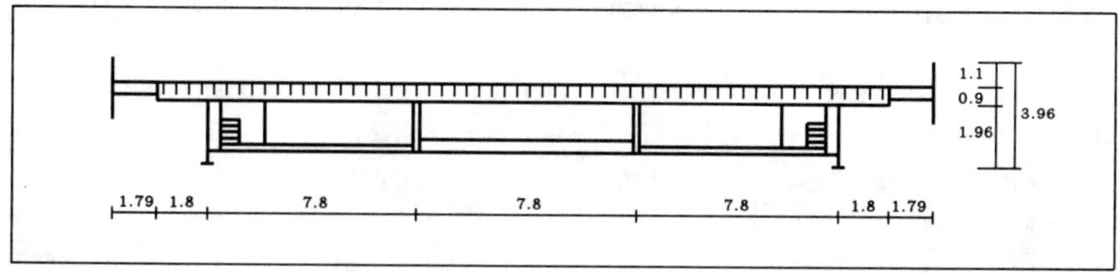

Figure 4: Cross-section

The FE-model and the following calculations are performed by the NASTRAN program. QUAD4 and TRIA elements are used for the roadway, main girders, ground plate and lateral disc. Beam elements are taken for the bracings, ribs, and pylons. Each of the 32 cables is represented by a rod. As is known, cables generally behave in a strongly nonlinear way. However, due to their high pre-stressing, the tension stresses will not change to compression during service, and only this is considered here. Consequently, linear dynamics, for example modal analysis, can be performed without restriction. A total of about 2000 elements with approximately 7000 static degrees of freedom are used. Fig. 5 shows the calculated eigenfrequencies with the modes.

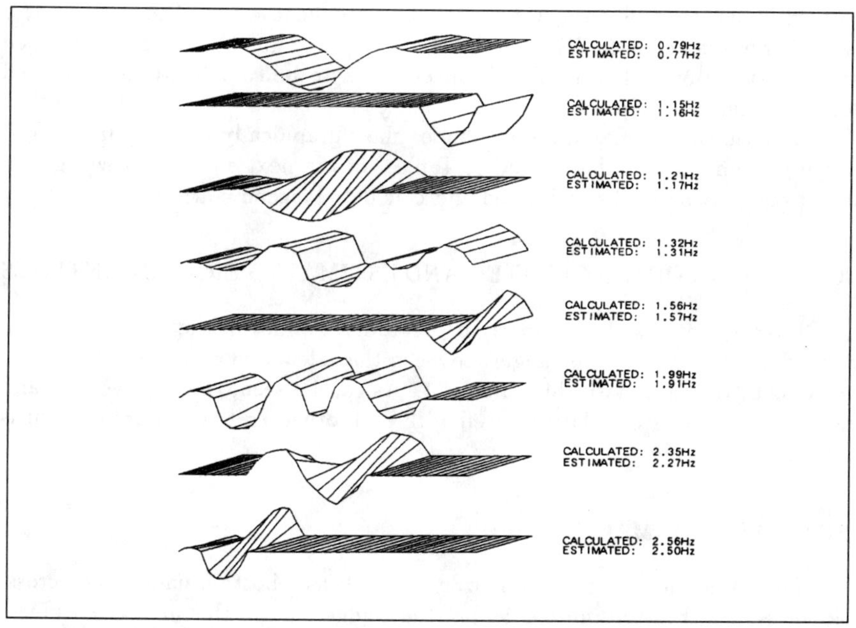

Figure 5: Calculated eigenfrequencies and normal modes

Table I: COMPARISON OF IDENTIFIED AND CALCULATED EIGENFREQUENCIES AND MODES

DESCRIPTION OF MODES	EIGENFREQUENCY [Hz]		DIFFERENCE %	MAC
	CALCULATED	MEASURED		
1.VERTICAL BENDING	0.79	0.756 ± 0.0021	3.3	0.997
2.VERTICAL BENDING	1.15	1.157 ± 0.0033	0.6	0.907
3.VERTICAL BENDING	1.32	1.307 ± 0.0051	1.3	0.887
4.VERTICAL BENDING	1.99	1.911 ± 0.0043	4.1	0.932
1.TORSION	1.21	1.173 ± 0.0054	3.2	0.996
2.TORSION	1.56	1.574 ± 0.0065	0.9	0.972
3.TORSION	2.35	2.274 ± 0.0069	3.3	0.832
4.TORSION	2.56	2.500 ± 0.0053	2.4	0.887

RESULTS OF EXPERIMENTAL MODAL ANALYSIS

It was impossible to stop the traffic in order to apply test signals, therefore the traffic was taken as excitation and modelled as broadband noise, knowing that heavy trucks should be modelled as interaction problems. Measurements of the dynamic responses ($\rightsquigarrow$ estimated frequency response functions) were performed for a measuring time of 60 minutes. Blocks of 8192 elements were therefore formed in the time domain, which lead to a frequency step of 0.0066 Hz at a sampling frequency of 54 Hz. This step was required due to the poorly damped steel construction. 1400 amplitude spectra had to be calculated and averaged for each reference point etc.

Because the estimated modal shapes show graphically no quantitative differences from the calculated modes, Table I of the next subsection will give a direct comparison of the estimated and calculated eigenquantities.

COMPARISON OF CALCULATED AND ESTIMATED EIGENQUANTITIES

Table I contains the comparison. The MAC-values in the last column are defined as the cosines of the angles between the calculated and estimated eigenvectors. In the ideal case this value is 1. As can be seen, the eigenvectors and eigenfrequencies agree relatively well with each other, so that an adjustment is not performed here.

SIMULATED DAMAGE

The lower chord of the main girders is built as a bottom flange. The cross-section is closed in the middle field of the bridge, and in this area the stiffness is mainly determined by the ground plate. However, in the end field, defined by the large deflection of the 2nd vertical bending mode (see Fig. 5), the bottom flanges of the longitudinal girders are essential. It is assumed that these flanges

Table II: COMPARISON OF THE EIGENQUANTITIES OF THE DAMAGED SYSTEM AND THE REFERENCE VALUES

DESCRIPTION OF MODES	EIGENFREQUENCY [Hz]		DIFFERENCE %	MAC
	REFERENCE	DAMAGED SYSTEM (SIMULATED)		
1.VERTICAL BENDING	0.765	0.79	3.3	0.985
2.VERTICAL BENDING	1.157	0.89	-23.0	0.860
3.VERTICAL BENDING	1.307	1.23	-5.9	0.562
4.VERTICAL BENDING	1.911	2.04	6.8	≈ 0
1.TORSION	1.173	1.23	4.9	0.994
2.TORSION	1.574	1.68	6.7	0.960
3.TORSION	2.274	2.37	4.2	0.818
4.TORSION	2.500		-5.2	0.120
5.TORSION	2.652	2.60	-2.0	0.938
6.TORSION	3.600	3.50	-2.8	0.922

are completely missing: the stiffnesses of the bottom flanges in the end field are removed in the related FE-model.

The damage simulated in this way leads to the eigenquantities as shown in Table II.

The effect is enormous. The second vertical bending shows a reduction of 23%. The related mode does not differ essentially in its shape. The number of degrees of freedom is reduced (modification of the model structure!). The MAC-values differ significantly from 1.

With respect to the assumed (large) modification, the eigenquantities are very sensitive to the simulated damage. They permit damage detection. Moreover, the large eigenfrequency deviation of the second bending from the reference value together with the mode shape, which shows only a main deflection in the end field (see Fig. 5), clearly gives the damage location.

If one wants to adjust the reference model to the model representing the damaged state, then an additional problem arises: model structure adjustment in addition to parameter adjustment, which should be solved by re-modelling.

OUTLOOK

Model-aided diagnosis is based on theoretical system analysis and measured quantities of the existing system. Taking into account the prior knowledge and the measurements with their uncertainties (by model building with the use of additional information (Natke, 1997)), then the best available knowledge base is obtained for diagnosis. This is illustrated by a real world system with simulated damage.

Additionally, it is mentioned that symptom-based diagnosis can complete the model-based one. The discussed model-aided diagnosis is based on linear time-invariant approximations. Time series analysis should take instationary processes into account. The respective transforms should then give information equivalent to the stationary cases (Natke and Cempel, 1997b). Nonlinear dynamic behaviour can be used for detection and localization. However, model adjustment has to include structure (of the model) adjustment (identification), which is generally an unsolved task. Here the Hilbert transform combined with the separation theorem can help to identify the model structure (Natke and Cempel, 1995b).

REFERENCES

Bendat, J.S., Piersol, A.G., 1986, *Random Data - Analysis and Measurement Procedures*, John Wiley & Sons New York, Chichester, Brisbane, Toronto, Singapore

Cempel, C., Natke, H.G., 1992, Application of transformed normal modes for damage location in structures; in: *Structural Integrity Assessment*, Ed. P. Stanley, Elsevier Appl. Sc., London, New York, 246-255

Friswell, M.I., Mottershead, J.E., 1995, *Finite Element Model Updating in Structural Dynamics*, Kluwer Academic Publ., Dordrecht, Boston, London

Natke, H.G., 1989, *Baudynamik - Einführung in die Dynamik mit Anwendungen aus dem Bauwesen*, B. G. Teubner Stuttgart

Natke, H.G., 1992a *Einführung in Theorie und Praxis der Zeitreihen- und Modalanalyse - Identifikation schwingungsfähiger elastomechanischer Systeme*, 3rd edition, Vieweg, Braunschweig, Wiesbaden

Natke, H.G., 1992b, On regularization methods within system identification; in: *Inverse Problems in Engineering Mechanics*, Eds. M. Tanaka, H.D. Bui, IUTAM Symposium Tokyo, Springer-Verlag Berlin, Heidelberg, New York, London, 3-20

Natke, H.G., Cempel, C. (eds.), 1995a, *Summer School on Systems Engineering, Collection of Lecture Notes*, Mitteilung des Curt-Risch-Instituts der Universität Hannover, CRI-K1/95

Natke, H.G., Cempel, C., 1995b, System analysis and identification - A holistic approach; in (Natke and Cempel, 1995a), 155-187

Natke, H. G., 1997, Uncertainties in mechanical system modelling: Definitions, models, measures, applications; in: *Uncertainties: Models and Measures*, Eds. H. G. Natke and Y. Ben-Haim, Akademie Verlag Berlin, 44-68

Natke, H.G., Cempel, C., 1997a, *Model-Aided Diagnosis of Mechanical Systems*, Springer-Verlag Berlin, Heidelberg, New York

Natke, H.G., Cempel, C., 1997b, Model-aided diagnosis based on symptoms; Structural Damage Assessment Using Advanced Signal Processing Procedures, EUROMECH 365, 30 June - 2 July 1997, Sheffield, UK (will be publ.)

The Use of Wave Propagation Models for Structural Damage Identification

D. J. PINES†

Abstract
This paper discusses the use of wave propagation models to detect damage in slender one-dimensional structural elements. This work is motivated by the successful application of the wave modeling approach to control of structures. Finite element analysis techniques have difficulty in detecting small changes in structural mass or stiffness. Wave models have higher fidelity because they are derived from the distributed nature of the structural dynamics. This paper studies this enhanced sensitivity on simple one-dimensional structures. The approach is compared qualitatively to a finite element model of comparable size to validate the improved performance in detecting small mass or stiffness changes. Simulated damage in the form of a changing wavenumber-frequency (reduced mass or stiffness) relation is used to locate the element with the damage. Experimental validation is carried out on a free-free rod structure with approximately collocated sensors and actuators.

1.0 Introduction

Interest in infrastructure health monitoring and damage detection has received a considerable amount of attention over the past two decades. Previous approaches to non-destructive evaluation of structures to assess their integrity typically involved some form of human interaction. Recent advances in smart materials and structures technology has resulted in a renewed interest in developing advanced self-diagnostic capability for assessing the state of a structure without any human interaction. The goal is to reduce human interaction while at the same time monitor the integrity of a structure. With this goal in mind, many researchers have made significant strides in developing damage detection methods for structures based on traditional modal analysis techniques [1-11]. These techniques are often well suited for structures which can be modeled by discrete lumped-parameter elements where the presence of damage leads to some low frequency change in the global behavior of the system. On the other hand small defects such as cracks are obscured by modal approaches since such phenomena are high frequency effects not easily discovered by examining changes in modal mass, stiffness or damping parameters. This is because at high frequency modal structural models are subject to uncertainty. This uncertainty can be reduced by increasing the order of the discrete

+Assistant Professor, Senior Member of AIAA, Member of AHS. email:djpterp@eng.umd.edu

model, however, this increases the computational effort of modal-based damage detection schemes [11]. On the other hand, wave propagation models of structures are well suited for detecting small defects since they are sensitive to changes in local dynamic impedance.

Recently, Lakshmanan et. al [12,13] have implemented a model-based wave propagation damage detection methodology on composite rotorcraft flexbeams to detect defects such as cracks and delaminations. This approach has proved to be mildly successful in detecting the location and size of these defects to within 15% for uncoupled composite flexbeams under rotation. This paper presents an overview of local and global wave propagation modeling and damage detection methods for determining the presence, location and magnitude of damage in structures composed of one-dimensional structural elements.

2.0 Wave Models of Structural Dynamics

Traveling wave descriptions begin with the modeling of continuum one-dimensional structural elements in terms of partial differential equations. These equations can then be transformed into ordinary differential equations by transforming the temporal variable into the frequency domain ω. Such an approach leads to a generic ordinary differential equation (ode) representation with constant coefficients given by

$$\frac{d^m u}{dx^m} + a_1 \frac{d^{m-1} u}{dx^{m-1}} + a_2 i\omega + \ldots + a_n u + b_1 i\omega u - b_2 \omega^2 u = 0 \qquad (1)$$

where m is the order of the system dynamics, u the displacement, and the a's are constant scalar coefficients. A suitable state space description in terms of a system of 1st order ordinary differential equations of order 2m is given by

$$\frac{d\bar{y}}{dx} = A(\omega)\bar{y} \qquad (2)$$

where $\bar{y}(x,\omega)$ represents a vector of member deflections and/or internal force variables. $A(\omega)$ is a frequency dependent matrix which characterizes homogeneous member dynamics.

2.1 Wave Mode Coordinates

Following the frequency domain formalism developed in references [14] and [15], the dynamic behaviour of each structural element can be obtained by superposing independently propagating traveling wave modes at every frequency. To expose this wave nature, equation (2) can be diagonalized to give the following de-coupled description of the spatial evolution of normalized wave mode variables, $\bar{w}(x,\omega)$

$$\frac{d\bar{w}}{dx} = \Gamma(\omega)\bar{w} \qquad (3)$$

where the entries of

$$\Gamma(\omega) = Y^{-1}(\omega)A(\omega)Y(\omega) = diag[k_1, \ldots, k_{2m}] \qquad (4)$$

are termed the propagation coefficients (with $k_j(\omega) = k_{rj}(\omega) + ik_{ij}(\omega)$) of traveling wave modes. The collection of these coefficients define the dispersion relations for a member which intimately connects the spatial and temporal domains. Depending on the 1-D structural element these coefficients can be very complex functions of the temporal frequency variable w. Structural elements which have linear dispersion relations, i.e. k vs. w, are termed non dispersive while members with non linear relations are termed dispersive. The difference between the two types of classifications lies in how fast information and energy flows at each frequency. The wave components $\vec{w}(x,\omega)$ thus appear in rightward and leftward pairs with the eigenvalues of $A(\omega)$ being restricted to the first and third quadrant of the complex spatial transform plane which adheres to the conservation of energy principle for a conservative system. The invertible matrix $Y(\omega)$ represents a frequency dependent set of complex eigenvectors which transform member deflections and internal forces into leftward and rightward traveling wave modes which propagate along the one-dimensional structure

$$\begin{aligned}\vec{y}(x,\omega) &= Y(\omega)\vec{w}(x,\omega) \\ &= \begin{bmatrix} Y_1 & Y_2 \\ Y_3 & Y_4 \end{bmatrix}\vec{w}(x,\omega)\end{aligned} \tag{5}$$

Each column of this matrix yields the relative magnitude and phase of the physical variables $\vec{y}(x,\omega)$ which are present in the corresponding wave type.

2.2 Scattering Dynamics

Traveling waves which propagate independently along each member can be scattered or generated at structural discontinuities or at locations where external excitations alter the homogeneous evolution of member dynamics. Locations at which the scattering or generation of traveling waves occur are referred to as junctions or boundaries. Associated with each member is a physical state vector $\vec{y}_m(x,\omega)$ defined at the interface of the member and the structural discontinuity. The junction boundary condition can then be described in terms of a composite state vector $\vec{y}_c^T = (\vec{y}_1,\ldots,\vec{y}_m)^T$, as ($B$ is a rectangular matrix with half as many rows as columns).

$$\begin{aligned}B(\omega)[\vec{y}] &= \vec{Q}(\omega) \\ B(\omega)Y(\omega)[\vec{w}] &= \vec{Q}(\omega)\end{aligned} \tag{6}$$

where the member transformation matrices have been used to substitute $\vec{w}_c$ for $\vec{y}_c$. Partitioning this equation into incoming and outgoing dynamics leads to

$$\begin{bmatrix} B_o(\omega) & B_i(\omega) \end{bmatrix}\begin{bmatrix} \vec{w}_o \\ \vec{w}_i \end{bmatrix} = \vec{Q}(\omega) \tag{7}$$

where the open loop description relating the generation of outgoing waves $\vec{w}_o(x,\omega)$ to the scattering of incoming waves $\vec{w}_i(x,\omega)$ and external excitation $\vec{Q}(\omega)$ is given by

$$\vec{w}_o(x,\omega) = S(\omega)\vec{w}_i + \Psi(\omega)\vec{Q}(\omega) \tag{8}$$

2.3 Transmission Dynamics

Wave modes at different locations in a structural member are related to wave modes at any other location of the same member through a frequency dependent transition matrix. This matrix is diagonal and consists of transcendental functions whose arguments depend on the path length between the two points in which the wave mode amplitudes are desired. A typical representation can be written as

$$\begin{aligned}\vec{w}(x_2,\omega) &= \xi(x_2,x_1,\omega)\vec{w}(x_1,\omega) \\ &= e^{\Gamma(\omega)(x_2-x_1)}\vec{w}(x_1,\omega) \\ &= \begin{bmatrix} \xi_1 & 0 \\ 0 & \xi_1^{-1} \end{bmatrix}\vec{w}(x_1,\omega)\end{aligned} \tag{9}$$

where $\vec{w}(x_1,\omega)$ and $\vec{w}(x_2,\omega)$ correspond to wave mode vectors at locations x_1 and x_2 respectively and $\xi(x_1,x_2,\omega)$ is the frequency dependent transition matrix.

2.4 Phase Closure

On finite structures waves generated by external point forces can circumnavigate the structure, interacting with structural discontinuities to close upon themselves in a constructive or destructive manner. This constructive behaviour characterizes the resonant dynamics of the member. Thus, the response at any location of a finite structure can be obtained by tracking the motion of individual leftward and rightward traveling wave components as they circulate a member and close upon themselves.

3.0 Global Structural Models Using Wave Propagation Analysis

Wave finite elements are derived by using the continuum mechanics description of structures developed above coupled with the phase closure principle on free-free one-dimensional structural elements. Therefore, instead of using polynomial weighting functions between displacement coordinates, complex spatial exponential weighting functions are used to couple the response between individual degrees of freedom. These spatial shape functions are used to weight the contribution of the individual wave-modes which can propagate along a structure.

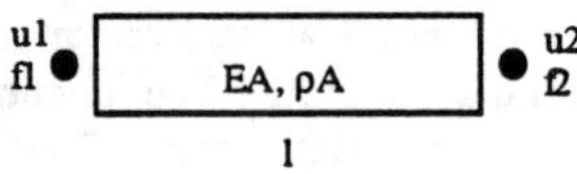

Fig. 1: Rod Element

3.1 Rod Wave Finite Elements (See Fig. 1)

Using equation (5), one can represent the deflection of a rod element in terms of complex spatial shape functions and the wave-mode coordinates as

$$u(x,\omega) = w_{rp}(\omega)e^{-ikx} + w_{lp}(\omega)e^{-ik(L-x)} \tag{10}$$

Using this relation and the principle of phase closure on a rod element, it is not difficult to derive the dynamic force-displacement relation given by

$$\begin{aligned}\vec{F} &= \begin{bmatrix} \dfrac{1+S_L\xi_l^2}{\psi_L(1-\xi_l^2)} & -\dfrac{\xi_l+S_L\xi_l}{\psi_L(1-\xi_l^2)} \\ -\dfrac{\xi_l+S_R\xi_l}{\psi_R(1-\xi_l^2)} & \dfrac{1+S_R\xi_l^2}{\psi_R(1-\xi_l^2)} \end{bmatrix} \begin{bmatrix} u_1 \\ u_2 \end{bmatrix} \\ &= \frac{EA}{l}G_e(kl)\vec{u}\end{aligned} \tag{11}$$

where

$$G_e(kl) = ikl \begin{bmatrix} \dfrac{1+e^{-i2kl}}{1-e^{-i2kl}} & \dfrac{-2e^{-ikl}}{1-e^{-i2kl}} \\ \dfrac{-2e^{-ikl}}{1-e^{-i2kl}} & \dfrac{1+e^{-i2kl}}{1-e^{-i2kl}} \end{bmatrix}; \quad S_R = S_L = 1; \quad \psi_R = \psi_L = \frac{1}{ikEA}; \quad \xi_l = e^{-ikl} \tag{12}$$

where the term, kl, is the non-dimensional wavenumber and l is the length of the rod element. The wavenumber-frequency relation is given by

$$k = \omega\sqrt{\frac{\rho A}{EA}} \tag{13}$$

where ρA represents the mass per unit length and EA represents the longitudinal stiffness of the element. Fig. 2 displays the components of the elemental rod dynamic stiffness matrix. Closer inspection of this figure reveals that the diagonal elements of G_e have both zeros and poles. However, the off-diagonal terms only have pole dynamics.

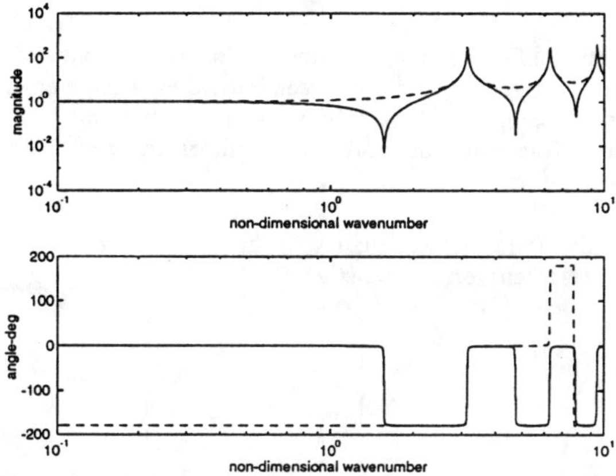

Fig. 2: Non-dimensional element dynamic stiffness matrix-Ge(kL)(2x2) versus non-dimensional wave number for a rod element. Ge(1,1) & Ge(2,2) are give by the solid curve. Ge(1,2) & Ge(2,1) are given by dashed curve.

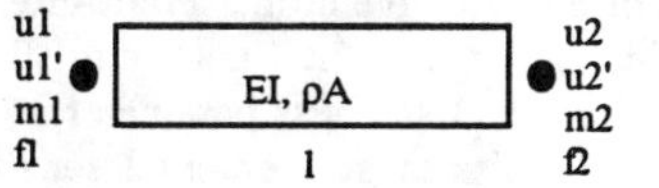

ul ul' ml fl EI, ρA l u2 u2' m2 f2

Fig. 3: Beam Element

3.2 Bernoulli-Euler Beam Wave Finite Elements (See Fig. 3)

Again using equation (5), the transverse deflection of a uniform Bernoulli-Euler beam can be written in terms of complex spatial shape functions and wave mode coordinates as

$$u(x,\omega)=w_{rp}(\omega)e^{-ikx}+w_{re}(\omega)e^{-kx}+ w_{lp}(\omega)e^{-ik(L-x)}+w_{le}(\omega)e^{-k(L-x)} \tag{14}$$

Again applying phase closure and the input output relation, the dynamic force-displacement relation becomes

$$\vec{F}=G_e(kl)\vec{u}=\begin{bmatrix}\Psi_L^{-1} & -\Psi_L^{-1}S_L\xi_1\\ -\Psi_R^{-1}S_R\xi_1 & \Psi_R^{-1}\end{bmatrix}\begin{bmatrix}Y_1 & Y_2\xi_1\\ Y_1\xi_1 & Y_2\end{bmatrix}^{-1}\begin{bmatrix}u_1\\ \phi_1\\ u_2\\ \phi_2\end{bmatrix}$$

$$S_R=S_L=\begin{bmatrix}-i & 1+i\\ 1-i & i\end{bmatrix};\quad \Psi_R=\Psi_L=\frac{-1}{2EIk^2}\begin{bmatrix}-1-i & \dfrac{-1-i}{k}\\ 1-i & \dfrac{-1-i}{k}\end{bmatrix};\quad \xi_1=\begin{bmatrix}e^{-ikl} & 0\\ 0 & e^{-kl}\end{bmatrix} \tag{15}$$

where the terms in G_e are much more involved than in the case of the rod. The wavenumber-frequency relation for the B-E beam is given by

$$k=\omega^{1/2}\sqrt{\sqrt{\frac{\rho A}{EI}}}$$

3.3 Global Assemblage

Similar to conventional finite element methods the overall response of the structure can be obtained by maintaining connectivity between individual wave elements in an analogous way to finite element methods. The only difference is that, this has to be done at every discrete frequency. This results in the following global structural response relation for the rod and beam formulations:

$$\bar{F} = G(kl)\bar{u} \tag{16}$$

where the global dynamic stiffness matrix, G, has the following banded form similar to conventional finite element formulations:

$$G = \begin{bmatrix} x & x & & & & & & \\ x & x & x & & & 0 & & \\ & x & x & x & & & & \\ & & x & x & x & & & \\ & & & x & x & x & & \\ & & & & x & x & x & \\ & 0 & & & & x & x & x \\ & & & & & & x & x \end{bmatrix}_{nxn} \tag{17}$$

<u>4.0 Dynamic Elemental Sensitivities</u>

An important component of many conventional Finite-Element based damage detection methods is that sensitivities of the elemental mass and stiffness matrices are computed with respect to the design mass and stiffness parameters. The use of wave descriptions of the structural dynamics permits these elemental sensitivities to be computed with respect to change in elemental wavenumber k along the respective element. The change in wavenumber is directly related to how the mass and stiffness properties change along the element.

Consider the non-dimensional dynamic stiffness matrix for a rod element which relates non-dimensional input forces (F_1/EA, F_2/EA) to displacements (U_1/L, U_2/L) at the free ends (See Figure 1). The sensitivity of these transfer functions to changing wavenumber is given by

$$\frac{dG_e(kl)}{dkl} = \begin{bmatrix} \frac{dG_{e11}}{dkl} & \frac{dG_{e12}}{dkl} \\ \frac{dG_{e21}}{dkl} & \frac{dG_{e22}}{dkl} \end{bmatrix} \tag{18}$$

Figure 4 plots the non-dimensional sensitivity with respect to wavenumber for a hypothetical free-free rod element. Notice that the sensitivity is slowly increasing as the wavenumber increases until $kl = n\pi$ ($n=1,2,.....$). These frequencies correspond to resonances of terms in the dynamic stiffness matrix. At these locations the sensitivity approaches infinity. Similarly, the sensitivity vanishes at the zeros of the terms in Ge. At low frequency ($kl<<1$) , the dynamic sensitivity approaches zero. This implies that the dynamics of the structure are more static than dynamic and the dynamic elemental stiffness approaches the non-dimensional static stiffness matrix k for a finite rod element given by

$$K_e = \begin{bmatrix} 1 & -1 \\ -1 & 1 \end{bmatrix}$$

Dynamic elemental sensitivities with respect EA and ρA can also be formulated by substituting the rod wavenumber-frequency relation into equation (18).

$$\frac{dG_e(kl)}{dEA} = \begin{bmatrix} \frac{dG_{e11}}{dEA} & \frac{dG_{e12}}{dEA} \\ \frac{dG_{e21}}{dEA} & \frac{dG_{e22}}{dEA} \end{bmatrix}; \frac{dG_e(kl)}{d\rho A} = \begin{bmatrix} \frac{dG_{e11}}{d\rho A} & \frac{dG_{e12}}{d\rho A} \\ \frac{dG_{e21}}{d\rho A} & \frac{dG_{e22}}{d\rho A} \end{bmatrix} \quad (19)$$

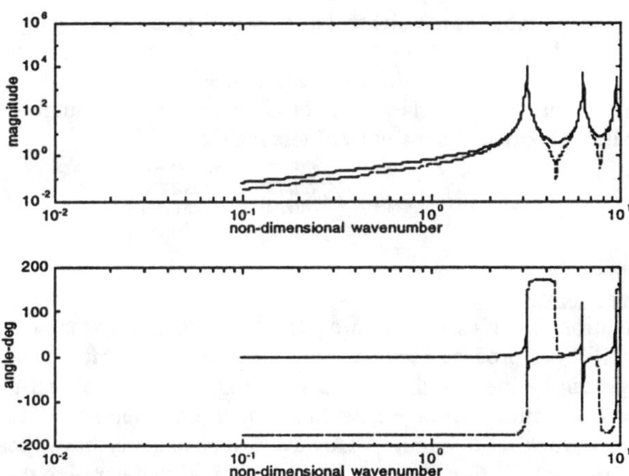

Fig. 4: Wave element sensitivities of the dynamic rod stiffness matrix $G_e(kl)$ versus kl. Elements (1,1) -solid, and (1,2)-dashed are plotted.

5.0 Modeling Structural Damage

Local defects are determined by modeling the scattering dynamics in a structure. Global defects are determined by modeling damage as a change in the local wavenumber-frequency relation in an individual structural element.

5.1 Local Models of Structural Damage

Using wave descriptions, structural discontinuities can be modeled to capture the salient features of the scattering dynamics of incident waves. This typically results in an undamaged homogeneous scattering equation given by:

$$w_o = S^u(\omega)w_i \quad (20)$$

When the structure becomes damaged, these homogeneous scattering dynamics are altered leading to a damaged frequency dependent scattering matrix approximated as

$$S^d(\omega) = S^u(\omega) + \Delta S(\omega) \quad (21)$$

where

$$\Delta S(\omega) = f(mass, stiffness, damping) \quad (22)$$

and characterizes how the change in scattering properties are affected by changes in the local structural properties. Thus, by computing the scattering matrix and examining the error difference between undamaged and damaged conditions, it is possible to estimate the local effect of damage on a structure.

5.2 Global Models of Structural Damage

Although, changes in local scattering dynamics will affect the global dynamics of a structure, these changes can also be modeled by assuming that the elemental wavenumber-frequency relation of a structural element has changed by some finite amount, α. This leads to the following representation of the damaged wavenumber-frequency relation for an element:

$$kl^d = kl^u + \Delta kl = kl^u + \alpha kl^u \tag{23}$$

Changes in local wavenumber at the elemental level will result in a change in the global dynamics approximated as

$$G^d(kl^d) = G^u(kl^u) + \Delta G(\alpha kl^u) \tag{24}$$

For a rod element, substitution of a damaged non-dimensional wavenumber results in the following change in the elemental dynamic stiffness matrix:

$$G_e(kl^d) = (ikl^u + i\alpha kl)\begin{bmatrix} \dfrac{1+e^{-i2(kl+\alpha kl)}}{1-e^{-i2(kl+\alpha kl)}} & \dfrac{-2e^{-i(kl+\alpha kl)}}{1-e^{-i2(kl+\alpha kl)}} \\ \dfrac{-2e^{-i(kl+\alpha kl)}}{1-e^{-i2(kl+\alpha kl)}} & \dfrac{1+e^{-i2(kl+\alpha kl)}}{1-e^{-i2(kl+\alpha kl)}} \end{bmatrix} \tag{25}$$

6.0 Damage Identification

Damage identification involves determining the location and extent of damage in a structure. Typically this process require some local or global model of the structure before the damage can be identified. Once a model is obtained, several techniques are available for global structural damage detection which lend themselves to global wave formulations of the structural dynamics. However, only recently have local and global based wave damage identification methods been developed to determine the location and extent of damage in a structure [12,13,16].

6.1 Local Damage Identification

Local damage identification relies on knowledge of the undamaged and damaged scattering matrix elements. Using this information, an optimization can be performed to determine properties of the damaged structure based on a local damage model. One detection strategy is to formulate a least squares cost function between elements of the scattering matrix

$$J = \sum_{i=1}^{4}\sum_{j=1}^{N}\left((S_i^u(\omega_j) - S_i^u(\omega_j))^H \cdot (S_i^u(\omega_j) - S_i^u(\omega_j))\right) \tag{26}$$

where subscripts u and d refer to the undamaged and damaged scattering matrix elements, i refers to the elements of the scattering matrix, j corresponds to the discrete frequency vector components and superscript H refers to the Hermitian operation on the error matrix. For a parameterized model of the local structural dynamics, this approach will yield local minima of equation (26). These solutions provide information about the location and extent of damage.

6.2 Global Damage Identification

Global descriptions of structures modeled in terms of wave propagation dynamics lend themselves to previously developed modal-based damage detection methods. Most of these approaches begin with the following representation of the structural dynamics:

$$M\ddot{x} + C\dot{x} + Kx = F\bar{u} \tag{27}$$

where M, C, and K are nxn mass, damping and stiffness matrices respectively, F is a nxm matrix of force inputs, and the vectors of displacements, x, and inputs, u, are nx1 and mx1 vectors respectively. Neglecting damping (C=0), the eigenvalue problem for this discrete representation is given by

$$(-\omega_i^2 M - K)\vec{\phi}_i = 0 \tag{28}$$

Similarly, the wave description of the global structural dynamics is given by

$$G\vec{u} = \vec{F} \tag{29}$$

where the global dynamic stiffness matrix, G, is nxn matrix and vectors u and F are of size nx1. This implies that the eigenvalue problem has the following form:

$$G(\omega_i)\vec{\phi}_i = 0 \tag{30}$$

Now incorporating structural damage via equation (24), the damaged eigenvalue problem becomes

$$G(\omega_i^d)\vec{\phi}_i^d = -\Delta G(\omega_i^d, \alpha)\vec{\phi}_i^d \tag{31}$$

where $\Delta G(\omega_i^d, \alpha)$ contains information regarding the location and extent of damage in a particular element. Notice that both mass and stiffness effects are coupled in the local wavenumber-relation of a structural element. With this representation of structural damage, it is not difficult to implement previously developed eigenstructure assignment and sensitivity based damage detection methods [16]. Such schemes are implemented on a simulated one-dimensional rod to illustrate the wave approach.

7.0 Simulation Results

To validate the wave propagation based damage detection strategy developed above, a mock structure was studied. This consisted of a simple three element free-free bar.

7.1 Free-Free Bar

Figure 5 displays a free-free bar with three rod wave finite elements corresponding to 4 physical displacements.

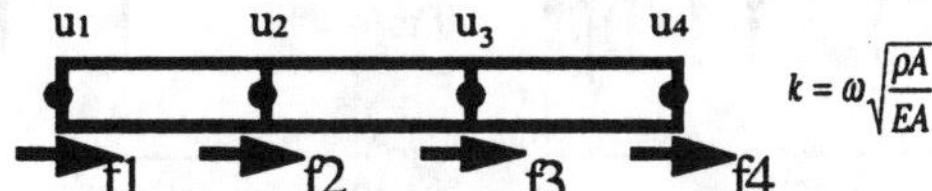

Fig. 5: Example-Free-Free Bar

The dynamic response of the bar can be obtained from the following expression relating the nodal forces with the displacements

$$\vec{F} = G\vec{u}$$

where the global undamaged dynamic stiffness matrix is given by

$$G = ikl\begin{bmatrix} \frac{1+e^{-i2kl}}{1-e^{-i2kl}} & \frac{-2e^{-ikl}}{1-e^{-i2kl}} & 0 & 0 \\ \frac{-2e^{-ikl}}{1-e^{-i2kl}} & 2\frac{1+e^{-i2kl}}{1-e^{-i2kl}} & \frac{-2e^{-ikl}}{1-e^{-i2kl}} & 0 \\ 0 & \frac{-2e^{-ikl}}{1-e^{-i2kl}} & 2\frac{1+e^{-i2kl}}{1-e^{-i2kl}} & \frac{-2e^{-ikl}}{1-e^{-i2kl}} \\ 0 & 0 & \frac{-2e^{-ikl}}{1-e^{-i2kl}} & \frac{1+e^{-i2kl}}{1-e^{-i2kl}} \end{bmatrix} \tag{32}$$

The characteristic values of equation (24) can be found by computing the determinant of the undamaged structure which leads to

$$1-e^{-i6kl}=0 \tag{33}$$

This implies that the undamped undamaged eigenvalues are given by

$$kl_i=(0,\frac{\pi}{3},\frac{2\pi}{3},\pi,\frac{4\pi}{3},....) \tag{34}$$

The corresponding normalized eigenvectors can be assembled into the following matrix

$$\phi=\begin{bmatrix} 0.5000 & -.6325 & .6325 & 0.5000 \\ 0.5000 & -.3162 & -.3162 & -0.5000 \\ 0.5000 & .3162 & -.3162 & -0.5000\cdots \\ 0.5000 & .6325 & .6325 & 0.5000 \end{bmatrix}_{4xN} \tag{35}$$

where N modes of the structure are considered.

7.2 Damage Identification Results

The non-dimensional transfer function response of the free-free bar can be found by inverting the dynamic stiffness matrix G to obtain a matrix H corresponding to the force to displacement transfer functions. A few of the elements of this matrix are plotted in Figure 6 for the case of no damage and the case of 10% damage (increase in wave number, reduction in stiffness) in element 1. Notice that there is a slight shift seen in the resonant dynamics of the transfer functions. The presence of damage causes a leftward shift of the resonant peaks, suggesting a softening of the bar.

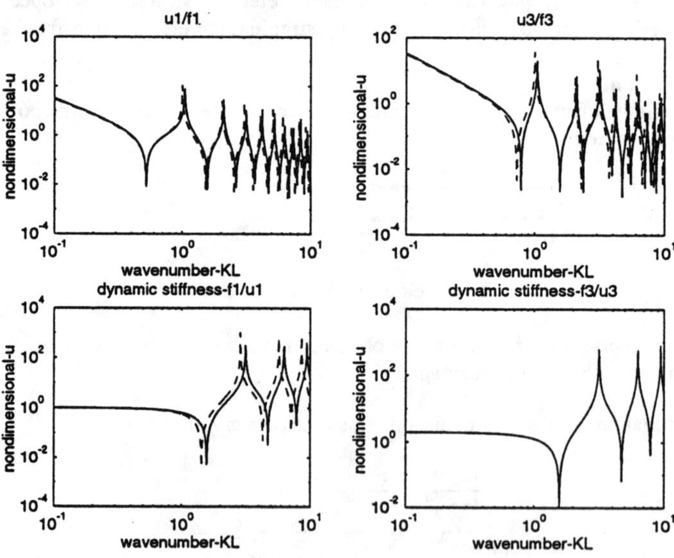

Fig. 6: Elements of global transfer function-H(ω) and dynamic stiffness-G(ω) matrices plotted versus undamaged non-dimensional wavenumber. 10%

Damage in wavenumber of element 1. Damaged curves (dashed), Undamaged curves (solid).

Using a Newton-Raphson iteration approach, it is not difficult to solve for the damaged eigenvalues from the transfer function responses shown in Fig. 6. This leads to the following damaged eigenvalues

$$kl_i^d = 0, 1.038, 2.0838, 3.0371 \tag{36}$$

The corresponding damaged eigenvectors can be found by substituting the damaged eigenvalues into a model of the system response. This leads to

$$\phi^D = \begin{bmatrix} 0.5000 & -.6046 & .6029 & 0.6087 & \\ 0.5000 & -.2752 & -.3761 & -0.2808 & \\ 0.5000 & .3597 & -.2855 & -.3497 & \cdots\cdots \\ 0.5000 & .6553 & .6430 & 0.6545 & \end{bmatrix}_{4xN} \tag{37}$$

Application of a sensitivity [16] based wave damage detection algorithm resulted in a 10% change in the local wavenumber of element 1 in the absence of noise.

7.3 Experimental Validation

Experimental validation of the global wave propagation based damage detection methodology was attempted on a free-free rectangular rod (See Table 1) displayed in Figure 7. Piezoelectric actuators were used to excite the structure in the longitudinal direction and collocated accelerometers were used to measure the response. To simulate damage a mass weighing approximately 10 grams was attached to the third element. Transfer Functions from force to acceleration were obtained for the undamaged and simulated damaged states. A three element wave model was constructed to detect damage using two dynamic modes of the structures. The rigid body mode was neglected. Figure 7 displays a comparison of the analytical and experimental undamaged and damaged transfer functions from piezoelectric actuator force, F1, to accelerometer sensor, u1. Notice that the three-element global wave model accurately predicts the response of the first two modes for both the damaged and undamaged transfer functions. The third mode is not as closely predicted because the model was tuned to mimic only the first two modes. Table 2 lists a comparison of the experimental resonant frequencies for the undamaged and damaged structure. Application of global based damage detection methods to this structure results in an estimate of 4.0% increase in the local wavenumber of element 3.

Table 1: Properties of Rod

	Value	Units
EA	11129010	N
ρA	0.435	kg/m

Table 2: Experimental Modal Frequencies

Freq. (Hz)	Undamaged	Damaged
ω_1	2604	2547
ω_2	5180	5149
ω_3	7477	7444

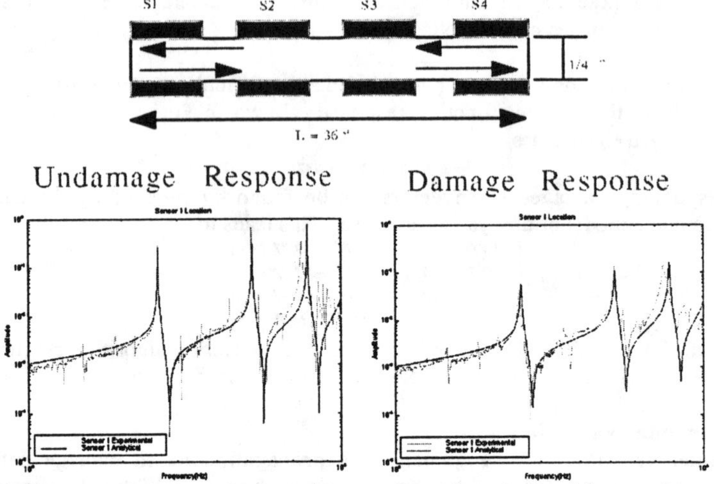

Figure 7: Transfer Functions from F_1 to u_1 for undamaged and damaged structure.

7.4 Qualitative Comparison with Discrete Modal Methods

Although a direct comparison of the above results with modal based techniques was not made, it is important to point out potential advantages and disadvantages of the wave propagation based damage detection approach developed in this work. The potential advantages in comparison to modal based damage detection techniques can be summarized as follows:

- fewer sensors to capture relevant structural dynamics.
- greater sensitivity in detecting small impedance changes caused by local defects.
- greater accuracy since no approximation is made of the spatial structural domain.
- damage detection scheme is computationally more efficient since it results in MxN modal eigenvectors instead of NxN where M < N (M is the number of sensors, N is the number of modes).

Some potential disadvantages of the approach include:

- unable to measure wave coordinate variables.
- difficulty working with complex exponentials.
- system identification of structural dynamics.
- the need for excitation and measurement at every node of the model.

Nevertheless, despite these shortcomings it appears that the local and global based wave propagation approaches may offer some significant advantages over current modal based approaches. Model reduction and expansion will need to be addressed for global wave descriptions.

8.0 Conclusions

This paper has presented a wave propagation approach to detecting damage in one-dimensional structures. Local and global damage detection methods have been presented to illustrate the advantages and disadvantages of the approach on a simulated one-dimensional rod structure. Results indicate that the approach is capable of detecting small changes in the local wavenumber-frequency relation of a structural element. Additional work is required to improve the algorithm and to validate against structures with more complex geometries. In addition, future will attempt to extend this approach to two-dimensional structures.

9.0 Acknowledgments

This work was supported in part by the National Science Foundation and the U.S. Army Research Office, under grants CMS9625004 and DAAL03-92-G-0121 respectively, with Dr.'s William Anderson and Gary Anderson serving as contract monitors.

10.0 References

[1] P. Cawley, and P.D. Adams, "The Location Of Defects In Structures From Measurements Of Natrual Frequencies," *Journal of Strain Analysis*, Vol. 14, No. 2, pp 49-57, 1979.

[2] M. Baruch, "Optimal Correction of Mass and Stiffness Matrices Using Measured Modes", *AIAA Journal*, Vol. 20, No. 11, 1982, pp. 1623-1626.

[3] A. Berman,. and E.J. Nagy., "Improvement of a Large Analytical Model Using Test Data", *AIAA Journal*, Vol. 21, No. 8, 1983, pp. 1168-1173.

[4] S.W. Smith, and S.L. Hendricks, "Damage Detection and Location in Large Space Trusses," *AIAA SDM Issues of the International Space Station,* A Collection of Technical Papers, AIAA, Washington, D.C., 1988, pp. 56-63.

[5] T. Kashangaki, "On-Orbit Damage Detection and Health Monitoring of Large Space Trusses - Status and Critical Issues," *AIAA Paper 91-1181*, April 1991.

[6] T. Kashangaki, S.W. Smith, and T.W. Lim, " Underlying Modal Data Issues for Detecting Damage in Truss Structures, " *AIAA Paper 92-2264*, April 1992.

[7] D.C. Zimmerman, and M. Kaouk, "Structural Damage Detection Using a Subspace Rotation Algorithm," *AIAA Paper 92-2521*, April 1992.

[8] F.J. Soeiro, and P. Hajela, "Damage detection in composite materials using identification techniques," *Journal of Aerospace Engineering*, Vol. 6, No. 4, pp 363-381).

[9] T.W. Lim, and T.A.L. Kashangaki, "Structural Damage Detection of Space Truss Structures using Best Achievable Eigenvectors", *AIAA Journal*, Vol. 32, No. 5, May 1994, pp. 1049 - 1057.

[10] J.M. Ricles, and J.B. Kosmatka, "Damage detection in elastic structures using vibratory residual forces and weighted sensitivity," *AIAA Journal*, Vol. 30, No. 9, pp. 2310-2317.

[11] C. Farrar and D. Jauregui, "Damage Detection Algorithms Applied to Experimental and Numerical Modal Data from I-40 Bridge", Report No. LA-13074-MS, Los Alamos National Laboratory, Los Alamos, NM, 1996.

[12] K.A. Lakshmanan and D.J. Pines, "Modeling Damage in Composite Rotorcraft Flexbeams Using Wave Mechanics", *Smart Materials and Structures*, Vol. 6, No. 3, June 1997, pp. 383-392.

[13] K.A. Lakshmanan and D.J. Pines, "Detecting Crack Size and Location in Composite Rotorcraft Flexbeams", *Proceedings of the SPIE Smart Structures and Materials*, March 3-6, 1997, San Diego, CA, Vo. 3041, pp. 408-416.

[14] D.W. Miller, "Modeling and Active Modification of Wave Scattering in Structural Networks," *PhD Dissertation*, Department of Aeronautics and Astronautics, *Massachusetts Institute of Technology*, Cambridge, Mass.,1988.

[15] A.H. von Flotow, "Disturbance Propagation in Structural Networks," *J. of Sound and Vibration*, 106(3), pp 433-450, 1986.

[16] D.J. Pines, "Damage Detection using Wave Element-by-Element Sensitivity Analysis on 1-dimensional structures", *AIAA/ASME/AHS Adaptive Structures Forum*, April 7-10, 1997, Kissimee, Florida.

Structural Damage Detection Using Modal Test Data

C. PAPADIMITRIOU, M. LEVINE-WEST and M. MILMAN

ABSTRACT

This study presents a methodology for updating the finite element model of a structure for damage detection purposes using an incomplete set of experimentally obtained modal frequencies and modeshapes. The proposed damage detection methodology involves a least squares minimization of the modal dynamic force balance residuals subject to quadratic inequality constraints introduced to properly account for the expected measurement and modeling errors. A three-step iterative procedure is proposed to iteratively predict the probable locations and size of significant damage by updating the properties of the finite element model of the structure at the element level. Simulated modal data obtained on a three-dimensional truss structure are used to assess the strengths, limitations, and overall performance of the proposed damage detection methodology in relation to the number and location of sensors, the location and magnitude of damage, as well as the number of measured modes.

INTRODUCTION

In past years, several studies have been devoted in reconciling finite element models with measured modal data. An important application of a model updating methodology is in the prediction of structural damage (see, for example, Natke and Yao, 1988; Stubbs, Broome and Osegueda 1990). The location and size of damage can be inferred by monitoring the reduction in stiffness and mass properties of the elements comprising the finite element model of the structure. The need for model updating also arises in the process of constructing a theoretical model of a structure. In most cases, in order to improve the accuracy in the

Costas Papadimitriou, Instructor, Division of Engineering & Applied Science 104-44, California Institute of Technology, Pasadena, CA 91125

Marie Levine-West, Technical Group Leader, Science and Technology Development Section, 4800 Oak Grove Dr., Jet Propulsion Laboratory, Pasadena, CA 91109-8099

Mark Milman, Member, Technical Staff, Automation and Control Section, MS 138-308, 4800 Oak Grove Dr., Jet Propulsion Laboratory, Pasadena, CA 91109-8099

model response predictions, the pre-test finite element model of the structure need to be updated to match available modal test data.

The general problem of model updating involves the selection of the best model from a parameterized class of models that best fits the modal data as judged by an appropriately selected measure of fit. The difficulties associated with this inverse problem are mainly due to the measurement error in the modal test data, the modeling error, and the incompleteness of the available modal data relative to the model complexity needed to produce physically meaninful models. As a result, the inverse problem leads to non-unique solutions and ill-conditioning (Berman, 1989; Beck, 1989; Mottershead and Friswell, 1993; Beck and Katafygiotis, 1997).

A literature review of existing finite element model updating and damage detection methods can be found in the survey by Mottershead and Friswell (1993). Each method has its own advantages and shortcomings and there is no acceptable methodology for successfully treating the model updating and damage detection problem. Most methods address the problem by choosing some mathematical criteria which creates a unique optimal model while neglecting other models that can give an equally good fit to the measured data. The present study is based on this class of methods. However, new methods (Beck, 1989; Beck and Katafygiotis, 1997) based on Bayesian statistical inference have been developed recently for properly addressing the non-uniqueness by computing and considering in the predictions all (finite or infinite) models that give acceptable fit to the data (Katafygiotis and Beck, 1997; Vanik, 1997; Beck and Vanik, 1996). The latter methods are powerful and have shown great promise in properly incorporating modeling and measurement errors, as well as properly addressing many of the difficulties encountered in the model updating problem, especially those associated with model and response prediction accuracy.

The problem of damage detection involves as a first step the identification of the locations of damage. Recently, various methods have been proposed for identifying probable damage locations in a structure using modal test data (Fahrat and Hemez, 1993; Levine-West, Milman and Kissil, 1996). They are based on modeshape expansion techniques and when combined with appropriately-defined localized error measures, they have shown promise in predicting the locations in the structure that are more likely to be damaged. In particular, the least-squares error measure subject to inequality constraints proposed by Levine-West, Milman and Kissil, (1996, 1997) properly accounts for the presence of measurement and modeling errors. The robustness and reliability of the resulting modeshape expansion technique for predicting the modeshape components at unmeasured points have been successfully evaluated in a previous study using actual experimental data obtained on the Jet Propulsion Laboratory micro-precision interferometer truss (Levine-West, Milman and Kissil, 1996). Compared with other modeshape expansion techniques, the least squares minimization technique with quadratic inequality constraints was found to provide the most reliable mode shape estimates and predictions of damage locations, even in adverse situations of significant measurement and model error.

Once the damage has been located, the magnitude of damage is predicted by updating the finite element model of the structure. The preferable methods of updating are usually the ones which preserve structural connectivity. The work by Farhat and Hemez (1993) combines modeshape expansion techniques with updating capabilities for predicting both the location and size of damage in a structure. It has been shown to work reasonably well for the cases which has been applied (Hemez and Farhat, 1995). In a recent work, Alvin (1997) has pointed out potential problems of this method and suggested several modifications which are found to provide a more robust technique for locating and sizing errors in the finite element model of a structure. In this work, a model updating and damage detection methodology is formulated based on the mode-shape expansion method presented by Levine-West, Milman and Kissil, (1996) to identify the probable locations of damage at the finite element level. The proposed measure of fit accounts for the expected measurement error in both modal frequencies and mode shape components, as well as the expected modeling errors.

STRUCTURAL MODELS AND PARAMETERIZATION

The structure is modeled by the following general class of linear models:

$$M(\theta)\,\ddot{x} + K(\theta)x = f(t) \tag{1}$$

where the global mass and stiffness matrices $M(\theta)$ and $K(\theta)$ are assembled, using a finite element analysis, from the element (or substructure) mass and stiffness matrices $M^e(\theta)$ and $K^e(\theta)$, respectively. The class of models has been parameterized using the parameter set θ which may represent mass and stiffness properties at the element or substructure level. The parameterization is chosen such that the undamaged finite element model of the structure corresponds to $\theta = 1$. Examples of finite element properties that can be included in the parameter set θ are: modulus of elasticity, cross-sectional area, thickness, moment of inertia and mass density of the finite elements comprising the model, as well as spring (translational or rotational) stiffnesses used to model fixity conditions at joints or boundaries.

In general, the system matrices $M(\theta)$ and $K(\theta)$ are nonlinear functions of θ. However, a parameterization which often arises in practical applications is the case for which both $M(\theta)$ and $K(\theta)$ are linear functions of θ, that is,

$$K(\theta) = K_0 + \sum_{i=1}^{p} K_i\theta_i \quad \text{and} \quad M(\theta) = M_0 + \sum_{i=1}^{p} M_i\theta_i \tag{2}$$

where K_0, K_i, M_0 and M_i are constant matrices independent of θ. Without loss of generality, the linear parameterization is used in this work. However, the incorporation of a general nonlinear parameterization is straightforward and will not be discussed further.

DAMAGE DETECTION METHODOLOGY

The objective in a modal-based damage detection methodology is to find the values of the parameter set θ so that the modal data generated by the linear class of models best matches, in some sense, the experimentally obtained modal data. A measure of fit that is explored herein is directly related to the modal dynamic force balance residuals $r(\omega, \phi, \theta)$, defined by

$$r(\omega, \phi, \theta) = [K(\theta) - \omega^2 M(\theta)]\phi \tag{3}$$

Note that the modal dynamic force balance residuals satisfy the equations $r(\omega_i(\theta), \varphi_i(\theta), \theta) = 0$, $i = 1, \cdots, m$, where $\omega_i(\theta)$ and $\varphi_i(\theta)$, $i = 1, \cdots, m$ are respectively the modal frequencies and mass-normalized mode shapes of the first m modes of the model (1).

For convenience, let the subsets a and o be the sets of measured and unmeasured model degrees of freedom, respectively. The set $[a, o]$ contains the total number of degrees of freedom of the structural model. Each mode shape vector ϕ_i can be partitioned in the form $\phi_i^T = [\phi_{ai}, \phi_{oi}]$, where ϕ_{ai} and ϕ_{oi} are the components of the mode shape ϕ_i at the measured and unmeasured degrees of freedom, respectively.

Let now $\tilde{\omega}_i$ and $\tilde{\phi}_{ai}$ be the experimentally obtained frequencies and modeshapes of the structure at the measured degrees of freedom. The proposed method for model updating searches for the optimal model parameters θ which minimize an appropriately selected norm of the modal dynamic force balance residuals $r(\omega_i, \phi_i, \theta)$ subject to conditions that reflect the fact that both the modal frequencies ω_i and modeshapes ϕ_i are sufficiently close, depending on the experimental error expected, to the measured modal frequencies and modeshape components. Mathematically, the model updating problem is stated as

$$\min_{\theta,\phi_i} \sum_{i=1}^{m} \|r(\tilde{\omega}_i, \phi_i, \theta)\|_{R_i} = \min_{\theta,\phi_i} \sum_{i=1}^{m} \left\|(K(\theta) - \tilde{\omega}_i^2 M(\theta))\phi_i\right\|_{R_i} \tag{4}$$

subject to

$$\left\|P\phi_i - \tilde{\phi}_{ai}\right\|^2 \leq \alpha_i \left\|\tilde{\phi}_{ai}\right\|^2, \quad i = 1, \cdots, m \tag{5}$$

where $\|x\|^2 = x^T x$ is the usual Eucledian norm, $\|x\|_R = x^T R x$, R_i is an appropriately selected weighting matrix which scales the contribution of each mode in the measure of fit (4), and P is a constant matrix of zeroes and ones such that $\phi_{ia} = P\phi_i$.

The unknown quantities involved in the proposed error measure (4) include the model properties θ, as well as the components of the vector ϕ_i of the contributing modes at both measured and unmeasured model degrees of freedom. The optimal vector ϕ_i resulting from the minimization can be viewed as the expanded modeshape consistent with the measured modal data.

Next, the measure of fit in (4) is further analyzed in order to determine a reasonable choice for the weights R_i. For this, consider first the contribution $J_i(\phi_i, \theta) = \|r(\tilde{\omega}_i, \phi_i, \theta)\|_{R_i}$ from the i-th modal term in the overall measure of fit (4). Using the model modeshapes φ_j, $j = 1, \cdots, N$ and expanding the vectors

ϕ_i in the form $\phi_i = \sum_{j=1}^{M} a_{ij}\varphi_j = \sum_{j=1}^{M}(\varphi_j^T M(\theta)\phi_i)\varphi_j$, the i-th modal residuals can be written in the form

$$J_i(\phi_i, \theta) = \sum_{j=1}^{N}\sum_{k=1}^{N} a_{ij}a_{ik}[\omega_j^2\omega_k^2 - \tilde{\omega}_i^2(\omega_k^2 + \omega_j^2) + \tilde{\omega}_i^4]\varphi_j^T M(\theta)R_i M(\theta)\varphi_k \qquad (6)$$

This expression can be simplified considerably by choosing R_i such that

$$\varphi_j^T M(\theta)R_i M(\theta)\varphi_k = \gamma_i^4 \delta_{jk} \qquad (7)$$

where δ_{jk} is the Kronecker delta, and γ_{ij} depends on the choice of R_i. Substitution of (7) into (6) gives

$$J_i(\phi_i, \theta) = \sum_{j=1}^{N}(\varphi_j^T M(\theta)\phi_i)^2 \frac{(\omega_j^2 - \tilde{\omega}_i^2)^2}{\tilde{\omega}_i^4}(\gamma_i\tilde{\omega}_i)^4 \qquad (8)$$

The choice (7) was preferred because it is the only one that results in positive terms in the modal measure of fit (6).

Note that for $\phi_i = \varphi_i$, $i = 1, \cdots, m$, i.e. the case of perfectly correlated expanded and model mode-shapes, all but the term corresponding to $j = i$ in the modal measure $J_i(\phi_i, \theta)$ disappear. The modal measure $J_i(\varphi_i, \theta)$ becomes proportional to the fractional difference between the squares of the model and measured modal frequencies for mode i, weighted by $(\gamma_i\tilde{\omega}_i)^4$. This equivalence between the measure $J_i(\varphi_i, \theta)$ and the more direct measure involving the difference between the squares of the model and measured modal frequencies was first reported in a recent study (Vanik, 1997). In the general case for which $\phi_i \neq \varphi_i$, all terms in the modal error measure (8) are present. In particular, the terms in the modal error measure (8) corresponding to $j \neq i$ involve the mass orthogonality condition between the expanded and model modeshapes, weighted by the factors $(\gamma_i\tilde{\omega}_i)^4(\omega_j^2 - \tilde{\omega}_i^2)^2/\tilde{\omega}_i^4$. Note that for a model which is well-correlated with the measured data, the factors $(\varphi_i^T M(\theta)\phi_i)^2 \approx 1$ and $(\varphi_j^T M(\theta)\phi_i)^2 \approx 0$ for $j \neq i$. Therefore, in the process of selecting the optimal model, the mass orthogonality conditions are also enforced through the terms in (8) corresponding to $j \neq i$.

The term in (8) corresponding to $j = i$ provides insight into the problem of choosing the weights R_i. From the aforementioned analysis, it becomes apparent that weighting the modal error measures $J_i(\phi_i, \theta)$ is equivalent to weighting the errors between the squares of the experimental and model modal frequencies. Thus, R_i can be selected such that errors between the experimental and model modal frequencies are weighted for each mode according to weights β_i. This is accomplished by choosing γ_i so that the factor $(\gamma_i\tilde{\omega}_i)^4$ is proportional to non-dimensional weights β_i. From a Bayesian statistical point of view, the weights β_i reflect the magnitude of the measurement errors expected between the experimental and model frequencies for each mode (Beck, 1989; Vanik, 1997). The size of these errors can be obtained from measurement data taken from repeated modal test analyses.

Several choices for the weights R_i can be made to satisfy condition (7) and, at the same time, guarantee that the factor $(\gamma_i\tilde{\omega}_i)^4$ is proportional to the non-dimensional quantity β_i. Attention is only given to the following two choices:

$$R_i = \beta_i M^{-1}(\theta)/\tilde{\omega}_i^4 \quad \Longrightarrow \quad (\gamma_i\tilde{\omega}_i)^4 = \beta_i \qquad (9)$$

$$R_i \quad = \quad \beta_i K^{-1}(\theta) M(\theta) K^{-1}(\theta) \qquad \Longrightarrow \qquad (\gamma_i \tilde{\omega}_i)^4 = \beta_i \tilde{\omega}_i^4 / \omega_j^4 \tag{10}$$

For the first choice to be applicable, it is required that $M^{-1}(\theta)$ is non-singular. Thus, it is not applicable for structures with zero mass degrees of freedom. However, this problem can easily be resolved by applying Guyan model reduction to eliminate the degrees of freedom corresponding to zero mass. For the second choice to be applicable, it is required that the matrix $K(\theta)$ is non-singular. Thus, it cannot be applied to structures that are not supported or they are partially supported such as those employed in space or tested in the lab by suspending them by very soft springs.

Finally, the inequality constraints (5) are introduced to account for the expected measurement error in the mode-shape components, with α_i controlling the expected magnitude of these errors. The value of α_i can be computed from a statistical analysis of measurement data taken from repeated modal test analyses. It is worth pointing out that the methodologies presented by Farhat and Hemez (1993) and Alvin (1997) are special cases of the measure (4) and condition (5). In particular, both methods correspond to values $\alpha_i = 0$, which fail to directly incorporate the expected measurement error. In contrast, the proposed inequality condition provides more flexibility in improving considerably the fit between model and measured modal data over the space of the parameter set θ.

The optimization in (4) and (5) can be performed using available inequality constaint optimization techniques. However, this is a complex and time-consuming nonlinear optimization problem. A more convenient three-step iterative procedure is proposed next which avoids some of the computational difficulties arising in the minimization of (4). In addition, the iterative procedure provides guidance in identifying the locations of damage and limiting the number of the parameters to be updating to only a few, thus reducing the problem of ill-conditioning and non-uniqueness expected when a large number of parameters is updated.

STEP 1: MODESHAPE EXPANSION

Given the current model of the structure at the k-th iteration step, corresponding to the value of the parameter set θ, designated by $\hat{\theta}^{(k)}$, expanded modeshapes are computed by solving the constrained minimization problem:

$$\min_{\phi_i} \sum_{i=1}^{m} \left\| (K(\hat{\theta}^{(k)}) - \tilde{\omega}_i^2 M(\hat{\theta}^{(k)})) \phi_i \right\|_{R_i} \tag{11}$$

subject to

$$\left\| P\phi_i - \tilde{\phi}_{ai} \right\|^2 \leq \alpha_i \left\| \tilde{\phi}_{ai} \right\|^2, \quad i = 1, \cdots, m \tag{12}$$

The minimization is performed with respect to the modeshape components at both measured and unmeasured degrees of freedom while holding the values of the model parameters θ fixed at their current values $\hat{\theta}^{(k)}$. Both the objective function and the inequality constraints are quadratic in the set of unknown parameters. It can be shown that a unique optimum exists (Levine-West, Milman and Kissil, 1997), denoted herein by $\hat{\phi}_i^{(k+1)}$, $i = 1, \cdots, m$. The algorithm for obtaining the unique solution is described in the work by Levine-West, Milman and Kissil (1996, 1997).

STEP 2: LOCATION OF DAMAGE

The expanded modeshapes predicted in the first step are used to identify possible locations of damage by examining, for each finite element (or substructure), the difference in element strain energy between the expanded mode-shapes and the model mode-shapes. The modal element (or substructure) strain energy for a finite element (or substructure) designated by $< e >$ is defined as

$$S^e(\phi_i) = (1/2)\phi_i^T K^e \phi_i \tag{13}$$

where K^e is the stiffness matrix assembled from the element or substructure. The following measure of modal strain energy error is used

$$\Delta S_i^e = \frac{S^e(\hat{\phi}_i^{(k+1)}) - S^e(\varphi_i^{(k)})}{S^e(\hat{\phi}_i^{(k+1)})} \tag{14}$$

where $\varphi_i^{(k)} = \varphi_i(\theta^{(k)})$ is the modeshape computed form the current structural model. It is expected that sufficiently large ΔS_i^e will be due to modeling errors in the particular element (or substructure) and will be indicative of probable damage in the element (or substructure). The properties of these elements (or substructures) are chosen to be updated if $|\Delta S_i^e| > tol_1$ for any mode i, where tol_1 is a user-specified threshold. These properties form an active subset of the parameter set θ, designated by $\theta_{act}^{(k+1)}$. The rest of the parameters in θ that are not included in the active set form the inactive set, designated by $\theta_{in}^{(k+1)}$. The properties of the finite element model included in the active set $\theta_{act}^{(k+1)}$ may differ from those in the set $\theta_{act}^{(k)}$ obtained form the previous iteration.

STEP 3: SIZE OF DAMAGE

In this step, the element stiffness, mass and geometrical properties at the identified probable locations of damage are updated using the latest estimates $\hat{\phi}_i^{(k+1)}$, $i = 1, \cdots, m$ of the complete modeshapes. Only the values of the active model parameters θ_{act} identified in step 2 are updated. The values of the inactive set $\theta_{in}^{(k+1)}$ are kept constant and equal to those in the set $\hat{\theta}^{(k)}$ determined in the previous iteration. The solution of the following unconstrained minimization problem:

$$\min_{\theta_{act}^{(k+1)}} J(\theta^{(k+1)}) = \min_{\theta_{act}^{(k+1)}} \sum_{i=1}^{m} \left\| (K(\theta^{(k+1)}) - \tilde{\omega}_i^2 M(\theta^{(k+1)}))\hat{\phi}_i^{(k+1)} \right\|_{R_i} \tag{15}$$

gives the optimal values $\hat{\theta}_{act}^{(k+1)}$. This is a nonlinear optimization problem which can be solved using available iterative schemes such as modified Newton method.

For the case for which $K(\theta)$ and $M(\theta)$ are linear function of θ and R_i is independent of θ, the objective function $J(\theta)$ is quadratic in θ and the unique solution $\hat{\theta}_{act}^{(k+1)}$ can be obtained without iterations by solving a linear algebraic system in θ.

The updated finite element model contains inaccuracies due to the fact that the expanded modeshapes are based on an inaccurate model obtained at the previous iteration step k. Thus, the three step procedure has to be repeated using the new updated finite element model until convergence is reached. Specifically, the iterative process is terminated when $\left\|\hat{\theta}^{(k+1)} - \hat{\theta}^{(k)}\right\| / \left\|\hat{\theta}^{(k+1)}\right\| < tol_2$ where tol_2 is a user-specified threshold level. Finally, The reduction in the values of θ_{act} of the parameter set θ is indicative of the probable locations and size of damage.

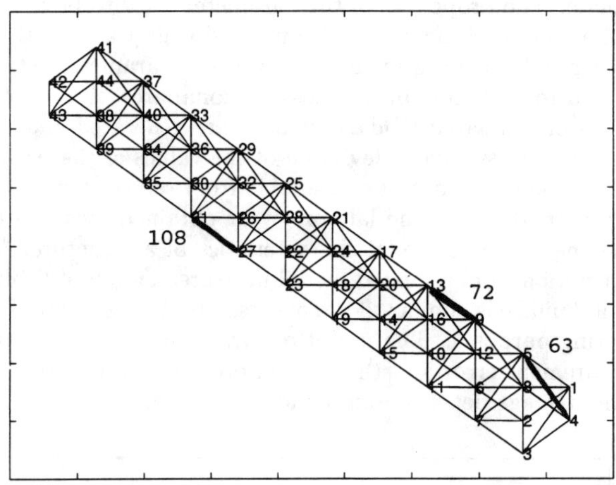

Figure 1: Three-dimensional truss structure

APPLICATION

The finite element model updating technique has been implemented in matlab to enhance the capabilities of the existing Integrated Modeling of Optical Systems (IMOS) software package developed at Jet Propulsion Laboratory. The model of the undamaged structure is a three-dimensional truss shown in Figure 1. It consists of 135 axial rod elements (1 per strut) with total of 120 degrees of freedom (3 per node). The structure is supported by restraining all degrees of freedom at nodes 1 to 4. The modulus of elasticity, cross-sectional area and mass density are the same for all members. The values are chosen such that the first eight modal frequencies of the model range from 10 Hz to 200 Hz.

Simulated modal data are used to assess the performance of the proposed damage detection methodology. The element 63, 72 and 108, located at different places on the structure as shown in Figure 1, are damaged by reducing the cross-sectional area of these element by 50%. The modal test data are produced by calculating the modal data for the lower eight modes of the damaged model and by adding Gaussian white noise to simulate measurement and model errors encountered in practice. The standard deviation of the noise is taken as 1% and 5% of the values of the modal frequencies and mode-shapes, respectively.

Two case studies are used to assess the performance of the method in relation to the number and location of sensors. In the first case, designated by Case A, a large number of 99 sensors are used. These sensors are places at nodes 5 through 37 to provide measurements at all three degrees of freedom for each node. For the second case, designated by Case B, only 15 sensors are used which are placed at nodes 5, 13, 21, 29 and 37 to provide measurements at all three degrees of freedom per node. The properties in the parameter set θ to be updated are the cross-sectional area of each member. The methodology was slightly modified to consider as acceptable changes in the values of the parameter set θ only those which correspond to reduction in the cross-sectional area of the members.

Multiple sets of simulated modal test data are generated and used to compute the mean and the standard deviation of the values of the parameter set θ. Multiple sets of modal test data are available from repeated modal test experiments usually carried out in the laboratoty or obtained by monitoring over a period of time the dynamic modal characteristics of a structure. This idea of using the information from repeated modal measurements to establish the location and size of damage is similar to the one used by Vanik (1997) for structural health monitoring purposes. The use of repeated measurement have shown to filter out measurement error and thus can improve substantially the predictive accuracy of the damage detection methodology.

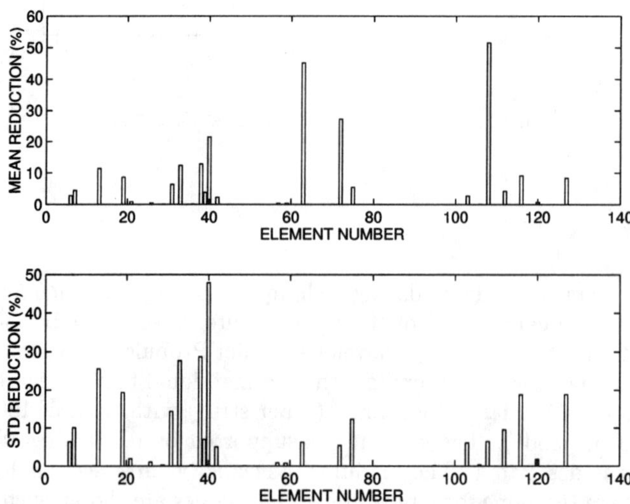

Figure 2: Predicted locations and magnitude of damage; Case A

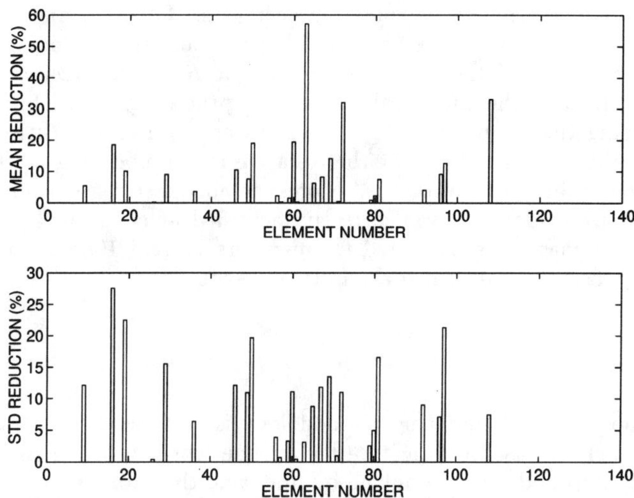

Figure 3: Predicted locations and magnitude of damage; Case B

Using the weights R_i given by (10) with $\alpha_i = 0.1$ and $\beta_i = 0.02$ for all modes, the predicted location and size of damage is shown in Figures 2 and 3 for the cases A and B, respectively. The results for the mean and the standard deviation of the predictions are based on five sets of simulated data. For case A, the predicted mean reductions in the cross-sectional area of members 63, 72 and 108 are 45% 28% and 52% with standard deviations 6%, 5% and 8%, respectively. These three members have correctly been identified as the damaged members with the highest mean reduction in cross-sectional area. The relatively small values of the standard deviation of these estimates is indicative of the relatively high confidence that damage has occurred in these members. In contrast, the standard deviation estimates of the rest of the members with non-zero mean reduction of cross-sectional area are relative large. This is due to the fact that only a small percentage of the data sets have resulted in non-zero reduction in cross-sectional area of these members. Specifically, 3 to 4 out of the 5 data sets predicted no reduction or almost insignificant reduction in the cross-sectional area for these members. The use of a small size of simulated modal data sets has resulted in relatively high mean reduction values. As the number of modal data and therefore the number of modal tests increases, the mean values and the standard deviation for these members decreases. The results for the case B show a similar pattern. The predicted mean reductions in the cross-sectional area of members 63, 72 and 108 are 58%, 32% and 33% with standard deviations 5%, 11% and 7%, respectively. It is worth noting that the resolution of the size of damage at element 108 is not as good as for the Case A because sensors are not directly placed in the vicinity of the member 108. However, the elements 63, 72 and 108 have been correctly identified as the damaged elements.

Extensive numerical studies has also been carried out which show that the accuracy of the predictions increases as the number of measured modes increases, or as the level of the measurement error decreases. Location and number of sensors also play a role in the resolution of location and size of damage.

Finally, the effect of the choice of the weight R_i on the results was also explored by repeating the numerical studies using the weight R_i defined by (9), as well as using the weight $R_i = I$ for all i, where I is the identity matrix. For both weights it was found that the iterative model updating methodology has much slower rate of convergence. Moreover, for most cases examined, the location and size of damage was correctly identified at element 63. However, no significant damage was predicted for members 72 and 108, although both members were correctly identified as faulty elements.

CONCLUSIONS

The proposed model updating methodology is suitable for damage detection purposes. It is based on an iterative scheme which provides estimates of probable locations and size of damage by updating the properties of the finite element model at the element level. The identification of the probable locations of damage is based on an element strain energy error measure between the expanded modeshapes predicted at a given iteration and model modeshapes predicted from the previous iteration. The size of damage is then updated using the predicted expanded mode-shapes. These estimates are iteratively updated until convergence is reached. A study using simulated data demonstrated that the methodology is promising for reliably predicting both the location and the size of damage in a structure. Measurement error was incorporated in the data by adding noise in the simulated data. The noise levels considered are similar to those expected in practical applications. Although the method suggested herein works well with simulated modal data and simulated measurement error, the practical use of this method with real data requires further study.

ACKNOWLEDGEMENT

The research described in this paper was carried out by the Jet Propulsion Laboratory, California Institute of Technology, under a contract with the National Aeronautics and Space Administration.

REFERENCES

1. Alvin, K.F., 1997. "Finite Element Model Update via Bayesian Estimation and Minimization of Dynamic Residuals." *AIAA J.*, 35(5):879-886.

2. Beck, J.L., 1989. "Statistical System Identification of Structures," in *Proc. 5th Int. Conf. on Structural Safety and Reliability, ASCE, II*, pp. 1395-1402.

3. Beck, J.L., and L.S. Katafygiotis, 1997. "Updating Structural Dynamic

Models and their Uncertainties - Statistical System Identification." *J. Engineering Mechanics, ASCE*, in print.

4. Beck, J.L, and M.W Vanik, 1996. "Structural Model Updating Using Expanded Modeshapes," in *in Proc. 11th Engineering Mechanics Conf.*, Y.K. Lin and T.C. Su, eds. ASCE, NY, pp. 152-155.

5. Berman, A., 1989. "Nonunique Structural System Identification," in *Proc. 7th Int. Modal Analysis Conference*, pp. 355-359.

6. Farhat, C., and F.M. Hemez, 1993. "Updating Finite Element Dynamics Models Using an Element-by-Element Sensitivity Methodology." *AIAA J.*, 31(9):1702-1711.

7. Hemez, F.M., and C. Farhat, 1995. "Structural Damage Detection via a Finite Element Model Updating Methodology." *The Int. Journal of Analytical and Experimental Modal Analysis*, 10(3):152-166.

8. Katafygiotis, L.S., and J.L. Beck, 1997. "Updating Structural Dynamic Models and their Uncertainties - Model Identifiability." *J. Engineering Mechanics, ASCE*, in print.

9. Levine-West, M.B., M. Milman, and A. Kissil, 1996. "Modeshape Expansion Techniques for Prediction: Experimental Evaluation." *AIAA J.*, 34(4):821-829.

10. Levine-West, M.B., M. Milman, and A. Kissil, 1997. "Modeshape Expansion Techniques for Prediction: Analysis." *AIAA J.*, submitted for publication.

11. Mottershead, J.E., and M.I. Friswell, 1993. "Model Updating in Structural Dynamics: A Survey." *J. Sound and Vibration*, 167(2):347-375.

12. Natke, H.G., and J.T.P. Yao, Eds, 1988. *Structural Safety Evaluation Based on System Identification Approaches*, Proc. of Workshop at Lambrecht/Pfatz, Frieder Vieweg and Sohn, Braunschweig, Germany.

13. Stubbs, N., T.H. Broome, and R. Osegueda, 1990. "Nondestructive Construction Error Detection in Large Space Structures." *AIAA J.*, 28(1):146-152.

14. Vanik, M.W, 1997. "A Bayesian Probabilistic Approach to Structural health Monitoring," EERL Report, Caltech, Pasadena, California.

Locating Structural Damage Using Frequency Response Reference Functions and Curvatures

M. J. SCHULZ,[1] P. F. PAI,[1] A. S. NASER,[1] S. K. THYAGARAJAN,[1] G. R. BRANNON[1] and J. CHUNG[2]

ABSTRACT

A technique to detect and locate damage by measuring the vibration response of a structure is presented. The technique uses measured frequency response functions from the healthy structure as reference data, and then monitors vibration measurements during the life of the structure to detect damage. Damage is located by defining a combined force vector (damage force vector + external force vector) and computing the cross-spectral densities between entries of this vector. The technique requires that the excitation forces are uniform, random, and uncorrelated, but the excitation force does not need to be measured and no structural model is needed. The accuracy of locating damage depends on the spatial density of the vibration measurements taken on the structure. A simpler technique that uses curvatures to detect damage is also presented. These two techniques combined are proposed as a health monitoring system for structures. A finite-element model of a beam is used to demonstrate this approach.

INTRODUCTION

Aircraft, rotorcraft, reusable launch vehicles and many other advanced structures are being built using lightweight composite materials/metals with design safety factors as low as 1.25 to 1.5. These advanced structures operate in uncertain and severe environments comprised of vehicle dynamic loads, unsteady aerodynamic loads, engine vibration, foreign object impact, lightning strikes, corrosion, and moisture absorption, and they are susceptible to damage such as delamination, fiber breaking/pullout, matrix cracking, and hygrothermal strain in composite materials, and fatigue and cracking in metals. To ensure human safety and load-

[1]Structural Mechanics and Control Laboratory, Dept. of Mechanical Engineering, North Carolina A&T State University, Greensboro, NC 27411, [2]Raytheon E-Systems, Greenville, Texas 75403

bearing integrity these structures must be inspected to detect and locate often invisible damage and faults as they occur, before becoming catastrophic.

Present techniques for non-destructive evaluation of structures include conventional methods such as ultrasonic scanning, radiographic inspection, visual inspection, dye penetrant, acoustic emissions, and smart paints, and new methods such as embedding fiber-optic wires, and vibration signature analysis methods. Conventional methods sometimes miss significant damage and are time consuming and expensive to perform, whereas embedding wires is difficult on a large scale and to retrofit to existing structures. In contrast, vibrometry or vibration signature techniques are a global method of structural integrity monitoring that potentially can efficiently detect damage on large structures, including damage that is away from sensor locations, and in the interior of structures. This can be done during maintenance inspections or during operation without affecting the integrity of the structure. Thus there is an enormous motivation behind the current intense research in vibrometry approaches to health monitoring [James, 1996; Doebling et al, 1996; Stevens et al, 1996; James et al, 1997]. Nonetheless, there is also controversy over how effective diagnostic structural vibrometry techniques can be [Friswell and Penny, 1997]. The central problems in developing health monitoring techniques are that Finite-Element-Models (FEMs) are not accurate for detecting small damage, the structural response must be measured at a large number of points to locate damage, the dynamic characteristics of a structure change due to environmental effects, the sensor system used for damage detection is sometimes more unreliable than the structure being monitored, and it is difficult to model the effects of damage, such as a crack, on the structural response. These issues are being addressed by many researchers and results and approaches are being presented that show progress and the potential to design practical health monitoring systems based on structural dynamics principles. Some background on emerging techniques for damage detection is given below, and then two new algorithms for health monitoring are presented.

Damage detection by vibration analysis can be done in the time, frequency, or modal domains, can be model-referenced or model-independent, and use either direct, optimization, or artificial neural network solution techniques. A time domain method [Banks et al, 1996] and an acoustic pinging method [Joshi and Bayard, 1995] have been applied to simple structural shapes. A Frequency Response Function Optimization technique [Schulz et al, 1996a] can relate damage to changes in physical properties of the structure, but the method is computationally intensive. A Frequency Response Function Assignment technique [Schulz et al, 1996b] can monitor critical sections of a structure to diagnose damage, but a model of the structure is required. Canceling the input force and environmental effects by normalization of responses is done in a Transmittance Function method [Schulz et al, 1997; Naser et al, 1997] for damage detection. Modal analysis methods [Lindner and Kirby, 1994; Kim and Bartkowicz, 1995; Zimmerman and Kaouk, 1992] provide simple damage detection algorithms based on the orthogonality of modes, proportional damping, and computing the Frequency Response Function (FRF) matrix using only one force input, but are inaccurate when damping is large/nonuniform or the modes are closely spaced [Lyon, 1995]. Small scale damage such as microcracking can be detected locally using a high frequency impedance technique [Sun et al, 1995]. Impedance techniques combine the sensor/actuator allowing more transfer functions to be determined, but they use high-frequency short-wave excitation and are disturbed by wave reflections and are sensitive to external noise, boundary, and temperature/loading conditions. Impedance methods also require large computer storage, peaks in the impedance function are due to resonance vibration of the structure and resonance of the electric circuit, and when they operate in the high-frequency range no FEM can be used as a guide in damage detection because models cannot accurately predict high-frequency dynamics. For in-situ damage detection, a direct solution [Zimmerman et al, 1995] approach reduces computations. An artificial neural network approach [Tsou and Shen, 1994] does not need a model but requires extensive training and may not extrapolate damage on large complex structures.

A Scanning Laser Doppler Vibrometer (SLDV) and holography technique have successfully detected damage to an aircraft fuselage section [James, 1996]. The holography technique uses sinusoidal excitation and the SLDV to produce images of the amplitude contours or fringe patterns of the vibrating structure on film. The amplitude contours can also be generated by computing FRFs between the force transducer and all the response points over the surface of the structure. The magnitude of the FRFs and the relative phase between response points are used to map a picture of vibration amplitude contours or operational deflections over the surface being interrogated. The fringe patterns or contours are examined visually and any abnormalities indicate damage (e.g. the fringes in a flexible skin structure extending into a rib section). The SLDV operational deflection and holography methods work well when there is relatively large surface damage to the structure wherein the low frequency vibration modes are changed. The SLDV can also be used to detect damage in composite materials by measuring the velocities of the surface vibration and damage is indicated where the velocities are low.

Surveys of current health monitoring techniques are given in [Doebling et al, 1996; James, 1996; James et al, 1997; Stevens et al, 1996] and it is apparent that no accepted universal technique exists. A frequency domain approach for damage detection is derived below. This method has the advantages that the excitation does not need to be measured, and no FEM is needed.

DAMAGE DETECTION USING FREQUENCY RESPONSE FUNCTIONS

In many structures it is impractical to use a finite-element model to diagnose damage because the model would not be accurate enough, and geometry changes due to cracks and nonlinear effects would be difficult to include. The Frequency Response Reference Function (FRRF) method uses full FRF measurements from the healthy structure to identify an input-output frequency domain model of the structure, where the number of model DOFs is equal to the number of accelerometers or sensors on the structure. This approach is accurate for detecting and locating damage, but the level and type of damage is not diagnosed.

The linear structure is assumed to be represented by:

$$M\ddot{\mathbf{x}} + C\dot{\mathbf{x}} + K\mathbf{x} = \mathbf{f}(t) \tag{1}$$

Taking the Fourier transform of (1) gives:

$$A(j\omega)\mathbf{x}(j\omega) = \mathbf{f}(j\omega) \tag{2}$$

where the system matrix $A(j\omega) = (K - \omega^2 M + j\omega C)$. Solving for the displacement vector gives:

$$\mathbf{x} = H\mathbf{f} \tag{3}$$

where $H = A^{-1}$ is the system FRF matrix and the $(j\omega)$ argument has been dropped for brevity. The damage location procedure to be developed requires that the H matrix be identified for the healthy structure in the bandwidth used to detect damage to the structure. The dimension of $\mathbf{x}$ is $nx1$ for n measurements on the structure (e.g. using accelerometers, PZT patches, strain gages, or a laser vibrometer), and thus the nxn H matrix must be determined experimentally for the healthy structure. The individual M, C, and K matrices do not need to be known.

To identify the H matrix for the healthy structure, a force, either random, sine, or impulse, is applied at each DOF one at a time. For example, for a random input the elements of the force vector are $f_i = 0, (i \neq k)$, and f_k is a random force. Then by multiplying both sides of (3) on the right by the complex conjugate force and taking expectations, the equation becomes:

$$\mathbf{h}_k = \mathbf{g}_{xf} / |f_k|^2 \quad k = 1,2,3,\ldots n \tag{4}$$

where $\mathbf{h}_k$ is the kth column of H, $\mathbf{g}_{xf}$ is the cross-spectral density vector between the displacement vector and the input force, and $|f_k|^2$ is the mean-square value of the random force. Thus from (4) each column and hence the full H matrix can be identified. Note that the H matrix is symmetric unless gyroscopic forces are present. A practical problem with computing the H matrix is that it is difficult to apply moments to obtain the FRFs for rotational DOFs. Offset fixtures to apply moments are currently being tested with a finite-difference approximation of angles to obtain rotational FRFs. Once H is determined, the A matrix is calculated as $A = H^{-1}$ at each frequency point in the bandwidth that is used to detect damage, and the A matrices are stored as historical data. Now a damage vector is defined as:

$$\mathbf{d} = A^h \mathbf{x}^d - \mathbf{f}^d \tag{5}$$

where h and d represent the healthy and damaged structure. If damage occurs the **d** vector will have non-zero forces only at the DOFs that are connected to the damaged elements. For example, if damage causes changes to the ith row of the stiffness matrix, only the ith entry of **d** will be nonzero. To locate damage using (5) requires that the excitation forces are known, and also the A matrix may change due to environmental effects such as temperature, pressure, or aging of the structure. A new damage indicator is developed to reduce these limitations by first defining a combined force vector (damage force + external force) as:

$$\mathbf{r} = (\mathbf{d} + \mathbf{f}^d) = A^h \mathbf{x}^d . \tag{6}$$

Multiplying **r** by its complex conjugate transpose and taking the expectation of both sides of (6) we obtain:

$$R = \mathrm{E}(\mathbf{r}\mathbf{r}^*) = \mathrm{E}(\mathbf{d}\mathbf{d}^* + (\mathbf{d}\mathbf{f}^{d*} + \mathbf{f}^d\mathbf{d}^*) + \mathbf{f}^d\mathbf{f}^{d*}) = A^h G^d_{xx} A^{h*} \tag{7}$$

where E is the expectation and $G^d_{xx}(j\omega)$ is the matrix of cross-spectral densities of the displacements. Using $\mathrm{E}(u+v) = \mathrm{E}(u) + \mathrm{E}(v)$, equation (7) becomes:

$$R = \mathrm{E}(\mathbf{d}\mathbf{d}^*) + \mathrm{E}(\mathbf{d}\mathbf{f}^{d*} + \mathbf{f}^d\mathbf{d}^*) + \mathrm{E}(\mathbf{f}^d\mathbf{f}^{d*}) = A^h G^d_{xx} A^{h*} . \tag{8}$$

We assume that the forces represented by **f** are random and uncorrelated, and approximately the same magnitude. The entries of R are defined as:

$$R_{ij} = \mathrm{E}(d_i d_j^*) + \mathrm{E}(d_i f_j^{d*} + f_i^d d_j^*) + \mathrm{E}(f_i^d f_j^{d*}) \tag{9}$$

where d_i is the i th entry of **d**. Damage to the structure can be located by the off diagonal terms of R. Thus we define a damage indicator matrix as:

$$D^2 = \frac{1}{\Delta f} \int_{f_1}^{f_2} R \circ R^* df, \quad (i \neq j), \quad D^2 = 0, \quad (i = j) \tag{10}$$

where f_1>0, $\circ$ denotes element by element multiplication, and $\Delta f = f_2 - f_1$. If the damage (change in stiffness, damping, or mass) is between DOFs i and j, then $R_{ij} \neq 0$ and $D^2_{ij} \neq 0$ and thus damage is located between DOFs (i.e. measurements) i and j on the structure.

During operation of the structure the damage indicator matrix $\overline{D}^2$ is actually computed with the R values given by (8). The elements of R can be written as:

$$R_{ij} = \sum_{p=1}^{n} \sum_{q=1}^{n} A^h_{ip} G^d_{pq} A^{h*}_{qj} \qquad (i \neq j) \tag{11}$$

where A^h_{ip} are the entries of the A^h matrix, and G^d_{pq} are entries of the G^d_{xx} matrix. Equation (11) shows that the damage indicator does not require measurement of the input forces. Global changes in the structural properties due to environmental effects such as changes in temperature and pressure will be interpreted as damage and this problem is being worked on considering normalization of (11) and frequency shifting. The advantage of this approach for damage detection is that, after surveying the healthy structure, only the random structural accelerations or velocities need to be measured to locate damage because $G_{xx} = G_{vv} / \omega^2 = G_{aa} / \omega^4$. The technique is also simple and suitable for continuous damage monitoring during operation of the structure, and for certain applications a scanning laser vibrometer can be used to locate single or multiple damage sites.

A limitation of the FRRF method, and all methods, is that it may not be possible to measure a large number of DOFs to detect damage, and thus the damage detection algorithm must use incomplete measurements or else approximate unmeasured coordinates. Typically, a model reduction procedure such as Guyan reduction or dynamic reduction/expansion is used to eliminate or approximate the unmeasured coordinates based on a model of the structure in the healthy condition. However, the reduction/expansion will almost always include DOFs near the damage and this causes an incorrect damage indication because the reduction/expansion based on the healthy model is incorrect for the damaged structure [Schulz et al, 1996a]. The approach taken here to the problem of incomplete measurements is to estimate the unmeasured DOFs from the measured transverse displacements of the damaged structure using finite-differences. This will provide a repeatable approximation that includes the effects of damage and eliminates the error caused by using information from the healthy model to perform a reduction on a damaged structure. The finite-difference approximation depends only on the geometry of deformation, not the model properties, and this improves the accuracy of damage detection.

DAMAGE DETECTION USING CURVATURES

The curvature at a point on a beam for small deflections is:

$$\rho = \partial^2 y / \partial x^2 \tag{12}$$

where x is the position along the beam, and y is the transverse displacement. For each element on the beam the curvature is computed approximately as:

$$\rho_i = (\theta_{i+1} - \theta_i) / l \tag{13}$$

where $(\theta_{i+1} - \theta_i)$ is the difference in angles at the ends of the element, and l is the element length. Damage is located by the percent change in curvature between the healthy and damaged structures, as:

$$d_i = |(\rho_i^d - \rho_i^h) / (\rho_i^h)| * 100 \tag{14}$$

It is required that the force is the same magnitude and location for testing both the healthy and damaged structures, and RMS curvatures from the healthy structure are stored as historical data.

DAMAGE DETECTION EXAMPLE USING A BEAM MODEL

An example is presented in which damage to a fixed-free and a fixed-fixed beam is determined. The finite-element model of a uniform fixed-free beam is shown in Fig. 1. A random force is applied at DOF 7, which is at the center of the beam. Damage, modeled as a 25% reduction in the values of the elemental stiffness matrix, is put in elements 1 through 8 and the damage indicators are plotted in Fig. 1. The damage location (peak value of indicator) is correct for each damage case, for both methods. Note that, for the FRRF method, the rotational DOFs are very sensitive to damage whereas the translational DOFs are relatively insensitive to damage. This has an implication for techniques that use mode shapes to detect damage. Since only translational DOFs are measured in modal analysis, very high wavenumber (1/wavelength) modes with large curvatures may be needed to detect small damage, and these are impractical to measure. The curvature method also located the damage by the peak amplitudes of the damage indicator, but the damage location is not as precise. Damage near the fixed end of the beam is more difficult to locate as it does not have much effect on the dynamics of the beam.

The finite-element model of a uniform fixed-fixed beam is shown in Fig. 2. The loading and damage conditions are the same as for the cantilever beam, and the results for damage detection are shown in Fig. 2. Again, both methods can predict the damage location. When lower levels of damage are considered the FRRF method indicates false damage at the force location, but still locates the damaged element. The curvature method works best at lower frequencies, and for small damage the method is sensitive to the frequency range used and sometimes the damage location is incorrect. Other results show that damage can also be predicted by both methods using finite-difference rotations, depending on the accuracy of the finite-difference approximation relative to the damage magnitude. Moreover, an improved approximation or a finer measurement pattern can improve the finite-difference results toward the exact solution. Conversely, the error always exists in model reduction procedures. Using rotational DOFs for health monitoring is an important area for continuing research. Although rotational DOFs are more sensitive to damage than translations, they are much smaller in magnitude. The small magnitudes make calculating rotational FRFs subject to numerical errors.

Advantages of the FRRF method are that; 1) the random force does not need to be measured; 2) the banding of the A matrix of the FEM model shows that only five simultaneous measurements are needed to check for damage using (10) at any single node on the beam; and 3) the accuracy of using finite-difference approximations can be improved by taking more measurement points using a SLDV. Damage will be accurately detected by measuring only translations and interpreting rotations. Based on numerical simulations, this will provide greater accuracy in damage detection than current methods that use the healthy model to eliminate or approximate unmeasured coordinates. Combining the FRRF and curvature methods provides a damage indicator system with a checking capability; when both methods agree, the damage prediction is likely correct.

Experiments to test the techniques are being performed and Fig. 3 shows the test beam with offset fixtures for rotational measurements, and the measured FRFs.

CONCLUSIONS

The FRRF method is potentially a useful health monitoring method for smaller structures where the full FRF matrix for the number of sensors used on the structure can be measured or approximated for the healthy structure. After the historical data is taken, only acceleration measurements from the on-line structure need to be measured to detect damage. Methods of measuring and approximating rotational FRFs, frequency shifting, and normalization to cancel environmental effects are under investigation for use with the FRRF method.

The curvature method is a useful health monitoring method because only curvatures from the healthy and damaged structure are needed to detect damage. The excitation must be the same for the before and after damage cases. A method that uses curvature transmittance functions to cancel the forces and environmental effects is also currently being developed. Use of a scanning laser vibrometer to compute finite-difference curvatures, and PZT patches to measure strain and hence curvatures, is also being investigated.

Combining the two damage detection methods presented, and the improvements discussed, may lead to a structural integrity monitoring system that provides reliability and confidence in the damage prediction.

REFERENCES

Banks, H.T., D.J. Inman, D.J. Leo, and Y. Wang, 1996. "An Experimentally Validated Damage Detection Theory In Smart Structures," *Journal of Sound and Vibration*, 191(5), 859-880.

Doebling, S.W., C.R. Farrar, M.B. Prime, and D.W. Shevitz, 1996. "Damage Identification and Health Monitoring of Structural and Mechanical Systems from Changes in their Vibration Characteristics: A Literature Review," *Los Alamos National Laboratory Report LA-13070-VA5*.

Friswell, M.J. and J.E. Penny, May 12-14, 1997. "The Practical Limits of Damage Detection and Location using Vibration Data," *11th Symp. on Structural Dynamics and Control, Blacksburg, VA*.

James, G. H., April 1996. "Development of Structural Health Monitoring Techniques Using Dynamics Testing," *SANDIA REPORT: SAND96-0810, UC-706*.

James, G.H., D.C. Zimmerman, C.R. Farrar, and S.W. Doebling, Jan. 30-Feb. 1, 1997. "Current Horizon for Structural Damage Detection Course," *Short Course Notes*, Orlando, Fla.

Joshi, S.S, and D.S. Bayard, March 1995. "Detecting Structural Failures Via Acoustic Impulse Responses," *JPL New Technology Report, NPO-19167.*

Kim, H.M., and T.J. Bartkowicz, 1995. "An Experimental Study For Damage Detection Using A Hexagonal Truss Structure," *AIAA-95-1116-CP*, pp. 3347-3356.

Lindner, D.K. and G. Kirby, April 18-22, 1994. "Location and Estimation of Damage In A Beam using Identification Algorithms," *35th Structural Dynamics and Materials Conference, Adaptive Structures Forum, Hilton Head, SC.*

Lyon, January, 1995. "Structural Diagnostics Using Vibration Transfer Functions," *Sound and Vibration Magazine*, pp. 28-31.

Naser, A.S., M.J. Schulz, P.F. Pai, P.F., W.N. Martin, D. Turrentine, and C. Wilkerson, July 6-12, 1997. "Health Monitoring of Composite Material Structures Using a Vibrometry Technique," *Fourth International Conference on Composites Engineering, Hawaii.*

Schulz, M.J., P.F. Pai, S.K. Thyagarajan, and J. Chung, April 18-19, 1996a. "Structural Damage Diagnosis by Frequency Response Function Optimization," *AIAA Dynamics Specialists Conference, Salt Lake City, Utah, AIAA-96-1223-CP,* pp. 187-199.

Schulz, M.J., P.F. Pai, and A.S. Abdelnaser, A.S., Feb. 12-15, 1996b. "Frequency Response Function Assignment Technique For Structural Damage Identification," *IMAC-XIV Conference, Dearborn Michigan*, pp. 105-111.

Schulz, M.J., P.F. Pai, A.S. Naser, M. Linville, and J. Chung, Feb. 3-7, 1997. "Detecting Structural Damage Using Transmittance Functions," *IMAC-XV Conference, Orlando Fla,* pp. 638-644.

Stevens, P.W., D.L. Hall, and E.C. Smith, June 4-6, 1996. "A Multidisciplinary Research Approach to Rotorcraft Health and Usage Monitoring," *American Helicopter Society 52nd Annual Forum, Washington, DC.*

Sun, F., C.A. Rogers, and C. Liang, 1995. "Structural Frequency Response Function Acquisition Via Electric Impedance Measurement Of Surface-Bonded Piezoelectric Sensor/Actuator", *AIAA-95-1127-CP*, pp. 3450-3458.

Tsou, P., and M.-H. Shen, January 1994. "Structural Damage Detection and Identification Using Neural Networks," *AIAA Journal,* 32(1), pp. 176-183.

Zimmerman, D.C., and M. Kaouk, 1992. "Structural Damage Detection Using A Subspace Rotation Algorithm," *AIAA-92-2521-CP*, pp. 2341-2350.

Zimmerman, D.C., T. Simmermacher, and M. Kaouk, 1995. "Model Correlation And System Health Monitoring Using Frequency Domain Measurements," *AIAA-95-1113-CP*.

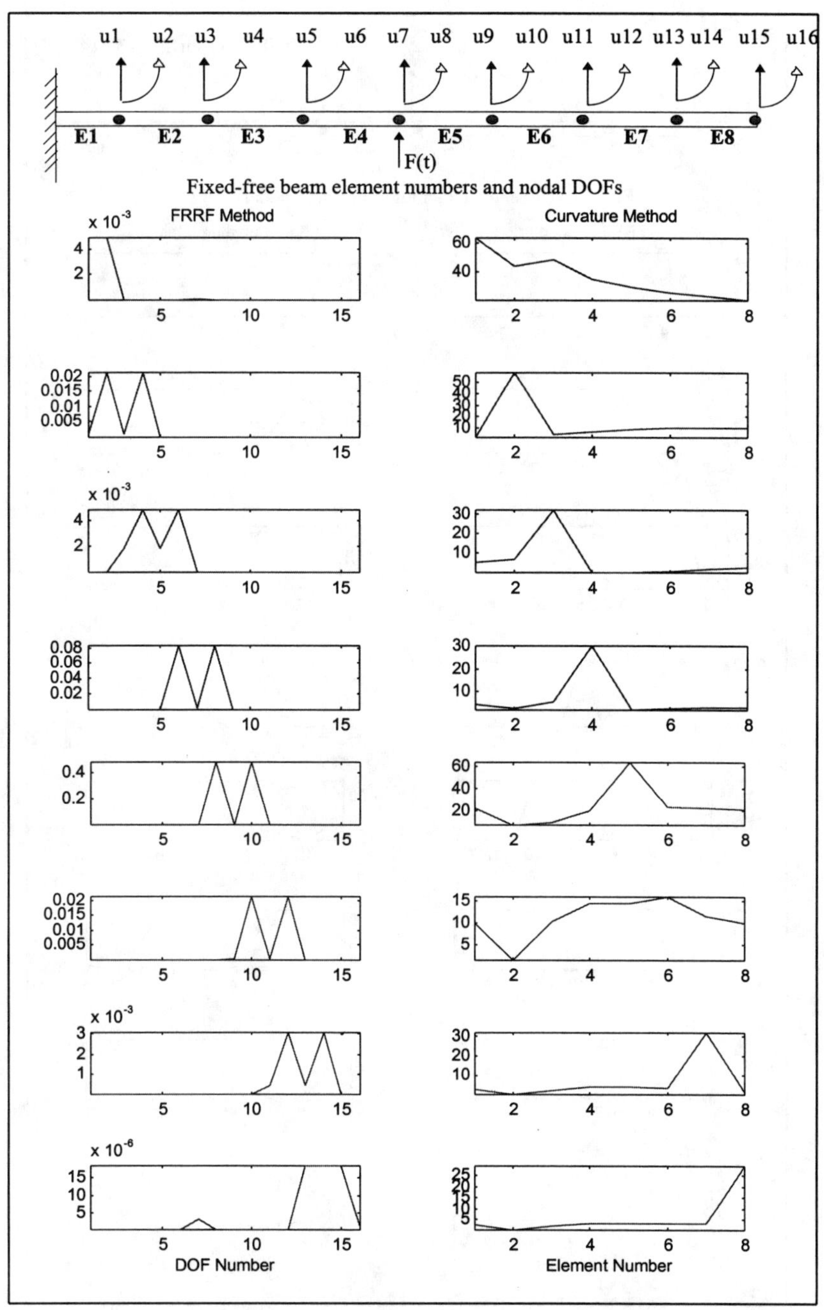

Figure 1. Damage indicators for fixed-free beam using exact response (for 25% damage to elements 1-8 (top to bottom), random force at beam center, 0.1-100 Hz)

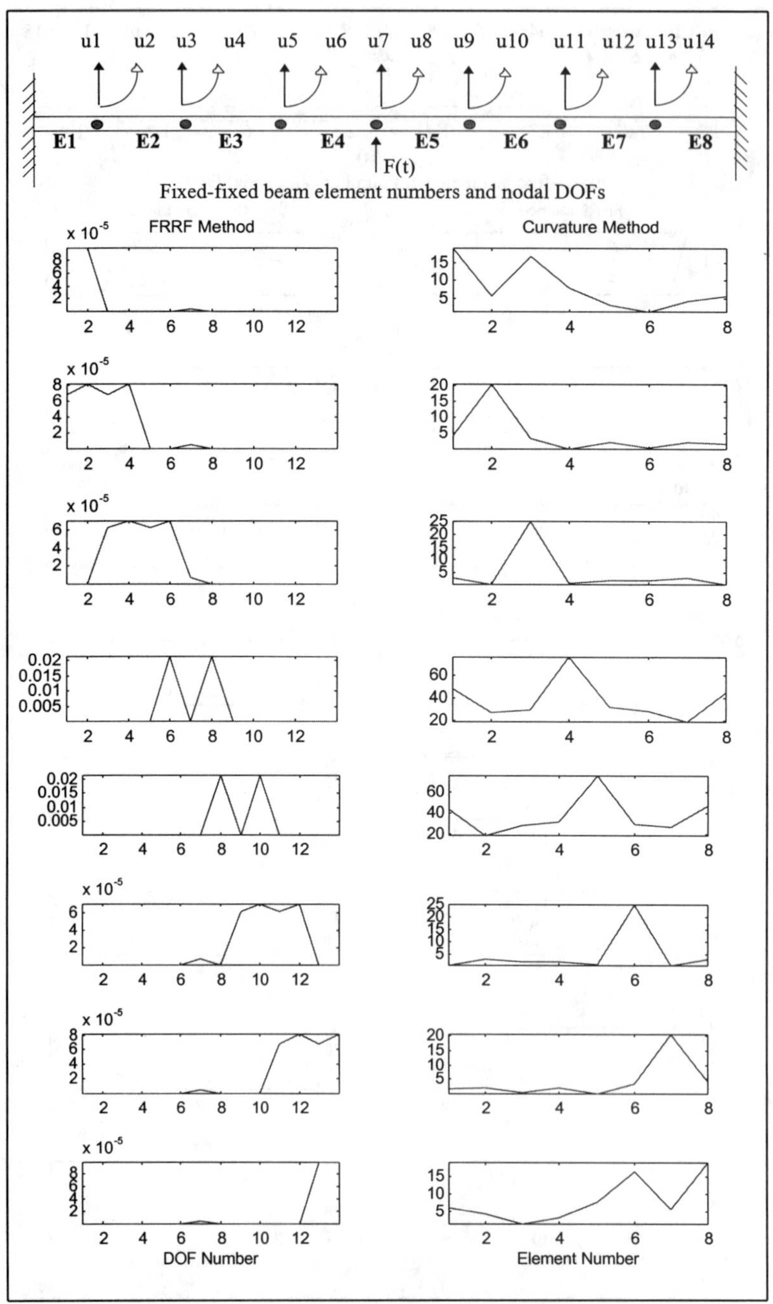

Figure 2. Damage indicators for fixed-fixed beam using exact response
(for 25% damage to elements 1-8 (top to bottom), random force at beam center, 0.1-210 Hz

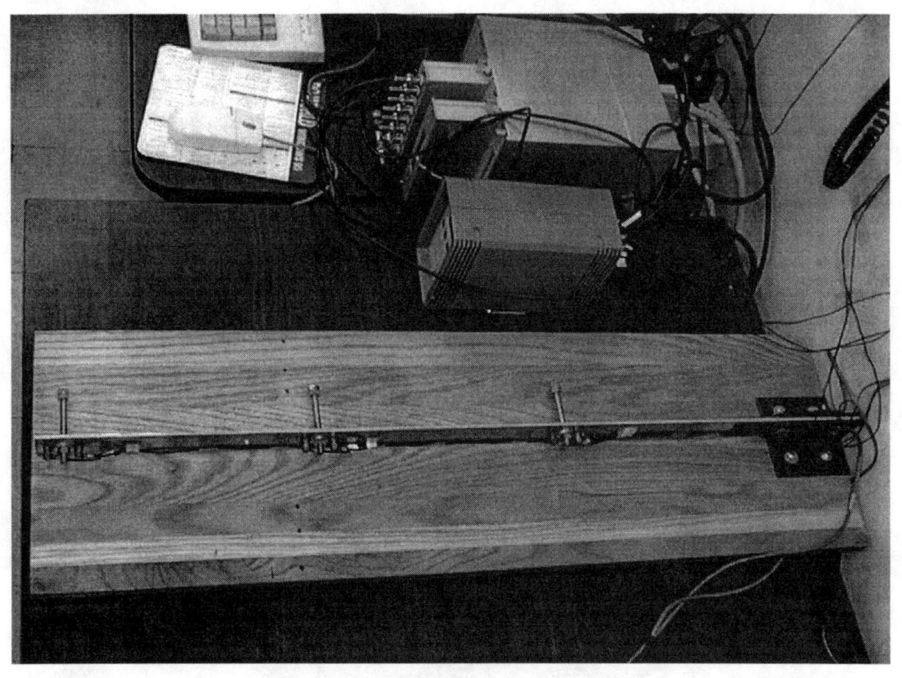

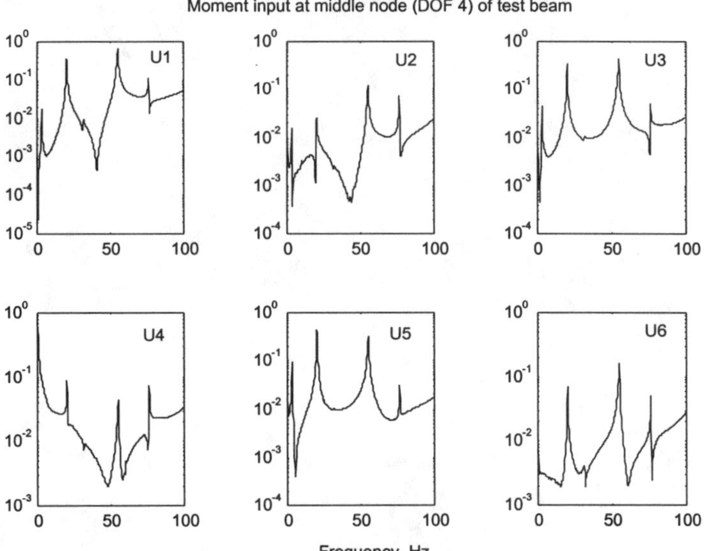

Figure 3. Measured FRFs for a fixed-free beam

KEYNOTE SPEAKERS

L. Melvin
S. Hanagud

Integrated Vehicle Health Monitoring (IVHM) for Aerospace Vehicles

L. MELVIN, B. CHILDERS, R. ROGOWSKI, W. PROSSER, J. MOORE, M. FROGATT, S. ALLISON, M. C. WU, J. BLY, C. AUDE, C. BOUVIER, E. ZISK, E. ENRIGHT, Z. CASSADABAN, R. REIGHTLER, J. SIRKIS, I. TANG, T. PENG, R. WEGREICH, R. GARBOS, W. MOUYOS, D. AIBEL and P. BODAN

ABSTRACT

The vehicle launch community is evaluating methods for reducing the excessive costs associated with access to space. System design strategies for future launch vehicles, like the X-33 Reuseable Launch Vehicle (RLV) demonstrator, focus on providing easy repair access for simplified servicing of infrastructures and expedited decision making from detected faults and anomalies. The Integrated Vehicle Health Monitoring (IVHM) system supports these strategies by providing reliable and low cost maintainability for the Single Stage to Orbit (SSTO) RLV, the planned replacement for the space shuttle. The objective of IVHM is to provide an automated collection and paperless health decision, maintenance and logistics system. This paper will discuss the sensor components, fiber optic and acoustic emission, of IVHM and their application to the X-33 demonstrator and their extension to the RLV.

Leland Melvin, Brooks Childers, William Prosser, Jason Moore Mark Froggatt, Robert Rogowski, NASA Langley Research Center M/S 231 Hampton, VA 23681
James Bly, Carl Aude Lockheed Martin 144 Research Dr. Hampton, VA 23666
Meng Chou Wu, AS&M, Inc. 107 Research Dr. Hampton, VA 23666
Edward Zisk, Eric Enright, Zip Cassadaban, Ron Reightler, Carl Bouvier, Lockheed Michoud Space Systems 13800 Old Gentilly Road New Orleans, Louisiana
James Sirkis, Iywu Tang, Tony Peng, Richard Wegreich, Department of Mechanical Engineering University of Maryland, College Park, Maryland 20742
William Mouyos, Dave Aibel, Patricia Bodan, Ray Garbos Sanders, A Lockheed Martin Company Mail Code: NCA1-6244 95 Canal ST Nashua, N.H. 03061

BACKGROUND

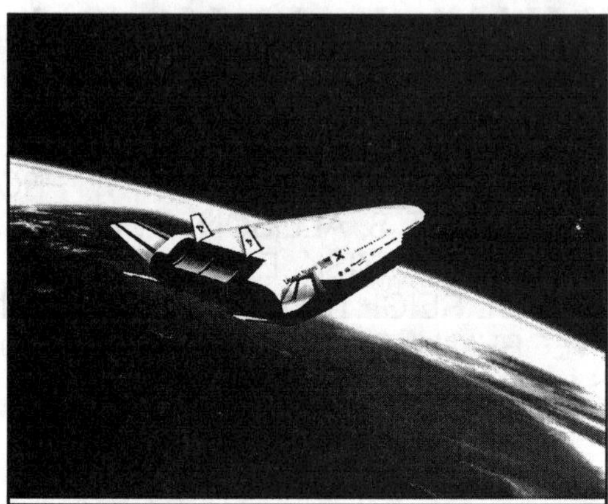

Figure 1 X-33 Concept

The X-33 (Figure 1) is a half-scale, sub-orbital experimental flight test vehicle that is a collaborative effort between NASA and Lockheed Martin. The unmanned X-33 flight test vehicle will be used to demonstrate many of the technologies necessary to make the follow-on Reusable Launch Vehicle (RLV) fleet successful. Approximately 15 missions are planned starting in 1999. One X-33/RLV program objective is to increase access to space by lowering on-orbit payload costs. A new RLV infrastructure is required to meet this objective. Reliability and efficient maintenance of the entire infrastructure is critical to success. The approach to System Health Management (SHM) must be integrated across the infrastructure to reduce costs and risk, and to meet the program objective. Therefore, the Lockheed Martin/NASA Team will strike the optimum balance between on-vehicle health management, prognostics, ground infrastructure, and operational cost.

To achieve on-time launch with near maintenance free operations, not only must the infrastructure be reliable and maintainable, it must also be able to determine the vehicle health by addressing faults enabling efficient, automated maintenance scheduling using paperless procedures. Lockheed Sanders, working with the Lockheed Martin IPT Teams, has developed a smart distributed Health Management System for both the vehicle and the ground systems that includes an automated, computerized, paperless ground computer environment. The team is demonstrating much of this technology during X-33 and is using the latest Commercial-Off-The-Shelf (COTS) workstation, software and internet tools to support the ground environments. This paper will focus on the distributed sensor systems for on-board IVHM implementation.

X-33 VEHICLE SENSOR SUITE

The X-33 will provide system and subsystem validation of distributed sensing technologies that utilize optical fibers and acoustic sensors. The optical fibers will employ Bragg gratings for strain and hydrogen, and Raman techniques for temperature. The acoustic emission sensors will be used to locate cracks and monitor their growth. The sensor suite will be utilized on the internal fuel tanking structure. It employs 16 single mode optical fibers with 20 to 25 Bragg gratings per fiber for strain and hydrogen detection on the hydrogen tank, two multimode temperature fibers per tank on all three tanks, and four acoustic transducers on the hydrogen tank. Once assembled, conventional inspection of the X-33 will be difficult making these built-in sensors a necessity. These sensors must operate in temperature ranges from -252 to 121°C and withstand launch and re-entry environments

DISTRIBUTED STRAIN SENSOR (DSS)

The DSS instrument consists of a wavelength tunable narrow linewidth laser, a fiber optic network containing a sensing fiber with Bragg grating sensors, light detection photodiodes, signal conditioning electronics, and a digital signal processor. The sensing fiber is routed from the VME chassis in the avionics bay to the cryogenic tanks for strain, and hydrogen measurement.

Bragg Grating Sensors

Bragg gratings written into an optical fiber by a UV laser can be used as a single point sensor along the length of an optical fiber. NASA Langley Research Center has developed a method for demodulation of over 50 gratings along a 7 m length of fiber for distributed measurement of strain, or hydrogen (Froggatt, 1996). The Bragg grating sensor reflects only a narrow band of light wavelengths propagating in the fiber. The equation governing the center wavelength of this reflected band is

$$2n\Lambda = \lambda \qquad (1)$$

where n is the index of refraction of the fiber core, λ is the reflected wavelength, and Λ is the grating spacing. Strain is defined by the change in length over length, which for a Bragg grating sensor is

$$\varepsilon \equiv \frac{\Lambda - \Lambda_B}{\Lambda_B} \qquad (2)$$

where the subscript B refers to a baseline grating spacing value obtained prior to flight. DSS measures the wavelength of light reflected by a grating to determine its spacing Λ via Equation (1). Therefore in terms of reflected wavelength strain can be written as

$$\varepsilon = \frac{\lambda - \lambda_B}{\lambda_B} \quad (3).$$

The form of the above equation, corrected for material properties, has been shown to be

$$\varepsilon = K\left(\frac{\lambda - \lambda_B}{\lambda_B} - \xi \Delta T\right) \quad (4)$$

where the constant K is a function of the refractive index, Poisson's ratio, and strain-optic constants of the fiber, ξ is the thermo-optic coefficient, and ΔT is change in temperature.

SENSOR DEMODULATION

A block diagram of the DSS demodulation technique is shown in Figure 2. DSS detects signals from the low finesse Fabry-Perot cavities formed between each grating and its reference reflector as the laser is tuned. Fourier methods are then used to recover the reflected center wavelength of each grating. The power spectrum

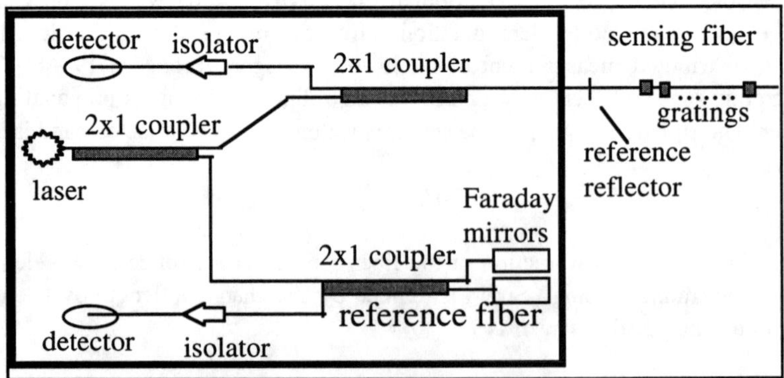

Figure 2 Schematic of demodulation system for optical network mounted on VME card(s). The sensing fiber is routed to tanks.

is computed from the forward Fourier transform of the sensing fiber signal to reveal the grating locations in the fiber. The grating locations in the associated forward transform are then windowed and the power spectra of the inverse Fourier transforms are computed to obtain the spectra of each grating.

DISTRIBUTED HYDROGEN SENSING

A Bragg grating is used as a sensor by imposing a strain on the fiber thereby shifting the grating period and its reflected spectrum. In order to measure hydrogen concentration, a Bragg grating sensor is bonded to palladium. Palladium is known to combine reversibly with hydrogen to form hydride and deform in the process. The state of strain of the grating is altered due to this deformation. The FOSS instrument measures this strain allowing the concentration of hydrogen to be inferred. NASA and Lockheed Martin are working with the University of Maryland to design, develop and implement a distributed hydrogen sensor network for X-33.

DISTRIBUTED TEMPERATURE SENSING (DTS)

Dakin (Dakin, et. al.,1985) demonstrated the use of Raman scattering in optical fiber for distributed temperature sensing in 1985. The method depends on the intensity ratio of the Stokes and anti-Stokes wavelengths in the Raman backscattered signal, which is temperature dependent. A commercial instrument is available which will measure temperature along an optical fiber up to four kilometers in length with a spatial resolution of one meter and a temperature accuracy of one degree Celsius.

ACOUSTIC EMISSION (AE)

Acoustic emission (AE) techniques detect fatigue crack initiation and growth by analysis of the elastic waves that are generated by the sudden release of strain energy as the crack propagates. The location of the crack site can be determined by triangulation using measurements of the time of the wave's arrival at a number of sensors at different locations, and a knowledge of the velocity of propagation. The technique is unlike other NDE methods in that it has the potential to provide global monitoring of a structure, under in-service loading conditions. It has been used in attempts to monitor a variety of structures including solid rocket motor casings (Green, Lockman, Steele, 1964), a reactor pressure vessel (Votava, Jax, 1979), and high pressure storage tanks.

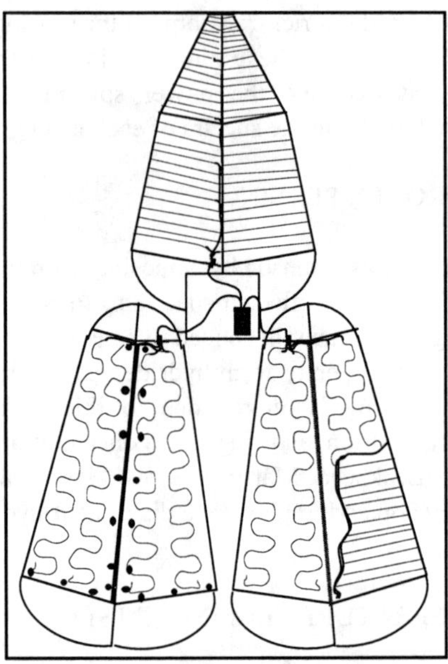

Figure 3 Distributed network of sensors for monitoring strain, temperature, and hydrogen on the X-33 tanks.

TESTING

X-33 REUSABLE CRYOGENIC TANK VHM USING DISTRIBUTED FIBER SENSING

The backbone of the X-33 Reusable Cryogenic Tank VHM system lies in the optical network of distributed strain, temperature and hydrogen sensors. This optical network will create a global strain and temperature map for monitoring the health of the tank structure, cryogenic insulation, and Thermal Protection System (TPS). Hydrogen sensors along the multilobe tank joint monitor for leak location. Figure 3 shows such a network of sensors on the LO2 and LH2 tanks of the X-33 vehicle.

DSS AND DTS IMPLEMENTATION:

Lockheed Martin Michoud Space Systems (LMMSS) is responsible for the IVHM system that monitors the main propulsion tanks of the X-33 vehicle shown in Figure 3. Significant progress towards developing techniques for installing and operating these systems has been made. Distributed Strain Sensor (DSS) and Distributed Temperature Sensor (DTS) fiber optic systems, provided by LaRC, were installed on a 70% scale dual-lobed composite Liquid Hydrogen (LH2) tank that was recently tested at NASA Stennis Space Center. The tank underwent two phases of testing for a total of 30 cryocycles up to 100 psi. These sensor systems were the first large scale application of fiber optic distributed sensor technology. Altogether over 600 feet of optical fiber sensors were bonded to the tank. The dual lobe tank utilized special techniques developed for installing optical fiber onto flight tankage: curved ramps, protective tubing, etc. The DSS system was applied to the composite substrate both at the bolted joint locations and along the barrels. The DTS system was installed on top of the foam in a repeating loop pattern to provide a thermal profile of the foam insulation.

Strain and temperature monitoring was performed during LH_2 filling and draining of the dual lobe composite tank at NASA Stennis. Temperature measurements were made with York Technologies DTS 80 Raman based distributed temperature sensor (DTS) to monitor insulation performance. Bragg gratings were written in the fiber bonded to the tank in Figure 4. Biaxial strain was measured by monitoring fibers placed at both the hoop and axial directions. The optical fiber temperature sensor network is shown in Figure 5. Successful measurement of both strain and temperature was made during tank testing. Figure 6 is a plot of Bragg grating sensor location along the fiber verses strain. The repeating pattern in the plot indicates orientation on the tank. The three curves show increasing tank pressure taken at ambient temperature. DSS and DTS showed good comparison with conventional strain gages and thermocouples during this test.

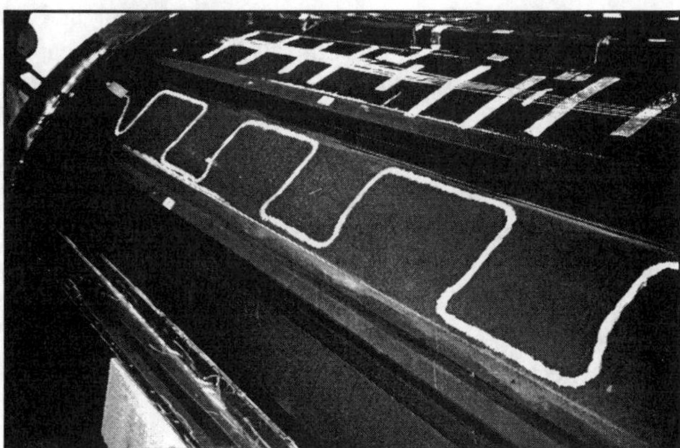

Figure 4 Distributed optical fiber strain sensor bonded to composite liquid hydrogen tank. The single fiber makes 14 biaxial strain measurements

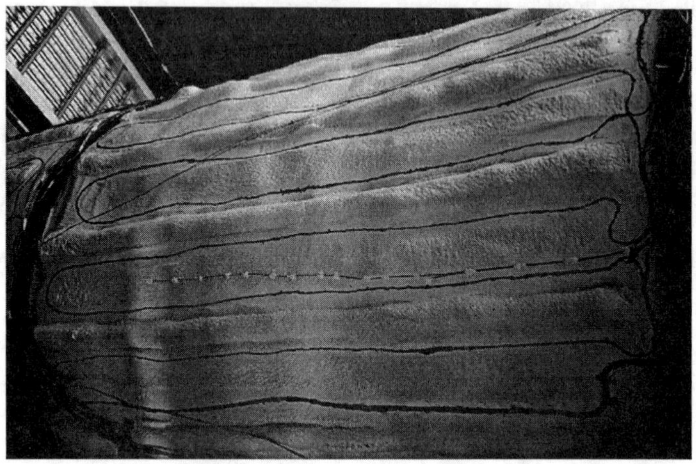

Figure 5 Distributed optical fiber temperature sensor bonded to cryogenic insulation. A single fiber provides temperature measurements at 50 locations along the fiber.

PRESSURE VESSEL TESTING WITH AE

Recent research (Gorman, 1990), (Prosser, 1991) using broad band, high-fidelity sensors, full waveform capture instrumentation, and wave propagation based analysis has lead to significant improvements in the capabilities of the AE technique. In plate and shell geometries, crack signals can be clearly discriminated from extraneous noise signals by their plate wave characteristics. Such geometries comprise many of the structures of interest for AE applications including air and space craft skins, pressure vessels, and piping. More accurate source location determination is also possible with this new approach (Ziola, Gorman, 1991). Figure 7 shows AE testing on subscale composite tanks. The wires (picture on left) are attached to a 6 channel AE system for recording waveforms of damaged or undamaged (impact) tanks during pressure cycling. Characterization of sprayed-on foam insulation effects on the acoustic signal was also studied.

SUMMARY

Cost effective fiber optic and acoustic emission sensors have been described as part of a system for performing Integrated Vehicle Health Monitoring on the X-33 demonstrator vehicle. These systems will be required to achieve the goals of on-time launch with near maintenance free operation of the RLV.

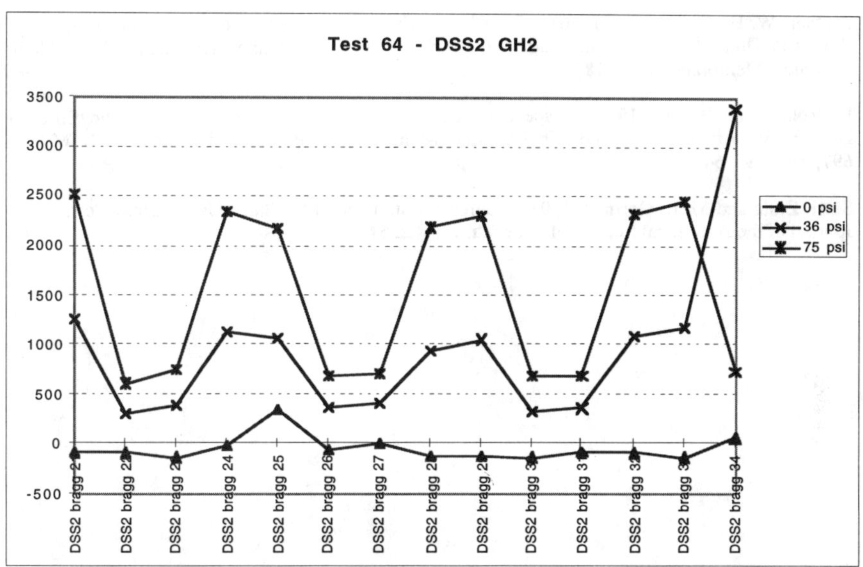

Figure 6 DSS dual Lobe tank data indicating grating sensor location verses strain.

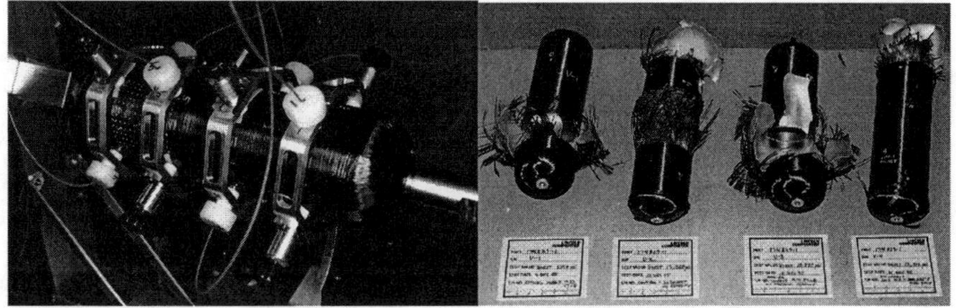

Figure 7 Composite pressure vessel testing with AE Sensors

REFERENCES

J. Dakin, D. Pratt, G. Bibby and J. Ross, 1985, "Distributed Optical Fibre Raman Temperature Sensing Using a Semiconductor Light Source and Detector," Electronic Letters, Vol. 21, pp. 569-570.

M. Froggatt, 1996, "Distributed measurement of the complex modulation of a photoinduced Bragg grating in an optical fiber", *Appl. Opt.*, 35, pp.5162-5164.

A. T. Green, C. S. Lockman, and R. K. Steele, 1964, "Acoustic Verification of Structural Integrity of Polaris Chambers," Modern Plastics, 41 p.137.

M. R. Gorman, 1990, "Plate Wave Acoustic Emission," J. Acoust. Soc. Am., 90(1), pp. 358-364.

Prosser, W. H., 1991, "The Propagation Characteristics of the Plate Modes of Acoustic Emission Waves in Thin Aluminum Plates and Thin Graphite/Epoxy Composite Plates and Tubes," NASA Technical Memorandum 104187.

E. Votava and P. Jax, 1979, "Inspection of Nuclear Reactors by Means of Acoustic Emission During Hydrostatic Test," Acoustic Emission Monitoring of Pressurized Systems, ASTM STP 697, pp. 149-164.

S. M. Ziola and M. R. Gorman, 1991, "Source Location in Thin Plates Using Cross-Correlation," Journal of the Acoustical Society of America, 90(5) 2551-2556.

Damage Detection and Health Monitoring Based on Structural Dynamics

S. HANAGUD and H. LUO

ABSTRACT

In this paper, we present a damage detection and health monitoring methodology that is based on the use of measured structural dynamic response. In a structural system that consists both metallic and layered composite sub-assemblies, the types of damages can include delaminations, transverse cracks, and impact damages. Usually, the dynamic response based damage detection methods deduce the information on damages by using the fact that the damages will alter the characteristics of the structure and thus the dynamic response. Existence of different types of damages in the same structure and the presence of multiple damages of the same type make the damage detection procedure more complicated.

In this paper, we have considered two different types of damages, delamination and stiffness loss due to impact damage and transverse cracks. Analytical models are proposed to predict the structural dynamic response of the damaged structure. A nonlinear dynamic response criterion is used to separate and identify the two types of damage by using the predicted and measured responses. We have also proposed a modified neural network technique to identify and separate damages by using the measured structural dynamic response. The developed damage detection procedures are validated by the use of specifically designed experiments.

INTRODUCTION

Structural damage detection and health monitoring is a topic which is of considerable interest in many fields. Applications of damage detection and health monitoring are found in many facets of engineering practice. Some examples include aerospace engineering, civil engineering, mechanical engineering, nuclear engineering, material engineering, chemical engineering, and automobile engineering. The health monitoring is an essential issue in enhancing the affordability as well as the safety of modern aircraft and spacecraft structures. One of the most common damage detection methods in the aircraft industry is through visual in-

---

S. Hanagud, Professor, School of Aerospace Engineering, Georgia Institute of Technology, Atlanta, GA 30332-0150

H. Luo, 301 Queens Drive, Schenectady, NY 12304

spection. Even with the aid of closed circuit television, visual inspection requires extensive amount of time. Other conventional damage detection methods are based on traditional nondestructive evaluation/testing (NDE/T) methods such as radiography, ultrasonics, acoustic emissions, optical methods, thermal methods, magnetic methods, and eddy current test. With the advances in computer science and technology, the integration of NDE/T instruments with microprocessors has enhanced the flexible signal processing capability and fast data storing and retrieving capability, thus making many of those traditional NDE/T methods more efficient. Despite the well-documented successes of many NDE/T methods, most of the conventional NDE/T methods have drawbacks and limitations when applied to real-time health monitoring. Usually, the conventional NDE/T methods are local in nature, passive, and labor intensive. Very often, special auxiliary instruments are required. Thus, it is difficult to implement these methods in on-board, automatic, real-time, and global health monitoring.

Structural dynamic response has been used for damage detection and health monitoring. In early research work, the frequency signature in the structural response function was used to identify the damages in the monitored structure. This kind of damage detection functions more like an expert system. The success relies on the complete collection of the "fingerprints" of the failure cases. Due to the complexity of the systems, a complete failure mode database is not possible.

Recently, much attention has been paid to the research of damage detection and health monitoring by using the measured structural dynamic response and the use of mathematical models. Successful applications, as well as their limitations, can be found in the literature. For example, experiment-based methods are simple in implementation and easy to understand, but empirical judgment is usually required. Systematic relations between the experimental data and the damage are not available. Model-based methods have a more rigorous mathematical background, but they suffer from some other drawbacks. For example, a precise numerical representation for the structure is hard to achieve and the physical meanings of the detected results are vague due to the complicated inverse operations. Furthermore, the realization of real-time detection is very difficult in the model-based methods because of the complicated mathematical manipulations.

Due to the complexity and high integrity of the aerospace structures, new concepts need to be introduced into the structural health monitoring system. These concepts include the smart/adaptive structure's concepts, global information acquisition, real-time signal processing, and automatic decision making.

The objective of this research was to develop a health monitoring system which uses the active global response information to identify and separate different types of damages like delaminations, impact damages, and transverse-crack-like defects. In this paper, it is assumed that only one type of the defect is present. However, *a priori*, we do not know the type of the defect, it location, or its magnitude. In the paper, first we have discussed the detection of delamination by the use of the nonlinear response. Next, we have discussed the stiffness loss detection by the use of curvature modes and an integral equation approach. The combined detection is discussed by the use of neural network and the mechanics of damaged structures.

MODELING OF COMPOSITE BEAMS WITH DELAMINATIONS

Delamination is one of the major failure modes in layered composite structures. Due to its importance, the modeling of composite with delaminations has been an active research topic for almost twenty years. One of the earliest models was proposed by Ramkumar *et al.* (1979) in the late seventy's. Later, Wang *et al.* (1982) proposed a model which included the delamination effects through the coupling between the flexural and longitudinal vibrations at the delamination boundaries. Mujumdar and Suryanarayan (1988) proposed a so-called constrained model, in opposite to the "free model" by Wang et al. Experimental work can be found in reports by Shen and Grady (1992), Hanagud and Luo (1994) and Luo and Hanagud (1995). The modeling of delaminations in layered anisotropic composite structures is attributed to Nagesh and Hanagud (1990). In this paper, we intend to include the nonlinear effects in the modeling. Shear deformation and rotary inertia terms, as well as the coupling between flexural and longitudinal vibrations, are also included in the modeling.

As shown in Figure 1, after delamination, a representative composite beam with width b, height H, and length L can be viewed as a combination of four beams connected at the delamination boundaries $x=L_1$ and $x=L_2$. In Figure 1, we denote m_i, D_i, S_i and A_i ($i=1,2,3,4$) the mass density per unit length, bending stiffness, cross sectional shear stiffness, and extensional stiffness of four beams, respectively. The notations H_2 and H_3 represent the distances between the neutral axis of the delaminated beam and the neutral axis of the intact part. For composite materials, it is worth noting that H_2+H_3 is not necessarily equal to $H/2$, since the geometric center and the neutral axis of a sublaminate, in general, do not coincide.

The effects between the delaminated surfaces depend on the relative position between the sublaminates during the vibrations. Here we assume that after delamination, partially intact matrix and fibers still fill the delamination crack. Some constraints between the upper and lower delamination surfaces still exist. Under a small amplitude vibration of the delaminated beam at a frequency corresponding to a delamination opening mode, the effect between delaminated sublaminates can be modeled as a distributed soft spring between them. When the amplitude exceeds a certain level, the spring effect becomes zero because the delamination opens beyond the small amplitude constraints. On the other hand, when the vibration mode does not tend to open the delamination, the delaminated sublaminates have the same flexural displacements and slopes. Thus, the exact behavior of the effects between the delaminated sublaminates may be described by a nonlinear spring model as shown qualitatively in Figure 2 by a dashed line.

To simplify the problem while keeping the nonlinear features of the delamination vibrations, we reduced the nonlinear model into a piecewise linear model based on the following observations: When the delamination tends to open in vibrations, that is, the relative displacement w_2-w_3 is positive, the distributed contact force is zero; When the delamination is completely closed during the vibrations, the relative displacement w_2-w_3 is a fixed value. Under such a circumstance, the spring model can be simplified by another straight line BC as shown in Figure 2; When the delamination beam is vibrating with relatively small displacement, for example $-d_0<w_2-w_3<0$, the spring model can be simplified by a linear spring model, as shown in Figure 2 by a solid straight line OB. For a given practical problem, the exact value for d_0 and spring constant in this region need to be determined by special experiments.

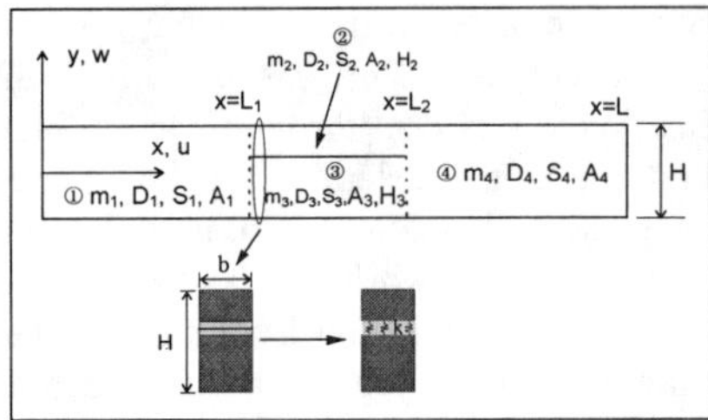

Figure 1. Modeling of Delamination Effects

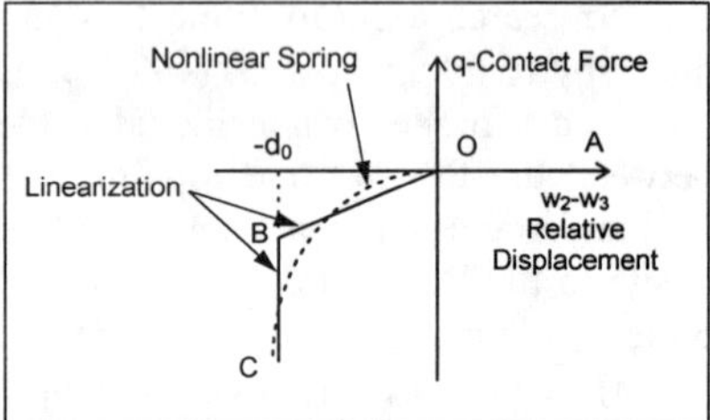

Figure 2. A Piecewise Linear Spring Model

The differential equations for beam segments are

$$A_i \frac{\partial^2 u_i}{\partial x^2} - m_i \frac{\partial^2 u_i}{\partial t^2} = P_i(x,t), \quad i=1, 2, 3, 4 \tag{1}$$

$$\begin{cases} \dfrac{\partial}{\partial x}(S_i\beta_i) - m_i \dfrac{\partial^2 w_i}{\partial t^2} = F_{1i}(x,t) \\ \dfrac{\partial}{\partial x}\left(D_i \dfrac{\partial \psi_i}{\partial x}\right) + S_i\beta_i - J_i \dfrac{\partial^2 \psi_i}{\partial t^2} = F_{2i}(x,t) \end{cases}, \quad i=1, 4 \tag{2}$$

$$\begin{cases} \dfrac{\partial}{\partial x}(S_i\beta_i) - m_i \dfrac{\partial^2 w_i}{\partial t^2} + q_i = F_{1i}(x,t) \\ \dfrac{\partial}{\partial x}\left(D_i \dfrac{\partial \psi_i}{\partial x}\right) + S_i\beta_i - J_i \dfrac{\partial^2 \psi_i}{\partial t^2} = F_{2i}(x,t) \end{cases}, \quad i=2,3 \tag{3}$$

In the equations, u_i and w_i denote the axial and flexural displacements, respectively; β_i is the angle of the shear at the beam segment neutral axis; ψ_i is the slope of the deflection curve caused by bending moment; m_i is the mass per unit length; J_i is the cross sectional mass moment of inertia; q_i is distributed lateral load. $P_i(x,t)$ is the external axial force.

The w_i, β_i, and ψ_i are connected by

$$\frac{\partial w_i(x,t)}{\partial x} = \psi_i(x,t) + \beta_i(x,t) \tag{4}$$

In our model, q_i has the form

$$\begin{cases} q_2(x,t) = k[w_3(x,t) - w_2(x,t)] \\ q_3(x,t) = k[w_2(x,t) - w_3(x,t)] \end{cases} \tag{5}$$

where k is the piecewise linear spring constant.

The eigensolutions of (1), (2), and (3) can be solved separately in the different vibration stages. The nonlinear dynamic response can be synthesized through a nonlinear modal analysis technique. Refer to Luo and Hanagud (1997b) for more details of the solution procedures. In Figure 3, we have shown a typical delamination mode, where the delamination opening is clearly seen. The delamination opening and closing during the vibration cause nonlinearity in the dynamic response. In Figure 4, the nonlinear dynamic response of a delaminated beam was detected by a PVDF film sensor, where the excitation was pure sinusoidal at a single frequency (24.3 Hz) that was provided by a PZT patch. The analytical prediction of the nonlinear response based on the proposed model is shown in Figure 5.

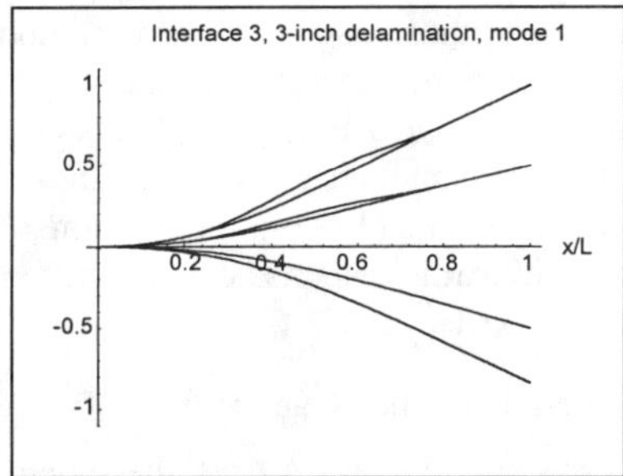

Figure 3 A Typical Vibration Mode of a Beam With Delamination

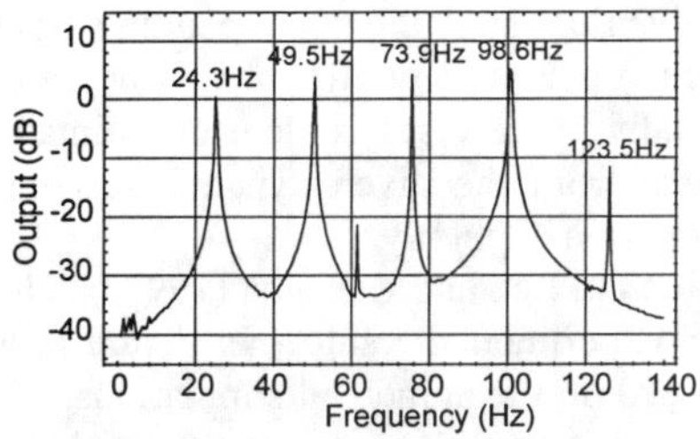

Figure 4 Experimentally Detected Nonlinear Response of a Delaminated Beam

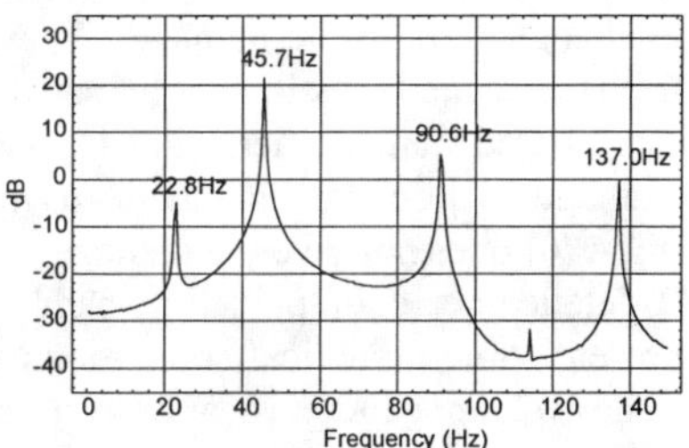

Figure 5 Analytically Predicted Nonlinear Response of a Delaminated Beam

A DAMAGE DETECTION SCHEME USING MEASURED DATA

Experiment-based damage detection methods use only the vibrational test data from intact and damaged structures to deduce the damage information in the structure. Typically, the measured data chosen are natural frequencies (Cawley and Adams, 1979; Pabst and Hagedorn, 1993), damping ratios (Lee *et al.* 1987; Griffin and Sun, 1991), mode shapes (Cempel *et al.* 1992), or curvature mode shapes (Pandey, 1991). The advantages of the experimental-based method lie in its simplicity and ease of implementation.

Even though some successful detection methods have been reported in the literature, most of the methods rely on empirical judgment. To get a complete damage information, one needs to detect the damage location and damage severity separately. Some of the methods are able to detect only partial information or detect the complete damage information, such as damage locations and damage severities, separately. For example, the curvature mode based method (Pandey *et al.*, 1991) and the flexibility based method (Pandey and Biswas, 1994) can only roughly detect the approximate damage location. The damage severity information could not be obtained. Pabst and Hagedorn (1993) provided a procedure to detect the location and severity of damage in a distributed system. The relationship among the changes in eigenfrequencies, the local stiffness loses and the mode shape functions were first set up. Cempel *et al.* (1992) discussed the diagnostic properties of the symptoms of a distributed system analytically. A symptom of damage is defined as a physical quantity covariable with damage. In both methods, an assumption was made that the mode shape changes before and after the structure damage was small. Unfortunately, small changes in displacement mode shapes of a structure due to damages do not necessarily mean small changes in curvature modes.

As discussed by Luo and Hanagud (1997a), the relationship between the damaged and the intact structural dynamic response can be written as

$$\int_{\Omega} \varphi_i(X)\Delta L[\overline{\varphi}_i(X)]d\Omega = \Delta\lambda_i \tag{6}$$

where φ_i is the eigen function of the intact system; $\overline{\varphi}_i = \varphi_i + \Delta\varphi_i$ is the eigen function of the damaged system; $\overline{\lambda} = \lambda + \Delta\lambda$ is the eigenvalue of the damaged system; λ is the eigenvalue of the intact system; ΔL is the differential operator difference due to damage. Equation (6) relates the damage information and the eigenvalue changes due to the damage. For a given damage, the corresponding eigenvalue changes can be estimated by equation (6) without any difficulty. But the experiment-based damage detection is essentially an inverse problem of estimating the damage location and damage magnitude from the given dynamic response. Equation (6) does not give the damage information explicitly.

A detailed detection procedure can be found in Luo and Hanagud (1997a). Here we present some numerical simulations and experimental results. In Figure 6, we have shown damage identification results based on the numerical simulations. The current detection method is accurate in the sense that the sensor accumulated stiffness loss is equal to the integration of the simulated damage (as shown in Figure 8). To examine the feasibility of the method in practical damage detection, we conducted experiments for different specimens with transverse crack and impact damage as shown in Figure 7. Results for transverse crack identification are shown in Figure 8.

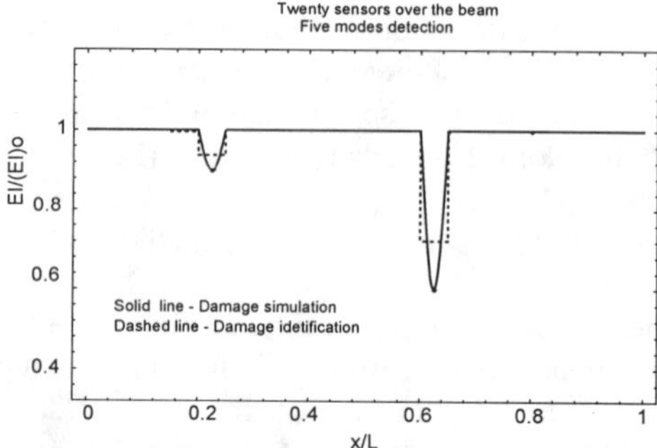

Figure 6 Double-Damaged Beam Iterative Detection

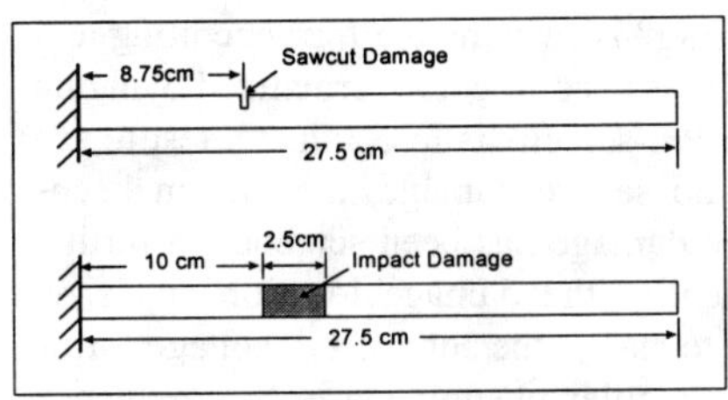

Figure 7 Saw-cut Damage and Hammer Impact Damage

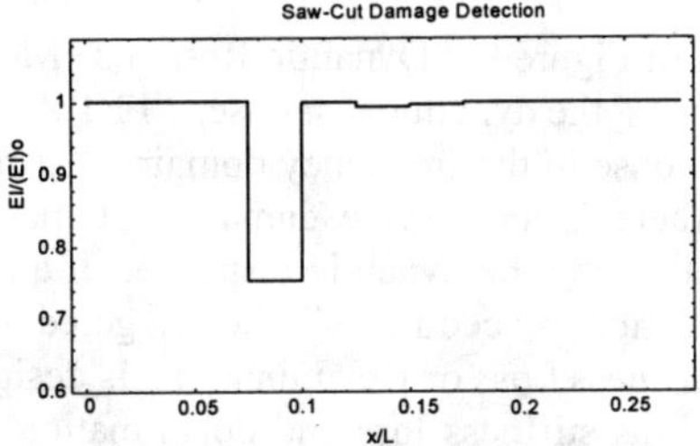

Figure 8 Detection Results of the Saw-cut Damaged Beam

INTEGRATED DAMAGE DETECTION AND HEALTH MONITORING

As shown the delamination modeling, the existence of delamination damage in a structure causes drastic change in the dynamic response. Nonlinearities appear in the dynamic response due to the delamination opening. A network design which covers the stiffness loss damage and delamination damage will increase the network topology complexity, training difficulty, and training data acquisition intricacy. A way to solve this problem is to design subnetworks undertaking the subtasks separately. Thus the objective is to use structural dynamic response and the design of subnetworks to separate delaminations and stiffness loss defects like impact damages and transverse cracks. The design flow chart is shown in Figure 9.

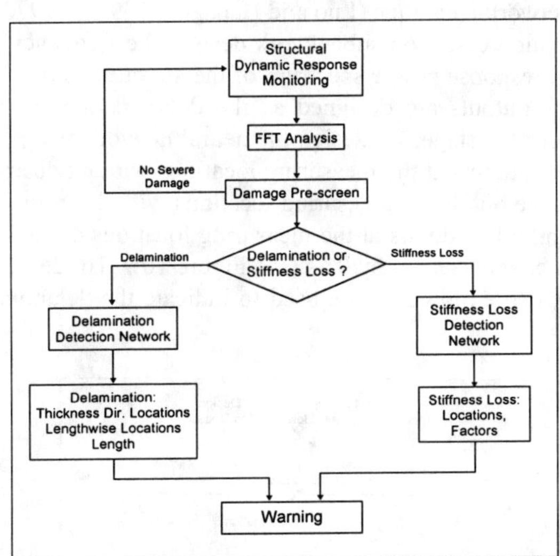

Figure 9 Health Monitoring System Design Flow Chart

In Figure 9, “Dynamic Response Monitoring” is responsible for generating and sensing the dynamic response. “FFT Analysis” is used to get a normalized dynamic response in the frequency domain. “Damage Pre-screen” is to conduct a test to see if there is any severe damage. If there is no severe damage, the system keeps monitoring the dynamic response. If a severe damage has been screened, a further detection procedure will be triggered to identify the damage location and size. “Stiffness Loss or Delamination” is designed to assign the subtasks. Damage forms such as stiffness loss and delamination share similar phenomena in the frequency domain, but there are also some significant difference between them. For example, in the delamination case, the opening mode causes nonlinearity in vibration, thus superharmonic components will appear in the frequency response function. This feature unique to delamination in composite structures can be used to determine which detection algorithm should be triggered in the next step. “Delamination Detection Network” is one of the subtasks. If a delamination damage has been detected, a further step will be switched to a network designed for delamination detection. The detection results include delamination location and delamination size. “Stiffness-loss Damage Detection Network” is the other subtask. If a stiffness-loss damage has been detected, a further step will be switched to a network designed for stiffness-loss damage detection. The detection results include the damage locations and damage severity. The detected damage information and a warning signal will be displayed in the “Warning."

In the following, we experiment the health monitoring system through a delamination detection subnetwork and a stiffness loss detection network based on a modified neural network technique (Luo and Hanagud, 1997c, 1997d).

In the delamination detection subnetwork design, the frequency response function or normalized response power spectrum of the structure is used as network input. The network outputs are designed as the delamination information of the structure. At the current stage, we design the neural network as a pattern classifier. We assume that the outputs at the measuring locations without delamination can be represented by 0s; the outputs at measuring locations with delamination can be represented by 1s; while the outputs at the measuring locations close to the delamination boundary can be represented by 0.5s (see Figure 10). To identify the thickness direction location, two extra codes are used to indicate the delamination thickness direction locations.

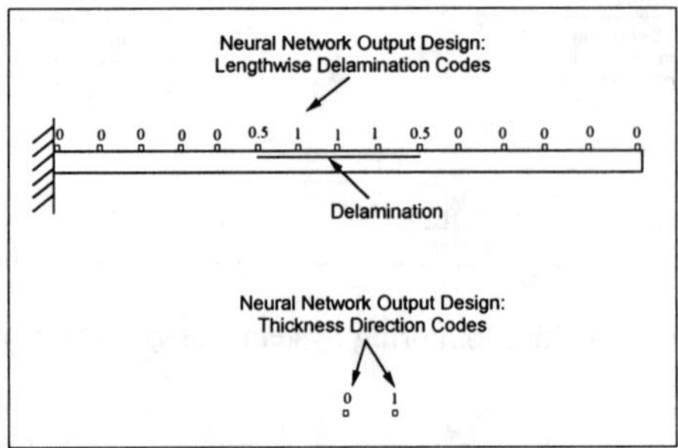

Figure 10 Delamination Detection Subnetwork Output Design

To verify the designed subnetwork, a delaminated composite beam experiment is designed. Specimens used to train the network were fabricated by 0/90 woven glass fiber cloth with the epoxy as matrix. Delaminations were simulated by placing very thin Teflon films between layers during the lay-up. All the specimens consist of 12 layers fiber glass cloth and were tailored in strips with dimension of 30cm×2.5cm. Delaminations were centered at the specimen length. Boundary conditions are simulated cantilever boundary by plant 2.5 cm of the specimens into a vice. Thus, the effective beam length is 27.5cm. The delamination information of the training set and the thickness direction delamination codes are listed in Table 1.

Table 1 Delamination Information of the Training Specimens

Specimen No.	Delamination Length(cm)	Thickness Location	Thickness Direction Code	
			Code1	Code2
1	0	0	0	0
2	3	6/6	0	1
3	10	6/6	0	1
4	18	6/6	0	1
5	3	4/8	1	0
6	10	4/8	1	0
7	18	4/8	1	0
8	3	2/10	1	1
9	10	2/10	1	1
10	18	2/10	1	1

In this example, the designed network was a three-layer feedforward network with 128 input nodes, 30 hidden nodes, and 18 output nodes. In Figure 11, we have shown the training iteration results of the SSD method and the FSD method (Luo and Hanagud, 1997c; 1997d). The superiority of the training performance of the FSD method over that of the SSD method is seen in the practical problems as well as benchmark problems.

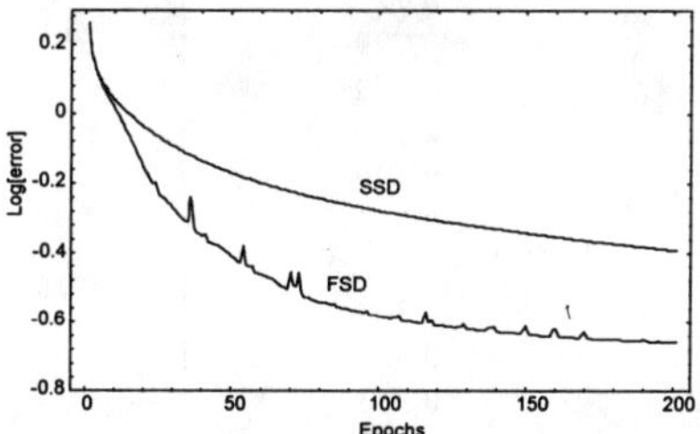

Figure 11 Training Performance of the Delamination Detection Subnetwork

Figure 12 3cm 4/8 Delamination Detection

After the network has been trained, another set of delamination specimens was passed through the network. An example of the detection is shown in Figure 12.

In the stiffness loss damage detection subnetwork, the frequency response functions are used as the damage information carriers. The frequency response function or normalized response power spectrum of the structure is used as network input. The network outputs are designed as the relative stiffness ratio, that is, the ratio of the stiffness of the damaged structure with respect to the stiffness at the intact state at the network output locations (see Figure 13).

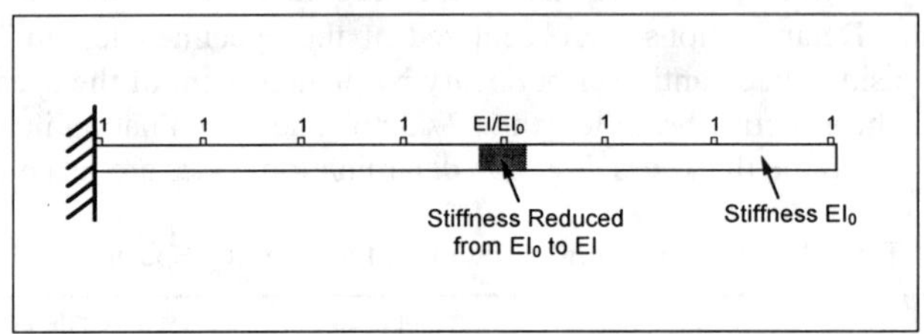

Figure 13 Stiffness Loss Damage Detection Subnetwork Output Design

For the purpose of design verification, only one stiffness loss location is assumed along the beam length. Numerical simulations were used to train the network.

To verify the designed subnetwork, different stiffness loss cases were numerically simulated for network training. The intact beam simulated in this subnetwork has the same equivalent stiffness and geometry as that of the intact specimen in the delamination damage detection subnetwork. Specimen boundary conditions were simulated with cantilever boundary conditions. Sixteen different stiffness loss cases were simulated. The descriptions of the stiffness loss cases are listed in Table 2.

Table 2 Stiffness Loss Damage Cases

Sample No.	Stiffness Loss Location	Stiffness Loss Ratio EI/EI_0
1	Intact	1
2	0.0075	0.729
3	0.0675	0.729
4	0.1275	0.729
5	0.1875	0.729
6	0.2475	0.729
7	0.0075	0.343
8	0.0675	0.343
9	0.1275	0.343
10	0.1875	0.343
11	0.2475	0.343
12	0.0075	0.125
13	0.0675	0.125
14	0.1275	0.125
15	0.1875	0.125
16	0.2475	0.125

In this simulation, the designed network was a three-layer feedforward network with 128 input nodes, 30 hidden nodes, and 5 output nodes. The network was trained by the 16 damage cases. In Figure 14, we have shown the training iteration results of the SSD method and the FSD method for this case. The trained network is able to detect the damage situations which have been seen in training accurately (Figure 15(a)). It can also conduct reasonable expansion. For example, as shown in Figure 15(b), the actual damages are located between two geometric output nodes of the trained network. Though these damage cases have not been seen by the network during the training, the network successfully detected the damage information at their neighboring locations.

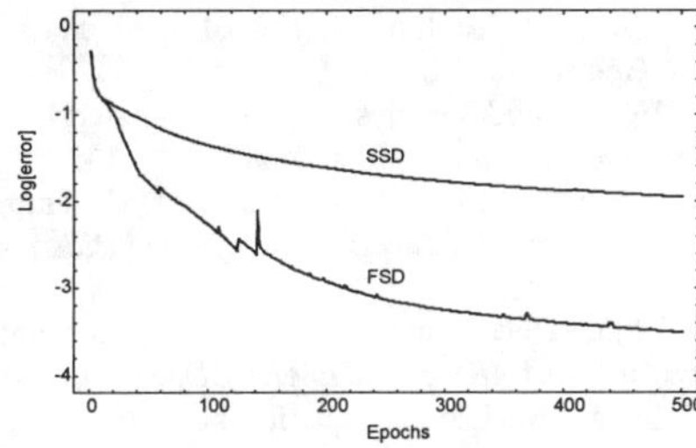

Figure 14 Training Performance of the Stiffness Loss Detection Subnetwork

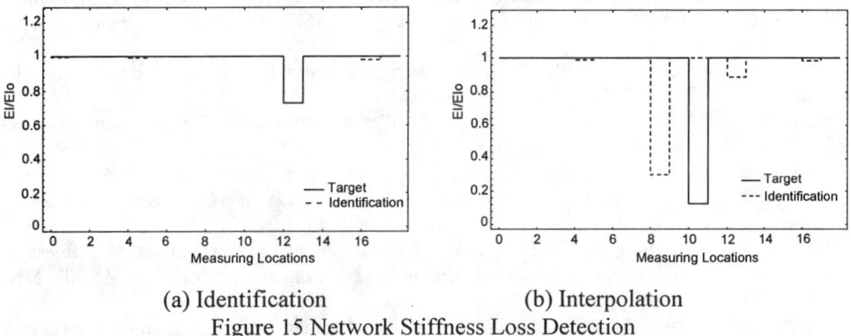

(a) Identification (b) Interpolation

Figure 15 Network Stiffness Loss Detection

CONCLUSIONS

In this paper, we have discussed a damage detection method that uses the structural dynamic response and the separation of different types of damages like delamination, and stiffness loss due to impact damage or transverse cracks. The method is based on the combined use of neural sub-networks and the information from the damaged structures. The information from the damaged structures includes the nonlinear response, curvature mode changes, and the integral relation between modes and changed natural frequencies.

REFERENCE

Anon, 1992. "Inspect of composite structures," *Aerospace Engineering* (Warrendale, Pennsylvania), 12(5): 9-13.

Cawley, P. and Adams, R. D., 1979. "The localization of defects in structure from measurements of natural frequencies," *Journal of Strain Analysis*, 14(2): 49-57.

Cempel, C. Natke, H. G. and Ziolkowski, A., 1992. "Application of transformed normal modes for damage location in structures," *Structural Integrity Assessment*, Edited by P. Stanley, Elsevier Applied Science, pp. 246-255.

Griffin, S. F. and Sun, C. T., 1991. "Health monitoring of dumb and smart structures," *28th Annual Technical Meeting of SES*, Gainesville, FL.

Hanagud, S. and Luo, H., 1994. "Modal analysis of a delaminated beam," *Proceedings of SEM Spring Conference on Experimental Mechanics*, pp. 880-887.

Lee, B. T., Sun, T. C. and Liu, D., 1987. "An assessment of damping measurements in the evaluation of integrity of composite beams," *Journal of Reinforced Plastics and Composites*, 6:114-125.

Luo, H. and Hanagud, S., 1995. "Delamination detection using dynamic characteristics of composite plates," *Proceedings of 36th AIAA/ASME/ASCE/AHS SDM Conference*, pp.129-139.

Luo, H and Hanagud, S. 1997a. "An integral equation between the structural dynamic characteristics of intact and damaged structures." To be published on the *International Journal of Solids and Structures*.

Luo, H. and Hanagud, S., 1997b. "Delaminated Beam Nonlinear Dynamic Response Calculation and Visualization," *Proceedings of the 38th AIAA/ASME/ ASCE/AHS SDM Conference*, pp.490-499.

Luo, H. and Hanagud, S. 1997c. "Fuzzy Learning Rate Neural Network Training And Structural Health Monitoring," *Proceedings of the 38th AIAA/ASME/ ASCE/AHS SDM Conference*, pp.634-644.

Luo, H. and Hanagud, S., 1997d, "Dynamic learning rate neural network training and composite structural damage detection," To appear, *AIAA Journal*.

Mujumdar, P. M. and Suryanarayan, S., 1988. "Flexural vibrations of beams with delaminations," *Journal of Sound and Vibrations*, 25(3): 441-461.

Nagesh Babu, G. L. and Hanagud, S., 1990. "Delamination in smart structures - A parametric study on vibrations", *Proceedings of 31th AIAA/ASME/ASCE/ AHS SDM Conference*, pp. 2417-2426.

Pabst, U. and Hagedorn, P., 1993. "On the identification of localized losses of stiffness in structures," *American Society of Mechanical Engineers, DE Vol. 59*, Published by ASME, New York, NY, USA, pp. 99-104.

Pandey, A. K., Biswas, M. and Samman, M. M., 1991. "Damage detection from changes in curvature mode shapes," *Journal of Sound and Vibration*, 145(2):321-332.

Pandey, A. K. and Biswas, M., 1994. "Damage detection in structures using changes in flexibility," *Journal of Sound and Vibration*, 169(1): 3-17.

Ramkumar, R. L., Kulkarni, S. V. and Pipes, R. B., 1979. "Free vibration frequencies of a delaminated beam," *34th Annual Technical Conference, 1979 Reinforced Plastics/Composites Institute*, The Society of the Plastics Industry, Inc., Section 22-E, pp. 1-5.

Shen, M,-H, H and Grady, J. E., 1992. "Free vibrations of delaminated beams," *AIAA Journal*, 30(5): 1361-1370.

Wang, J. T. S., Liu, Y. Y. and Gibby, J. A., 1982. "Vibration of split beams," *Journal of Sound and Vibrations*, 84(4): 491-502.

SESSION 13

MODELING AND DIAGNOSTIC METHODS V

Dynamic Response of Smart Composites with Delaminations

A. CHATTOPADHYAY, D. DRAGOMIR-DAESCU and H. GU

ABSTRACT

A refined higher order theory-based finite element model is presented for modeling the dynamic response of delaminated smart composite beam. The refined displacement field accurately accounts for transverse shear deformations through the thickness and all traction-free boundary conditions are satisfied at all free surfaces including delamination interfaces. New formulations are presented to include nonlinear induced strain effects. Vibration control is accomplished by piezoelectric layers incorporating the composite beam. The finite element model is shown to agree well with published experimental data and the results show significant improvements compared to the existing analytical solutions. Numerical results presented in the paper indicate significant changes in the natural frequencies, mode shapes, static and dynamic responses due to delaminations.

INTRODUCTION

Recently, smart composite structures have received considerable attention due to the potential for designing adaptive structures which are both light in weight and possessing adaptive control capabilities for shape correction and vibration control. In designing with composites, it is important to take into consideration imperfections, such as delamination, that are often pre-existing or are generated by external impact forces during the service life. The existence of delamination can significantly alter the dynamic response of composite structures. Several mathematical models have been reported in the literature for the analysis of beams and plates with piezoelectric sensing/actuation. The classical theory-based approach (Crawley and Anderson, 1989; Lee, 1990) was first introduced to investigate such problem with thin plates. This was followed by the first order Mindlin type analyses (Chandrashekhara and Agarwal, 1993; Tzou and Zhong, 1993) and the potentially expensive layer-wise theories (Robbins and Reddy, 1991; Lee and Saravanos, 1995). An hybrid theory has also been reported later (Mitchell and Reddy, 1995). It is well known that refined higher order theories are capable of capturing the transverse shear

Aditi Chattopadhyay, Associate Professor, Associate Fellow AIAA, Member SPIE, ASME, AHS, ISSMO, AAM, Department of Mechanical and Aerospace Engineering, Arizona State University, Tempe, AZ 85287-6106
Dan Dragomir-Daescu, Graduate Research Assistant
Haozhong Gu, Postdoctoral Fellow, Member AIAA, ASME

deformation through the thickness quite accurately (Chattopadhyay and Gu, 1994). These theories are applicable to laminates of thicker construction and have been shown to be useful in modeling smart composite laminates (Reddy, 1990; Chattopadhyay and Seeley, 1996). Finite element based solution procedures (Chandrashekhara and Agarwal, 1993; Robbins and Reddy, 1991; Chattopadhyay and Seeley, 1996) are practical since real geometries and boundary conditions can be investigated.

A significant amount of research has also been performed in modeling delamination in composites. Although three dimensional approaches (Yang et al., 1994; Whitcomb, 1989) are more accurate than two dimensional theories (Pavier and Clarke, 1996; Kardomateas and Schmueser, 1988; Gummadi and Hanagud, 1995), their implementation can be very expensive for practical applications. The layer-wise approach (Barbero and Reddy, 1991) is an alternative since it is capable of modeling displacement discontinuities. However, the computational effort increases with the number of plies. Recently, a refined higher order theory, developed by Chattopadhyay and Gu (1994), was shown to be both accurate and efficient for modeling delamination in composite plates and shells of moderately thick construction.. This theory has also been shown to agree well with both elasticity solutions (Chattopadhyay and Gu, 1996) and experimental results (Chattopadhyay and Gu, 1996).

Preliminary research (Keilers and Chang 1995) has been conducted on the use of smart materials for detecting pre-existing delamination. However, the mathematical models used in those work are simply classical theory based approaches, which exclude the transverse shear effects. The importance of transverse shear deformation in composite laminates have been identified by many researchers. As much as 50% deviations in structural response has been reported in thick constructions (Barbero and Reddy, 1991; Chattopadhyay and Gu, 1994). Therefore, the objective of the current research is modeling dynamic response of delaminated smart composite structures using a refined order theory.

A higher order theory based finite element model is presented for delaminated smart composite beams. The model properly takes into account the distributed nature of the presence of discrete delaminations, actuators and sensors. Since the relationships between the induced strain due to actuation and the applied electric field are nonlinear in nature (Crawley and Lazarus, 1989), new formulations are presented in the current research to include these nonlinear induced strain effects. A control algorithm is implemented for vibration reduction.

HIGHER ORDER THEORY BASED FORMULATION

A general higher order displacement field is extended to model composite beams with induced strain actuations due to piezoelectric materials. The inplane displacements are assumed to be effectively expressed by a cubic function through the thickness (z) and the transverse displacement is assumed to be independent of z. To model delamination in such structures, the general displacement field is defined as

$$\begin{aligned} U(x,z) &= u(x) + (z-c)\left[\psi(x) - \frac{\partial w}{\partial x}\right] + (z-c)^2\phi(x) + (z-c)^3\varphi(x) \\ W(x,z) &= w(x) \end{aligned} \tag{1}$$

where, U and W are the inplane and outplane displacements at a point (x,z), u and w denote the midplane displacements, ψ represents the rotation of the midplane, ϕ and φ are the higher order terms.

For anisotropic elastic body, the constitutive relations among the stress, strain, charge and electric field can be derived from the electric enthalpy density function given as follows.

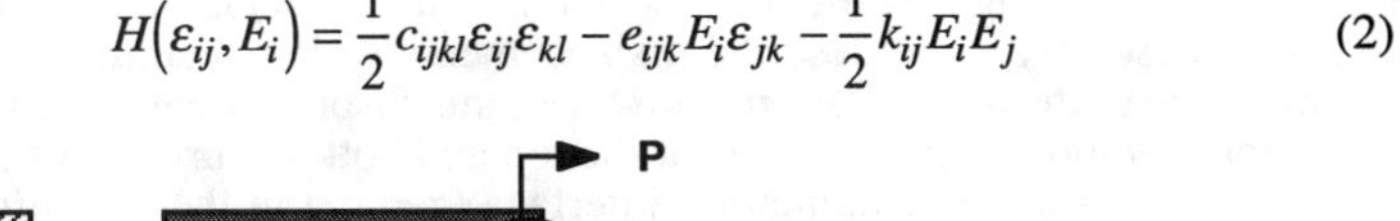

$$H(\varepsilon_{ij},E_i)=\frac{1}{2}c_{ijkl}\varepsilon_{ij}\varepsilon_{kl}-e_{ijk}E_i\varepsilon_{jk}-\frac{1}{2}k_{ij}E_iE_j \tag{2}$$

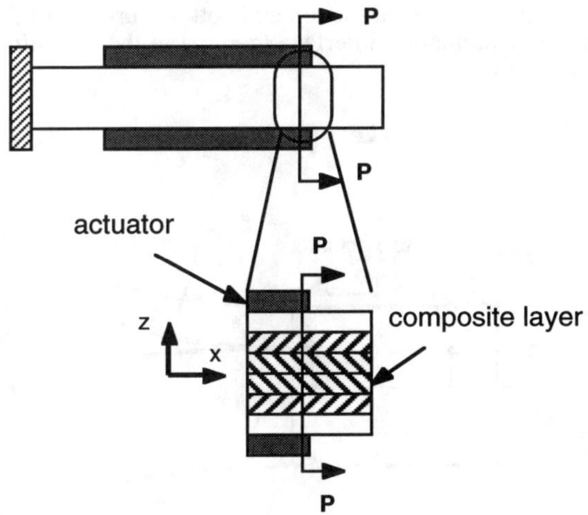

Fig. 1 Cantilever composite laminate with piezoelectric actuators

where ε_{ij} and E_i are components of the strain tensor and electric field vector, respectively and c_{ijkl}, e_{ijk}, and k_{ij} are the elastic, piezoelectric and dielectric permitivity constants, respectively. The stress and charge are then determined as

$$D_i=-\frac{\partial H}{\partial E_i} \tag{3}$$

$$\sigma_{ij}=\frac{\partial H}{\partial \varepsilon_{ij}} \tag{4}$$

For an orthotropic composite beam with piezoelectric layers, the constitutive relationships can be simplified as

$$\begin{bmatrix}\sigma_x\\ \tau_{xz}\end{bmatrix}_k=\begin{bmatrix}\overline{Q}_{11} & 0\\ 0 & \overline{Q}_{55}\end{bmatrix}\begin{bmatrix}\varepsilon_x-\Lambda_x\\ \gamma_{xz}\end{bmatrix} \tag{5}$$

$$D_{3_k}=\overline{Q}_{11}d(\varepsilon)\varepsilon_{x_k} \tag{6}$$

where Λ_x are the induced strains ($\Lambda_x = d(\varepsilon)\, E_3$).

To account for the delamination effects, it is necessary to partition the laminate into several different regions as shown in Fig. 2. These regions include the nondelaminated region Ω^u, the region above the delamination Ω^a and the region below the delamination Ω^b. The interface between the nondelaminated region and the delaminated region, indicated by the dashed line in Fig. 2, is denoted S. The general form of the higher order displacement field (Eq. 1) is independently applied to each of these regions to describe displacements which account for slipping and separation due to the delamination. However, this displacement field need to satisfy the traction-free boundary condition at the top and bottom surfaces of the plate ($z = \pm h/2$) as well as on the delamination interface ($z = h_1$) in the delaminated region. That is,

$$\tau_{xz}(x,z^*) = 0, \qquad x \in \Omega^r \quad (r = u,a,b, \qquad z^* = \pm h, h_1) \tag{7}$$

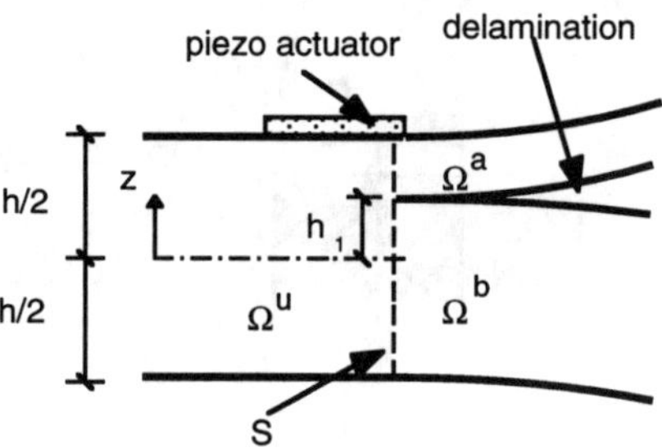

Fig. 2 Laminate cross section.

in which the superscript r corresponds to either the nondelaminated region (u), or the regions above and below the debonding (a and b), respectively. For orthotropic plates, these conditions are equivalent to the requirement that the corresponding transverse shear strains be zero on these surfaces. A refined displacement field is then obtained by applying these boundary conditions in each region as follows.

$$U^r = u^r + \left(z - c^r\right)\left(\psi^r - \frac{\partial w^r}{\partial x}\right) - \frac{4\left(z - c^r\right)^3}{3\left(d^r\right)^2}\psi^r \qquad (r = u,a,b) \tag{8}$$

$$W^r = w^r$$

where c^r is the local midplane and d^r is the local thickness of the region. With the identification of the higher order terms, the variables in the generalized displacement field are reduced. It is important to note that the total thickness of the beam, h, may vary due to the presence of surface bonded sensors/actuators.

Additional continuity conditions must be imposed to ensure the continuity of displacements at the interface of the nondelaminated and the delaminated regions (S) as shown in Fig. 2. The continuity conditions at the interface of the nondelaminated and the delaminated regions are imposed as follows.

$$\left.\begin{aligned} U^u &= U^a \\ W^u &= W^a \\ U^u &= U^b \\ W^u &= W^b \end{aligned}\right\} \quad x \in S \tag{9}$$

The above equations can be exactly satisfied with the classical theory since it assumes a linear displacement distribution through the thickness. However, the displacement distribution using the refined theory is nonlinear and therefore Eqs. 9 can be approximated at the local midplane as follows.

$$\begin{array}{lll} U^u = U^a, & W^u = W^a & \\ \dfrac{\partial U^u}{\partial z} = \dfrac{\partial U^a}{\partial z}, & \dfrac{\partial^2 U^u}{\partial z^2} = \dfrac{\partial^2 U^a}{\partial z^2} & z = c^a \\ U^u = U^b, & W^u = W^b & \\ \dfrac{\partial U^u}{\partial z} = \dfrac{\partial U^b}{\partial z}, & \dfrac{\partial^2 U^b}{\partial z^2} = \dfrac{\partial^2 U^b}{\partial z^2} & z = c^b \end{array} \quad x \in S \tag{10}$$

Using Eqs.10, the relations of displacement fields defined in different regions at region interface S can now be derived.

$$\begin{aligned} & u^a = u^u + \frac{1}{2}\left(\frac{h}{2} + h_1\right)\left(\psi^u - \frac{\partial w^u}{\partial x}\right) - \frac{\left(\frac{h}{2} + h_1\right)^3}{6h^2}\psi^u \\ & w^a = w^u, \qquad \psi^a = \psi^u, \qquad \frac{\partial w^a}{\partial x} = \frac{\partial w^u}{\partial x} \\ & u^b = u^u + \frac{1}{2}\left(\frac{h}{2} - h_1\right)\left(\psi^u - \frac{\partial w^u}{\partial x}\right) - \frac{\left(\frac{h}{2} - h_1\right)^3}{6h^2}\psi^u \\ & w^b = w^u, \qquad \psi^b = \psi^u, \qquad \frac{\partial w^b}{\partial x} = \frac{\partial w^u}{\partial x} \end{aligned} \tag{11}$$

NONLINEAR PIEZOELECTRIC ACTUATION

It is assumed that the piezoelectric actuators are surface bonded on the composite laminate. Therefore, the piezoelectric material can be treated as an additional surface layer. The electro-mechanical coupling between the applied electric field and the induced strain in the piezoelectric material is governed by the coupling coefficients, d, which represent piezoelectric material properties. According to Crawley and Lazarus (1991), the coefficient depends on the actual strain in the actuator as well. That is

$$\Lambda = d(\varepsilon)E_3 \tag{12}$$

where E_3 is the applied electric field through the thickness of the actuator. However, due to the weak nonlinearity, Eq. 12 can be rewritten using first order Taylor series expansion. That is

$$\begin{aligned}\Lambda &= (d_0 + d_1\varepsilon + \frac{1}{2}\varepsilon^2 + ...)E_3 \\ &= d_0E_3 + d_1\varepsilon E_3\end{aligned} \tag{13}$$

The coefficients d_0 and d_1 can be identified using functional relationships of the strain versus electric field obtained from experimental data of an unconstrained piezoelectric actuator.

CONTROL ALGORITHM

Since the piezoelectric sensor can accurately detect the strain rate when they are connected with a current amplifier and proportional feed beck is the simplest to apply for structural vibration control, a piezoelectric sensor model is developed here for monitoring the proportional variables which are necessary for control feed back. When the kth piezoelectric sensor layer is deformed, it accumulates an electric charge on its surface electrode which is given by

$$q_k(t) = \int\limits_A\int\limits_z D_{3_k} dzdA = \int\limits_A\int\limits_z \overline{Q}_{11} d_{31}\varepsilon_{x_k} dzdA \tag{14}$$

The current developed from the electric charge is obtained by differentiating the charge equation (Eq. 14) with respect of time.

$$i_k(t) = \frac{dq_k(t)}{dt} = \int\limits_A\int\limits_z \overline{Q}_{11} d(\varepsilon)\frac{d\varepsilon_{x_k}}{dt} dzdA \tag{15}$$

With a design of a control law using Lyapunov direct method, the beam stability in the Lyapunov sense is guaranteed if the feedback actuator voltage is selected as

$$E_{3k}(t) = Gi_k(t) = G\int\limits_A\int\limits_z \overline{Q}_{11} d(\varepsilon)\frac{d\varepsilon_{x_k}}{dt} dzdA \tag{16}$$

with gain $G > 0$.

FINITE ELEMENT MODELING

The finite element method (FEM) is used to implement the refined higher order theory since it allows for the analysis of practical geometries and boundary conditions. The continuity conditions presented in Eqs. 11, between the nondelaminated region (Ω^u) and delaminated regions (Ω^a, Ω^b), are applied to the finite element degrees of freedom at the interface of the nondelaminated and delaminated regions (S) by first presenting these discretized conditions in matrix form as follows.

$$\mathbf{u}_s^k = \mathbf{T}^k \mathbf{u}_s^u \qquad k = a, b \tag{17}$$

where

$$\mathbf{u}_s^k = \left[u^k, \psi^k, w^k, \frac{\partial w^k}{\partial x} \right]_s^T, \qquad \mathbf{u}_s^u = \left[u^u, \psi^u, w^u, \frac{\partial w^u}{\partial x} \right]_s^T$$

$$\mathbf{T}_k = \begin{bmatrix} 1 & h_k - \dfrac{4h_k^3}{3h^2} & 0 & -h_k \\ 0 & 1 & 0 & 0 \\ 0 & 0 & 1 & 0 \\ 0 & 0 & 0 & 1 \end{bmatrix}, \qquad \begin{array}{ll} h_k = \dfrac{1}{2}\left(\dfrac{h}{2} + h_1 \right), & \textit{when } k = a \\ h_k = \dfrac{1}{2}\left(\dfrac{h}{2} - h_1 \right), & \textit{when } k = b \end{array} \tag{18}$$

Using Eqs.18, the node variables defined in the delaminated region can easily be expressed in terms of those defined in the nondelaminated region at the region interface.

The finite element equations are derived using the discretized form of Hamilton's principle, which is stated as follows.

$$\delta\Pi = \int_{t_1}^{t_2} \sum_{e=1}^{N_e} \left[\delta K^e - \delta U^e + \delta W^e \right] dt = 0 \tag{19}$$

where t_1 and t_2 are the initial and the final times, respectively and δK^e, δU^e and δW^e represent the elemental variations in the kinetic, the strain and the potential energies, respectively. The finite element matrices are formulated as follows.

$$\int_{t_1}^{t_2} \sum_{e=1}^{N_e} \left[\delta\mathbf{u}^e \mathbf{M}^e \ddot{\mathbf{u}}^e + \delta\mathbf{u}^e \mathbf{K}^e \mathbf{u}^e - \delta\mathbf{u}^e \left(\mathbf{F}^e + \mathbf{F}_P^e \right) \right] = 0 \tag{20}$$

where N_e is the number of elements, an overdot indicates a derivative with respect to time and $\mathbf{u}^e$.represent the nodal generalized displacement vector for each element. The mass matrix, $\mathbf{M}^e$, is formulated from the density and the elemental shape functions. The stiffness matrix, $\mathbf{K}^e$, includes the stiffness of smart beams and the additional stiffness modification due to nonlinear actuation. Two force vectors are formulated for the distributed load, $\mathbf{F}^e$, and the piezoelectric forces $\mathbf{F}_P^e$. The global finite element equations of motion are then expressed as follows.

$$\mathbf{M}\ddot{\mathbf{u}} + \mathbf{K}\mathbf{u} = \mathbf{F} + \mathbf{F}_P \tag{25}$$

where the quantities **M**, **K** , **F** and **u** denote the mass and the stiffness matrices, the force vector due to a distributed load and the nodal displacement vector, respectively. The quantity $\mathbf{F}_p$ is the force vector due to piezoelectric actuation. Linear shape functions are used for the inplane displacements and rotations (u, ψ) while a 4 terms cubic polynomial is used for the transverse displacements (w). The resulting two noded beam elements are computationally efficient and contain 4 degrees of freedom at each node.

NUMERICAL RESULTS

Numerical results are presented first to compare the higher order theory with published experimental and analytical results to ascertain its validity. Next, dynamic response and vibration control using piezoelectric actuation are presented for a cantilever composite plate with varying delamination length. Differences between the developed higher order theory and the first order theory are also discussed. The $[0°/90°/0°/90°]_S$ Graphite/Epoxy laminates with surface bonded piezoelectric actuator pairs are used in the analysis. The material properties used for the smart composite beam are listed in Table 1, where E is the Young's modulus and υ is the Poisson's ratio. Beams with length $L = 0.127\ m$, thickness $h = 0.001016\ m$ and width $b = 0.0127\ m$, are considered. The piezoelectric actuator pairs have dimensions of $L_p = 0.0127\ m$, $h_p = 0.000254m$ and $b_p = 0.0127\ m$, located different positions (Fig.1).

Table 1. Material properties.

	E_1 (GPa)	E_2 (GPa)	υ_{12}	G_{12}, G_{13} (GPa)
Graphite/Epoxy	134.4	10.3	0.33	5
PZT	63	63	0.33	24.2
	G_{23} (GPa)	ρ(x10^3 Kg/m^3)	d_0 (x10^{-12} m/V)	d_1 (x10^{-7} m/V)
Graphite/Epoxy	2	1.477		
PZT	24.2	7.6	247	8.38

Table 2. First natural frequency for delamination along the midplane.

Delamination length (mm)	Experiment 1 (Hz)	Experiment 2 (Hz)	Experiment 3 (Hz)	Existing model (Hz)	Current model (Hz)
0	79.875	79.875	79.750	82.042	81.859
25.4	78.376	79.126	77.001	80.133	77.644

The results from the current approach are compared to finite element results and experimental data obtained by Shen and Grady (1991) as show in Tab. 1. The finite element model in that research is based on the Timoshenko beam theory. Delamination lies on the midplane of the beam and starts from the clamped end. The natural frequencies obtained from higher order theory based finite element approach show much better correlations with experimental data, compared to the existing finite element results, especially in the delaminated case. This indicates that the higher order theory-based model provides improved accuracy while maintaining same computational efficiency since the number of degrees of freedom per node is the same as the Timoshenko beam theory based FEM.

Next, the nonlinear induced strain effect on smart composite beams is considered. The static response under actuation only is analyzed and it is assumed that all of the beam surfaces are covered with the piezoelectric actuators. The finite element model comprises 10 equal-sized elements with 44 total degrees of freedom. Comparisons of beam deflections with and without nonlinear actuation effect are presented in Figs. 3 and 4 for different beam stiffness along the beam span. The results from the $[0°/90°/0°/90°]_S$ Graphite/Epoxy laminate are presented in Fig. 3 and those of the $[90°]_8$ Graphite/Epoxy laminate used in Fig. 4. Less than 1 percent deviation is observed for beams with higher stiffness along the span in Fig. 3. However, significant deviation (about 11 percent) is obtained for beams with lower stiffness along the span as shown in Fig. 4. This implies that the nonlinear actuation effect on smart composite beams is closely relate to the composite beam stiffness. The lower the stiffness, the higher is the nonlinear actuation effect.

The effect of closed loop control due to piezoelectric actuation is investigated. The beam is released from a real bended position initially, which is produced by the piezoelectric actuations under 127 Volts. Ten equal-sized elements are used and are numbered from clamped end to the free end. The actuator pairs are located at the top and the bottom surfaces of the second, third, fifth, sixth, eighth and ninth elements (Fig.1). The internal structural damping is assumed to be 1% from critical damping for each degree of freedom in following cases. Figures 5 and 6 present the dynamic responses of nondelaminated smart composite beams with and without closed loop control, respectively. Compared to Fig. 5, Fig. 6 shows the actuators effectively damp out the beam vibrations with the use of the closed loop control. The dynamic response of delaminated smart composite beam are also presented in Fig. 7 and 8 for cases with and without feed back control, respectively. Same beam configurations

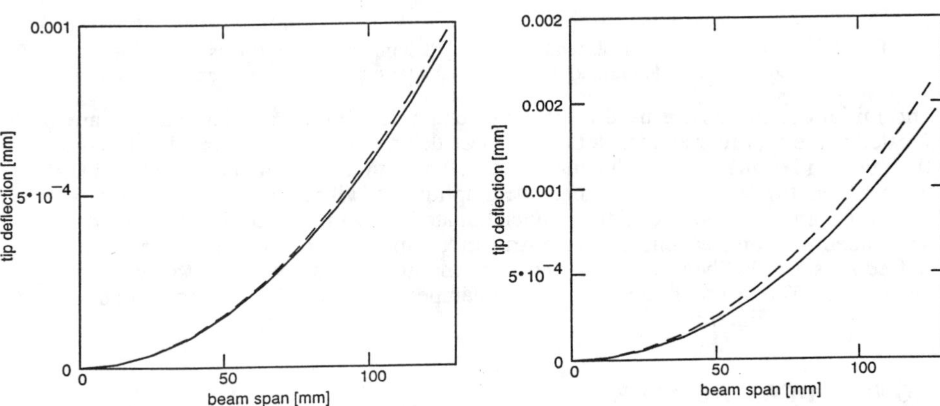

Fig. 3 Nonlinear actuation effect on tip deflection of $[0°/90°/0°/90°]_S$ beam

Fig. 4 Nonlinear actuation effect on tip deflection of $[0°]_8$ beam

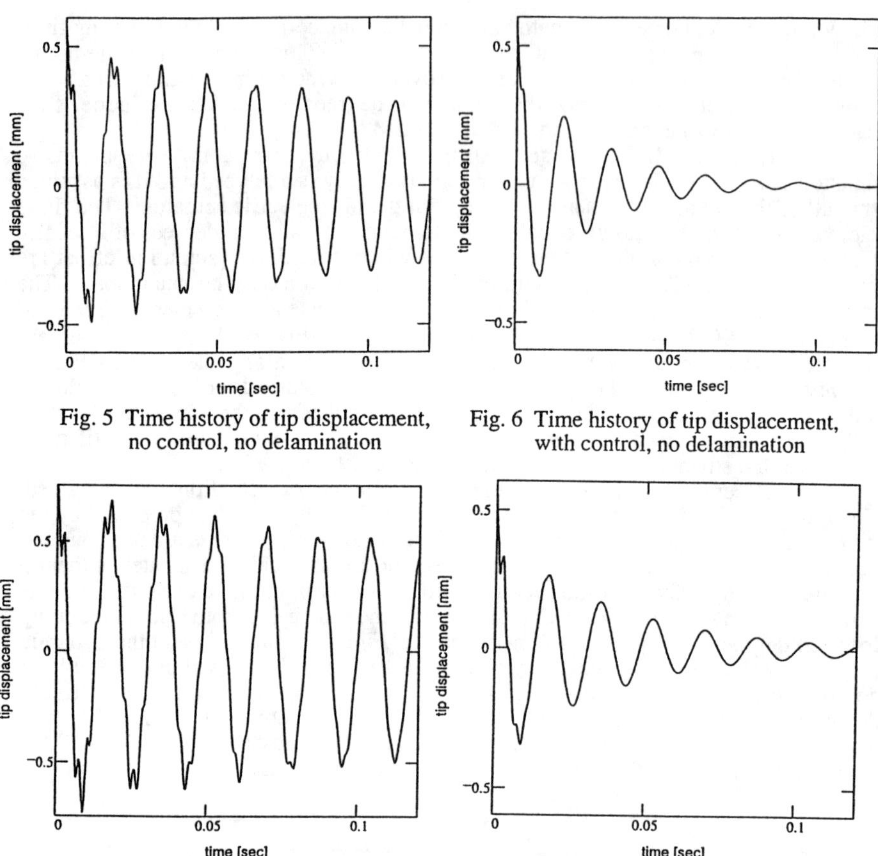

Fig. 5 Time history of tip displacement, no control, no delamination

Fig. 6 Time history of tip displacement, with control, no delamination

Fig. 7 Time history of tip displacement, no control, with delaminations

Fig. 8 Time history of tip displacement, with control, with delaminations

and initial conditions are used as previous example. Multiple delaminations are located at the first four elements. Each of these delaminations are assumed to lies on the midplane of the beam with equal-sized delamination of length equal to half the element length. With delaminations, the amplitude of vibrations become larger as shown in Fig. 7, compared to the nondelaminated cases in Fig. 5. In other words, since delamination weakens the beam structure, same load condition produces more deflections of the beam. Moreover, delaminations also weaken the control authorities. The beam vibrations are being damped out slower in Fig.8 compared to that shown in Fig. 6.

CONCLUDING REMARKS

A general framework has been developed for the analysis of smart composite beams with surface bonded piezoelectric actuators and sensors in the presence of delamination. A refined third order theory has been used which accurately captures the transverse shear deformation through the thickness of the smart composite laminate while satisfying the stress free boundary conditions on the free surfaces, including the delaminated region. The presence of pre-existing delaminations in the composite laminate at the layer interface was studied. The following important observations were made from this study.

1) The developed theory provides an accurate, computationally efficient analysis tool for the study of smart composite laminates with piezoelectric sensing and actuation in the presence of delamination.

2) The theory is implemented using the finite element method to allow the incorporation of practical geometries, boundary conditions and the presence of discrete piezoelectric transducers.

3) The developed theory correlates well with experimental results.

4) The magnitude of nonlinear actuation effect depends on the composite beam stiffness. The lower the stiffness, the higher is the nonlinear actuation effect.

5) Delaminations weaken the beam structures, produce larger vibration amplitude and reduce the closed loop control authority.

ACKNOWLEDGMENTS

The research was supported by the Air Force Office of Scientific Research, grant number: F49620-96-0195, technical monitor Dr. Brian Sanders. The authors would like to thank Dr. Charles E. Seeley, for helpful discussions.

REFERENCES

Barbero, E. J. and Reddy, J. N. 1991. "Modeling of Delamination in Composite Laminates Using a Layer-wise Plate Theory" *International Journal of Solids and Structures*, Vol. 28, No. 3, pp. 373-388.

Chandrashekhara, K. and Agarwal, A. N. 1993. "Active Vibration control of Laminated Composite Plates Using Piezoelectric Devices: A Finite Element Approach" *Journal of Intelligent Material Systems and Structures*, Vol. 4, Oct., pp. 496-508.

Chattopadhyay, A. and Gu, H. 1994. "A New Higher-Order Plate Theory in Modeling Delamination Buckling of Composite Laminates" *AIAA Journal*, Vol. 32, No. 8, Aug., pp. 1709-1718.

Chattopadhyay, A. and Gu, H. 1996. "An Experimental Investigation of Delamination Buckling and Postbuckling of Composite Laminates" *Proc.* ASME International Mechanical Engineering Congress and Exposition, Atlanta, GA, Nov. 17-22,.

Chattopadhyay, A. and Gu, H. 1996. "Elasticity Solution for Delamination Buckling of Composite Plates", *Proc.* 37th AIAA/ASME/ASCE/AHS/ASC Structures, Structural Dynamics and Materials Conference, Salt Lake City, UT, Apr. 15-18.

Chattopadhyay, A. and Seeley, C. E. 1996. "A Higher Order Theory for Modeling Composite Laminates with Induced Strain Actuators" *Composites Part B: Engineering*, Feb., *accepted for publication.*

Chattopadhyay, A., and Seeley, C. E. 1997. "Experimental Investigation of Composites with Piezoelectric Actuation and Debonding" *Proc.* International Mechanical Engineering Congress and Exposition Winter Annual Meeting of the ASME: Adaptive Structures and Material Systems Symposium, Dallas, TX, Nov. 16-21.

Crawley, E. F. and Anderson, E. H. 1989. "Detailed Models of Piezoelectric Actuation of Beams" *Proc.* of the 30th AIAA/ASME/ASCE/AHS/ASC Structures, Structural Dynamics and Materials Conference, Mobile, Al., April, pp. 2000-2010.

Gummadi, L. N. B. and Hanagud, S. 1995. "Vibration Characteristics of Beams with Multiple Delaminations" *Proc.* 36th AIAA/ASME/ASCE/AHS/ASC Structures, Structural Dynamics and Materials Conference - Adaptive Structures Forum, New Orleans, LA, Apr. 10-14, pp. 140-150.

Kardomateas, G. A and Schmueser, M. 1988. "Buckling and Postbuckling of Delaminated Composites Under Compressive Loads Including Transverse Shear Effects", *AIAA Journal*, Vol. 26, No. 3, 1988, pp. 337-343.

Keilers, C. H. and Chang F. K. 1995, "Identifying Delamination in Composite Beams Using Built-In Piezoelectric: Part I -- Experiments and Analysis; Part II -- An identification Method, *Journal of Intelligent Material System and Structures*, Vol.5, No.5, pp.649-663.

Lee C. K. 1990. "Theory of Laminated Piezoelectric Plates for the Design of Distributed Sensors/Actuators. Part 1: Governing Equations and Reciprocal Relationships" *Journal of the Acoustical Society of America*, Vol. 87, No. 3, March, pp. 1144-1158.

Lee, H. J. and Saravanos, D. A. 1995. "Coupled Layerwise Analysis of Thermopiezoelectric Smart Composite Beams" *AIAA Journal*, Aug., submitted for possible publication.

Lee, J. E. and Fassois, S. D. 1990. "A Stochastic Suboptimum Maximum Likelihood Approach to Structural Dynamics Identification" *Proc.* of the 8th International Model Analysis Conference, Kissimmee, FL, Jan. 29-Feb. 1, pp. 1424-1433.

Mitchell, J. A. and Reddy, J. N., 1995. "A Refined Hybrid Plate Theory for Composite Laminates with Piezoelectric Laminae", *International Journal of Solids and Structures*, Vol. 32, No. 16, pp. 2345-2367.

Pavier, M. J. and Clarke, M. P. 1996. "A Specialized Composite Plate Element for Problems of Delamination Buckling and Growth" *Composite Structures*, Vol. 35, 45-53.

Reddy, J. N. 1990. "A General Non-Linear Third-Order Theory of Plates with Moderate Thickness" *International Journal of Non-Linear Mechanics*, Vol. 25, No. 6, pp. 677-686.

Robbins, D. H. and Reddy, J. N. 1991. "Analysis of Piezoelectrically Actuated Beams Using a Layerwise Displacement Theory" *Computers and Structures*, Vol. 41, No. 2, pp. 265-279.

Seeley, C. E. and Chattopadhyay, A. 1996. "Modeling Delaminations in Smart Composite Laminates" *Proc.* of the 37th AIAA/ASME/ASCE/AHS/ASC Structures, Structural Dynamics and Materials Conference and Adaptive Structures Forum, Salt Lake City, UT, Apr. 15-19, pp. 109-119.

Seeley, C. E., Chattopadhyay, A. 1997. "Modeling of Smart Composite Laminates Including Debonding: A Finite Element Approach" *Proc.* of the 38th AIAA/ASME/ASCE/AHS/ASC Structures, Structural Dynamics and Materials Conference - Adaptive Structures Forum, Kissimmee, FL, April 7-11.

Shen, M. H. and Grady, J.E. 1991. "Free Vibration of Delaminated Beams" *Proc.* of the 31th AIAA/ASME/ASCE/AHS/ASC Structures, Structural Dynamics and Materials Conference, Baltimore, MD, April 8-10.

Tzou, H. S. and Zhong, J. P. 1993. "Electromechanics and Vibrations of Piezoelectric Shell Distributed Systems" *Journal of Dynamic Systems, Measurement, and Control*, Vol. 115, Sept., pp. 506-517.

Wang, J. and Lin, C. 1995. "Vibration of Beam-Plates Having Multiple Delaminations", *Proc.* 36th AIAA/ASME/ASCE/AHS/ASC Structures, Structural Dynamics and Materials Conference - Adaptive Structures Forum, New Orleans, LA, Apr. 10-14.

Whitcomb, J. D., 1989. "Three Dimensional Analysis of a Postbuckled Embedded Delamination", *Journal of Composite Materials*, Vol. 23, pp. 862-889.

Yang, H. T. Y. and He, C. C. 1994. "Three-Dimensional Finite Element Analysis of Free Edge Stresses and Delamination of Composite Laminates", *Journal of Composite Materials*, Vol. 28, No. 15, pp. 1394-1412.

Identification of Hole Defect for GFRP Woven Laminates Using Neural Network Scheme

S. T. JENQ and W. D. LEE

ABSTRACT

The purpose of the current study is to identify the various hole defect sizes and locations on composite beam structures utilizing neural networks. A series of vibration tests have been conducted on the woven GFRP laminated beam structures in order to measured the corresponding frequency response using impact hammer test equipment. The measured frequency response is then compared with the corresponding ANSYS finite element analysis solution in order to assure the accuracy of test result. Ninty finite element simulations are then performed to compute the vibration response of the structures containing various size of hole defect at different location. These simulated results are utilized as a knowledge base for further neural network analysis. The back-propagation neural network with an adaptive learning rate scheme is adopted. The trained network is capable of predicting the defect size and location for arbitrary untrained cases closely.

INTRODUCTION

In order to monitor the health condition of structure in question, it would be crucial to develop an appropriate method to keep track of the structural damage/defect status. The damage detection of structure elements using neural networks has been reported in the past five years by a number of researchers. Previous studies employing different methods for detection and qualification of structural damage were briefly summarized by Chaudhry et al. [1] A number of examples of (1) damage detection using neural networks with analytical data and (2) experimental studies performed using neural networks to detect the severity and/or location of specific damage have also been summarized by Jones et al. [2] .

As we understand that the ability to sense and monitor structural health condition is an important issue in the development of intelligent material / structure

S. T. Jenq and W. D. Lee, Institute of Aeronautics & Astronautics, National Cheng Kung University, Tainan, Taiwan, 701, R.O.C.

system. The identification of the location and size of defects in advanced composite structures using neural networks is certainly an important topics. Cawley and Adams [3] reported a method of non-destructively assessing the integrity of structures using measurements of the structural natural frequencies. They showed how measurements made at a single point in the structure can be used to detect and locate and roughly quantify damage on an aluminum plate and a cross-ply carbon-fibre reinforced plastic plate. It was reported [3,4] that frequency change is related to the damage state and the size and location of hole cut-out defect of structures, and it would be interesting to identify the aforementioned defects from measured frequencies via an appropriate neural network identification process.

Present work is focused on determining the size and location of the hole defect on the GFRP woven laminate utilizing neural network scheme. In order to reduce the experimental testing effort, the commercial finite element code ANSYS is used to generate the training data base. In order to check against the neural network identified result, a series of vibration tests and the corresponding numerical simulations have been performed with prescirbed defects different from those used in the training set.

BACKPROPAGATION NEURAL NETWORK

The mathematical form of the output of a neuron j (y_j) can be described by the input signals (y_i, i=1 to p), the synaptic weights (w_{ji}) of neuron j, the threshold or bias (θ_j),the net internal activity level (v_j) and the activation function ($\varphi(\cdot)$) as:

$$y_j = \varphi_j \left(\sum_{i=1}^{p} w_{ji} y_i - \theta_j \right) = \varphi_j \left(\sum_{i=0}^{p} w_{ji} y_i \right) = \varphi_j (v_j(n)) . \tag{1}$$

The error signal of the output of neuron j at the presentation of the nth training pattern is defined by

$$e_j(n) = d_j(n) - y_j(n) \tag{2}$$

where d_j is the desired (or target) response for neuron j at time (or iteration) n. The corresponding instantaneous sum of square errors of all neurons in the output layer can written as

$$E(n) = \frac{1}{2} \sum_{j \in C} e_j^2(n) , \tag{3}$$

where the set C includes all the neurons in the output layer of the network. Note that if there are N patterns (examples) contained in the training set, the averaged squared error is obtained by summing E(n) over all n (n=1 to N) and the normalized with respect to the size N of the training set. The simple back-propagation algorithm [5] applies a correction Δw_{ji} to the linking weight $w_{ji}(n)$ by the delta rule:

$$\Delta w_{ji} = -\eta \frac{\partial E(n)}{\partial w_{ji}} , \tag{4}$$

where η is called the learning-rate parameter. One can also determine [5] that the synaptic weight correction Δw_{ji} connecting neuron i to neuron j is equal to multiplication of learning rate parameter (η), local gradient ($\delta_j(n)$) and input signal of neuron j ($y_i(n)$). The local gradient $\delta_j(n)$ equals the product of error signal and the derivative of activation function for the output node; the product of the associated derivative of activation function and weighted sum of the local gradient computed for the neurons in the next hidden or output layer connected.

For present application, the batch mode of pack-propagation learning is used to perform weight updating after the presentation of all the training examples that constitute an epoch. In general, we may define the cost function as the average squared error by summing the instantaneous sum of squared errors of the network E(n) given in eqn. (3) and then normalizing with respect to the total number of patterns N contained in the training set , as

$$E_{avg} = \frac{1}{2N} \sum_{n=1}^{N} \sum_{j \in C} e_j^2(n). \tag{5}$$

In order to accelerate the convergence of the back-propagation method, we use the modified Rumelhart algorithm [6,7, 8] as:

$$\Delta w_{ji}(m+1) = \eta(1-\alpha)\Sigma \delta_{pi} y_{pi} + \alpha \Delta w_{ji}(m), \tag{6}$$

where m represents the iteration number rather than the presentation number n.. When momentum constant α is set to zeor, then eqn. (6) is equivalent to eqn. (4). During computation, if an update results in reduced total error, η is multiplied (accelerated) by a factor greater than one for the next iteration. If a step produces a network with a total error more than a few (usually one to five) percent above the previous value, all changes to the weights are rejected and the learning rate (η) is multiplied by a factor less than unity. In addition, the momentum constant (α) is set equal to zero and the step is repeated. When a successful step is obtained , the momentum constant is reset to its original value.

VIBRATION TESTS AND NUMERICAL SIMULATIONS

Both 6061 aluminum samples and plain weaved GFRP laminates (200 x 25.4 x 4 mm^3) are prepared for the present study. The composite samples are made of woven glass fiber (TGFW-600, Taiwan Glass Co.) and epoxy resin system (Aradite LY564 and HY2954, CibaGeigy). Detailed description of the composite curing process is referred to previously work in reference [9]. The mechanical properties of aluminum specimen are summarized in Table 1. The test determined averaged fiber volume fraction of the composite laminates is 52.3%. Apply the average stiffness matrix method [10] the cross-ply GFRP laminate's elastic mechanical properties can be estimated and the result is tabulated in Table 2.

The vibration tests performed on the cantilever beam type structure utilizes an impact hammer (PCB 086B01) and its associated accelerometer (PCB309A). In addition, the spectrum analyzer (HP 3567) and a PC based STAR vibration analysis software (Structural Measurement Systems, Milpitas, CA) are also used to analyze the vibration frequencies and the corresponding mode shapes. The first four vibration frequencies for metallic and composite samples without hole cut-out defect are presented in the 2nd and the 5th column of Table 3. In order to examine the effect of circular hole cut-out on the natural frequency shift for the cantilever beam structure made of aluminum or GFRP composite, a series of tests using aforementioned equipment have been conducted. Table 4 reports the measured first four vibration frequencies for composite specimens with cut-outs. Similar result has also been obtained for aluminum specimens but not reported here.

The commercial finite element code ANSYS is used to simulate the present problem with or without hole defect for cantilever beam structures. A total of 1,434 10-node tetrahedral solid elements are generated in an intact sample for vibration simulations. The code simulated first four natural frequencies for aluminum and GFRP composite using the material properties given in Tables 1 and 2 are summarized in the 3rd and 6th columns of Table 3, respectively. It is found that the difference between the test determined average first three bending natural frequencies and code computed results is within 1.3% and 6.5% for aluminum and composite specimens, respectively. For composite specimens, the difference seems greater especially for the torsional mode.

Ninety set of numerical simulations have been performed for both aluminum and composite cantilever structures containing circular hole cut-out with diameters (2R shown in fig. 1) and defect locations (x shown in fig. 1) ranging from 9 to 14 mm and 3 to 17 cm, respectively. The effect of the circular hole cut-out (i.e. 11mm. diameter) on the frequency shift of the first four fundamental modes is presented in figure 1 for glass/epoxy laminated cantilever structures. Figure 2 shows the relationship between the natural frequency shift versus the diameter of the circle hole cut-out when the defect is located 13 cm from the clamped end of GFRP samples. Notice that the aluminum samples also preserve a similar trend as that found for composite specimens in figures 1 and 2. In addition, figures 1 and 2 maintain specific patterns for each mode, and these information will to converted to the natural frequency shift used as input for the subsequent neural network identification process in order to locate the size and location of defect in the range of study.

DEFECT IDENTIFICATION USING NEURAL NETWORK

The architecture of the present problem is consisted of four inputs (i.e. the percent of frequency shift for mode one to four) and two outputs (i.e. the defect size and its location). Therefore, the output linear layer only have two neurons. Two layers network is adopted in this work and the hypertangent sigmoid hidden layer is consisted of fifteen neurons. A schematic drawing the neural network is presented in figure 3. In order to attain better learning result, the previous mentioned finite element ANSYS simulated results (90 runs) are randomly accessed in order to generate an input sample space containing 500 cases. The sample space consists of the vibration frequency shift (compared with that of an intact sample) information for the first four modes when defect diameter and location are ranged from 9 to 14 mm and 3 to 17 cm, respectively. The batch mode of back-propagation learning process and the modified Rumelhart algorithm (eqn. (6)) are adopted to perform weight updating after the presentation of all the training examples that constitutes an epoch. After one hundred thousand learning epochs, the initial sum of square error with an order of 10^5 initially is decreased to an order of 10^1, as shown in figure 4. Figure 5 schematically presents the batch mode neural network identification process for the current problem.

The trained neural network is then used to identify the cut-out defect size and location with the given test determined first four vibration frequencies for composite samples containing either a 9 mm or 11 mm diameter of defect located at 3.5, 6.5, 9.5 12.5 or 15.5 cm from the clamped end of the cantilever beam structure. Table 5 shows the network predicted defect location and size for ten GFRP composite specimens. The averaged error found for diameter of cut-out identification for 10 samples shown in Table 5 is within 15%. In addition, the average error is within 9% when we predict the location of cut-out for composite specimens. It is noted that the defect location of these samples are different from those used in training the weighting parameters of neural network based on ANSYS finite element computed first four vibration frequencies. Moreover, a series of ANSYS finite element simulations are also performed with defect diameter and location shown in Table 6 as the target values. These simulations are selected so that their defects are different from those specified in training the present neural network model. Upon examining Table 6, it reveals that the averaged error for predicting the diameter of cut-out defect and the location of defect from the clamped end is calculated to be less than 7% and 6%, respectively.

CONCLUSION

Two-layer neural network with four vibration frequency shifts (corresponding to the first four vibration mode) as input is used to predict the location and size of circular hole cut-out defect for plain weaved GFRP cantilever structures. The batch mode of back-propagation learning process and the modified Rumelhart algorithm are used to train the network. The ANSYS finite element analysis is used to simulate the vibration response of current problem with various defect location and size. Numerical computed results are then randomly used to generate a sample space containing 500 elements for network training process. The trained neural network is used to identify the defect location and size for composite cantilever structures, and good agreement is found. Note that the defect conditions specified for identification is different from those specified for the network training cases.

ACKNOWLEDGEMENT

The author like to thanks the National Science of Council, R.O.C. for sponsoring the present project in part under contract no. NSC86-2212-E006-062 and the Ministry of Education.

REFERENCES

1. Chaudhry, Z. And A. J. Ganino, 1994, "Damage detection using neural networks: an initial experimental study on debonded beams", J. Intelligent Material Systems & Structures, 5(4), pp. 585-589.
2. Jones, R. T., J. S. Sirkis and E. J. Friebele, 1997, "Detection of impact location and magnitude for isotropic plate using neural networks", J. Intelligent Material

Systems & Structures, 8(1), pp. 90-99.
3. Cawley, P. And R. D. Adams, 1979, "The location of defects in structures from measurements of natural frequencies", J. Strain Analysis, 14(2), pp. 49-57.
4. Jenq, S. T., G. C. Hwang, and S. M. Yang, 1993, "The effect of square cut-outs on the natural frequencies and mode shapes of GRP cross-ply laminates", Composites Science Technology, 47, pp. 91-101.
5. Haykin, Simon, 1994, Neural Networks, Macmillan Pub. Co., NJ, USA.
6. Vogl, T. P., J. K. Mangis, A. K. Rigler, W. T. Zink, and D. L. Alkon, 1983, "Accelerating the convergence of the back-propagation method", Biological Cybernetics, 59, pp. 257-263.
7. Rumelhart, D. E., G. E. Hinton and R. J. William, 1986, "Learning internal representation by error propagation", in Parallel distributed processing, Rumelhart, D. E., McClelland J. L. and the PDP Research Group (eds), v. 1, chap 8, MIT Press, Cambridge, MA.
8. Neural Network Toolbox, 1993, the MathWorks, Inc., Natick, MA.
9. Jenq, S. T. And J. J. Mo, 1996, "Ballistic impact response for two-step braided three-dimensional textile composites", AIAA, 34(2). Pp. 375-383
10. Chou, T. W., 1992, Microstructural Design Of Fiber Composites, Cambridge University Press, Cambridge, U.K.

Table 1 The mechanical properties of aluminum specimens

Young's modulus (E)	70 GPa
shear modulus (G)	26 GPa
Poisson's ratio (ν)	0.33
density (ρ)	2,700 kg/m^3

Table 2 Model predicted elastic properties for laminated specimens

	computed value	Note
E_{xx}	15.64 GPa	test determined value is 17.14 GPa
E_{yy}	15.64 GPa	
E_{zz}	6.33 GPa	
ν_{xy}	0.12	
ν_{yz}	0.17	
ν_{zx}	0.17	
G_{xy}	1.93 GPa	
G_{yz}	1.87 GPa	
G_{zx}	1.87 GPa	

Table 3 The first four vibration frequencies for metallic and composite samples without hole cut-out.

	For aluminum specimen			For composite specimens		
	Test result*(Hz)	ANSYS result(Hz)	Difference (%)	Test result*(Hz)	ANSYS result(Hz)	Difference (%)
Mode 1 (bending)	62.1	62.42	0.48	40.1	37.89	5.6

Mode 2 (bending)	391.9	390.71	0.3	250.	236.86	5.2
Mode 3 (torsion)	915.	921.51	0.71	422.6	331.19	21.6
Mode 4 (bending)	1108.4	1094.0	1.3	706.7	660.78	6.5

* represents for the test averaged result

Table 4 The measured natural frequencies for glass/epoxy laminated cantilever beam with or without circular hole.

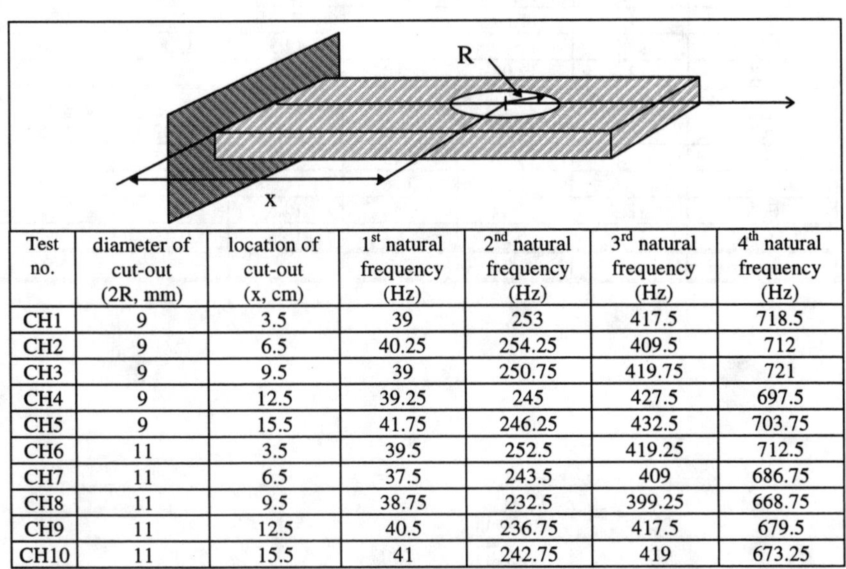

Test no.	diameter of cut-out (2R, mm)	location of cut-out (x, cm)	1st natural frequency (Hz)	2nd natural frequency (Hz)	3rd natural frequency (Hz)	4th natural frequency (Hz)
CH1	9	3.5	39	253	417.5	718.5
CH2	9	6.5	40.25	254.25	409.5	712
CH3	9	9.5	39	250.75	419.75	721
CH4	9	12.5	39.25	245	427.5	697.5
CH5	9	15.5	41.75	246.25	432.5	703.75
CH6	11	3.5	39.5	252.5	419.25	712.5
CH7	11	6.5	37.5	243.5	409	686.75
CH8	11	9.5	38.75	232.5	399.25	668.75
CH9	11	12.5	40.5	236.75	417.5	679.5
CH10	11	15.5	41	242.75	419	673.25

Table 5 The network predicted defect location and size for GFRP composites using test determined frequencies as input in neural network analysis

sample	diameter of cut-out (mm)			location of defect (cm)		
no.	target value	predicted value	error (%)	target value	predicted value	error (%)
TC1	9	10.76	19.6	3.5	3.48	0.6
TC2	9	9.27	3.0	6.5	5.38	17.3
TC3	9	10.13	12.6	9.5	10.56	11.1
TC4	9	8.68	3.5	12.5	13.1	4.8
TC5	9	12.13	34.7	15.5	16.97	9.5
average error			14.68			8.66
TC6	11	10.69	2.8	3.5	4.36	19.8
TC7	11	12.28	11.7	6.5	6.58	1.2
TC8	11	12.22	11.1	9.5	10.47	9.3
TC9	11	13.82	25.7	12.5	12.56	0.44
TC10	11	12.90	17.3	15.5	15.88	2.4
average error			13.7			6.62

Table 6 The network predicted defect location and size for GFRP composites using ANSYS code simulated frequencies as input in neural network analysis

sample no.	diameter of cut-out (mm)			location of defect (cm)		
	target value	predicted value	error (%)	target value	predicted value	error (%)
FC1	9.5	8.91	6.3	3.5	3.05	12.9
FC2	10.5	10.37	1.3	6.5	6.98	7.4
FC3	11.5	11.04	4.0	9.5	9.74	2.5
FC4	12.5	11.49	8.1	12.5	12.86	2.86
FC5	13.5	12.03	10.9	15.5	14.95	3.56
average error			6.1			5.83
FC6	9.5	9.21	3.0	3.5	3.09	11.6
FC7	10.5	10.46	0.43	6.5	6.97	7.25
FC8	11.5	11.59	0.76	9.5	9.82	3.42
FC9	12.5	12.44	0.49	12.5	12.03	3.72
FC10	13.5	15.22	12.7	15.5	16.04	3.52
average error			3.5			5.9

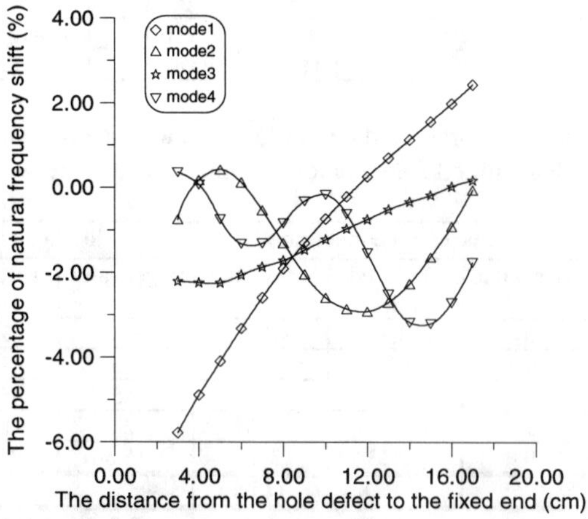

Figure 1 Effect of of the hole cut-out (11mm dia.) on the natural frequency shift for glass/epoxy laminated cantilever structure.

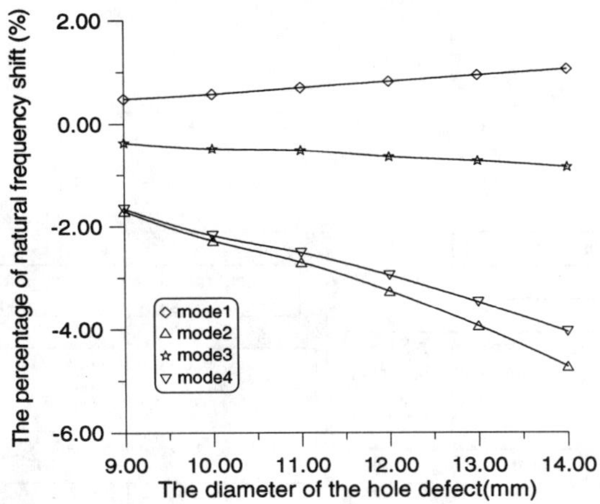

Figure 2 A plot of the natural frequency shift versus the diameter of hole cut-out when the hole defect is located 13 cm from the clamped end for GFRP specimen.

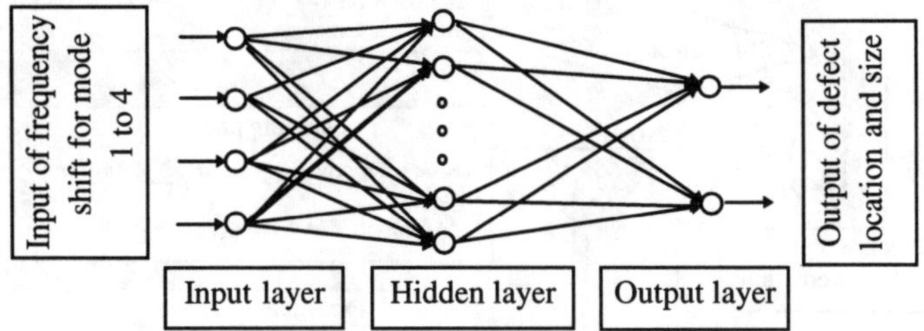

Figure 3 A schematic drawing the neural network studied.

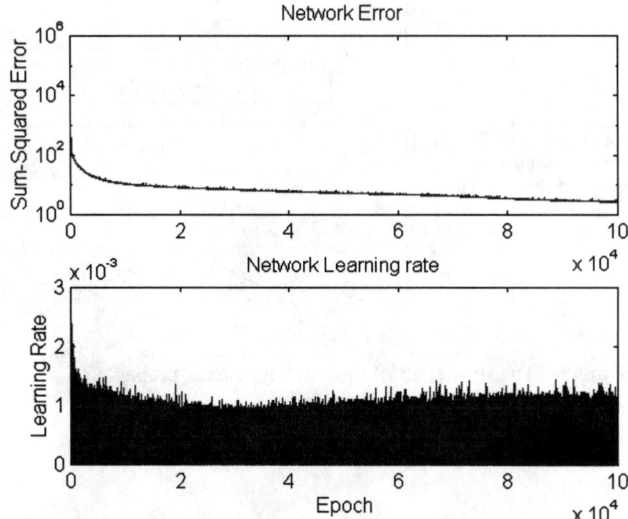

Figure 4 The sum-square error and learning rate of the network versus the number of epoch for GFRP specimens.

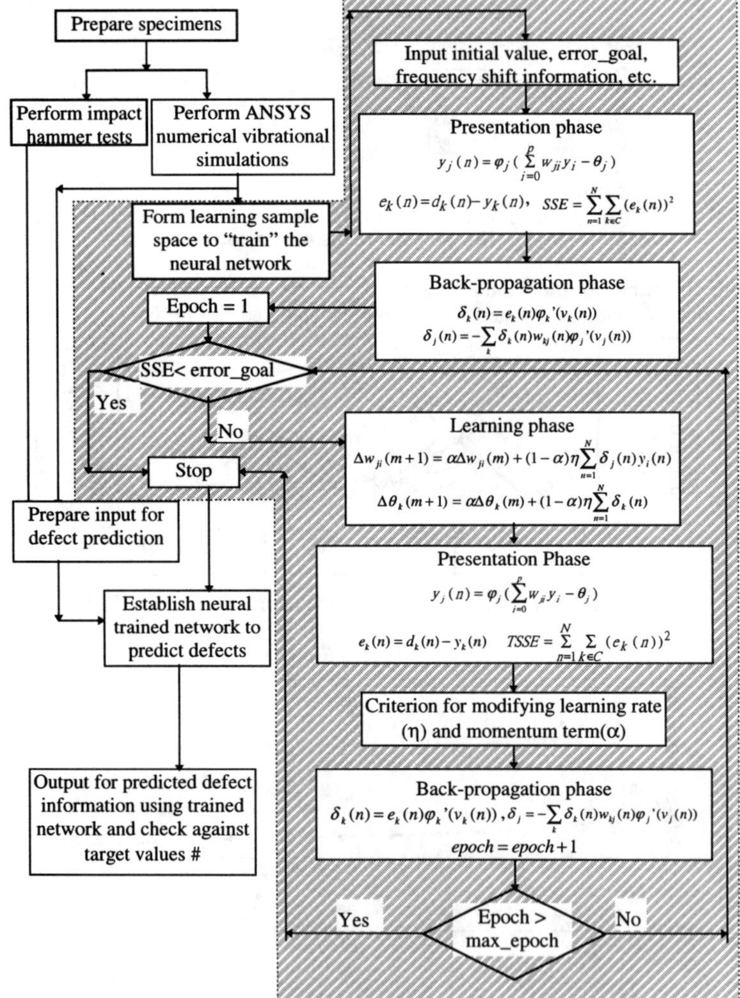

Figure 5 The flow chart of the batch type back-propagation identification process.

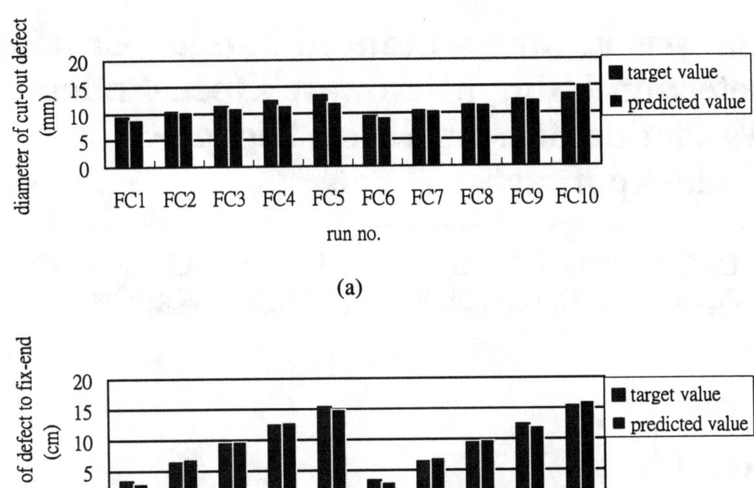

Figure 6 The neural network predicted (a) diameter of cut-out defect and (b) distance of cut-out defect to structure fix-end. The ANSYS computed frequencies are used as input to trained network identification.

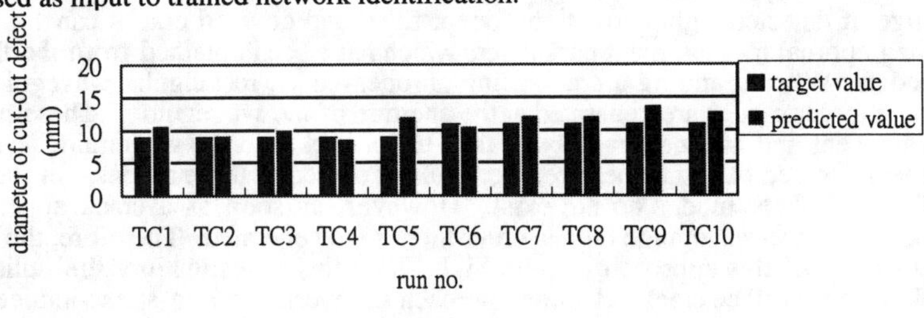

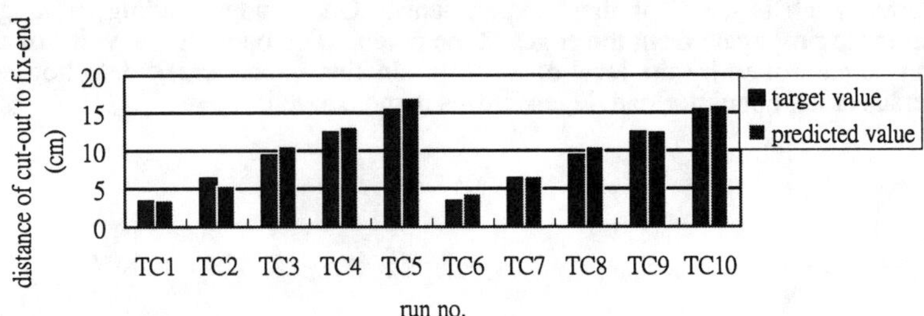

Figure 7 The neural network predicted (a) diameter of cut-out defect and (b) distance of cut-out defect to structure fix-end. The test determined frequencies are used as input to trained network identification.

Analysis of Stress Induced Fatigue Crack Detection Using Microwave Open-Ended Waveguide Sensors Using Higher-Order Mode Approach

N. QADDOUMI, R. MIRSHAHI, E. RANU, V. OTASHEVICH, R. ZOUGHI, J. D. McCOLSKEY and R. LIVINGSTON

ABSTRACT

Surface crack detection in metals is of utmost importance to ensure the integrity of structures such as bridges, buildings, airplane wings, etc. In the microwave regime, open-ended waveguide sensors have shown the potential for surface crack detection in metals. Some of the applications in which open-ended waveguide sensors may be used include detection of cracks in clean steel, cracks under dielectric coatings, cracks filled with dielectric materials such as rust or paint and remote crack detection. The development of theoretical models allows for optimization of measurement parameters for a given range of crack dimensions. Thus, the dynamic range of detected signals for tight, deep, filled and covered cracks can be increased using optimal measurement parameters which have been obtained from the theoretical models. When scanning a crack using an open-ended rectangular waveguide probe, higher order modes are generated at the aperture of the waveguide. These modes are evanescent and attenuate rapidly as they travel back into the waveguide. A pick up sensor, located near the aperture, is capable of detecting these modes. In the absence of a crack these modes do not exist. However, as soon as a crack appears in the aperture of the waveguide higher-order modes are generated. Therefore, the detection sensitivity of this approach is quite high. For this investigation, the utility of this microwave surface crack detection approach is investigated on stress-induced fatigue cracks under various loads. A specially prepared (at the National Institute of Standards and Technology, NIST, in Boulder, CO) crack specimen using a sawcut starter notch is used for these experiments. Once under loading, a fatigue crack begins to propagate from the edges of the notch. The opening or width of this crack can be controlled by the level of loading. In this paper, crack detection sensitivity versus crack opening (load dependent) is demonstrated.

INTRODUCTION

Successful detection of tight cracks in metallic structures subject to loading, in particular steel bridges, is an important concern. Presently, there are several conventional nondestructive inspection techniques that are used for this purpose (Bovig, 1989). In addition to the fact that each of these techniques possesses some advantages and disadvantages, in some cases they may be the only applicable technique and not necessarily the optimal one. Some microwave nondestructive testing methods, for surface crack detection in metals, have also been devised (Feinstein and Hruby, 1967; Bahr, 1981; Robinson and Gysel, 1972; Bahr and Watjen, 1981; Ash and Husain, 1973; Auld, 1978). However, no activity in this area took place for a long time since these early investigations until 1992, during which a novel microwave method using open-ended rectangular waveguides was developed at the Applied Microwave Nondestructive Testing Laboratory (*amntl*) by Yeh and Zoughi (Yeh and Zoughi, 1994). This method has since been expanded in its applicability to filled and covered cracks (Zoughi , et al., 1995; Huber, et al., 1997; Huber and Zoughi, 1995). Furthermore, two electromagnetic models have also been developed which describe the interaction of a rectangular waveguide aperture with a metallic surface perturbed by a tight crack (Yeh and Zoughi, 1994; Huber, et al., 1997). These models may be used to optimize measurement parameters such as the frequency of operation for a given set of crack dimensions.

This particular microwave approach possesses several important practical features, such as:

1. the method is fast, reliable and relatively inexpensive,
2. the sensor may or may not be in contact with the surface under examination offering the possibility of remote inspection,
3. cracks may be filled with dielectric materials such as paint, dirt, rust, etc.,
4. cracks under dielectric coatings such as paint can also be detected and their properties evaluated without the need for removing the paint (which is time consuming, expensive and environmentally hazardous for lead paint),
5. cracks may be on non-ferromagnetic as well as ferromagnetic metals or alloys,
6. the detected signal is only due to surface defects and not to interior flaws (i.e. easier signal interpretation),
7. the dimensions of a crack as well as orientation, edge and tip locations may be evaluated,
8. no special operator skills or surface preparation are required,
9. the technique is environmentally compliant, operator friendly and safe (low microwave power),
10. theoretical modeling provides for prior-to-detection optimization of measurement parameter(s).

One of the features of this approach is the ability to take advantage of the generated higher-order modes, when a crack appears inside a waveguide aperture, by detecting them as an indication of crack presence. This has been shown to be a very sensitive approach for the detection of tight cracks (Yeh, Ranu and Zoughi, 1994). The utility of this approach for detecting stress-induced fatigue cracks under loading is presented in this paper.

BACKGROUND

Figures 1a-b show the side and plan views of an open-ended waveguide aperture scanning a metal surface in which there is a crack (slot). When the crack is outside of the waveguide aperture the waveguide is considered to be short circuited. But when the crack appears in the aperture the reflection properties of the metal surface change. This is due to the fact that the surface currents induced in the metal are now perturbed which give rise to this change in the reflection properties. Considering the incident electric field to be that of the dominant TE_{10} mode, the presence of a crack generates an infinite number of higher-order modes. These modes are evanescent and attenuate rapidly as a function of the distance away from the waveguide aperture. Therefore, to detect their presence one must use a probe very close to the waveguide aperture. Figure 2 shows the schematic of the higher-order probe (Yeh, Ranu and Zoughi, 1994). It has also been shown that the location of the probe on the narrow dimension of the waveguide plays a major role in the detection sensitivity (Yeh, Ranu and Zoughi, 1994). This is expected since the effect of these higher-order modes is detected as a coherent addition of those modes that are generated. It has been shown that when $\mathcal{T}$ is 0.17b to 0.2b away from the waveguide wall, most optimal detection takes place given a waveguide band of frequencies (Yeh, 1994). The magnitude of higher-order modes diminishes rapidly as a function of increasing ℓ'. Thus, ℓ' is usually no more than a millimeter. Figure 3 shows the calculated power (proportional to the detector output voltage) associated with the higher-order modes as a function of scanning distance for $\ell' = 0.2$ mm (0.008"), $\mathcal{T} =$ 017b, at a frequency of 12.4 GHz for a slot with a width of 0.144 mm (0.005") and a depth of 1.2 mm (0.05"). The results show that as the crack enters the waveguide aperture a relatively large signal is detected indicating its presence. As the crack leaves the waveguide aperture a similar but lower in magnitude signal is detected again. Figure 3 is referred to as the *higher-order mode crack characteristic signal.* Since detection of a crack is the objective, the probe should be located at around 0.17b from the waveguide wall. It must be noted that this numerical result was obtained assuming that only TM (transverse magnetic)higher-order modes are generated.

One of the most important attributes of this approach is its high sensitivity to the presence of a crack. In addition, tight cracks may be detected at relatively low microwave frequencies (Yeh, Ranu and Zoughi, 1994). This is due to the fact that in the absence of a crack there are no higher-order modes, however as a crack appears in the waveguide aperture a finite amount of signal is detected. Therefore, theoretically

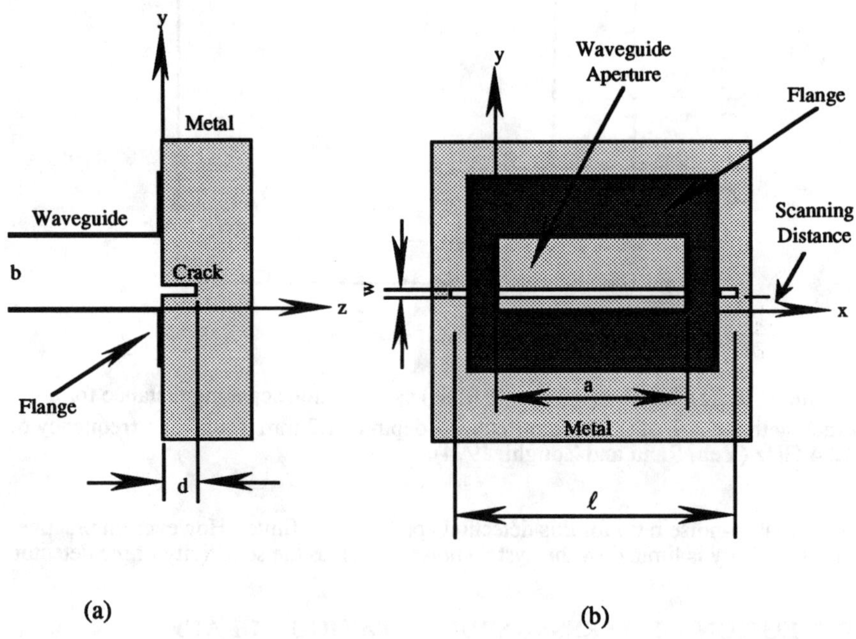

Figure 1: a) Side view, b) plan view of a rectangular waveguide aperture scanning a surface crack (Yeh, Ranu and Zoughi, 1994).

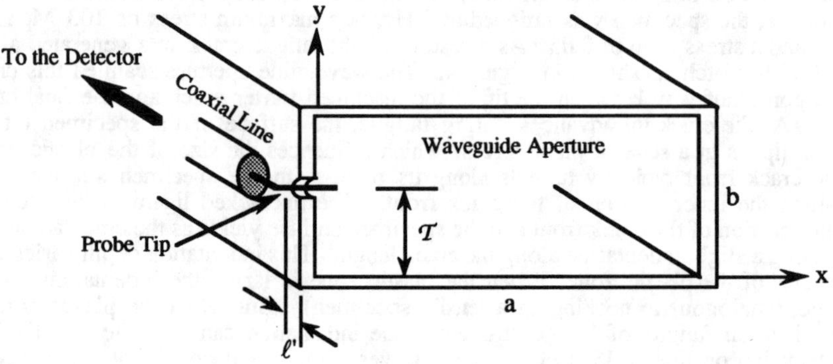

Figure 2: The schematic of the higher-order mode probe (Yeh, Ranu and Zoughi, 1994).

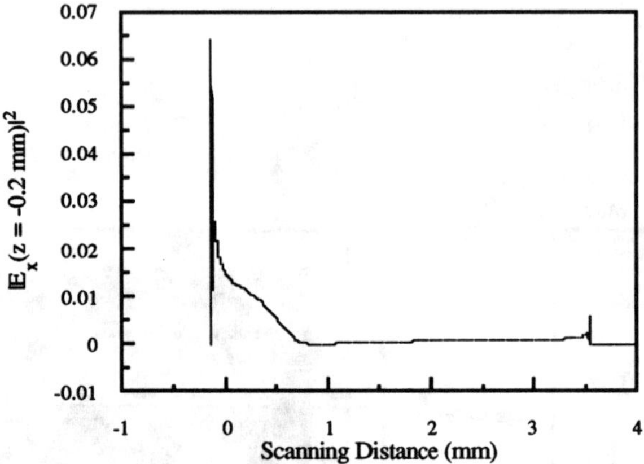

Figure 3: $|E_x(z)|^2$ at $\ell' = -0.2$ mm (0.008") as a function scanning distance for a crack with a width of 0.144 mm (0.005"), depth of 1.2 mm (0.05")at a frequency of 12.4 GHz (Yeh, Ranu and Zoughi, 1994).

the signal-to-noise ratio for this detection approach is infinite. However, in practice this sensitivity is limited by the system noise as well as the sensitivity of the detector

DETECTION OF STRESS-INDUCED FATIGUE CRACKS

To illustrate the utility of this microwave approach for tight crack detection, a fatigue crack was generated in an A-36 steel specimen with a thickness of 12.7 mm (0.5"). A small hole was drilled through the specimen and a sawcut starter notch was introduced on both sides of the hole. Using a closed loop servo-hydraulic fatigue machine, the specimen was fatigued at 8 Hz, at a maximum stress of 103 Mpa (15 ksi) and a stress ratio of 0.05. As a result, a tight fatigue crack was generated at the end of the notch, as shown in Figure 4. The waveguide aperture scanned this crack at a point half way between the tip of the machined starter notch and the final crack tip. As the crack tip advances during fatigue, the surface of the specimen (at the crack tip) is in a state of plane stress, which influences the size of the plastic zone. The crack front typically tunnels along its path, with the specimen surface crack trailing the inner portion of the crack front. The uncracked ligament between the inner portion of the crack front and the specimen surface yields as the crack advances, creating a slight indentation along the crack length. This indentation depth varies with the size of the plastic zone. When the plastic zone is large, the indentation can be large, (analogous to necking in a tensile specimen). But when the plastic zone is small, as in fatigue of bridge structures, the indentation can be quite small. The indentation on this A-36 steel sample was very small, and could not accurately be measured without destroying the sample. An estimate of the indentation, using a replication technique, showed the depth to be less than 0.040 mm (0.0016"). Table 1 shows the crack opening as a function of load level.

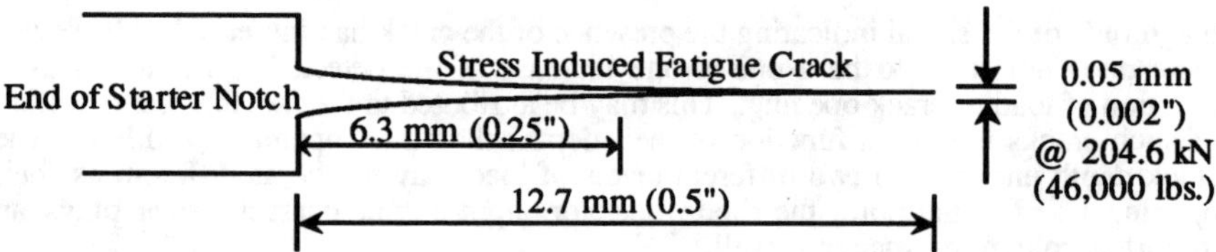

Figure 4: Schematic of a stressed-induced fatigue crack under 204.6 kN (46,000 lb.) of load.

Table 1: Crack opening as a function of load.

Load (kN , kips)	**Crack Opening (micrometers/inches)**
8.9 , 2	1.3/0.00005
26.7 , 6	5.1/0.0002
44.5 , 10	7.6/0.0003
62.3 , 14	12.7/0.0005
80.0 , 18	17.8/0.0007
97.9 , 22	20.3/0.0008
115.7 , 26	25.4/0.0010
133.4 , 30	30.5/0.0012
151.2 , 34	35.6/0.0014
169.0 , 38	40.6/0.0016
186.8 , 42	47.0/0.00185
195.7 , 44	50.8/0.0020
204.6 , 46	50.8/0.0020

A probe was constructed operating in the Ka-band frequency range (26.5-40 GHz), and was subsequently used to scan this crack as a function of different load levels at a frequency of 35.5 GHz. Figure 5 shows the higher-order mode crack characteristic signal for the fatigue crack under no load. The results show that the crack is indeed detected even though it has zero opening. The numerical code would not be able to re-produce the results of Figure 5 since it requires a finite opening or crack width. The reason that the crack is detected is partially due the fact that while being produced, during the cyclical loading of the steel specimen, shear lips were generated on the edges of the crack. Although their dimensions are quite small, nevertheless they generate some perturbation in the surface of the steel specimen. The indentations created by the shear lips and the crack itself must be enough of a surface perturbation for the probe to detect them. Furthermore, the results shown in Figure 5 follow the same trend of that shown in Figure 3 where, as the crack enters the waveguide aperture, a larger signal is detected than that when it leaves the aperture. The width of these two signals, generated by the crack, are however wider than one would expect. This is thought to be directly related to the presence of the shear lips which, in effect, present a wider crack (surface perturbation) to the scanning probe. Clearly, more analysis of the results are required to completely understand the results shown in Figure 5. However, pertaining to the goal at hand, the crack is indeed *detected.* Figure 6 shows the higher-order mode crack characteristic signal for the same crack while under 71 kN (16,000 lb.) of load. Clearly, the crack is detected and when compared to the results of Figure 5 the magnitude of the signal indicating the presence of the crack has increased. However, there does not seem to be a one-to-one increase in the detected signal levels as a function of load or crack opening. This may be attributed to the fact that the detection of such cracks is more a function of their depths than their opening (widths). The crack depth under these two different levels of load may not be as different as their opening is. Furthermore, the diode detector input-output characteristics plays an important role in this matter as well.

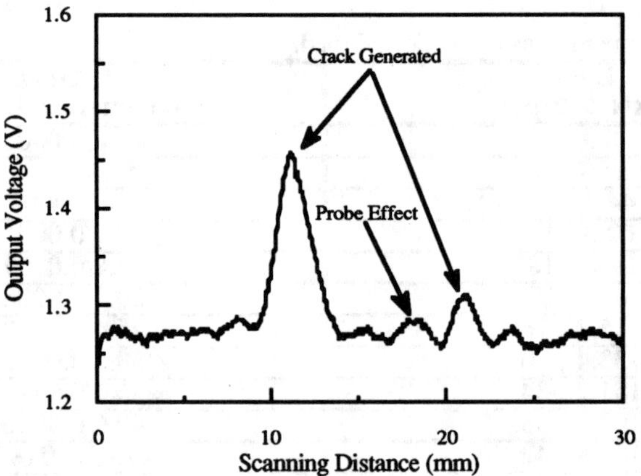

Figure 5: Higher-order mode crack characteristic signal for the fatigue crack under no load.

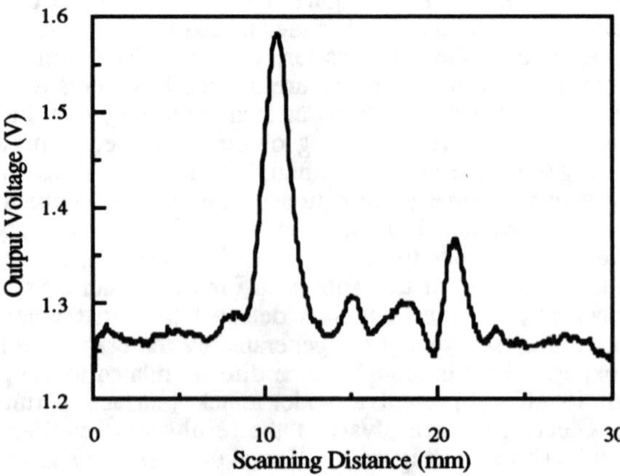

Figure 6: Higher-order mode crack characteristic signal for the fatigue crack under 71 kN (16,000 lb.) of load.

The results further showed that as the load level increased, the difference between the two signals, indicating the entering and exiting of the crack from the aperture, continuously decreased and beyond a load level of 169 kN (38,000 lb.) the signal in the right portion of Figure 6 became larger than the one in the left portion. Figure 7 shows the results for 187 kN (42,000 lb.) of load. This is a phenomenon that had not been seen before using rectangular slots cut out of metal specimens in the laboratory investigations. It is again believed that the combination of crack opening and crack depth dictates the detected signal magnitude. Currently, our recent electromagnetic code is being used to investigate this hypothesis.

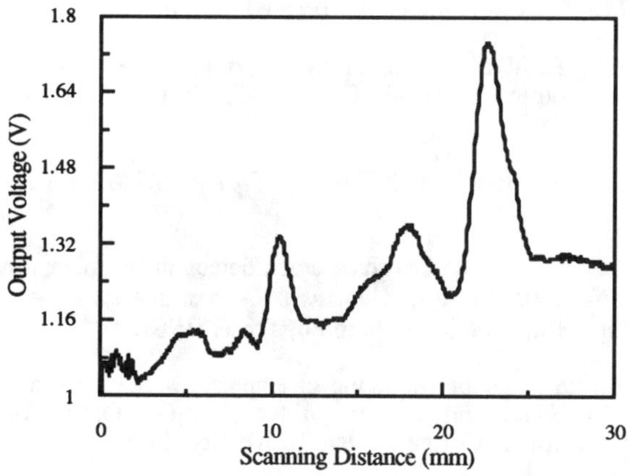

Figure 7: Higher-order mode crack characteristic signal for the fatigue crack under 187 kN (42,000 lb.) of load.

SUMMARY

The utility of the higher-order mode rectangular waveguide approach was shown for detecting stress-induced fatigue cracks in steel. It was also shown that the dynamic range of the detected signal increases as a function of increasing crack opening or width (due to loading). Some new issues have risen as a result of these investigations which must be further investigated. Currently, it is believed that the properties of the higher-order mode crack characteristic signal may be a function of crack width as a well as crack depth. If shown to be true, this may lead to a technique by which crack dimensions may also be evaluated using this technique. These issues will be looked into during the coming months.

Acknowledgment: This research has been funded by the Federal Highway Administration (FHWA) grant no. DTFH61-94-X-00023.

REFERENCES

1. Ash, E.A. and A. Husain, "Surface examination using a superresolution scanning microwave microscope," *Proc. of the 3rd European Microwave Conf.*, pp. c.15.2, 1973.

2. Auld, B.A., "Theory of ferromagnetic resonance probes for surface cracks in metals," G.L. Report 2839, E.L. Ginzton Laboratory, Stanford University, Stanford, CA July 1978.

3. Bahr, A.J., "Microwave eddy-current techniques for quantitative nondestructive evaluation," in *Eddy-Current Characterization of Materials and Structures*, ASTM STP 722, G. Birnbaum and G. Free editors, pp. 311-331, 1981.

4. Bahr, A.J. and J.P. Watjen, "Novel eddy-current probe development," Semi-annual Report, Contract F33615-80-C-5025, SRI Project 1908, Menlo Park, CA, Feb. 1981.

5. Bovig, K.G., "NDE Handbook," *Teknisk Forlag A/S (Danish Tech. Press)*, 1989.

6. Feinstein, L. and R.J. Hrby, "Surface crack detection by microwave methods," *6th Symp. on Nondestructive Evaluation of Aerospace and Weapons Systems Components and Materials*, San Antonio, Texas, 1967.

7. Huber, C., "Electromagnetic Modeling of Exposed and Covered Surface Crack Detection Using Open-Ended Waveguides," Ph.D. Dissertation, Electrical Engineering Department, Colorado State University, 1996.

8. Huber, C., H. Abiri, S. Ganchev and R. Zoughi, "Modeling of Surface Hairline Crack Detection in Metals Under Coatings Using Open-Ended Rectangular Waveguides," To appear in the *IEEE Transactions on Microwave Theory and Techniques*, 1997.

9. Huber, C., H. Abiri, S. Ganchev and R. Zoughi, "Modeling of Surface Hairline Crack Detection in Metals Under Coatings Using Open-Ended Rectangular Waveguides," *IEEE Transactions on Microwave Theory and Techniques*, 1997.

10. Huber, C. and R. Zoughi, "Higher Order Modes as Indicators of Surface Cracks Under Stratified Dielectric Coatings Using an Open-Ended Waveguide," *Proceedings of the Review of Progress in Quantitative NDE*, vol. 14A, pp. 637-642, 1995.

11. Robinson, L.A. and U.H. Gysel, "Microwave coupled surface crack detector," Final report, Contract DAAG-46-72-C-0019, SRI Project 1490, Stanford Research Institute, Menlo Park, CA, Aug. 1972.

12. Yeh, C., "Detection and Sizing of Surface Cracks in Metals Using Open-Ended Rectangular Waveguides," Ph.D. Dissertation, Electrical Engineering Department, Colorado State University, 1994.

13. Yeh, C. and R. Zoughi, "A Novel Microwave Method for Detection of Long Surface Cracks in Metals," *IEEE Transactions on Instrumentation and Measurement*, vol. 43, no. 5, pp. 719-725, October, 1994.

14. Yeh, C., E. Ranu and R. Zoughi, "A Novel Microwave Method for Surface Crack Detection Using Higher Order Waveguide Modes," *Materials Evaluation*, vol. 52, no. 6, pp. 676-681, June, 1994.

15. Zoughi, R., S. Ganchev, C. Huber, E. Ranu and R. Runser, "A Novel Microwave Method for Filled and Covered Surface Crack Detection in Steel Bridges Including Crack Tip Identification," Final (Fourth Quarterly) Report, Federal Highway Administration (FHWA), Grant no. DTFH61-94-X-00023, p. 187, September, 1995.

SESSION 14

CIVIL INFRASTRUCTURES IV

Optical Fiber Crack Sensor for Concrete Structures

C. K. Y. LEUNG, N. ELVIN, N. OLSON, T. MORSE and Y. F. HE

ABSTRACT

The condition of concrete structures can be assessed through the monitoring of crack openings. Researchers at MIT and Brown University have recently developed a novel concept for the sensing of cracks in concrete structures. The sensing capability is based on the light loss as microbending occurs in a fiber bridging the crack. To use the sensor, only the crack plane orientation in the structure (rather than the exact crack locations) needs to be known. With the use of OTDR, distributed sensing is possible. In this paper, the novel crack sensing concept is first introduced. To guide the design of sensors for various performance requirements, a theoretical model is developed to relate signal loss and crack opening. Prediction from the model is compared with experimental results. Using laboratory-sized specimens, the detection of both surface and interior cracks are demonstrated.

INTRODUCTION

The condition of many important concrete structures can be assessed through the detection and monitoring of cracks. For example, in concrete bridge decks, crack openings beyond 0.15 to 0.2 mm will allow excessive penetration of water and chloride ions, leading to corrosion of steel reinforcements. Crack opening of the order of mm's, which may occur after a major earthquake, is a sign of severe structural damage. As another example, hazardous waste sites may be confined by closely spaced concrete columns grouted in drilled holes. With the grout penetrating into the soil between the columns, a continuous barrier for the waste can be formed. However, if cracking in the grouted columns is left unnoticed, leakage occurs and the barrier will lose its purpose.

Conventionally, crack detection and monitoring for bridges have been carried out by eye inspection. The procedure is time consuming, expensive and yet unreliable. For buried structures like grouted columns, no reliable techniques is available for the detection of slightly opened cracks. Recently, various researchers has developed fiber optics based crack sensors for concrete structures. Compared with other

Christopher K.Y. Leung, Niell Elvin, Noah Olson, Dept. of Civil and Environmental Engineering, MIT, Cambridge, MA 02139

Theodore F. Morse and Yi-Fei He, Laboratory for Lightwave Technology, Division of Engineering, Brown University, Providence, RI 02912

sensors and transducers, optical fibers have several advantages including non-conductivity (hence not vulnerable to lightning), immunity to electromagnetic interference and low weight. Also, with the parallel developments in the telecommunications industry, the cost of fibers and opto-electronics equipment can potentially drop significantly with time.

Existing optical crack sensors are, however, very limited in their applications. Sensing based on fiber breakage (Rossi and LaMaou, 1989, Huston et al, 1992) can distinguish between the presence or absence of cracking but cannot provide information on gradual structural degradation. The sensors developed by Ansari et al (1993) and Voss and Wanser (1994) are 'point' sensors, which means that they can detect and monitor the opening of a crack only if cracking occurs in a small region that is known in a-priori. Thus, while the sensor is useful in experimental fracture mechanics, where crack location can be controlled, its application to real concrete structures is very limited. Wolff and Meisseler (1992) developed an optical fiber sensing system which can be attached to concrete structures for the monitoring of surface displacement. If the attachment points are too close, a crack may not pass through the sensing system and if the points are too far apart, sensitivity will be low and it is not possible to distinguish between the presence of one single widely opened crack or several narrower ones. In summary, one can see that a new approach needs to be developed for reliable crack monitoring in real structures.

A NOVEL CRACK DETECTION AND MONITORING CONCEPT

Researchers at MIT and Brown University are currently developing a novel sensor for the detection and monitoring of cracks in a concrete structure. The principle of the sensor is illustrated in Fig.1, which shows a 'zig-zag' sensor at the bottom of a bridge deck. The backscattered power is measured as a function of time (with Optical Time Domain Reflectometry or OTDR). Before the formation of cracks, the backscattered signal vs time should follow a relatively smooth curve (the upper line in Fig.1b). The loss in signal power with time is due to the increasing loss with distance traveled (which is directly proportional to the traveling time). In the straight portions of the fiber, the loss is due to absorption and scattering (Allard, 1990). In the curved portion (where the fiber turns in direction), bending loss may occur depending on the radius of curvature. A simple physical explanation for the bending loss is as follows. The optic fiber consists of a core surrounded by a cladding with a lower refractive index. For a straight fiber, a light ray launched into the core at a low angle to its axis will undergo total internal reflection at the core/cladding interface and will always stay guided in the core. However, when the fiber is bent, the increase in curvature may reduce the incident angle at the core/cladding interface to a value below the critical angle (Ansari et al, 1993). Some light energy will then move into the cladding and get dissipated.

When a crack opens in the structure, a fiber intersecting the crack at an angle other than 90° has to bend to stay continuous (Fig.1c). The sudden bending of an optical fiber at the crack results in a sharp drop in the optical signal (lower line, Fig.1b). From the times on the OTDR record corresponding to the sharp signal drops, the location of each crack in the structure can be deduced. Also, from the magnitude of the drop, the crack opening can be obtained if a calibration relation is available. The theoretical development of such a calibration curve will be discussed in the next section. The angle with which the fiber intersects the crack is critical,

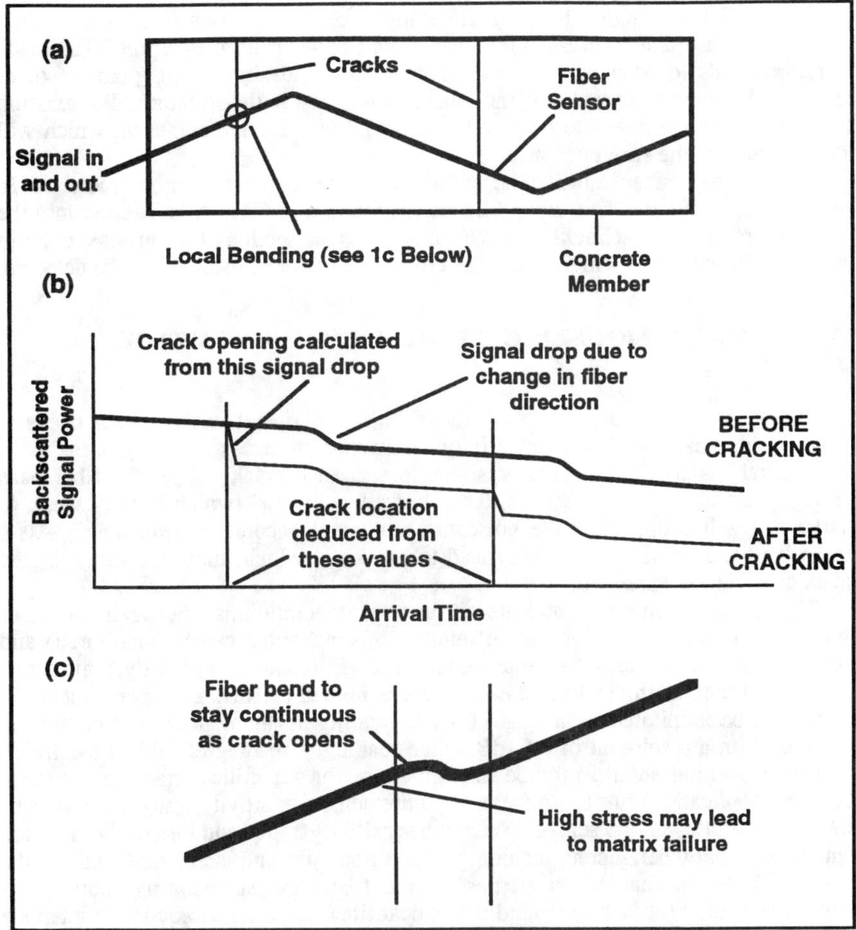

Figure 1 A Novel Concept for Crack Sensing

since this determines not only the number of cracks we may sense, but also the "bandwidth" of the measurement. Note that the signal loss at the fiber turning point is shown for a general case. If the sensor geometry is designed to have a small curvature at the turning point, the loss will be negligible.

The proposed technique does not require prior knowledge of the crack locations, which is a significant advancement over existing crack monitoring techniques. Moreover, a number of cracks can be detected and monitored with a single fiber. For the sensor to work, however, crack directions need to be known. An ideal application of the sensor is in the monitoring of flexural cracks in bridges, which may form at arbitrary locations along the deck, but essentially perpendicular to the spanning direction. To sense cracks effectively, several sensors should be employed. With a single fiber, if a crack intersects the 'zig-zag' fiber at a location where the fiber direction is changing, results will be difficult to interpret. The crack therefore needs to be picked up by another fiber with its turning point at other sections. For new constructions, the sensor can be cast in the structure. For existing structures, the fiber needs to be first incorporated into a polymeric sheet which will then be glued to the structure surface.

The sensing concept can also be applied to grouted concrete barrier structures. In such a case, the fiber is first wound helically around a rod which is inserted into the holes before grouting. Cracks which occurs across the section of the grouted column (as a result of bending) will then intersect the optical fiber at an angle and be detected.

THEORETICAL MODELING OF SIGNAL LOSS VS CRACK OPENING

In order to detect the presence of cracks with small openings, the sensor needs to be designed to have sufficient sensitivity. However, an excessively high sensitivity is undesirable since it implies large signal loss at each crack. After the fiber pass through a small number of cracks, the backscattered signal (which is very weak to start with) will drop below the noise threshold and become undetectable. Also, excessive bending of the fiber may lead to breakage, which limits the range of the crack opening measurement.

The above discussions indicate an important compromise between the fiber sensitivity, loss (which governs the number of cracks that can be monitored) and range. To give an example, assume we have an OTDR system with a dynamic range of 50dB. If the maximum loss at each crack is limited to 1dB, an upper limit of 50 cracks can be monitored by a single fiber (assuming negligible loss at fiber turning points). With a resolution of 0.01dB in the reading, a reading of 1dB at maximum allowable opening should provide enough room for sub-critical crack monitoring. Different applications impose different requirements (sensitivity, range, number of cracks to monitor) on the sensor. Since the sensitivity, range and loss of the fiber are functions of many parameters including (i) the fiber size and inclination angle to the crack, (ii) the mechanical properties of the fiber, the surrounding matrix, the protective coating (or jacket) around the optical fiber and the matrix/coating interface and (iii) the optical properties of the fiber core and cladding, design of sensors with specific performance from purely empirical testing is almost impossible. To provide a guideline for sensor design, a theoretical model relating signal loss to crack opening needs to be developed. Such a model involves two major steps (i) micromechanical modeling of fiber bending as the crack opens, and (ii) modeling of electromagnetic wave propagation in a deformed waveguide.

MICROMECHANICAL MODELING OF FIBER BENDING

Crack opening leads to bending of the optical fiber. Due to anti-symmetry, there is an inflection point in the middle of the crack. In the analysis, the fiber is first cut at the inflection point to give two parallel but separate segments. A displacement is then applied to each of the cut ends to bring them back together. The deformation along the fiber corresponding to the applied displacement is obtained through a three-dimensional finite element analysis. In the analysis, 8-node solid elements are employed to model the fiber, the matrix as well as the polymeric coating around the fiber. The boundary of the matrix is placed at 5 fiber diameters from the fiber center. This is found to be sufficient to simulate a matrix of infinite extent. A typical plot of deformation along the fiber is shown in Fig.2. Note that most of the deformation is absorbed in the plastic coating (or jacket) around the glass fiber. As shown in Fig.2, when the fiber is bent upward, part of it may separate from the matrix at the lower side. In the finite element model, material separation is allowed by removing a layer of elements at the jacket/matrix interface. The number of elements to be removed is determined from an analysis with perfectly bonded fiber and matrix. Elements at the interface that are under radial tension will be removed.

With the finite element analysis, fiber displacement is only obtained at a given number of points. To obtain an accurate estimate of the second derivative of the displacement, which governs the signal loss, a reliable interpolation technique is required. It has been shown (Leung and Li, 1992) that a fiber in a cementitious matrix can be modeled as a beam on an elastic foundation, with the foundation stiffness increasing with distance from the crack face. Here, we assume the foundation stiffness to be a linear function of distance from the crack face ($k = \beta^4 x$, where β is a constant to be obtained). A fourth order differential equation can then be set up and solved numerically. In reality, the stiffness will reach a constant value rather than increasing continuously. However, since the foundation reaction decreases rapidly with distance from the crack face, the assumptions of a continuously increasing stiffness and one with stiffness reaching a plateau give no noticeable difference in the results. By varying β, the best fit to the finite element result can be obtained (Fig.3). From the best fit curve, the curvature along the fiber is deduced for the calculation of signal loss.

MODELING OF SIGNAL LOSS IN A CURVED OPTICAL FIBER

Bending loss along an optical fiber with arbitrary curvature can be obtained with the beam propagating method (Feit and Fleck, 1978). For a time harmonic Electric field (E field), one can separate the time variable from the wave equation to obtain the Helmholtz equation in space variables. Instead of solving the Helmholtz equation in curvilinear coordinates, the curved fiber is first transformed into an equivalent straight fiber with a modified refractive index profile (Heiblum and Harris, 1975). A numerical solution of the Helmholtz equation is then obtained through the spectral decomposition technique. As the initial condition, the E field corresponding to that in a straight fiber is imposed at a point far away from the crack (where no loss has yet occurred). The E field is decomposed into its spectral components (in the space variables) through the use of discrete Fourier Transform. Each spectral component is then propagated separately and recombined by carrying out inverse Fourier Transform. With the availability of Fast Fourier Transform algorithms, spectral decomposition allows a much more efficient solution scheme than the more direct finite difference methods. The analysis is carried out along the fiber until it becomes straight again at the other side of the crack. The ratio of power at the input and output sections allows the calculation of dB loss as a result of fiber bending.

THEORETICAL PREDICTIONS AND COMPARISON TO PRELIMINARY EXPERIMENTAL RESULTS

Typical signal loss vs crack opening curves for a given set of optical and mechanical properties are shown for three different fiber inclination angles in Fig.4.

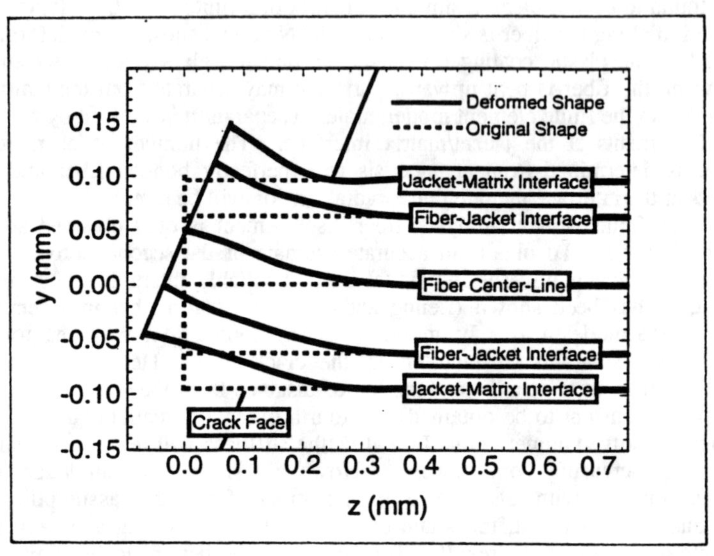

Figure 2 Typical Deformation Results from Finite Element Analysis

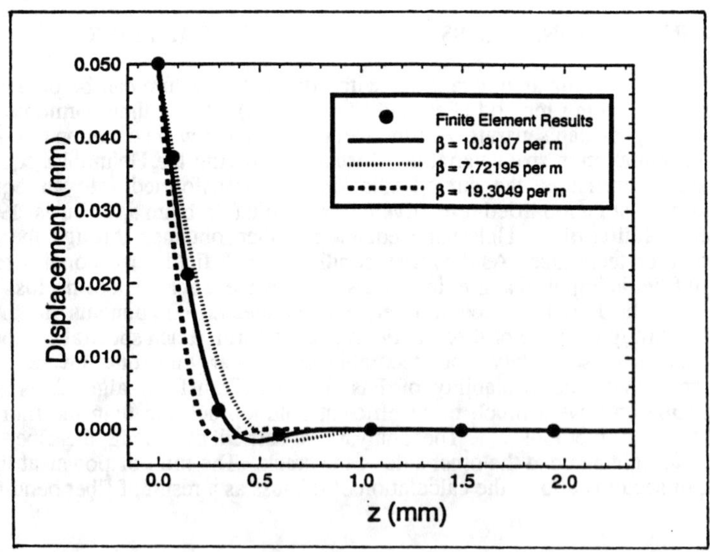

Figure 3 Fitting of Finite Element Results with Analytical Expressions

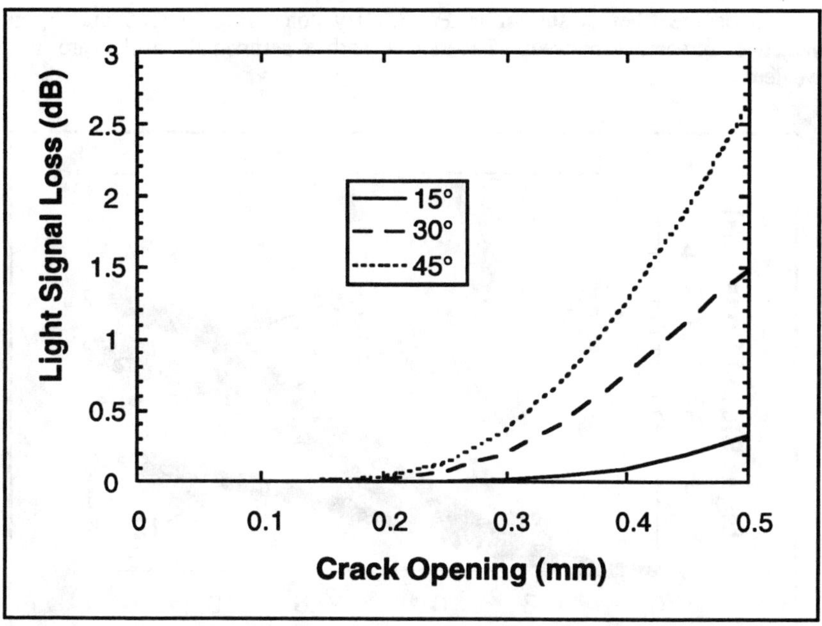

Figure 4 Typical Results from Theoretical Model

The detailed verification of the theoretical results is still on the way. However, a preliminary set of experiments has been carried out to study the qualitative behavior of the sensor. A rectangular epoxy block is cast with an embedded steel fiber of a size slightly large than that of the optical fiber. Before the epoxy hardens, the steel fiber is removed. After hardening, the block is cut into two parts and an optical fiber is carefully inserted into the hole formed by the steel fiber. The two blocks are then pushed together and gripped in a loading fixture. By pulling the blocks apart and simultaneously recording the block separation and the optical signal loss, a plot of signal attenuation vs crack opening is obtained. The result for three different fiber orientation angles are shown in Fig.5. The experimental results up to about 0.5mm show good qualitative agreement with the theoretical prediction. The sensitivity is initially very low but increases drastically after a certain amount of opening. At large crack openings, up to 1mm, the experimental results show that the loss is approaching an asymptotic value. This is due to the fact that as the crack opening becomes very large, all the curvature changes will concentrate at the locations where the fiber enter the crack. No matter how far apart the crack faces are, the changes stay essentially the same. The theoretical simulations in Fig.4 has not reached this regime. Since the asymptotic value represents the maximum possible loss at a crack, it is an important design parameter governing the minimum number of cracks that can be monitored with a single fiber. To theoretically predict the asymptotic value, we have extended the analysis to consider large displacement of the fiber. Also, since the beam propagation method breaks down for long propagation length, signal loss is calculated with a different approach (Mustieles et al, 1993). An example simulation for a 45 degree fiber is shown in Fig.6. By comparing Figs. 5 and 6, good qualitative agreement between theoeretical and experimental results are clearly revealed.

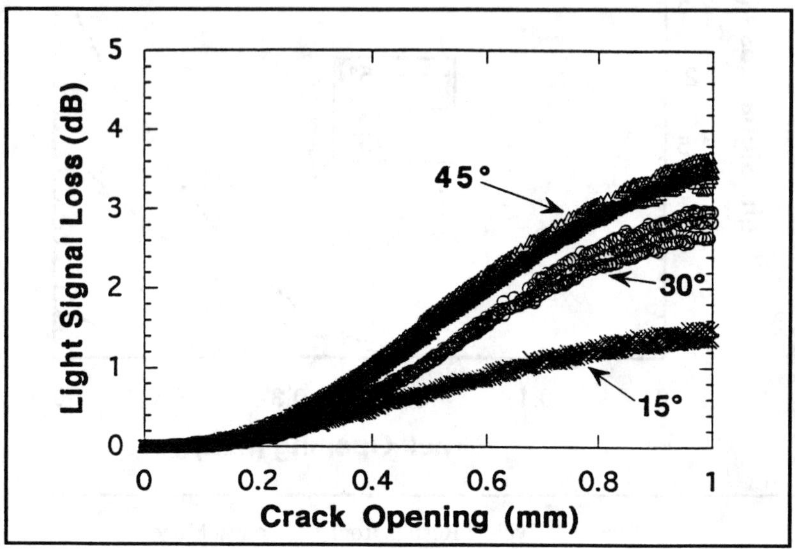

Figure 5 Preliminary Experimental Results for Various Fiber Inclination Angles

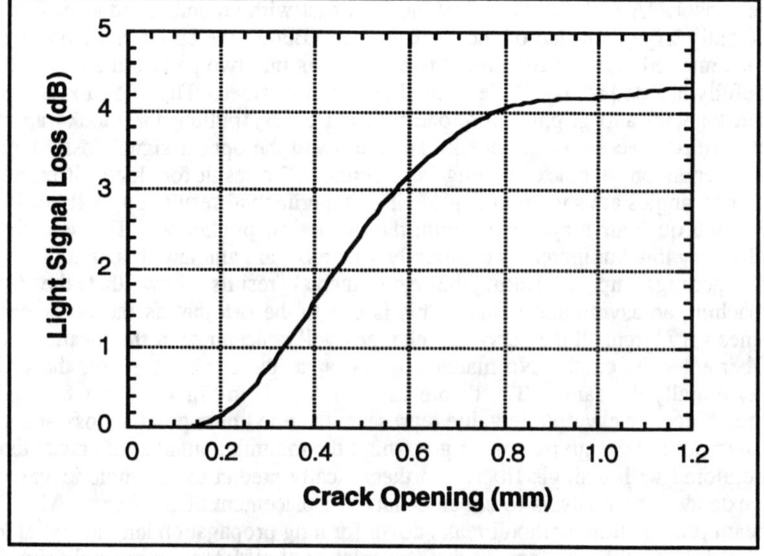

Figure 6 Example Simulation of Signal Loss for Large Crack Opening

DEMONSTRATION OF SENSOR APPLICATIONS

MONITORING OF SURFACE CRACKING

When surface cracks form on concrete structures, the penetration of water and salt may cause corrosion of the steel bars. Moreover, excessive surface cracking is a sign of severe damage which requires immediate repair. To monitor surface cracking, we have developed a polymeric sensor sheet with an embedded optical fiber (Fig.7a). To detect cracks, the sheet is glued onto the surface of the structure. This technique is applicable to both new and existing structures. An ideal application is in the monitoring of flexural cracks under bridge decks, which are currently assessed manually by eye inspection. For successful crack sensing, the polymer sheet has to bond strongly to the concrete. Also, the sheet has to be brittle enough for cracks in the concrete structure to penetrate into it. Epoxy is chosen for its well-known bond compatibility with concrete (epoxy has been used as crack sealant or bonding agent in concrete structures for years).

Crack sensing relies on the bending of optical fiber to induce signal loss. For the fiber to bend at the crack, free sliding in the polymer matrix has to be allowed. A releasing agent is therefore applied to the fiber surface before it is embedded into the polymer matrix. To demonstrate the applicability, the sensing sheet is glued to a beam specimen (Fig. 7b). A pre-crack is cut on each side of the beam so the crack position is known. A displacement transducer is placed across the crack. When loading is applied, the change in optical signal and transducer output are both recorded. Figure 7c shows a plot of signal loss vs crack opening (measured by the displacement transducer). The signal loss is detectable at small crack openings below 0.2mm. Also, the shape of the curve (at small crack opening) is in good agreement with the theoretical prediction (Fig. 6). This simple experiment clearly demonstrates the feasibility of surface crack monitoring with the proposed approach.

MONITORING OF INTERNAL CRACKING

In buried structures formed in-situ, such as grouted concrete barriers or diaphragm walls, it is not feasible to install a crack sensor on the surface of the structure. In such a case, cracking can be monitored with a sensing rod cast into the structure. The sensing rod is a rod with an optical fiber wrapped around it in a helix. When cracking occurs in a direction perpendicular to the axis of the rod, it will intersect the fiber at an angle, inducing bending and signal loss. To verify this sensing approach, a sensing rod is made with mortar and cast into a mortar beam. The beam is then tested in three point bending as shown in Fig.8a. Also shown in the figure is the pre-notched cross section as well as the location of the sensing rod and the optical fiber. A displacement transducer is placed across the notch near the bottom of the beam. A plot of light signal loss vs displacement is shown in Fig. 8b. Since the displacement transducer is much closer to the crack mouth than the optical fiber, the signal loss is 'lagging' behind the measured displacement. Also, the optical measurements seem to be affected by noise. Despite these shortcomings, the result clearly indicates the ability of the sensing rod to detect and monitor internal cracks.

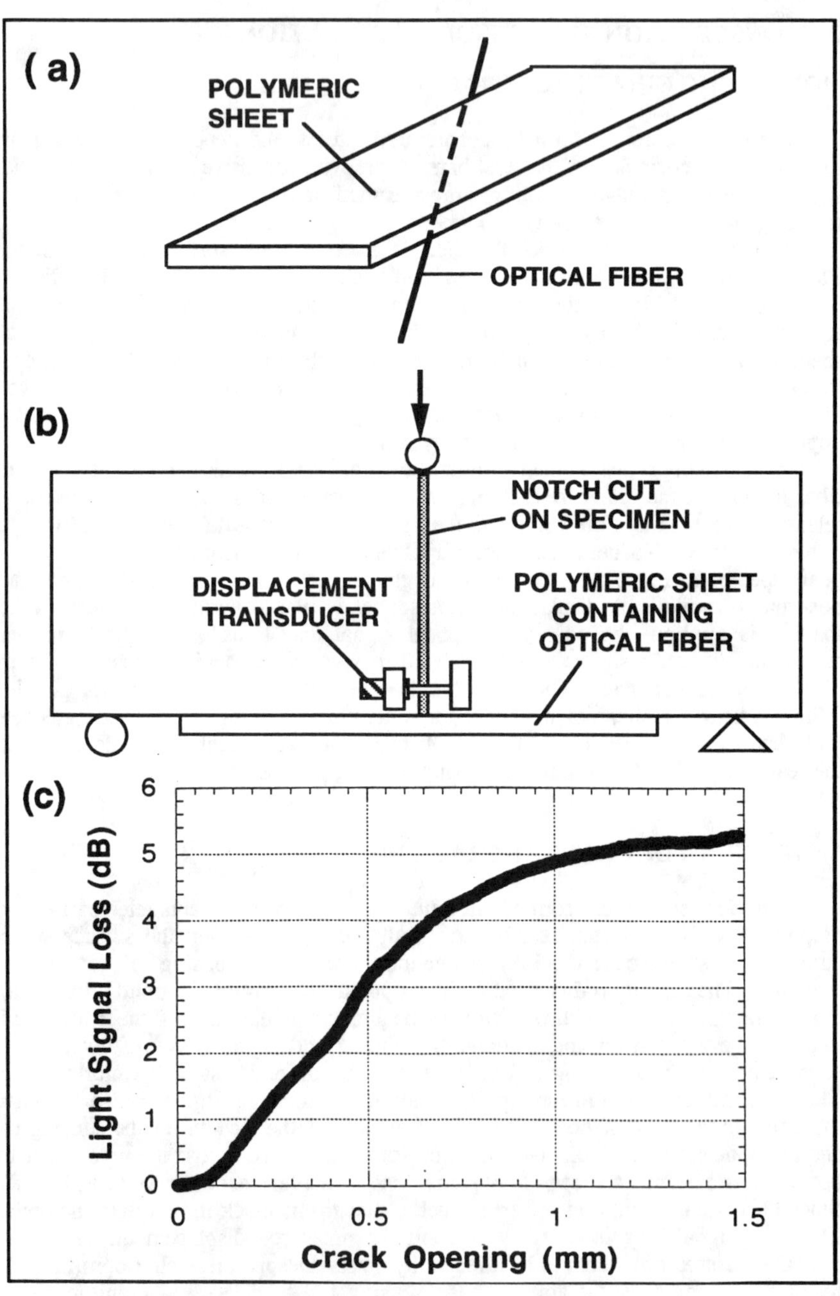

Figure 7 (a) Sensing Sheet with Optical Fiber
(b) Beam Specimen with Sensing Sheet at the Bottom
(c) Signal Loss at Optical Fiber vs Crack Opening

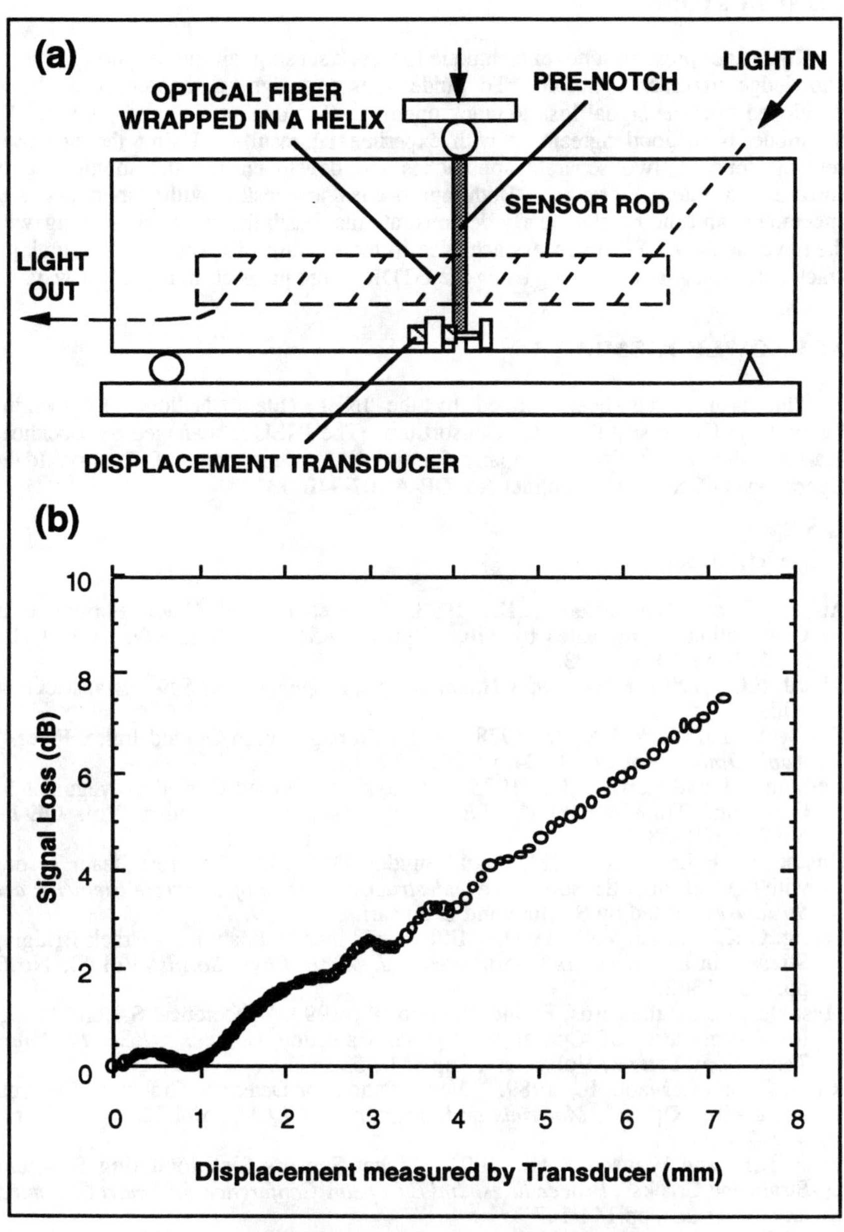

Figure 8 (a) Beam Specimen containing Sensing Rod
(b) Signal Loss in Embedded Optical Fiber vs Displacement

CONCLUSIONS

This paper presents a novel technique for crack sensing without requiring a-priori knowledge of crack location. To guide sensor design, a theoretical model is developed to relate signal loss to crack opening. The qualitative trend predicted by the model is in good agreement with experimental results. Using the proposed sensing concept, two separate approaches are developed for the monitoring of surface and internal cracks. Both approaches are tested with laboratory size specimens, and the results clearly demonstrate the feasibility of crack sensing with the novel sensors. With success achieved in the sensing of a single crack, multiple crack detection and monitoring using the OTDR technique is currently under way.

ACKNOWLEDGEMENTS

This project has been funded by the INEL (Idaho National Engineering Laboratory) University Research Consortium. The INEL is managed by Lockheed Martin Idaho technologies Company for the U.S. Department of Energy, Idaho Operations Office, under Contract No. DE-AC07-94ID13223.

REFERENCES

Ansari, F. and Navalurkar, R.K., 1993. "Kinematics of Crack Formation in Cementitious Composites by Fiber Optics", *ASCE J. Eng. Mech.*, Vol.119, No.5, 1048-1061, 1993.

Allard, F.C., 1990. *Fiber Optics Handbook for Engineers and Scientists*, McGraw Hill.

Feit, M.D. and Fleck, J.A., Jr., 1978. "Light Propagation in Graded-Index Fibers", *Appl. Optics*, Vol.17, No.24, pp.3990-3998.

Heiblum, M. and Harris, J.H., 1975. "Analysis of Curved Optical Waveguides by Conformal Transformation", *IEEE J. of Quantum Electronics*, Vol. QE-11, No.2, pp.75-83.

Huston, D., Fuhr, P., Kajenski, P. and Snyder, D., 1992. "Concrete Beam Testing with Optical Fiber Sensors", in *Nondestructive Testing of Concrete Elements and Structures*, edited by S. Sture and F. Ansari, pp. 60-69.

Leung, C.K.Y. & Li, V.C., 1992. "Effects of Fiber Inclination on Crack Bridging Stresses in Brittle Matrix Composites", *J. Mech. Phys. Solids*, Vol.40, No.6, pp.1333-1362.

Mustieles, F.J., Ballesteros, E. and baquero, P., 1993. "Theoretical S-Bend Profile for Optimization of Optical Waveguide Radiation Losses." *IEEE Photonics Technology Letters*, Vol.5, No.5, pp.551-553.

Rossi, P. and LaMaou, F., 1989. "New Method for Detecting Cracks in Concrete using Fiber Optics", *Materials and Structures (RILEM)*, Vol.22, No.132, pp. 437-442.

Voss, K.F. and Wanser, K.H., 1994. "Fiber Sensors for Monitoring Structural Strain and Cracks", *Proceedings, 2nd European Conference on Smart Structures and Materials*, pp.144-147.

Wolff, R. and Miesseler, H., 1992. "Monitoring of Prestressed Concrete Structures with Optical Fiber Sensors", *Proceedings, 1st European Conference on Smart Structures and Materials*, pp. 23-29.

Realtime Seismic Monitoring of Buildings and Bridges in Taiwan

W. H. K. LEE and T. C. SHIN

ABSTRACT

A realtime seismic monitoring system was designed using commercially available hardware components and software published by the International Association of Seismology and Physics of the Earth's Interior (IASPEI) in collaboration with the Seismological Society of America (SSA). Since 1992, about 50 such systems were installed in selected buildings and bridges in Taiwan to monitor their structural health under strong ground motions caused by large earthquakes. During the past six years in operation, these systems performed extremely well and recorded hundreds of earthquakes with maximum ground motion exceeding the threshold level (usually set to be about 0.05 g in acceleration). We will present the monitoring results of a bridge in Taipei during a magnitude 6 earthquake located about 60 km away. This realtime monitoring system can be easily adapted for other applications and had been used for monitoring volcanoes, explosions, and missle flights, etc. worldwide.

INTRODUCTION

Taiwan is an island (about 300 km by 100 km) located in southeastern Asia, where the Philippine Sea plate is subducting under the Eurasian plate at a rate of about 7 cm per year. Earthquakes are frequent in Taiwan and many disastrous earthquakes had occurred in the past.

In 1992, the Central Weather Bureau (CWB) of Taiwan began a six-year program (with a budget of about US$ 40 millions) to greatly expand the seismic monitoring capabilities in Taiwan. At present, there are about 1,000 digital seismic stations/arrays in operation in Taiwan, or about 5,000 digital seismic channels. In comparison, California (which is about 12 times larger in area) has only about 2,500 digital seismic channels in operation.

William H. K. Lee, U.S. Geological Survey, MS-977, Menlo Park, CA 94025.
Tzay Chyn Shin, Central Weather Bureau, 64 Kung Yuan Rd., Taipei, Taiwan.

One major component of the Taiwan earthquake program is to monitor the structural health of buildings and bridges under strong ground motions caused by large earthquakes. In this paper, we will present the design and implementation of a realtime seismic system for structural health monitoring in Taiwan. We will also present monitoring results of a bridge in Taipei for illustration purposes.

SYSTEM DESIGN

The system design for seismic monitoring of buildings and bridges in Taiwan was developed under a joint United States and Taiwan cooperation program. It was based on an earlier design developed by the senior author and his associates at the U.S. Geological Survey for monitoring explosions and earthquakes in the 1980's. The origin design was first released in Lee, Tottingham and Ellis (1988), and was subsequently published (Lee, Tottingham and Ellis, 1989; Lee, 1994). The system design is characterized by using only commercially available hardware, and by writing software only if it is not commercially available.

There are four basic elements in a realtime monitoring system: (1) sensors, (2) telemetry, (3) data acquisition, and (4) data processing and analysis. For seismic monitoring of strong ground motion, the force-balance accelerometers are the usual choice for sensors. Telemetry is dictated by actual application and cost. In a building or a bridge, the simplest telemetry is to use cables, which link the sensors at various locations to a central data acquisition site, i.e., analog telemetry. After signal conditioning (SC), data acquisition is then performed using a personal computer (PC) with an appropriate analog-to-digital (A/D) board. Data processing and analysis can also be performed either on the same PC or on another PC networked to the A/D PC. Figure 1 illustrates the hardware components used in such a system, and Figure 2 illustrates the software components for data acquisition and processing.

Alternatively, SC and A/D can be performed at the sensors, resulting in signals for digital telemetry. Although there are some advantages in this design, but it is more expensive to implement and difficult to achieve time synchronization among all the sensor signals. Furthermore, it is much easier to service the electronics, vary the signal conditioning, or vary the sampling rate in a system with analog telemetry than with digital telemetry. Actually, our system design also supports digital telemetry with the appropriate digital accelerographs, and the software (and information on the necessary hardware) was also published in Lee (1994).

TAIWAN IMPLEMENTATION

Because the hardware components are commercially available and the soft-

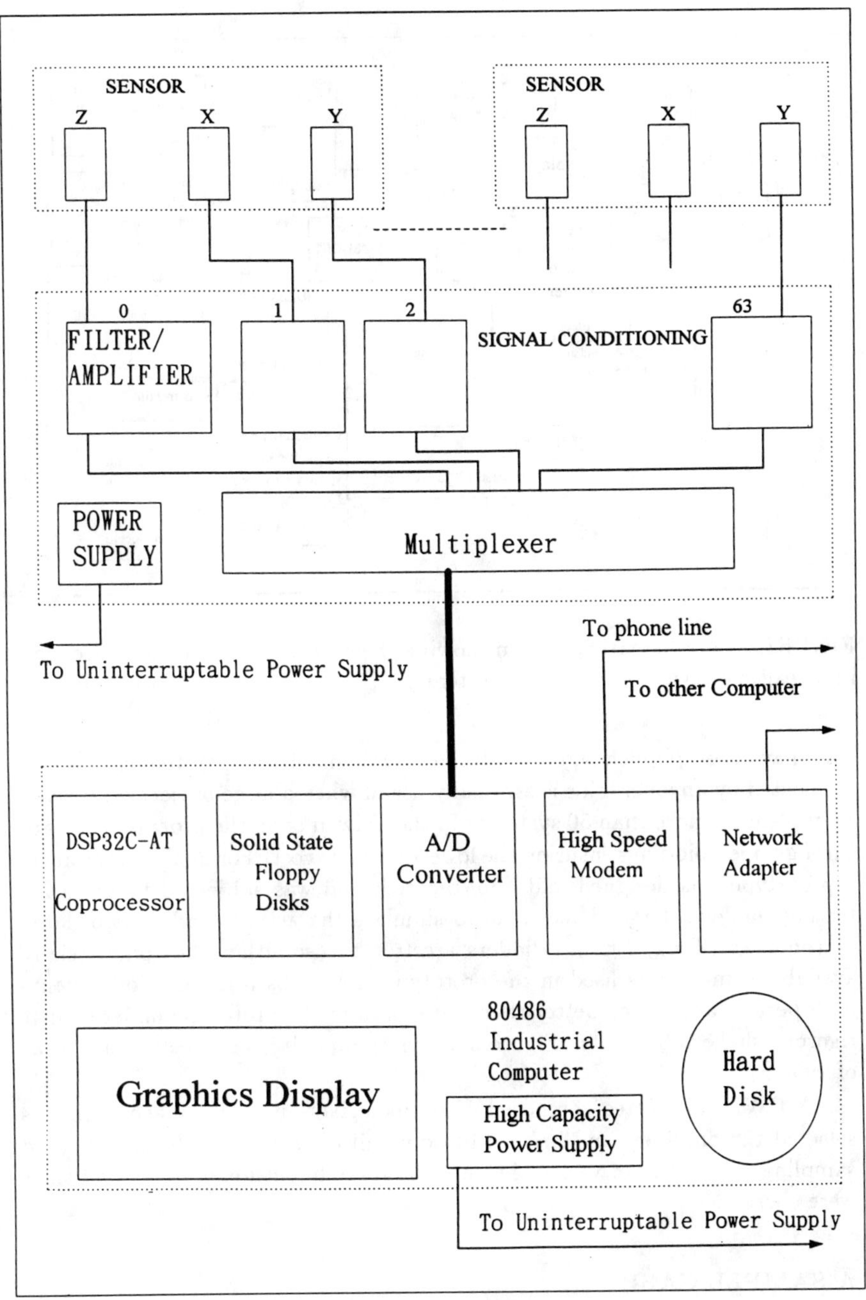

FIGURE 1. A schematic diagram showing the hardware components in a realtime seismic monitoring system.

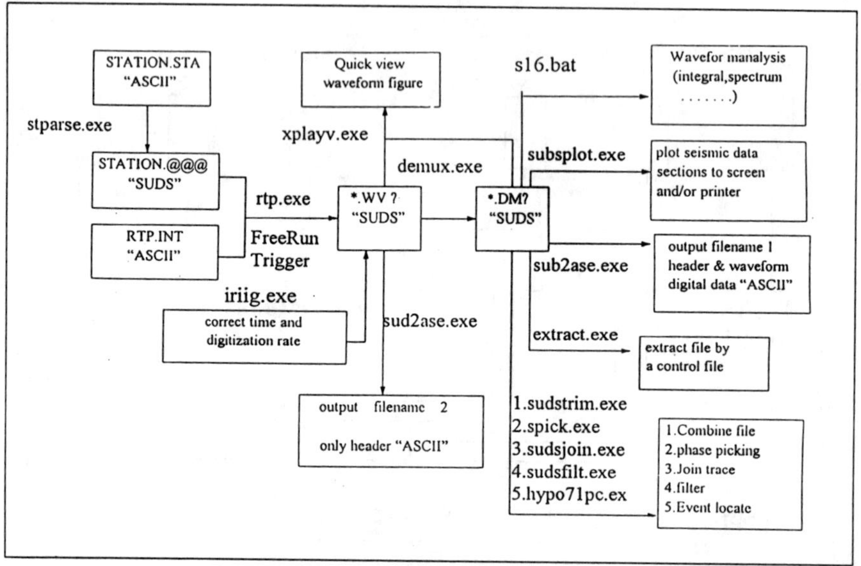

FIGURE 2. A schematic diagram showing the data acquisition and processing in a realtime seismic monitoring system.

ware have been published, our realtime seismic monitoring system can be implemented by anyone. This is especially useful when a large project is involved to implement more than 50 systems. In the Taiwan case, the procurements are through open bidding, ensuring the lowest possible cost. For quality assurance, a prototype was designed and implemented, and was subjected to extensive testings under a large shake table to simulate the actual conditions of large earthquakes (Teng, 1992). Bidders are free to use either the commercially available components used in the prototype or to substitute any components with performance at or better than those used in the prototype under similar rigorous shake table tests. Figure 3 shows the monitoring sites in Taiwan as of 1996.

A CWB committee composed of seismologists and earthquake engineers selected the buildings and bridges to be monitored. In order to have a good sampling, structures of various heights and on different foundation conditions were chosen.

A SAMPLE CASE

Since the program began in 1992, about 10 structures were instrumented per year. So far, hundreds of earthquakes had been recorded. However, since no large earthquakes occurred nearby any of the instrumented structures in

Taiwan, we must wait. Nevertheless, the records obtained so far can be shown to be useful in finding out the resonance frequencies of the structure under moderate shaking. We will present a sample case.

Figure 4 shows the sensor locations of a bridge that was instrumented in 1994 with 54 components of accelerometers. This bridge is located in Taipei and is heavily used. On June 25, 1995, a moderate earthquake ($M_L = 6.4$; $M_S = 5.5$) occurred about 60 km southeast from this bridge. Figures 5 and 6 show the actual accelerograms recorded in realtime for the acceleration components along the X-axis (along the bridge), and for the acceleration components along the Y-axis (perpendicular to the bridge), respectively.

Along the bridge, the highest acceleration occurred on the bridge tower (S8 and S9 in Fig. 4) and as expected with a maximum acceleration of about 0.37 g. The free-field site (located just off the bridge, i.e., S18) recorded a maximum acceleration of about 0.03 g in the same direction. Perpendicular to the bridge, the highest acceleration also occurred on the bridge tower (S8 and S9) with a maximum acceleration of about 0.11 g. The free-field site (S18) recorded a maximum acceleration of about 0.02 g in the same direction.

The resonance frequencies of this bridge can be found by computing the Fourier amplitude spectra of the recorded accelerograms. In Figure 7, the upper graph shows the Fourier spectrum of the free-field acceleration (S18); it is relatively flat between 1 to 4 Hz. The lower graph of Figure 7 shows the Fourier spectrum of the bridge tower acceleration (S8) along the X-axis; it has a large spectral peak at 3 Hz, about 30 times larger than the corresponding value for the free-field acceleration.

Similarly, Figure 8 shows the spectral comparison in the direction perpendicular to the bridge. The spectral peak at the bridge tower is at 0.9 Hz and is about 20 times larger than the free-field acceleration in the same direction.

CONCLUSION

The realtime seismic monitoring program of structures in Taiwan is the most extensive one in the world. It is based on a well-proven design, using commercially available components and published software. The system operates automatically and requires little maintenance. We hope that these systems will provide useful data when large earthquakes occurred.

REFERENCES

Lee, W. H. K. (Editor), 1994. *Realtime Seismic Data Acquisition and Processing.* IASPEI Software Library Volume 1 (2nd Edition), Seismological Society of America, El Cerrito, CA.

Lee, W. H. K., D. M. Tottingham and J. O. Ellis, 1988. "A PC-based Seismic Data Acquisition and Processing System." Open-File Report 88-751, U. S. Geological Survey,

Menlo Park, CA.

Lee, W. H. K., D. M. Tottingham and J. O. Ellis, 1989. "Design and Implementation of a PC-based Seismic Data Acquisition, Processing, and Analysis system." *IASPEI Software Library*, 1:21-46.

Teng, T. L., 1992. "Design and Implementation of a Prototype Realtime Strong-Motion Array System," Technical Report, Southern California Earthquake Center, University of Southern California, Los Angeles, CA.

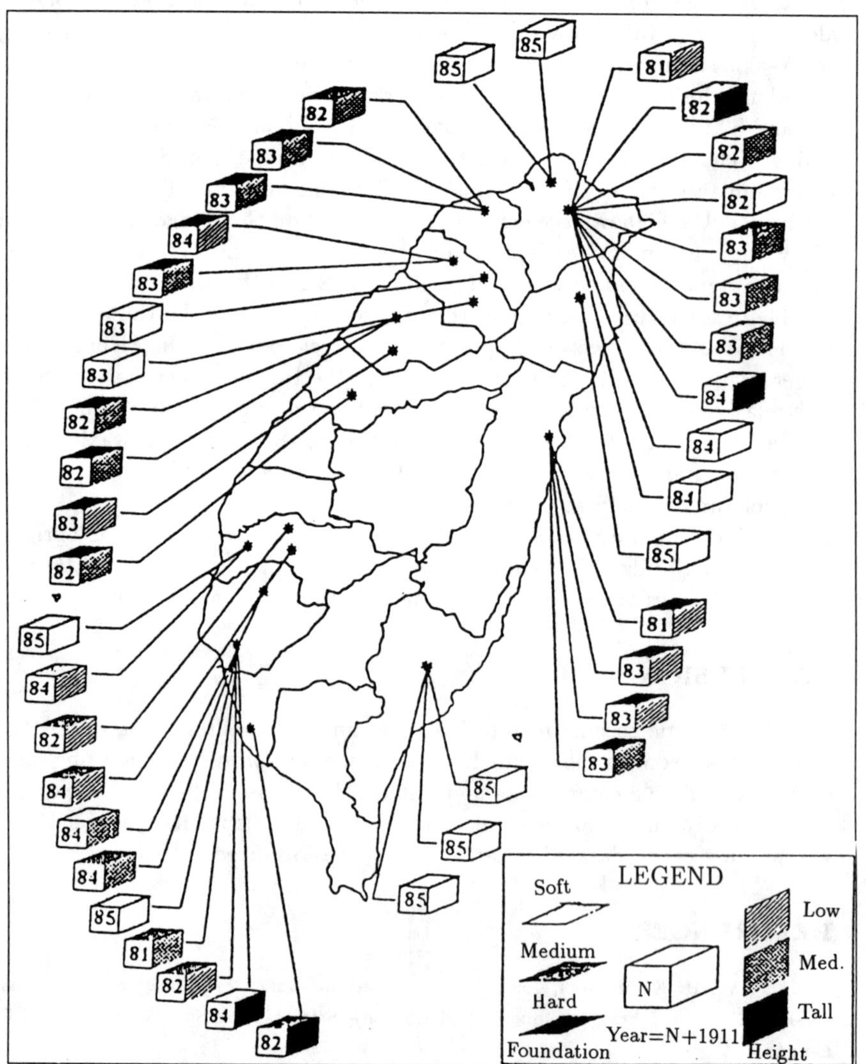

FIGURE 3. Geographic distribution of instrumented structures in Taiwan. Bridges have blank top and blank side.

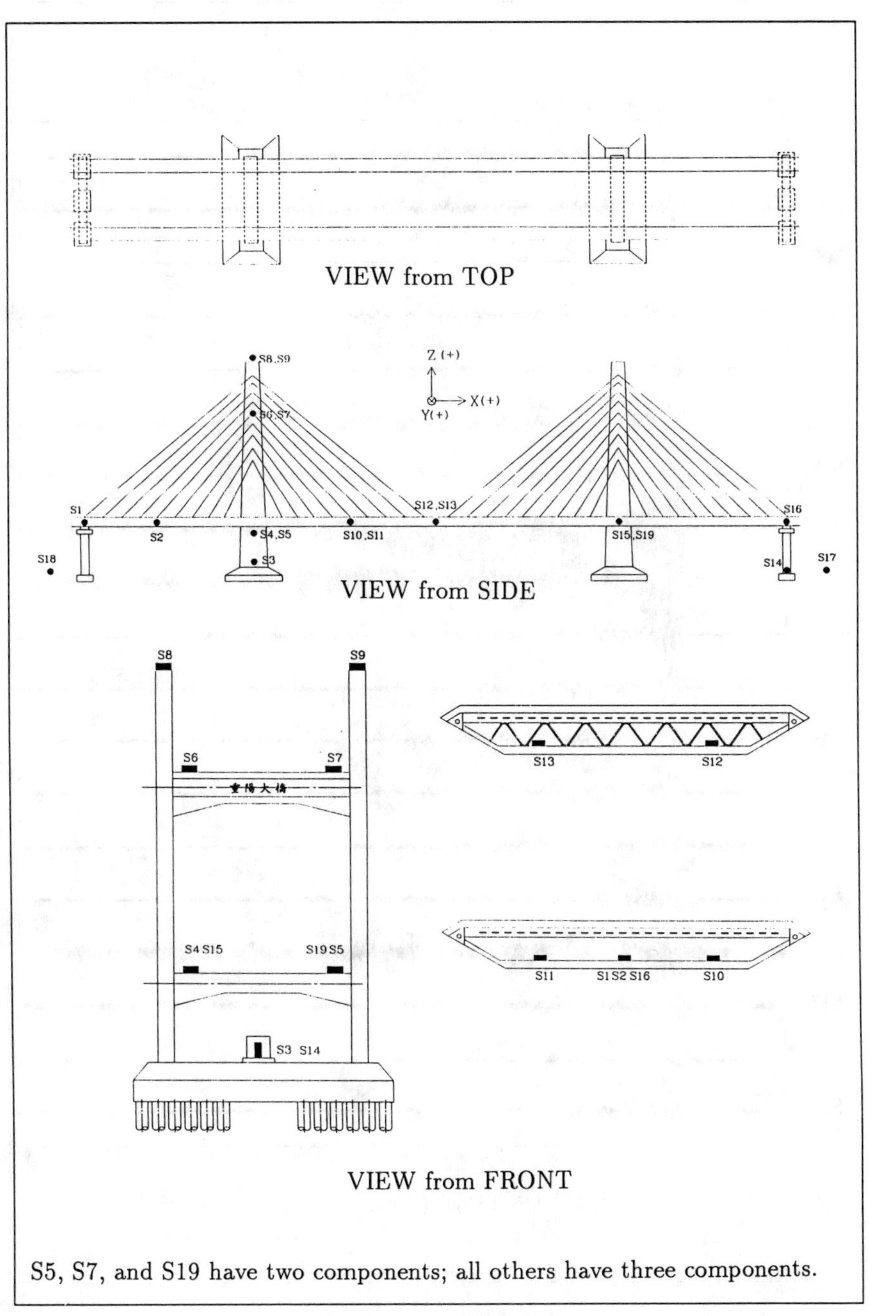

FIGURE 4. Sensor locations of an instrumented bridge in Taipei.

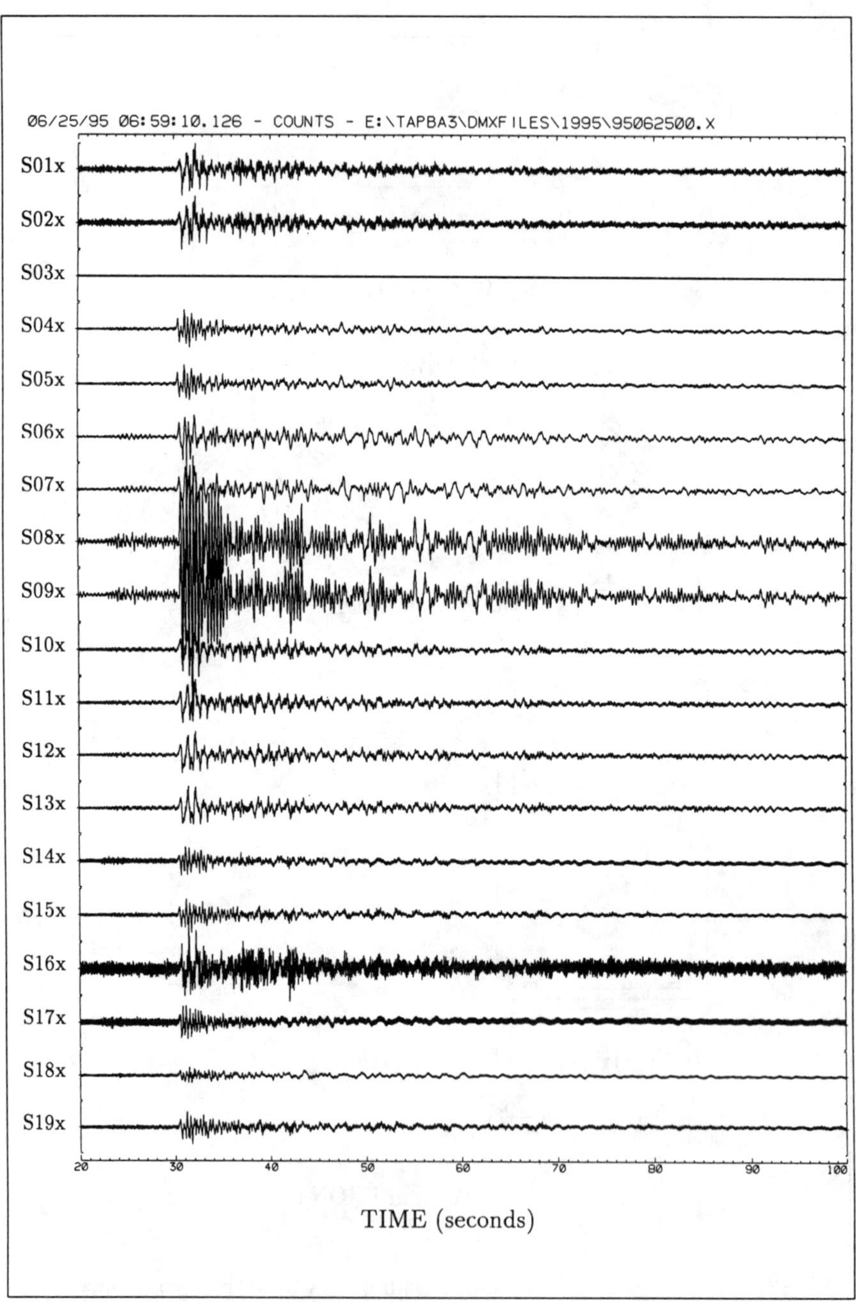

FIGURE 5. Recorded accelerations (along the X-axis) at a bridge site in Taipei for the June 25, 1995 earthquake.

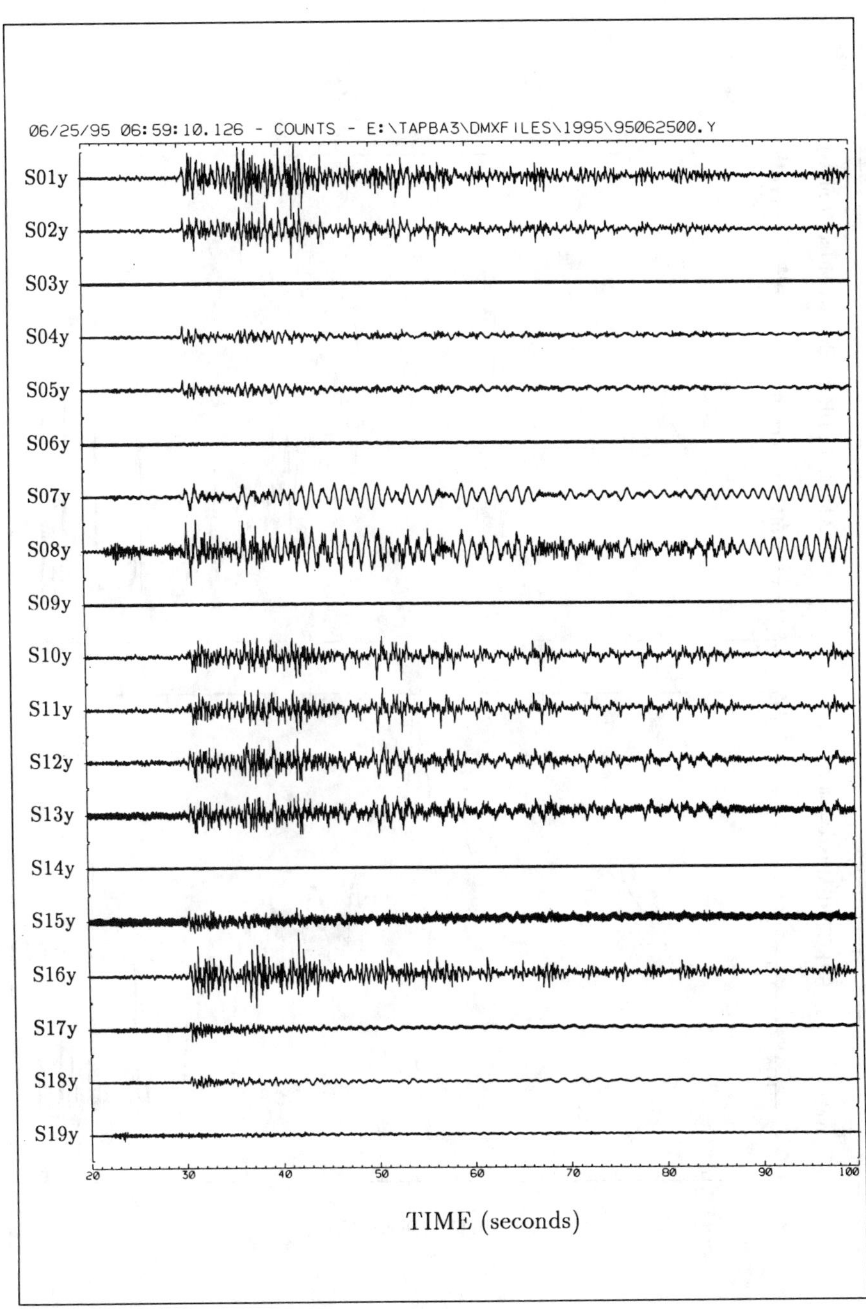

FIGURE 6. Recorded accelerations (along the Y-axis) at a bridge site in Taipei for the June 25, 1995 earthquake.

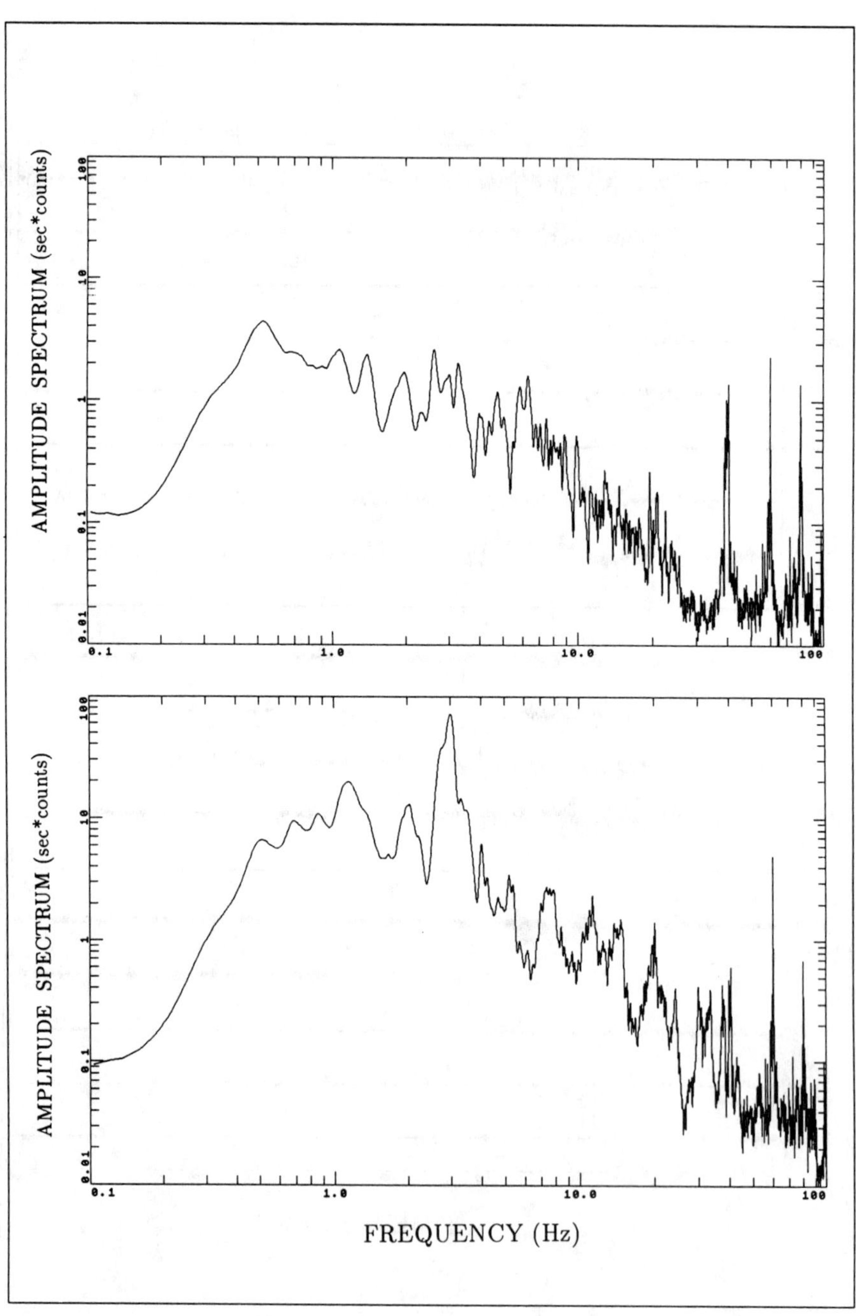

FIGURE 7. Fourier acceleration spectrum (along the X-axis) at a Taipei bridge site (Top: free-field; Bottom: bridge tower).

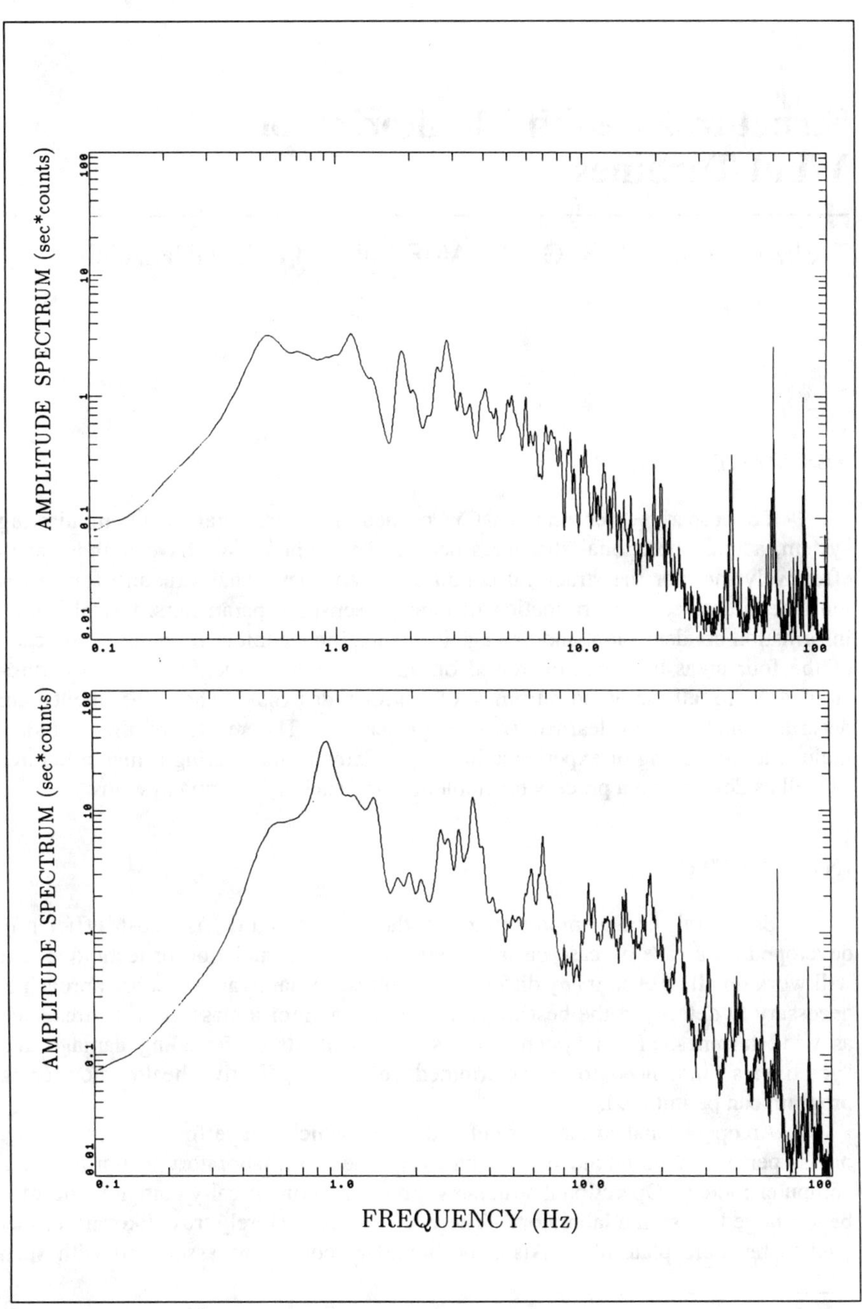

FIGURE 8. Fourier acceleration spectrum (along the Y-axis) at a Taipei bridge site (Top: free-field; Bottom: bridge tower).

Structural Health Monitoring of Wind Turbines

T. SIMMERMACHER, G. H. JAMES III and J. E. HURTADO

ABSTRACT

To properly determine what is needed in a structural health monitoring system, actual operational structures need to be studied. We have found that to effectively monitor the structural condition of an operational structure four areas must be addressed: determination of damage-sensitive parameters, test planning, information condensation, and damage identification techniques. In this work, each of the four areas has been exercised on an operational structure. The structures studied were all be wind turbines of various designs. The experiments are described and lessons learned will be presented. The results of these studies include a broadening of experience in the problems of monitoring actual structures as well as developing a process for implementing such monitoring systems.

INTRODUCTION

Structural health monitoring, at the present time, is most efficiently developed on a case by case basis. There exists no formulation or technique that will work on all or even many different structures. A hardware specific approach is necessary to determine the best methodology to monitor a class of structures such as wind turbines. Each system has its own methods of revealing damage and "gotch ya's" that need to be determined before an effective health monitoring program can be initiated.

An operational structure is defined as one which can perform, is performing, or has performed its intended function as opposed to a laboratory test article or a computer model. Operational structures are often geometrically complex and may be too large to test in a laboratory. These structures are rarely truss-like and in fact tend to be more plate-like. Also, the boundary conditions associated with such

Todd Simmermacher, Postdoctoral Research Assistant, twsimme@sandia.gov
George H. James, Technical Staff Member, ghjames3@aol.com
John E. Hurtado, Technical Staff Member, jehurta@sandia.gov
Sandia National Laboratories, Mail Stop 0557, Albuquerque, NM 87185-0557

structures are not as well known as a laboratory test structure or a computer model. Finally, the environment associated with an operational structure (e.g. weather, traffic patterns, or location) is usually changing and has a serious impact on the measured structural response. Therefore, it is challenging but desirable to perform health monitoring research and development on structures possessing such characteristics.

One structure that has received much attention at Sandia National Laboratory is the wind turbine. It is a good representative structure to describe the systematic approach described in this paper. Also, the wind turbine is a structurally simple structure where the dynamics can easily be physically interpreted. In addition, a failure of a wind turbine blade can have severe consequences, as that can damage other blades, the tower, internal mechanical systems, other wind turbines, or to workers. The failure means loss of revenue, loss of equipment, and usually negative public relations. Therefore there is much interest in the renewable energy community to develop reliable and quick health monitoring systems for wind turbines.

There are two basic types of wind turbines that will be studied in this work. They are identified by the orientation of their axis. The first is the Vertical Axis Wind Turbine (VAWT). This turbine has its axis of rotation perpendicular to the ground, like an eggbeater (Fig. 1). The VAWT is not dependent on the wind direction for its performance. The more traditional propeller style Horizontal Axis Wind Turbine (HAWT), also shown in Fig 1, in contrast, requires the ability to yaw in response to a change in wind direction. The two structures can be studied together since both are basically beam-like, have similar maintenance procedures, have the same operating environments, and are rotating structures.

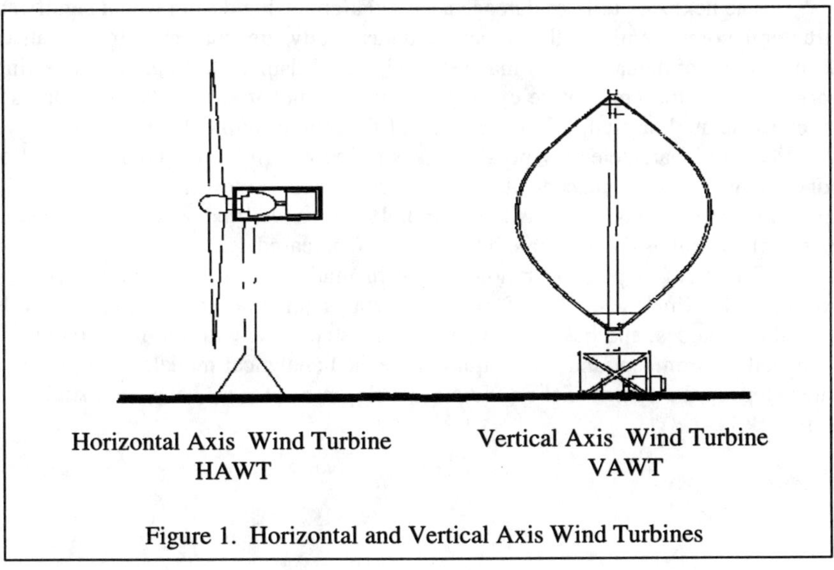

Figure 1. Horizontal and Vertical Axis Wind Turbines

This paper is separated into four sections. The first section will summarize the concept of operational evaluation. The next section will describe laboratory mock-ups that were used for fatigue and Laser Doppler Vibrometer (LDV) studies. These studies provided information on modes of failures and also gave some insight on some of the modal characteristics of a blade. In the third section, two field experiments will be described. These tests provided experience on testing large structures with low frequencies. In the final section, a field test on a HAWT is described that was performed to study the effectiveness of an LDV in the field and the observablity of simulated damage on a blade. This test utilizes much of the experience developed in the other sections. Taken together, these sections address each of the four technology areas which are necessary for implementing a structural health monitoring system: determination of damage-sensitive parameters, test planning, information condensation, and damage identification techniques.

OPERATIONAL EVALUATION

The operation evaluation development has centered around techniques needed to answer two questions in the implementation of a structural health monitoring system:

1) What data needs to be acquired to track important structural changes?
2) How is this data to be collected in the operational environment?

The answer to these questions can be found by a three step process. The first step utilizes engineered flawed specimens to develop an initial understanding of which parameters are sensitive to the expected damage and to validate the diagnostic measurements. The use of analytical tools such as experimentally-validated Finite Element Models (FEM) can be a great asset in this process.

The next step utilizes damage accumulation testing during which significant structural components of the structure under study are subjected to a realistic accumulation of damage. This may require induced-damage testing, fatigue testing, corrosion growth, temperature cycling, etc. to accumulate certain types of damage in an accelerated fashion. Hence, a study of the relationship between damage level and measured parameters is possible as well as determining initial information concerning sensor placement, data acquisition interval, and, possibly, environmental effects. As with the initial step, a verified analytical model is extremely useful as the available information is increased.

The final step is operational implementation. This step in the process concerns the final selection of sensors, data acquisition, monitoring intervals, excitation sources, and baseline data set. This step deals with the full structure in its actual environment and may require a verified analytical model. Aspects of all three steps in the operational evaluation development process have been studied in this work.

LABORATORY STUDIES

Before any health monitoring methodology can be implemented, laboratory studies must be performed. Studies on specimens of the actual structure provide insight into failure modes of the structure, help to develop an intuition about the vibrational characteristics of the structure, as well as to help determine what parameters are most sensitive to damage. Studies on engineered specimens, which may or may not be exactly representative of the actual structure, can provide valuable information on instrumentation issues, signal processing algorithms, and damage detection procedures that will be of use on the operational structure. Both types of experiments are described in this work.

ENGINEERED SPECIMENS

The engineered specimens described in this work were developed for aircraft damage detection. They were used to understand and refine the use of the LDV. Issues that were addressed were data condensation, pointing issues, and also gaining basic experience with this new technology. This work is directly applicable to wind turbines as a potential non-contacting measurement system. Instrumenting an operational wind turbine is a non-trivial task. Through the use of a LDV, a large time savings may be realized. The use of the LDV on an operational wind turbine will be described in the last section of this paper.

The LDV is a non-contacting measurement system that infers the velocity of the target (test article) by calculating the Doppler shift between the incident and reflected laser beam. This device can be used at a larger distance than most other non-contacting measurement systems such as Nearfield Acoustic Holography (Veronesi and Maynard, 1989) or fiber-optic or capacitance displacement sensors. The laser is pointed at the desired measurement points by controlling the voltage signals into the pointing mirrors. This may be done either by a computer or by manually controlling power supplies. By using a computer controller, automatic scanning is possible. The LDV can allow a higher spatial resolution to be realized than would be possible with accelerometers. A disadvantage of the LDV system is that a stationary reference is necessary in order to determine phase information for the Frequency Response Function (FRF). This is typically an accelerometer but could be another LDV system that does not rove over the structure.

Two LDV tests contributed significantly to the health monitoring of wind turbines. The first was a simulated aircraft panel with a cut stringer (Doebling, 1993). This test provided a look at the capabilities of the LDV system and where it differs from traditional modal analysis tools, i.e. accelerometers. The test compared output from both the LDV system and accelerometers. A study was also performed on the mass loading effects of accelerometers. Comparisons were also made between two damage detection algorithms. It was found that the LDV was capable of accurately estimating natural frequencies, although signal drop-out problems with the LDV prevented accurate mode shapes to be determined. Since the damage detection algorithms studied required mode shape information, accelerometer data was used for that study. This test initiated later tests on full scale aircraft structures (Robinson et. al. 1996 and Meza, et. al. 1997).

The second LDV experiment to be describe here was performed on five composite plates (James, et. al 1996). The purpose of this study was to try to determine if the LDV was capable of identifying debonded regions of the plates. The debonds were engineered into the construction of the plates. The plates were 24 inches by 24 inches constructed of a 0.5 inch Nomex honeycomb core sandwiched between four ply T300 plain weave graphite cloth panels. The simulated flaws were typical of what is normally seen in composite aerospace structures. This test utilized a white dye penetrant to decrease the occurrence of signal drop-out as was observed in the test described above. This allowed damage detection algorithms to be evaluated directly using LDV data (Meza, et. al. 1996).

These studies provided many lessons in the use of the LDV. Among these lessons are the need for surface treatment to reduce drop-outs, the need for a reference accelerometer to determine phase information, and the need for efficient data reduction and damage identification algorithms to handle the large amount of data that can be gathered using the LDV system (Robinson, et. al. 1996).

WIND TURBINE LABORATORY FATIGUE STUDIES

Fatigue studies can demonstrate some of the probable modes of failure for a system and provide information on which parameters should be monitored in an operational structure. They also provide data on the history of the propagation of the failure. This information may be of use in predicting remaining life of a structure as fatigue studies provide a direct connection to the mechanics of the materials used in the structure. Predicting remaining life is ultimately the capability desired of any health monitoring system.

There were two fatigue tests performed on composite wind turbine blades. The first experiment was a resonant fatigue test, during which the test article was excited at its fundamental frequency for an extended period of time. This has the effect of increasing the speed of fatigue (which decreases the length of the test) because the large amplitude strains can be induced at a high rate. A FloWind Corporation blade joint from the 17EHD VAWT was the subject of this initial resonant fatigue test (Rodemen and Gregory, 1994). The test article was a 14 foot long section of pultruded fiberglass blade bonded to steel attachment hardware that would bolt to the tower on the actual turbine. Strain gages were placed at 20 locations to monitor stress concentrations and load transfer characteristics. Thirty-four accelerometers were also used for the structural health monitoring.

A modal test was performed to obtain an initial damping estimate. A difficult task in performing the resonant fatigue test was the selection of a proper excitation source. The final configuration had the blade mounted on a vibration slip table and driven by an UnHoltz-Dickie Model T-4000 electrodynamic shaker. The test article was excited at the first resonant frequency (initially at 4.3 Hz). Failure occurred after 22,000 cycles as opposed to the 100,000 estimated. A design flaw was found to be contributing to the premature failure. This was subsequently corrected by the company. There was a simple analytical model of this structure which assisted in test planning and analysis.

The final test was also a resonant fatigue test on a second pultruded fiberglass VAWT blade obtained from FloWind. This 16 foot section was of a newer and lighter design and did not include the root joint. A resonant fatigue test was planned and performed on this specimen. A free-free configuration was used on this test. The difficult issue in the design of this test was the load transfer fixture. The blade was instrumented at seven strain gage locations and with 70 accelerometers. The excitation frequency of 25 Hz resonated the blade at its fundamental frequency. During the course of the test it became obvious that a large-area non-contacting transducer such as an LDV would have been much more efficient than traditional accelerometers which tended to break off of the structure during the high-level excitation. The blade failed after 15.5 hours of testing and 1.325 million cycles. The failure was an axial crack that was not in the highest stress location. The first two bending modes were unaffected by damage while the first torsion mode decreased in frequency all the way to failure. This would be expected for the observed failure. This test also had the advantage of an analytical model for test planning and analysis.

These studies showed the necessity of maintaining a consistent set of boundary and environmental conditions during laboratory studies. This phase of the process was designed to focus on material, structural, signal processing, and data acquisition issues. More specifically, these tests showed that the damage accumulation and failure does not necessarily manifest itself in the forcing frequencies. For resonant fatigue tests, the use of non-contact sensors and long-stroke shakers would be most beneficial. In all cases, the availability of a simple analytical model assisted in test planning, interpretation of results, and damage detection.

FIELD TESTING OF WIND TURBINES

Obviously, an important aspect of developing a health monitoring procedure is developing an ability to test the structure in the field. Wind turbines present a set of special problems that need to be addressed. One problem is that they typically have very low resonant frequencies, with potentially many modes below 10 Hz. consequently the problem becomes one of excitation. Three methods of excitation have been used: step relaxation, wind excitation, and impact. Another problem is one of instrumentation. Instrumenting a structure which can be as tall as 110 m with accelerometers and all the required cabling can be difficult at best. Two tests are described in this section. The first is a research two meter VAWT which was developed by and tested at Sandia National Laboratories (Carne, et. al., 1988). This turbine was tested both in a parked and in an operating condition. The other turbine is the 110 m EOLE VAWT (Carne, et. al, 1989). This test was performed in a parked condition.

There were two objectives in testing the Sandia two meter research turbine. The first was to verify a Finite Element (FE) code that was developed to calculate mode shapes and natural frequencies of a rotating VAWT. The FE code included the tension stiffening, centrifugal, and Coriolis terms necessary for a flexible body in a rotating reference. The code was developed in the NASTRAN DMAP (MSC/NASTRAN, 1981) language which allowed these effects to be included in a NASTRAN model. This simplified code development by eliminating the need to write a completely new code.

The second objective of this test was to explore techniques of testing rotating VAWTs. The turbine was tested in a parked and in a rotating configuration. The parked test consisted of 22 accelerometers and was excited using an impact excitation on the tower and on the blades. The modes were lightly damped and measured up to 60 Hz. The rotating test was performed using two accelerometers and seven strain gages. The signals were transmitted through slip rings on the tower. To excite the rotating blades, a method known as step-relaxation was used. This method consists of a pretensioned cable between the tower and a blade with a force transducer and a quick release mechanism on the cable. When the cable is released, a step function is imparted on the system which is measured by the force transducer. Although this method applied force only in the plane of the blades, out of plane modes were excited due to coupling from the Coriolis force. The blade was tested at different rotational speeds from 100 to 600 rpm.

These experiments developed procedures to field test operational wind turbines. The testing of rotating structures required the development of specific experimental and signal processing tools. As an example, follow-on work to these projects developed a procedure called the Natural Excitation Technique (NExT) (James, et. al. 1992) to allow the extraction of modal parameters from an operating wind turbine using only the wind excitation. Also, a significant effort in model updating allowed the development of system identification and model correlation tools and experience during these activities. Several follow-on projects used the lessons learned from these experiments to more fully understand VAWT technology; however, the typical VAWT requires only a single rotating coordinate system and is therefore reasonable to model. In contrast, the HAWT problem requires several coordinate systems to model the various components. More recent work (as discussed in the next section) has concentrated on understanding and monitoring HAWT's.

FIELD TESTING AND EVALUATION OF A WIND TURBINE

The work up until this point has described tests that are not directly related to health monitoring, but developed critical supporting technologies. The fatigue studies gave information on what failures to expect and some insight on what parameters may indicate impending failure. The plate studies provided experience on the LDV and on information condensation. The full scale modal tests provided insight on successful methods of field testing wind turbines. The next step is to test a HAWT with the objective of health monitoring. This was performed recently in Bushland, Texas on a 15 meter diameter Atlantic Orient Corporation turbine owned by the United States Department of Agriculture (USDA). The hub of the blade was approximately 25 meters from the ground. The objectives of this test are presented. The data has not been fully analyzed as of yet and therefore will not be discussed.

The primary objective of this test was to see if an LDV could be used in the field for distances that would be expected for use on a wind turbine. To test this, the HAWT was instrumented with accelerometers and retro-reflective tape was placed near the accelerometers. The LDV system was set approximately 210 ft from the measurement points on the blade of the structure. From this setup, the response measured from the accelerometers and the LDV can be compared. Impact testing was performed, so that the input could be measured and a true FRF could be measured.

Another objective of the test was to see if accurate NExT data could be gathered by the LDV. This is requiring the development of scaling algorithms and new test procedures as the sequential nature of LDV data acquisition is somewhat incompatible with traditional NExT. For the Bushland tests, the data from the accelerometers was saved as a comparison with the LDV data. Another check on the quality of the NExT data, two simulated damage cases were inflicted on the blade. The damage was inflicted by loosening the bolts at the root of the blade various amounts to simulate a root failure. If the damage is visible using current state-of-the-art damage detection algorithms and the NExT data, then the NExT data is useful for damage detection.

Preliminary results show that the LDV system performed well in the field at the distance that it was used. The acceleration data, integrated to produce velocities, corresponded well with the LDV velocity data. Figure 2 shows an overlay of the FRF from the impact test for an accelerometer and the LDV for a point on the tip of the blade. The success of the other objectives will be presented elsewhere (Rumsey, et. al, 1997).

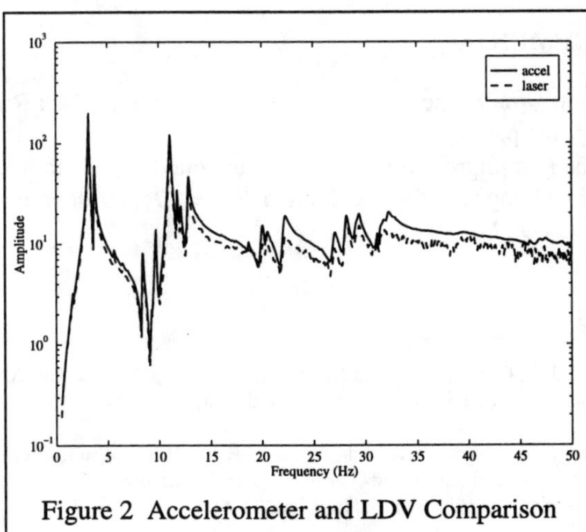

Figure 2 Accelerometer and LDV Comparison

CONCLUSIONS

This paper has presented a broad collection of work performed at Sandia National Laboratories over the last several years which is developing the capability to perform structural health monitoring on operational structures. The electricity generating wind turbine has been used as a target structure for much of this work. This class of structures has allowed all aspects of structural health monitoring technology to be exercised and developed: determination of damage-sensitive parameters, test planning, information condensation, and damage identification techniques. Although different structures may possess different damage-sensitive parameters, the work presented herein has shown that modal frequencies, mode shapes, modal damping, static flexibility, and static stiffness provide a reasonably inclusive set of parameters to chose from. The ongoing full-scale HAWT tests should provide a specific set of parameters for at least one example structure.

Test planning activities have included the development of procedures such as step relaxation testing of fielded structures and NExT for operational structures. Also, techniques which use finite element models for planning and pre-analysis have been developed and proven to be extremely valuable. Information condensation developments have produced new techniques for reducing large data sets rapidly into manageable information sources. This is extremely valuable with devices such as the LDV which can access a great deal of information in a semi-automated fashion. Damage identification techniques which are appropriate for operational structures have been exercised. These include fingerprinting of dynamic parameters but have included experimentally based comparisons of flexibility shapes or matrices as well as dynamic residual formulations using analytical models. Taken as a whole, this work represents a significant body of expertise needed to implement structural health monitoring on operational structures.

ACKNOWLEGEMENTS

The authors would like to thank Thomas Carne and Mark Rumsey for their help in preparing this paper.

Sandia is a muliprogram laboratory operated by Sandia Corporation, a Lockheed Martin Company, for the United States Department of Energy under Contract DE-ACO4-94AL85000.

REFERENCES

Carne, T.G., Lauffer, J.P., Gomez, A.J., and Benjannet, H., 1989, "Natural and Artificial Excitation for Modal Testing of Large Structures," Sound and Vibration, Nov.

Carne, T.G., Lobitz, D.W., Nord, A.R., and Watson, R.A., 1988, "Finite-Element Analysis and Modal Testing of a Rotating Wind Turbine," The International Journal of Analytical and Experimental Modal Analysis, Vol. 3, No. 1, pp. 32-41.

Doebling, S.W., 1993 "Evaluation of a Laser Doppler Measurement System and Techniques for Damage Detection," Final Report for Sandia National Laboratories OSSP Program, also, Appendix A of "Development of Structural Health Monitoring Techniques Using Dynamics Testing," SAND96-0810, Sandia National Laboratories, Albuquerque, NM, April, 1996.

James, G.H., Carne, T.G., and Lauffer, J.P., "The Natural Excitation Technique (NExT) for Modal Parameter Extraction from Operating Wind Turbines," SAND 92-1666, Sandia National Laboratories, Albuquerque, NM, 1992.

James, G. H., Hansche, B., Meza, R., and Robinson, N., 1996, "Health Monitoring Studies on Composite Structures for Aerospace Applications," Proceedings of the 5th ASCE International Conference on Engineering, Construction, and Operations in Space - SPACE '96, Albuquerque, NM.

James, G. H., Mayes, R., Carne, T., Simmermacher, T., and Goodding, J., 1995. "Health Monitoring of Operational Structures - Initial Results," Proceedings of the AIAA/ASME/ASCE/AHS/ASC Structures, Structural Dynamics, and Materials Conference, New Orleans, LA.

MCS/NASTRAN User's Manual, 1981, I and II, The MacNeal-Schwendler Corporation, Los Angeles, CA,

Meza, R., Carrasco, C.J., Osegueda, R.A., James, G. H., and Robinson, N.A., "Damage Detection in a DC-9 Fuselage Using Laser Doppler Velocimetry," Proceedings of the 15th International Modal Analysis Conference, Orlando, FL., February 3-6, 1997.

Meza, R., Osegueda, R.A., Carrasco, C.J., James, G.H., "Modal Tests of Composite Honeycomb Panels Using Laser Doppler Velocimetry for Damage Assessment," Report FAST 96-02, University of Texas at El Paso, El Paso, TX, September 1996.

Robinson, N.A. Peterson, L.D., James, G.H., and Meza, R. "Health Monitoring Studies for Aircraft Applications," Proceedings of the 1996 AIAA Adaptive Structures Forum in conjunction with the 1996 Structures, Structural Dynamics, and Materials Conference, Salt Lake City, UT., April 18-19, 1996.

Rodeman, R. and Gregory, D., 1994, "Wind Turbine Blade Joint Fatigue Test," Internal Memo, Sandia National Laboratories, also Appendix C of "Development of Structural Health Monitoring Techniques Using Dynamics Testing," SAND96-0810, Sandia National Laboratories, Albuquerque, NM, April, 1996.

Rumsey, M., Hurtado, J., Hansche, B., Simmermacher, T., Carne, T.G., James, G.H.,1997 "The In-Field Use of Laser Doppler Vibrometer to Monitor the Health of a Wind Turbine Blade," Proceedings of the ASME/AIAA Wind Energy Symposium, Reno, NV.

Veronesi, M. and Maynard, J. D., 1989, "Digital Holographic Reconstruction of Sources with Arbitrarily Shaped Surfaces," Journal of the Acoustical Society of America, Vol. 33, No. 2, pp. 588-598.

Author Index